Lecture Notes in Computer Science 7626

Commenced Publication in 1973
Founding and Former Series Editors:
Gerhard Goos, Juris Hartmanis, and Jan van Leeuwen

Georgy Gimel'farb   Edwin Hancock
Atsushi Imiya   Arjan Kuijper
Mineichi Kudo   Shinichiro Omachi
Terry Windeatt   Keiji Yamada (Eds.)

Structural, Syntactic, and Statistical Pattern Recognition

Joint IAPR International Workshop
SSPR & SPR 2012
Hiroshima, Japan, November 7-9, 2012
Proceedings

 Springer

Volume Editors

Georgy Gimel'farb
University of Auckland, Department of Computer Science, New Zealand
E-mail: g.gimelfarb@auckland.ac.nz

Edwin Hancock
University of York, Department of Computer Science, UK
E-mail: erh@cs.york.ac.uk

Atsushi Imiya
Chiba University, Institute of Media and Information Technology, Japan
E-mail: imiya@faculatiy.chiba-u.jp

Arjan Kuijper
Technische Universität Darmstadt/Fraunhofer IGD, Germany
E-mail: arjan.kuijper@igd.fraunhofer.de

Mineichi Kudo
Hokkaido University, Graduate School of Information Science and Technology
Sapporo, Hokkaido, Japan
E-mail: mine@main.eng.hokudai.ac.jp

Shinichiro Omachi
Tohoku University, Graduate School of Engineering, Sendai, Miyagi, Japan
E-mail: machi@ecei.tohoku.ac.jp

Terry Windeatt
University of Surrey, Centre for Vision, Speech and Signal, Guilford, UK
E-mail: t.windeatt@surrey.ac.uk

Keiji Yamada
NEC Corporation, C&C Innovation Research Laboratories, Ikoma, Nara, Japan
E-mail: kg-yamada@cp.jp.nec.com

ISSN 0302-9743   e-ISSN 1611-3349
ISBN 978-3-642-34165-6   e-ISBN 978-3-642-34166-3
DOI 10.1007/978-3-642-34166-3
Springer Heidelberg Dordrecht London New York

Library of Congress Control Number: 2012948959

CR Subject Classification (1998): I.2, H.3, H.4, I.4-5, F.1, H.2.8

LNCS Sublibrary: SL 6 – Image Processing, Computer Vision, Pattern Recognition, and Graphics

Typesetting: Camera-ready by author, data conversion by Scientific Publishing Services, Chennai, India

Printed on acid-free paper

Springer is part of Springer Science+Business Media (www.springer.com)

Preface

The joint IAPR International Workshops on Structural and Syntactic Pattern Recognition (SSPR 2012) and Statistical Techniques in Pattern Recognition (SPR 2012) were held at Miya-jima Itsukushima, Hiroshima, during November 7–9, 2012. These were, respectively, the 14th and 9th editions of the SSPR and SPR workshops. This joint event is biannually organized by Technical Committee 1 (Statistical Pattern Recognition Technique) and Technical Committee 2 (Structural and Syntactical Pattern Recognition) of the International Association of Pattern Recognition (IAPR), and held in conjunction with the International Conference on Pattern Recognition (ICPR). This year the 21st ICPR was held during November 11–15, 2012, at the Tsukuba International Congress Center, Tsukuba, Japan. As is now tradition, during the SPR workshop the Pierre Devijver Award recipient presents an invited lecture. This year the Pierre Devijver Award winner was Professor George Nagy from Rensselaer Polytechnic Institute in Troy, USA. The workshop also contained invited talks by Ales Leonardis from the University of Birmingham and Kenichi Kanatani from Okayama University.

In 2012 the joint SSPR and SPR Workshops were co-hosted by the pattern recognition research groups from four Japanese universities, namely, Hokkaido University, Tohoku University, Hiroshima University, and Chiba University. The Special Interest Group of Pattern Recognition and Media Understanding (SIG PRMU, formerly SIG PR) of the Institute of Electronic Information and Communication Engineers (IEICE) of Japan offered formal support for this event. SIG PRMU(PR) is one of the oldest communities for pattern recognition in the world, dating back to the 1960s. Interestingly, the origins of the Principal Component Analysis technique now universally used in pattern recognition can be traced back to independent early work by Taizo Iijima in 1963 at the former Electrotechnical Laboratory of MITI and Satosi Watanabe in 1962, from the University of Hawaii. Today PCA is an indispensable tool in pattern recognition that has recently been extended to give both sparse and kernel methods, providing powerful new tools for data reduction. In the 1970s basic methodology from structural and syntactical pattern recognition was used in a national project concerned with "Kanji" (Chinese characters used in Japanese context) character recognition, and the results presented and discussed at a historically significant meeting of SIG PR. Based on this long tradition of pattern recognition in Japan, we welcomed SS+SPR 2012 at Miyajima.

There were 120 papers submitted to the joint workshops, of which we accepted 80 papers from 18 countries. We thank the members of the international Program Committee for their thoughtful reviews, which led to the interesting and varied set of papers contained within this volume.

We gratefully acknowledge the financial support from the Institute of Media and Information Technology, Chiba University, and from Chiba University. We also acknowledge valuable support from Hokkaido University, Tohoku University, Hiroshima University, and the Special Interest Group of Pattern Recognition and Media Understanding in Institute of Electronic Information and Communication Engineers of Japan. We gratefully extend our thanks to Takio Kurita and Toru Tamaki for their help with the local organization in Hiroshima. Without their assistance, the workshops at Hiroshima would not have been possible. Finally, we thank Hayato Itoh, Shun Inagaki, Fumiki Sekiya, and Ken Nobehara for their invaluable help in assembling this volume.

November 2012

Georgy L. Gimel' farb
Edwin Hancock
Atsushi Imiya
Mineichi Kudo
Arjan Kuijper
Shinichiro Omachi
Terry Windeatt
Keiji Yamada

Organization

Co-organizers

Atsushi Imiya	IMIT, Chiba University, Japan
Mineichi Kudo	Hokkaido University, Japan
Georgy L. Gimel'farb	University of Auckland, New Zealand
Keiji Yamada	NEC Corporation, Japan

Program Co-chairs

Arjan Kuijper	Fraunhofer IGD and TU Darmstadt, Germany
Edwin Hancock	University of York, UK
Shinichiro Omachi	Tohoku University, Japan
Terry Windeatt	University of Surrey, UK

Invited Speakers

George Nagy	Rensselaer Polytechnic Institute, USA
Kenichi Kanatani	Okayama University, Japan
Ales Leonardis	University of Ljubljana, Slovenija

SSPR

Program Committee

Gady Agam	Illinois Institute of Technology, USA
Juan Andrade-Cetto	CSIC-UPS, Spain
Luc Brun	ENSICAEN, France
Tiberio Caetano	NICTA, Australia
Francisco Escolano	University of Alicante, Spain
Ana Fred	IST-Technical University of Lisbon, Portugal
Francois Jacquenet	University of Saint-Etienne, France
Xiaoyi Jiang	University of Münster, Germany
Yukiko Kenmochi	LIGM, CNRS, France
Koichi Kise	Osaka Prefecture University, Japan
Walter G. Kropatsch	Vienna University of Technology, Austria
Josep Llados	Universitat Autònoma de Barcelona, Spain
Marcello Pelillo	Ca' Foscari University of Venice, Italy
José Manuel Iñesta Quereda	University de Alicante, Spain
Antonio Robles-Kelly	NICTA, Australia
Tomoya Sakai	Nagasaki University, Japan

SPR

Program Committee

Local Organizers

Kazuhiko Kawamoto	IMIT, Chiba University, Japan
Takio Kurita	Hiroshima University, Japan
Ken-ichi Maeda	Toshiba, Japan
Yoshihiko Mochizuki	Waseda University, Japan
Tomoya Sakai	Nagasaki University, Japan
Toru Tamaki	Hiroshima University, Japan
Yukihiko Yamashita	Tokyo Institute of Technology, Japan
Heitoh Zen	IMIT, Chiba University, Japan
Hayato Itoh	Chiba University, Japan
Shun Inagaki	Chiba University, Japan
Fumiki Sekiya	Chiba University, Japan
Keiko Morita	Chiba University, Japan
Ken Nobehara	Chiba University, Japan

Sponsoring Institutions

International Association of Pattern Recognition
Technical Committe 1. Statistical Pattern Recognition Techniques
Technical Committe 2. Structural and Syntactical Pattern Recognition
The Institute of Electronics, Information and Communication Engineers
Hokkaido University
Tohoku University
Hiroshima University
Institute of Media and Information Technology, Chiba University

Table of Contents

Randomized Methods and Image Analysis

Kernel Methods in Structural and Syntactical Pattern Recognition

Applications of Structural and Syntactical Pattern Recognition

Clustering

Learning

Kernel Methods in Statistical Pattern Recognition

Applications of Statistical Pattern Recognition

Applications of Structural, Syntactical, and Statistical Methods

Estimation, Learning, and Adaptation: Systems That Improve with Use

George Nagy

RPI, Troy, NY 12180, USA

Abstract. The accuracy of automated classification (labeling) of single patterns, especially printed, hand-printed, or handwritten characters, has leveled off. Further gains in accuracy require classifying sequences of patterns. Linguistic context, already widely used, relies on 1-D lexical and syntactic constraints. Style-constrained classification exploits the shape-similarity of sets of same-source (isogenous) characters of either the same or different classes. For understanding tables and forms, 2-D structural and relational constraints are necessary. Applications of pattern recognition that do not exceed the limits of human senses and cognition can benefit from green interaction wherein operator corrections are recycled to the classifier.

Keywords: Devijver, adaptive classification, style consistency, tables, green interaction.

1 Introduction

Pierre Devijver and I shared several interests – nearest neighbors, Delaunay triangulation, clustering, connected components, error estimation, and context. I have dabbled in computational geometry, computer-aided design, remote sensing, and geographical information systems. However, most of my studies – and those of my students – have been devoted to document image analysis and to one of its fundamental components, character recognition.

The fact that research on character recognition has contributed so much to pattern recognition and machine learning cannot be attributed mainly to our desire to live in a paperless world. Character recognition is a limitless field of research in SPR because of the wealth of relationships induced by messages conveyed through sequences of visually recognizable patterns characterized by multi-dimensional feature vectors to be classified into possibly hierarchical classes with minimum error or cost. One can investigate and model the statistical distributions of individual features, of all the features of a single sample, and the relationships between the features of multiple patterns and class variables. Any of the above patterns may consist of a single stroke, a single letter or numeral, part of a word, or a whole word, page, or document. The various models used or proposed in character recognition can be represented concisely by Bayesian networks [1].

G.L. Gimel' farb et al. (Eds.): SSPR & SPR 2012, LNCS 7626, pp. 1–10, 2012.
© Springer-Verlag Berlin Heidelberg 2012

If this paper fails to conform to some rules, I claim the Senior Citizens' Exemption. First I shall reminisce about Pierre Devijver and his technical legacy. Then I shall ruminate about some of my own hobby horses, including a few whose connection with statistical pattern recognition may be less than obvious. Table processing, for example, may fit better under "syntactic and structural." The conclusions mention some developments that I did not anticipate.

The reader will find no formulas or experimental results herein. Since all of what I recount has already been published – some of it more than once – I keep to a bird's eye view. Details can be found in the references cited, or in the references cited in those references.

2   Pierre Devijver

Pierre Devijver was a man of many accomplishments, and I am honored to have a chance to talk about my favorite topic, statistical pattern recognition, under his aegis. I had occasion to meet Pierre at several conferences where our similar technical interests and a predilection for lunch-time walks promoted conversation. (If I had known that he was a marathon runner I would have worried about those long walks with him.)

I missed the 1973 ICPR in Washington DC where Pierre first introduced his ideas about the relationship of the Bayes Risk to the Mean Square Error [2]. He published an article on error bounds the following year in the *IEEE Transactions on Computers* (where the best papers on pattern recognition were published before *PAMI* came along) [3]. So even before I ever met him I had studied some of his work. Later, when I taught pattern recognition and document image processing at RPI, I benefited a great deal from his and Joseph Kittler's rigorous text [4], which includes some of their results on the bias and variance of k-NN based error estimates. I still return to their lucid presentation of probabilistic distance measures.

I know that Pierre and I interacted at the 1980 Pattern Recognition in Practice workshop organized by Edzard Gelsema† and Laveen Kanal in Amsterdam. We both presented papers at the ICPR earlier that year in Miami Beach, but I don't recall any conversation between us. I also cannot remember any specific discussion at the 1984 ICPR in Montreal. By then Pierre was vice-president of IAPR, so he must have spent any time left after the technical sessions at committee meetings. In spite of his conscientious work for IAPR, the following examples show that he found time for remarkable technical contributions.

2.1   Connected Components

For many years the Devijver-Ronse monograph on Connected Component Detection was the only book on the subject. It expounded efficient disk access while tracing a CC [5]. Pierre foresaw that CC detection would be a cornerstone of document image analysis even though at the time only a small swath of a scanned page would fit into primary memory.

2.2   Markov Algorithms

In 1966, working under first-order Markov Chain assumptions, Joseph Raviv devised an iterative algorithm to convert the information from feature vectors of preceding patterns into the prior probability for the current pattern by using bigram and trigram class-transition probability tables. At the 1984 ICPR, Pierre extended this to take into account the information from any number of *succeeding* patterns by adding an iterative backward pass [6,]. He subsequently improved the numerical stability of Baum's HMM training algorithm by computing joint rather than conditional probabilities [7].

Already then his interest lay in Markov meshes and MRFs for image processing, so he avoided any assumptions that applied only to 1-D. I wish that Pierre had completed his studies of MRFs a few years earlier, when my UNL students and I were struggling to set appropriate constraints on causal generation of 2-D Markov fields. Pierre's deep insights would have been invaluable.

2.3   Nearest Neighbors

As Godfried Toussaint repeatedly demonstrated, statistical pattern recognition and computational geometry mix well. The decision boundaries of a nearest neighbor classifier are a subset of the edges of its Voronoi diagram in feature space. The Voronoi diagram, in turn, can be computed rapidly from its dual Delaunay triangulation. One way to speed up nearest-neighbor classification is by removing from the reference set all the patterns with same-class Voronoi Neighbors. However, removing a pattern changes both the Voronoi diagram and the underlying Delaunay triangulation. Pierre devised fast algorithms for *dynamic* Delaunay triangulation in high-dimensions [8]. He also derived bounds on the fraction of the training set that can be harmlessly edited out.

Pierre's research included methods of estimating the error rate on the test set from results on the training set (still a hot topic), feature extraction, the relationship between clustering and mixture identification, and applications ranging from tumor detection to astronomy. This sketch of his contributions is far too superficial to do justice to his pervasive and persuasive ideas.

3   My Own Trail

Before I get into even a modicum of technicalities, I wish to acknowledge what good fortune I had in my collaborators and co-authors, and how much I learned from my students (some regrettably already retired). One of the best things about the field of pattern recognition is that it has attracted such bright, generous, and convivial scholars.

3.1   Feature Extraction

My first graduate student, at the Université de Montréal, was Kamal Abdali. I set him to solve the optimal feature extraction problem because nobody knew yet about

NP-completeness. My last student, Xiaoli Zhang, confirmed my belief that the class-conditional statistical dependence structure (e.g., the covariance matrix) of features depends far more on the chosen feature set than on the data itself [9]. NP-complete it may be, but there is more to be done.

3.2   Unsupervised Classification

There is no such thing as unsupervised learning. Children learn without being explicitly taught, but only because they emulate the behavior of respected teachers (grown-up or other children). They often get to surpass the abilities of those whom they imitate. In 1965 we programmed a classifier to trust the labels assigned to scanned printed characters by an off-the-shelf journeyman classifier, and to use them for its own training set. On the data that it was retrained on, this apprentice classifier turned out to be better than the original classifier, and so we used it as a role model for still another classifier. To our surprise and delight, the error rate kept dropping during several iterations and then flattened out [10,11]. Almost three decades later, these results impressed Henry Baird, so we tried "mean adaptation" with his own features and his then humungous 100-font dataset [12]. Henry concluded that the expected gain is considerable, while the downside risk was small.

Ho and Agrawala had pointed out earlier that we were lucky because the many datasets on which we had experimented all fell under restrictive constraints [13,14]. With features crafted by Hiromichi Fujisawa and Cheng-Lin Liu of the Hitachi Central Research Laboratory, Harsha Veeramachaneni used Expectation Maximization to re-estimate both the means and the class-covariance matrices using classifier-assigned labels. This turned out even luckier (at least on NIST hand-printed digits) than just mean adaptation [15].

We have not yet found necessary and sufficient conditions that would guarantee that adaptation will reduce the error on a set of same-source samples. Is there a principled way to predict the results of adaptation? A place to start might be Castelli's and Cover's insights on the relative values of labeled and unlabeled samples.

3.3   Prototype Based Text-Image Compression

Any clustering method can be viewed as data compression with each cluster prototype serving as surrogate for all the patterns in that cluster. In 1970 or thereabouts, Pete Welch, my boss's boss at IBM Research, suggested that we apply the bitmap clustering methods that we had developed earlier for Chinese character recognition [16] to image data compression. It worked like a charm! We could not patent it because of a Government anti-monopoly suit, so IBM waited four years before letting us publish it [17].

Our method eventually resurfaced in DjVu and JBIG, but with a critical improvement that we had missed. We transmitted only run-length coded prototypes, or their (compressed) id and position. Subsequent researchers encoded the difference between the prototype and the actual glyph, thereby rendering the scheme lossless. Current methods are nearing the theoretical limits.

3.4   Decision Trees

Although I worked at IBM on several OCR projects, including the three million dollar reader for the Social Security Administration, the only algorithm that made it into a product was a probabilistic decision tree for isolated bitmapped characters [18]. Dick Casey developed most of the theory, and I programmed it up in APL during a summer at the IBM San Jose Research Center. After everything was reprogrammed efficiently in Japan it became the IBM TextReader. I still have copies of the shrink-wrapped floppies.

3.5   Language Context

Both children and adults expand their vocabulary by guessing and refining the meaning of unknown words or phrases according to what makes sense time after time. If in a foreign land most street signs end in a particular string, it is likely to mean "street" or "avenue". Early proponents of the use of language context in pattern recognition include Allen Hanson, Ed Riseman†, Joe Raviv†, Godfried Toussaint, and Ching Suen.

Meanings can be assigned to unknown alphabetic glyphs so that they form words that are part of the language. Substitution ciphers have been solved this way since at least the days of the Roman Empire. I have participated in three initiatives to automate this process and apply it to scanned text.

In a first attempt, Dick Casey and I clustered bitmaps of scanned single-case English text in one of four different typefaces. We solved the resulting cryptograms by matching the frequencies of cluster numbers and bigrams of cluster numbers to the known letter unigram and bigram frequencies of the English language [19]. We were very pleased when Scientific American asked us to describe our methods in laymen's terms [20]. Twenty years later at the University of Nebraska, we improved the scheme by recursive matching of trial assignments against a lexicon of a few hundred words instead of letter n-grams [21]. In another ten years, Tin Ho at Bell Labs used a larger lexicon and improved the matching scheme. We demonstrated "OCR with no shape training" on Spitz glyphs at the Barcelona ICPR [22].

3.6   Style

At the 1992 ICPR I proposed exploiting the family resemblance of same-font letters and numerals for recognizing individually ambiguous characters when they appeared in their usual company [23]. I called this notion *spatial context*. During his doctoral research Prateek Sarkar dubbed the distinction between feature distributions originating from patterns from isogenous typeface, printer, writer or speaker, and distributions from patterns from heterogeneous sources, as *style*. Harsha Veeramachaneni explained style as follows: "the way Alice writes 1 helps predict the way she will write 7." Applying these concepts to fields of same-source patterns may not be so difficult, but defining them formally requires a lot of notation and subscripts.

A critical property of style context defined by Harsha is *order independence*, known in probability theory as *exchangeability* [24]. Pure style implies that the probability of any pattern field given the field class is equal to the probability of any permutation of that pattern field given the field class subjected to the same permutation. Order independence vitiates most types of language context, but style and multi-pattern language context can still be combined.

Another useful distinction can be made between *intra-class style* and *inter-class style*. In intra-class style, an "e" in a field of patterns to be recognized is always in the same font, or was written by the same person. One might, however, find this "e" next to a "c" of a different font or by a different writer. So there is some statistical dependence between the feature distributions of all the patterns of the same class, but no way to tell anything about a pattern of class E from a pattern of class C. The inter-class constraint is more rigid: samples from all the classes must be isogenous. Therefore the features of patterns even of different classes exhibit observable statistical dependence. Most of the adaptive classifiers discussed above require only intra-class style. The style classifiers described below make use of inter-class style.

Prateek Sarkar derived algorithms for optimal classification of style-consistent fields of arbitrary length [25]. He posited that the features of each pattern, while dependent on the features of other patterns in the field because of the same-style constraint, were independent of the *classes* of the other patterns. In other words, every "e" in the field looks the same, regardless of whether the field spells "element" or "dependent". He formulated several ways of combining Gaussians as mixture distributions to model the class-and-style-conditional probabilities via weighting factors that depend on both class and style. In terms of hand print, his method can classify fields of never-before-seen hybrid Ann-Jen script after training only on separate fields of Ann's, Bonnie's, Dave's and Jen's writing.

Because the computation of the optimal maximum likelihood assignment requires lengthy sum-of-products-of-sums computations, Prateek devised a *top-label* approximation equivalent to selecting from a set of style-specific feature classifiers the one that yields the highest field-feature likelihood. He trained his classifiers with a mixture of isogenous (*isofont*) fields, and tested them on isogenous fields of lengths different from those of the training set.

Harsha Veeramachaneni considered a continuous distribution of Gaussians instead of mixtures of a predetermined fixed number. His insight was that the posterior distribution of a field of any length can be determined from the cross-covariance matrices of only *pairs* of same-source pattern feature vectors. This led to quadratic field classification with computation proportional rather than exponential with field length [26].

From the perspective of style-constrained field classification, the field of an adaptive classifier encompasses the entire set of isogenous data rather than a fixed number of patterns. This observation may explain why some adaptive classifiers exploit only intra-class consistency. On short fields, on the other hand, more powerful and more computation-intensive classifiers can take full benefit of inter-class style consistency.

In practical OCR applications, style-constrained classification aims at scenarios similar to font or writer recognition. Both of these are effective tools for decreasing the error rate by substituting a single-font or single-writer classifier for a more error-prone omnifont or omni-writer classifier. In theory, however, style classifiers should achieve a lower error rate because they do not "waste" any statistical information on font or writer identification. In some applications, however, it is desirable to identify font or writer in addition to producing a transcription. We have also pursued, with mixed, success, variations of style classification based on nearest-neighbors and support-vector machines.

4 Tables

We began our studies of tables twenty years ago with foreign language tables that gave us a chance to see how much information can be derived from table structure without lexical help. Since then mainstream table recognition has progressed from scanned paper tables to computer-generated HTML and PDF tables. All of this work has been part of a long-standing and most enjoyable collaboration with Dave Embley (BYU), Sharad Seth (UNL), Moorthy Krishnamoorthy (RPI) and Dan Lopresti (Lehigh), often under the aegis of TANGO [27]. We have written far too many surveys and reports, especially considering how often our views have shifted, so rather than reciting progressive steps I just list some articles of faith (for which I take sole responsibility and which I may retract next year).

- The underlying grid of a table reveals a 2-D indexing scheme. This geometric indexing is interwoven with possibly higher-dimensional, logical "Wang" categories which can be interpreted as geometric indexing in a higher-dimensional space.
- The essential task of table analysis is to establish the relationship of column and row headers to individual data cells. This is trickier than might first appear because of the possible occurrence of hierarchical headers, spanning cells and headers in the row stub, and because the appearance of a table depends on the rendering program as well as the file containing the table Additional tasks require extracting metadata (table caption, title, footnote references, footnotes, aggregates, units, ...).
- Tables are distinguished from *forms* because tables are meant to disseminate information rather than collect it. The distinction is often obvious, but a filled-out spreadsheet might be either a table or a form. In most forms individual field captions take the place of 2-D indexing. Their structure can be represented by graphs.
- Tables are distinguished from *lists* by 2-D indexing of data cells by row and column headers.. Even ordered lists like telephone books require a search to locate a cell. The table vs. list question arises only when nested lists of uniform length are laid out on a grid, or when table ill-formed table headers preclude unique indexing.
- Tables prepared for human readers are different from relational tables. The designer of a relational table must determine what is an attribute and what is a key, and orient the table accordingly, with attributes on top. In contrast, the orientation of tables prepared for hardcopy or web publication is usually determined by matching the number of row and column headers to the page or display format. Therefore in

tables of Canadian statistics the column headers are often provinces, while in tables of US statistics the states are usually row headers. Visual tables are essentially symmetric with respect to rows and columns, but relational tables are not. This does not preclude the transformation of visual tables into relational tables.

5   Green Interaction

CAVIAR (*Computer Assisted Visual InterActive Recognition*) for flowers is an attempt at efficient human-computer interaction [28]. When a flower image is presented, the program extracts visually verifiable features (like the shape or number of petals), and classifies the flower into one of a hundred or so classes. If the user is unsatisfied, she or he can edit the features and reject or approve the classification according to the resemblance of the flower to reference samples. The classifier, in turn, adapts its parameters using what it learns from the user. Both the computer and the user improve with time. A most enjoyable part of this project was collecting over 600 wildflower samples.

In OCR, a perennial problem is obtaining large-enough *labeled* training sets. Studies have shown that classification on the test set improves even after tens or hundreds of thousands of training samples. The output of an OCR system, especially in error intolerant applications like medical or financial form entry, is often routed to operators for verification or correction. The best possible training set is the stream of data encountered during actual operation. Therefore all final corrected labels should be associated with the scanned patterns and routed back periodically to retrain the classifier.

Green interaction means that expensive and time-consuming human effort devoted to approving or correcting the output of any pattern recognition system should not be wasted. More on this at ICPR 2012.

6   Conclusion

We miss Pierre Devijver. We are fortunate that he left behind so much to think about.

Claims made ever since Pierre and I were starting out, to the effect that OCR was essentially a solved problem, turned out to be uninformed. I had the good luck to work on a variety of related problems with perspicacious colleagues and students. The study of each problem revealed new problems begging for solution. Writing surveys has shown me many more. It is like a Garden of Eden where the quandary is which fruit to taste first. Fortunately there are still some sweet low-hanging fruit left.

Progress in some areas surprises me. In 1968, when we worked on hand-printed numeral recognition, I was sure that writer-independent cursive script recognition was a pipe dream. Speech and face recognition also work better than I expected. I underestimated the scope and power of Expectation Maximization (but eventually made a strenuous effort to understand it more thoroughly). While I did predict in a 1983 scanner survey that camera-based systems would be right along, I never dreamed that so much image processing and recognition software would be crammed into a

KitKat-sized camera cell phone. I was skeptical that Wikipedia could become a useful and tolerably reliable source of information about pattern recognition. I hope that there are many more equally pleasant surprises coming down the pike!

One of my most agreeable duties from 1967 till 2007 was teaching a graduate course in pattern recognition. I always offered students completing the course with at least a B a lifetime guarantee to make myself available for any technical question that they might want to discuss. Some have taken me up on it. Now that I am retired, I have gone back to being a full time student. In addition to the occasional déjà vu, I look forward to learning much new material during SSSPR and ICPR 2012.

References

1. Veeramachaneni, S., Sarkar, P., Nagy, G.: Modeling Context as Statistical Dependence. In: Dey, A.K., Kokinov, B., Leake, D.B., Turner, R. (eds.) CONTEXT 2005. LNCS (LNAI), vol. 3554, pp. 515–528. Springer, Heidelberg (2005)
2. Devijver, P.A.: Relationship between statistical risk and the least-mean-square error criterion in pattern recognition. In: Proceedings of the First International Joint Conference on Pattern Recognition, Washington, DC (1973)
3. Devijver, P.A.: On a new class of bounds on Bayes risk in multihypothesis pattern recognition. IEEE Trans. on Computers C-23(1), 70–80 (1974)
4. Devijver, P.A., Kittler, J.: Pattern recognition: A statistical approach. Prentice/Hall (1982)
5. Devijver, P.A., Ronse, C.: Connected Components in Binary Images: The Detection Problem. John Wiley & Sons, Inc., New York (1984)
6. Devijver, P.A.: Classification in Markov Chains for minimum symbol error rate. In: Proceedings of International Conference on Pattern Recognition, pp. 1334–1336 (1984)
7. Devijver, P.A.: Baum's forward-backward algorithm revisited. Pattern Recognition Letters 3(6), 369–373 (1985)
8. Devijver, P.A., Dekesel, M.: Computing multidimensional Delaunay tessellations. Pattern Recognition Letters 4(5-6), 311–316
9. Nagy, G., Zhang, X.: Simple statistics for complex features spaces. In: Basu, M., Ho, T.K. (eds.) Data Complexity in Pattern Recognition, pp. 173–195. Springer (2006)
10. Nagy, G., Shelton, G.L.: Self-Corrective Character Recognition System. IEEE Trans. Information Theory 12(2), 215–222 (1966)
11. Nagy, G.: The Application of Nonsupervised Learning to Character Recognition. In: Kanal, L. (ed.) Pattern Recognition, pp. 391–398. Thompson Book Company, Washington (1968)
12. Baird, H.S., Nagy, G.: A Self-correcting 100-font Classifier. In: Proceedings of SPIE Conference on Document Recognition, San Jose, CA. SPIE, vol. 2181, pp. 106–115 (February 1994)
13. Ho, Y.C., Agrawala, A.K.: On the self-learning scheme of Nagy and Shelton. Proceedings of the IEEE 55, 1764–1765 (1967)
14. Nagy, G., Tuong, N.G.: On a Theoretical Pattern Recognition Model of Ho and Agrawala. Proceedings of the IEEE 56(6), 1108–1109 (1968)
15. Veeramachaneni, S., Nagy, G.: Classifier Adaptation with Non-representative Training Data. In: Lopresti, D.P., Hu, J., Kashi, R.S. (eds.) DAS 2002. LNCS, vol. 2423, pp. 123–133. Springer, Heidelberg (2002)

16. Casey, R.G., Nagy, G.: Recognition of Printed Chinese Characters. IEEE Trans. Electronic Computers 15(1), 91–101 (1966)
17. Ascher, R.N., Nagy, G.: A Means for Achieving a High Degree of Compaction on Scan-Digitized Printed Text. IEEE Trans. Computers 23(11), 1174–1179 (1974)
18. Casey, R.G., Nagy, G.: Decision Tree Design Using a Probabilistic Model. IEEE Trans. Information Theory 30(1), 93–99 (1984)
19. Casey, R.G., Nagy, G.: Autonomous Reading Machine. IEEE Trans. Computers 17(5), 492–503 (1968)
20. Casey, R.G., Nagy, G.: Advances in Pattern Recognition. Scientific American 224(4), 56–71 (1971)
21. Nagy, G., Seth, S., Einspahr, K.: Decoding Substitution Ciphers by means of Word Matching with Application to OCR. IEEE Trans. Pattern Analysis and Machine Intelligence 9(5), 710–715 (1987)
22. Ho, T.K., Nagy, G.: OCR with no shape training. In: Proceedings of International Conference on Pattern Recognition-XV, Barcelona, Spain, vol. 4, pp. 27–30 (September 2000)
23. Nagy, G.: Teaching a computer to read. In: Proceedings of the Eleventh International Conference on Pattern Recognition, vol. 2, pp. 225–229, The Hague (1992)
24. Veeramachaneni, S., Nagy, G.: Analytical results on style-constrained Bayesian classification of pattern fields. IEEE Trans. Pattern Analysis and Machine Intelligence 29(7), 1280–1285 (2007)
25. Sarkar, P., Nagy, G.: Style consistent classification of isogenous patterns. IEEE Trans. Pattern Analysis and Machine Intelligence 27(1), 88–98 (2005)
26. Veeramachaneni, S., Nagy, G.: Style context with second order statistics. IEEE Trans. Pattern Analysis and Machine Intelligence 27(1), 14–22 (2005)
27. Tijerino, Y.A., Embley, D.W., Lonsdale, D. W., Nagy, G.: Towards ontology generation from tables. World Wide Web Journal 6(3) (2005)
28. Zou, J., Nagy, G.: Visible models for interactive pattern recognition. Pattern Recognition Letters 28, 2335–2342 (2007)

Optimization Techniques
for Geometric Estimation: Beyond Minimization

Kenichi Kanatani

Department of Computer Science, Okayama University, Okayama 700-8530, Japan
kanatani@suri.cs.okayama-u.ac.jp

Abstract. We overview techniques for optimal geometric estimation from noisy observations for computer vision applications. We first describe estimation techniques based on minimization of given cost functions: least squares (LS), maximum likelihood (ML), which includes reprojection error minimization (Gold Standard) as a special case, and Sampson error minimization. We then formulate estimation techniques not based on minimization of any cost function: iterative reweight, renormalization, and hyper-renormalization. Showing numerical examples, we conclude that hyper-renormalization is robust to noise and currently is the best method.

1 Introduction

One of the most important tasks of computer vision is to compute the 2-D and 3-D shapes of objects exploiting *geometric constraints*, by which we mean properties that can be described by relatively simple equations such as the objects being lines or planes, their being parallel or orthogonal, and the camera imaging geometry being perspective projection. We call the inference based on such geometric constraints *geometric estimation*. In the presence of noise, however, the assumed constraints do not exactly hold. To do geometric estimation "optimally" in the presence of noise, a lot of efforts have been made since 1980s by many researchers. This paper summarizes that history and reports the latest results.

2 Preliminaries

2.1 Definition of Geometric Estimation

The geometric estimation problem we consider here is defined as follows. We observe some quantity x (a vector), which is assumed to satisfy in the absence of noise an equation

$$F(x; \theta) = 0, \tag{1}$$

parameterized by unknown vector θ. This equation is called the *geometric constraint*. Our task is to estimate the parameter θ from noisy instances x_α, $\alpha = 1$, ..., N, of x. Many computer vision problems are formulated in this way, and we

G.L. Gimel' farb et al. (Eds.): SSPR & SPR 2012, LNCS 7626, pp. 11–30, 2012.

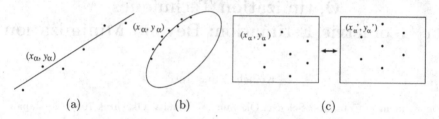

Fig. 1. (a) Line fitting. (b) Ellipse fitting. (c) Fundamental matrix computation.

can compute from the estimated $\boldsymbol{\theta}$ the positions, the shapes, and the motions of the objects we are viewing. In many problems, we can reparameterize the problem so that the constraint is liner in the parameter $\boldsymbol{\theta}$ (but generally nonlinear in the data $\boldsymbol{x}$). Then, Eq. (1) has the form

$$(\boldsymbol{\xi}(\boldsymbol{x}), \boldsymbol{\theta}) = 0, \tag{2}$$

where $\boldsymbol{\xi}(\boldsymbol{x})$ is a vector-valued nonlinear function of $\boldsymbol{x}$. In this paper, we denote the inner product of vectors $\boldsymbol{a}$ and $\boldsymbol{b}$ by $(\boldsymbol{a}, \boldsymbol{b})$. Equation (2) implies that the scale of $\boldsymbol{\theta}$ is indeterminate, so we hereafter normalize $\boldsymbol{\theta}$ to unit norm: $\|\boldsymbol{\theta}\| = 1$.

Example 1. (Line fitting) To a given point sequence (x_α, y_α), $\alpha = 1, ..., N$, we fit a line

$$Ax + By + C = 0. \tag{3}$$

(Fig. 1(a).) If we define

$$\boldsymbol{\xi}(x, y) \equiv (x, y, 1)^\top, \qquad \boldsymbol{\theta} \equiv (A, B, C)^\top, \tag{4}$$

the line equation is written as

$$(\boldsymbol{\xi}(x, y), \boldsymbol{\theta}) = 0. \tag{5}$$

Example 2. (Ellipse fitting) To a given point sequence (x_α, y_α), $\alpha = 1, ..., N$, we fit an ellipse

$$Ax^2 + 2Bxy + Cy^2 + 2(Dx + Ey) + F = 0. \tag{6}$$

(Fig. 1(b).) If we define

$$\boldsymbol{\xi}(x, y) \equiv (x^2, 2xy, y^2, 2x, 2y, 1)^\top, \qquad \boldsymbol{\theta} \equiv (A, B, C, D, E, F)^\top, \tag{7}$$

the ellipse equation is written as

$$(\boldsymbol{\xi}(x, y), \boldsymbol{\theta}) = 0. \tag{8}$$

Example 3. (Fundamental matrix computation) Corresponding points (x, y) and (x', y') in two images of the same 3-D scene taken from different positions satisfy the *epipolar equation* [8]

$$\left(\begin{pmatrix} x \\ y \\ 1 \end{pmatrix}, F \begin{pmatrix} x' \\ y' \\ 1 \end{pmatrix} \right) = 0, \tag{9}$$

where F is called the *fundamental matrix*, from which we can compute the camera positions and the 3-D structure of the scene [8] (Fig. 1(c)). If we define

$$\xi(x, y, x', y') \equiv (xx', xy', x, yx', yy', y, x', y', 1)^\top, \tag{10}$$

$$\theta \equiv (F_{11}, F_{12}, F_{13}, F_{21}, F_{22}, F_{23}, F_{31}, F_{32}, F_{33})^\top, \tag{11}$$

the epipolar equation is written as

$$(\xi(x, y, x', y'), \theta) = 0. \tag{12}$$

2.2 Modeling of Noise

In the context of image analysis, "noise" means *uncertainty of image processing operations*, rather than random fluctuations over time or space as commonly understood in physics and communications. It reflects the fact that standard image processing operations such as feature extraction and edge detection are not perfect and do not necessarily output exactly what we are looking for. We model this uncertainty in statistical terms: the observed value x_α is regarded as a perturbation from its true value $\bar{x}_\alpha$ by an independent random Gaussian variable Δx_α of mean $\mathbf{0}$ and covariance matrix $V[x_\alpha]$. Furthermore, $V[x_\alpha]$ is assumed to be known *up to scale*. Namely, we write it as

$$V[x_\alpha] = \sigma^2 V_0[x_\alpha] \tag{13}$$

for some unknown constant σ, which we call the *noise level*. The matrix $V_0[x_\alpha]$, which we call the *normalized covariance matrix*, describes the orientation dependence of uncertainty in relative terms and is assumed to be known. The separation of $V[x_\alpha]$ into σ^2 and $V_0[x_\alpha]$ is merely a matter of convenience; there is no fixed rule. This convention is motivated by the fact that optimal estimation can be done, as shown shortly, only from the knowledge of $V_0[x_\alpha]$.

If the observation x_α is regarded as a random variable, its nonlinear mapping $\xi(x_\alpha)$, which we write ξ_α for short, is also a random variable. Its covariance matrix $V[\xi_\alpha] = \sigma^2 V_0[\xi_\alpha]$ is evaluated to a first approximation in terms of the Jacobi matrix $\partial\xi/\partial x$ of the mapping $\xi(x)$ as follows:

$$V_0[\xi_\alpha] = \left.\frac{\partial\xi}{\partial x}\right|_{x=\bar{x}_\alpha} V_0[x_\alpha] \left.\frac{\partial\xi}{\partial x}^\top\right|_{x=\bar{x}_\alpha}. \tag{14}$$

This expression contains the true value $\bar{x}_\alpha$, which is replaced in actual computation by the observation x_α. It has been confirmed by experiments that this replacement does not practically affect the final result. It has also been confirmed that upgrading the first approximation to higher orders does not have any practical effect.

2.3   Geometric Models for Geometric Estimation

One of the most prominent distinctions of the geometric estimation from the traditional statistical estimation is that the starting equation, Eq. (1) (or Eq. (2)), which we call the *geometric model*, only specifies the necessary constraint and does not explain the mechanism as to how the data x_α are generated. Hence, we cannot express x_α in terms of the parameter θ as an explicit function.

Another big difference is that while the traditional statistical estimation is based on *repeated* observations regarded as sampled from the statistical model (= probability density), and hence accuracy vs. the number N of observations in the asymptotic limit $N \to \infty$ is a major concern, geometric estimation is done from *one* set of data $\{x_1, ..., x_N\}$. Naturally, the estimation accuracy increases with less observation uncertainty. Hence, accuracy vs. the noise level σ in the limit of $\sigma \to 0$ is a major concern [14].

In computer vision applications, the asymptotic analysis of $N \to \infty$ does not have much sense, because the number of data obtained by image processing operations is limited in number. Usually, the output of an image processing operation is accompanied by its reliability index, and we select only those data that have high reliability indices. If we want to increase the number of data, we necessarily need to include those with low reliability, but they are often misdetections. Despite the basic differences, however, two approaches exist in both statistical and geometric estimation domains:

Minimization Approach. We choose the value θ that minimizes a specified cost function. This is regarded as the standard for computer vision applications.

Non-minimization Approach. We compute the value θ by solving a set of equations, called *estimating equations* [6], which need not be derivatives of some function. Hence, the solution does not necessarily minimize any cost function. In traditional statistical estimation domains, this approach is regarded as more general and more flexible with a possibility of yielding better solutions than the minimization approach, but it is not widely recognized in computer vision research.

2.4   KCR Lower Bound

For minimization or non-minimization approaches, there exists a theoretical accuracy limit. We assume that the true values $\bar{\xi}_\alpha$ of the observations ξ_α satisfy the constraint $(\bar{\xi}_\alpha, \theta) = 0$ for some θ. If it is estimated from the observation $\{\xi_\alpha\}_{\alpha=1}^N$ by some means, the estimate $\hat{\theta}$ is as a function $\hat{\theta}(\{\xi_\alpha\}_{\alpha=1}^N)$ of $\{\xi_\alpha\}_{\alpha=1}^N$, called an *estimator* of θ. Let $\Delta\theta$ be its error, i.e., write $\hat{\theta} = \theta + \Delta\theta$, and define the covariance matrix of $\hat{\theta}$ by

$$V[\hat{\theta}] = E[\Delta\theta\Delta\theta^\top], \tag{15}$$

where $E[\cdot]$ denotes expectation over data uncertainty. If we can assume that

- each $\boldsymbol{\xi}_\alpha$ is perturbed from its true value $\bar{\boldsymbol{\xi}}_\alpha$ by independent Gaussian noise
  of mean $\mathbf{0}$ and covariance matrix $V[\boldsymbol{\xi}_\alpha] = \sigma^2 V_0[\boldsymbol{\xi}_\alpha]$, and
- the function $\hat{\boldsymbol{\theta}}(\{\boldsymbol{\xi}_\alpha\}_{\alpha=1}^N)$ is an *unbiased estimator*, i.e., $E[\hat{\boldsymbol{\theta}}] = \boldsymbol{\theta}$ identically
  holds for whatever $\boldsymbol{\theta}$,

then the following inequality holds [3,11,12,14].

$$V[\hat{\boldsymbol{\theta}}] \succ \frac{\sigma^2}{N}\Big(\frac{1}{N}\sum_{\alpha=1}^{N}\frac{\bar{\boldsymbol{\xi}}_\alpha\bar{\boldsymbol{\xi}}_\alpha^\top}{(\boldsymbol{\theta}, V_0[\boldsymbol{\xi}_\alpha]\boldsymbol{\theta})}\Big)^{-}. \tag{16}$$

Here, $\boldsymbol{A} \succ \boldsymbol{B}$ means that $\boldsymbol{A} - \boldsymbol{B}$ is a positive semidefinite symmetric matrix, and $(\cdot)^{-}$ denotes the pseudo inverse. Chernov and Lesort [3] called the right side Eq. (16) *Kanatani-Cramer-Rao* (KCR) *lower bound*.

3 Minimization Approach

First, we overview popular geometric estimation techniques for computer vision that are based on the minimization approach.

3.1 Least Squares (LS)

Since the true values $\bar{\boldsymbol{\xi}}_\alpha$ of the observations $\boldsymbol{\xi}_\alpha$ satisfy $(\bar{\boldsymbol{\xi}}_\alpha, \boldsymbol{\theta}) = 0$, we choose the value $\boldsymbol{\theta}$ that minimizes

$$J = \frac{1}{N}\sum_{\alpha=1}^{N}(\boldsymbol{\xi}_\alpha, \boldsymbol{\theta})^2 \tag{17}$$

for noisy observations $\boldsymbol{\xi}_\alpha$ subject to the constraint $\|\boldsymbol{\theta}\| = 1$. This can also be viewed as minimizing $\sum_{\alpha=1}^{N}(\boldsymbol{\xi}_\alpha, \boldsymbol{\theta})^2/\|\boldsymbol{\theta}\|^2$. Equation (17) can be rewritten in the form

$$J = \frac{1}{N}\sum_{\alpha=1}^{N}(\boldsymbol{\xi}_\alpha, \boldsymbol{\theta})^2 = \frac{1}{N}\sum_{\alpha=1}^{N}\boldsymbol{\theta}^\top\boldsymbol{\xi}_\alpha\boldsymbol{\xi}_\alpha^\top\boldsymbol{\theta} = (\boldsymbol{\theta}, \underbrace{\frac{1}{N}\sum_{\alpha=1}^{N}\boldsymbol{\xi}_\alpha\boldsymbol{\xi}_\alpha^\top}_{\equiv \boldsymbol{M}}\boldsymbol{\theta}) = (\boldsymbol{\theta}, \boldsymbol{M}\boldsymbol{\theta}), \tag{18}$$

which is a quadratic form of $\boldsymbol{M}$. As is well known, the unit vector $\boldsymbol{\theta}$ that minimizes this form is given by the unit eigenvector of $\boldsymbol{M}$ for the smallest eigenvalue.

Since the sum of squares is minimized, this method is called *least squares* (*LS*). Equation (17) is often called the *algebraic distance*, so this method is also called *algebraic distance minimization*. Because the solution is directly obtained without any search, LS is widely used in many applications. However, it has been observed that the solution has a large statistical bias. For ellipse fitting in *Example 2* (Sec. 2.1), for instance, the fitted ellipse is almost always smaller than the true shape. For this reason, LS is not suited for accurate estimation. However, LS is convenient for rough estimation for guiding image processing, for the outlier-detection voting, and for initializing iterative optimization schemes.

3.2   Maximum Likelihood (ML)

If the noise in each $\boldsymbol{x}_\alpha$ is an independent Gaussian variable of mean $\boldsymbol{0}$ and co-variance matrix $V[\boldsymbol{x}_\alpha] = \sigma^2 V_0[\boldsymbol{x}_\alpha]$, the *Mahalanobis distance* of the observations $\{\boldsymbol{x}_\alpha\}$ from their true values $\{\bar{\boldsymbol{x}}_\alpha\}$ is

$$J = \frac{1}{N} \sum_{\alpha=1}^{N} (\boldsymbol{x}_\alpha - \bar{\boldsymbol{x}}_\alpha, V_0[\boldsymbol{x}_\alpha]^{-1}(\boldsymbol{x}_\alpha - \bar{\boldsymbol{x}}_\alpha)), \tag{19}$$

and the likelihood of $\{\boldsymbol{x}_\alpha\}$ is written as $Ce^{-NJ/2\sigma^2}$, where C is a normalization constant that does not depend on $\bar{\boldsymbol{x}}_\alpha$ or $\boldsymbol{\theta}$. Thus, *maximum likelihood (ML)* is equivalent to minimizing Eq. (19) subject to the constraint

$$(\boldsymbol{\xi}(\bar{\boldsymbol{x}}_\alpha), \boldsymbol{\theta}) = 0. \tag{20}$$

As a special case, if the noise is homogeneous, i.e., independent of α, and isotropic, i.e., independent of orientation, we can write $V_0[\boldsymbol{x}_\alpha] = \boldsymbol{I}$ (the identity), which reduces Eq. (19) to the *geometric distance*

$$J = \frac{1}{N} \sum_{\alpha=1}^{N} \|\boldsymbol{x}_\alpha - \bar{\boldsymbol{x}}_\alpha\|^2. \tag{21}$$

Minimizing this subject to Eq. (20) is called *geometric distance minimization* by computer vision researchers and *total least squares (TLS)* by numerical analysis researchers[1]. If $\bar{\boldsymbol{x}}_\alpha$ represents the projection of the assumed 3-D structure onto the image plane and $\boldsymbol{x}_\alpha$ is its actually observed positions, Eq. (21) is called the *reprojection error*. Minimizing it subject to Eq. (20) is often called *reprojection error minimization*.

Geometrically, ML can be interpreted to be fitting to N points $\boldsymbol{x}_\alpha$ in the data space the parameterized hypersurface $(\boldsymbol{\xi}(\boldsymbol{x}), \boldsymbol{\theta}) = 0$ by adjusting $\boldsymbol{\theta}$, where the discrepancy of the points from the surface is measured not by the Euclid distance but by the Mahalanobis distance of Eq. (19), which inversely weights the data by their covariances, thereby imposing heavier penalties on the points with higher certainty. In the field of computer vision, this approach is widely regarded as the ultimate method and often called the *Gold Standard* [8]. However, this is a highly nonlinear optimization problem and difficult to solve by a direct means. The difficulty stems from the fact that Eq. (20) is an implicit function of $\bar{\boldsymbol{x}}_\alpha$. If we could solve Eq. (20) for $\bar{\boldsymbol{x}}_\alpha$ to express it as an explicit function of $\boldsymbol{\theta}$, we could substitute it into Eq. (19) to obtain an unconstrained optimization problem for $\boldsymbol{\theta}$ alone, but this is generally not possible. In *Examples 1* (line fitting), *2* (ellipse fitting), and *3* (fundamental matrix computation) in Sec. 2.1, for instance, we cannot express (x, y) or (x, y, x', y') in terms of $\boldsymbol{\theta}$.

---

[1] If the data $\boldsymbol{x}_\alpha$ are 2-D positions $\boldsymbol{x}_\alpha = (x_\alpha, y_\alpha)$ and the y-coordinate alone undergoes noise, we only need to minimize $(1/N) \sum_{\alpha=1}^{N} (y_\alpha - \bar{y}_\alpha)^2$. In general, if only some components of the data $\boldsymbol{x}_\alpha$ contain noise, the problem is called *partial least squares (PLS)*.

3.3   Bundle Adjustment

A standard technique for minimizing Eq. (19) subject to Eq. (20) is to introduce a problem-dependent auxiliary variable to each $\boldsymbol{X}_\alpha$ and express $\bar{\boldsymbol{x}}_\alpha$ in terms of $\boldsymbol{X}_\alpha$ and $\boldsymbol{\theta}$ in the form

$$\bar{\boldsymbol{x}}_\alpha = \bar{\boldsymbol{x}}_\alpha(\boldsymbol{X}_\alpha, \boldsymbol{\theta}). \tag{22}$$

Then, we substitute this into Eq. (19) and minimize

$$J(\{\boldsymbol{X}_\alpha\}_{\alpha=1}^N, \boldsymbol{\theta}) = \frac{1}{N} \sum_{\alpha=1}^N (\boldsymbol{x}_\alpha - \bar{\boldsymbol{x}}_\alpha(\boldsymbol{X}_\alpha, \boldsymbol{\theta}), V_0[\boldsymbol{x}_\alpha]^{-1}(\boldsymbol{x}_\alpha - \bar{\boldsymbol{x}}_\alpha(\boldsymbol{X}_\alpha, \boldsymbol{\theta}))) \tag{23}$$

over the joint parameter space of $\{\boldsymbol{X}_\alpha\}_{\alpha=1}^N$ and $\boldsymbol{\theta}$.

A typical example of this approach is 3-D reconstruction from multiple images, for which $\boldsymbol{x}_\alpha$ has the form of $\boldsymbol{x}_\alpha = (x_\alpha, y_\alpha, x'_\alpha, y'_\alpha, ..., x''_\alpha, y''_\alpha)$, concatenating the projections (x_α, y_α), (x'_α, y'_α), ..., (x''_α, y''_α) of the αth point in the scene onto the images. The unknown parameter $\boldsymbol{\theta}$ specifies the state of all the cameras, consisting of the extrinsic parameters (the positions and orientations) and the intrinsic parameters (the focal lengths, the principal points, the aspect ratios, and the skew angles). If we introduce the 3-D position $\boldsymbol{X}_\alpha = (X_\alpha, Y_\alpha, Z_\alpha)$ of each point in the scene as the auxiliary variable, the true value $\bar{\boldsymbol{x}}_\alpha$ of $\boldsymbol{x}_\alpha$ can be explicitly expressed in the form $\bar{\boldsymbol{x}}_\alpha(\boldsymbol{X}_\alpha, \boldsymbol{\theta})$, which describes the image positions of the 3-D point $\boldsymbol{X}_\alpha$ that should be observed if the cameras have the parameter $\boldsymbol{\theta}$. Then, we minimize the *reprojection error*, i.e., the discrepancy of the observed projections $\boldsymbol{\xi}_\alpha$ from the predicted projections $\bar{\boldsymbol{x}}_\alpha(\boldsymbol{X}_\alpha, \boldsymbol{\theta})$. The minimum is searched over the entire parameter space of $\{\boldsymbol{X}_\alpha\}_{\alpha=1}^N$ and $\boldsymbol{\theta}$. This process is called *bundle adjustment* [23,32], a term originated from photogrammetry, meaning we "adjust" the "bundle" of lines of sight so that they pass through the observed points in images. The package program is available on the Web [23]. The dimension of the parameter space is $3N +$ 'the dimension of $\boldsymbol{\theta}$', which becomes very large when many points are observed.

This bundle adjustment approach is not limited to 3-D reconstruction from multiple images. In *Examples 1* (line fitting) and *2* (ellipse fitting) in Sec. 2.1, for example, if we introduce the arc length s_α of the true position $(\bar{x}_\alpha, \bar{y}_\alpha)$ along the line or the ellipse from a fixed point as the auxiliary variable, we can express (x_α, y_α) in terms of s_α and $\boldsymbol{\theta}$. Then, we minimize the resulting Mahalanobis distance J over the entire parameter space of $s_1, ..., s_N$ and $\boldsymbol{\theta}$. Instead of the arc length s_α, we can alternatively use the argument ϕ_α measured from the x-axis [30]. A similar approach can be done for fundamental matrix computation [2].

The standard numerical technique for the search of the parameter space is the *Levenberg-Marquardt* (*LM*) *method* [27], which is a hybrid of the Gauss-Newton iterations and the gradient descent. However, depending on the initial value of the iterations, the search may fall into a local minimum. Various global optimization techniques have also been studied [7]. A typical method is *branch and bound*, which introduces a function that gives a lower bound of J over a given region and divides the parameter space into small cells; those cells which have lower bounds that are above the tested values are removed, and other cells are recursively subdivided [7,9]. However, the evaluation of the lower bound involves a complicated technique, and searching the entire space requires a significant amount of computational time.

3.4   Gaussian Approximation of Noise in the ξ-Space

The search in a high-dimensional parameter space of the bundle adjustment approach can be avoided if we introduce Gaussian approximation to the noise distribution in the ξ-space. If the noise in the observation x_α is Gaussian, the noise in its nonlinear transformation $\xi_\alpha = \xi(x_\alpha)$ is not strictly Gaussian, although it is expected to have a Gaussian-like distribution if the noise is small. If it is approximated to be Gaussian, the optimization computation becomes much simpler. Suppose ξ_α has noise of mean $\mathbf{0}$ with the covariance matrix $V[\xi_\alpha] = \sigma^2 V_0[\xi_\alpha]$ evaluated by Eq. (14). Then, the ML computation reduces to minimizing the Mahalanobis distance

$$J = \frac{1}{N} \sum_{\alpha=1}^{N} (\xi_\alpha - \bar{\xi}_\alpha, V_0[\xi_\alpha]^{-1}(\xi_\alpha - \bar{\xi}_\alpha)) \tag{24}$$

in the ξ-space subject to the *linear* constraint

$$(\bar{\xi}_\alpha, \theta) = 0. \tag{25}$$

Geometrically, this is interpreted to be fitting to N points ξ_α in the ξ-space the parameterized "hyperplane" $(\xi, \theta) = 0$ by adjusting θ, where the discrepancy of the points from the plane is measured by the Mahalanobis distance of Eq. (24) inversely weighted by the covariances of the data in the ξ-space. Since Eq. (25) is now "linear" in $\bar{\xi}_\alpha$, this constraint can be eliminated using Lagrange multipliers, reducing the problem to unconstrained minimization of

$$J = \frac{1}{N} \sum_{\alpha=1}^{N} \frac{(\xi_\alpha, \theta)^2}{(\theta, V_0[\xi_\alpha]\theta)}. \tag{26}$$

Today, Eq. (26) is called the *Sampson error* [8] after the ellipse fitting scheme introduced by P. D. Sampson [29].

3.5   Sampson Error Minimization

Various numerical techniques have been proposed for minimizing the Sampson error in Eq. (26). The best known is the *FNS* (*Fundamental Numerical Scheme*) of Chojnacki et al. [5], which goes as follows:

1. Let $W_\alpha = 1$, $\alpha = 1, ..., N$, and $\theta_0 = \mathbf{0}$.
2. Computer the matrices

$$M = \frac{1}{N} \sum_{\alpha=1}^{N} W_\alpha \xi_\alpha \xi_\alpha^\top, \qquad L = \frac{1}{N} \sum_{\alpha=1}^{N} W_\alpha^2 (\theta_0, \xi_\alpha)^2 V_0[\xi_\alpha]. \tag{27}$$

3. Solve the eigenvalue problem $(M - L)\theta = \lambda\theta$, and compute the unit eigenvector θ for the smallest[2] eigenvalue λ.

4. If $\theta \approx \theta_0$ up to sign, return θ and stop. Else, let

$$W_\alpha \leftarrow \frac{1}{(\theta, V_0[\xi_\alpha]\theta)}, \qquad \theta_0 \leftarrow \theta, \qquad (28)$$

and go back to Step 2.

The background of FNS is as follows. At the time of convergence, the matrices M and L have the form

$$M = \frac{1}{N} \sum_{\alpha=1}^{N} \frac{\xi_\alpha \xi_\alpha^\top}{(\theta, V_0[\xi_\alpha]\theta)}, \qquad L = \frac{1}{N} \sum_{\alpha=1}^{N} \frac{(\theta, \xi_\alpha)^2 V_0[\xi_\alpha]}{(\theta, V_0[\xi_\alpha]\theta)^2}. \qquad (29)$$

It is easily seen that the derivative of the Sampson error J in Eq. (26) is written in terms of these matrices in the form

$$\nabla_\theta J = 2(M - L)\theta. \qquad (30)$$

It can be shown that if the above iterations converge, the eigenvalue λ must be 0. Hence, the returned value θ is the solution of $\nabla_\theta J = 0$.

Other methods exist for minimizing Eq. (26) including the *HEIV* (*Heteroscedastic Errors-in-Variables*) of Leedan and Meer [22] and Matei and Meer [24], and the *projective Gauss-Newton iterations* of Kanatani and Sugaya [18]; all compute the same solution. Note that the "initial solution" obtained in the beginning by letting $W_\alpha = 1$ coincides with the LS solution described in Sec. 3.1.

3.6 Computation of the Exact ML Solution

Since the Sampson error of Eq. (26) is obtained by approximating the non-Gaussian noise distribution in the ξ-space by a Gaussian distribution, the solution does not necessarily coincide with the ML solution that minimizes the Mahalanobis distance in Eq. (19). However, once we have obtained the solution θ that minimizes Eq. (26), we can iteratively modify Eq. (26) by using that θ so that Eq. (26) coincides with Eq. (19) in the end. This means that we obtain the exact ML solution. The procedure goes as follows [21]:

1. Let $J_0^* = \infty$ (a sufficiently large number), $\hat{x}_\alpha = x_\alpha$, and $\tilde{x}_\alpha = 0$, $\alpha = 1, ...,$ N.

2. Evaluate the normalized covariance matrices $V_0[\hat{\xi}_\alpha]$ by replacing x_α by $\hat{x}_\alpha$ in their definition.

3. Compute the following ξ_α^*:

$$\xi_\alpha^* = \xi_\alpha + \frac{\partial \xi}{\partial x}\bigg|_{x=x_\alpha} \tilde{x}_\alpha. \qquad (31)$$

---

[2] We can alternatively compute the unit eigenvector θ for the smallest eigenvalue λ in absolute value, but it has been experimentally confirmed that convergence is faster for computing the smallest eigenvalue [18].

4. Compute the value $\boldsymbol{\theta}$ that minimizes the *modified Sampson error*

$$J^* = \frac{1}{N} \sum_{\alpha=1}^{N} \frac{(\boldsymbol{\xi}_\alpha^*, \boldsymbol{\theta})^2}{(\boldsymbol{\theta}, V_0[\hat{\boldsymbol{\xi}}_\alpha]\boldsymbol{\theta})}. \tag{32}$$

5. Update $\tilde{\boldsymbol{x}}_\alpha$ and $\hat{\boldsymbol{x}}_\alpha$ as follows:

$$\tilde{\boldsymbol{x}}_\alpha \leftarrow \frac{(\boldsymbol{\xi}_\alpha^*, \boldsymbol{\theta})V_0[\boldsymbol{x}_\alpha]}{(\boldsymbol{\theta}, V_0[\hat{\boldsymbol{\xi}}_\alpha]\boldsymbol{\theta})} \frac{\partial \boldsymbol{\xi}}{\partial \boldsymbol{x}}\bigg|_{\boldsymbol{x}=\boldsymbol{x}_\alpha}^{\top} \boldsymbol{\theta}, \qquad \hat{\boldsymbol{x}}_\alpha \leftarrow \boldsymbol{x}_\alpha - \tilde{\boldsymbol{x}}_\alpha. \tag{33}$$

6. Evaluate J^* by

$$J^* = \frac{1}{N} \sum_{\alpha}^{N} (\tilde{\boldsymbol{x}}_\alpha, V_0[\boldsymbol{x}_\alpha]\tilde{\boldsymbol{x}}_\alpha). \tag{34}$$

If $J^* \approx J_0$, return $\boldsymbol{\theta}$ and stop. Else, let $J_0 \leftarrow J^*$ and go back to Step 2.

Since the modified Sampson error in Eq. (32) has the same form as the Sampson error in Eq. (26), we can minimize it by FNS (or other methods). According to numerical experiments, this modification converges after four or five rounds, yet in many practical problems the first four or five effective figures remain unchanged [19,20]. In this sense, we can practically identify the Sampson error minimization with the ML computation.

3.7   Hyperaccurate Correction of ML

It has been widely recognized that the Sampson error minimization solution, which can be practically identified with the ML solution, has very high accuracy. However, it can be shown by detailed error analysis that the solution has statistical bias of $O(\sigma^2)$ and that the magnitude of the bias can be theoretically evaluated [14]. This implies that the accuracy can be further improved by subtracting the theoretically expected bias. This process is called *hyperaccurate correction* and goes as follows [13,14]:

1. Estimate the square noise level σ^2 from the computed solution $\boldsymbol{\theta}$ and the corresponding matrix M in Eq. (29) by

$$\hat{\sigma}^2 = \frac{(\boldsymbol{\theta}, M\boldsymbol{\theta})}{1 - (n-1)/N}, \tag{35}$$

where n is the dimension of the vector $\boldsymbol{\theta}$.
2. Compute the correction term[3]

$$\Delta_c \boldsymbol{\theta} = -\frac{\sigma^2}{N} M_{n-1}^- \sum_{\alpha=1}^{N} W_\alpha(\boldsymbol{e}_\alpha, \boldsymbol{\theta})\boldsymbol{\xi}_\alpha + \frac{\hat{\sigma}^2}{N^2} M_{n-1}^- \sum_{\alpha=1}^{N} W_\alpha^2(\boldsymbol{\xi}_\alpha, M_{n-1}^- V_0[\boldsymbol{\xi}_\alpha]\boldsymbol{\theta})\boldsymbol{\xi}_\alpha, \tag{36}$$

where $\boldsymbol{e}_\alpha$ is a vector that depends on individual problems, and M_{n-1}^- is the pseudoinverse of M with truncated rank $n-1$ (the smallest eigenvalue is replaced by 0 in its spectral decomposition).

---

[3] The first term of Eq. (36) is omitted in [13,14].

3. Correct the ML solution $\boldsymbol{\theta}$ in the form

$$\boldsymbol{\theta} \leftarrow \mathcal{N}[\boldsymbol{\theta} - \Delta_c \boldsymbol{\theta}], \tag{37}$$

where $\mathcal{N}[\cdot]$ is the normalization operator into unit norm ($\mathcal{N}[\boldsymbol{a}] \equiv \boldsymbol{a}/\|\boldsymbol{a}\|$).

The vector $\boldsymbol{e}_\alpha$ is $\boldsymbol{0}$ for many problems including line fitting (*Example 1* in Sec. 2.1) and fundamental matrix computation (*Example 3* in Sec. 2.1). It is generally $\boldsymbol{0}$ if multiple images are involved. A typical problem of nonzero $\boldsymbol{e}_\alpha$ is ellipse fitting (*Example 2* in Sec. 2.1), for which $\boldsymbol{e}_\alpha = (1, 0, 1, 0, 0, 0)^\top$. However, the effect is negligibly small, and the solution is practically the same if $\boldsymbol{e}_\alpha$ is replaced by $\boldsymbol{0}$.

The above bias correction concerns geometric estimation based on the geometric model of Eq. (2). In statistics, on the other hand, it is known that ML entails statistical bias in the presence of what is known as "nuisance parameters", and various studies exist for analyzing and removing bias in the ML solution. Okatani and Deguchi [25,26] applied them to vision problems by introducing auxiliary variables in the form of Eq. (22). They analyzed the relationship between the bias and the hypersurface defined by the constraint [25] and introduced the method of projected scores [26].

For those computer vision researchers who regarded reprojection error minimization as the ultimate method, or the *Gold Standard* [8], the fact that the accuracy of ML can be improved by the above hyperaccurate correction was rather surprising. For hyperaccurate correction, however, one first needs to obtain the ML solution by an iterative method such as FNS and also estimate the noise level σ. Then, a question arises. Is it not possible to directly compute the corrected solution from the beginning, say, by modifying the FNS iterations? We now show that this is possible if we adopt the non-minimization approach of geometric estimation.

4   Non-minimization Approach

4.1   Iterative Reweight

The oldest method that is not based on minimization is the following *iterative reweight*:

1. Let $W_\alpha = 1$, $\alpha = 1, ..., N$, and $\boldsymbol{\theta}_0 = \boldsymbol{0}$.
2. Computer the following matrix $\boldsymbol{M}$:

$$\boldsymbol{M} = \frac{1}{N} \sum_{\alpha=1}^{N} W_\alpha \boldsymbol{\xi}_\alpha \boldsymbol{\xi}_\alpha^\top. \tag{38}$$

3. Solve the eigenvalue problem $\boldsymbol{M}\boldsymbol{\theta} = \lambda\boldsymbol{\theta}$, and compute the unit eigenvector $\boldsymbol{\theta}$ for the smallest eigenvalue λ.

4. If $\boldsymbol{\theta} \approx \boldsymbol{\theta}_0$ up to sign, return $\boldsymbol{\theta}$ and stop. Else, let

$$W_\alpha \leftarrow \frac{1}{(\boldsymbol{\theta}, V_0[\boldsymbol{\xi}_\alpha]\boldsymbol{\theta})}, \qquad \boldsymbol{\theta}_0 \leftarrow \boldsymbol{\theta}, \tag{39}$$

and go back to Step 2.

The motivation of this method is the *weighted least squares* that minimizes

$$\frac{1}{N}\sum_{\alpha=1}^{N} W_\alpha(\boldsymbol{\xi}_\alpha, \boldsymbol{\theta})^2 = \frac{1}{N}\sum_{\alpha=1}^{N} W_\alpha \boldsymbol{\theta}^\top \boldsymbol{\xi}_\alpha \boldsymbol{\xi}_\alpha^\top \boldsymbol{\theta} = (\boldsymbol{\theta}, \underbrace{\frac{1}{N}\sum_{\alpha=1}^{N} W_\alpha \boldsymbol{\xi}_\alpha \boldsymbol{\xi}_\alpha^\top}_{\equiv M} \boldsymbol{\theta}) = (\boldsymbol{\theta}, M\boldsymbol{\theta}).$$

$$\tag{40}$$

This is minimized by the unit eigenvector $\boldsymbol{\theta}$ of the matrix M for the smallest eigenvalue. As is well known in statistics, the optimal choice of the weight W_α is the inverse of the variance of that term. Since $(\bar{\boldsymbol{\xi}}_\alpha, \boldsymbol{\theta}) = 0$, we have $(\boldsymbol{\xi}_\alpha, \boldsymbol{\theta}) = (\Delta\boldsymbol{\xi}_\alpha, \boldsymbol{\theta}) + \cdots$, and hence the leading term of the variance is

$$E[(\Delta\boldsymbol{\xi}_\alpha, \boldsymbol{\theta})^2] = E[\boldsymbol{\theta}^\top \Delta\boldsymbol{\xi}_\alpha \Delta\boldsymbol{\xi}_\alpha^\top \boldsymbol{\theta}] = (\boldsymbol{\theta}, E[\Delta\boldsymbol{\xi}_\alpha \Delta\boldsymbol{\xi}_\alpha^\top]\boldsymbol{\theta}) = \sigma^2(\boldsymbol{\theta}, V_0[\boldsymbol{\xi}_\alpha]\boldsymbol{\theta}). \tag{41}$$

Hence, we should choose

$$W_\alpha = \frac{1}{(\boldsymbol{\theta}, V_0[\boldsymbol{\xi}_\alpha]\boldsymbol{\theta})}, \tag{42}$$

but $\boldsymbol{\theta}$ is unknown. So, we do iterations, determining the weight W_α from the value of $\boldsymbol{\theta}$ in the preceding step. The "initial solution" computed with $W_\alpha = 1$ coincides with the LS solution, minimizing Eq. (17) in Sec. 3.1.

If Eq. (42) is substituted, Eq. (40) coincides with the Sampson error in Eq. (26). With the iterative update in Eq. (39), it appears that Eq. (26) is minimized. However, we are computing at each step the value of $\boldsymbol{\theta}$ that minimizes the numerator part for the fixed value of the denominator terms determined in the preceding step. Hence, at the time of the convergence, the resulting solution $\boldsymbol{\theta}$ is such that

$$\frac{1}{N}\sum_{\alpha=1}^{N} \frac{(\boldsymbol{\xi}_\alpha, \boldsymbol{\theta})^2}{(\boldsymbol{\theta}, V_0[\boldsymbol{\xi}_\alpha]\boldsymbol{\theta})} \leq \frac{1}{N}\sum_{\alpha=1}^{N} \frac{(\boldsymbol{\xi}_\alpha, \boldsymbol{\theta}')^2}{(\boldsymbol{\theta}, V_0[\boldsymbol{\xi}_\alpha]\boldsymbol{\theta})} \tag{43}$$

for any $\boldsymbol{\theta}'$, but this does not mean

$$\frac{1}{N}\sum_{\alpha=1}^{N} \frac{(\boldsymbol{\xi}_\alpha, \boldsymbol{\theta})^2}{(\boldsymbol{\theta}, V_0[\boldsymbol{\xi}_\alpha]\boldsymbol{\theta})} \leq \frac{1}{N}\sum_{\alpha=1}^{N} \frac{(\boldsymbol{\xi}_\alpha, \boldsymbol{\theta}')^2}{(\boldsymbol{\theta}', V_0[\boldsymbol{\xi}_\alpha]\boldsymbol{\theta}')}. \tag{44}$$

The fact that iterative reweight does not minimize a particular cost function has not been well recognized by vision researchers.

The perturbation analysis in [14] shows that the covariance matrix $V[\boldsymbol{\theta}]$ of the resulting solution $\boldsymbol{\theta}$ agrees with the KCR lower bound (Sec. 2.4) up to $O(\sigma^4)$. Hence, it is practically impossible to reduce the variance any further. However, it has been widely known that the iterative reweight solution has a large bias [11]. Thus, the following strategies were introduced to improve iterative reweight:

– Remove the bias of the solution.
– Exactly minimize the Sampson error in Eq. (26).

The former is Kanatani's renormalization [10,11], and the latter is the FNS of Chojnacki et al. [5] and the HEIV of Leedan and Meer [22] and Matei and Meer [24].

4.2 Renormalization

Kanatani's renormalization [10,11] goes as follows[4]:

1. Let $W_\alpha = 1$, $\alpha = 1$, ..., N, and $\theta_0 = 0$.
2. Compute the following matrices M and N:

$$M = \frac{1}{N} \sum_{\alpha=1}^{N} W_\alpha \xi_\alpha \xi_\alpha^\top, \qquad N = \frac{1}{N} \sum_{\alpha=1}^{N} W_\alpha V_0[\xi_\alpha]. \qquad (45)$$

3. Solve the generalized eigenvalue problem $M\theta = \lambda N\theta$, and compute the unit eigenvector θ for the smallest eigenvalue λ in absolute value.
4. If $\theta \approx \theta_0$ up to sign, return θ and stop. Else, let

$$W_\alpha \leftarrow \frac{1}{(\theta, V_0[\xi_\alpha]\theta)}, \qquad \theta_0 \leftarrow \theta, \qquad (46)$$

and go back to Step 2.

The motivation of renormalization is as follows. Let $\bar{M}$ be the true value of the matrix M in Eq. (45). Since $(\bar{\xi}_\alpha, \theta) = 0$, we have $\bar{M}\theta = 0$. Hence, θ is the eigenvector of $\bar{M}$ for eigenvalue 0, but $\bar{M}$ is unknown. So, we estimate it. Since $E[\Delta\xi_\alpha] = 0$ to a first approximation, the expectation of M is

$$E[M] = E[\frac{1}{N} \sum_{\alpha=1}^{N} W_\alpha(\bar{\xi}_\alpha + \Delta\xi_\alpha)(\bar{\xi}_\alpha + \Delta\xi_\alpha)^\top] = \bar{M} + \frac{1}{N} \sum_{\alpha=1}^{N} W_\alpha E[\Delta\xi_\alpha \Delta\xi_\alpha^\top]$$

$$= \bar{M} + \frac{\sigma^2}{N} \sum_{\alpha=1}^{N} W_\alpha V_0[\xi_\alpha] = \bar{M} + \sigma^2 N. \qquad (47)$$

Thus, $\bar{M} = E[M] - \sigma^2 N \approx M - \sigma^2 N$, so instead of $\bar{M}\theta = 0$ we solve $(M - \sigma^2 N)\theta = 0$, or $M\theta = \sigma^2 N\theta$. Assuming that σ^2 is small, we regard it as the smallest eigenvalue λ in absolute value. As in the case of iterative reweight, we iteratively update the weight W_α so that it approaches Eq. (42).

Kanatani's renormalization [10,11] attracted much attention because it exhibited higher accuracy than any other then known methods. However, questions

---

[4] This is slightly different from the description in [10], in which the generalized eigenvalue problem is reduced to the standard eigenvalue problem, but the resulting solution is the same [11].

were repeatedly raised as to what it minimizes, perhaps out of the deep-rooted preconception that optimal estimation should minimize something. Chojnacki et al. [4] argued that renormalization can be "rationalized" if viewed as approximately minimizing the Sampson error. However, the renormalization process is not minimizing any particular cost function.

Note that the initial solution with $W_\alpha = 1$ solves $\left(\sum_{\alpha=1}^{N} \boldsymbol{\xi}_\alpha \boldsymbol{\xi}_\alpha^\top\right)\boldsymbol{\theta} = \lambda\left(\sum_{\alpha=1}^{N} V_0[\boldsymbol{\xi}_\alpha]\right)\boldsymbol{\theta}$, which is nothing but the method of Taubin [31], known to be very accurate algebraic method without requiring iterations. Thus, *renormalization is an iterative improvement of the Taubin solution*. According to many experiments, renormalization is shown to be more accurate than the Taubin method with nearly comparable accuracy with the FNS and the HEIV. The accuracy of renormalization is analytically evaluated in [14], showing that the covariance matrix $V[\boldsymbol{\theta}]$ of the solution $\boldsymbol{\theta}$ agrees with the KCR lower bound up to $O(\sigma^4)$ just as iterative reweight, but the bias is much smaller. That is the reason for the high accuracy of renormalization.

4.3   Analysis of Covariance and Bias

Since the covariance matrix $V[\boldsymbol{\theta}]$ of the renormalization solution $\boldsymbol{\theta}$ agrees with the KCR lower bound up to $O(\sigma^4)$, the covariance of the solution cannot be substantially improved any further. Very small it may be, however, the bias is not 0. Note that the renormalization procedure reduces to iterative reweight if the matrix $\boldsymbol{N}$ is replaced by the identity $\boldsymbol{I}$. This means that the reduction of the bias is attributed to the matrix $\boldsymbol{N}$. This observation implies the possibility of further reducing the bias by *optimizing* the matrix $\boldsymbol{N}$ in the form

$$N = \frac{1}{N} \sum_{\alpha=1}^{N} W_\alpha V_0[\boldsymbol{\xi}_\alpha] + \cdots , \tag{48}$$

so that *the bias is zero up to high order error terms*. Using the perturbation analysis in [14], Al-Sharadqah and Chernov [1] actually did this for ellipse fitting, and Kanatani et al. [15] extended it to general geometric estimation. Their analysis goes as follows. We write the observation $\boldsymbol{x}_\alpha$ as the sum $\boldsymbol{x}_\alpha = \bar{\boldsymbol{x}}_\alpha + \Delta\boldsymbol{x}_\alpha$ of the true value $\bar{\boldsymbol{x}}_\alpha$ and the noise term $\Delta\boldsymbol{x}_\alpha$. Substituting this into $\boldsymbol{\xi}_\alpha = \boldsymbol{\xi}(\boldsymbol{x}_\alpha)$ and expand it in the form

$$\bar{\boldsymbol{\xi}}_\alpha + \Delta_1\boldsymbol{\xi}_\alpha + \Delta_2\boldsymbol{\xi}_\alpha + \cdots , \tag{49}$$

where and hereafter the bar denotes the noiseless value and Δ_k denotes terms of $O(\sigma^k)$. We similarly expand $\boldsymbol{M}$, $\boldsymbol{\theta}$, $\boldsymbol{\lambda}$, and $\boldsymbol{N}$ and express the generalized eigenvalue problem in the form

$$(\bar{\boldsymbol{M}}+\Delta_1\boldsymbol{M}+\Delta_2\boldsymbol{M}+\cdots)(\bar{\boldsymbol{\theta}}+\Delta_1\boldsymbol{\theta}+\Delta_2\boldsymbol{\theta}+\cdots)$$
$$= (\bar{\lambda}+\Delta_1\lambda+\Delta_2\lambda+\cdots)(\bar{\boldsymbol{N}}+\Delta_1\boldsymbol{N}+\Delta_2\boldsymbol{N}+\cdots)(\bar{\boldsymbol{\theta}}+\Delta_1\boldsymbol{\theta}+\Delta_2\boldsymbol{\theta}+\cdots).$$

Equating the terms of the same order in σ, we obtain

$$\Delta_1\boldsymbol{\theta} = -\bar{\boldsymbol{M}}^-\Delta_1\boldsymbol{M}\bar{\boldsymbol{\theta}}, \tag{50}$$

$$\Delta_2^{\perp}\boldsymbol{\theta} = \bar{\boldsymbol{M}}^-\Big(\frac{(\bar{\boldsymbol{\theta}}, \boldsymbol{T}\bar{\boldsymbol{\theta}})}{(\bar{\boldsymbol{\theta}}, \bar{\boldsymbol{N}}\bar{\boldsymbol{\theta}})}\bar{\boldsymbol{N}}\bar{\boldsymbol{\theta}} - \boldsymbol{T}\bar{\boldsymbol{\theta}}\Big), \tag{51}$$

where $\bar{\boldsymbol{M}}^-$ is the pseudoinverse of $\bar{\boldsymbol{M}}$; since $\bar{\boldsymbol{M}}$ has the eigenvector $\bar{\boldsymbol{\theta}}$ of eigenvalue 0, its rank is $n - 1$ (n is the dimension of $\boldsymbol{\theta}$). The symbol $\Delta_2^{\perp}\boldsymbol{\theta}$ denotes the component of the second order noise term orthogonal to $\bar{\boldsymbol{\theta}}$; since $\boldsymbol{\theta}$ is a unit vector, it has no error in the direction of itself, so we are interested in the error orthogonal to it. The matrix $\boldsymbol{T}$ in Eq. (51) is defined to be

$$\boldsymbol{T} \equiv \Delta_2\boldsymbol{M} - \Delta_1\boldsymbol{M}\bar{\boldsymbol{M}}^-\Delta_1\boldsymbol{M}. \tag{52}$$

From Eq. (50), we can show that the leading term of the covariance matrix of $\boldsymbol{\theta}$ has the following form [14].

$$V[\boldsymbol{\theta}] \equiv E[\Delta_1\boldsymbol{\theta}\Delta_1\boldsymbol{\theta}^{\top}] = \frac{\sigma^2}{N}\bar{\boldsymbol{M}}^-. \tag{53}$$

From this we observe:

  - The covariance matrix $V[\boldsymbol{\theta}]$ is $O(\sigma^2)$.
  - The right side of Eq. (16) agrees with the KCR lower bound.
  - Eq. (53) does not contain the matrix $\boldsymbol{N}$.

Thus, we cannot change the value of Eq. (53) by adjusting the matrix $\boldsymbol{N}$. However, the root-mean-square (RMS) error of $\boldsymbol{\theta}$ is the sum of the covariance term and the bias term, and the bias term is also $O(\sigma^2)$ (the expectation of odd order noise terms is 0, so the first order bias is $E[\Delta_1\boldsymbol{\theta}] = \boldsymbol{0}$). Since the second order bias term contains the matrix $\boldsymbol{N}$, we can reduce it by adjusting $\boldsymbol{N}$. From Eq. (51), the second order bias has the following expression:

$$E[\Delta_2^{\perp}\boldsymbol{\theta}] = \bar{\boldsymbol{M}}^-\Big(\frac{(\bar{\boldsymbol{\theta}}, E[\boldsymbol{T}\bar{\boldsymbol{\theta}}])}{(\bar{\boldsymbol{\theta}}, \bar{\boldsymbol{N}}\bar{\boldsymbol{\theta}})}\bar{\boldsymbol{N}}\bar{\boldsymbol{\theta}} - E[\boldsymbol{T}\bar{\boldsymbol{\theta}}]\Big). \tag{54}$$

4.4   Hyper-renormalization

Equation (54) implies that if we can choose an $\boldsymbol{N}$ such that

$$E[\boldsymbol{T}\bar{\boldsymbol{\theta}}] = c\bar{\boldsymbol{N}}\bar{\boldsymbol{\theta}} \tag{55}$$

for some constant c, we will have

$$E[\Delta_2^{\perp}\boldsymbol{\theta}] = \bar{\boldsymbol{M}}^-\Big(\frac{(\bar{\boldsymbol{\theta}}, c\bar{\boldsymbol{N}}\bar{\boldsymbol{\theta}})}{(\bar{\boldsymbol{\theta}}, \bar{\boldsymbol{N}}\bar{\boldsymbol{\theta}})}\bar{\boldsymbol{N}}\bar{\boldsymbol{\theta}} - c\bar{\boldsymbol{N}}\bar{\boldsymbol{\theta}}\Big) = \boldsymbol{0}, \tag{56}$$

i.e., *the second order bias is completely eliminated.* Kanatani et al. [15] showed that if the matrix $\bar{N}$ is defined by

$$\bar{N} = \frac{1}{N} \sum_{\alpha=1}^{N} \bar{W}_\alpha \left(V_0[\bar{\xi}_\alpha] + 2S[\bar{\xi}_\alpha e_\alpha^\top] \right)$$

$$- \frac{1}{N^2} \sum_{\alpha=1}^{N} \bar{W}_\alpha^2 \left((\bar{\xi}_\alpha, \bar{M}^- \bar{\xi}_\alpha) V_0[\bar{\xi}_\alpha] + 2S[V_0[\bar{\xi}_\alpha] \bar{M}^- \bar{\xi}_\alpha \bar{\xi}_\alpha^\top] \right), \quad (57)$$

then $E[T\bar{\theta}] = \sigma^2 \bar{N}\bar{\theta}$ holds, where e_α is a vector that depends on individual problems (the same vector as that in Eq. (36)), and $S[\cdot]$ denotes symmetrization ($S[A] = (A + A^\top)/2$). In actual computation, the true values in Eq. (57) are replaced by computed values. This entails errors of $O(\sigma)$, but since the expectation of odd order noise terms is 0, Eq. (56) is $O(\sigma^4)$. Thus, we obtain the following procedure of *hyper-renormalization*:

1. Let $W_\alpha = 1$, $\alpha = 1, ..., N$, and $\theta_0 = 0$.
2. Compute the following matrices M and N:

$$M = \frac{1}{N} \sum_{\alpha=1}^{N} W_\alpha \xi_\alpha \xi_\alpha^\top, \quad (58)$$

$$N = \frac{1}{N} \sum_{\alpha=1}^{N} W_\alpha \left(V_0[\xi_\alpha] + 2S[\xi_\alpha e_\alpha^\top] \right)$$

$$- \frac{1}{N^2} \sum_{\alpha=1}^{N} W_\alpha^2 \left((\xi_\alpha, M_{n-1}^- \xi_\alpha) V_0[\xi_\alpha] + 2S[V_0[\xi_\alpha] M_{n-1}^- \xi_\alpha \xi_\alpha^\top] \right). \quad (59)$$

Here, M_{n-1}^- is the pseudoinverse of M with truncated rank $n - 1$ (cf. Eq. (36)).
3. Solve the generalized eigenvalue problem $M\theta = \lambda N\theta$, and compute the unit eigenvector θ for the smallest eigenvalue λ in absolute value.
4. If $\theta \approx \theta_0$ up to sign, return θ and stop. Else, let

$$W_\alpha \leftarrow \frac{1}{(\theta, V_0[\xi_\alpha]\theta)}, \qquad \theta_0 \leftarrow \theta, \quad (60)$$

and go back to Step 2.

It turns out that the initial solution with $W_\alpha = 1$ coincides with what is called *HyperLS* [16,17,28], which is derived to remove the bias up to second order error terms within the framework of algebraic methods without iterations[5]. Thus, *hyper-renormalization is an iterative improvement of HyperLS.*

Standard linear algebra routines for solving the generalized eigenvalue problem $M\theta = \lambda N\theta$ assume that N is positive definite, but the matrix N in Eq. (59)

---

[5] The expression of Eq. (59) with $W_\alpha = 1$ lacks one term as compared with the corresponding expression of HyperLS, but the same solution is produced.

has both positive and negative eigenvalues. For renormalization, the matrix N is positive semidefinite, having eigenvalue 0. This, however, causes no trouble, because the problem can be rewritten as

$$N\theta = \frac{1}{\lambda} M\theta. \tag{61}$$

The matrix M is positive definite for noisy data, so we can use a standard routine to compute the eigenvector θ for the eigenvalue $1/\lambda$ with the largest absolute value. If the matrix M happens to have eigenvalue 0, it indicates that the data are all exact, so the unit eigenvector for the eigenvalue 0 is the exact solution.

5   Numerical Examples

We define 30 equidistant points on the ellipse shown in Fig. 1(a). The major and minor axis are set to 100 and 50 pixels, respectively. We add random Gaussian noise of mean 0 and standard deviation σ to the x and y coordinates of each point independently and fit an ellipse to the noisy point sequence using : 1) LS, 2) iterative reweight, 3) the Taubin method, 4) renormalization, 5) HyperLS, 6) hyper-renormalization, 7) ML, and 8) hyperaccurate correction of ML.

Since the computed θ and its true value $\bar{\theta}$ are both unit vectors, we measure the discrepancy between them by the orthogonal component $\Delta^{\perp}\theta = P_{\bar{\theta}}\theta$, where $P_{\bar{\theta}}$ ($\equiv I - \bar{\theta}\bar{\theta}^{\top}$) is the projection matrix along $\bar{\theta}$. We generated 10000 independent noise instances and evaluated the bias B (Fig. 1(b)) and the RMS (root-mean-square) error D (Fig. 1(c)) defined by

$$B = \left\| \frac{1}{10000} \sum_{a=1}^{10000} \Delta^{\perp}\theta^{(a)} \right\|, \quad D = \sqrt{\frac{1}{10000} \sum_{a=1}^{10000} \|\Delta^{\perp}\theta^{(a)}\|^{2}}, \tag{62}$$

where $\theta^{(a)}$ is the solution in the ath trial. The dotted line in Fig. 1(c) indicates the KCR lower bound. The interrupted plots in Fig. 2(a) for iterative reweight, ML, and hyperaccurate correction of ML indicate that the iterations did not converge beyond that noise level. Our convergence criterion is $\|\theta - \theta_0\| < 10^{-6}$ for the current value θ and the value θ_0 in the preceding iteration; their signs are adjusted before subtraction. If this criterion is not satisfied after 100 iterations, we stopped. For each σ, we regarded the iterations as not convergent if any among the 10000 trials does not converge.

We can see from Fig. 2(a) that LS and iterative reweight have very large bias, in contrast to which the bias is very small for the Taubin method and renormalization. The bias of HyperLS and hyper-renormalization is still smaller and even smaller than ML. Since the leading covariance is common to iterative reweight, renormalization, and hyper-renormalization, the RMS error reflects the magnitude of the bias as shown in Fig. 2(b). Because the hyper-renormalization solution does not have bias up to high order error terms, it has nearly the same

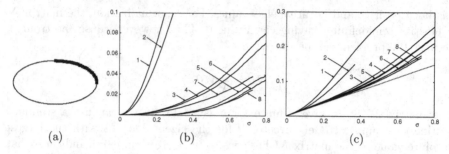

  (a)   (b)   (c)

Fig. 2. Thirty points on an ellipse (a). The bias (a) and the RMS error (b) of the fitted ellipse for the standard deviation σ of the added noise over 10000 independent trials. 1) LS, 2) iterative reweight, 3) the Taubin method, 4) renormalization, 5) HyperLS, 6) hyper-renormalization, 7) ML, 8) hyperaccurate correction of ML. The dotted line in (c) indicates the KCR lower bound.

accuracy as ML, or reprojection error minimization. A close examination of the small σ part reveals that hyper-renormalization outperforms ML. The highest accuracy is achieved, although the difference is very small, by hyperaccurate correction of ML. However, it first requires the ML solution, and the FNS iterations for its computation may not converge above a certain noise level, as shown in Figs. 2(a), (b). On the other hand, hyper-renormalization is very robust to noise. This is because the initial solution is HyperLS, which is itself highly accurate already as shown in Fig. 2. For this reason, we conclude that it is the best method for practical computations.

6 Concluding Remarks

We have overviewed techniques for optimal geometric estimation from noisy observations for computer vision applications. We first described minimization-based approaches: LS, ML, which includes reprojection error minimization (Gold Standard) as a special case, and Sampson error minimization. We then formulated non-minimization approaches: iterative reweight, renormalization, and hyper-renormalization, which can be viewed as iterative improvement of LS, the Taubin method, and HyperLS, respectively (Table 1). Showing numerical examples, we conclude that hyper-renormalization is robust to noise and currently is the best method.

Table 1. Summary of non-minimization approaches

initial	weight update	final
LS	$\longrightarrow$	iterative reweight
Taubin	$\longrightarrow$	renormalization
HyperLS	$\longrightarrow$	hyper-renormalization

Acknowledgments. A major part of this paper resulted from the author's collaboration with Prasanna Rangarajan of Southern Methodist University, U.S.A., Ali Al-Sharadqah of the University of Mississippi, U.S.A., Nikolai Chernov of the University of Alabama at Birmingham, U.S.A., and Yasuyuki Sugaya of Toyohashi University of Technology, Japan. This work was supported in part by JSPS Grant-in-Aid for Challenging Exploratory Research (24650086).

References

1. Al-Sharadqah, A., Chernov, N.: A doubly optimal ellipse fit. Comp. Stat. Data Anal. 56(9), 2771–2781 (2012)
2. Bartoli, A., Sturm, P.: Nonlinear estimation of fundamental matrix with minimal parameters. IEEE Trans. Patt. Anal. Mach. Intell. 26(3), 426–432 (2004)
3. Chernov, N., Lesort, C.: Statistical efficiency of curve fitting algorithms. Comp. Stat. Data Anal. 47(4), 713–728 (2004)
4. Chojnacki, W., Brooks, M.J., van den Hengel, A.: Rationalising the renormalization method of Kanatani. J. Math. Imaging Vis. 21(11), 21–38 (2001)
5. Chojnacki, W., Brooks, M.J., van den Hengel, A., Gawley, D.: On the fitting of surfaces to data with covariances. IEEE Trans. Patt. Anal. Mach. Intell. 22(11), 1294–1303 (2000)
6. Godambe, V.P. (ed.): Estimating Functions. Oxford University Press, New York (1991)
7. Hartley, R., Kahl, F.: Optimal algorithms in multiview geometry. In: Proc. 8th Asian Conf. Comput. Vis., Tokyo, Japan, vol. 1, pp. 13–34 (November 2007)
8. Hartley, R., Zisserman, A.: Multiple View Geometry in Computer Vision, 2nd edn. Cambridge University Press, Cambridge (2004)
9. Kahl, F., Agarwal, S., Chandraker, M.K., Kriegman, D., Belongie, S.: Practical global optimization for multiview geometry. Int. J. Comput. Vis. 79(3), 271–284 (2008)
10. Kanatani, K.: Renormalization for unbiased estimation. In: Proc. 4th Int. Conf. Comput. Vis., Berlin, Germany, pp. 599–606 (May 1993)
11. Kanatani, K.: Statistical Optimization for Geometric Computation: Theory and Practice. Elsevier, Amsterdam (1996); reprinted, Dover, New York (2005)
12. Kanatani, K.: Cramer-Rao lower bounds for curve fitting. Graphical Models Image Process. 60(2), 93–99 (1998)
13. Kanatani, K.: Ellipse fitting with hyperaccuracy. IEICE Trans. Inf. & Syst. E89-D(10), 2653–2660 (2006)
14. Kanatani, K.: Statistical optimization for geometric fitting: Theoretical accuracy analysis and high order error analysis. Int. J. Comput. Vis. 80(2), 167–188 (2008)
15. Kanatani, K., Al-Sharadqah, A., Chernov, N., Sugaya, Y.: Renormalization Returns: Hyper-renormalization and Its Applications. In: Fitzgibbon, A., Lazebnik, S., Perona, P., Sato, Y., Schmid, C. (eds.) ECCV 2012, Part III. LNCS, vol. 7574, pp. 384–397. Springer, Heidelberg (2012)
16. Kanatani, K., Rangarajan, P.: Hyper least squares fitting of circles and ellipses. Comput. Stat. Data Anal. 55(6), 2197–2208 (2011)
17. Kanatani, K., Rangarajan, P., Sugaya, Y., Niitsuma, H.: HyperLS and its applications. IPSJ Trans. Comput. Vis. Appl. 3, 80–94 (2011)
18. Kanatani, K., Sugaya, Y.: Performance evaluation of iterative geometric fitting algorithms. Comp. Stat. Data Anal. 52(2), 1208–1222 (2007)

19. Kanatani, K., Sugaya, Y.: Compact algorithm for strictly ML ellipse fitting. In: Proc. 19th Int. Conf. Patt. Recog., Tampa, FL, U.S.A (December 2008)
20. Kanatani, K., Sugaya, Y.: Compact fundamental matrix computation. IPSJ Tran. Comput. Vis. Appl. 2, 59–70 (2010)
21. Kanatani, K., Sugaya, Y.: Unified computation of strict maximum likelihood for geometric fitting. J. Math. Imaging Vis. 38(1), 1–13 (2010)
22. Leedan, Y., Meer, P.: Heteroscedastic regression in computer vision: Problems with bilinear constraint. Int. J. Comput. Vis. 37(2), 127–150 (2000)
23. Lourakis, M.I.A., Argyros, A.A.: SBA: A software package for generic sparse bundle adjustment. ACM Trans. Math. Software 36(1), 2, 1–30 (2009)
24. Matei, J., Meer, P.: Estimation of nonlinear errors-in-variables models for computer vision applications. IEEE Trans. Patt. Anal. Mach. Intell. 28(10), 1537–1552 (2006)
25. Okatani, T., Deguchi, K.: On bias correction for geometric parameter estimation in computer vision. In: Proc. IEEE Conf. Comput. Vis. Patt. Recog., Miami Beach, FL, U.S.A, pp. 959–966 (June 2009)
26. Okatani, T., Deguchi, K.: Improving accuracy of geometric parameter estimation using projected score method. In: Proc. Int. Conf. Comput. Vis., Kyoto, Japan, pp. 1733–1740 (September/October 2009)
27. Press, W.H., Teukolsky, S.A., Vetterling, W.T., Flannery, B.P.: Numerical Recipes in C: The Art of Scientific Computing, 2nd edn. Cambridge University Press, Cambridge (1992)
28. Rangarajan, P., Kanatani, K.: Improved algebraic methods for circle fitting. Electronic J. Stat. 3, 1075–1082 (2009)
29. Sampson, P.D.: Fitting conic sections to "very scattered" data: An iterative refinement of the Bookstein algorithm. Comput. Graphics Image Process. 18(1), 97–108 (1982)
30. Sturm, P., Gargallo, P.: Conic fitting using the geometric distance. In: Proc. 8th Asian Conf. Comput. Vis., Tokyo, Japan, vol. 2, pp. 784–795 (November 2007)
31. Taubin, G.: Estimation of planar curves, surfaces, and non-planar space curves defined by implicit equations with applications to edge and range image segmentation. IEEE Trans. Patt. Anal. Mach. Intell. 13(11), 1115–1138 (1991)
32. Triggs, B., McLauchlan, P.F., Hartley, R.I., Fitzgibbon, A.W.: Bundle Adjustment – A Modern Synthesis. In: Triggs, B., Zisserman, A., Szeliski, R. (eds.) ICCV-WS 1999. LNCS, vol. 1883, pp. 298–375. Springer, Heidelberg (2000)

Hierarchical Compositional Representations of Object Structure

Aleš Leonardis*

[1] Visual Cognitive Systems Laboratory
Faculty of Computer and Information Science
University of Ljubljana
Tržaška 25, SI-1001 Ljubljana
Slovenia
[2] School of Computer Science
University of Birmingham
Edgbaston, Birmingham, B15 2TT
United Kingdom
ales.leonardis@fri.uni-lj.si, a.leonardis@cs.bham.ac.uk

Abstract. Visual categorisation has been an area of intensive research in the vision community for several decades. Ultimately, the goal is to efficiently detect and recognize an increasing number of object classes. The problem entangles three highly interconnected issues: the internal object representation, which should compactly capture the visual variability of objects and generalize well over each class; a means for learning the representation from a set of input images with as little supervision as possible; and an effective inference algorithm that robustly matches the object representation against the image and scales favorably with the number of objects.

In this talk I will present our approach which combines a learned compositional hierarchy, representing (2D) shapes of multiple object classes, and a coarse-to-fine matching scheme that exploits a taxonomy of objects to perform efficient object detection. Our framework for learning a hierarchical compositional shape vocabulary for representing multiple object classes takes simple contour fragments and learns their frequent spatial configurations. These are recursively combined into increasingly more complex and class-specific shape compositions, each exerting a high degree of shape variability. At the top-level of the vocabulary, the compositions represent the whole shapes of the objects. The vocabulary is learned layer after layer, by gradually increasing the size of the window of analysis and reducing the spatial resolution at which the shape configurations are learned. The lower layers are learned jointly on images of all classes, whereas the higher layers of the vocabulary are learned incrementally, by presenting the algorithm with one object class after another.

However, in order for recognition systems to scale to a larger number of object categories, and achieve running times logarithmic in the number of classes, building visual class taxonomies becomes necessary. We propose

---

This is a joint work with Sanja Fidler and Marko Boben.

G.L. Gimel' farb et al. (Eds.): SSPR & SPR 2012, LNCS 7626, pp. 31–32, 2012.
© Springer-Verlag Berlin Heidelberg 2012

an approach for speeding up recognition times of multi-class part-based object representations. The main idea is to construct a taxonomy of constellation models cascaded from coarse-to-fine resolution and use it in recognition with an efficient search strategy. The structure and the depth of the taxonomy is built automatically in a way that minimizes the number of expected computations during recognition by optimizing the cost-to-power ratio. The combination of the learned taxonomy with the compositional hierarchy of object shape achieves efficiency both with respect to the representation of the structure of objects and in terms of the number of modeled object classes. The experimental results show that the learned multi-class object representation achieves a detection performance comparable to the current state-of-the-art flat approaches with both faster inference and shorter training times.

Information Theoretic Prototype Selection for Unattributed Graphs

Lin Han[1], Luca Rossi[2], Andrea Torsello[2],
Richard C. Wilson[1], and Edwin R. Hancock[1]

[1] Department of Computer science,University of York,UK
[2] Department of Environmental Science, Informatics and Statistics,
Ca' Foscari Univerisity of Venice, Italy

Abstract. In this paper we propose a prototype size selection method for a set of sample graphs. Our first contribution is to show how approximate set coding can be extended from the vector to graph domain. With this framework to hand we show how prototype selection can be posed as optimizing the mutual information between two partitioned sets of sample graphs. We show how the resulting method can be used for prototype graph size selection. In our experiments, we apply our method to a real-world dataset and investigate its performance on prototype size selection tasks.

Keywords: Prototype Selection, Mutual information, Importance Sampling, Partition function.

1 Introduction

Relational graphs provide a convenient means of representing structural patterns. Examples include the arrangement of shape primitives or feature points in images, molecules and social networks. Recently, there has been considerable interest in learning prototype graphs which can capture the structural variations given a set of sample graphs [1,12]. These approaches are frequently sample-based, having several candidate prototype graphs in hand, we are confronted with the problem of selecting the best one. This problem falls into the category of model selection, which is one of the fundamental tasks in pattern analysis. A good model should be able to summarize the observed data well. Moreover, it should have good predictive capabilities. There are a wealth of principles in the literature for selecting the best model [9,11,10]. Generally speaking, although the principles are motivated from different viewpoints, most of them employ penalizing the parameters (or complexity) of the model in order to generalize well on a new dataset. For example, the Akaike information criterion(AIC) penalizes the model using the value of twice the number of free parameters of the model [13], while the minimum description length criterion uses a universal coding [14].

The main drawback of these approaches is that they cannot be easily extended from the vector domain to the graph domain. On the other hand, other frameworks such as the approximate set coding [3] can be transformed to the graph domain with the help of sampling techniques such as Importance Sampling.

G.L. Gimel' farb et al. (Eds.): SSPR & SPR 2012, LNCS 7626, pp. 33–41, 2012.

In this paper we present an approach to selecting the optimal prototype size for a set of sample graphs. Our method is an extension of the theory of the approximate set coding to graph data. The prototype of optimal size is that which maximizes the mutual information between the two partitioned sets of the sample graphs. To measure the mutual information, we need to compute the partition functions of the two partitioned sets and their joint partition function. The computation of the partition function involves exploring the complete hypothesis space and this is a NP hard problem for graphs. We locate an approximate solution to this problem by using the importance sampling approach.

The remainder of the paper is organized as follows. We first briefly introduce the theory of the approximate set coding [3]. Then we describe how we extend the theory on model selection in vector domain to the graph domain. This includes how to characterize the sample sets using the partition function and how to approximate the value of the partition function using the importance sampling approach. In the last part we provide some preliminary experimental results.

2   Approximation Set Coding

In this section we briefly review the theory of the approximate set coding proposed in [3,4]. In this context, a *hypothesis* is a solution to our pattern recognition problem. In this specific case, a hypothesis c is a mapping (matching) of all of our sample graphs to a prototype graph. We also have a *cost function* $R(c)$ which evaluates the quality of a particular matching. Naturally $R(c)$ depends on the prototype graph proposed for the data samples.

Given a prototype graph drawn from set of possible prototypes (usually of different sizes or complexity), we can find the best matching and prototype configuration by optimizing $R(c)$. We denote the best hypothesis as $c^{\perp}$. As usual, we cannot use $R(c)$ to select the best prototype from the set, as the more complex prototypes have lower costs (they fit the samples better) but do not generalize well.

Approximation set coding uses the observation that there are a set of transformations which alter the sample data without essentially changing the prototype in any way. For example, if we consider the sample graphs in a different order, or their nodes are permuted in some way, then the structure of the recovered prototype should be the same (although the prototype graph nodes may also be in a different order). We can use this fact to measure how good our prototype is at recovering these transformations when they are coded using the prototype graph and sent through a noisy channel. To do this, we split the sample data into two partitions. The first partition is used to code the transformation, and the second partition provides a prototype graph to decode the transformation. We then attempt to maximize the amount of information transmitted. The analysis in [3] shows that the mutual information between sender and receiver is

$$I_\gamma = \frac{1}{n} \log \left(\frac{|T||\Delta C_{\gamma,12}|}{|C_{\gamma,1}|C_{\gamma,2}} \right) \tag{1}$$

where $|C_{\gamma,1}|$ is the number of hypotheses which are within a cost γ of the best cost in set 1 (and likewise for $|C_{\gamma,2}|$). The quantity $|\Delta C_{\gamma,12}|$ is the number of hypotheses *on set 2* which are within a cost γ of the best cost *in set 1*. To calculate this, we need a way of transferring hypotheses from set 2 to set 1. For more details of these techniques, the reader is referred to [3,4].

3   Prototype Selection for Graphs

In this section, we extend the methodology of the approximate set coding from the vector domain to the graph domain. Our main contribution here is that we redefine three important ingredients in approximate set coding (i.e. hypothesis, cost function and partition function), and generalize them from the vector domain to the graph domain. In the following, we commence by introducing our problem and then give formal definitions of the ingredients.

Given a set of sample graphs, our aim is to select the optimal size of the prototype graph for the sample graphs. To ensure that the optimal prototype graph generalizes well on a new dataset, we adopt a two-sample set scenario and partition the sample graphs into two sets of the same size $\mathcal{G}_1 = \{G_1^{(1)}, G_2^{(1)}, ..., G_n^{(1)}\}$, $\mathcal{G}_2 = \{G_1^{(2)}, G_2^{(2)}, ..., G_n^{(2)}\}$. Here the superscripts indicate different sample-set and the subscripts indicate the graph indices. The best prototype graph is determined according to its generalization capability on the two sets.

3.1   Hypothesis

The hypotheses originally proposed in the clustering problem (where approximate set coding was first used) are the assignments of data points to clusters [4]. Here in our problem the hypotheses consist of a set of mappings of each of the sample graphs onto its corresponding prototype graph. By direct analogy with the clustering problem, each mapping is equivalent to an assignment of a point to a cluster; the prototype graph here is equivalent to the cluster centroid. For each dataset $\mathcal{G}_q(q \in \{1, 2\})$ a hypothesis is $c_q = \{S_1^{(q)}, S_2^{(q)}, ..., S_n^{(q)}\}$ where $S_i^{(q)}$ ($i \in \{1, ..., n\}$) is the assignment matrix between graph i from set q and its corresponding prototype graph $G_M^{(q)}$. The set of all possible hypotheses is $\mathcal{C}$, which consists of all the possible mappings between all samples and the prototype graph.

3.2   Cost Function

To proceed, we require a cost function $R_q(c_q)$ to quantify the effectiveness of a particular hypothesis c_q. The cost function measures how consistent the given mappings are with the prototype graph. Here the cost function of a hypothesis is the negative logarithm of the matching probability between the sample graph and the prototype graph under the hypothesis modelled using the technique described in [15].

$$R_q(c_q) = -\log P(\mathcal{G}_q|G_M^{(q)}, c_q)$$

$$= \sum_{G_i^{(q)}} \sum_{a \in V_i^{(q)}} -\log \sum_{a \in V_M^{(q)}} K_a^i \exp\left[\mu \sum_{b \in V_i^{(q)}} \sum_{\beta \in V_M^{(q)}} D_{iab}^{(q)} M_{\alpha\beta}^{(q)} S_{ib\beta}^{(q)}\right]. \quad (2)$$

In the above, $D_i^{(q)}$ is the adjacency matrix for the sample graph G_i from set q and $M^{(q)}$ is the adjacency matrix for the prototype graph $G_M^{(q)}$. The matrix $S_i^{(q)}$ is the assignment matrix between the two graphs. If nodes a and b of the sample graph $G_i^{(q)}$ are connected, their corresponding element $D_{iab}^{(q)}$ in $D_i^{(q)}$ has a unit value otherwise it is zero. This is same for the nodes α, β of the prototype graph $G_M^{(q)}$. The elements of the assignment matrix $S_{ia\alpha}^{(q)}$ are unit if node a in graph $G_i^{(q)}$ is matched to node α in graph G_M. The cost function above is a natural choice in our problem because it is also involved in measuring the similarity between the sample graphs and the prototype graph during the learning procedure of the prototype graph.

In order to normalize the minimum cost of the hypotheses to zero, we define the relative cost of hypothesis. Suppose the optimal hypothesis (i.e., the hypothesis yielding the lowest costs between the sample graphs and their prototype graph) is $c_q^{\perp}$, the relative cost of hypothesis c_q is $\Delta R_q(c_q) = R_q(c_q) - R_q(c_q^{\perp})$.

3.3   Partition Function

The measurement of the mutual information of the two sample-sets requires counting the number of hypotheses which are within a certain cost of the optimal solution. However, this is hard to do since it involves exploring all the hypotheses. Fortunately, this value can be estimated using concepts from statistical physics. Considering the hypotheses as microcanonical ensembles in statistical mechanics, their number can be estimated by calculating the partition function [4]

$$Z_q = \sum_{c_q \in C_q} \exp[-\beta \Delta R_q(c_q)] \qquad (3)$$

where β is a positive scaling parameter known as the inverse computational temperature. Essentially, β coarsens the precision of the partition function approximating the number of hypotheses that fit the sample set [3]. When β is zero, the partition function is equal to the number of all the possible hypotheses. When β is very large, the partition function only counts the number of optimal hypotheses. Because β controls the number of hypotheses fitting the sample set, we will call these β-optimal hypotheses. In our case, the hypothesis space is the set of all the possible mappings between the sample graphs and their prototype graph. The hypothesis space is very large and the computation of the partition function will be expensive. Later we show how we use the importance sampling approach to sample the mapping between the sample graphs and their prototype graph and approximate the partition function.

To measure how well the hypotheses generalize for the two sample sets, we count the number of β-optimal hypotheses in the first set which also exist in the second set, when transferred to the first set. We therefore need a way of transferring hypotheses from the second dataset to the first. We denote the cost of the hypothesis c_2 between the transferred graphs and prototype graph $G_M^{(2)}$ as $R_t(c_2)$. This is the cost of making hypothesis c_2 for the graphs $\mathcal{G}_2$ when evaluated against the data in $\mathcal{G}_1$. The following procedure may be used to find the transfer.

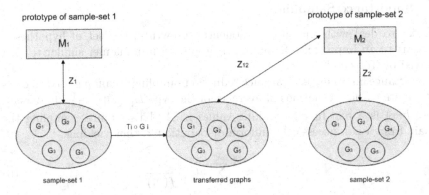

Fig. 1. A diagram illustrates the procedure of computing the three partition functions. When we compute the partition function Z_{12}, we need to count how many of our hypotheses are β-optimal when we use the prototype from set 2 and the data graphs from set 1. We therefore need a way of transferring hypotheses from the second set to the first.

For each $G_i^{(1)}$ graph in $\mathcal{G}_1$, we find the most similar graph in $\mathcal{G}_2$ and the mapping between T_i between the two. $T_i \circ G_i^{(i)}$ is then the image of this graph in the second set. From these images, we compute the cost of c_2 by comparing the images to the prototype graph $\mathcal{G}_M^{(2)}$ under the mappings in c_2. Finally, the joint partition function is formulated as

$$Z_{12} = \sum_{c_2 \in \mathcal{C}_2} \exp[-\beta(\Delta R_t(c_2) + \Delta R_2(c_2))] \ . \tag{4}$$

The quantity $\Delta R_t(c_2)$ is the relative cost of hypothesis c_2 between the image graphs of $\mathcal{G}_1$ in the second set and the prototype graph $G_M^{(2)}$. It is equivalent to the cost of hypothesis c_2 between the image graphs and $G_M^{(2)}$ minus their minimum cost. Figure 1 illustrates the procedure of computing partition functions Z_1, Z_2 and the joint partition function Z_{12}.

Prototype graphs with different sizes are ranked according to their mutual information between the two sets

$$I_\beta = \frac{1}{n} \log \left(\frac{k Z_{12}}{Z_1 Z_2} \right) \ . \tag{5}$$

In the above equation, Z_1 and Z_2 are the respective partition functions of two sample sets, and Z_{12} is their joint partition function. The constant k is a normalization factor which keeps the value of the mutual information equal to zero when β is zero. The value of the mutual information can be interpreted as the maximum generalization capacity of prototype graphs. Hence our problem is posed as that of finding the prototype graph that maximizes this mutual information.

3.4   Importance Sampling

In order to deal with the super-exponential growth of the set of hypotheses, we resort to an Importance Sampling [7] approach in a manner similar to that reported by Torsello [2].

Importance Sampling is a variance reduction sampling technique used to compute Monte Carlo estimations of averages of the type $E[h(x)] = \frac{1}{||A||}\int_A h(x)dx$, where $h(x)$ is a real function taking values in A. This requires sampling the domain from a non necessarily uniform distribution f, thus yielding

$$E_f[h(x)] \approx \frac{1}{k}\sum_{i=1}^{k} h(x_i)\frac{\frac{1}{||A||}}{f(x_i)} \tag{6}$$

where $\frac{\frac{1}{||A||}}{f(x_i)}$ is the *importance factor* used to correct the bias introduced when sampling from the distribution f. Note that in the limit if $f = \frac{h(x)}{\int_A h(x)dx}$ then the variance of the estimator is zero. In practice then, we would like choose f as close as possible to $\frac{h(x)}{\int_A h(x)dx}$.

In this paper, we need to approximate the value of the partition functions $\mathcal{Z}_1$, $\mathcal{Z}_2$ and $\mathcal{Z}_{12}$. Since the approximation procedure is going to be the same in all the three cases, we simply review the equations for $\mathcal{Z}_1$. In this case, $||A|| = n!$ and $h(x) = \exp[-\beta\Delta R_1(c_1)]$, and thus

$$\mathcal{Z}_1 = E_{c_1}\Big[\exp[-\beta\Delta R_1(c_1)]\Big]n! \approx \frac{1}{|\mathcal{C}_1|}\sum_{c_1\in\mathcal{C}_1}\frac{\exp[-\beta\Delta R_1(c_1)]}{P(c_1)} \tag{7}$$

To implement the importance sampler along the lines suggested in [2], recall that $\Delta R_q = R_q(c_q) - R_q(c_q^{\perp})$ and $R_q(c_q) = -\log P(\mathcal{G}_q|G_M^{(q)}, c_q)$, where G_q is the observed graph and $G_M^{(q)}$ is the prototype graph. We aim to sample a mapping $c_q \in C_q$ with probability close to $\frac{P(\mathcal{G}_q|G_M^{(q)},c_q)}{\sum_{c_q\in c_q}P(\mathcal{G}_q|G_M^{(q)},c_q)}$. The procedure is as follows. Assume that we know the node-correspondence matrix $\bar{M} = (m_{a\alpha})$ giving the probability that graph node a was generated by prototype node α. Then we can first sample a correspondence for the prototype node 1 with probability m_{1a_1}. The next step is to condition the matrix to the current match by taking into account the structural information between the sampled nodes and all the remainder. Finally we project the conditioned matrix onto a double-stochastic matrix by using the Sinkhorn process [16], yielding the matrix $\bar{M}_1^{a_1}$. We repeat this procedure for each node of the prototype graph, until we have sampled a mapping c_q with probability $P(c_q) = (\bar{M})_{1,a_1} \cdot (\bar{M}_1^{a_1})_{2,a_2} \cdot \ldots \cdot (\bar{M}_{1,\ldots,n-1}^{a_1,\ldots,a_{n-1}})_{n,a_n}$.

4   Experiments

In this section, we report some experimental results of the application of our prototype size selection method on real-world dataset. The dataset used is the

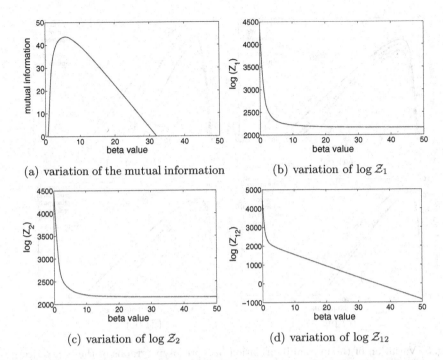

(a) variation of the mutual information   (b) variation of $\log \mathcal{Z}_1$

(c) variation of $\log \mathcal{Z}_2$ (d) variation of $\log \mathcal{Z}_{12}$

Fig. 2. How the mutual information and logarithm of partition functions change as β increases from 0 to 50

COIL [5] which consists of images of different objects, with 72 views of each object obtained from equally spaced viewing directions over 360°. We extract corner features from each image and use the detected feature points as nodes to construct sample graphs by Delaunay triangulation.

We first investigate how the value of the mutual information and the three partition functions vary as the value of β increases. To do this, we randomly partition the graphs from a given object, e.g. the cat images, into a training set and a test set which are of the same size. The bijective mapping of the graphs between the two sets is located by minimizing the sum of the edit distances between the mapped graphs. We learn two prototype graphs of the same size for the two sets using the method in [1]. Given this setting, we compute the value of the mutual information and the logarithms of the three partition functions $\log \mathcal{Z}_1$, $\log \mathcal{Z}_2$ and $\log \mathcal{Z}_{12}$. Figure 2 shows how these quantities vary as we increase the value of β from 0 to 50. From the plot in Figure 2(a), we observe that the mutual information initially increases and achieves the highest value around $\beta = 8$, and afterwards it begins to decrease. To maintain the non negativity of the mutual information, we set its value to zero when it falls below zero. Figure 2(b) and 2(c) respectively show the value of the logarithms of partition functions $\log \mathcal{Z}_1$ and $\log \mathcal{Z}_2$. From the plots it is clear that these two quantities converge to a horizontal asymptote. The reason for this is that while the relative cost of the optimal hypothesis is zero and thus its contribution to the partition function is a constant positive value, the

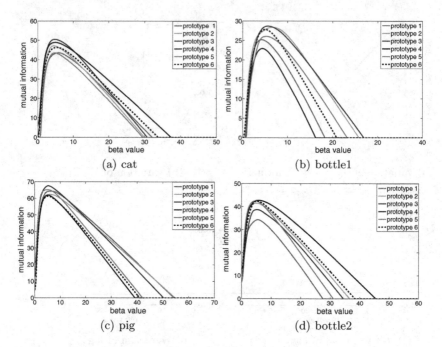

Fig. 3. Variation of the mutual information of 6 prototype graphs of the four objects

exponential of the relative costs given by the non optimal hypotheses converges to zero as β increases, thus yielding the observed horizontal asymptote. On the other hand, the logarithm of the joint partition function $\log \mathcal{Z}_{12}$ in Figure 2(d) continues to decrease as β increases. This indicates that the optimal hypotheses of the graphs in the test set do not necessarily generalize to the optimal hypotheses of their mapped graphs in the training set. For this reason the relative costs of all the hypotheses in the joint partition function are positive values. As a result their exponentials converge to zero as β increases. Consequently, the joint partition function converges to zero and its corresponding logarithm becomes both large and negative.

Our second experimental goal is to select the optimal size of the prototype graph for several objects from the COIL dataset. Here the objects we used are the cat, pig and two bottles. To perform these tasks, for each object we learn 6 prototype graphs of different size using the method in [1] and then compute the mutual information of these prototype graphs. The optimal size of the prototype graph is that which gives the highest mutual information as β is varied. Figure 3 shows plots of the mutual information versus β for the four objects. From the plots it is clear that for each objet there is a prototypes size that gives optimized performance. Finally, note that unlike what is expected using other standard model complexity selection methods, which may choose the model with the smallest size, in our experiments we observe that in 3 out of 4 objects the proposed method favours some value between the largest and the smallest size.

5 Conclusion

In this paper we have developed a method for selecting the optimal size of a prototype graph used to represent a set of sample graphs. The optimal size of the prototype graph is selected so as to maximize the mutual information of the two partitioned sets of the sample graphs. To compute the mutual information, we extend the theory of approximate set coding from the vector domain to the graph domain. Experimental results show that our method works well for prototype graph selection in object recognition. Future work will concentrate on validating our prototype graph size selection method. Moreover, while the prototype selection step is currently a separate post processing step which takes place after the learning procedure, we intend to investigate how to integrate the two together, so as to reduce the overall complexity.

Acknowledgement. Edwin Hancock was supported by a Royal Society Wolfson Research Merit Award.

References

1. Han, L., Wilson, R.C., Hancock, E.R.: A Supergraph-based Generative Model. In: ICPR, pp. 1566–1569 (2010)
2. Torsello, A.: An Importance Sampling Approach to Learning Structural Representations of Shape. In: CVPR, pp. 1–7 (2008)
3. Buhmann, J.M., Chehreghani, M.H., Frank, M., Streich, A.P.: Information Theoretic Model Selection for Pattern Analysis. JMLR: Workshop and Conference Proceedings 7, 1–15 (2011)
4. Buhmann, J.M.: Information Thereotic Model Validation for Clustering. In: International Symposium on Information Theory, pp. 1398–1402 (2010)
5. Nene, S.A., Nayar, S.K., Murase, H.: Columbia Object Image Library(COIL100). Columbia University (1996)
6. National Center for Biotechnology Information, http://www.ncbi.nlm.nih.gov
7. Hammersley, J.M., Handscomb, D.C.: Monte Carlo Methods. Wiley, New York (1964)
8. Han, L., Hancock, E.R., Wilson, R.C.: Learning Generative Graph Prototypes Using Simplified von Neumann Entropy. In: GbRPR, p. 4251 (2011)
9. Rissanen, J.: Modelling by Shortest Data Description. Automatica 14, 465–471 (1978)
10. Schwarz, G.E.: Estimating the dimension of a model. Annals of Statistics 6, 461–464 (1978)
11. Foster, D.P., George, E.I.: The Risk Inflation Criterion for Multiple Regression. Annals of Statistics 22, 1947–1975 (1994)
12. White, D., Wilson, R.C.: Parts based generative models for graphs. In: ICPR, pp. 1–4 (2008)
13. Akaike, H.: A new look at the statistical model identification. IEEE Transactions on Automatic Control 19, 716–723 (1974)
14. Grnwald, P.D., Myung, I.J., Pitt, M.A.: Advances in Minimum Description Length: Theory and Applications. The MIT Press (2005)
15. Luo, B., Hancock, E.R.: Structural graph matching using the EM alogrithm and singular value decomposition. IEEE Transactions on PAMI 23, 1120–1136 (2001)
16. Sinkhorn, R.: A Relationship Between Arbitrary Positive Matrices and Doubly Stochastic Matrices. The Annals of Mathematical Statistics 35, 876–879 (1964)

Graph Kernels: Crossing Information from Different Patterns Using Graph Edit Distance

Benoit Gaüzère[1], Luc Brun[1], and Didier Villemin[2]

[1] GREYC CNRS UMR 6072, Caen, France
[2] LCMT CNRS UMR 6507, Caen, France

Abstract. Graph kernels allow to define metrics on graph space and constitute thus an efficient tool to combine advantages of structural and statistical pattern recognition fields. Within the chemoinformatics framework, kernels are usually defined by comparing the number of occurences of patterns extracted from two different graphs. Such a graph kernel construction scheme neglects the fact that similar but not identical patterns may lead to close properties. We propose in this paper to overcome this drawback by defining our kernel as a weighted sum of comparisons between all couples of patterns. In addition, we propose an efficient computation of the optimal edit distance on a limited set of finite trees. This extension has been tested on two chemoinformatics problems.

1 Introduction

Chemoinformatics aims to predict molecule's properties from their structural similarity. Most of existing methods are based on fingerprints defined as collections of descriptors such as boiling point, logP, molar refractivity, etc. An alternative strategy consists to extract a set of descriptors directly from the molecular graph $G = (V, E, \mu, \nu)$, where the unlabeled graph (V, E) encodes the structure of the molecule while μ maps each vertex to an atom's label and ν characterizes a type of bond between two atoms (single, double, triple or aromatic). Considering this representation, similarity between molecules can be deduced from the similarity of their molecular graphs.

Graph kernels can be understood as symmetric graph similarity measures. Using a semi definite positive kernel, the value $k(G, G')$, where G and G' encode two graphs, corresponds to a scalar product between two vectors $\psi(G)$ and $\psi(G')$ in an Hilbert space. Graph kernels provide thus a natural connection between structural and statistical pattern recognition fields. A large family of kernels is based on bags of patterns. These methods extract a bag of patterns from each graph and deduce graph's similarity from bag's similarity by comparing the number of occurrences of each pattern within both graphs. Most of existing methods are defined on linear patterns [6]. Such methods have generally a low complexity but are limited by the lack of expressivity of linear patterns on graphs. In order to use more structural information, some methods are based on non linear patterns, such as the tree-pattern kernel [7]. This last method is

G.L. Gimel' farb et al. (Eds.): SSPR & SPR 2012, LNCS 7626, pp. 42–50, 2012.
© Springer-Verlag Berlin Heidelberg 2012

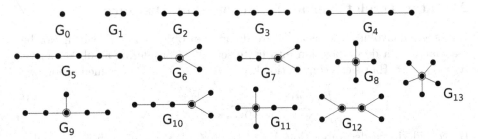

Fig. 1. Set of sub structures enumerated from the graphs

based on an implicit enumeration of tree patterns, ie. trees where a node can appear more than once.

Another approach, called treelet kernel [4], computes an explicit enumeration of a limited set of subtrees. Treelet kernel is a graph kernel defined as a kernel between two sets of patterns extracted from both graphs to be compared. The set of extracted patterns, called treelets and denoted $\mathcal{T}$, is composed of all labeled trees with a number of nodes less than or equal to 6 (Figure 1). Based on the enumeration of this set of substructures, each graph G is associated to a vector $f(G)$. Each component of this vector $f_t(G)$ equals the number of occurrences of a given treelet t in G:

$$f(G) = (f_t(G))_{t \in \mathcal{T}(G)} \text{ with } f_t(G) = |(t \trianglelefteq G)| \tag{1}$$

where $\mathcal{T}(G)$ denotes the set of treelets extracted from G and $\trianglelefteq$ the sub graph isomorphism relationship. Using this vector representation, similarity between treelet distributions is computed using a sum of sub kernels between treelet's frequencies:

$$K_{\mathcal{T}}(G, G') = \sum_{t \in \mathcal{T}(G) \cap \mathcal{T}(G')} k(f_t(G), f_t(G')) \tag{2}$$

where $k(.,.)$ defines any positive definite kernel between real numbers such as linear kernel, Gaussian kernel or intersection kernel. Unfortunately, similarity of occurrences is only computed between isomorphic patterns and not between similar patterns. From a mathematical point of view, computing similarities only between isomorphic patterns relies to consider that each axis encoding a pattern is orthogonal with all other axis. This assumption is dubious since large patterns are composed by smaller ones, hence encoding partially the same information. Moreover, from a chemical point of view, two sub structures may have a similar influence on a chemical property if they slightly differ, hence showing the interest of crossing information collected from differents treelets.

In order to capture this similarity, we propose to extend treelet kernel by adding comparisons of non isomorphic treelets. In Section 2, we propose to weigh the influence of any pair of treelets by their edit distance. In Section 2.1, we propose an efficient way to compute an exact edit distance between treelets. Then, in Section 3, this treelet kernel extension is tested and discussed on an experimental comparison involving two chemoinformatics problems.

2   Inter Treelet Kernel Based on Edit Distance

Haussler's convolution kernels [5] are defined on objects $x \in \mathcal{X}$ which can be associated to a decomposition into finite sets $\mathcal{X}_x$. Considering a sub kernel $k : \mathcal{X}_x \times \mathcal{X}_x \to \mathbb{R}$, Haussler's convolution kernel $K : \mathcal{X} \times \mathcal{X} \to \mathbb{R}$ is defined as follows:

$$K(x, y) = \sum_{(x', y') \in \mathcal{X}_x \times \mathcal{X}_y} k(x', y') \tag{3}$$

By considering a decomposition $\mathcal{X}_G = \{(t, f_t(G)) | t \trianglelefteq G\}$ of each graph and a tensor product $(k \otimes k')$ of two kernels $k' : \mathcal{T} \times \mathcal{T} \to \mathbb{R}$ and $k : \mathbb{R} \times \mathbb{R} \to \mathbb{R}$, treelet kernel (Equation 2) can be reformulated as a convolution kernel:

$$K(G, G') = \sum_{\substack{(t, f_t(G)) \in \mathcal{X}_G \\ (t', f_{t'}(G')) \in \mathcal{X}_{G'}}} (k' \otimes k)(t, f_t(G), t', f_{t'}(G'))$$

$$K(G, G') = \sum_{\substack{(t, f_t(G)) \in \mathcal{X}_G \\ (t', f_{t'}(G')) \in \mathcal{X}_{G'}}} k'(t, t') k(f_t(G), f_{t'}(G')) \tag{4}$$

where $k(f_t(G), f_{t'}(G'))$ is defined as in Equation 2 and $k'(t, t') = 1 \iff t \simeq t', 0$ otherwise. Note that $k'(t, t')$ is equal to 1 only if treelet t in $\mathcal{T}(G)$ is isomorphic to t' and thus belongs simultaneously to $\mathcal{T}(G)$ and $\mathcal{T}(G')$ (Equation 2). Such a definition of $k'(t, t')$ restricts comparison of occurrences to isomorphic treelets. In order to relax this restriction and based on the assumption that similar structures should have a similar chemical activity, we propose to define $k'(t, t')$ in Equation 4 as a measure of similarity between t and t'. This similarity measure is based on the graph edit distance defined as the sequence of operations transforming G into G' with a minimal cost [8]. Such a sequence, called an edit path, may include vertex or edge addition, removal and relabeling. Given a cost function $c(.)$ associated to each operation, the cost of a sequence of operations is defined as the sum of each elementary operation's costs. A high edit distance indicates a low similarity between two graphs while a small one indicates a strong similarity. Unfortunately, trivial kernels defined on graph edit distance are not always semi definite positive and thus do not define valid kernels. In order to define semi definite positive kernels, we apply a regularization scheme as defined by [4, 8]. According to [8], the computational cost of the exact edit distance grows exponentially with the size of graphs. To overcome this problem, Fankhauser and al. [3] propose a method to compute an approximate edit distance in $O(n^3)$ where n is equals to the number of nodes and to the maximal degree of both graphs. Such an edit distance computation provides an efficient way to compute an approximate edit distance between graphs at the cost of a lower precision.

2.1   Exact Treelet Edit Distance

Exact edit distance is hard to compute when considering the whole set of possible graphs. Given a finite set of n structures $B = \{(V_1, E_1), \ldots, (V_n, E_n)\}$, we thus restrict our study to sets of graphs D such that for any $G = (V, E, \mu, \nu) \in D$ we have $(V, E) \in B$. We show in the remaining of this section that within this framework, exact edit distance may be computed within a reasonable computational time using ad hoc methods. In order to present such methods, let us introduce some common definitions. A graph $G' = (V', E', \mu', \nu')$ is a structural sub graph of $G = (V, E, \mu, \nu)$, denoted $G' \trianglelefteq_s G$, iff $V' \subseteq V$ and $E' \subseteq E \cap (V' \times V')$. In addition, if $\mu'_{|V'} = \mu$ and $\nu'_{|E'} = \nu$, $f_|$ denoting the restriction of function f to a particular domain, then G' is a sub graph of G, denoted $G' \trianglelefteq G$. A graph $G = (V, E, \mu, \nu)$ is structurally isomorphic to a graph $G' = (V', E', \mu', \nu')$, denoted $G \simeq_s G'$ iff there exists a bijective function $f : V \to V'$ such that $(u, v) \in E \Leftrightarrow (f(u), f(v)) \in E'$. If $\mu' \circ f = \mu$ and $\nu' \circ f = \nu$, then G is isomorphic to G', denoted $G \simeq G'$. If $G = G'$ then f is called an automorphism. If f is only injective then it exists a sub graph isomorphism between G and G'. A graph $\hat{G}$ is a maximal common sub graph of G_1 and G_2 if it is a sub graph of G_1 and G_2 and if it is not a sub graph of any other common sub graph of G_1 and G_2. A graph $\hat{G}$ is called a maximum common sub graph of G_1 and G_2 if it is a common sub graph of G_1 and G_2 with a maximal number of nodes. The notions of maximal structural sub graph and maximum structural sub graph are defined the same way using the notion of structural sub graph.

Under mild assumptions [1], the sequence of edit operations encoding an edit path can be ordered into a sequence of deletions, substitutions and additions as illustrated in Figure 2(a). The first sequence transforms the initial graph G_1 into one of its sub graphs $\hat{G}_1$ by deleting a set of nodes corresponding to $V_1 - \hat{V}_1$ and a set of edges corresponding to $E_1 - \hat{E}_1$. The second sequence represents the set of substitutions transforming $\hat{G}_1$ into $\hat{G}_2$. This set of substitutions defines a one to one matching between $\hat{V}_1$ and $\hat{V}_2$ on the one hand and between $\hat{E}_1$ and $\hat{E}_2$ on the other. Substitutions matching two elements having a same label are denoted as identical substitutions. Finally, the last sequence corresponds to the addition of a set of nodes and edges in order to transform $\hat{G}_2$ into G_2. Note that the set of operations transforming $\hat{G}_1$ into $\hat{G}_2$ is only composed of substitutions

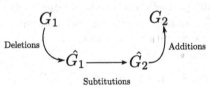

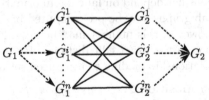

(a) Different steps describing an edit path.

(b) Possible edit paths passing through maximum common structural sub graphs $\{\hat{G}_1^1, \dots, \hat{G}_1^n\}$ and $\{\hat{G}_2^1, \dots, \hat{G}_2^n\}$. Dashed lines correspond to structural operations, other to substitutions.

Fig. 2. General edit path scheme and edit paths passing through maximum common structural sub graphs

which do not modify the structure of graphs. Therefore, $\hat{G}_1$ and $\hat{G}_2$ have the same structure and correspond to two structurally isomorphic sub graphs of G_1 and G_2. We define costs on edit operations as non negative constant functions for edges (c_{e*}) and vertex (c_{v*}) deletions (c_{*d}), insertions (c_{*i}) or substitutions (c_{*s}). In addition, the cost associated to an identical substitution is equals to 0 since such an operation does not modify the graph. Using the representation described in Figure 2(a) and cost functions previously defined, the cost of any edit path is equal to:

$$\gamma(P) = |V_1 - \hat{V}_1|c_{vd} + |E_1 - \hat{E}_1|c_{ed} + V_f c_{vs} + E_f c_{es} + |V_2 - \hat{V}_2|c_{vi} + |E_2 - \hat{E}_2|c_{ei} \quad (5)$$

with V_f, resp. E_f, denoting the number of non identical substitutions on nodes, resp. edges, required to transform $\hat{G}_1$ into $\hat{G}_2$. Bunke has shown that under some slightly different conditions on edge operations, constraining the costs to $c_{vd} + c_{vi} < c_{vs}$ and $c_{es} < c_{vs}$ induces that $\hat{G}_1 \simeq \hat{G}_2$ correspond to a maximum common sub graph of G_1 and G_2 [2]. However, maximum common sub graph of two graphs depends both on structure and labels. This last point does not allow us to use efficiently our assumption that the number of different structures of any set of graphs is bounded and known a priori. We propose to study if different conditions between costs can lead to a possible efficient algorithm to compute an exact edit distance.

Proposition 1. *Given two graphs G_1, G_2, let us denote by δ_v the number of vertices of their maximum structural common sub graph and by δ_e, the maximal number of edges, of their structural common sub graphs. If $\frac{c_{vd} + c_{vi}}{c_{vs}} \geq \delta_v + \frac{c_{es}}{c_{vs}}\delta_e$ and $\frac{c_{ed} + c_{ei}}{c_{es}} \geq \delta_e + \frac{c_{vs}}{c_{es}}\delta_v$, then $\hat{G}_1$ is a maximal common structural sub graph of G_1 and G_2.*

Proof. Can be found in [1]

Considering two graphs G_1 and G_2, this first proposition ensures that sequences of structural operations transform G_1 and G_2 into one of their maximal common structural sub graphs. Since the maximal common structural sub graph does not depend on labeling information, the set of maximal common structural sub graphs may be pre computed between any pair of structure belonging to B. However, this number may be large hence forbidding an efficient pre computation of the exact edit distance. By restricting conditions on costs, we obtain a relationship leading to a reduced set of sub structures:

Proposition 2. *Let us suppose that $c_{es} = c_{ed} + c_{ei}$. Given two graphs G_1 and G_2, let us further denotes by δ_v the number of vertices of their maximum common structural sub graphs and by δ_e the maximal number of edges of all maximal common structural sub graphs. If $c_{vd} + c_{vi} \geq c_{vs}\delta_v + c_{es}\delta_e$, then $\hat{G}_1$ is a maximum common structural sub graph of G_1 and G_2.*

Proof. Can be found in [1]

Proposition 2 states that under some hypothesis on the costs c_{*d}, c_{*i} and c_{*s} any optimal edit path between two graphs G_1 and G_2 should pass through one of their maximum common structural sub graphs. Let us consider two graphs G_1 and G_2 and without loss of generality let us suppose that these two graphs share only one maximum common structural sub graph $\hat{G} = (\hat{V}, \hat{E})$. Let us denote as $\{\hat{G}_1^0, \ldots, \hat{G}_1^i, \ldots, \hat{G}_1^{\hat{n}_1}\}$ and $\{\hat{G}_2^0, \ldots, \hat{G}_2^j, \ldots, \hat{G}_2^{\hat{n}_2}\}$ the sets of sub graphs of G_1 and G_2 structurally isomorphic to G (Figure 2(b)). By Proposition 2, any optimal edit path P between G_1 and G_2 should pass through one $\hat{G}_1^i$ and $\hat{G}_2^j$. The cost associated to P can be decomposed into two parts: a structural cost $\gamma_{struc}(P)$, corresponding to insertion and deletion operations, and a substitution cost $\gamma_{label}(P)$, corresponding to the label substitutions required to transform $\hat{G}_1^i$ into $\hat{G}_2^j$:

$$\gamma(P) = \gamma_{struc}(P) + \gamma_{label}(P) \tag{6}$$

Following Equation 5, we have:

$$\begin{cases} \gamma_{struc}(P) = |V_1 - \hat{V}_1|c_{vd} + |E_1 - \hat{E}_1|c_{ed} + |V_2 - \hat{V}_2|c_{vi} + |E_2 - \hat{E}_2|c_{ei} \\ \gamma_{label}(P) = V_f c_{vs} + E_f c_{es} \end{cases} \tag{7}$$

For any $i \in \{1, \ldots, n_1\}$, since $\hat{G}_1^i \trianglelefteq G_1$, we have $\hat{V}_1^i \subseteq V_1$ and $\hat{E}_1^i \subseteq E_1$ and thus:

$$\begin{cases} |\hat{V}_1^i - V_1| = |V_1| - |\hat{V}_1^i| = |V_1| - |\hat{V}| \\ |\hat{E}_1^i - E_1| = |E_1| - |\hat{E}_1^i| = |E_1| - |\hat{E}| \end{cases} \tag{8}$$

Similarly, the same holds for G_2 and $\hat{G}_2^j$ for any $j \in \{1, \ldots, n_2\}$. Structural cost corresponding to edit path P is thus equals to:

$$\begin{aligned} \gamma_{struct}(P) = &|V_1|c_{vd} + |V_2|c_{vi} + |E_1|c_{ed} + |E_2|c_{ei} \\ &- |\hat{V}|(c_{vd} + c_{vi}) - |\hat{E}|(c_{ed} + c_{ei}) \end{aligned} \tag{9}$$

Computing substitution cost $\gamma_{label}(P)$ (Equation 7) relies on computing the number of non identical node substitutions V_f and edge substitutions E_f transforming $\hat{G}_1^i$ into $\hat{G}_2^j$. Let $\Phi(\hat{G})$ denotes the set of structural automorphisms of $\hat{G}$. Given both sub graphs $\hat{G}_1^i$ and $\hat{G}_2^j$, each automorphism $\phi \in \Phi(\hat{G})$ induces a mapping of $\hat{G}_1^i$ onto $\hat{G}_2^j$ and thus a substitution of the label of each vertex v (resp. edge e) of $\hat{G}_1^i$ onto the label of $\phi(v)$ (resp. $\phi(e)$) in $\hat{G}_2^j$. More precisely, let us denote by $P_{i,j,\phi}$ the edit path associated to the triplet $(\hat{G}_1^i, \hat{G}_2^j, \phi)$. The number of non identical substitutions V_f and E_f induced by $P_{i,j,\phi}$ is equals to:

$$\begin{aligned} V_f(P_{i,j,\phi}) &= |\{v \in \hat{V}_1 \mid \hat{\mu}_1^i(v) \neq \hat{\mu}_2^j(\phi(v))\}| \\ E_f(P_{i,j,\phi}) &= |\{(v, v') \in \hat{E}_1 \mid \hat{\nu}_1^i(v, v')) \neq \hat{\nu}_2^j(\phi(v), \phi(v'))\}| \end{aligned} \tag{10}$$

Substitution cost of edit path $P_{i,j,\phi}$ is thus equals to $\gamma_{label}(P_{i,j,\phi}) = V_f(P_{i,j,\phi})c_{ns} + E_f(P_{i,j,\phi})c_{es}$. Let us denotes by P_{opt} the edit path minimizing the substitution cost:

$$P_{opt} = P_{i_0,j_0,\phi_0} \text{ with } (i_0, j_0, \phi_0) = \underset{(i,j,\phi)\in\{1,...,n_1\}\times\{1,...,n_2\}\times\Phi(\hat{G})}{\text{argmin}} \gamma_{label}(P_{i,j,\phi})$$

$$(11)$$

Since $\gamma_{struct}(P_{i,j,\phi})$ is the same for any $(i, j, \phi) \in \{1, \ldots, n_1\} \times \{1, \ldots, n_2\} \times \Phi(\hat{G})$ (Equation 9), P_{opt} is an edit path having a minimal cost. Therefore, under our assumptions, the edit path associated to the edit distance is the one which passes through the pair of maximum common structural sub graphs and which minimizes the number of substitutions (Equation 11). This exact edit distance computation algorithm can be applied to treelets since the set of treelets is composed of 14 different structures. In addition, by restricting the set of edit paths to the ones which preserve the connectedness of intermediate graphs [1], we can obtain a lower bound on the ratio between substitutions and insertion/deletion costs.

Proposition 3. *Considering edit paths preserving connectedness and given two trees T_1, $T_2 \in \mathcal{T}$, if $\frac{c_{vd}+c_{vi}}{c_{vs}} \geq \delta_v$ and $\frac{c_{ed}+c_{ei}}{c_{es}} \geq \delta_v - 1$, then $\hat{G}_1$ is a maximum common structural sub tree of T_1 and T_2.*

Proof. Can be found in [1]

When computing tree edit distance on the set of treelets, δ_v is bounded by 6 and if we define costs as symmetric, i.e. $c_{vd} = c_{vi}$ and $c_{ed} = c_{ei}$, bounds on costs lead to: $c_{vd} > 3c_{vs}$ and $c_{ed} > 2.5c_{es}$. Since the set of treelets represents all trees having a size less than or equal to 6, the maximum common structural sub tree of two treelets T_1 and T_2 is a treelet. The set of possible sub graphs and automorphisms for any pair of treelets can be easily pre computed since we have to consider only 14 patterns. Therefore, computing exact edit distance between two treelets consists in comparing at most $\max_{i,j\in\{0,...,13\}}(n_i * n_j * |\Phi_{ij}|)$ label sequences where Φ_{ij} denotes the set of automorphisms of the maximum common structural subtree $\hat{T}$ of treelets T_i and T_j and n_i, n_j the numbers of sub trees of T_i and T_j isomorphic to $\hat{T}$. The value of this product on the set of treelets is bounded by 120, hence inducing a constant time complexity for the computation of the exact tree edit distance. Note that, without our restriction to a set of specific tree structures, the complexity of the edit distance calculation between labeled unordered unrooted trees is NP-Complete [9]. In addition, given a trainset D, our kernel is defined as the 0-extension of the kernel defined by the regularization of matrix $\left(e^{-d(t_i,t_j)/\sigma}\right)_{(i,j)\in\{1,...,n\}^2}$, where n is the number of different treelets extracted from D. Note that this regularisation has to be performed only once since this kernel only operates on treelets and not directly on graphs.

3 Experiments

Our first experiment evaluates our inter treelet kernel on a regression problem which consists in predicting molecule's boiling points[1]. This dataset is composed

---

[1] All databases are available on the IAPR TC15 Web page:
   http://www.greyc.ensicaen.fr/iapr-tc15/links.html#chemistry

Table 1. Boiling point prediction

Method	RMSE ($^\circ C$)	Learning Time (s)
1 Random Walks Kernel	18.72	19.10
2 Gaussian edit distance	10.27	1.35
3 Tree Pattern Kernel	11.02	4.98
4 Treelet Kernel	8.10	0.07
5 Treelet Kernel with backward selection	6.75	10363
6 Inter Treelet Kernel with approximate edit distance	6.09	0.70
7 Inter Treelet Kernel with exact edit distance	5.89	0.50

of 183 acyclic molecules and prediction is performed using a 10-fold cross validation. Table 1 shows prediction accuracy and time required to compute the Gram matrix and to perform linear regressions for each method.

Due to the limited expressivity of linear patterns, random walks kernel [6] (Table 1, Line 1) does not permit to predict correctly molecule's boiling points. Line 2 shows results obtained by a Gaussian kernel applied on graph edit distance [8]. This last method based on global similarity of graphs obtains a better result than kernel based on linear patterns. In the same way, tree pattern [7] and treelet kernels (Table 1, Lines 3 and 4) improve the accuracy of prediction model based on linear patterns by including information encoded by non linear patterns. Then, Line 5 shows results obtained by combining treelet kernel with a variable selection step [4] which leads to a better prediction accuracy (Table 1, Line 5) than Treelet Kernel without variable selection step (Table 1, Line 4), at the price of an high computation time. Lines 6 and 7 show results obtained using our inter treelet kernel. First, inter treelet kernel obtains a better prediction accuracy than using treelet kernel restricted to the comparison of similar treelets, hence showing the relevance of including pairs of non isomorphic treelets within kernel computation. Second, we can note that the use of an exact edit distance provides a slightly more accurate weighting than using an approximate edit distance (Table 1, Lines 6 and 7).

Our second experiment is defined as a classification problem on the monoamine oxidase (MAO) dataset which is composed of 68 molecules divided into two classes: 38 molecules inhibit the monoamine oxidase (antidepressant drugs) and 30 do not. Classification accuracy is measured for each method using a leave one out

Table 2. Classification accuracy on the monoamine oxidase (MAO) dataset

Method	Classification Accuracy	Learning Time (s)
1 Random Walks Kernel	82% (56/68)	58.80
2 Gaussian edit distance	90% (61/68)	1.03
3 Tree Pattern Kernel	96% (65/68)	7.18
4 Treelet Kernel	91% (62/68)	0.3
5 Inter Treelet Kernel with approximate edit distance	93% (63/68)	0.67
6 Inter Treelet Kernel with exact edit distance	94% (64/68)	0.54

procedure with a two-class SVM. This classification scheme is made for each of the 68 molecules of the dataset. In this experiment, best results are obtained using a Tree Pattern Kernel (Table 2, Line 3). Methods based on non linear patterns (Table 2, Lines 3 to 6) outperform methods based on linear patterns (Table 2, Line 1) and graph edit distance (Table 2, Line 2). In addition, the better accuracy obtained by methods crossing information from differents patterns (Table 2, Lines 5 and 6) shows the relevance of the proposed extension. As highlighted on our first experiment, difference between the two methods may be explained by the better accuracy provided by the exact edit distance. Note that the 3 molecules misclassified by Tree Pattern Kernel method are also misclassified by all others methods.

4   Conclusion

In this article, we have presented an extension of the Treelet Kernel which consists in crossing information encoded by non isomorphic treelets according to their structural similarities. In addition, we have defined a new relation between edit distance and maximum common structural sub graphs which leads to an efficient computation of edit distance between treelets. One major perspective of this work is to define the weighting of non isomorphic treelet pairs using their relevance according to a property to predict and no more by an a priori similarity measure such as edit distance.

References

1. Brun, L., Gaüzère, B., Fourey, S.: Relationships between graph edit distance and maximal common unlabeled subgraph. Technical report, CNRS UMR 6072 GREYC (2012), http://hal.archives-ouvertes.fr/hal-00714879
2. Bunke, H.: Error correcting graph matching: On the influence of the underlying cost function. IEEE Transactions on Pattern Analysis and Machine Intelligence 21(9), 917–922 (1999)
3. Fankhauser, S., Riesen, K., Bunke, H.: Speeding Up Graph Edit Distance Computation through Fast Bipartite Matching. In: Jiang, X., Ferrer, M., Torsello, A. (eds.) GbRPR 2011. LNCS, vol. 6658, pp. 102–111. Springer, Heidelberg (2011)
4. Gaüzère, B., Brun, L., Villemin, D.: Two new graph kernels and applications to chemoinformatics. Pattern Recognition Lett. (in Press, 2012)
5. Haussler, D.: Convolution kernels on discrete structures. Technical report, Dept. of Computer Science, University of California at Santa Cruz (1999)
6. Kashima, H., Tsuda, K., Inokuchi, A.: Kernels for graphs, ch. 7, pp. 155–170. MIT Press (2004)
7. Mahé, P., Vert, J.-P.: Graph kernels based on tree patterns for molecules. Machine Learning 75(1), 3–35 (2009)
8. Neuhaus, M., Bunke, H.: Bridging the gap between graph edit distance and kernel machines. World Scientific Pub. Co. Inc. (2007)
9. Zhang, K., Statman, R., Shasha, D.: On the editing distance between unordered labeled trees. Information Processing Letters 42(3), 133–139 (1992)

Mode Seeking Clustering by KNN
and Mean Shift Evaluated

Robert P.W. Duin[1], Ana L.N. Fred[2], Marco Loog[1], and Elżbieta Pękalska[3]

[1] Pattern Recognition Laboratory,
Delft University of Technology, The Netherlands
r.duin@ieee.org,m.loog@tudelft.nl
[2] Department of Electrical and Computer Engineering, Instituto Superior Técnico
(IST - Technical University of Lisbon), Portugal
afred@lx.it.pt
[3] School of Computer Science,
University of Manchester, United Kingdom
pekalska@cs.man.ac.uk

Abstract. Clustering by mode seeking is most popular using the mean shift algorithm. A less well known alternative with different properties on the computational complexity is kNN mode seeking, based on the nearest neighbor rule instead of the Parzen kernel density estimator. It is faster and allows for much higher dimensionalities. We compare the performances of both procedures using a number of labeled datasets. The retrieved clusters are compared with the given class labels. In addition, the properties of the procedures are investigated for prototype selection.

It is shown that kNN mode seeking is well performing and is feasible for large scale problems with hundreds of dimensions and up to a hundred thousand data points. The mean shift algorithm may perform better than kNN mode seeking for smaller dataset sizes.

1 Introduction

The mean shift clustering procedure is based on a 1975 paper by Fukunaga [7]. It has been made most popular by Cheng [2] and by Comaniciu and Meer [3]. They showed how the idea of finding the modes of a non-parametrically estimated probability density function based on the Parzen kernel could be implemented sufficiently fast such that it can be used for segmenting images. As for reliable estimates many data points are needed, this became only feasible when sufficiently large memories and fast CPUs entered the market after 2000. The mean shift algorithm however suffers from the fact that determining and tracking the gradient in high dimensional spaces is still computationally heavy. Its use is thereby mainly restricted to applications with small sets of features, e.g. three color images.

There exists an interesting alternative for the Parzen kernel in mode seeking: the k-Nearest Neighbor (kNN) rule. This has also been shown by Fukunaga and his colleagues Koontz and Narendra in 1976 [9], but it did not receive much attention. An early version of the algorithm has been included in the PRTools

G.L. Gimel' farb et al. (Eds.): SSPR & SPR 2012, LNCS 7626, pp. 51–59, 2012.
© Springer-Verlag Berlin Heidelberg 2012

Matlab package [4] about 15 years ago. We renewed this implementation to make it feasible for 10^4 - 10^5 data points and 10^2 - 10^3 dimensions. It is the purpose of this paper to compare the two mode-seeking algorithms with each other. We will show that the performance of kNN mode seeking is reasonable, sometimes worse, sometimes better than mean shift, but it has the advantage of a significantly better computational efficiency. This will be shown and explained.

Comparing procedures for data analysis in general and for cluster analysis in particular is a mining field. It is very difficult to make general statements and one can easily be deceived. For that reason, we include a discussion (Section 2) on our philosophy on benchmarking cluster procedures and give arguments for the choices we have made. In Section 3, the algorithms will be discussed and our version of kNN mode seeking will be specified. In Section 4, the algorithms are compared on a number of datasets for two performance criteria and their computing times. The paper is finished with a discussion, summarizing the main properties and differences of both algorithms, see Section 5.

2   Comparing Cluster Procedures

The aim of clustering is to find an interesting structure in data, e.. sensor data collected in some scientific study. What is interesting is usually not pre-defined. Any structure that makes sense for building an understanding of the observations may give a hint to the researcher to think in a particular direction. For this reason, clustering is necessarily ill defined. Attempts to specify 'interesting' in terms of numerically well defined criteria may limit the analyst in his exploration. Consequently, a vast number of procedures has been developed, partially as diversity is essential, but also because there is no way inside clustering to estimate performances.

A way out of this dilemma is to use datasets that have already been analyzed by experts and for which they defined labels to identify objects that belong to the same structure (classes). This may be done by inspecting the data itself, or by other means, outside the given data. It is thereby possible to find in retrospect the cluster procedures yielding results that are consistent with expert supplied class labels. In this way, examples of procedures can be found that make sense in real world problems.

Cluster procedures give the following two types of results. They may group the objects, i.e. they define subsets of objects that are in one way or another similar. Some procedures determine in addition, or sometimes just instead, small sets of prototypes or examples of objects that are representative for the whole set. Procedures that don't deliver both can be extended with an additional step to determine the missing information. If just clusters are obtained and no prototypes, then for every cluster its centre (or medoid) will define a prototype (i.e. the object for which the maximum (mean) distance to the other objects in the cluster is minimized). If just prototypes are found, clusters can be defined by applying the nearest mean classifier trained by the prototypes to all objects.

These two results give two different ways to determine how consistent a cluster result is with a given class labeling. It depends on the number of clusters or prototypes that have been found. Almost every clustering procedure has a parameter that controls this number. In many studies, attempts are made to determine an optimal number of clusters. In comparing clustering results with given class labeling, this is not always appropriate. The expert may have used a varying resolution in distinguishing and naming classes. To remove this problem, we decided to use in this study a series of clusterings obtaining $k = 1$ up to (e.g.) $k = 50$ clusters. Next a performance measure is computed relating the result of all these clusterings with the given, true class labels. We used two performance measures, one that focuses on the obtained clusters and one on the prototypes.

An obtained clustering differs from a given class labeling in two ways. Objects in the same cluster may belong to different classes and objects in different clusters may belong to the same class. Let the cluster index of an object x be given by $C(x) \in \{1, 2, ..., k\}$ and let its object label be $\lambda(x) \in \{1, 2, ..., n\}$. The following two probabilities for an arbitrarily selected pair of objects $\{x_i, x_j\}$ are a measure for the consistency of the clustering with the true class labeling:

$$\epsilon_1 = Prob(C(x_i) \neq C(x_j)|\lambda(x_i) = \lambda(x_j)) \tag{1}$$

$$\epsilon_2 = Prob(C(x_i) = C(x_j)|\lambda(x_i) \neq \lambda(x_j)) \tag{2}$$

A clustering is consistent with the class labelling if $\epsilon_1 = 0$ in case $k <= n$ and $\epsilon_2 = 0$ in case $k >= n$. Both should be zero if $k = n$ in case of consistency. For a set of clusterings with varying k a set of $(\epsilon_1^k, \epsilon_2^k)$ pairs is obtained for every value of k, constituting a curve in the (ϵ_1, ϵ_2)-plane. An example is given in Fig. 1 for the two procedures, kNN mode seeking and mean shift, applied to the Iris dataset, see Section 3 and Section 4 for more details.

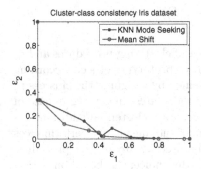
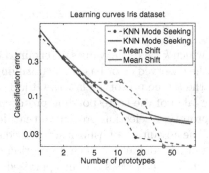

Fig. 1. Cluster-class consistency plot for the Iris dataset

Fig. 2. Learning curves for the Iris dataset. Dashed curves as estimated, solid ones are approximated by Eq. 4.

Points on the horizontal axis relate to clusterings in which every cluster contains just objects that belong to the same class. Points on the vertical axis relate to clusterings in which all objects of the same class are taken by a single cluster.

A point in the origin corresponds to a clustering that fully coincides with the class labeling. The previous two types of clustering are still consistent with this one as either a merge or a split of clusters may produce this origin clustering.

As the two curves in Fig. 1 describing the two clustering procedures cross each other it is not clear whether one of the two is better than the other for the Iris dataset. A solution to judge this can be found be reducing such a curve to a single number. A possible option is to find the point on the curve that is most close to the origin, i.e. use $\epsilon_1 + \epsilon_2$ as a criterion. As this depends on just a single point, we prefer the following. Like in judging ROC curves that describe the trade-off between two types of classification errors [1], we may use the Area Under the Curve (AUC) for judging it. In this case, the lower the better. If the AUC is zero the set of clusterings is perfect as all clusterings are consistent with the class labels.

In case the clustering procedure results in prototypes, another way of judging its performance is possible. By simulating an active learning scenario, the labels of these prototypes may be used to label the clusters they belong to. By repeating such a procedure as a function of the number of prototypes, a learning curve, is created, see Fig. 2 for an example. To convert such a curve into a single performance measure, the learning curve, the classification error ϵ as a function of the number of training samples k, can be approximated by the following function:

$$\epsilon(k) = \epsilon_0 * k^{-\alpha} + \epsilon_\infty \tag{3}$$

in which ϵ_0 is the starting value and ϵ_∞ is the asymptotic value that equals here the nearest neighbor error for an infinite training set. The important parameter is the *learning speed* α that can be considered as a measure for the quality of the prototypes. We estimate α by

$$\alpha = argmin_\alpha \sum_k (\hat{\epsilon}_k - \epsilon(k))^2 \tag{4}$$

in which k takes the values determined by the set of clustering procedures and $\hat{\epsilon}_k$ is the observed error based on classifying the out-of-prototypes objects according to the k true prototype labels. Objects are classified by assigning the clusters to the label of the corresponding prototype, or by the nearest mean rule in case of a prototype selection procedure that does not generate clusters in addition.

The advantage of using a supervised criterion for judging cluster performances is that, by definition, such a criterion cannot be used by any cluster procedure itself. This contrasts the unsupervised criteria. Any choice, e.g. based on within and/or between cluster distances would result in a bias towards specific cluster algorithms.

Before discussing algorithms and datasets, we like to emphasize that a comparative study is not a match between procedures. The target is not to find a winner in one way or the other. As algorithms differ, they are good for different datasets. Any cluster algorithm can be considered as an estimator of some statistics of the data defined by the performance measure. Different algorithms relate to different estimators. Any estimator is biased, as it is based on some assumptions or a model. It will be better if these assumptions hold for a considered dataset. What is done in a comparative study over a collection of datasets is that one tries to find out for which problems, which estimators are better, or at least, whether for different estimators different datasets can be found for which they are useful. Any voting or averaging of results over a collection is arbitrary, unless one is sure that the collection is very representative for the problems to be studied in future.

3   The Algorithms

Mode seeking clustering can be considered as an agglomerative approach. First, a density function is estimated for the dataset. In general, it has several local maxima: the modes. In the clustering phase, for every object it is decided to which mode it belongs by following the density gradient from that object until a mode is found. Objects that end up in the same mode are considered to belong to the same cluster.

By this procedure, the number of clusters is identical to the number of modes. Here this approach differs from the Mixture-of-Gaussians (MoG) procedures as for these every component may be related to a cluster, but not every component constitutes its own mode in the mixture. The two mode seeking procedures discussed in this paper are not based on a mixture of Gaussians, but on nonparametric density estimates based on the Parzen kernel (the mean shift procedure) and the k-nearest-neighbor estimator (kNN mode seeking). Both have a width parameter that influences the number of modes in the density estimate. This number may thereby vary between one, for a very wide kernel or a large k, and m for a narrow kernel or $k = 1$. We do not try to optimize the width parameter, but instead consider the clustering results as a function of this parameter.

The use of the Parzen kernel for mode seeking clustering can be traced back to a 1975 paper by Fukunaga [7]. It resulted in the mean shift algorithm, which uses the observation that the shift of the mean of a kernel of a single object after weighing it with the neighboring objects inside the kernel points into the direction of the gradient. It has been made most popular by Cheng [2] and by Comaniciu and Meer [3]. They showed how this idea could be implemented sufficiently fast such that it can be used for segmenting images. For our study, we used the Matlab implementation by Bart Finkston [5]. It is not deterministic as it depends on the order in which objects are considered. Instead of the kernel width we used the number of nearest neighbors like in the kNN mode seeking (see below). The kernel width was set to the average distance to the k-th nearest neighbor over the entire dataset.

kNN mode seeking is originally described by Koontz [9]. It is related to an algorithm studied by Kittler [8] for which Shaffer et al. [10] stated that although it is based on another idea, its results may be very similar to single-linkage hierarchical clustering. Our experiments have shown that this is not true for our version. In our definition of kNN mode seeking, the density of every object is proportional to the distance to its k-th neighbor. We define for every object a pointer to the object with the highest density in its neighborhood. Finally, these pointers are followed to the object that points to itself. It stands for a mode in the density as it is itself the object with the highest density in its neighborhood.

The main difference between the two mode seeking procedures, kNN and mean shift, is that the latter uses a kernel with a neighborhood size that, in terms of distances, is constant over the entire space. kNN on the other hand uses a fixed neighborhood size in terms of the number of objects and thereby adapts itself to areas with higher or lower densities. In addition, from the implementation point of view, it is much faster to jump from object to object than to compute and follow a gradient.

The below description of kNN mode seeking is good for very large data set sizes m (e.g. $m = 10^5$) objects. All pairwise distances are needed twice and as $m^2 = 10^{10}$ distances cannot be stored, they are computed twice. This is done for a set of n neighborhood sizes k (e.g. $n = 25$) in parallel, by which it is needed to store mn densities and mn pointers. These are used to compute mn cluster indices for the n clusterings.

1. Define a set of target neighborhood sizes $K = k_1, k_2, ..., k_n$.
2. Repeat for all m objects x_i.
3.   Compute its distances d_{ij} to all other objects x_j.
4.   Sort them: $s_{ir} = sort_j(d_{ij})$
5.   Store density estimates $\forall k \in K : f_{ij} = 1/s_{ik}$.
6. Next i
7. Repeat for all m objects x_i.
8.   Compute its distances d_{ij} to all other objects x_j.
9.   Rank them: $q_{ir} = argsort_j(d_{ij})$
10.   Store for all $k \in K$ a pointer $p_i = argmax_{r=1,k}(q_{ir})$
11. Next i
12. Repeat for all neighborhood sizes $k \in K$
13.   Repeat until no change $\forall i : p_i = p_{p_i}$
14.   Store clustering for neighborhood size k: $C_k = [p_1, p_2, ...p_m]$
15. Next k

4   Experiments

In this section it is shown that the two procedures defined in Section 3 are both useful. For both of them, datasets can be found for which one is better than the other. The datasets belong to the standard distribution of PRTools [4]. Most of them originate from the UCI repository [6]. Table 1 presents the area-under-the-curve values for the cluster-class consistency plots as defined is Section 2 and Eq. 1-2. The underlined values are the best for a dataset.

In Table 2, the learning speeds (Equation 4) are shown for the same datasets and algorithms. They show how valuable the procedures are for selecting proto-types to be used for labeling an entire dataset. As the mean shift algorithm does not find modes exactly on the position of objects, we used the medoids of the clusters it finds. In comparing Tables 1 and 2 it can be concluded that although

Table 1. The cluster-class-consistency AUC values for a collection of datasets (the lower the better). Underlined are the best results per dataset. The datasets Aviris* and MNIST* are sampled versions (10%) of the originals.

Dataset	m	d	c	kNN	MS
Hepatitis	155	19	2	0.52	0.47
Wine	178	13	3	0.31	0.28
Biomed	194	5	2	0.43	0.28
Glass	214	9	4	0.40	0.34
Malaysia	291	8	20	0.45	0.46
Ecoli	336	7	8	0.20	0.08
Auto-mpg	398	6	2	0.25	0.39
Arrhythmia	420	278	13	0.47	0.29
Breast	699	9	2	0.33	0.05
Diabetes	768	8	2	0.49	0.47
Car	1728	6	4	0.50	0.50
mfeat-fou	2000	76	10	0.27	0.16
Aviris*	2109	200	17	0.37	0.39
MNIST*	6006	784	10	0.35	0.29
Satellite	6435	36	6	0.21	0.17
Ringnorm	7400	20	2	0.50	0.34
Twonorm	7400	20	2	0.10	0.07
ChromoBands	12000	30	24	0.28	0.23

Table 2. The Learning Speed values of the cluster algorithms (the higher the better)

Dataset	kNN	MS
Hepatitis	1.73	0.09
Wine	0.66	0.29
Biomed	0.62	0.47
Glass	0.34	0.16
Malaysia	0.07	0.09
Ecoli	0.32	0.48
Auto-mpg	0.84	0.99
Arrhythmia	1.99	-0.70
Breast	0.76	27.27
Diabetes	0.75	0.26
Car	0.14	0.36
mfeat-fou	0.59	0.47
Aviris*	-0.03	-0.09
MNIST*	0.30	-0.05
Satellite	0.58	0.44
Ringnorm	0.00	0.37
Twonorm	51.76	0.71
ChromoBands	0.32	0.17

our collection of datasets contains more problems for which mean shift has a better AUC value than the kNN procedure, the latter has more problems with a lower learning speed. So for these examples mode shift finds clusters that are more consistent with the classes, but kNN finds better prototypes.

If the modes of the density function have to be used to define clusters, good density estimates should be available. This points to the direction of large datasets in comparison with the (intrinsic) dimensionality. The mean shift algorithm can handle large datasets for low dimensional spaces, but has problems for tracking the density gradient in high dimensions. Our implementation of kNN mode seeking can handle both, large numbers of objects and high dimensions.

It is interesting that the kNN procedure is significantly faster than mean shift, see Table 3. kNN mode seeking can easily handle datasets that are even an order larger than the ones studied here. The mean shift algorithm then fails, either due to intolerable computing times or because of the need to handle too large distance matrices.

5   Discussion

Mode seeking seems a natural procedure for cluster analysis. It is, however, necessary to have a sufficiently large dataset to obtain good density estimates.

Table 3. The total computing times (over about 25 clusterings per problem) in seconds needed for the various experiments

Dataset	m	d	c	kNN	MS
Aviris*	2109	200	17	2	381
ChromB	12000	30	24	72	5976
MNIST*	6006	784	10	43	12974
Ringnorm	7400	20	2	28	3014
Twonorm	7400	20	2	29	1469
Satellite	6435	36	6	22	886

Formally the mean shift procedure should be better able to handle this as the Parzen kernel takes care for a smooth estimate. The kNN procedure, however, offers interesting possibilities for a feasible and fast implementation.

Except for speed the restriction to use densities located in the objects offers a few additional advantages. As it is well defined, there is no inaccuracy in the exact position of the mode. In the mean shift algorithm, thresholds have to be set to determine whether a newly found mode is really new. We studied this in a small experiment based on the mfeat-fou dataset that has 2000 objects. In Tables 4 and 5, the sizes of the largest clusters are given for the two procedures for a set of values for k between 25 and 200. The mean shift procedure has the tendency to select one or a few large clusters and many with a size of one or two objects. kNN mode seeking finds more balanced cluster sizes.

Table 4. The number of objects in the 7 largest clusters found by kNN mode seeking with $k = \{25, ..., 200\}$ for the mfeat-fou dataset (2000 objects)

k							
25	316	245	207	180	159	146	146
50	381	339	303	209	188	142	141
75	595	486	355	253	214	97	
100	746	580	448	226			
125	704	560	502	234			
150	752	577	433	238			
175	741	621	398	240			
200	1086	914					

Table 5. The number of objects in the 7 largest clusters found by mean shift with $k = \{25, ..., 200\}$ for the mfeat-fou dataset (2000 objects)

k							
25	447	327	294	189	184	166	144
50	1410	323	165	67	2	1	1
75	1658	324	2	1	1	1	1
100	1739	252	1	1	1	1	1
125	1734	261	1	1	1	1	1
150	1752	244	1	1	1	1	
175	1996	1	1	1	1		
200	1997	1	1	1			

Another advantage of the kNN mode seeking procedure is that it is based on the object distances only. No computations in the feature space are needed. Consequently, it may operate on given distance matrices and after conversion on similarity matrices as well.

Finally, we showed that for the given collection of datasets kNN mode seeking is a better prototype selector, while the mean shift algorithm found often clusters that are more consistent with given class labelings.

References

1. Bradley, A.P.: The use of the area under the ROC curve in the evaluation of machine learning algorithms. Pattern Recognition 30(7), 1145–1159 (1997)
2. Cheng, Y.: Mean shift, mode seeking, and clustering. IEEE Trans. Pattern Anal. Mach. Intell. 17(8), 790–799 (1995)
3. Comaniciu, D., Meer, P.: Mean shift: A robust approach toward feature space analysis. IEEE Trans. Pattern Anal. Mach. Intell. 24(5), 603–619 (2002)
4. Duin, R., Juszczak, P., de Ridder, D., Paclík, P., Pękalska, E., Tax, D., Verzakov, S.: PRTools 4.1, a Matlab toolbox for pattern recognition, http://prtools.org
5. Finkston, B.: Mean shift clustering
6. Frank, A., Asuncion, A.: UCI machine learning repository (2010), http://archive.ics.uci.edu/ml
7. Fukunaga, K., Hostetler, L.D.: The estimation of the gradient of a density function, with applications in pattern recognition. IEEE Trans. Information Theory 21(1), 32–40 (1975)
8. Kittler, J.V.: A locally sensitive method for cluster analysis. Pattern Recognition 8(1), 23–33 (1976)
9. Koontz, W.L.G., Narendra, P.M., Fukunaga, K.: A graph-theoretic approach to nonparametric cluster analysis. IEEE Trans. Computer 25, 936–944 (1976)
10. Shaffer, E., Dubes, R.C., Jain, A.K.: Single-link characteristics of a mode-seeking clustering algorithm. Pattern Recognition 11(1), 65–70 (1979)

Learning Sparse Kernel Classifiers in the Primal

Zhouyu Fu[1], Guojun Lu[2], Kai-Ming Ting[2], and Dengsheng Zhang[2]

[1] School of Computing, University of Western Sydney, Penrith, NSW 2750, Australia
[2] Gippsland School of IT, Monash University, Churchill, VIC 3842, Australia
z.fu@uws.edu.au, {guojun.lu,kaiming.ting,dengsheng.zhang}@monash.edu

Abstract. The increasing number of classification applications in large data sets demands that efficient classifiers be designed not only in training but also for prediction. In this paper, we address the problem of learning kernel classifiers with reduced complexity and improved efficiency for prediction in comparison to those trained by standard methods. A single optimisation problem is formulated for classifier learning which optimises both classifier weights and eXpansion Vectors (XVs) that define the classification function in a joint fashion. Unlike the existing approach of Wu et al, which performs optimisation in the dual formulation, our approach solves the primal problem directly. The primal problem is much more efficient to solve, as it can be converted to the training of a linear classifier in each iteration, which scales *linearly* to the size of the data set and the number of expansions. This makes our primal approach highly desirable for large-scale applications, where the dual approach is inadequate and prohibitively slow due to the solution of *cubic-time* kernel SVM involved in each iteration. Experimental results have demonstrated the efficiency and effectiveness of the proposed primal approach for learning sparse kernel classifiers that clearly outperform the alternatives.

1 Introduction

Kernel classifiers have been widely used in pattern classification applications due to its superior predictive performance. The major issue with kernel classifiers is the heavy computational cost involved in both training and prediction. While existing methods have mainly focused on reducing the training cost for kernel classifiers especially the Support Vector Machine (SVM) [2], one should not overlook the issue of prediction cost. The reason is that a kernel classifier takes a linear expansion of kernel evaluations for prediction, where the number of expansion terms is usually determined by the size of training data. This makes kernel classifiers quite inefficient for large-scale applications where prediction speed is a main concern, such as classifying pictures, music and web documents in a large collection.

A few methods have been proposed in the literature for learning sparse kernel classifiers with fewer expansion terms in the learned classifiers [3,4,1,5]. The fewer expansions, the sparser the classifier, the smaller number of kernel function evaluations needed, and hence the more efficient the prediction stage is. While early methods such as the Reduced Set (RS) [4] and Reduced SVM (RSVM) [3] focused

G.L. Gimel' farb et al. (Eds.): SSPR & SPR 2012, LNCS 7626, pp. 60–69, 2012.

on pre- and post-processing steps for building sparse kernel classification models, they do not explicitly take into account label information and can result in the loss of discriminant information in fitting the kernel expansions. More recent methods [1,5] adopted a discriminant approach by searching for the expansion vectors (XV) which form the kernel function terms in the resulting classifier so as to maximise the margin and minimise the mis-classification cost. Specifically, Wu et al [1] proposed the Sparse Kernel Learning Algorithm (SKLA), a direct approach for building kernel classifiers with significantly smaller number of XVs and comparable predictive performance to the standard SVM. Compared to the greedy incremental algorithm in [5], SKLA formulates sparse kernel learning in a single optimisation problem and is able to select XVs at arbitrary locations, making it possible for further reduction of the classifier.

A major issue with SKLA is its training complexity. It is an iterative algorithm which involves training a full kernel SVM model in each iteration. This makes it extremely inefficient for large data sets, where SVM training becomes quite costly. On the other hand, it is more likely to have a over-complex kernel classifier with thousands of SVs when we apply standard SVM methods to large data sets. Hence producing sparser classifiers to improve prediction efficiency becomes a real issue for problems with large data sets. For these problems, one needs to have an effective algorithm for sparse kernel classifier learning with low computational cost in *both training and testing*.

The main contribution of this paper is a solution to the above problem that scales well to large data sets and produces sparse prediction models for testing. A similar formulation to [1] is developed using a differentiable loss function. This allows us to tackle the resulting optimisation problem directly in its primal form instead of converting it to the dual form for solution as in [1]. Moreover, with transformation of variables, we are able to convert the primal problem into a standard linear SVM. An iterative technique [6] is then employed to solve the formulated problem, where each iteration only involves solving a linear SVM with significant computational savings particularly for large data. The resulting algorithm, dubbed Primal Sparse Kernel Classifier (PSKC), is shown to perform competitively with SKLA and SVM while being more efficient.

2   Learning Sparse Kernel Classifiers

We focus on kernel SVM in this paper, but the proposed algorithm can easily be adapted to other kernel classifiers. Consider a binary classification problem with data set $(\mathcal{X}, \mathcal{Y}) = \{(\mathbf{x}_i, y_i) | i = 1, \ldots, N, x_i \in \mathbb{R}^d, y_i \in \{-1, 1\}\}$, kernel SVM training can be cast as the following optimisation problem

$$\min_{\mathbf{w}, b} \quad \frac{1}{2} \|\mathbf{w}\|^2 + C \sum_i^N \ell(f(\mathbf{x}_i), y_i) \tag{1}$$

$$f(\mathbf{x}_i) = \mathbf{w}^T \varphi(\mathbf{x}_i) + b$$

where φ specifies an implicit feature mapping function. The exact form of φ is unknown but the inner product between two feature maps $\varphi(\mathbf{x}_i)$ and $\varphi(\mathbf{x}_j)$ is well defined by the kernel function κ such that $\kappa(\mathbf{x}_i, \mathbf{x}_j) = \langle \varphi(\mathbf{x}_i), \varphi(\mathbf{x}_j) \rangle$ holds for any $\mathbf{x}_i, \mathbf{x}_j \in \mathbb{R}^d$. And $\ell(f_i, y_i) = \max(0, 1 - f_i y_i)$ is the Hinge loss for SVM and varnishes whenever the margin is greater than 1 ($f_i y_i \geq 1$).

Despite the simple form, it is difficult to directly solve problem (1) without knowing φ in closed form. Hence the following dual problem is solved instead

$$\min_{\alpha} \frac{1}{2} \sum_{i,j} \alpha_i \alpha_j \kappa(\mathbf{x}_i, \mathbf{x}_j) - \sum_i \alpha_i y_i \tag{2}$$

$$s.t. \sum_i \alpha_i = 0 \quad \text{and} \quad 0 \leq \alpha_i y_i \leq C$$

where α_i is the dual variable for input example i and the primal variable $\mathbf{w}$ and the classifier f in (1) can be expressed in terms of dual variables

$$\mathbf{w} = \sum_{i, \alpha_i \neq 0} \alpha_i \varphi(\mathbf{x}_i) \tag{3}$$

$$f(\mathbf{x}) = \sum_{i, \alpha_i \neq 0} \alpha_i \kappa(\mathbf{x}_i, \mathbf{x}) + b \tag{4}$$

Note that only examples with nonzero dual variables are included in the expansion above. These are called Support Vectors (SV), referring to examples incurring positive Hinge losses. The complexity of kernel SVM depends on the number of SVs in the expansion (4). With larger training data set, it is likely to produce a classifier with a large number of SVs. The purpose for sparse kernel classifiers is to reduce the number of expansion terms in (3) and (4) without affecting the performance. The weight vector takes a similar form below

$$\mathbf{w} = \sum_{j=1}^{m} \beta_j \varphi(\mathbf{z}_j) \tag{5}$$

where $\mathbf{z}_i$'s are the eXpansion Vectors (XV) that form the bases of $\mathbf{w}$. Unlike SVs in kernel SVM, XVs do not necessarily overlap with input data and thus provide more flexibility in fitting the weight $\mathbf{w}$. The number of XVs m is much smaller than the number of SVs in standard SVM, making predictions more efficient.

Various strategies can be used for selecting XVs here, such as random selection from the input examples (RSVM, [3]) or fitting a trained SVM classifier with a fixed number of XVs that minimises the reconstruction error (RS, [4]). A more principled approach would account for the cost function to be minimised and embed XV selection into the optimisation process. Wu et al [1] added Equation (5) as an explicit constraint into the SVM formulation in (1). They then showed that a new dual problem can be formulated for sparse SVM, resembling the standard formulation in (2) with a modified kernel function

$$\hat{\kappa}_z(\mathbf{x}_i, \mathbf{x}_j) = \psi_i^T (\mathbf{K}^z)^{-1} \psi_j \tag{6}$$

$$\psi_i = [\kappa(\mathbf{x}_i, \mathbf{z}_1), \dots, \kappa(\mathbf{x}_i, \mathbf{z}_m)]^T \tag{7}$$

where $\mathbf{K}^z$ is a $m \times m$ Gram matrix whose (i,j)th entry is given by the kernel evaluation $\kappa(\mathbf{z}_i, \mathbf{z}_j)$. Each entry $\hat{\kappa}_z(\mathbf{x}_i, \mathbf{x}_j)$ depends on the XVs $\mathbf{z}_i$'s. This suggests the use of a perturbed optimisation technique [6] to solve the formulated problem iteratively. In each iteration, the gradient of the dual objective function with respect to z_i is computed by solving the dual SVM problem with modified kernel (6) and fixing the dual variables to their optimal values in gradient computation as if they do not depend on the XVs. Then a line search is pursued in the direction of the negative gradient for sufficient decrease of cost function value. Each linear search updates the values of z_i's and hence involves retraining of the kernel SVM. The validity of this approach is established by a theorem for optimal value functions in [6].

3   Primal Sparse Kernel Classifier Learning

Despite the effectiveness of SKLA [1] for learning sparse kernel classifiers, it is extremely expensive in training, making it impractical for large-scale applications. The complexity of SKLA arises mainly in two aspects. Firstly, being an algorithm of iterative nature, SKLA involves repeatedly retraining of the kernel SVM problem for each function evaluation involved in gradient computation and line search. Note the complexity of kernel SVM training is at best cubic in the number of SVs, which is roughly proportional to the training data size. In addition, for difficult problems, there is a high probability of failure for line searches. This would greatly increase the number of times for SVM retraining and the overall training time. Secondly, the gradient computation is also very costly, as it involves taking the derivative of each $\hat{\kappa}_z(\mathbf{x}_i, \mathbf{x}_j)$ in the modified kernel w.r.t. $\mathbf{z}_j$'s. The time complexity for computing the gradient of a single XV is $O(N^2 m^2 d)$, quadratic in both the training data size and the number of XVs. This is also undesirable for large-scale applications.

The main hurdle to the efficiency of SKLA is the optimisation of dual variables, which are then used to compute the β variables that define the weight vector in (5). A more direct approach would aim to solve β_i's directly. This motivates the development of the PSKC algorithm, which provides a primal optimisation framework for sparse kernel classifier learning. By substituting weight vector $\mathbf{w}$ in (5) into the primal problem in (1) and rewriting the cost function in terms of β and $\mathbf{z}_j$'s, we have

$$\min_{\beta, b; \mathbf{Z}} \quad f(\beta, b; \mathbf{Z}) = \frac{1}{2}\beta^T \mathbf{K}^z \beta + C \sum_i^N \ell(\beta^T \psi_i + b, y_i) \tag{8}$$

Here $\beta = [\beta_1, \ldots, \beta_m]^T$ is the vector of expansion coefficients, ψ_i is defined in (7), and $\mathbf{Z}$ is a concatenation of XVs $\mathbf{z}_j$'s. Moreover, the squared Hinge loss is used here $\ell(f_i, y_i) = \max(0, 1 - f_i y_i)^2$ instead of the Hinge loss for standard SVM. The reason for utilising the squared Hinge loss will become apparent shortly.

Two sets of variables need to be optimised for the above problem, the expansion coefficient vector β and the XVs $\mathbf{Z}$. Hence, we adopt an efficient approach to solve this joint optimisation problem. Specifically, we convert the original problem in (8) into the following problem which depends on variable $\mathbf{Z}$ only

$$\min_{\mathbf{Z}} g(\mathbf{Z}) \quad \text{with} \quad g(\mathbf{Z}) = \min_{\beta, b} f(\beta, b; \mathbf{Z}) \tag{9}$$

The new cost function $g(\mathbf{Z})$ is special because itself is the optimal value of f minimised over variables (β, b). In our case, $g(\mathbf{Z})$ not only exists, but is also differentiable at each $\mathbf{Z}$. This result can be established by applying Theorem 4.1 of [6], which provides sufficient conditions for the existence of derivatives for optimal value functions like $g(\mathbf{Z})$. According to the theorem, $g(\mathbf{Z})$ is differentiable if $f(\beta, b; \mathbf{Z})$ is differentiable w.r.t. β and b and has unique optimal value over variables β and b for each given $\mathbf{Z}$. The first condition is guaranteed by the use of squared Hinge loss discussed before, which is differentiable everywhere. Note this condition is not true for the standard Hinge loss, as it is non-differentiable if $y_i f_i = 1$. The uniqueness condition is true because f is a quadratic function over β with positive definite Hessian $\mathbf{K}^z$. Thus f is convex in β and ensures the optimal solution is unique. Let $(\overline{\beta}, \overline{b}) = \arg\min f(\beta, b; \mathbf{Z})$ be the minimiser of f at given $\mathbf{Z}$, the derivative of $g(\mathbf{Z})$ w.r.t. each XV $\mathbf{z}_j$ can then be computed by substituting $\overline{\beta}$ and $\overline{b}$ into (8) and taking the corresponding derivative as if $g(\mathbf{Z})$ does not depend on $\overline{\beta}$ and $\overline{b}$

$$\frac{\partial g}{\partial \mathbf{z}_j} = \sum_{i=1} m \overline{\beta}_i \frac{\partial \kappa(\mathbf{z}_i, \mathbf{z}_j)}{\partial \mathbf{z}_j} \overline{\beta}_j + 2C \sum_{i \in \mathcal{S}} (\beta^T \psi_i + b - y_i) \beta_j \frac{\partial \kappa(\mathbf{x}_i, \mathbf{z}_j)}{\partial \mathbf{z}_j} \tag{10}$$

where $\mathcal{S} = \{i | \ell(y_i, f_i) > 0\}$ denotes the index set of examples with positive loss terms. The partial derivative of the kernel function depends on the choice of the kernel. In this paper, we have adopted the following Gaussian kernel, but the algorithm works for all differentiable kernel functions

$$\kappa(\mathbf{x}, \mathbf{z}) = \exp\left(-\gamma \|\mathbf{x} - \mathbf{z}\|^2\right)$$
$$\frac{\partial \kappa(\mathbf{x}, \mathbf{z})}{\partial \mathbf{z}} = 2\gamma \kappa(\mathbf{x}, \mathbf{z})(\mathbf{x} - \mathbf{z}) \tag{11}$$

With the derivatives of $g(\mathbf{Z})$, we can use a gradient descent algorithm with back-tracing line search to iteratively optimise the values of $\mathbf{Z}$. The only problem left is how to evaluate the value of $g(\mathbf{Z})$ for each $\mathbf{Z}$, which equals solving the minimisation problem in (8) for β and b with fixed $\mathbf{Z}$. This problem is in fact equivalent to a linear SVM with the one-to-one mapping of variables below.

$$\vartheta = (\mathbf{K}^z)^{\frac{1}{2}} \beta \qquad \Longleftrightarrow \qquad \beta = (\mathbf{K}^z)^{-\frac{1}{2}} \vartheta \tag{12}$$

By substituting β in (8), we can rewrite it as a cost function over ϑ

$$f'(\vartheta, b; \mathbf{Z}) = \vartheta^T \vartheta + C \sum_i^N \ell(\vartheta^T (\mathbf{K}^z)^{-\frac{1}{2}} \psi_i + b, y_i) \tag{13}$$

The above is the same as the cost function for a linear SVM. ϑ is the weight vector of the linear SVM, and $(\mathbf{K}^z)^{-\frac{1}{2}} \psi_i$'s are the input vectors. Let $(\overline{\vartheta}, \overline{b})$ be the solution of the linear SVM trained with transformed feature vectors, the minimiser $(\overline{\beta}, \overline{b})$ for $f(\beta, b; \mathbf{Z})$ can be easily obtained by mapping the solution $\overline{\vartheta}$ back to $\overline{\beta}$ via (12).

Same as the case of SKLA, the complexity of PSKC depends mainly on function evaluation and gradient computation. We have shown above that function evaluation is equal to solving a linear SVM with transformed features. Both linear SVM training and feature transformation has linear time complexity $O(N)$ with input data size N. This is much better than SKLA with time complexity of $O(N^3)$ for kernel SVM training. Since m is a small number compared to N, the computation of $(\mathbf{K}^z)^{-\frac{1}{2}}$ and map from ϑ to β is negligible. From (10), we can see that the cost of gradient computation for each XV $\mathbf{z}_j$ is roughly $O(Nmd)$, a factor-$O(Nm)$ saving compared to that in SKLA.

4 Experimental Results

We first tested our PSKC algorithm on a synthetic example to showcase its interesting properties. The synthetic data set shown in Figure 1. has four classes, each occupying a separate cluster generated from a Gaussian distribution. Points from different classes are marked with different symbols. For this example, a minimum of 4 XVs overlapping with the centroid of each cluster is sufficient to distinguish the four classes. We deliberately initialise PSKC with poor initial locations of the XVs far from their respective cluster centroids, as denoted by squares in the top-left plot. By running PSKC and recording the locations of XVs over each iteration, a trajectory is created for each XV which keeps track of its evolution during optimisation. It can be seen that eventually all XVs have converged to locations close to the cluster centroids, as denoted by the circles in the same plot. The improvement on XVs locations is a natural consequence of the reduction in cost function values over each iteration, as shown in the top-right plot. These plots have empirically demonstrated the effectiveness of PSKC in finding good XVs for sparse kernel classifiers.

We have obtained similar results with the dual SKLA algorithm [1]. However, SKLA is much more costly than PSKC in training. To show this, we conducted two experiments on the same data distributions. The first experiment compares the training speed of PSKC and SKLA by increasing the size of the training data while the second one focuses on the effect of increasing feature dimensions. The results are shown on the second row of Figure 1, with the number of seconds spent on training over different training data sizes in the left plot and increasing feature dimensions on the right for both PSKC ((in solid lines) and SKLA (in broken lines). It can be easily seen that PSKC has superior scalability in comparison to SKLA in both cases. This is especially true with large training data sizes. Whereas SKLA has cubic time complexity with the sample size, PSKC scales linearly and is well suited for large-scale applications.

We now turn our attention to real-world data sets. 12 data sets from the UCI Machine Learning Repository[1] were used in our experiment. A summary of data sets used is given in Table 1, including the size of the training (Ntr) and testing (Nts) sets, feature dimension (Dim), number of classes (Cls). We then applied the

---

[1] http://archive.ics.uci.edu/ml/

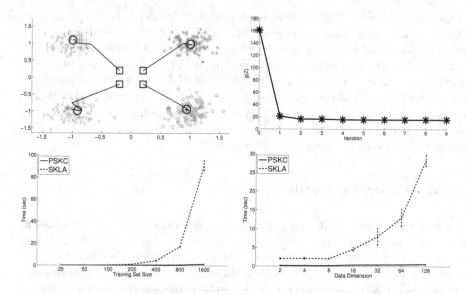

Fig. 1. Results on synthetic data. Top row: trajectories for the locations of each XV (left) and cost function values over the iterations (right); Bottom row: training speed for PSKC and SKLA with increasing data sizes (left) and feature dimensions (right).

standard SVM with the Gaussian kernel on each data set and compared the results of SVM with four candidate sparse kernel classifiers - RS [4], RSVM [3], the dual SKLA [1] and the proposed PSKC. Input data were scaled to have unit variance for each coordinate. The parameter γ for the Gaussian kernel was empirically set to the inverse of the feature dimension and the SVM parameter C was 10 for all methods under comparison. We have used the LibSVM package[2] for training kernel SVM classifiers and the LibLinear package[3] for training linear SVMs for RSVM and PSKC. The algorithms was implemented in Octave on a MacBook Pro with Intel Core i5 CPU and 4Gb memory. We have repeated the experiment 10 times over random partitions of training and testing data. The classification performance in terms of average accuracy values and their standard deviations are reported in Table 2.

For the SVM classifier, we have also recorded the number of SVs produced on average for each data set, which are the numbers in the brackets on the second column of Table 2. For each sparse kernel classifier, we have used a fixed number of expansions equal to 1% of the training data size capped at a minimum value of 10 and a maximum value of 100. The exact number for each data set is shown inside the brackets following the names on the first column. For each data set, we have highlighted the accuracy value corresponding to the best-performing sparse kernel classifier. More than one values could be highlighted in cases of ties, which are determined by the results of paired t-tests at the confidence interval of 95%.

---

[2] http://www.csie.ntu.edu.tw/~cjlin/libsvm/
[3] http://www.csie.ntu.edu.tw/~cjlin/liblinear/

Table 1. Statistics of the real-world data sets used in the comparison experiment

Data	Ntr	Nts	Dim	Cls	Data	Ntr	Nts	Dim	Cls
australian	346	344	14	2	letter	15000	5000	16	26
breast-cancer	342	341	10	2	connect-4	33780	33777	126	3
dna	1001	999	180	3	shuttle	43500	14500	9	7
segment	1155	1155	19	7	ijcnn1	49990	91701	22	2
satimage	4435	2000	36	6	mnist	60000	10000	784	10
usps	7291	2007	256	10	SensIT	78823	19705	100	3

We can clearly see from Table 2 that PSKC and SKLA are the most competitive sparse kernel classifiers. Their performances could approach that of kernel SVM albeit with a significantly smaller number of kernel expansions. In contrast, RS and RSVM do not perform as well as PSKC and SKLA. Specifically, among the four sparse kernel classifiers under comparison, PSKC is the exclusive winner for 6 out of 12 data sets, and winning 10 of them including tied cases. Moreover, for the majority of data sets, PSKC beats RSVM by a large margin in performance. This is clear evidence for the effectiveness of the PSKC optimisation algorithm, as RSVM is the special case of PSKC without any optimisation.

Table 2. Comparison of classification performance for SVM and various sparse kernel classifiers on real-world data sets

	SVM	RS	RSVM	SKLA	PSKC
australian (10)	84.2 ± 1.4(156)	82.2 ± 3.1	77.9 ± 2.6	**85.7 ± 2.1**	**85.8 ± 1.6**
breast-cancer (10)	95.2 ± 0.8(56)	94.6 ± 1.6	**96.4 ± 0.7**	96.1 ± 0.9	96.0 ± 0.8
dna (10)	95.3 ± 0.6(821)	**94.7 ± 0.4**	67.0 ± 2.5	94.3 ± 0.6	94.2 ± 0.5
segment (12)	94.4 ± 0.6(384)	71.3 ± 7.7	85.7 ± 2.5	91.7 ± 1.1	**92.9 ± 0.9**
satimage (44)	91.0 ± 0.7(1425)	**89.8 ± 0.8**	88.5 ± 1.0	90.2 ± 0.8	**90.4 ± 0.6**
usps (73)	97.8 ± 0.8(2175)	96.1 ± 1.0	94.4 ± 1.0	**96.9 ± 0.8**	**97.0 ± 0.9**
letter (100)	96.5 ± 0.4(6260)	76.9 ± 1.9	88.3 ± 0.3	-	**94.6 ± 0.3**
connect-4 (100)	83.7 ± 0.1(16721)	**80.3 ± 0.7**	74.5 ± 0.3	-	80.1 ± 0.5
shuttle (100)	99.8 ± 0.0(605)	98.8 ± 1.1	99.8 ± 0.0	-	99.8 ± 0.0
ijcnn1 (100)	99.0 ± 0.1(2683)	**98.5 ± 0.2**	96.4 ± 0.2	-	98.5 ± 0.2
mnist (100)	97.7 ± 0.1(9883)	94.3 ± 0.4	90.5 ± 0.4	-	**95.8 ± 0.2**
SensIT (100)	84.0 ± 0.2(21249)	82.0 ± 0.8	80.0 ± 0.6	-	**83.6 ± 0.4**

Though the performances of PSKC and SKLA are quite comparable, there is a huge difference in training cost. SKLA is very inefficient and scales poorly to large data, hence we only have the results for SKLA in Table 2 for the first six data sets. For the remaining data sets with over 10,000 training examples, the training cost is prohibitive for SKLA, which involves many iterations of kernel SVM training and heavy gradient computations. The training times for each method are reported in Table 3. It is evident that SKLA is very inefficient in training. On the other hand, PSKC scales much better with increasing training sizes. After all,

Table 3. Comparison of training time for different methods

	SVM	RS	RSVM	SKLA	PSKC
australian	0.02 sec	0.35 sec	0.01 sec	1.0 sec	0.2 sec
breast-cancer	0.01 sec	0.39 sec	0.01 sec	0.4 sec	0.1 sec
dna	1.2 sec	2.9 sec	0.01 sec	$\approx$ 4.2 min	0.5 sec
segment	0.2 sec	2.5 sec	0.02 sec	31.1 sec	0.6 sec
satimage	2.5 sec	27.9 sec	0.2 sec	$\approx$ 47.0 min	9.0 sec
usps	32.9 sec	$\approx$ 4.7 min	0.9 sec	> 3 hour	38.4 sec
letter	52.1 sec	$\approx$ 7.0 min	6.2 sec	-	$\approx$ 3.7 min
connect-4	$\approx$ 13.4 min	$\approx$ 21.0 min	3.0 sec	-	$\approx$ 4.7 min
shuttle	9.7 sec	$\approx$ 1.0 min	5.3 sec	-	$\approx$ 2.3 min
ijcnn1	31.2 sec	$\approx$ 1.3 min	1.9 sec	-	$\approx$ 1.3 min
mnist	$\approx$ 1.5 hour	$\approx$ 2 hour	12.3 sec	-	$\approx$ 12.8 min
SensIT	$\approx$ 2 hour	$\approx$ 2.5 hour	18.6 sec	-	$\approx$ 11.4 min

it only involves repeated training of linear SVM during optimisation. Although it is not as efficient as SVM for small and median data sets, the asymptotic complexity for PSKC is better. This is empirically justified by the less amount of time spent for training on two largest data sets - "mnists" and "SensIT" in Table 3. In addition, PSKC uses only 10% to 1% of XVs compared to the number of SVs in SVM. This reduces prediction time by a factor of 10 to 100, which depends linearly on the number of XVs in the classifier. The improvement in efficiency for PSKC is more pronounced for large data sets compared to SVM. This makes PSKC particularly suited for large-scale applications.

5   Conclusions

In this paper we presented PSKC, an efficient and effective algorithm for learning sparse kernel classifiers. Training PSKC is quite efficient and only involves solving linear classifiers repeatedly. Experiments show that PSKC outperforms other sparse kernel classifiers and is comparable with kernel SVM in predictive accuracy at lower training and prediction costs for large data sets.

Acknowledgment. The work was partly supported by the Australian Research Council under the Discovery Project "Automatic music feature extraction, classification and annotation" (DP0986052).

References

1. Wu, M., Scholkopf, B., Bakir, G.: A direct method for building sparse kernel learning algorithms. Journal of Machine Learning Research 7, 603–624 (2006)
2. Platt, J.: Fast training of support vector machines using sequential minimal optimization. In: Advances in Kernel Methods - Support Vector Learning (1998)

3. Lee, Y.J., Mangasarian, O.L.: Rsvm: Reduced support vector machines. In: Siam Data Mining Conf. (2001)
4. Scholkopf, B., Smola, A.: Learning with Kernels: Support Vector Machines, Regularization, Optimization, and Beyond. MIT Press (2002)
5. Keerthi, S., Chapelle, O., DeCoste, D.: Building support vector machines with reduced classifier complexity. Journal Machine Learning Res. 7, 1493–1515 (2006)
6. Bonnans, J.F., Shapiro, A.: Optimization problems with pertubation: A guided tour. SIAM Review 40(2), 202–227 (1998)

Evolutionary Weighted Mean Based Framework for Generalized Median Computation with Application to Strings

Lucas Franek and Xiaoyi Jiang

Department of Mathematics and Computer Science
University of Münster, Germany
{lucas.franek,xjiang}@uni-muenster.de

Abstract. A new general framework for generalized median approximation is proposed based on the concept of weighted mean of a pair of objects. It can be easily adopted for different application domains like strings, graphs or clusterings, among others. The framework is validated for strings showing its superiority over the state-of-the-art.

1 Introduction

The concept of median is widely used in order to estimate a single representative of a set of objects. Another motivation of the median concept is to eliminate some erroneous objects by averaging over all objects. Further, the median concept is also motivated by the results received from supervised classifier combination: It is well known that by averaging the results of several classifiers a more reliable classification can be achieved [9].

While finding the Euclidean median in vector space was originally posed by Fermat in the 17th century and is referred to as the Fermat-Weber problem, in the last years the median problem was also formulated for more general spaces and objects like strings [7], graphs [3], clusterings [12], and segmentations [11], among others. In most cases, however, the computation of generalized median turns out to be very demanding, partly even of $\mathcal{NP}$-completeness [10,12]. This fact motivates the design of approximate approaches.

There is very little work on general frameworks for generalized median approximation. The embedding approach is based on embedding the objects into the vector space, in which the Weiszfeld algorithm [14] can be applied to find the median point. Then, an inverse transformation to the original object domain is performed by using the weighted mean of a pair of objects (to be discussed later). This framework has been adopted to strings [7] and graphs [3]. However, the transformation in vector space and back into object (string, graph, etc.) domain is not trivial and such an embedding may cause undesired distortions.

In this work we propose a new framework for generalized median approximation. It is formulated for objects in general spaces and can be adopted to different application domains, such as strings, graphs, and clusterings, among

G.L. Gimel' farb et al. (Eds.): SSPR & SPR 2012, LNCS 7626, pp. 70–78, 2012.

others. The proposed framework is motivated by the lower bound for the generalized median [6]. It is observed that in case of a tight lower bound generalized median is received by computing the weighted mean of a pair of objects. This motivates us to formulate an algorithm for generalized median approximation by using the concept of weighted mean. The definition of weighted mean is directly motivated by the weighted mean of two numbers (or vectors) and it has already been adopted to the domain of strings [2], graphs [1] and clusterings [5].

In the experimental part of this work the proposed framework is adopted to the domain of strings. Further, a comparison with the embedded based approach [7] is provided.

The rest of the paper is organized as follows. In the next section we introduce the fundamentals of this work (formal definition of the generalized median problem and weighted mean). Our new general framework for generalized median approximation is presented in Section 3. The application to the domain of strings and experiments are shown in Section 4. We conclude in Section 5.

2   Fundamentals

We first define the problem of generalized median formally for a set of general objects.

Definition 1. *Let $X = \{x_1, \ldots, x_n\}$ be a set of objects in a general space U and $d : U \times U \to \mathbb{R}_0^+$ a distance function defined on U. Then, the generalized median $\hat{x}$ is defined by*

$$\hat{x} = \arg \min_{x \in U} \sum_{i=1}^{n} d(x, x_i) = \arg \min_{x \in U} SOD(x), \tag{1}$$

where the summation will be called sum of distances (SOD) of object x.

It intends to infer a representative sample out of the ensemble X. If the minimizer $\hat{x}$ is restricted to be within the ensemble X, then the corresponding solution is called set median.

Further, for the proposed algorithm we need the concept of weighted mean of a pair of objects. Consider two points in the n-dimensional real space, $x, y \in \mathbb{R}^n$. The weighted mean of x and y is defined as

$$z = \alpha x + (1 - \alpha)y, \ 0 \leq \alpha \leq 1. \tag{2}$$

If $\alpha = \frac{1}{2}$, then z is the (normal) mean of x and y. Clearly, z is a point on the line segment between x and y and the distance between z to x and y is controlled by the parameter α.

Generally, the weighted mean of two objects can be defined as follows.

Definition 2. *Let x_1 and x_2 denote two objects in a space U of all objects, and $d : U \times U \to \mathbb{R}_0^+$ a distance function defined on U which measures the*

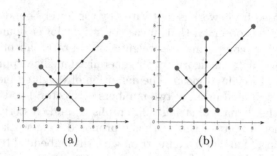

Fig. 1. Generalized median in two-dimensional vector space. Red points: initial points. Small black points: weighted means. Green point: generalized median. (a) Generalized median is located at the intersection point of the line segments of opposite point pairs. (b) Line segments do not intersect in one point. Further iterations are necessary to approach the generalized median.

dissimilarity of two objects. The weighted mean of x_1 and x_2 is an object x_w such that

$$d(x_1, x_w) = \alpha \tag{3}$$
$$d(x_1, x_2) = \alpha + d(x_w, x_2) \tag{4}$$

where α is a constant with $0 \leq \alpha \leq d(x_1, x_2)$.

The concept of weighted mean has been brought to pattern recognition for strings [2], graphs [1], and clusterings [5].

3 Evolutionary Weighted Mean Based Framework for Generalized Median Computation

In order to motivate our new method for generalized median approximation let us first consider the lower bound for the generalized median [6]. Let $\mathcal{P}$ denote a partition of the objects X into $m = \frac{n}{2}$ pairs (n even for convenience):

$$\mathcal{P} = \{(x_{11}, x_{12}), (x_{21}, x_{22}), \dots, (x_{m1}, x_{m2})\}, \quad x_{i,j} \in X, \quad \bigcup_{i=1,\dots,m, j=1,2} \{x_{ij}\} = X.$$

Further, $\mathfrak{P}$ denotes the set of all such partitions $\mathcal{P}$. If d is a metric, then it can be shown that a lower bound Γ on the SOD of the generalized median $\hat{x}$, i.e. $0 \leq \Gamma \leq SOD(\hat{x})$, can be computed by estimating the optimal set of pairs $\hat{\mathcal{P}} \in \mathfrak{P}$ such that the sum of distances of the pairs (x_{i1}, x_{i2}) is maximal (see [6] for a proof):

$$\Gamma = \max_{\mathcal{P} \in \mathfrak{P}} \sum_{i=1}^{m} d(x_{i1}, x_{i2}). \tag{5}$$

This lower bound formulation motivates an approximation algorithm for the generalized median. By approaching the lower bound the generalized median is obviously also approached. In the ideal case, where the lower bound is tight, i.e. $SOD(\hat{x}) = \Gamma$, even the true generalized median could be found by approaching Γ. In this case it directly follows from the metric property $d(x_{i1}, \hat{x}) + d(\hat{x}, x_{i2}) = d(x_{i1}, x_{i2})$, i.e. the generalized median is the weighted mean of each pair $(x_{i1}, x_{i2}) \in \hat{\mathcal{P}}$. Since the lower bound is estimated by an optimum partition of pairs, the generalized median is approached by computing the weighted means of such pairs of objects.

For each pair the optimum weight is unknown a priori. But a condition for optimality is obvious: The optimum weighted means of all pairs are estimated such that they match in one point, namely the generalized median.

The idea is depicted in Fig. 1 for a set of points. From a geometrical point of view, the median point in an Euclidean space is ideally located in the exact intersection point between the opposite pairs of points (Fig. 1(a)). Thus, the generalized median is received by computing for each of the pairs a set of weighted means (represented by the smaller points on the line segments). The generalized median is then located at the point where the weighted means match. Note that in general the lower bound will not be tight. To say it another way, in geometric space the weighted means of all pairs will not match in one point as depicted in Fig. 1(b). Thus, we resort to an iterative algorithm:

Step 1: Compute the optimum (opposite) pairs of objects.
Step 2: Estimate for each pair the optimum weighted mean in terms of SOD and add it to the current set of objects.
Step 3: Select the optimum objects in the current set of objects.

These steps are detailed in the following.

Step 1: Optimum Pairs of Objects

The question arises how to estimate the optimum set of pairs of objects such that the sum of distances of pairs (Eq. (5)) is maximized. To handle this problem it is proposed to build a graph, where each object corresponds to a vertex and each edge between two vertices is weighted by the distance between the corresponding objects. Then, finding the optimum pairs is equipollent to solving the maximum weighted graph matching problem. A solution to this problem is provided by [8]. Note that in the situation of an odd number of input objects one vertex remains unmatched (see also [6]). The unmatched point is stored in the current set of objects which is processed in the third step.

Step 2: Optimum Weighted Mean in Terms of SOD

After having computed the optimum set of pairs the weighted mean for each pair of objects is computed. In this situation, however, the problem arises that the weight may vary for each pair in the range $[0, 1]$ (after normalization) and the optimum weight yielding the generalized median is unknown a priori. To find the optimum weight a search procedure is applied. First, several weighted means are computed and in the next step it is decided which weighted means are suitable and which are not, i.e. a kind of fitness function is needed. Considering that the generalized median aims at minimizing the SOD in Eq. (1) it is proposed to use this SOD as fitness function. To estimate the best weight α the whole range is sampled a fixed number of times in an equidistant way. The weighted means are evaluated by the SOD and the weighted mean with the lowest SOD is selected. Note that this kind of search procedure is a linear search. Further search procedures could be also adopted. Optionally, the two weighted means with the lowest SOD may be selected. This is discussed in the next step.

Step 3: Selecting the Optimum Set of Objects

In an ideal situation the optimum weighted mean for each pair of objects in terms of SOD (as computed in Step 2) would be equal to the generalized median of the initial set. In this case the algorithm should terminate. In a non-ideal situation, however, it cannot be expected that the computed optimum weighted means of different pairs are equal. Probably, some weighted means will be more suitable than other ones. Again, it is proposed to resort to the SOD as a fitness function in order to distinguish between suitable weighted means and less suitable ones.

More specifically, in the proposed algorithm the optimum weighted mean for each pair of objects is added to the current set of objects. Then, the optimum set of objects is estimated by selecting the best n_{max} objects from the current set of objects, where n_{max} is a parameter of the algorithm. Hereby, the best objects are again selected by evaluating their SOD with respect to the input set of objects and selecting the objects with the lowest SOD. The parameter n_{max} is fixed such that the size of the set of optimum objects is limited during the iteration process. Note that optionally, the second best weighted mean from Step 2 may also be added to the set of objects because in a non-ideal case it may contain valuable information as well.

The process is now iterated beginning with the first step using the current optimum set of objects. The algorithm may finish when either the lower bound is reached or when it converges to some solution.

Evolutionary Weighted Mean Based Framework

The proposed framework can now be formulated as follows. Given a set of objects $O = \{o_1, \ldots, o_n\}$.

1. Consider all pairs of objects and compute their weights by the distance between the corresponding objects. Save them into the distance matrix D.
2. Determine the optimal set of pairs using maximum weighted graph matching on D. Let $E = (e_1, \ldots, e_m)$ denote the corresponding optimum set of edges.
3. For each edge $e_i \in E$ consider the corresponding pair (o_{i1}, o_{i2}) and:
   (a) Compute w weighted means by using $\alpha = \frac{i \cdot d(o_{i1}, o_{i2})}{w+1}, i = 1, \ldots, w$.
   (b) Evaluate the w weighted means by the fitness function SOD and select the best weighted mean o^*. Update the current set of objects by adding the best weighted mean: $O = O \cup \{o^*\}$.

4. Evaluate all objects in O by SOD and delete the worst instances such that the resulting current set O consists of a maximum number n_{max} of objects.
5. It is checked if the lower bound is matched or if convergence is achieved. Otherwise, the procedure starts from step 1 with the current O. A maximum number of iterations I_{max} prevents the algorithm from getting inefficient.

Obviously, the algorithm is convergent, because the values of the fitness function (SOD) can only decrease. However, I_{max} is introduced in the last step for efficiency reasons.

It is emphasized that the only requirement for this framework is that the weighted mean is well defined for the space under consideration. In the next section this framework will be adopted and validated for the median string problem.

4 Application to Strings

Strings are a fundamental representation in structural pattern recognition. Here, our framework is adopted to the domain of strings. As a comparison method the embedding based generalized median computation proposed by Jiang et al. [7] will be used, which has been demonstrated to outperform the related algorithms from the literature.

In order to be able to apply our proposed approach to the domain of strings two requirements have to be fulfilled. First, a suitable distance function is needed in order to compare strings. Secondly, based on this distance function the weighted mean has to be defined. Here, we use the popular Levenshtein edit distance [13].

The weighted mean of a pair of strings $(\mathfrak{S}_1, \mathfrak{S}_2)$ for the edit distance was introduced in [2]. It is defined analogously to Definition 2 as a string $\mathfrak{S}_w$ with

$$d(\mathfrak{S}_1,\ \mathfrak{S}_w) = \alpha,\ d(\mathfrak{S}_1,\mathfrak{S}_2) = \alpha + d(\mathfrak{S}_w,\mathfrak{S}_2),\ 0 \le \alpha \le d(\mathfrak{S}_1,\mathfrak{S}_2).$$

$\mathfrak{S}_w$ is constructed by selecting a subsequence of all edit operations used for transforming $\mathfrak{S}_1$ into $\mathfrak{S}_2$, such that $d(\mathfrak{S}_1,\mathfrak{S}_\mathfrak{w}) = \alpha$. Applying this subsequence to $\mathfrak{S}_1$ yields $\mathfrak{S}_w$.

The evolutionary weighted mean algorithm can now be directly applied to the domain of strings. In the implementation we set $w = 3$ and $n_{max} = 10$, i.e. the set of strings will consist of a maximum number of 10 strings after the first iteration. Further, the iteration is stopped after $I_{max} = 5$ iterations.

4.1 Experimental Settings

In order to be able to compare the proposed approach with the embedding based approach the experimental settings from [7] are used. A synthetic dataset is generated for test purpose by distorting an initial string p times. Hereby, each symbol of the initial string is distorted with a fixed probability $p_{distort}$. If a symbol is distorted, then the three elementary operations substitution, deletion, and insertion are chosen by a fixed probability. Five strings (Scotland, Birmingham, Philadelphia, TristanDaCunha, WesternPatagonia) are used. 100 datasets are generated for each initial string and the average performance measures are reported. The same parameter values as in [7] are used:

- $p = 40$: Number of strings in the initial set $\mathcal{S}$.
- $p_{distort} = 12\%$: Distortion probability for each symbol.
- The probabilities of the three basic operations: $p_{substitute} = 87\%$, $p_{delete} = 9\%$, $p_{insert} = 4\%$.
- $c(s \rightarrow \bar{s}) = c(\epsilon \rightarrow s) = c(s \rightarrow \epsilon) = 1$: Equal costs for the edit operations.

We also summarize some settings of the embedding based method [7]. It uses prototype selection in order to reduce computational efforts. In our experiments the K-medians prototype selector is used as it was proposed in [7]. The number of prototypes is chosen to be 25% of the original set size. A comparison with all variants of the embedding based approach (concerning the inverse transformation to the original object domain) is provided, namely linear, triangulation, and recursive. Additionally, set median is also included into the comparison.

4.2   Experimental Results

In order to evaluate the obtained median strings it has to be taken into account that both the generalized median $\hat{\mathfrak{S}}$ and its $\mathrm{SOD}(\hat{\mathfrak{S}})$ are unknown. Consequently, we have to resort to the lower bound Γ (see Eq. (5)). Then, the quality of the obtained median $\tilde{\mathfrak{S}}$ is evaluated by $\Delta = \mathrm{SOD}(\tilde{\mathfrak{S}})/\Gamma$. If $\Delta \approx 1$, it is a strong hint that $\tilde{\mathfrak{S}}$ is an accurate approximation of the generalized median.

Probability of symbol distortion. In the first experiment the robustness against distortions in the input strings is investigated (see Fig. 2). The distortion probability is varied ($p_{distort} \in [2, 50]$). As observed in [7], our results confirm that the recursive approach dominates the other embedding based approaches up to 20%. For higher distortion probabilities the deviation Δ increases. The proposed evolutionary weighted mean algorithm clearly outperforms all variants for all distortion probabilities. Even for a distortion probability of 35% the deviation Δ is less than 1.05, making our method suitable also for high distortion levels.

Number of strings. Now we study the algorithm behavior with respect to the number of strings in the initial set of strings, i.e. we vary $p = 10, 12, \ldots, 40$. The results are shown in Fig. 3. While for example the recursive approach needs 40 initial strings in order to yield a deviation $\Delta \leq 1.05$ the evolutionary weighted mean algorithm performs very well already for more than 10 input strings with a deviation $\Delta \leq 1.01$.

Length of initial string. In Fig. 4 the performance is plotted for different string lengths. The evolutionary weighted mean algorithm clearly outperforms the embedding based approach for all string lengths. The SOD of the obtained median is very close to the lower bound indicating that the obtained median of the proposed algorithm is very close to the generalized median.

Time complexity. The time complexity for one iteration depends on the number of weighted means computed for one pair of strings as well as on the maximum

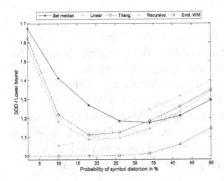

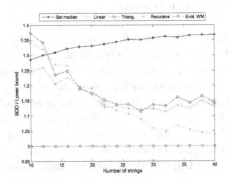

Fig. 2. Performance as a function of distortion probability

Fig. 3. Performance as a function of number of strings

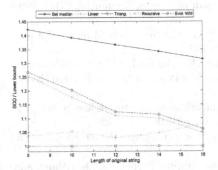

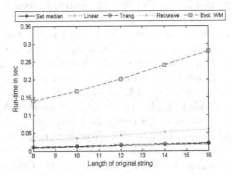

Fig. 4. Performance as a function of string length

Fig. 5. Computational time for test series "String length"

number of pairs in the current set of objects, i.e. $\mathcal{O}(w \cdot \frac{n_{max}}{2})$. Consequently, for a maximum number of iterations I_{max} the overall time complexity is $\mathcal{O}(I_{max} \cdot w \cdot \frac{n_{max}}{2})$. The computational time was measured on an Intel Core i7 2.80 GHz with 6 GB RAM. The result with respect to the string length is shown in Fig. 5. Note that the proposed approach is slightly more complex than the embedding based approach because of a higher number of necessary weighted mean computations. Nevertheless, the computational time is absolutely negligible (less than half a second in all cases).

Discussion. The performed experiments have shown that the evolutionary algorithm clearly outperforms the embedding based approach in all cases. It can handle high distortion levels very well and it works also quite well for a small number of initial input strings, whereas for example the embedding based recursive approach needs a significantly higher number of input strings in order to yield comparable results. Moreover, the results have shown that in many cases the performance is less than 2-3% compared to the lower bound, indicating that the obtained result is very close to the unknown generalized median.

5   Conclusion

A new algorithm for generalized median computation was formulated based on the concept of weighted mean. Experimental results were shown for strings. The proposed algorithm clearly outperforms the embedding based approach (and thus the related algorithms from the literature). The main advantage of the proposed framework is that it can be easily adopted for every application domain, in which the weighted mean is defined. Recently, the framework was adopted for ensemble clustering and its superiority with respect to several state-of-the-art ensemble clustering methods has been shown [4]. In future we will consider the application of the framework to further domains such as graphs and image segmentation.

References

1. Bunke, H., Günter, S.: Weighted mean of a pair of graphs. Computing 67(3), 209–224 (2001)
2. Bunke, H., Jiang, X., Abegglen, K., Kandel, A.: On the weighted mean of a pair of strings. Pattern Anal. Appl. 5(1), 23–30 (2002)
3. Ferrer, M., Valveny, E., Serratosa, F., Riesen, K., Bunke, H.: Generalized median graph computation by means of graph embedding in vector spaces. Pattern Recognition 43(4), 1642–1655 (2010)
4. Franek, L.: Ensemble Algorithms with Applications to Clustering and Image Segmentation. Ph.D. thesis, University of Münster (2012)
5. Franek, L., Jiang, X.: Weighted mean of a pair of clusterings. Pattern Anal. Appl. (under revision)
6. Jiang, X., Münger, A., Bunke, H.: On median graphs: Properties, algorithms, and applications. IEEE Trans. Pattern Anal. Mach. Intell. 23(10), 1144–1151 (2001)
7. Jiang, X., Wentker, J., Ferrer, M.: Generalized median string computation by means of string embedding in vector spaces. Pattern Recognition Letters 33(7), 842–852 (2012)
8. Munkres, J.: Algorithms for the assignment and transportation problems. Journal of the Society of Industrial and Applied Mathematics 5(1), 32–38 (1957)
9. Rokach, L.: Pattern classification using ensemble methods. World Scientific Pub. Co. Inc. (2010)
10. Sim, J.S., Park, K.: The consensus string problem for a metric is NP-complete. J. Discrete Algorithms 1(1), 111–117 (2003)
11. Singh, V., Mukherjee, L., Peng, J., Xu, J.: Ensemble clustering using semidefinite programming with applications. Mach. Learn. 79(1-2), 177–200 (2010)
12. Vega-Pons, S., Ruiz-Shulcloper, J.: A survey of clustering ensemble algorithms. Int. J. Pattern Recognition and Artificial Intelligence 25(3), 337–372 (2011)
13. Wagner, R.A., Fischer, M.J.: The string-to-string correction problem. J. ACM 21(1), 168–173 (1974)
14. Weiszfeld, E., Plastria, F.: On the point for which the sum of the distances to n given points is minimum. Annals of Operations Research 167, 7–41 (2009)

Graph Complexity from the Jensen-Shannon Divergence

Lu Bai and Edwin R. Hancock*

Department of Computer Science
University of York, UK
Deramore Lane, Heslington, York, YO10 5GH, UK

Abstract. In this paper we aim to characterize graphs in terms of structural complexities. Our idea is to decompose a graph into substructures of increasing layers, and then to measure the dissimilarity of these substructures using Jensen-Shannon divergence. We commence by identifying a centroid vertex by computing the minimum variance of its shortest path lengths. From the centroid vertex, a family of centroid expansion subgraphs of the graph with increasing layers are constructed. We then compute the depth-based complexity trace of a graph by measuring how the Jensen-Shannon divergence varies with increasing layers of the subgraphs. The required Shannon or von Neumann entropies are computed on the condensed subgraph family of the graph. We perform graph clustering in the principal components space of the complexity trace vector. Experiments on graph datasets abstracted from bioinformatics and image data demonstrate effectiveness and efficiency of the graphs complexity traces.

1   Introduction

Graph based relational representations have proven to be both powerful and flexible in pattern recognition. Compared to vector based pattern recognition, a major drawback with graph representations is the lack of a natural correspondence order. This limits the direct application of standard machine learning algorithms for problems such as graph clustering. One way to overcome this problem is to embed the graph data into a vector space, where standard machine learning techniques can be deployed. There have been several successful solutions which include a) embedding graph into vector space using the dissimilarity embedding [5], b) representing graph structure using permutation invariant polynomials computed from the eigenvectors of the Laplacian matrix using algebraic graph theory [10], and c) computing permutation-invariant graph features via the Ihara zeta function [7]. The limitations of the existing methods is that they usually depend on the graph topology or size, and as a result they tend to be computationally burdensome or can not be efficiently computed in an algebraic manner.

* Edwin R. Hancock is supported by a Royal Society Wolfson Research Merit Award.

G.L. Gimel' farb et al. (Eds.): SSPR & SPR 2012, LNCS 7626, pp. 79–88, 2012.

To overcome the limitations of existing methods, we propose a novel framework for characterizating graphs based on computing complexity traces. Depth-based representations of undirected graph structures have proved powerful for characterizing their topological structure in terms of intrinsic complexity [1,2]. One approach is to gauge information content flow through K layer subgraphs of a graph (e.g.subgraphs around a vertex having a maximum topology distance or minimal path length K) of increasing size and to use the flow as a structural signature. This approach allows a complexity trace to be defined which gauges how the complexity of the graph varies as a function of depth [2]. Unfortunately, to construct such a trace requires a measure of the intrinsic structural complexity, and this requires burdensome computations. In this paper we focus on developing an efficient depth-based signature, that can both capture fine structure and can be evaluated relatively efficiently. To compute a complexity trace of a graph G, we identify the centroid vertex v_C in G by selecting the vertex with minimum variance of shortest path lengths. Based on v_C, we derive a family of expansion subgraphs from v_C with in increasing layer size K. Then we construct a complexity trace of G by measuring how the dissimilarity between the K layer subgraph and G varies on the expansion subgraphs with the increasing layer K. To compute the proposed depth-based complexity trace efficiently, we turn to the Jensen-Shannon divergence as the dissimilarity measure. This is a nonextensive information theoretic measure derived from the mutual information between probability distributions over different structures. Here the required entropies of the Jensen-Shannon divergence are computed using the Shannon entropy or von Neumann entropy on the (sub)graphs. We empirically demonstrate that our Jensen-Shannon complexity trace can easily scale to large graphs. The performance of our framework is competitive to the state of the art methods in the literature.

2 Centroid Expansion Subgraphs

In this section, we introduce a set of subgraphs which we refer to as centroid expansion subgraphs of a given graph. We first describe how to identify the centroid vertex for a graph and explain how to extract the centroid expansion subgraphs from the graph with regard to the centroid vertex. Then we describe how to compute entropies on these centroid expansion subgraphs.

2.1 Centroid Vertex

The shortest path for a pair of vertices v_i and v_j in an undirected graph $G(V, E)$ can be obtained by using Dijkstra algorithm. We refer to the matrix S_G whose elements $S_G(i, j)$ represents the shortest path length between vertices v_i and v_j as shortest path matrix for $G(V, E)$. The average-shortest-path vector S_V for $G(V, E)$ is a vector with the same vertex sequence as S_G, with each element $S_V(i) = \sum_{j=1}^{|V|} S_G(i, j)/|V|$ representing the average shortest path length from vertex v_i to the remaining vertices. We then locate the centroid vertex v_i for $G(V, E)$ as follows

$$\hat{i} = \arg\min_i \sum_{j=1}^{|V|} [S_G(i, j) - S_V(i)]^2. \tag{1}$$

The centroid vertex v_i of $G(V, E)$ is located through selecting a vertex with a minimum variance of shortest path lengths out of all vertices in $G(V, E)$. Therefore, the shortest paths starting from the centroid vertex v_i form a *steady* path set that exhibits less length variability than those path sets starting from other vertices. For a graph $G(V, E)$ with the centroid vertex v_C, the K-layer centroid expansion subgraph $\mathcal{G}_K(\mathcal{V}_K; \mathcal{E}_K)$ is

$$\begin{cases} \mathcal{V}_K := \{u \in V | S(v_C, u) \le K, K \ge 1\}; \\ \mathcal{E}_K := \{\{v, u\} \subseteq \mathcal{V}_K | \{v, u\} \in E\}. \end{cases} \tag{2}$$

The number of centroid expansion subgraphs is equal to the greatest length of the shortest path from the centroid vertex to the other vertices of the graph.

2.2 Entropies on K-Layer Centroid Expansion Subgraphs

The definition of steady state random walks and entropy on a subgraph is similar to that for a graph. Given the K-layer centroid expansion subgraph $\mathcal{G}_K(\mathcal{V}_K; \mathcal{E}_K)$ of a graph $G(V, E)$, the adjacency matrix A_K for $\mathcal{G}_K(\mathcal{V}_K; \mathcal{E}_K)$ has elements

$$A_K(i, j) = \begin{cases} 1 \text{ if} (v_i, v_j) \in \mathcal{E}_K; \\ 0 \text{ otherwise.} \end{cases} \tag{3}$$

The vertex degree matrix of $\mathcal{G}_K(\mathcal{V}_K; \mathcal{E}_K)$ is a diagonal matrix D_K whose elements are given by $D_K(v_i, v_i) = d_K(i) = \sum_{v_i, v_j \in \mathcal{V}_K} A_K(i, j)$. From the matrixes D_K and A_K we can construct the Laplacian matrix $L_K = D_K - A_K$. The normalized Laplacian matrix is given by $\hat{L}_K = D_K^{-1/2} L_K D_K^{-1/2}$. The spectral decomposition of the normalized Laplacian matrix is $\hat{L}_K = \hat{\Phi}_K \hat{\Lambda}_K \hat{\Phi}_K^T$ where $\hat{\Lambda}_K = diag(\hat{\lambda}_{K_1}, \hat{\lambda}_{K_2}, ..., \hat{\lambda}_{K_{|\mathcal{V}_K|}})$ is a diagonal matrix with the ordered eigenvalues as elements $(0 = \hat{\lambda}_{K_1} < \hat{\lambda}_{K_2} < ... < \hat{\lambda}_{K_{|\mathcal{V}_K|}})$ and $\hat{\Phi}_K = (\hat{\phi}_{K_1} | \hat{\phi}_{K_2} | ... | \hat{\phi}_{K_{|\mathcal{V}_K|}})$ is a matrix with the corresponding ordered orthonormal eigenvectors as columns. The normalized Laplacian matrix is positive semi-definite and so has all eigenvalues non-negative. The number of zero eigenvalues is the number of connected components in $\mathcal{G}_K(\mathcal{V}_K; \mathcal{E}_K)$. In [12], the von Neumann entropy of $\mathcal{G}_K(\mathcal{V}_K; \mathcal{E}_K)$ associated with the normalized Laplacian eigenspectrum is defined as $H_{VN} = -\sum_{i=1}^{|\mathcal{V}_K|} \frac{\hat{\lambda}_{K_i}}{2} \log \frac{\hat{\lambda}_{K_i}}{2}$. Since the computation of the von Neumann entropy requires cubic number of vertices operations, Han et al. [3] have shown how the computation can be rendered quadratic in the number of the vertices. By approximating the von Neumann entropy by its quadratic counterpart, the approximated von Neumann entropy for $\mathcal{G}_K(\mathcal{V}_K; \mathcal{E}_K)$ is given by

$$H_{VN}(\mathcal{G}_K) = \frac{|\mathcal{V}_K|}{4} - \sum_{(v_i, v_j) \in \mathcal{E}_K} \frac{1}{4 \, d_K(i) d_K(j)} \tag{4}$$

Furthermore, the probability of a steady state random walk on $\mathcal{G}_K(\mathcal{V}_K; \mathcal{E}_K)$ visiting vertex v_i is $P_K(i) = d_K(i)/\sum_{v_j \in \mathcal{V}_K} d_K(j)$. The Shannon entropy of $\mathcal{G}_K(\mathcal{V}_K; \mathcal{E}_K)$ with the probability distribution $P\{\mathcal{G}_K\} = P_K$ is then given by

$$H_S(P\{\mathcal{G}_K\}) = H_S(P_K) = -\sum_{i=1}^{|\mathcal{V}_K|} P_K(i) \log P_K(i). \qquad (5)$$

3   Jensen-Shannon Complexity Traces of Graphs

In this section, we investigate how to use the Jensen-Shannon divergence as a means of constructing a depth-based complexity trace of graph-structure.

3.1   Jensen-Shannon Divergence Measure

The Jensen-Shannon divergence is a nonextensive mutual information measure. It is defined on probability distributions over structured data [4]. The Jensen-Shannon divergence $JSD(P_m, P_n)$ between probability distributions P_m and P_n is given by:

$$JSD(P_m, P_n) = H_S(\frac{P_m + P_n}{2}) - \frac{H_S(P_m) + H_S(P_n)}{2} \qquad (6)$$

where $H_S(P_m)$ is the Shannon entropy for the probability distribution P_m.

3.2   Composite Structure of Subgraphs

Before we use the Jensen-Shannon divergence as a means of constructing a complexity trace of a graph, we required a composite structure graph of a pair of (sub)graphs. For a pair of subgraphs $\mathcal{G}_K(\mathcal{V}_K, \mathcal{E}_K)$ and $\mathcal{G}_{K'}(\mathcal{V}_{K'}, \mathcal{E}_{K'})$, their composite structure graph $\mathcal{G}_K \oplus \mathcal{G}_{K'}$ has vertex and edge sets $\mathcal{V}_K \oplus \mathcal{V}_{K'}$ and $\mathcal{E}_K \oplus \mathcal{E}_{K'}$ respectively. The most common algorithms to create a composite structure graph of two initial (sub)graphs are formed by taking graph product and graph union. For reason of the efficient computation here we take the (sub)graph union. To construct an union graph $\mathcal{G}_U(\mathcal{V}_U, \mathcal{E}_U)$ of $\mathcal{G}_K(\mathcal{V}_K, \mathcal{E}_K)$ and $\mathcal{G}_{K'}(\mathcal{V}_{K'}, \mathcal{E}_{K'})$, we perform pairwise correspondence matching. Details of the construction are outside the scope of this paper. Our approach follows that of Han et.al's work in [11].

3.3   Complexity Characterisation of Graph Structure

We define a depth-based Jensen-Shannon complexity trace for a graph. For a graph $G(V, E)$ the full set of its centroid expansion subgraphs is $G_C^{v_C} = \{\mathcal{G}_1, ..., \mathcal{G}_K, ..., \mathcal{G}_L\}$ where v_C is the centroid vertex of G, L is the greatest length of shortest paths from the centroid vertex v_C to the remaining vertices in $G(V, E)$, and $\mathcal{G}_K$ is the K-layer centroid expansion subgraph of $G(V, E)$. The essentiality of the L layer subgraph is the graph $G(V, E)$ itself. Suppose we have probability distributions resulting from steady state random walks on each of the K layer centroid expansion subgraph $\mathcal{G}_K$ denoted by $P\{\mathcal{G}_1\}, ..., P\{\mathcal{G}_K\}, ..., P\{\mathcal{G}_L\}$. The complexity trace is computed as

$$CT = [JSD(P\{\mathcal{G}_1\}, P\{\mathcal{G}_L\}), ..., JSD(P\{\mathcal{G}_K\}, P\{\mathcal{G}_L\}), ..., JSD(P\{\mathcal{G}_L\}, P\{\mathcal{G}_L\})]^T \tag{7}$$

where $JSD(P\{\mathcal{G}_K\}, P\{\mathcal{G}_L\})$ is the Jensen-Shannon divergence between the K layer centroid expansion subgraph and the L layer centroid expansion subgraph (i.e. graph $G(V, E)$). This complexity trace encapsulates an mutual information based interior dissimilarity transformation between the graph $G(V, E)$ and its K, which is from 1 to L, layer centroid expansion subgraphs with their steady state random walk probability distributions. The Jensen-Shannon divergence $JSD(P\{\mathcal{G}_K\}, P\{\mathcal{G}_L\})$ is defined as:

$$JSD(P\{\mathcal{G}_K\}, P\{\mathcal{G}_L\}) = H_S(\frac{P\{\mathcal{G}_K\} \oplus P\{\mathcal{G}_L\}}{2}) - \frac{H_S(P\{\mathcal{G}_K\}) + H_S(P\{\mathcal{G}_L\})}{2} \tag{8}$$

where $\frac{P\{\mathcal{G}_K\} \oplus P\{\mathcal{G}_L\}}{2}$ represents the probability distribution of the steady state random walk over the union graph $\mathcal{G}_U(\mathcal{V}_U, \mathcal{E}_U)$ of $\mathcal{G}_K(\mathcal{V}_K, \mathcal{E}_K)$ and $\mathcal{G}_L(\mathcal{V}_L, \mathcal{E}_L)$. As the L layer expansion subgraph $\mathcal{G}_L(\mathcal{V}_L, \mathcal{E}_L)$ contains the full structure of the K layer expansion subgraph $\mathcal{G}_K(\mathcal{V}_K, \mathcal{E}_K)$, using the graph union mentioned in Section 3.3, $\mathcal{G}_U(\mathcal{V}_U, \mathcal{E}_U)$ can be represented by $\mathcal{G}_L(\mathcal{V}_L, \mathcal{E}_L)$. As a result (8) can be rewritten as:

$$JSD(P\{\mathcal{G}_K\}, P\{\mathcal{G}_L\}) = \frac{H_S(P\{\mathcal{G}_L\}) - H_S(P\{\mathcal{G}_K\})}{2} \tag{9}$$

Since we also use the von Neumann entropy in (4) to construct the complexity trace CT, then CT in (7) can also be written as

$$CT = [JSD(\mathcal{G}_1, \mathcal{G}_L), ..., JSD(\mathcal{G}_K, \mathcal{G}_L), ..., JSD(\mathcal{G}_L, \mathcal{G}_L)]^T \tag{10}$$

where $JSD(\mathcal{G}_K, \mathcal{G}_L)$ is given by

$$JSD(\mathcal{G}_K, \mathcal{G}_L) = \frac{H_{VN}(\mathcal{G}_L) - H_{VN}(\mathcal{G}_K)}{2} \tag{11}$$

3.4 Graphs of Different Size

The L layer expansion subgraph is the undirected graph itself, and the dimension of a Jensen-Shannon complexity trace vector is thus equal to greatest layer L. However, the complexity trace vectors for graphs of different sizes may exhibit various lengths. To compare these graphs by using complexity trace vectors, we need to make vector lengths uniform. This is achieved by padding out the dimensions of the complexity trace vectors. Hence, for complexity trace vectors CT_m and CT_n of two graphs G_m and G_n with dimensions L_m and L_n respectively, where $L_m > L_n$, we use the L_n-th element value of CT_n as the added padding value for the extended $L_n + 1$-th to L_m-th elements of CT_n. Since the L_n-th element Jensen-Shannon divergence value is 0, the padding values are 0.

3.5   Computational Complexity Evaluation

The computational complexity of proposed complexity trace is governed by four computational step. Consider a sample graph $G(V, E)$ with size s and highest shortest path length L for the centroid vertex. The Dijkstra shortest path calculation requires $O(s^2)$ operations. The processing of centroid expansion subgraphs requires $O(Ls^2)$ operations. Since the L layer centroid expansion subgraph possesses the full structures of any K layer centroid expansion subgraphs, the union graph is the L layer centroid expansion subgraph. As a result the union graph construction approximately requires $O(s)$ operations. The Jensen-Shannon divergence calculation approximately requires $O(\sqrt[3]{s^2})$ operations. The L is approximated equal to $\sqrt[3]{s}$.

4   Experimental Evaluation

4.1   Interior Complexity Evaluation

We commence by illustrating how the representational power of the proposed complexity traces of graphs, and demonstrate that these can be used to distinguish different objects. The evaluation utilizes graphs extracted from images of a box and a house, taken respectively from the ALOI and CMU databases. For each object we use 18 images captured from different viewpoints. The graphs are the Delaunay triangulations of feature points extracted from the different images. For each graph, we identify the centroid vertex and construct centroid expansion subgraphs. The interior complexity values are computed using (9) or (11). Figs.1(a)(b) and (c)(d) show the sets of complexity histograms of complexity traces using Shannon or von Neumann entropy (18 per object) for each object in turn respectively. The main features to note are that the distributions from the same object are similar to each other, whereas those from different objects are dissimilar.

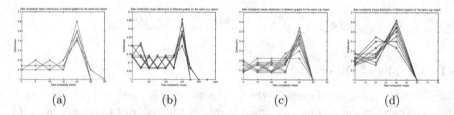

(a)   (b)   (c)   (d)

Fig. 1. Complexity histograms of complexity traces of different graphs

4.2   Stability Evaluation on Centroid Vertex

To evaluate the stability of our proposed complexity trace from the centroid vertex, we explore the relationship between graph edit distance and the pattern vectors resulting from our complexity trace vectors of graphs. The evaluation utilizes two randomly generated seed graphs. The two seed graphs have 500 vertices and

300 vertices respectively. For each seed graph, we first identify its centroid vertex as the original centroid vertex, then we apply random edit operations of edges fraction addition to simulate the effects of noise. The feature distance of the original seed graph G_O and its noise corrupted counterpart G_E is defined as their Euclidean distance $d_{G_O,G_E} = \sqrt{(CT_O - CT_E)^T (CT_O - CT_E)}$ where CT_O and CT_E are complexity traces of G_O and G_E from the same centroid vertex, i.e. the original centroid vertex. The experimental results are shown in Fig.2. Fig.2(a)(b) and (c)(d) show the feature distance between pattern vectors using Shannon or von Neumann entropy for the two seed graphs and their edited graphs respectively. In each subfigure, the x-axis shows the 1% to 35% of edges randomly added, and the y-axis shows the value of the Euclidean distance d_{G_O,G_E} between G_O and G_E. From Fig.2 it is clear that when less than 5% are added the fluctuation is small, and when around 20% are added the fluctuation becomes moderate. This implies that the proposed complexity trace from the centroid vertex is robust even when the seed graph structures undergo relatively large perturbations. As a whole, there's an approximately linear relationship between the graph edit distance and the Euclidean distance. This implies that the proposed method possesses the ability to distinguish graphs under controlled structural error.

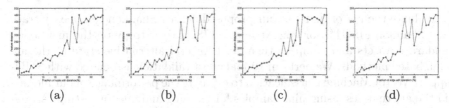

(a)   (b)   (c)   (d)

Fig. 2. Distance distribution between feature vectors versus graph edit operation

4.3 Real-World Datasets

We compare our proposed complexity trace method with several state of the art methods. The methods for comparison include 1) Jensen-Shannon graph kernel (JSGK) [6], 2) von-Neumann thermodynamic depth complexity (VNTD) [2,3], 3) information functionals f^V (FV) (e=1) and f^P (FP) (e=1) [8], and 4) Ihara coefficients for graphs (CIZF) [7]. We use three standard graph based datasets abstracted from bioinformatics datasets [9,2] for experimental evaluation. For the FV and FP, we set the parameters α as 2, and c_k and b_k as $\rho - k + 1$ [8].

MUTAG: The MUTAG benchmark is based on graphs representing 188 chemical compounds, and aims to predict whether each compound possesses mutagenicity. The maximum and average number of vertices are 28 and 17.93 respectively. As the vertices and edges of each compound are labeled with a real number, we transform these graphs into unweighted graphs.

PPIs: The PPIs dataset consists of protein-protein interaction networks (PPIs). The graphs describe the interaction relationships between histidine kinase in different species of bacteria. Histidine kinase is a key protein in the development of signal transduction. If two proteins have direct (physical) or indirect (functional) association, they are connected by an edge. There are 219 PPIs in this dataset and they are collected from 5 different kinds of bacteria. We select *Proteobacteria*40 PPIs and *Acidobacteria*46 PPIs as the second group test graphs. The maximum, minimum and average number of vertices of selected graphs are 238, 6 and 109.60 respectively.

D&D: The D&D dataset contains 1178 protein structures. Each protein is represented by a graph, in which the nodes are amino acids and two nodes are connected by an edge if they are less than 6 Angstroms apart. The prediction task is to classify the protein structures into enzymes and non-enzymes. The maximum and average number of virtices are 5748 and 284.32 respectively.

ENZYMES: The ENZYMES dataset is a dataset based on graphs representing protein tertiary structures consisting of 600 enzymes from the BRENDA enzyme database. In this case the task is to correctly assign each enzyme to one of the 6 EC top-level classes. The maximum and average number of vertices are 126 and 32.63 respectively.

4.4   Performance Comparison

We evaluate the performance of our proposed Jensen-Shannon complexity trace using Shannon (JSCTS) or von Neumann (JSCTV) entropy on the mentioned standard datasets and compare them with several alternative state of the art graph based methods. We perform 10-fold cross-validation associated with SMO-Support Vector Machine Classification to evaluate the performance of our method and the alternatives, using nine samples for training and one for testing. All parameters of the SVMs were optimized. The codes of our previous work in [6] and the other methods were also re-optimized. We report the average prediction accuracies and runtime of each method in Table 1(-:infeasible runtime; =: over computing), the runtime were measured under Matlab R2011a running on a ThinkPad T61p with an Intel 2.2GHz 2-Core processor and 2GB RAM.

Table 1. Experimental Comparison on Bioinformatics Datasets

Datasets	JSCTS	JSCTV	JSGK	VNTD	FV	FP	CIZF
MUTAG	85.63	82.44	87.76	83.51	84.57	85.63	80.85
PPIs	76.74	77.90	69.85	67.44	70.93	70.93	70.93
Enzymes	29.00	32.16	27.05	30.50	24.17	23.33	32.00
D&D	75.32	76.15	78.00	–	=	=	– =

Datasets	JSCTS	JSCTV	JSGK	VNTD	FV	FP	CIZF
MUTAG	1"	1"	2"	19'21"	1"	1"	1"
PPIs	1"	1"	2"	52'27"	1"	1"	55"
Enzymes	1"	1"	19"	4h37'	1"	1"	11"
D&D	42"	44"	14'59"	–	=	=	– =

In terms of the runtime and graph size, our method can efficiently compute graph complexity traces even for graphs with thousands of vertices, while VNTD and CIZF prove computationally burdensome or can not be finished in one day on D&D dataset. Our method outperforms all the alternatives on classification accuracy, only the JSGK is competitive to ours on D&D and MUTAG datasets. Our method outperforms all the alternatives on runtime for datasets of large graphs. Compare to depth-base complexity measures VNTD, FV and FP, our depth-based Jensen-Shannon complexity trace using Shannon or von Neumann entropy outperforms all of them on classification and runtime.

5 Conclusion

In this paper, we have shown how to construct a depth-based Jensen-Shannon complexity trace for a graph. Our method is based on the graph decomposition and Jensen-Shannon divergence. For a graph, we have identified a centroid vertex by computing the minimum variance of its shortest path lengths, and thus obtained a family of expansion subgraphs with increasing layers. The proposed complexity trace of a graph has been constructed by measuring how the Jensen-Shannon divergence varies with increasing layers of the subgraphs. We use the Shannon entropy or von Neumann entropy to calculate the required entropies in the Jensen-Shannon divergence. Experiments on graph datasets abstracted from bioinformatics demonstrate effectiveness and efficiency of the proposed complexity trace.

References

1. Crutchfield, J.P., Shalizi, C.R.: Thermodynamic depth of causal states: Objective complexity via minimal representations. Physical Review E 59, 275–283 (1999)
2. Escolano, F., Hancock, E.R., Lozano, M.A.: Heat diffusion: Thermodynamic depth complexity of networks. Physical Review E 85, 206–236 (2012)
3. Han, L., Escolano, F., Hancock, E.R.: Graph characterizations from von Neumann entropy. To appear in Pattern Recognition Letter (2012)
4. Martins, A.F.T., Smith, N.A., Xing, E.P., Aguiar, P.M.Q., Figueiredo, M.A.T.: Nonextensive information theoretic kernels on measures. Journal of Machine Learning Research 10, 935–975 (2009)
5. Bunke, H., Riesen, K.: Improving vector space embedding of graphs through feature selection algorithms. Pattern Recognition 44, 1928–1940 (2011)
6. Bai, L., Hancock, E.R.: Graph Clustering Using the Jensen-Shannon Kernel. In: Real, P., Diaz-Pernil, D., Molina-Abril, H., Berciano, A., Kropatsch, W. (eds.) CAIP 2011, Part I. LNCS, vol. 6854, pp. 394–401. Springer, Heidelberg (2011)
7. Ren, P., Wilson, R.C., Hancock, E.R.: Graph Characterization via Ihara Coefficients. IEEE Transactions on Neural Networks 22, 233–245 (2011)
8. Dehmer, M., Mowshowitz, A.: A history of graph entropy measures. Proceedings of Information Sciences 181, 57–78 (2011)
9. Shervashidze, N., Schweitzer, P., Leeuwen, E.J., Mehlhorn, K., Borgwardt, K.M.: Weisfeiler-Lehman graph kernels. Journal of Machine Learning Research 1, 1–48 (2010)

10. Wilson, R.C., Hancock, E.R., Luo, B.: Pattern vectors from algebraic graph theory. IEEE Transactions Pattern Analysis and Machine Intelligence 27, 1112–1124 (2005)
11. Han, L., Hancock, E.R., Wilson, R.C.: Learning Generative Graph Prototypes Using Simplified von Neumann Entropy. In: Jiang, X., Ferrer, M., Torsello, A. (eds.) GbRPR 2011. LNCS, vol. 6658, pp. 42–51. Springer, Heidelberg (2011)
12. Passerini, F., Severini, S.: Quantifying complexity in networks: the von Neumann entropy. International Journal of Agent Technologies and Systems 1, 58–67 (2009)

Complexity of Computing Distances between Geometric Trees

Aasa Feragen

Department of Computer Science, University of Copenhagen, Universitetsparken 1,
2100 Copenhagen, Denmark
aasa@diku.dk

Abstract. Geometric trees can be formalized as unordered combinatorial trees whose edges are endowed with geometric information. Examples are skeleta of shapes from images; anatomical tree-structures such as blood vessels; or phylogenetic trees. An inter-tree distance measure is a basic prerequisite for many pattern recognition and machine learning methods to work on anatomical, phylogenetic or skeletal trees. Standard distance measures between trees, such as tree edit distance, can be readily translated to the geometric tree setting. It is well-known that the tree edit distance for unordered trees is generally NP complete to compute. However, the classical proof of NP completeness depends on a particular case of edit distance with integer edit costs for trees with discrete labels, and does not obviously carry over to the class of geometric trees. The reason is that edge geometry is encoded in continuous scalar or vector attributes, allowing for continuous edit paths from one tree to another, rather than finite, discrete edit sequences with discrete costs for discrete label sets. In this paper, we explain why the proof does not carry over directly to the continuous setting, and why it does not work for the important class of trees with scalar-valued edge attributes, such as edge length. We prove the NP completeness of tree edit distance and another natural distance measure, QED, for geometric trees with vector valued edge attributes.

1 Introduction

Trees are basic structures in mathematics and computer science, as well as in nature. Tree-structures appear, for instance, as airway trees in the lungs [20,21], as blood vessel trees [13], or as skeleta of more general shapes [4,9,10,15,17,19]. Anatomical and biological trees carry information about the organ or organism that contains them, and many pattern recognition algorithms, e.g., in computer vision and medical image analysis, require a distance measure between tree-structures as input [5,10,15]. Tree edit distance (TED) is a classical distance measure between trees, which has been used in many applications [9,10,14,15, 17]. Anatomical trees are geometric trees, in the sense that they carry useful geometric information about their branches' shape, size and position. TED is readily translated to handle geometric properties, but anatomical trees are often

G.L. Gimel' farb et al. (Eds.): SSPR & SPR 2012, LNCS 7626, pp. 89–97, 2012.
© Springer-Verlag Berlin Heidelberg 2012

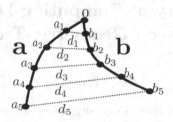

Fig. 1. For many applications, each edge e is represented by an edge attribute consisting of a set of n equidistant landmark points $a_i \in \mathbb{R}^m$, where $m = 2, 3$, giving a point $a = (a_i)_{i=1}^{n} \in \mathbb{R}^{mn}$. We typically assume that the first landmark point a_1 is translated to the origin. The cost of deforming one edge attribute, or shape, a into another edge attribute, or shape, b is the Euclidean norm $\|a - b\|_2 = \sqrt{d_1^2 + \ldots + d_n^2}$.

not adorned with a natural branch labeling or order. This means that we need to be able to compare unordered trees.

Tree edit distance for unordered trees is generally NP complete to compute [1, 22]. However, the classical proof of NP completeness is made for a particular case of edit distance with integer edit costs for trees with discrete labels, and it does not obviously carry over to the class of geometric trees. This is because the geometric trees have branch descriptors that are vectors or scalars, which thus form a path-connected set of branch attributes, with continuous edit costs.

1.1   Geometric Trees

By a *tree* we shall mean a rooted combinatorial tree $\mathscr{T} = \{V, E, r\}$ where V is a set of vertices, $E \subset V \times V$ is a set of edges, and $r \in V$ is a designated root vertex. By *geometric tree* we shall mean a pair $(\mathscr{T}, x)$ where $\mathscr{T}$ is a combinatorial tree and $x \colon E \to A$ is a map from the edge set of $\mathscr{T}$ into a space A of geometric attributes, which attaches an edge attribute $x_e \in A$ to every edge $e \in E$. The space A of geometric attributes could, for instance, be a space of edge lengths, $(\mathbb{R}_{\geq 0})$, a space of edge embeddings into plane or space ($\{f \colon [0, 1] \to \mathbb{R}^m\}$, $m = 2, 3$), or, as a discretization of the latter, a space of landmark point sets that describe the shape of the edge in plane or space ($(\mathbb{R}^m)^n$, where n is the number of landmark points per edge, and $m = 2, 3$), see fig. 1. In this paper, we shall consider situations where the attribute space is $\mathbb{R}_{\geq 0}$ or $\mathbb{R}^N$ for some $N \in \mathbb{N}$.

1.2   Related Work

Tree edit distance, or TED [1,11,16,22], is defined as the minimal total sum of costs of edit operations needed in order to turn the first tree into the second. In its most general form, TED is formulated for combinatorial trees $\mathscr{T} = (V, E, r)$ endowed with edge (or vertex) *labels* given by a mapping $x \colon E \to \mathscr{L}$, where $\mathscr{L}$ is a space of labels. The set of labels could be a vector space, as in the case of geometric trees, but in many applications previously studied, the set of labels is a finite dictionary. The set of edit operations typically consists of *deletion* of edges, *insertion* of edges, and *relabeling* of edges[1] (although extra edit operations have been introduced in some cases [9]). Note that insertion and deletion can also happen to edges which are not leaves. A sequence of edit operations that turn one tree into another is called an *edit path* between the two trees. When restricted to classes of trees with additional assumptions, such as *edge order* or *bounded size*, there exist a number of polynomial time algorithms [3, 11, 22] for computing the tree edit distance, and with further restrictions on the allowed complexity of the edit paths, there are linear time algorithms available [18]. However, for general, unordered trees, the edit distance computation problem has been shown to be NP complete by Zhang, Statman and Shasha [22]. Their proof can be simplified to the trick explained in section 2 below, originally used by Matousek and Thomas to prove NP completeness of the subtree problem [12]. However, as we shall see below, this proof does not automatically transfer to geometric trees with continuous edge labels, and in fact it fails for trees with scalar valued labels. The same is true for the original, slightly more complicated proof in [22]. Using a construction similar to the NP completeness proof of [12], we prove that the computation of tree edit distance is NP complete for the space of geometric trees with vector valued edge labels.

From a statistical point of view, TED is not an optimal distance between geometric trees, as it does not define unique geodesics [6]. Feragen et al. [7] have defined a metric on geodesic trees called *the QED metric* and showed that it has better statistical properties [6]. In section 3 we give a brief account of this metric and prove that it, too, is NP complete to compute for geometric trees with vector edge attributes.

2 Tree Edit Distance

The original proof of NP completeness for edit distance between unordered, rooted trees, is formulated for the class of rooted trees $\mathscr{T} = (V, E, r)$ with edge labels $x \colon E \to \mathscr{L}$ where $\mathscr{L}$ is a discrete set of labels. The available edit operations are *edge deletion*, *edge insertion* and *edge relabeling*, which have cost 1 each. This measure is called *integer TED*.

The Exact 3-Cover Problem. The NP completeness proof for integer TED [12, 22] is based on the exact 3-cover problem. Let $L = \{l_1, \ldots, l_{3q}\}$ be a set, and let $\mathscr{S} = \{C_i | i = 1 \ldots N\}$ be a cover[2] of L by sets $C_i \subset L$, all with 3 elements. The exact 3-cover problem is the problem of deciding whether there is an exact

---

[1] In some papers, e.g. [22] the edit operations (delete, add, edit) are performed on vertices rather than edges. This is equivalent to the approach taken here: Represent the branches of an anatomical tree as attributed nodes, joined together in the obvious tree structure. By defining edit operations on nodes, we would get exactly the same definition as the one used here. We, however, prefer to represent branches in geometric trees as edges, as this is more intuitive, and also quite standard [9, 15].

[2] A *cover* of L is a family of subsets $C_i \subset L$ such that $L \subset \bigcup_{i=1}^{N} C_i$.

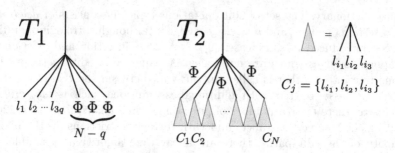

Fig. 2. Any instance of the exact 3-cover problem can be solved by computing the edit distance between these two trees

subcover[3] of $\mathscr{S}$ (and identifying such a subcover). The exact 3-cover problem is a classical NP-complete problem [8].

TED and the Exact 3-Cover Problem. We first review the original proof of NP completeness for integer TED [12,22]. Assume given an instance of the exact 3-cover problem, i.e. assume given a finite set $L = \{l_1, \ldots, l_{3q}\}$ with a cover $\mathscr{S} = \{C_i | i = 1 \ldots N\}$ by sets C_i that have 3 elements each. Build the edge-labeled trees T_1 and T_2 with labels from $L \cup \{\Phi\}$, as in fig. 2, where Φ is some label not in L. We shall see that

i) by computing the TED distance between T_1 and T_2, we can determine whether there exists an exact subcover of $\mathscr{S}$, and
ii) if there is an exact subcover, we can retrieve it from the optimal edit path from T_1 to T_2.

Let us ignore the tree-structure of T_1 and T_2 for a second and only consider the two sets of attributed edges. To find the minimal total cost of editing one set to become the other, note that the set of edge attributes $L_1 = x_1(E_1)$ in T_1 is contained in the set of edge attributes $L_2 = x_2(E_2)$ in T_2. There are $N + 2q + 1$ edge attributes in L_1 and $4N + 1$ edge attributes in L_2, so in order to transform $L_1 \subset L_2$ into L_2, we only need to insert $3N - 2q$ edges, at a total cost of

$$b_l = (4N + 1) - (N + 2q + 1) = 3N - 2q.$$

This number b_l is a lower bound for the edit distance between T_1 and T_2.

If there is a solution to the exact 3-cover problem on S, consisting of a set $\mathscr{S}' = \{C_i | i = 1 \ldots q\}$ of 3-sets, then the following edit path from T_1 to T_2 actually *has* length b_l:

– insert an edge with attribute Φ above each triple of elements in some $C_i \in \mathscr{S}'$, $i = 1 \ldots q$.

---

[3] An *exact subcover* of $\mathscr{S}$ is a sub-family $\mathscr{S}' = \{C_{i_j}\}_{j=1}^{M}$ of $\mathscr{S}$ such that $\mathscr{S}'$ is a cover of L and $C_{i_{j_1}} \cap C_{i_{j_2}} = \emptyset$ for all $j_1 \neq j_2$.

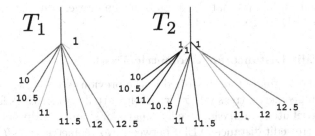

Fig. 3. The edit path indicated by the colored edge matches has minimal length, even though the corresponding exact 3-cover problem does not have a solution. Thus, the proof from integer TED does not carry over to the case of continuous edge attributes.

– insert the remaining $3(N-q)$ edges with attributes belonging to the remaining $N - q$ yellow subtrees below each of the Φ branches in T_1.

Since this edit path has length b_l, which is also the lower bound for the length, the TED distance between T_1 and T_2 is b_l. Thus, a solution to the exact 3-cover problem yields a) a solution to the TED problem and b) a total distance b_l between T_1 and T_2. If we can show that edit paths that do *not* yield solutions to the exact 3-cover problem are *longer* than b_l, then we have proven our claim. But this is easy, since any mapping from T_1 to T_2 which does not correspond to a solution to the exact 3-cover problem must involve either changing some label l_i to another label l_j, or deleting edges, or both. This has to cost more than b_l.

It follows that the computation of integer TED is NP complete.

2.1 Example: Geometric Trees with Scalar Branch Attributes

To see that the same idea of proof does not carry directly over to geometric trees, consider the following set $L = \{10, 10.5, 11, 11.5, 12, 12.5\}$ and the following cover of L by 3-sets: $\mathscr{S} = \{\{10, 10.5, 11\}, \{10.5, 11, 11.5\}, \{11, 12, 12.5\}\}$. Clearly, $\mathscr{S}$ does not have an exact subcover. As above we form trees T_1 and T_2 as in fig. 3, where the lengths of edges labeled by elements in L are the corresponding real numbers, and the lengths of edges labeled with Φ are, say, 1.

A lower bound b_l for the edit distance between T_1 and T_2 is, just like above, found by just considering sets of edge attributes, forgetting about tree topology for a second, matching the sets of edge attributes up, and adding the costs of the entire matching process. Again, all edge attributes from T_1 can be matched to an identical edge attribute from T_2, so the only nonzero matching costs come from the additional edges in T_2, namely $2 * \|\Phi\| + 10.5 + 11 + 11 = 34.5$.

We already know that there is no exact 3-cover of $\mathscr{S}$; nevertheless, we can, in fact, find an edit path from T_1 to T_2 of length b_l, where the branches indicated by colors in fig. 3 are matched (deformed to match) and all branches appearing in black in T_2 are inserted. The total cost of deformation edits is 1 and the total cost of insertion edits is 33.5, giving an edit distance of $34.5 = b_l$ between T_1 and T_2, although the edit path does not correspond to a solution of the exact

3-cover problem. It follows that the proof from the integer edit distance does not carry over to TED for geometric trees.

2.2   Tree Edit Distance for Geometric Trees

Building on the original proof described in the previous section, we consider the class $\mathscr{X}$ of all geometric trees $(\mathscr{T}, x)$ with edge attributes $x \colon E \to \mathbb{R}^N$, $N \in \mathbb{N}$. These edge attributes could, e.g., be edge length, or shape descriptors as in fig. 1.

Define the tree edit distance (TED) between two geometric trees T_1 and T_2 in $\mathscr{X}$ as the smallest possible total cost of transforming T_1 into T_2 through a finite sequence of edit operations, which belong to the following three categories:

i) *Delete an edge* $e \in E$ (and correspondingly a vertex from V), which costs $\|x(e)\|$, where $\|\cdot\|$ is the Euclidean norm,

ii) *Insert an edge* e to E (and correspondingly a vertex from V), which costs $\|x(e)\|$, and

iii) *Deform an edge* $e \in E$ by changing its attribute from $x(e)$ to a new value a; this costs $\|a - x(e)\|$.

Theorem 1. *If $N \geq 2$, then computing tree edit distance in $\mathscr{X}$ is NP-complete.*

Proof. As for the combinatorial edit distance, this is proven by reducing an arbitrary instance of the exact 3-cover problem to an instance of the edit distance problem. We prove the theorem for $N = 2$; the proof trivially generalizes to $N \geq 2$. Denote by T_1, T_2 the trees in fig. 2, labeled with elements from L and an additional label Φ, where the l_i and Φ represent *distinct* vector edge attributes of length 1.

As before, we can forget about the tree structure and only consider sets of edges. The set of edge attributes in T_1 is, again, contained in the set of edge attributes in T_2, and the minimal total edit cost of transforming the set of $N + 2q + 1$ edge attributes in T_1 to the set of $4N + 1$ edge attributes in T_2 is the cost of inserting the rest of the edge attributes from T_2, which all cost 1 each. This gives us total cost

$$b_l = (4N + 1) - (N + 2q + 1) = 3N - 2q.$$

Again, we need to prove that any edit path that does not correspond to a solution to the exact 3-cover problem must have length $> b_l$. An edit path that does not correspond to a solution to the exact 3-cover problem will have to either:

a) map some edge with (nonzero) attribute l_i to an edge with (nonzero) attribute l_j which is *not* l_i, or

b) delete some edge, or

c) map the edges from T_1 into edges in more than q subtrees C_i in T_2.

Note that

a) the cost of mapping l_i to some $l_j \neq l_i$ has cost $\|l_i - l_j\|$, which is > 0 since the l_i are distinct. This cost comes in addition to inserting at least $3N - 2q$ branches, which gives total cost $> 3N - 2q = b_l$.

b) this means we have to insert more than $3N - 2q$ branches, giving total cost $> 3N - 2q = b_l$.

c) this means we will have to delete some of the branches with attribute Φ from T_1, and thus we have to grow out more than $3N - 2q$ branches, giving total cost $> 3N - 2q = b_l$.

Thus, any edit path which does not correspond to a solution to the exact 3-cover problem has length $> b_l$. This concludes the proof of theorem 1.   □

Remark 2. *In a), the crucial part is, in fact, that the vectors l_i are not parallel. This is to avoid examples like the scalar attribute case in section 2.1.*

3   NP Completeness for Quotient Euclidean Distance

In order to use geometric tools for statistical analysis of geometric trees, e.g., use of geodesics in the spirit of manifold statistics, it is useful to construct a space of geometric trees, and endow it with a geodesic metric. The non-uniqueness of TED geodesics disqualifies TED as a metric of choice in such a framework. A more suitable metric is the QED metric on the space of tree-like shapes as defined by Feragen et al. [6, 7], which has been used to study the shapes of airway trees from human lungs.

By a *tree-shape*, we shall mean a tree which is embedded in $\mathbb{R}^d$, where d is typically 2 or 3. In this paper, we are mainly concerned with the case $d = 3$, since planar trees ($d = 2$) typically induce a canonical edge ordering. The space of tree-like shapes is constructed as follows: Consider a combinatorial rooted, binary tree $T = (V, E, r, <)$ which is sufficiently large to span all the tree-like shapes of interest (T could be infinitely large). The space

$$X = \prod_{e \in E} (\mathbb{R}^m)^n, \ m = 2, 3, \tag{3}$$

contains representatives of all tree-shapes spanned by T, whose edges are represented by landmark point shape descriptors as in fig. 1. That is, a point $x \in X$ corresponds to a map $x \colon E \to (\mathbb{R}^m)^n$. Trees with fewer edges are represented by collapsing (contracting) redundant branches, and higher-order vertices are represented in a similar fashion, also using collapsed branches, as in fig. 4. Some tree-shapes will have more than one representative in X, also shown in fig. 4. In the space of tree-like shapes, these representations are all identified through an equivalence relation. That is, whenever two points $x_1, x_2 \in X$ represent the same tree-shape, they are said to be equivalent: $x_1 \sim x_2$. The space of tree-like shapes $\bar{\bar{X}}$ is defined as the quotient space of X by the equivalence $\sim$:

$$\bar{\bar{X}} = X / \sim .$$

The induced tree-shape space $\bar{\bar{X}}$ is highly nonlinear, and has self-intersections that stem from the identifications made by the equivalence. From the Euclidean metric on X, Feragen et al. work with the *quotient metric* on $\bar{\bar{X}}$, which in this

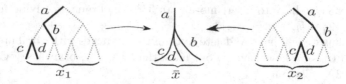

Fig. 4. Higher-order vertices can be represented by the binary tree by collapsing internal branches, shown as dotted lines

case is called the *QED metric*. The quotient metric is a standard mathematical construction [2], which here creates a piecewise Euclidean metric on $\bar{\bar{X}}$. Note that $\bar{\bar{X}}$ geometrically corresponds to a folded Euclidean space. This construction is actually closely related to the TED metric: If the Euclidean metric on X is replaced with an l_1 product of Euclidean metrics on $(\mathbb{R}^m)^n$ in the product in (3), the geometric TED metric studied in section 2 above is retrieved as quotient metric on $\bar{\bar{X}}$ [7].

It turns out that computing the QED distance is generally also NP complete:

Theorem 4. *Computing QED distances in $\bar{\bar{X}}$ is NP-complete.*

Proof. Just as for TED, the QED shortest paths consist of deleting, inserting and deforming edges. Using the same two trees T_1 and T_2 shown in fig. 2, we see that again, if we disregard the tree structure, the lower bound b_l for the QED distance from T_1 to T_2 is given by $b_l = \sqrt{3N - 2q}$, which can be obtained as a shortest QED path length if and only if there exists a solution to the exact 3-cover problem, using the same matchings as in the TED case. Again, the non-parallel property as noted in Rem. 2 is essential. □

Remark 5. *As in section 2.1 the proof would not hold if we replaced the edge shape space $(\mathbb{R}^m)^n$ by scalar edge descriptors $\mathbb{R}$, because the proof depends on the non-parallel assumption on attributes.*

4 Discussion and Conclusion

In this paper we see that the most common distances between unlabeled, unordered geometric trees with vector edge attributes are generally NP complete to compute, just like the edit distance between purely combinatorial unordered, unlabeled trees. NP completeness is a result of the exponential search space which arises when there is no or little formal limitation to the possible mappings between the trees. For trees with scalar edge attributes, such as edge length, the proofs of NP completeness do not hold, and we conjecture that computing these distances is, in fact, also NP complete.

Acknowledgements. The author would like to thank Sean Skwerer and Scott Provan for valuable discussions on complexity of tree algorithms.

References

1. Bille, P.: A survey on tree edit distance and related problems. Theor. Comput. Sci. 337(1-3), 217–239 (2005)
2. Bridson, M.R., Haefliger, A.: Metric spaces of non-positive curvature. Springer (1999)
3. Demaine, E.D., Mozes, S., Rossman, B., Weimann, O.: An optimal decomposition algorithm for tree edit distance. ACM Trans. Algorithms 6, 2:1–2:19 (2009)
4. Demirci, F., Shokoufandeh, A., Dickinson, S.J.: Skeletal shape abstraction from examples. TPAMI 31(5), 944–952 (2009)
5. Demirci, M., Platel, B., Shokoufandeh, A., Florack, L., Dickinson, S.: The representation and matching of images using top points. JMIV 35, 103–116 (2009)
6. Feragen, A., Hauberg, S., Nielsen, M., Lauze, F.: Means in spaces of tree-like shapes. In: ICCV (2011)
7. Feragen, A., Lauze, F., Lo, P., de Bruijne, M., Nielsen, M.: Geometries on Spaces of Treelike Shapes. In: Kimmel, R., Klette, R., Sugimoto, A. (eds.) ACCV 2010, Part II. LNCS, vol. 6493, pp. 160–173. Springer, Heidelberg (2011)
8. Garey, M.J., Johnson, D.S.: Computers and Intractability: A Guide to the Theory of NP-completeness. Freeman (1979)
9. Klein, P., Tirthapura, S., Sharvit, D., Kimia, B.: A tree-edit-distance algorithm for comparing simple, closed shapes. In: SODA, pp. 696–704 (2000)
10. Klein, P.N., Sebastian, T.B., Kimia, B.B.: Shape matching using edit-distance: an implementation. In: SODA, pp. 781–790 (2001)
11. Klein, P.N.: Computing the Edit-Distance between Unrooted Ordered Trees. In: Bilardi, G., Pietracaprina, A., Italiano, G.F., Pucci, G. (eds.) ESA 1998. LNCS, vol. 1461, p. 91. Springer, Heidelberg (1998)
12. Matoušek, J., Thomas, R.: On the complexity of finding iso- and other morphisms for partial k-trees. Discrete Mathematics 108(1-3), 343–364 (1992)
13. Metzen, J.H., Kröger, T., Schenk, A., Zidowitz, S., Peitgen, H.-O., Jiang, X.: Matching of anatomical tree structures for registration of medical images. Im. Vis. Comp. 27, 923–933 (2009)
14. Riesen, K., Bunke, H.: Approximate graph edit distance computation by means of bipartite graph matching. Im. Vis. Comp. 27(7), 950–959 (2009)
15. Sebastian, T.B., Klein, P.N., Kimia, B.B.: Recognition of shapes by editing their shock graphs. TPAMI 26(5), 550–571 (2004)
16. Tai, K.-C.: The tree-to-tree correction problem. J. ACM 26, 422–433 (1979)
17. Torsello, A., Robles-Kelly, A., Hancock, E.R.: Discovering shape classes using tree edit-distance and pairwise clustering. IJCV 72(3), 259–285 (2007)
18. Touzet, H.: A Linear Tree Edit Distance Algorithm for Similar Ordered Trees. In: Apostolico, A., Crochemore, M., Park, K. (eds.) CPM 2005. LNCS, vol. 3537, pp. 334–345. Springer, Heidelberg (2005)
19. Trinh, N., Kimia, B.: Skeleton search: Category-specific object recognition and segmentation using a skeletal shape model. IJCV, 1–26 (2011)
20. Tschirren, J., McLennan, G., Palágyi, K., Hoffman, E.A., Sonka, M.: Matching and anatomical labeling of human airway tree. TMI 24(12), 1540–1547 (2005)
21. Weibel, E.R.: What makes a good lung? Swiss Med. Weekly 139(27-28), 375–386 (2009)
22. Zhang, K., Statman, R., Shasha, D.: On the editing distance between unordered labeled trees. Inf. Process. Lett. 42(3), 133–139 (1992)

Active Graph Matching Based on Pairwise Probabilities between Nodes*

Xavier Cortés, Francesc Serratosa, and Albert Solé-Ribalta

Universitat Rovira i Virgili, Departament d'Enginyeria Informàtica i Matemàtiques, Spain
{xavier.cortes,francesc.serratosa,albert.sole}@urv.cat

Abstract. We propose a method to perform active graph matching in which the active learner queries one of the nodes of the first graph and the oracle feedback is the corresponding node of the other graph. The method uses any graph matching algorithm that iteratively updates a probability matrix between nodes (Graduated Assignment, Expectation Maximisation or Probabilistic Relaxation). The oracle's feedback is used to update the costs between nodes and arcs of both graphs. We present and validate four different active strategies based on the probability matrix between nodes. It is not needed to modify the code of the graph-matching algorithms, since our method simply needs to read the probability matrix and to update the costs between nodes and arcs. Practical validation shows that with few oracle's feedbacks, the algorithm finds the labelling that the user considers optimal because imposing few labellings the other ones are corrected automatically.

Keywords: Machine Learning, Active Graph Matching, Interactive Graph Matching, Least Confident, Maximum Entropy, Expected Model Change.

1   Introduction

Generally speaking, *machine learning* is a discipline concerned with the design and development of algorithms that allow computers to evolve behaviours based on examples [1]. In this discipline, a learner can take advantage of examples to capture characteristics of interest from the data respect of their class and to be able to deduct the class that new examples could belong to. *Error-tolerant graph matching* [2] is another discipline that aims to find the best labelling between nodes of both graphs such that the cost of this optimal labelling is the minimum among all possible labellings. If we put together machine learning and error-tolerant graph matching disciplines, we can define a model in which examples are composed by the set of nodes of one of the graphs and classes are the nodes of the other graphs. Therefore, what we want to learn is which is the matching between two graphs that is considered to be the best.

The key idea behind *active learning* [3, 7, 20] is that a machine learning algorithm can achieve a greater accuracy with fewer classified training examples if it is allowed to

---

* This research is supported by Consolider Ingenio 2010: project CSD2007-00018 & by the CICYT project DPI 2010-17112.

G.L. Gimel' farb et al. (Eds.): SSPR & SPR 2012, LNCS 7626, pp. 98–106, 2012.

choose the data from which it learns. Active learning is well motivated in many modern machine-learning problems, where unclassified examples may be abundant but finding the class is difficult, time-consuming or expensive to obtain [1, 21]. Active learning has been applied in several fields such as speech recognition [6, 21], information extraction [10, 13, 15, 17], robotics [12], transcription of text images [22, 28] or object classification in general [4, 5, 6, 9, 14, 18]. And in general, for parameter selection [8].

We present a model in which we have put together the active learning and graph matching concepts. In this case, the learner queries the node that it is supposed to produce a greater impact on the labelling between both graphs. In our case, the active learner may ask queries in the form of graph nodes and asks which are the nodes of the other graph that they have to be matched. The answerer of the query might be another automatic system or a human annotator (in general, it is called an *oracle*).

Active learning scenarios are usually classified in three classes. The active learner generates a query de novo [18], the active learner receives a stream of objects and decides whether query them or not [19] and the pool-based active learning, in which active learner decides to query an element from a sub-set of unlabelled objects [4, 11], or with sub-sampling [5]. In our case, we always have nodes of both graphs, for this reason, our scenario can be classified as the third scenario: *pool-based active learning*.

All active learning scenarios are involved in evaluating the informativeness of unclassified examples, which can either be generated de novo or sampled from a given distribution. It has been proposed many ways of formulating such query strategies in the literature [3]. We have considered two main strategies: *Uncertainly Sampling* and *Expected Model Change*. We have modelled the first one in three different strategies and the second one in one strategy.

The rest of the paper is organised as follows. In the next section, we present the graph-matching problem. In section 3, we present four different active strategies applied to the problem of finding the best labelling. In section 4, we show the algorithm to compute the active graph matching. Finally, in section 5 we show the practical evaluation and we conclude the paper in section 6.

2   Graph Matching and Isomorphism between Graphs

Let g^1 and g^2 be two attributed graphs. We suppose that g^1 and g^2 have the same number of nodes n since they have been enlarged enough to incorporate null nodes. We define nodes in g^1 and g^2 as $v_i^1 \in \Sigma_v^1$ and $v_a^2 \in \Sigma_v^2$ and we define arcs as $e_{ij}^1 \in \Sigma_e^1$ and $e_{ab}^2 \in \Sigma_e^2$, $\forall i, j, a, b \in \{1, ..., n\}$. Moreover, let f be a bijective labelling between nodes of both graphs. The cost of matching graphs g^1 and g^2, given this isomorphism f, is represented by

$$C_f(g^1, g^2) = \sum_{v_i^1 \in \Sigma_v^1} c_v(v_i^1, v_a^2) + \sum_{e_{ij}^1 \in \Sigma_e^1} c_e(e_{ij}^1, e_{ab}^2) \tag{1}$$

where $f(v_i^1) = v_a^2$ and $f(v_j^1) = v_b^2$. That is, the cost is defined as the addition of the pairwise costs of matching nodes and arcs [10]. These local costs can be represented through two matrices $C_v \in \mathbb{R}^{+^2}$, $C_v[i, a] = c_v(v_i^1, v_a^2)$ and $C_e \in \mathbb{R}^{+^4}$, $C_e[i, j, a, b] = c_e(e_{ij}^1, e_{ab}^2)$ and their definition depends on the application. Usual examples are the Euclidean distance, when attributes have the position of the node in the image or the distance between local features such as SIFTs or HOGs.

There are several error-tolerant graph-matching algorithms that, using a minimization criteria, such as eq. (1), return the best isomorphism f between two graphs. For instance: probabilistic relaxation [23], Graduated-Assignment [24] or Expectation-Maximisation [25]. In fact, the input of these algorithms can be matrices C_v and C_e instead of graphs g^1 and g^2 since matrices capture all the differences between graphs and the minimisation cost is defined through these matrices (eq. 1). Considering that the involved graphs have a degree of disturbance and also the exponential complexity of the problem, these algorithms do not return exactly the isomorphism f but a probability matrix related to it. We represent this matrix by P where each cell contains $P[v_i^1, v_a^2] = Prob(f(v_i^1) = v_a^2)$. Thus, given the probability matrix P it is necessary to derive the final labelling f by a discretization process. There are several techniques to perform this discretization, e.g. [30]. Figure 1 represents the probabilistic graph-matching paradigm.

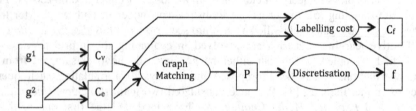

Fig. 1. Probabilistic graph matching framework

In the next section, we present four different strategies that, with the information of the probability matrix P, derive the node that has to be queried. Besides in section 4, we show how to use the interactive algorithm presented in [31] that modifies matrices C_v and C_e to consider the oracle feedback.

3   Active Learning Strategies

In this section, we present four strategies to select a node v^{1^*} of g^1 that have to be queried to an oracle since the model assumes the exact knowledge of its mapping will increase the accuracy of the system. Therefore, given the selected node v^{1^*} we ask for $f(v^{1^*})$ and the oracle feedback is v^{2^*}. In all strategies, the pool of nodes to be queried is composed by the nodes that have never been queried before. The logical function $Q(i)$ shows if node v_i^1 has been queried. When the active algorithm is initialised, $Q(i)$ takes the *False* value for all nodes of g^1 and this value is changed to *True* in each query. This logical function is used to assure a node is not queried several times. Note that in the case that $Q(i) = True$ for all nodes of g^1 then the following strategies return a null value. Nevertheless, in this case, the active algorithm (section 4) has to stop since the whole nodes have been queried. Besides, the computation of the following strategies is performed at function *Oracle_Feedback* in the algorithm.

The four strategies we present are classified on *Uncertainly Sampling* and *Expected Model Change* [3].

Uncertainly Sampling. The active learner queries the instances about which it is least certain how to classify. We define three different strategies.

Least Confident (LC): This strategy queries the element that its highest probability of belonging to a class is the lower one between all the elements. In our model, the learner queries the node $v_{LC}^{1^*}$ of g^1 that has not been previously queried and whose maximum probability given the nodes of g^2 is the lower. Node $v_{LC}^{1^*}$ is obtained in two steps. Firstly, we obtain the set of nodes in g^2: $\left\{ v^{2^{(1)}}, ..., v^{2^{(i)}}, ..., v^{2^{(n)}} \right\}$ such that,

$$v^{2^{(i)}} = \underset{\forall j=\{1,..,n\}}{\text{argmax}}\, P\left[v_i^1, v_j^2\right]; \; \forall i = \{1,..,n\} \tag{2}$$

note that some of the nodes in this set can appear several times, $v^{2^{(i)}} = v^{2^{(j)}}; i \neq j$.

And secondly, we select the node in g^1 such that its respective node in the set obtains the minimum probability,

$$v_{LC}^{1^*} = \underset{\forall i=\{1,..,n\}|Q(i)=\,False}{\text{argmin}}\, P\left[v_i^1, v^{2^{(i)}}\right] \tag{3}$$

Least Confident given the Current Labelling (LCCL): The aim of this strategy is to query the nodes that are matched through the current labelling but they have not been queried. Therefore, it could be seen as the method tries to minimise the hamming distance between the current labelling and the ideal labelling (the labelling that would have been predicted by the oracle if all the nodes were queried). The learner queries node $v_{LCCL}^{1^*}$ of g^1 that has not been previously queried and it has the minimum probability given the current labelling f. Formally,

$$v_{LCCL}^{1^*} = \underset{\forall i=\{1,..,n\}|Q(i)=\,False}{\text{argmin}}\, P[v_i^1, f(v_i^1)] \tag{4}$$

Maximum Entropy (ME): This strategy queries the element with maximum Shanon Entropy given the probabilities. The main idea of the method is to query the elements that they are more difficult to be classified. In our model, the selected node $v_{ME}^{1^*}$ is,

$$v_{ME}^{1^*} = \underset{\forall i=\{1,..,n\}|Q(i)=\,False}{\text{argmax}}\, -\sum_{j=1}^{n} P[v_i^1, v_j^2] \cdot log\big(P[v_i^1, v_j^2]\big) \tag{5}$$

Expected Model Change (EMC): An active learner queries the instances that would impart the greatest change to the current model if we knew its class. A possible query strategy could be the "expected gradient length". Since graph-matching probabilistic models are usually trained using gradient-based optimisation; the change imparted to the model can be measured by the length of the training gradient. In other words, the learner should query the instance that if changed its labelling, the gradient between the current labelling and the new one would have the largest magnitude. Considering this aim, we propose to query the node $v_{EMC}^{1^*}$ defined through the following equation,

$$v_{EMC}^{1^*} = \underset{\forall i=\{1,..,n\} \wedge Q(i)=\,False}{\text{argmax}}\, \{R_i\} \tag{6}$$

The value R_i shows the maximum magnitude of any possible change of the current labelling at node v_i^1.

$$R_i = \max_{\forall j=\{1,..,n\}} \{P[v_i^1, v_j^2]\} - P[v_i^1, f(v_i^1)] \tag{7}$$

If $R_i > 0$, the current labelling of v_i^1 is not the ideal one, considering only probabilities $P[v_i^1, v_j^2]$, $\forall j = \{1,..,n\}$. On the contrary, if $R_i = 0$ then the current labelling is the one that obtains the maximum probability, so, it is the ideal case. Note that $R_i < 0$ is not possible.

4   Active Algorithm

Algorithm *Interactive Graph Matching* presented in [31] obtains a labelling between nodes of attributed graphs g^1 and g^2 considering the human feedback. That is, it computes several times a sub-optimal graph-matching algorithm (for instance [23, 24, 25]), but in each step, the cost matrices C_v and C_e are modified through the current user feedback. In fact, we assume the input of the algorithm is not both graphs but matrices C_v and C_e. The feedback of the user is introduced into the algorithm through a vector of simple actions w. One of the actions is $w_q = Set(v_i^1, v_a^2)$, in which the user imposes that the labelling has to be $v_a^2 = f(v_i^1)$. The matrix costs C_v and C_e are updated considering the human feedback through functions *Interactive_Node_Costs* and *Interactive_Edge_Costs*. See [31] for more details.

The active algorithm we present has a similar structure. We have only added function *Active_Query* and the expression $w_1 = Set(v^{1^*}, v^{2^*})$ with the aim of functions *Interactive_Node_Costs* and *Interactive_Edge_Costs* being compatible with the interactive algorithm in [31]. The algorithm stops when the oracle returns a special node (for instance a negative value) or all the nodes of g^1 have been queried. The final labelling cost obtained at the end of the algorithm is computed through the original costs, C_v^0 and C_e^0.

```
Algorithm Active Graph Matching
Input: Attributed Graphs g¹ and g²
Output: Labelling f and Cost Cf
```
$C_v^0, C_e^0 = Initialise\_Cost(g^1, g^2);\ C_v = C_v^0;\ C_e = C_e^0.$
$f = Graph\_Matching(C_v, C_e).$
```
Do
```
    $v^{1^*} = Active\_Query(P, f).$
    $v^{2^*} = Oracle\_Feedback(g^1, g^2, v^{1^*}, f).$
    $w_1 = Set(v^{1^*}, v^{2^*}).$
    $C_v = Interactive\_Node\_Costs(w, C_v).$
    $C_e = Interactive\_Edge\_Costs(w, C_e).$
    $f = Graph\_Matching(C_v, C_e).$
```
Since Stop
```
Compute $C_f(C_v^0, C_e^0)$
```
End Algorithm
```

Figure 2 shows the probabilistic graph-matching framework with active learning. Dashed lines connect the active modules that do not appear in the classical framework shown in figure 1. Moreover, we have added the original costs C_v^0 and C_e^0, since the labelling cost C_f is computed through these costs.

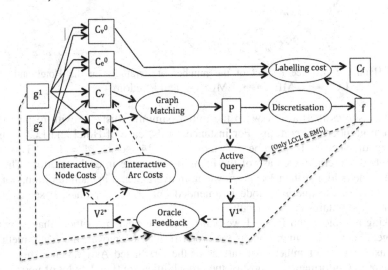

Fig. 2. Probabilistic graph matching framework with active learning

5   Practical Evaluation

We have used the CMU "house" and "castle" sequences. There are two datasets consisting of 111 frames of a toy house and a castle. Each frame in these sequences has been hand-labelled, with the same 30 landmarks identified in each frame [29]. From each landmark, we have only considered their bidirectional position in the image. From each frame, we have defined an attributed graph of 30 nodes using the 3-nearest neighbour technique. Nodes represent these landmarks and arcs represent proximity. Attributes on nodes are the position of the landmarks and arcs do not have attributes. The cost between nodes $C_v[v_i^1, v_a^2]$ is the Euclidean distance of their image positions. The cost between arcs $C_e[e_{ij}^1, e_{ab}^2]$ is 0 if both arcs exist or do not exist and 1 if only on of the arcs exists. We have used all pairs of graphs that have been extracted from images that the separation between frames is 60. The final result values are the average of these experiments. We have used the Graduated Assignment algorithm [24] and, in each iteration of the active algorithm, we only permitted a maximum of 30 iterations of the external loop and 20 iterations of the internal loop.

We assess the quality of the current labelling through the Hamming distance between the current labelling and the hand-made labelling [29]. Figure 3 shows this Hamming distance throughout the number of iterations of the active algorithm and using the four previously commented active strategies and also a random strategy. In this random strategy, the *Active_Learning* sequentially returns the nodes of g^1. The algorithm stops when all the 30 nodes have been queried.

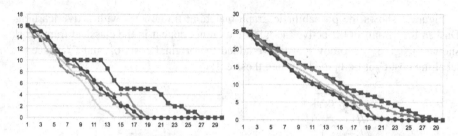

Fig. 3. Hamming distance respect of the number of iterations on the Hotel and House. LCCL: ▬▬▬, LC: ▬●▬, ME: ▬●▬, EMC: ▬▲▬ and Random: ▬■▬.

Table 1 shows the ratio between the initial hamming distance with respect to the maximum number of iterations. For instance, in the case of Hotel and LCCL, initial hamming distance = 16, number of iterations = 14, so 16/14 = 1.14. This value represents the average decrease of the hamming distance in each iteration. The case that the value is higher than 1 appears when, in average and in each iteration, not only the labelling of the queried node is amended but also other ones. It seems as the LCCL method obtains the best results.

Looking at table 1 and figure 1 we realise that the random method, that is, without "intelligent" active learning, obtains the worst results. Moreover, the EMC method is very sensitive to the number of iterations of the Graduated Assignment. This method can deduct few information if most of the probabilities are 0 or 1 (a lot of iterations) or if most of the iterations are near to 1/n being n the number of nodes (few iterations).

Table 1. Ratio between hamming distance and iterations of the Hotel and House

	LCCL	LC	ME	EMC	Random
Hotel	1.14	0.84	0.88	0.94	0.59
House	1.10	1.10	1.06	0.87	0.85
Average	1.12	0.97	0.97	0.90	0.72

Figure 4 shows the evolution of the current labelling cost C_f. As described in the algorithm, this cost is computed through the original costs. All methods tend to obtain the optimal labelling, for this reason, at the end, the cost is similar for all of them. Note that in some steps, the cost increases. This means that forcing some nodes to be mapped, the sub-optimal graph-matching algorithm finds a worse labelling.

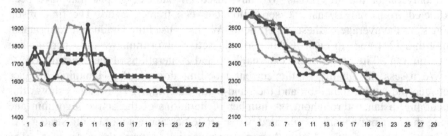

Fig. 4. Labelling cost respect of the number of iterations on the Hotel and House. LCCL: ▬▬▬, LC: ▬●▬, ME: ▬●▬, EMC: ▬▲▬ and Random: ▬■▬.

6   Conclusions and Future Work

We have presented four different strategies to be applied on an active graph-matching algorithm. These strategies are based on classical active machine learning but they are applied to the case of searching for the best labelling between nodes. Moreover they are based on the probability matrix between nodes that some sub-optimal algorithms use to iteratively find the best labelling. Due to the active algorithm only updates the costs between nodes and arcs and reads the probability matrix; it is not needed to modify the code of these well-known algorithms. Experimental validation shows that the Least Confident method that uses the current labelling (LCCL) tends faster to find the optimal labelling.

We have planned to apply this method to Expectation-Maximisation and Probabilistic Relaxation algorithms. Moreover, we have obtained different results while using different number of iterations of the two main loops of the Graduated Algorithm. These results have not been shown due to space problems and we have shown in this paper the ones we considered being the best ones. We wish to publish in a journal the whole method presented in this paper together with [31] and the other experiments commented before.

References

1. Mitchell, T.: Machine Learning. Mcgraw-Hill International Edit (1997) ISBN-13: 978-0071154673
2. Conte, D., Foggia, P., Sansone, C., Vento, M.: Thirty Years Of Graph Matching In Pattern Recognition. International Journal of Pattern Recognition and Artificial Intelligence 18(3), 265–298 (2004)
3. Settles, B.: Active Learning Literature Survey. Computer Science Technical Report 1648, University of Wisconsin-Madison
4. Wang, R., Kwong, S., Chen, D.: Inconsistency-based active learning for support vector machines. Pattern Recognition 45(10), 3751–3767 (2012)
5. Gorisse, D., Cord, M., Precioso, F.: SALSAS: Sub-linear active learning strategy with approximate k-NN search. Pattern Recognition 44(10-11), 2343–2357 (2011)
6. Yu, D., Varadarajan, B., Deng, L., Acero, A.: Active learning and semi-supervised learning for speech recognition: A unified framework using the global entropy reduction maximization criterion. Computer Speech & Language 24(3), 433–444 (2010)
7. Zhang, Q., Shiliang: Sun Multiple-view multiple-learner active learning. Pattern Recognition 43(9), 3113–3119 (2010)
8. Wang, Z., Yan, S., Zhang, C.: Active learning with adaptive regularization. Pattern Recognition 44(10-11), 2375–2383 (2011)
9. Patra, S., Bruzzone, L.: A cluster-assumption based batch mode active learning technique. Pattern Recognition Letters 33(9), 1042–1048 (2012)
10. Zhang, D., Wang, F., Shi, Z., Zhang, C.: Interactive localized content based image retrieval with multiple-instance active learning. Pattern Recognition 43(2), 478–484 (2010)
11. Kanamori, T.: Pool-based active learning with optimal sampling distribution and its information geometrical interpretation. Neurocomputing 71(1-3), 353–362 (2007)

12. Baranes, A., Oudeyer, P.-Y.: Active learning of inverse models with intrinsically motivated goal exploration in robots. Robotics and Autonomous Systems (available online, 2012)
13. Cord, M., Gosselin, P.H., Philipp-Foliguet, S.: Stochastic exploration and active learning for image retrieval. Image and Vision Computing 25(1), 14–23 (2007)
14. Lughofer, E.: Hybrid active learning for reducing the annotation effort of operators in classification systems. Pattern Recognition 45(2), 884–896 (2012)
15. Cheng, J., Wang, K.: Active learning for image retrieval with Co-SVM. Pattern Recognition 40(1), 330–334 (2007)
16. da Silva, A.T., Falcão, A.X., Magalhães, L.P.: Active learning paradigms for CBIR systems based on optimum-path forest classification. Pattern Recognition 44(12), 2971–2978 (2011)
17. Gosselin, P.H., Precioso, F., Philipp-Foliguet, S.: Incremental kernel learning for active image retrieval without global dictionaries. Pattern Recognition 44(10-11), 2244–2254 (2011)
18. King, R.D., et al.: Functional genomic hypothesis generation and experimentation by a robot scientist. Nature 427(6971), 247–252 (2004)
19. Yu, H.: SVM selective sampling for ranking with application to data retrieval. In: Proceedings of the International Conference on Knowledge Discovery and Data Mining (KDD), pp. 354–363. ACM Press (2005)
20. Kotsiantis, S.B.: Supervised Machine Learning: A Review of Classification Techniques. Informatica 31, 249–268 (2007)
21. Sanchís, A., Juan, A., Vidal, E.: A Word-Based Naïve Bayes Classifier for Confidence Estimation in Speech Recognition. IEEE Transactions on Audio, Speech & Language Processing 20(2), 565–574 (2012)
22. Toselli, A.H., Romero, V., Pastor, M., Vidal, E.: Multimodal interactive transcription of text images. Pattern Recognition 43(5), 1814–1825 (2010)
23. Fekete, G., Eklundh, J.O., Rosenfeld, A.: Relaxation: Evaluation and Applications. IEEE Transactions on Pattern Analysis and Machine Intelligence 3(4), 459–469 (1981)
24. Gold, S., Rangarajan, A.: A Graduated Assignment Algorithm for Graph Matching. IEEE Transactions on Pattern Analysis and Machine Intelligence 18(4), 377–388 (1996)
25. Luo, B., Hancock, E.R.: Structural graph matching using the EM algorithm and singular value decomposition. IEEE Transactions on Pattern Analysis and Machine Intelligence 23(10), 1120–1136 (2001)
26. Sanromà, G., Alquézar, R., Serratosa, F., Herrera, B.: Smooth Point-set Registration using Neighbouring Constraints. Available on line Pattern Recognition Letters (2012)
27. Sanromà, G., Alquézar, R., Serratosa, F.: A New Graph Matching Method for Point-Set Correspondence using the EM Algorithm and Softassign. Computer Vision and Image Understanding 116(2), 292–304 (2012)
28. Romero, V., Toselli, A.H., Vidal, E.: Multimodal interactive handwritten text transcription. Word Scientific (2012)
29. Caetano, T.S., Caelli, T., Schuurmans, D., Barone, D.A.C.: Graphical Models and Point Pattern Matching. IEEE Trans. Pattern Analysis and Machine Intelligence 28(10), 1646–1663 (2006)
30. Kuhn, H.W.: The Hungarian method for the assignment problem Export. Naval Research Logistics Quarterly 2(1-2), 83–97 (1955)
31. Serratosa, F., Cortés, X., Solé-Ribalta, A.: Interactive Graph Matching by means of Imposing the Pairwise Costs. In: International Conference on Pattern Recognition, ICPR 2012, Tsukuba, Japan (accepted for publication, 2012)

On the Relation between the Common Labelling and the Median Graph*

Nicola Rebagliati[1], Albert Solé-Ribalta[2],
Marcello Pelillo[1], and Francesc Serratosa[2]

[1] Università Ca' Foscari Venezia, Italy
[2] Universitat Rovira i Virgili, Spain
{rebagliati,pelillo}@dsi.unive.it,
{albert.sole,francesc.serratosa}@urv.cat

Abstract. In structural pattern recognition, given a set of graphs, the computation of a Generalized Median Graph is a well known problem. Some methods approach the problem by assuming a relation between the Generalized Median Graph and the Common Labelling problem. However, this relation has still not been formally proved. In this paper, we analyse such relation between both problems. The main result proves that the cost of the common labelling upper-bounds the cost of the median with respect to the given set. In addition, we show that the two problems are equivalent in some cases.

1 Introduction

In many pattern recognition applications, we are given a set of different representations of the same object and the goal is to summarize these representations into a single one. The resulting representation should capture the important features of the object and discard noisy or unexpected variations. When the representation is made using attributed graphs, this graph is identified as the Generalized Median Graph [1], or simply the Median Graph. Given a training set of graphs, the Median Graph is formally defined as a graph which minimizes the sum of costs to all other graphs in the set.

If we assume that vertices are not uniquely labelled, like in [2], the problem of finding the Median Graph is, in its general form, at least as difficult as the problem of matching two graphs under a particular cost function, e.g. the Graph Edit Distance, which is a NP-Hard problem [3]. Indeed, the Median Graph cannot be computed in closed form since its synthesis depends on the matchings between itself and the given graphs and the matchings to the Median Graph clearly require having the Median Graph. A usual way to deal with this chicken-egg problem is using an incremental approach where the Median Graph is coarsely constructed and then iteratively refined until all graphs in the training set are considered.

* We acknowledge financial support from the FET programme within the EU FP7, under the SIMBAD project (contract 213250), from Consolider Ingenio 2010 project (CSD2007-00018) and from the CICYT project (DPI2010-17112).

G.L. Gimel' farb et al. (Eds.): SSPR & SPR 2012, LNCS 7626, pp. 107–115, 2012.

Several approaches address the problem in this fashion [4–9]. A completely different approach to compute the Median Graph is to decouple the matchings and the synthesis process. This approach relies on the assumption that given the vertex labellings that compute the Median Graph, its computation can be, in most applications, done efficiently in polynomial time, e.g. averaging the vertices and edge attributes. This approach can be summarized in two steps. In the first step, we obtain a Common Labelling between the given graphs. The objective of the Common Labelling, initially defined in [10, 11], is to minimize the pair-wise labellings among a set of graphs with some transitivity restrictions. Once we know this information, we can easily compute an Approximated Median Graph. Figure 1 illustrates the complete process to generate a Median Graph using a Common Labelling. Note the given set of graphs is labelled to a virtual node set and Median Graph is not computed until the end of the process. The main advantage of using a Common Labelling approach for approximating the Median Graph relies on the fact that the Median Graph does not need to be computed until the end of the process. In this way, labellings of the initial graphs to the Median Graph are not needed and the initial chicken-egg problem disappears.

Several works exist in the literature which decouple the problem of the Median Graph computation. The first method to completely decouple the matching process from the synthesis process was presented by Hlaoui and Wang [7]. Another recent method, based on linear programming, has been proposed in [12] and [13]. But possibly the most complete work on these kind of methods is presented in [14]. Experiments in [14] show that using the Common Labelling for computing the Median Graph gives satisfactory results, but up to now a formal relation between the Common Labelling problem and the Median Graph synthesis was missing. In this work, we show that, if the cost for matching graphs is a metric, the two problems are tightly connected because we can bound the Median Graph error using the Common Labelling value. The obtained bounds show that, when the error of the Common Labelling is low, the obtained graph median is close to the real one. In addition, in the specific case of unattributed graphs with the squared Euclidean distance as cost function, the two problems are equivalent.

2   Definitions

Let $\mathcal{H}$ be a set of attributed graphs representing the input/output space of our problems. Each graph is represented as a tuple $G = (V, E, A_V, A_E)$, where $V = \{v_1, ..., v_n\}$ represents the vertex set, $E \subseteq \{e_{a,b}, \forall a, b \in 1..n\}$ the edge set, and functions $A_V : V \to D_V$ and $A_E : E \to D_E$ assign attributes to vertex and edges respectively.

Given a set of m attributed graphs $S = \{G_1, ..., G_m\}, G_i = (V_i, E_i, A_V, A_E) \in \mathcal{H}$, we assume that each of these graphs have the same number of vertices n. If this is not the case, several solutions have been proposed to extend the size of the graphs [9, 15]. However, the most common approach is to include null vertices [15] which represent deletions and insertion of vertices in the resulting labelling. In the general graph matching setting, vertices of each graph are not

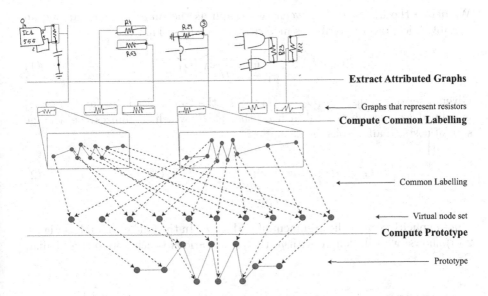

Fig. 1. The process for computing an Approximated Median Graph with the Common Labelling. After representing the given objects, which in this figure are sketches of electrical circuits, with attributed graphs we look for labelling the nodes of each graph. The virtual set of nodes does not have structure and is used to compare the labelling for each graph and evaluate their pairwise matching cost. After choosing a labelling for each graph we convert the virtual set of nodes into the actual graph prototype. See section 3 for a formal definition of the Common Labelling problem.

uniquely identified by their index, i.e. we cannot assign or identify $v_3 \in V_1$ with $v_3 \in V_2$ only because they have the same index 3. Indeed, the difficult part of comparing a pair of graphs relies in finding a suitable bijection π of vertices which provides the right ordering. In the following, given a bijection π, the notation G^π means that $v_i(\pi) = v_{\pi(i)}$, so that vertices of V and edges E are permuted accordingly to π. The bijection $id \in \Pi$ represents the identity $v_i(id) = v_i$. Figure 2 (a) shows how graphs are permuted to a common reference system with permutations π_i and ρ_i. The function $c : \mathcal{H} \times \Pi \times \mathcal{H} \times \Pi \to R^+$ is a user-defined cost between two graphs whose vertices have a fixed bijection. We assume that $c(\cdot, \cdot, \cdot, \cdot)$ can be computed efficiently in polynomial time because the vertex to vertex correspondence is fixed, and consequently also the edge to edge correspondence and their attributes. We use the shorthand $c(G_i^{\pi_i}, G_j^{\pi_j}) = c(G_i, \pi_i, G_j, \pi_j)$ and $c(G_i, G_j) = c(G_i, id, G_j, id)$. If the cost function is a metric we denote it as c_M in this case, given a fixed set of bijections $\pi_{1,...,m}$, the following axioms hold:

identity $c_M(G_i, G_j) = 0 \Leftrightarrow G_i = G_j$,
positivity $c_M(G_i, G_j) \geq 0$,
symmetry $c_M(G_i, G_j) = c_M(G_j, G_i)$,
triangle inequality $c_M(G_i, G_j) \leq c_M(G_i, G_k) + c_M(G_k, G_j)$.

We define the distance d between two graph as the minimum cost among all possible bijections of attributes in vertices and edges. That is,

$$d(G_1, G_2) := \min_{\pi_1, \pi_2 \in \Pi} c_M(G_1^{\pi_1}, G_2^{\pi_2}) \qquad (1)$$

Given a set of graphs $S = (G_1, ..., G_m) \subseteq \mathcal{H}$, the Generalized Median Graph [1] is defined as a graph G^*, taken from the set $\mathcal{H}$, which minimizes the average sum of costs to all graphs in S:

$$GM^*(\mathcal{H}) := \min_{\substack{\rho_1, ..., \rho_m \in \Pi \\ G \in \mathcal{H}}} \frac{1}{m} \sum_{i=1}^{m} c(G_i^{\rho_i}, G) \qquad (2)$$

If not explicitly stated the argument of GM^* is $\mathcal{H}$. In the following, and as Figure 2 (a) shows, we will denote with ρ_i the permutations which obtain the Median Graph.

3   The Common Labelling Problem

Given a set of graphs $S = (G_1, ..., G_m) \subseteq \mathcal{H}$, the Common Labelling problem aims at finding a, possibly low cost, consistent multiple isomorphism between the graphs, such that for every three mappings $\pi_{i,j}, \pi_{j,r}$ and $\pi_{i,r}$ we have $\pi_{i,j} \circ \pi_{j,r} = \pi_{i,r}$. Equivalently, we look for m consistent bijections assigning vertices of the graph of a virtual vertex set and that minimize the average sum of pairwise distances between graphs in S. Its normalized objective function is the following:

$$CL^* := \min_{\pi_1, ..., \pi_m \in \Pi} \frac{1}{m^2} \sum_{i=1}^{m} \sum_{j=1}^{m} c(G_i^{\pi_i}, G_j^{\pi_j}) \qquad (3)$$

Once the Common Labelling and the m bijections $\pi_{1,...,m}$ that computes the value are obtained, we assume that we can efficiently estimate a median graph $\overline{G}$:

$$\overline{G} \in \underset{G \in \mathcal{H}}{\operatorname{argmin}} \sum_{i=1}^{m} c(G_i^{\pi_i}, G) \qquad (4)$$

which we call Approximated Median Graph. In the following, and as Figure 2 (a) shows, we will denote with π_i the permutations which obtain the Approximated Median Graph through the Common Labelling.

4   Relating the Common Labelling with the Generalized Median Graph

In this section, we show two main results of this work. The first theorem shows the relationship between the objective function of the Common Labelling, CL^*, and the objective function of the Median Graph, GM*. The second theorem shows that, if the functional of the Common Labelling CL* has a low value, the Approximated Median Graph $\overline{G}$ is close to the Median Graph G^*.

Theorem 1. *Let $\mathcal{H}$ be a set of graphs and $S = \{G_1, \ldots, G_m\}$ a subset of $\mathcal{H}$. In addition, let $\overline{G}$ be the Approximated Median Graph computed considering S and the bijections obtained by the Common Labelling, $\pi_{1,\ldots,m}$. Let the cost function c_M be a metric. Then*

$$CL^* \geq GM^*(\{\overline{G}\}) \geq GM^* \geq \frac{1}{2}CL^* \tag{5}$$

Proof. We start with the left hand side of (5):

$$
\begin{aligned}
CL^* &= \frac{1}{m^2} \sum_{i=1}^{m} \sum_{j=1}^{m} c_M(G_i^{\pi_i}, G_j^{\pi_j}) \\
&\geq \frac{1}{m^2} \sum_{i=1}^{m} \sum_{j=1}^{m} c_M(\overline{G}, G_j^{\pi_j}) \\
&\geq \frac{1}{m} \sum_{j=1}^{m} c_M(\overline{G}, G_j^{\pi_j}) \\
&= GM^*(\{\overline{G}\}) \\
&\geq GM^*
\end{aligned}
\tag{6}
$$

The second step comes from optimality of the Approximated Median Graph, see (4).

The right hand side of (5) follows from:

$$
\begin{aligned}
GM^* &= \frac{1}{m} \sum_{i=1}^{m} c_M(G_i^{\rho_i}, G^*) \\
&= \frac{1}{2m^2} \sum_{i=1}^{m} \sum_{j=1}^{m} c_M(G_i^{\rho_i}, G^*) + c_M(G^*, G_j^{\rho_j}) \\
&\geq \frac{1}{2m^2} \sum_{i=1}^{m} \sum_{j=1}^{m} c_M(G_i^{\rho_i}, G_j^{\rho_j}) \\
&\geq \frac{1}{2m^2} \sum_{i=1}^{m} \sum_{j=1}^{m} c_M(G_i^{\pi_i}, G_j^{\pi_j}) \\
&= \frac{1}{2}CL^*
\end{aligned}
\tag{7}
$$

The third step uses the triangle inequality and the forth step comes from considering the optimality of π_i and π_j.

Theorem 2. *Let $\mathcal{H}$ be a set of graphs and $S = \{G_1, \ldots, G_m\}$ be a subset of $\mathcal{H}$. In addition, let $\overline{G}$ be the Approximated Median Graph computed considering S and G^* the Generalized Median Graph. Then,*

$$d(\overline{G}, G^*) \leq 2CL^* \leq 4GM^* \tag{8}$$

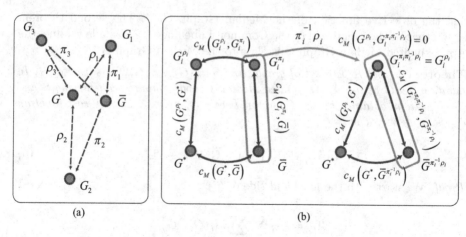

Fig. 2. (a) Notation for Theorems 1 and 2. (b) Graphical representation of (9), which is the basic inequality for proving Theorem 2.

Proof. Let $\pi_{1,\ldots,m}$ be the bijections obtained by the Common Labelling and $\rho_{1,\ldots,m}$ the bijections related to G^* and c_M a metric cost function. Since c_M is a metric we have for each single graph G_i:

$$c_M(G^*, \overline{G}) \leq c_M(G^*, G_i^{\rho_i}) + c_M(\overline{G}, G_i^{\rho_i})$$
$$\leq c_M(G^*, G_i^{\rho_i}) + c_M(G_i^{\rho_i}, G_i^{\pi_i}) + c_M(\overline{G}, G_i^{\pi_i}) \quad (9)$$

since ρ_i and π_i may be different $c_M(G_i^{\rho_i}, G_i^{\pi_i}) \neq 0$. However, applying bijection $\pi_i^{-1}\rho_i$ to $G_i^{\pi_i}$ and $\overline{G}$ costs are preserved $c_M(\overline{G}, G_i^{\pi_i}) = c_M(\overline{G}^{\pi_i^{-1}\rho_i}, G_i^{\rho_i})$ and $c_M(G_i^{\rho_i}, G_i^{\pi_i \pi_i^{-1}\rho_i}) = 0$. This reasoning is visualized in Figure 2 (b). Hence,

$$c_M(\overline{G}^{\pi_i^{-1}\rho_i}, G^*) \leq c_M(\overline{G}^{\pi_i^{-1}\rho_i}, G_i^{\rho_i}) + c_M(G_i^{\rho_i}, G^*). \quad (10)$$

In (10), vertices and edges of $G_1, \ldots, G_m$ and $\overline{G}$ have been permuted accordingly to G^*. To ease notation, assume that π_i correspond to the identity. Consequently,

$$d(\overline{G}, G^*) \leq c_M(\overline{G}^{\rho_i}, G^*) \leq c_M(\overline{G}^{\rho_i}, G_i^{\rho_i}) + c_M(G_i^{\rho_i}, G^*). \quad (11)$$

Then, adding inequality (11) for the different G_i's we get:

$$d(\overline{G}, G^*) \leq \frac{1}{m} \sum_{i=1}^{m} c_M(\overline{G}^{\rho_i}, G_i^{\rho_i}) + c_M(G_i^{\rho_i}, G^*)$$
$$= \text{GM}^*(\{\overline{G}\}) + \text{GM}^* \quad (12)$$
$$\leq 2\text{CL}^*$$

A desirable output for the user is that the Approximated Median Graph is an ϵ approximation of the given objects. The following corollary shows that, in this case, this Approximated Median Graph is close to the actual Median Graph.

Corollary 1. *Let $S = \{G_1, \ldots, G_m\}$ admit an Approximated Median Graph $\overline{G}$ such that $\text{GM}^*(\{\overline{G}\}) \leq \epsilon$. Then $d(\overline{G}, G^*) \leq 3\epsilon$.*

The proof is based on equation (12) of theorem 2 and is left to the reader.

Theorem 1 and 2 are proven considering the optimal computation of CL^*. If we relax this assumption with a suboptimal computation we get the following corollary.

Corollary 2. *Let $\mathcal{H}$ be a set of graphs and $S = \{G_1, \ldots, G_m\}$ be a subset of $\mathcal{H}$. In addition, let $\overline{G'}$ be the Approximated Median Graph computed considering S and the, possibly suboptimal, bijections obtained by Common Labelling whose value is CL. Let the cost function c_M be a metric. Then $\mathrm{CL} \geq \mathrm{GM}(\overline{G'}) \geq \mathrm{GM}^*$ and $d(\overline{G'}, G^*) \leq 2\mathrm{CL}$.*

5   Median Graph of Weighted Graphs

Clearly, the notion of Median Graph can be used with a large set of different cost functions. In this section, we will show how using the original proposed cost [1] between graphs and restricting to weighted graphs, the Median Graph problem reduces exactly to the Common Labelling problem. Let $A_V : V \to [0, 1]$ and $A_E : E \to [0, 1]$ be the domain of vertices and edges attributes. In this case, the value "1" indicates that the graph vertex, or edge, exists and value "0" that the vertex, or edge, does not exist. We use a vector/matrix representation, so that $\mathbf{V_i}(r) = A_V(v_r)$ and $\mathbf{E_i}(r, s) = A_E(e_{r,s})$ where $v_r \in V_i$ and $e_{r,s} \in E_i$ and bijections π_i are represented as permutation matrices $\mathbf{p_i}$. In case no vertex position is indicated, $\mathbf{V_i}$, we refer to the complete vector.

As cost function we use the squared Euclidean distance, $c(v_r, v_s) = \|\mathbf{V_i}(r) - \mathbf{V_j}(s)\|^2$ where $v_r \in V_i$ and $v_s \in V_j$. The edge cost function is defined in an equivalent form. This cost was also used in the genetic algorithm of [1] where authors proved the best prototype for a set of graphs, with fixed labellings, is the average of attributes

$$
\begin{cases}
\overline{\mathbf{V}}(r) = \dfrac{1}{m} \displaystyle\sum_{k=1}^{m} \mathbf{V_k}(r) \\[2ex]
\overline{\mathbf{E}}(r, s) = \dfrac{1}{m} \displaystyle\sum_{k=1}^{m} \mathbf{E_k}(r, s)
\end{cases}
\tag{13}
$$

Under these considerations, we can state the following theorem:

Theorem 3. *Let $\mathcal{H}$ be a set of weighted graphs, $S = \{G_1, \ldots, G_m\}$ a given subset of $\mathcal{H}$. and $p_{1,\ldots,m} \in \mathbb{R}^{N \times N}$ m permutation matrices. Considering the cost given by the squared Euclidean distance, we have:*

$$
\frac{1}{2}\mathrm{CL}^* = \mathrm{GM}^*.
\tag{14}
$$

Proof. The scalar product of two vectors is:

$$
\langle \mathbf{V_i}, \mathbf{V_j} \rangle = \sum_{r=1}^{n} \mathbf{V_i}(r)\mathbf{V_j}(r).
$$

The proof follows the lines of the Huygens theorem [16].

$$
\begin{aligned}
\tfrac{1}{2}\mathrm{CL}^* &= \frac{1}{2m^2} \sum_{i=1}^{m} \sum_{j=1}^{m} \|\mathbf{p_i V_i}\|^2 - 2\langle \mathbf{p_i V_i}, \mathbf{p_j V_j}\rangle + \|\mathbf{p_j V_j}\|^2 \\
&= \frac{1}{m^2} \sum_{i=1}^{m} \sum_{j=1}^{m} \|\mathbf{p_i V_i}\|^2 - \langle \mathbf{p_i V_i}, \mathbf{p_j V_j}\rangle \\
&= \frac{1}{m} \sum_{i=1}^{m} \|\mathbf{p_i V_i}\|^2 + \sum_{i=1}^{m}\sum_{j=1}^{m} -\frac{2}{m^2}\langle \mathbf{p_i V_i}, \mathbf{p_j V_j}\rangle + \frac{1}{m^2}\langle \mathbf{p_i V_i}, \mathbf{p_j V_j}\rangle \\
&= \frac{1}{m} \sum_{i=1}^{m} (\|\mathbf{p_i V_i}\|^2 - \frac{2}{m}\langle \mathbf{p_i V_i}, \sum_{j=1}^{m} \mathbf{p_j V_j}\rangle) + \langle \frac{1}{m}\sum_{i=1}^{m}\mathbf{p_i V_i}, \frac{1}{m}\sum_{j=1}^{m}\mathbf{p_j V_j}\rangle \\
&= \frac{1}{m} \sum_{i=1}^{m} \|\mathbf{p_i V_i}\|^2 - 2\langle \mathbf{p_i V_i}, \overline{\mathbf{V}}\rangle + \|\overline{\mathbf{V}}\|^2 \\
&\geq \mathrm{GM}^*
\end{aligned}
\tag{15}
$$

The converse inequality is similarly proved and the process is equivalent for the edge costs.

As an immediate consequence of theorem 3 we have that the Approximated Median Graph error is the same as the Generalized Median Graph. The proof is based on theorem 3 and is left to the reader.

Corollary 3. *Under the hypothesis of theorem 3 we have:*

$$
\mathrm{GM}^*(\{\overline{G}\}) = \mathrm{GM}^* \tag{16}
$$

By exploiting the particular properties of the squared Euclidean distance, which is not a metric, we get a much stronger result than theorem 1.

6   Discussion

In this paper we analysed the relation between two structural pattern recognition problems, the Median Graph and the Common Labelling. We proved that these problems are closely related and in some special cases they are in fact equivalent, thereby formalising a connection which up to now was unknown. This connection confirms that algorithms based on the Common Labelling, to compute the Median Graph, are theoretically sound. In addition, the proposed bounds are useful in practice, when the Common Labelling is computed using non-exact algorithms, like in [11].

References

1. Jiang, X., Müunger, A., Bunke, H.: On median graphs: Properties, algorithms, and applications. IEEE Trans. Pattern Anal. Mach. Intell. 23(10), 1144–1151 (2001)
2. Dickinson, P.J., Bunke, H., Dadej, A., Kraetzl, M.: Matching graphs with unique node labels. Pattern Anal. Appl. 7(3), 243–254 (2004)

3. Zeng, Z., Tung, A.K.H., Wang, J., Feng, J., Zhou, L.: Comparing stars: on approximating graph edit distance. In: Proc. VLDB Endow., vol. 2(1), pp. 25–36 (August 2009)
4. Ferrer, M., Serratosa, F., Sanfeliu, A.: Synthesis of Median Spectral Graph. In: Marques, J.S., Pérez de la Blanca, N., Pina, P. (eds.) IbPRIA 2005, Part II. LNCS, vol. 3523, pp. 139–146. Springer, Heidelberg (2005)
5. Ferrer, M., Serratosa, F., Valveny, E.: Evaluation of Spectral-Based Methods for Median Graph Computation. In: Martí, J., Benedí, J.M., Mendonça, A.M., Serrat, J. (eds.) IbPRIA 2007, Part II. LNCS, vol. 4478, pp. 580–587. Springer, Heidelberg (2007)
6. Ferrer, M., Valveny, E., Serratosa, F., Riesen, K., Bunke, H.: Generalized median graph computation by means of graph embedding in vector spaces. Pattern Recognition 43(4), 1642–1655 (2010)
7. Hlaoui, A., Wang, S.: Median graph computation for graph clustering. Soft Computing - A Fusion of Foundations, Methodologies and Applications 10, 47–53 (2006)
8. Jain, B.J., Wysotzki, F.: Central clustering of attributed graphs. Machine Learning 56, 169–207 (2004)
9. Jain, B., Obermayer, K.: Elkan's k-Means Algorithm for Graphs. In: Sidorov, G., Hernández Aguirre, A., Reyes García, C.A. (eds.) MICAI 2010, Part II. LNCS, vol. 6438, pp. 22–32. Springer, Heidelberg (2010)
10. Solé-Ribalta, A., Serratosa, F.: Graduated Assignment Algorithm for Finding the Common Labelling of a Set of Graphs. In: Hancock, E.R., Wilson, R.C., Windeatt, T., Ulusoy, I., Escolano, F. (eds.) SSPR&SPR 2010. LNCS, vol. 6218, pp. 180–190. Springer, Heidelberg (2010)
11. Solé-Ribalta, A., Serratosa, F.: Models and algorithms for computing the common labelling of a set of attributed graphs. Comput. Vis. Image Underst. 115(7), 929–945 (2011)
12. Justice, D., Hero, A.: A binary linear programming formulation of the graph edit distance. IEEE Transactions on Pattern Analysis and Machine Intelligence 28, 1214 (2006)
13. Mukherjee, L., Singh, V., Peng, J., Xu, J., Zeitz, M., Berezney, R.: Generalized median graphs and applications. Journal of Combinatorial Optimization 17, 21–44 (2009)
14. Solé-Ribalta, A.: Multiple graph matching and applications. PhD thesis, Universitat Rovira i Virgili (2012)
15. Wong, A.K.C., You, M.: Entropy and distance of random graphs with application to structural pattern recognition. IEEE Transactions on Pattern Analysis and Machine Intelligence (5), 599–609 (1985)
16. Edwards, A., Cavalli-Sforza, L.: A method for cluster analysis. Biometrics 21, 362–375 (1965)

A Hierarchical Image Segmentation Algorithm Based on an Observation Scale*

Silvio Jamil F. Guimarães[1,2], Jean Cousty[2],
Yukiko Kenmochi[2], and Laurent Najman[2]

[1] PUC Minas - ICEI - DCC - VIPLAB
sjamil@pucminas.br
[2] Université Paris-Est, LIGM, ESIEE - UPEMLV - CNRS
{j.cousty,y.kenmochi,l.najman}@esiee.fr

Abstract. Hierarchical image segmentation provides a region-oriented scale-space, *i.e.*, a set of image segmentations at different detail levels in which the segmentations at finer levels are nested with respect to those at coarser levels. Most image segmentation algorithms, such as region merging algorithms, rely on a criterion for merging that does not lead to a hierarchy. In addition, for image segmentation, the tuning of the parameters can be difficult. In this work, we propose a hierarchical graph based image segmentation relying on a criterion popularized by Felzenszwalb and Huttenlocher. Quantitative and qualitative assessments of the method on Berkeley image database shows efficiency, ease of use and robustness of our method.

Keywords: hierarchical segmentation, edge-weighted graph, saliency map.

1 Introduction

Image segmentation is the process of grouping perceptually similar pixels into regions. A hierarchical image segmentation is a set of image segmentations at different detail levels in which the segmentations at coarser detail levels can be produced from simple merges of regions from segmentations at finer detail levels. Therefore, the segmentations at finer levels are nested with respect to those at coarser levels. Hierarchical methods have the interesting property of preserving spatial and neighboring information among segmented regions. Here, we propose a hierarchical image segmentation in the framework of edge-weighted graphs, where the image is equipped with an adjacency graph and the cost of an edge is given by a dissimilarity between two points of the image.

Any hierarchy can be represented with a minimum spanning tree. The first appearance of this tree in pattern recognition dates back to the seminal work

* The authors are grateful to FAPEMIG and CAPES, which are Brazilian research funding agencies, and also to Agence Nationale de la Recherche through contract ANR-2010-BLAN-0205-03 KIDICO, which is a French research funding agency.

G.L. Gimel' farb et al. (Eds.): SSPR & SPR 2012, LNCS 7626, pp. 116–125, 2012.

of Zahn [1]. Lately, its use for image segmentation was introduced by Morris *et al.* [2] in 1986 and popularized in 2004 by Felzenszwalb and Huttenlocher [3]. However the region-merging method [3] does not provide a hierarchy. In [4,5], it was studied some optimality properties of hierarchical segmentations. Considering that, for a given image, one can tune the parameters of the well-known method [3] for obtaining a reasonable segmentation of this image. We provide in this paper a hierarchical version of this method that removes the need for parameter tuning.

The algorithm of [3] is the following. First, a minimum spanning tree (MST) is computed, and all the decisions are taken on this tree. For each edge linking two vertices x and y, following a non-decreasing order of their weights, the following steps are performed:

(i) Find the region X that contains x.
(ii) Find the region Y that contains y.
(iii) Merge X and Y according to a certain criterion.

The criterion for region-merging in [3] measures the evidence for a boundary between two regions by comparing two quantities: one based on intensity differences across the boundary, and the other based on intensity differences between neighboring pixels within each region. More precisely, in step (iii), in order to know whether two regions must be merged, two measures are considered. The *internal difference* $Int(X)$ of a region X is the highest edge weight among all the edges linking two vertices of X in the MST. The *difference* $Diff(X, Y)$ between two neighboring regions X and Y is the smallest edge weight among all the edges that link X to Y. Then, two regions X and Y are merged when:

$$Diff(X,Y) \leq \min\{Int(X) + \frac{k}{|X|}, Int(Y) + \frac{k}{|Y|}\} \qquad (1)$$

where k is a parameter allowing to prevent the merging of large regions (*i.e.*, larger k forces smaller regions to be merged).

The merging criterion defined by Eq. (1) depends on the scale k at which the regions X and Y are observed. More precisely, let us consider the *(observation) scale* $S_Y(X)$ of X *relative to* Y as a measure based on the difference between X and Y, on the internal difference of X and on the size $|X|$ of X:

$$S_Y(X) = (Diff(X,Y) - Int(X)) \times |X|. \qquad (2)$$

Then, the *scale* $S(X, Y)$ is simply defined as:

$$S(X,Y) = \max(S_Y(X), S_X(Y)). \qquad (3)$$

Thanks to this notion of a scale, Eq. (1) can be written as:

$$k \geq S(X,Y). \qquad (4)$$

In other words, Eq.(4) states that the neighboring regions X and Y merge when their scale is less than the threshold parameter k.

Even if the image segmentation results obtained by the method proposed in [3] are interesting, the user faces two major issues:

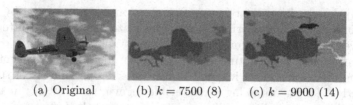

(a) Original   (b) $k = 7500$ (8)   (c) $k = 9000$ (14)

Fig. 1. A real example illustrating the violation of the causality principle by [3]: the number of regions (in parentheses) increases from 8 to 14, instead of decreasing when the so-called "scale of observation" increases

- first, the number of regions may increase when the parameter k increases. This should not be possible if k was a true scale of observation: indeed, it violates the *causality principle* of multi-scale analysis, that states in our case [6] that a contour present at a scale k_1 should be present at any scale $k_2 < k_1$. Such unexpected behaviour of missing causality principle is demonstrated on Fig. 1.
- Second, even when the number of regions decreases, contours are not stable: they can move when the parameter k varies, violating a *location principle*. Such a situation is illustrated on Fig. 2.

Given these two issues, the tuning of the parameters of [3] is a difficult task.

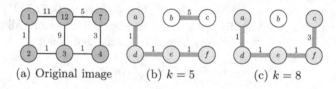

(a) Original image   (b) $k = 5$ (c) $k = 8$

Fig. 2. An example illustrating the violation of the location principle by [3]: the contours are unstable from one "scale" to another

Following [6], we believe that, in order for k to be a true scale-parameter, we have to satisfy both the causality principle and the location principle, which leads to work with a hierarchy of segmentations. Reference [7] is the first to propose an algorithm producing a hierarchy of segmentations based on [3]. However, this method is an iterative version of [3] that uses a threshold function, and requires a tuning of the threshold parameter.

The main result of this paper is an efficient hierarchical image segmentation algorithm based on the dissimilarity measure of [3]. Our algorithm has a computational cost similar to [3], but provides all scales of observations instead of only one segmentation level. As it is a hierarchy, the result of our algorithm satisfies both the locality principle and the causality principle. Namely, and in contrast with [3], the number of regions is decreasing when the scale parameter increases, and the contours do not move from one scale to another.

Figure 3 illustrates the results obtained by applying our method to the same image of Fig. 1(a), with segmentations at two different scales of observations, as

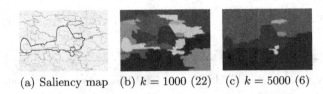

(a) Saliency map   (b) $k = 1000$ (22)   (c) $k = 5000$ (6)

Fig. 3. A real example illustrating the saliency map of Fig. 1(a) computed with our approach. We display in (b) and (c) two image segmentations extracted from the hierarchy at scales 1000 and 5000, together with their numbers of regions (in parentheses).

well as a saliency map [8,4,5] (a map indicating the disparition level of contours and whose thresholds give the set of all segmentations).

This work is organized as follows. In Section 2, we present our hierarchical method for color image segmentation. Some experimental results performed on Berkeley image database are given in Section 3. Finally, in Section 4, some conclusions are drawn and further works are discussed.

2   A Hierarchical Graph Based Image Segmentation

In this section, we describe our method to compute a hierarchy of partitions based on observation scales as defined by Eq. (3). Let us first recall some important notions for handling hierarchies [2,4,5].

To every tree T spanning the set V of the image pixels, to every map $w : E \to \mathbb{N}$ that weights the edges of T and to every threshold $\lambda \in \mathbb{N}$, one may associate the partition $\mathcal{P}_\lambda^w$ of V induced by the connected components of the graph made from V and the edges of weight below λ. It is well known [2,5] that for any two values λ_1 and λ_2 such that $\lambda_1 \geq \lambda_2$, the partitions $\mathcal{P}_{\lambda_1}^w$ and $\mathcal{P}_{\lambda_2}^w$ are *nested* and $\mathcal{P}_{\lambda_1}^w$ is *coarser* than $\mathcal{P}_{\lambda_2}^w$. Hence, the set $\mathcal{H}^w = \{\mathcal{P}_\lambda^w \mid \lambda \in \mathbb{N}\}$ is a *hierarchy of partitions induced by the weight map w.*

Our algorithm does not explicitly produce a hierarchy of partitions, but instead produces a weight map L (scales of observations) from which the desired hierarchy $\mathcal{H}^L$ can be inferred on a given T. It starts from a minimum spanning tree T of the edge-weighted graph built from the image. In order to compute the scale $L(e)$ associated with each edge of T, our method iteratively considers the edges of T in a non-decreasing order of their original weights w. For every edge e, the new weight map $L(e)$ is initialized to ∞; then for each edge e linking two vertices x and y the following steps are performed:

(i) Find the the region X of $\mathcal{P}_{w(e)}^w$ that contains x.
(ii) Find the the region Y of $\mathcal{P}_{w(e)}^w$ that contains y.
(iii) Compute the hierarchical observation scale $L(e)$.

At step (iii), the *hierarchical scale* $S'_Y(X)$ of X relative to Y is needed to obtain the value $L(e)$. Intuitively, $S'_Y(X)$ is the lowest observation scale at which some sub-region of X, namely X^*, will be merged to Y. More precisely, using an internal parameter v, this scale is computed as follows:

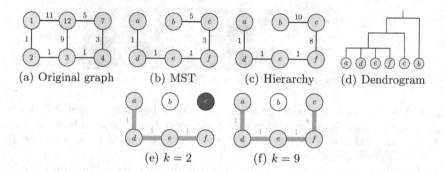

(a) Original graph   (b) MST   (c) Hierarchy   (d) Dendrogram

(e) $k = 2$ (f) $k = 9$

Fig. 4. Example of hierarchical image segmentations. In contrast to example in Fig. 2, the contours are stable from a scale to another, providing a hierarchy.

(1) Initialize the value of v to ∞.
(2) Decrement the value of v by 1.
(3) Find the the region X^* of $\mathcal{P}_v^L$ that contains x.
(4) Repeat steps 2 and 3 while $S_Y(X^*) < v$
(5) Set $S'_Y(X) = v$.

With the appropriate changes, the same algorithm allows $S'_X(Y)$ to be computed. Then, the hierarchical scale $L(e)$ is simply set to:

$$L(e) = \max\{S'_Y(X), S'_X(Y)\}. \tag{5}$$

Figure 4 illustrates the result of our method on a pedagogical example. Starting from the graph of Fig. 4(a), our method produces the hierarchical observation scales depicted in Fig. 4(c). As for the method of [3], our algorithm only considers the edges of the minimum spanning tree (see Fig. 4(b)). The whole hierarchy is depicted as a dendrogram in Fig. 4(d), whereas two levels of the hierarchy (at scales 2 and 9) are shown in Fig. 4(e) and (f).

2.1   Implementation Issues

To efficiently implement our method, we use some data structures similar to the ones proposed in [5]; in particular, the management of the collection of partitions is due to Tarjan's union find and Fredman and Tarjan's Fibonnacci heaps. Furthermore, we made some algorithmic optimizations to speed up the computations of the observation scales. In order to illustrate an example of computation time, we implemented all our algorithm in C++ on a standard single CPU computer under windows Vista, we run it in a Intel Core 2 Duo, 4GB. For the image illustrated in Fig. 1(a) (with size 321x481), the hierarchy is computed in 2.7 seconds, and the method proposed in [3] spent 1.3 seconds.

3   Experimental Results

In this section, we present a quantitative and a qualitative assessments in order to better compare our method to the method proposed in [3] (called method FH

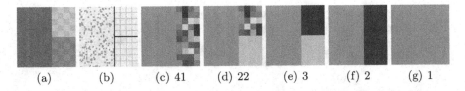

<table>
<tr><td>(a)</td><td>(b)</td><td>(c) 41</td><td>(d) 22</td><td>(e) 3</td><td>(f) 2</td><td>(g) 1</td></tr>
</table>

Fig. 5. An example of a hierarchical image segmentations of a synthetic image containing three perceptually big regions. The saliency map of the image (a) is shown in (b). The number of regions of the segmented images is written under each figure.

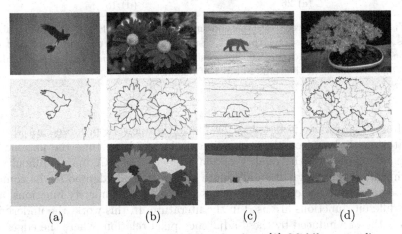

(a)   (b)   (c)   (d)

Fig. 6. Top row: some images of the Berkeley database [9]. Middle row: saliency maps of these images according to our hierarchical method. The numbers of scales of these hierarchies are (a) 240, (b) 443, (c) 405 and (d) 429. Bottom row: according to our subjective judgment, the best segmentations extracted from the hierarchies. The numbers of regions are (a) 3, (b) 18, (c) 6 and (d) 16.

Table 1. Performances of our method and the method FH [3] using two different measures: Ground-truth Covering (GT Covering) and Probabilistic Rand Index. The presented scores are optimal considering a constant scale parameter for the whole dataset (ODS) and a scale parameter varying for each image (OIS). See [9] for more details on the evaluation method.

Area	GT Covering				Prob. Rand. Index			
	ODS		OIS		ODS		OIS	
	Ours	FH	Ours	FH	Ours	FH	Ours	FH
20	0.42	0.43	0.52	0.52	0.75	0.75	0.81	0.79
50	0.44	0.43	0.52	0.52	0.76	0.75	0.81	0.79
500	0.46	0.43	0.53	0.53	0.76	0.76	0.81	0.79
1000	0.46	0.44	0.53	0.53	0.76	0.76	0.80	0.80
1500	0.46	0.44	0.52	0.54	0.76	0.76	0.80	0.80

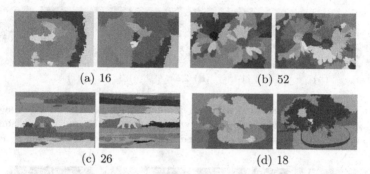

(a) 16   (b) 52

(c) 26   (d) 18

Fig. 7. Comparison between our method and the method FH [3]. For each pair of images, the right image shows the best result (according to our judgment and our experiments) from [3] and the left image shows a segmentation extracted from our hierarchical result, with the same number of regions.

hereafter). The former is based on evaluation framework proposed in [9], and the later one is based on three experiments in which we tune the parameters to (visually) evaluate the quality of the segmentations. A major difficulty of experiments is the design of an adequate edge-cost, well adapted to the content to be segmented. A practical solution is to use some dissimilarity functions, and many different functions are used in the literature. In this work, the underlying graph is the one induced by the 4-adjacency pixel relation, where the edges are weighted by a simple color gradient computed by the Euclidean distance in the RGB space. Before presenting the quantitative and qualitative assessments, we illustrate some results of our method.

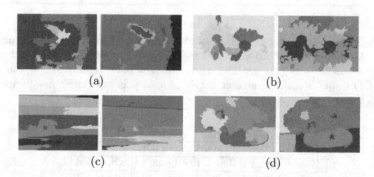

(a)   (b)

(c)   (d)

Fig. 8. Examples of image segmentation where the number of regions has been set to 15. For each pair of images, the left one shows a segmentation extracted from our hierarchy; and the right one shows the result obtained with [3] by varying the parameter k until the desired number of regions is found.

(a)   (b)   (c)

Fig. 9. Examples of segmentations for images corrupted by a random salt noise. The corrupted images (at different levels - 70% and 90%) are shown on the first column. The results of our method and [3] are illustrated in the second and third columns, respectively.

In Fig. 5, we present some results on an artificial image containing three perceptually large regions. With this example, one can easily verify the hierarchical property of our method by looking at the segmentations at scales *resp.* 1000, 2000, 5000, 140000 and 224000 (*resp.* Fig. 5(c), (d), (e), (f) and (g)). Since the resulting segmentations are nested, the whole hierarchy can be presented in a saliency map (see Fig. 5(b)). Figure 6 illustrates the performance of our method when applied to some images of the Berkeley's database [9]. Note that, as in [3], an area filtering is applied to eliminate small regions (smaller than 500 pixels).

In the sequel, we present the quantitative assessment followed by the qualitative one. Table 1 assesses the equivalent performances of our method and of the method FH [3], according to the evaluation framework proposed in [9], in terms of Ground-truth Covering and Probabilistic Rand Index, when applied on 200 test images of the Berkeley's database [9]. For this experiment, an area filtering is applied to eliminate small regions varying from 20 to 1500 pixels.

For the qualitative assessment, we made three experiments. First, we try to set the reasonable parameter for [3], *i.e.* the parameter that produces the best (subjective) visual result (Fig. 7). We can compare this result with the segmentation result at a scale in our hierarchy such that it contains the same number of regions that of [3] (Fig. 7). In a second experiment, we fixed the number of regions to 15 for all images, and tune the parameter for [3] to obtain this number of regions. For our method, we use breadth-first traversal in the hierarchy (tree structure) to find the scale that givens 15 regions. We compare those segmentations in Fig. 8. The last illustration (Fig. 9) is designed to assess the robustness to random impulse noise. From these experiments, we observe that in general, our method produces "objects" (or regions) better defined with respect to the results obtained by the method FH. Moreover, the contours are stable, *i.e.*, the contours do not move from one scale to another, and the number of regions is decreasing when the scale parameter increases.

4   Conclusions

This paper proposes an efficient hierarchical segmentation method based on the observation scales of [3]. In contrast to [3], our method produces the complete set of segmentations at every scale, and satisfies both the causality and location principle defined by [6]. An important practical consequence of these properties is to ease the selection of a scale level adapted to a particular task. We assess our method and the method of [3] on the Berkeley database following the methodology introduced in [9]. We visually assessed our method on some real images by comparing our segmentations to those of [3]. From theses quantitative and qualitative assessments, the produced segmentations are promising, in particular w.r.t. robustness. As future work, we will investigate using more information into the definition of observation scale as well as learning which information is pertinent for a given practical task. Moreover, we will investigate theoretical properties of our method.

References

1. Zahn, C.T.: Graph-theoretical methods for detecting and describing gestalt clusters. IEEE Trans. Comput. 20, 68–86 (1971)
2. Morris, O., Lee, M.J., Constantinides, A.: Graph theory for image analysis: an approach based on the shortest spanning tree. Communications, Radar and Signal Processing, IEE Proceedings F 133(2), 146–152 (1986)
3. Felzenszwalb, P.F., Huttenlocher, D.P.: Efficient graph-based image segmentation. IJCV 59, 167–181 (2004)
4. Najman, L.: On the equivalence between hierarchical segmentations and ultrametric watersheds. JMIV 40, 231–247 (2011)
5. Cousty, J., Najman, L.: Incremental Algorithm for Hierarchical Minimum Spanning Forests and Saliency of Watershed Cuts. In: Soille, P., Pesaresi, M., Ouzounis, G.K. (eds.) ISMM 2011. LNCS, vol. 6671, pp. 272–283. Springer, Heidelberg (2011)

6. Guigues, L., Cocquerez, J.P., Men, H.L.: Scale-sets image analysis. IJCV 68(3), 289–317 (2006)
7. Haxhimusa, Y., Kropatsch, W.: Segmentation Graph Hierarchies. In: Fred, A., Caelli, T.M., Duin, R.P.W., Campilho, A.C., de Ridder, D. (eds.) SSPR&SPR 2004. LNCS, vol. 3138, pp. 343–351. Springer, Heidelberg (2004)
8. Najman, L., Schmitt, M.: Geodesic saliency of watershed contours and hierarchical segmentation. PAMI 18(12), 1163–1173 (1996)
9. Arbelaez, P., Maire, M., Fowlkes, C., Malik, J.: Contour detection and hierarchical image segmentation. PAMI 33, 898–916 (2011)

A Discrete Scale Space Neighborhood for Robust Deep Structure Extraction

Martin Tschirsich[1] and Arjan Kuijper[1,2]

[1] Technische Universität Darmstadt, Germany
[2] Fraunhofer IGD, Darmstadt, Germany

Abstract. Linear or Gaussian scale space is a well known multi-scale representation for continuous signals. The exploration of its so-called deep structure by tracing critical points over scale has various theoretical applications and allows for the construction of a scale space hierarchy tree. However, implementational issues arise, caused by discretization and quantization errors. In order to develop more robust scale space based algorithms, the discrete nature of computer processed signals has to be taken into account. Aiming at a computationally practicable implementation of the discrete scale space framework, we investigated suitable neighborhoods, boundary conditions and sampling methods. We show that the resulting discrete scale space respects important topological invariants such as the Euler number, a key criterion for the successful implementation of algorithms operating on its deep structure. We discuss promising properties of topological graphs under the influence of smoothing, setting the stage for more robust deep structure extraction algorithms.

1 Introduction

In the field of computer vision, deriving information from observed images is a central problem. Various strategies have been invented to do so in a performant manner, usually by applying some kind of operator. Their performance depends on the *inner scale*, the sampling density or resolution of the image they operate on. To overcome this dependence between operator and inner scale, various strategies of multi-scale representations have been proposed. Almost all those strategies consist of transforming the given images into a scale independent representation first before applying an operator on this representation. A common requirement for such preliminary transformations is to mask as little information present in the original image as possible. The Gaussian scale space satisfies these requirements and can be thought of as the natural generalization of the lowpass pyramid. It is also popular for its theoretical foundation. A Gaussian scale space representation of a given signal is a family of derived signals, progressively smoothed with a Gaussian filter.

Its *deep structure* consists of critical points or zerocrossings traced over scale. An implementation of important scale space based algorithms can be found in the software tool ScaleSpaceViz [1]. ScaleSpaceViz has, according to the authors, "proven to be useful in exploring the deep structure of images and constructing applications involving scale space interest points, such as reconstruction and matching". Admittedly, this

G.L. Gimel' farb et al. (Eds.): SSPR & SPR 2012, LNCS 7626, pp. 126–134, 2012.
© Springer-Verlag Berlin Heidelberg 2012

holds true under certain conditions. Nevertheless, ScaleSpaceViz suffers from robustness problems. As it is the case with many such scale space applications, its implementation is based on a discretized continuous scale space.

The discrete scale space proposed by Lindeberg [2] takes the discrete nature of computer processed signals into account. It is based on equivalent assumptions and axioms that have been used to derive the continuous Gaussian scale space adapted to discrete signals. It is our belief that porting scale space algorithms from a discretized continuous to the discrete scale space will eventually lead to more accurate, robust and possibly faster implementations. The discrete scale space formalized by Lindeberg however *does not* respect important topological invariants such as the Euler number. Since most algorithms that operate on the deep structure of the Gaussian scale space require this topological invariant to hold, we present in this paper a modified definition of the discrete scale space respecting the Euler number. A subsequent investigation of various properties of this discrete scale space then results in a fast and robust sampling algorithm [3]. We propose the application of topological graphs together with adaptive sampling in order to reliably extract the deep structure of the discrete scale space.

2 Discrete Signals

Important discrete operators, including those needed to build a scale space over f, operate on functions of infinite domain $\mathbb{Z}^2$. Therefore, signals with bounded domain must be expanded to cover the whole range of $\mathbb{Z}^2$. Imposing a Dirichlet or zero-border condition implies assuming the signal to be of constant or zero-value outside the originally bounded domain. A zero-border condition may reduce the computational complexity of certain operations, namely discrete convolution with infinitely large kernels. However, a zero-border has drawbacks concerning the computation of a scale space over f. These can be avoided by assuming f to be periodic instead with period M for the first and N for the second dimension, thus imposing a periodic boundary condition.

Image structures or features are often defined in terms of limited subsets of image points, so called interest points. These include stationary points such as minima, maxima and saddles in the two-dimensional case. If the discrete signal f is interpreted as lattice height data, these critical points are then called peaks, pits and passes. They reveal important topological characteristics. Critical point detection has to be performed in order to find their positions.

Provided that f is a sampled continuous signal and the sampling operator is known, it is often possible to reconstruct and perform critical point detection on a functional representation of the original signal using the well known gradient or slope based definitions of critical points on continuous data [4]. In the general case however, this is not possible and we need a separate definition of critical points on discrete data.

2.1 Euler Number

A well known topological invariant stating a stable relation between the number of extrema and saddle points on closed continuous surfaces is the Euler number. The Euler number, also called Euler formula or mountaineer's equation [5] for a closed continuous surface is given by $m^+ - s + m^-$, where m^+, s and m^- denote the number of local maxima, saddles and minima on that surface.

Various algorithms that exploit the deep structure of the continuous scale space rely on a stable relation between the number of these critical points. This invariant should therefore equally hold true for discrete critical points on discrete signals. Firstly, we need a definition of the Euler number for a discrete signal f and then a definition of discrete critical points that respects this invariant. An important aspect of this definition is the neighborhood of each point in $D(f)$. The neighborhood of a point consists of all those points it is adjacent to. This relation is usually expressed as a mapping from a point to the set of its neighbors $N : \mathbb{Z}^2 \to \mathcal{P}(\mathbb{Z}^2)$. In order to reason about critical points and the relations between the number of critical points in a given neighborhood, we resort to a graph representation of f.

The discrete signal f and a symmetric neighborhood N can be represented as an undirected graph $G_f = (V, E)$ with vertices $V = \{v_{x,y} : (x, y) \in D(f)\}$, edges $E = \{\{v_{x,y}, v_{x',y'}\} \in \mathcal{P}(V) : (x', y') \in N(x, y)\}$ and a function $f_G : V \to \mathbb{R}_+$ relating each vertex to a value $f_G(v_{x,y}) = f(x, y)$. A bijection between $D(f)$ and V relates each point $(x, y) \in D(f)$ to a vertex $v_{x,y} \in V$. For each pair of points (x, y), (x', y') that is adjacent given the neighborhood N, E contains an edge $\{v_{x,y}, v_{x',y'}\}$. Thus, the domain of f is represented via V, the neighborhood via E and the mapping f via the mapping f_G over the set of vertices.

If the graph G_f can be embedded into a closed surface such that its edges do not intersect, the graph genus g then denotes the minimal genus of the closed surface the graph can be embedded in [6]. For closed continuous surfaces, the Euler characteristic of that surface $\chi = 2 - 2g$ coincides with the Euler number [4]. The value of the Euler characteristic and thus the Euler number depends on the chosen neighborhood. A graph with no edges for example can always be embedded into a sphere, a closed surface with genus 0. By contrast, the complete graph K_5 on 5 vertices is not planar and can only be embedded into a torus, a closed surface with genus 1. Intuitively, the 4-neighborhood $N_4(x, y) = \{(x', y') \in \mathbb{Z}^2 : \|(x - x', y - y')\|_1 = 1\}$ seems to be a natural choice given the square lattice f is defined upon, but we will later see that a 6-neighborhood N_6 resulting from a Delaunay triangulation of the square lattice is a more practical choice. For bounded signals, G_f also has to reflect the chosen boundary condition. For positive signals, a zero-border condition can be modeled by introducing a virtual pit $v_o \in V$ with $f_G(v_0) = 0$ connected to all boundary points (Fig. 1a) [5]. In this case, the neighborhood of boundary points might deviate from N, but this does not matter as long as we are only interested in topological properties. Modeling a periodic boundary condition however requires a strict correspondence between the chosen neighborhood and the graph representation, modeled by duplication of $D(f)$ in all directions and connecting opposing boundary points through additional edges (Fig. 1b). A periodic replication of the square grid $D(f)$ does not affect the Euler number.

If we impose a zero-border and not a periodic boundary condition, the graph representation of a signal f with bounded domain $[1, 3] \times [1, 3] \subseteq D(f) \subset \mathbb{Z}^2$ and both 4-neighborhood N_4 or 6-neighborhood N_6 can be embedded into a sphere, i.e. G_f is planar (Fig. 1a). Intersection-free embedding into a sphere is not possible for periodic f.

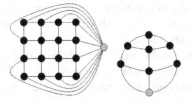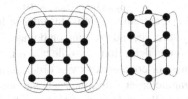

(a) 4-neighborhood with virtual pit (gray), graph genus 0, embedded in a sphere (spherical polyhedron).

(b) periodic 4-neighborhood, graph genus 1, embedded in a torus (toroidal polyhedron).

Fig. 1. Embedding an undirected graph with 4-neighborhood N_4 into a closed surface

Proof. Using Kuratowski's theorem stating that a finite graph is not planar if it does contain a subgraph that is homeomorphic to K_5, it can be shown that the graph representation G_f of a periodic signal f with $D(f) = [1,3] \times [1,3] \subset \mathbb{Z}^2$ is not planar. Since the graph representation of a periodic signals f whose domain is a superset of $D(f)$ is a supergraph of G_f, it is homeomorphic to K_5 as well. The same holds for arbitrary neighborhoods N with $N_4(x,y) \subseteq N(x,y)$, since their graph representation is also a supergraph of G_f.   □

The surface with the next higher genius is a torus. From Fig. 1b it becomes clear that it is possible to embed G_f into a torus for periodic f. For the Euler characteristic and therefore the Euler number then holds $\chi = 2 - 2g = 0 \Leftrightarrow m^+ + m^- = s$.

Now that we have transfered the invariant stated by the Euler number for closed continuous surfaces to the graph representation G_f and thus the periodic discrete signal f for the N_4 neighborhood and its triangulations N_6, we can reduce the number of reasonable discrete critical point definitions upon these neighborhoods to those that respect the Euler number. Eventually, such a critical point definition allows for easier adaption of algorithms that work well in the continuous case to discrete signals.

2.2 Detecting All Critical Points

There are principally two different ways to define critical points in the continuous case, gradient and local path based definitions [4]. This translates to the discrete case if proper definitions of gradient and local paths are given, both depending on the chosen neighborhood. According to Kovalevsky [7], a stationary (homogeneous) neighborhood suitable to formulate a definition of discrete critical points respecting the Euler number in the discrete n-dimensional space must contain $2(2^n - 1)$ points. Any stationary 6-neighborhood given by a triangulation of the square lattice meets this condition. Later, it was shown by Takahashi [5] that any arbitrary triangulation of the lattice is equally topologically consistent. However, a stationary neighborhood is inherently simpler and introduces no perturbations in the position of critical points, thus we will use a stationary hexagonal neighborhood $N_6(x,y) = \{(x \pm 1, y), (x, y \pm 1), (x \pm 1, y \mp 1)\}$ as proposed by Kuijper [8].

We can now give a definition of discrete critical points respecting the Euler number on the hexagonal neighborhood N_6: $(x_0, y_0) \in \mathbb{Z}^2$ is said to be a local *maximum (minimum)* for a signal $f : \mathbb{Z}^2 \to \mathbb{R}$ within neighborhood $N(x_0, y_0) \subseteq \mathbb{Z}^2$ iff

$\forall (x, y) \in N : (x_0, y_0) > (<) (x, y)$. This definition is only valid as long as there are no local extremal regions. In order to incorporate these plateaus, we introduce an arbitrary strict second ordering on points of equal value.

A common definition of saddle points is via sign-changes in the neighborhood, given a clockwise or counterclockwise order of the neighbors. Given a neighborhood N, the lexicographic order with minor adaptations to the boundary provides such an arrangement. We will use the definition given by Takahashi [5] who differs between regular and degenerated saddles. Four sign changes indicate a regular, six sign changes a degenerated saddle. At degenerated saddle points, more than two contour lines intersect. Degenerated saddles on N_6 can be split into two regular saddles.

Provided a neighborhood and a suitable critical point definition as outlined above, we can extract the surface network and other topological graphs such as the Reeb graph from f. This can be done using the algorithms outlined by Takahashi [5] and Scott [4] with minor adaptations to periodic signals whose representative graph is of genus 1.

3   Neighborhood and Discrete Scale Space

In this section, we will give a concise definition of the Gaussian scale space [9, 10] for two-dimensional discrete signals comprising a continuous scale parameter. The scale space representation of a discrete signal $f : \mathbb{Z}^2 \rightarrow \mathbb{R}$ is a one-parameter family of derived signals $L : \mathbb{Z}^2 \times \mathbb{R}^+$ defined by a set of discrete scale space axioms. It was first described by Lindeberg [2], who introduced a discrete version of the *non enhancement of local extrema* axiom based on *weak* local extrema within the 8-neighborhood (Fig. 2).

An equivalent definition via convolution with a scale space kernel yields $L (\cdot, \cdot; t) = k (\cdot, \cdot; t) * f (\cdot, \cdot)$ where $k : \mathbb{Z}^2 \times \mathbb{R}_+ \rightarrow \mathbb{R}$ is the discrete analogue of the continuous Gaussian scale space kernel.

$$L (x, y; t) = \sum_{m=-\infty}^{\infty} k (m; t) \sum_{n=-\infty}^{\infty} k(n; t) f(x - m, y - n)$$

The scale space representation L of a signal f is the solution of the diffusion equation. For the axioms chosen by Lindeberg, it takes the form

$$\partial_t L = \alpha \nabla_2^5 L + \beta \nabla_\times^2 L \qquad (1)$$

with $L (\cdot, \cdot; 0) = f (\cdot, \cdot)$ as initial condition and some $\alpha, \beta \geq 0$. The five-point operator ∇_2^5 and the cross operator $\nabla_\times^2$ approximate the continuous Laplacian operator. They correspond to convolution of $L (\cdot, \cdot; t)$ with kernel $\begin{bmatrix} & 1 & \\ 1 & -4 & 1 \\ & 1 & \end{bmatrix}$ respectively $\begin{bmatrix} \frac{1}{2} & 0 & \frac{1}{2} \\ 0 & -2 & 0 \\ \frac{1}{2} & 0 & \frac{1}{2} \end{bmatrix}$.

For $\alpha = 1$ and $\beta = 0$ Eq. (1) simplifies to $\partial_t L = \nabla_2^5 L$ and results in k being separable.

If one wishes to choose a strict local extrema definition and another neighborhood because of topological reasons, the axiomatic scale space definition may change. Only the strict local extrema definition within a 6-neighborhood has proven to respect the Euler number. However, not every spatial extremum in L within an arbitrary 6-neighborhood satisfies the aforementioned non-enhancement axiom based on weak local extrema

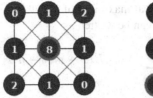 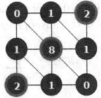

Fig. 2. Different neighborhoods (left: N_8, right: N_6) over the same signal. Only the central maximum satisfies the non-enhancement axiom $\partial_t L \leq 0$ for both neighborhoods and arbitrary $\alpha \geq 0$, $\beta > 0$.

within the 8-neighborhood chosen by Lindeberg. Thus, in order to incorporate the Euler number as topological invariant into the discrete scale space, either L and therefore the discretized diffusion equation has to be modified or the 6-neighborhood has to be chosen in a way that the extrema within such neighborhood represent a subset of the weak extrema within the 8-neighborhood.

3.1 Discretized Scale Space for N_6

We want to show that the discrete scale space over f with $N = N_6$ is defined by the solution of a discretized version of the differential equation as it is the case for $N = N_8$.

Necessity. Let $N = N_6$. Then, following the approach given by Lindeberg [2], it becomes evident that the scale space representation $L : \mathbb{Z}^2 \times \mathbb{R}_+ \to \mathbb{R}$ of signal $f : \mathbb{Z}^2 \to \mathbb{R}$ satisfies the differential equation $\partial_t L = \alpha \nabla_5^2 L$ for some $\alpha \geq 0$ with initial condition $L(\cdot, \cdot; 0) = f(\cdot, \cdot)$.

Proof. Strictly following the proof given by Lindeberg and resulting from his set of proposed discrete scale space axioms *without* the non-enhancement of local extrema axiom and thus not depending on the chosen neighborhood, $\partial_t L = AL$ holds for all $t \geq 0$ and a linear shift-invariant operator A. Because of the shift-invariance, this can be written as

$$(AL)(x, y; t) = \sum_{m=-\infty}^{\infty} \sum_{n=-\infty}^{\infty} a_{m,n} L(x - m, y - n; t).$$

Let $N_6^+(x, y) = N_6(x, y) \cup \{(x, y)\}$. Now, from using the non-enhancement of local extrema axiom for extrema within N_6, it follows that $a_{m,n} = 0$ for $(m, n) \in N_6^+$. Otherwise, assuming that $a_{\tilde{m}, \tilde{n}} \neq 0$ for one $(\tilde{m}, \tilde{n}) \in N_6^+$, we could define a function $f_1 : \mathbb{Z}^2 \to \mathbb{R}$ with

$$f_1(x, y) = \begin{cases} \epsilon > 0 & \text{if } (x, y) = (0, 0) \\ 0 & \text{if } (x, y) \in N_6(0, 0) \\ 1 & \text{if } (x, y) = (\tilde{m}, \tilde{n}) \\ 0 & \text{otherwise} \end{cases}$$

and $\partial_t L\left(0,0;0\right) = \partial_t f\left(0,0\right) = \epsilon a_{o,o} + a_{\tilde{m},\tilde{n}} \geq 0$ for well-chosen ϵ, violating the non-enhancement axiom since $(0,0)$ is a local maximum of f within N_6. The same holds for local minima and $\epsilon < 0$. Thus, $\partial_t L$ can be written as

$$(AL)\left(x,y;t\right) = \sum_{m,n \in N_6^+} a_{m,n} L\left(x-m,y-n;t\right).$$

Imposing symmetry conditions in analogy to Lindeberg [2], AL equals a convolution of L with kernel $\begin{bmatrix} 0 & b & 0 \\ b & c & b \\ 0 & b & 0 \end{bmatrix}$ for some b and c. Considering another function $f_2\left(\cdot,\cdot\right) = 1$ with $f_2\left(0,0\right) = 1 + \left(-\right)\epsilon$ According to the strict definition of local extrema given earlier, the point $(0,0)$ is a local maximum (minimum), thus $\partial_t L\left(0,0;0\right) = \partial_t f_2\left(0,0\right) = 4b + c = 0$. Therefore, $\begin{bmatrix} 0 & b & 0 \\ b & c & b \\ 0 & b & 0 \end{bmatrix} = \alpha \begin{bmatrix} 0 & 1 & 0 \\ 1 & -4 & 1 \\ 0 & 1 & 0 \end{bmatrix}$ for some α. Finally, from $\partial_t L\left(0,0;0\right) = \partial_t f_1\left(0,0\right) = \alpha\epsilon c \leq 0$ it becomes evident that $\alpha \geq 0$. $\qquad\square$

Following this proof, it becomes apparent that not only the hexagonal 6-neighborhood, but arbitrary 6-neighborhoods (arbitrary *triangulations*) as well as the 4-neighborhood N_4 result in the same differential equation $\partial_t L = \alpha \nabla_5^2 L$. For simplicity, the parameter α, which only affects the scaling of the scale parameter, is usually set to 1.

Sufficiency. The solution $L : \mathbb{Z}^2 \times \mathbb{R}_+ \to \mathbb{R}$ of the differential equation $\partial_t L = \alpha \nabla_5^2 L$ for some $\alpha \geq 0$ with initial condition $L\left(\cdot,\cdot;0\right) = f\left(\cdot,\cdot\right)$ for $f : \mathbb{Z}^2 \to \mathbb{R}$ is a scale space representation of f.

Proof. Let $\tilde{L}$ be a scale space representation of f. According to the previous definition, $\tilde{L}$ is then also the solution of $\partial_t L = \tilde{\alpha} \nabla_5^2 L$ for some $\tilde{\alpha} \geq 0$ with initial condition $L\left(\cdot,\cdot;0\right) = f\left(\cdot,\cdot\right)$. Since $\tilde{\alpha}$ is only a linear scaling parameter along the direction of t, $\tilde{\tilde{L}}\left(\cdot,\cdot;t\right) = \tilde{L}\left(\cdot,\cdot;\frac{\alpha}{\tilde{\alpha}}t\right)$ is the solution of $\partial_t L = \alpha \nabla_5^2 L$ and therefore $\tilde{\tilde{L}} = L$. It is obvious that $\tilde{\tilde{L}}$ still satisfies all the discrete scale scape axioms, thus L is a scale space representation of f. $\qquad\square$

Linking. In order to extract critical curves, the discrete scale space has to be sampled along its continuous scale parameter, since we do not know how to compute the exact occurrences of the zerocrossings in the neighborhood as stated above. The known way to extract critical curves is to compute several scale images of L at preselected scales, then detecting spatially critical points on these scales and finally linking spatially close critical points on subsequent scales into critical curves. We need a criterion that tells us when further subsampling is required and when the sampling density is high enough to guarantee a correct result. Such a criterion is found in topological graphs or more precisely in the difference of topological graphs of subsequent scales. Tracking changes in the surface network over scale is a promising approach, since the set of possible changes between scale space events such as creations or annihilations of critical points is strictly limited.

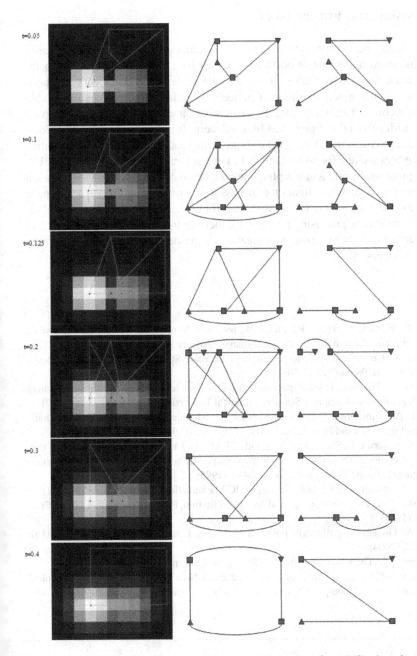

Fig. 3. A sampled discrete scale space. From the critical points alone (left), the subsequent annihilation and creation between scale $t_1 = 0.05$ and $t_2 = 0.125$ is not visible and might lead to incorrect linking. However, the surface networks (middle, incomplete) of these scales differ, thus providing a criterion whether further subsampling might be necessary. The Reeb graphs (right, incomplete) are identical, though.

4   Conclusion and Future Work

The discrete scale space as an equivalent to the two-dimensional Gaussian scale space has been discussed and some important properties have been derived. A computationally practicable implementation of the discrete scale space framework has been outlined. The regular 6-neighborhood, a periodic boundary condition and a suitable critical point definition respecting the Euler number have led to discrete scale space axioms that differ from those proposed by Lindeberg. It has been shown that the discretized diffusion equation inducing the discrete scale space derived from the modified axioms resembles the one found by Lindeberg but has one less degree of freedom. Using our computationally efficient sampling method, based on properties of the Laplacian kernel [3], we gave a first investigation of the deep structure of the discrete scale space illustrating the need for a more robust algorithm for critical curve extraction. Topological graphs have shown promising properties under the influence of changes in scale. However, further and more formal investigation of the deep structure of the discrete scale space is necessary.

References

1. Kanters, F., Florack, L., Duits, R., Platel, B., ter Haar Romeny, B.: Scalespaceviz: a-scale spaces in practice. Pattern Recognition and Image Analysis 17, 106–116 (2007)
2. Lindeberg, T.: Discrete Scale-Space Theory and the Scale-Space Primal Sketch. PhD thesis, Royal Institute of Technology (1991)
3. Tschirsich, M.: The discrete scale space as a base for robust scale space algorithms. Technical report, Department of Computer Science, Technical University of Darmstadt (June 2012)
4. Scott, P.J.: An algorithm to extract critical points from lattice height data. International Journal of Machine Tools and Manufacture 41(13-14), 1889–1897 (2001)
5. Takahashi, S., Ikeda, T., Shinagawa, Y., Kunii, T.L., Ueda, M.: Algorithms for extracting correct critical points and constructing topological graphs from discrete geographical elevation data. Computer Graphics Forum 14(3), 181–192 (1995)
6. Perez, A.: Determining the genus of a graph. HC Mathematics Review 1(2), 4–13 (2007)
7. Kovalevsky, V.A.: Discrete topology and contour definition. Pattern Recognition Letters 2(5), 281–288 (1984)
8. Kuijper, A.: On detecting all saddle points in 2d images. Pattern Recognition Letters 25(15), 1665–1672 (2004)
9. Koenderink, J.J.: The structure of images. Biological Cybernetics 50, 363–370 (1984)
10. Weickert, J., Ishikawa, S., Imiya, A.: Scale-space has been discovered in japan. Technical report, Department of Computer Science, University of Copenhagen (August 1997)

On the Correlation of Graph Edit Distance and L_1 Distance in the Attribute Statistics Embedding Space

Jaume Gibert[1], Ernest Valveny[1], Horst Bunke[2], and Alicia Fornés[1]

[1] Computer Vision Center, Universitat Autònoma de Barcelona,
Edifici O Campus UAB, 08193 Bellaterra, Spain
{jgibert,ernest}@cvc.uab.es

[2] Institute for Computer Science and Applied Mathematics, University of Bern,
Neubrückstrasse 10, CH-3012 Bern, Switzerland
bunke@iam.unibe.ch

Abstract. Graph embeddings in vector spaces aim at assigning a pattern vector to every graph so that the problems of graph classification and clustering can be solved by using data processing algorithms originally developed for statistical feature vectors. An important requirement graph features should fulfil is that they reproduce as much as possible the properties among objects in the graph domain. In particular, it is usually desired that distances between pairs of graphs in the graph domain closely resemble those between their corresponding vectorial representations. In this work, we analyse relations between the edit distance in the graph domain and the L_1 distance of the attribute statistics based embedding, for which good classification performance has been reported on various datasets. We show that there is actually a high correlation between the two kinds of distances provided that the corresponding parameter values that account for balancing the weight between node and edge based features are properly selected.

1 Introduction

The comparison of relational structures has been widely studied over the past years [3]. Graph edit distance constitutes a major paradigm due to its ability to handle arbitrary graph structures [2,10]. It is defined as the minimum amount of distortion that is needed to transform one graph into another. This distance measure is very intuitive in nature since the edit path it looks for is based on substituting, deleting and inserting nodes and edges such that the source and the target graph become isomorphic.

Graph edit distance, however, has a high computational complexity. Modern ways for graph matching try to avoid this high complexity. Extracting graph features and building up pattern vectors for the analysis of graphs —known as graph embedding— is a common way to reduce the computational complexity and make efficient learning algorithms available for the domain of graphs. A desired property of any generic graph embedding scheme is that it should be

G.L. Gimel' farb et al. (Eds.): SSPR & SPR 2012, LNCS 7626, pp. 135–143, 2012.

able to approximate the original distribution of patterns in the graph domain. In other words, distances between objects in the graph domain should be similar to their corresponding distances in the embedding space. For instance, for the dissimilarity space embedding proposed in [9] it has been shown that the graph edit distance between two graphs is an upper bound of the Euclidean distance between the corresponding vectorial maps. Similarly, in [6], the Ihara coefficients have been experimentally shown to be a set of features with distances that correlate linearly with the edit distance.

In this paper, we investigate how the edit distance is related to a discrete version of the embedding methodology proposed in [5]. The features under this embedding methodology account for the number of nodes with a certain label that appear in a graph, and the number of edges with a given label that exist between two nodes with certain labels. In other words, this kind of embedding is based on occurrence and co-occurrence statistics of labels in the underlying graph. Absolute differences between node-based features indicate how many nodes with a certain label exist in one graph that are not present in the other graph. This is, in fact, exactly the same situation that occurs when performing the edit distance computation between graphs with discrete attributes under a cost function that disregards substitution of nodes with different labels and forces node deletions and insertions instead. This observation is one of the main motivations of our work.

In particular, we express both ways of computing graph distances —the edit distance and the L_1 distance for the embedding methodology— in terms of a weighting parameter balancing the impact of nodes and edges in the resulting distance values. We investigate how distances are correlated as a function of these two parameters, and also how corresponding distance-based classifiers behave.

The rest of the article is organized as follows. Graph edit distance is reviewed in the next section, and the edit cost function used for the case of discretely attributed graphs is specified. Section 3 describes the embedding methodology based on statistics of labelling information. Correlation experiments of both ways of comparing graphs and a discussion of the results are presented in Section 4. Finally, Section 5 draws conclusions from this work.

2   Graph Edit Distance

A graph $g = (V, E, \mu, \nu)$ is a 4-tuple where V is the set of nodes, $E \subseteq V \times V$ the set of edges, and $\mu : V \to L_V$ and $\nu : E \to L_E$ are the labelling functions of nodes and edges, respectively. In this work, we use undirected graphs where the labels come from finite discrete domains.

As already stated above, the main idea of graph edit distance is to define a dissimilarity measure between graphs by the minimum amount of distortion that is needed to transform one graph into the other [2,10]. Distortions are defined in terms of edit operations between two graphs, such as node and edge deletion, insertion and substitution. A sequence of edit operations transforming the source graph into the target graph is called an edit path. Edit costs define whether a

Table 1. Edit cost function

	Deletion / Insertion	Substitution
Nodes	$c(u \to \epsilon) = c(\epsilon \to v) = 1 - \rho$	$c(u \to v) = \begin{cases} 0, & \text{if } \mu(u) = \mu(v) \\ 2 \cdot (1 - \rho), & \text{otherwise} \end{cases}$
Edges	$c(e_1 \to \epsilon) = c(\epsilon \to e_2) = \rho$	$c(e_1 \to e_2) = \begin{cases} 0, & \text{if } \nu(e_1) = \nu(e_2) \\ 2 \cdot \rho, & \text{otherwise.} \end{cases}$

given operation constitutes a large deformation between the two involved graphs or not. Between similar graphs there should exist an inexpensive edit path, while dissimilar graphs are characterized by an edit path with high cost. The edit distance between two graphs is thus defined as the cost of the edit path with the minimum cost among all possible edit paths between two graphs.

The exact computation of the edit distance is a computationally hard task and many approximations have been proposed in the literature. In this work, we use the suboptimal approach of [8] where an approximate solution of graph edit distance is provided by means of solving the assignment problem of nodes of one graph to nodes of the other. A cost matrix regarding the substitution of the local structure of every node of the source graph by the local structure of every other node in the target graph is built. Then the optimal assignment is extracted by the Munkres' algorithm, and an edit path can be inferred from this assignment.

As a prerequisite, we need to assign costs to every edit operation between graph elements, i.e., nodes and edges. In particular, in this work we focus on the same cost function used in [1], where substitutions of nodes and edges with different labels are heavily penalized, forcing the (sub)optimal path to, first, delete the source node (or edge) and then insert the target node (or edge). Formally, deleting or inserting a given node (or edge) has a constant cost c, while substituting it has at least twice that cost if the corresponding labels are different. Without loss of generality, we set $c = 1$. Furthermore, we assume null cost of substituting two nodes (or edges) with the same label. In order to weight the node operations against those on the edges we introduce a parameter $\rho \in [0, 1]$ and multiply the node costs by $1 - \rho$ and the edge costs by ρ. The resulting cost function is summarized in Table 1.

3   Attribute Statistics Based Embedding

Consider a set of graphs $\mathcal{G} = \{g_1, \ldots, g_N\}$, with $g_i = (V_i, E_i, \mu_i, \nu_i)$ being the ith graph in the set with labelling alphabet L_{V_i} for the nodes and L_{E_i} for the edges. We assume that all graphs in $\mathcal{G}$ have the same labelling alphabets, this is $L_{V_i} = L_{V_j}$ and $L_{E_i} = L_{E_j}$ for all $i, j \in \{1, \ldots, N\}$. We do not assume, however, that each node and edge label necessarily occurs in each graph. Let $L_V = \{\alpha_1, \ldots, \alpha_p\}$ and $L_E = \{\omega_1, \ldots, \omega_q\}$ be the discrete common labelling alphabets.

For each graph $g = (V, E, \mu, \nu) \in \mathcal{G}$, we define p unary features measuring the number of times each label in L_V appears in the graph, this is

$$U_i = \#(\alpha_i, g) = |\{v \in V \mid \alpha_i = \mu(v)\}|, \quad \forall i \in \{1, \ldots, p\}. \tag{1}$$

We also define $\frac{1}{2} \cdot q \cdot p \cdot (p+1)$ binary features counting the frequency of an edge with a specific label (ω_k) between two nodes with two given labels (α_i and α_j). Formally,

$$\begin{aligned} B_{ij}^k &= \#([\alpha_i \leftrightarrow \alpha_j]_{\omega_k}, g) \\ &= |\{e = (u, v) \in E \mid \alpha_i = \mu(u) \wedge \alpha_j = \mu(v) \wedge \omega_k = \nu(e)\}| \end{aligned} \tag{2}$$

where $k \in \{1, \ldots, q\}$ and $1 \leq i \leq j \leq p$. These features describe the local structure of every graph in terms of how frequently a simple substructure —an edge with a given label between two given node labels— occurs in a given graph.

These two sets of features can be combined in order to give a more global structural representation of the graphs by bringing together various pieces of local information. Formally, we define the embedding of graphs in the following way.

Definition 1 (Graph Embedding). *Given a graph* $g \in \mathcal{G}$, *let* $\varphi_n(g)$ *and* $\varphi_e(g)$ *be the vectors*

$$\varphi_n(g) = \left(\{U_i\}_{1 \leq i \leq p} \right) \tag{3}$$

$$\varphi_e(g) = \left(\{B_{ij}^k\}_{1 \leq i \leq j \leq p}^{1 \leq k \leq q} \right) \tag{4}$$

where U_i *and* B_{ij}^k *are defined in Eqs. (1) and (2), respectively. The embedding of graph* g *is defined as the concatenation of these two vectors,*

$$\varphi(g) = [\varphi_n(g) \; \varphi_e(g)]. \tag{5}$$

The above definition has been proved successful in our previous work [5]. In the current paper, we go one step further and assign a different weight to the node related vector $\varphi_n(g)$ and the edge related vector $\varphi_e(g)$. This leads to a generalized distance between the map of two graphs, where the information included in the nodes can be weighted differently from the information included in the edges. Given two graphs g_1 and g_2, we define the vectorial distance between them by

$$D(g_1, g_2) = (1 - \alpha) \cdot d_{L_1}(\varphi_n(g_1), \varphi_n(g_2)) + \alpha \cdot d_{L_1}(\varphi_e(g_1), \varphi_e(g_2)), \tag{6}$$

where $\alpha \in [0, 1]$ and $d_{L_1}(\cdot, \cdot)$ is the L_1 distance $d_{L_1}(x, y) = \sum_{i=1}^n |x_i - y_i|$. Clearly, the case $\alpha = 0.5$ is identical to the scenario in [5]. Now parameter α of Eq. (6) can be related to parameter ρ of the edit distance introduced in Section 2. As a matter of fact, Eq. (6) emulates the edit operations defined by the cost function of the edit distance. As described above, the cost function maintains all nodes and edges with identical labels, but deletes and subsequently inserts all nodes and edges with different labels. Concerning the features we have defined, these operations translate into checking the absolute differences between vector coordinates. Note that the distance of Eq. (6) can alternatively be obtained if we would first weight both components of vector in Eq. (5) with $1 - \alpha$ and α, respectively, and then compute the L_1 distance between the weighted vectors.

The parameter α measures the strength we give to the components of $\varphi_n(g)$ relative to $\varphi_e(g)$. In this way, there is a clear resemblance with ρ which weights the cost of operations on the nodes relative to the cost of operations on the edges. In Section 4.2, we experimentally check for the correlation of these parameters. From the definitions given above, it follows that the pair $(\rho, \alpha) = (0, 0)$ will result in a correlation coefficient equal to 1.

4 Experiments

4.1 Databases

We work with four datasets of discretely attributed graphs. These datasets are divided into two categories: object image datasets and molecule datasets. The object images are subsets of the ALOI and ODBK collections [4,11]. Images are segmented and a region adjacency graph is built, where nodes are labelled with a color name of the color naming theory and edges are labelled according to whether the common border of two adjacent regions is *short*, *medium* or *long*.

The molecule datasets are the AIDS and MUTAG collections from the IAM repository [7]. Nodes correspond to atoms labelled with the corresponding chemical element and edges represent chemical bonds with the corresponding covalent number.

All four dataset are divided into a training, a validation and a test sets. In the following, we will use the training and the validation sets for computing pairwise distances. The test sets are not used.

4.2 Distance Correlation

Given a pair of values (ρ, α), we compute the sets of all pairwise graph distances X_ρ and Y_α between all graphs in the training set and all graphs in the validation set, using parameter value ρ for the edit distance and parameter value α for the embedding distance. For these two sets of distance values, X_ρ and Y_α, we compute the correlation coefficient by

$$C_{(\rho,\alpha)} = \frac{cov(X_\rho, Y_\alpha)}{\sigma_{X_\rho} \sigma_{Y_\alpha}}, \tag{7}$$

where $cov(X_\rho, Y_\alpha)$ is the covariance between distributions X_ρ and Y_α, and σ_{X_ρ} and σ_{Y_α} are the corresponding standard deviations. We compute such a coefficient for all pairs $(\rho, \alpha) \in [0, 1]^2$ and plot the corresponding 3D functions and correlation maps. Results can be seen in Fig. 1 (because of limited space we omit the ALOI and MUTAG cases but their behavior is very similar to that of ODBK and AIDS, and thus all discussions are valid for them as well).

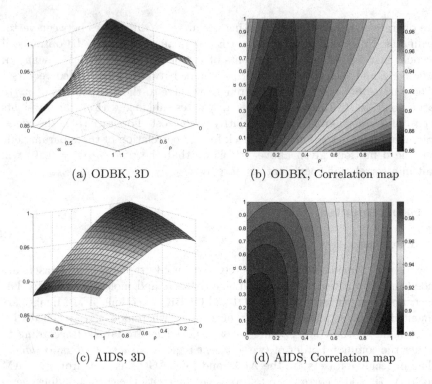

(a) ODBK, 3D   (b) ODBK, Correlation map

(c) AIDS, 3D   (d) AIDS, Correlation map

Fig. 1. Correlation values as a function of the weighting parameters

First of all, we note how values close to $(\rho, \alpha) = (0, 0)$ have, both in the object and molecule datasets, a high correlation coefficient. This confirms that the embedding features under the L_1 metric replicate the edit distances when node information is considered as more relevant than edge information. If this is the case in the underlying application, we suggest to use the attribute statistics based graph embedding rather than working with graph edit distance because, first, the relative graph distribution is well maintained and, second, the computation efficiency is much higher.

In Fig. 1, we can also observe the biased effect of the correlation values with respect to the ideal case, where a diagonal behavior should be observed. The explanation for the biased relation is the fact that the edge-based embedding features still keep quite some information of the node labels. In particular, the co-occurrence of a certain pair of node labels at the end of an edge tells us that these particular node labels do appear in the graph. Therefore, it is clear that considering edge-based features only, the embedding representation still keeps information about the node attributes. As a consequence of this phenomenon, the correlation of the embedding distances for $\alpha = 1$ is maximized by values $\rho \simeq 0.2$, suggesting that 80% of the node information in the graph domain is still included in the embedding representation when only edge-based features are considered.

In Fig. 1, when both ρ and α tend to 1, low correlation values result. This might be explained by the fact that the edit distance computation looks for an edit path that completely disregards the information of the nodes. Thus, since edge-based embedding features still keep some of this information, the behavior of distances in both domains becomes different.

Also worth noting is the shape of the correlation regions, which is more ellipse-like in the molecule datasets than in the object datasets. This observation has an interpretation in terms of how important the actual structural configuration of graphs is in each dataset. In the molecule datasets edge information is more salient than in the object datasets. The more weight we put on the edge-based features ($\alpha \to 1$) the faster the correlation values for $\rho \simeq 0.2$ descend, which means that edge-based features are less correlated with the node information in the graph domain and thus we should put more attention on the edges.

4.3   Classifier Correlation

Another way to check how well the edit distances are reproduced in the embedding space is to see how a distance based classifier performs. In this paper, we use a kNN classifier with both ways of computing the distances between graphs and look for the difference in performance. In particular, we compare the performance of the classifier based on the distances in the vector space for all values of α with the performance of the classifier using the edit distances in the graph domain for those values of ρ that maximize the correlation for every value of α. We indicate by

$$\bar{\rho}_\alpha = \operatorname*{argmax}_{\rho \in [0,1]} C_{(\rho,\alpha)} \tag{8}$$

the ρ value that maximizes the correlation coefficient for a given α value. In Fig. 2 we show the corresponding classification curves on the validation sets. The x-axis shows the range of parameter α and the y-axis the classification rates. In particular, the results' curves for the embedding distances are stretched in such a way that the value $\alpha = 1$ coincides with the ρ values maximizing its correlation, and the corresponding intermediate values of α are maximally correlated with the respective ρ values in the curve. This is, for each value of α, the corresponding result of the edit distance curve is that of $\bar{\rho}_\alpha$. We also show the corresponding scatter plots of the accuracies of both classifiers, for all pairs $(\alpha, \bar{\rho}_\alpha)$ and give the correlation coefficient for these scatter plots on top of Figs. 2(b) and 2(d).

With regards to the correlation of the classifiers, we observe a great degree of similarity of both curves, supporting the hypothesis of a high correlation. We notice that the classifiers' correlation is higher for the object datasets than for the molecule ones. This result is explained by the same reason we have been discussing before. The fact that the embedding features correlate with edit distance whenever the node information of graphs is actually relevant makes the classifiers perform in similar ways. On the other hand, the molecules need some more attention on the edge structure and therefore the edit distance and the embedding based distance differ more.

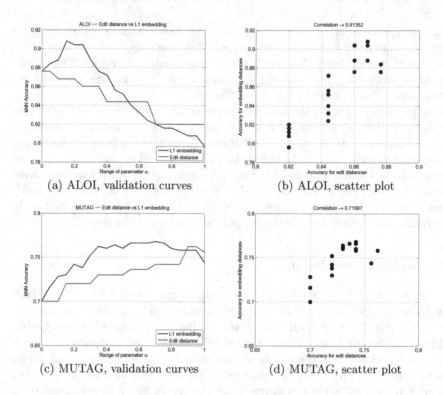

(a) ALOI, validation curves   (b) ALOI, scatter plot

(c) MUTAG, validation curves   (d) MUTAG, scatter plot

Fig. 2. Classifiers performance: L_1 embedding as a function of α, edit distance as a function of $\overline{\rho}_\alpha$, and scatter plots of the accuracies of all pairs $(\alpha, \overline{\rho}_\alpha)$

With respect to the performance of the classifier on the molecules dataset, we observe how the embedding curve obtains its highest result for an intermediate value of the parameter, thus confirming that here edges have higher importance than in the objects case, where the highest result is obtained by a value of the parameter closer to 0. Another point supporting this idea is the fact that for the object datasets the case $\alpha = 0$ gives a better result than that of $\alpha = 1$, and for the molecule datasets this is the other way around.

As a final comment, we note how in most of the cases the performance of the embedding classifier for a given α outperforms that of the edit distance classifier for $\overline{\rho}_\alpha$. This suggests that whenever both ways of computing distances are regarding the same type of information, the embedding distances are more capable to distinguish among graph categories. Because of this observation and because of its much higher efficiency, the use of the embedding methodology for graph comparison is recommended.

5   Conclusions

In this work we have established a relation between graph edit distance and the L_1 vectorial distance in the attribute statistics embedding space. It has been shown that under a special class of cost functions, where node and edge label insertions and deletions are favored over substitutions, there is a close relation between the graph edit distance and the L_1 distance of the corresponding vectors obtained through graph embedding. Our formal analysis has been confirmed in a series of experiments. We have experimentally shown that there exists a high correlation between both types of graph distances and between the corresponding classifiers, provided that corresponding parameter values are chosen for both distances. The analysis provided in this paper may help in developing a better understanding of label statistics based embedding [5], which has been demonstrated to perform very well in practice but has been lacking, until now, a more rigorous formal investigation of its properties.

The current paper is limited to graphs with discrete labels. However, in future, a similar study for the case of continuous attributed graphs is planned. In addition, it would be interesting to exploit the embedding features to derive necessary conditions for subgraph isomorphism in terms of component-wise relations between the corresponding vectorial representations of graphs.

References

1. Bunke, H.: On a relation between graph edit distance and maximum common subgraph. Pattern Recognition Letters 18(8), 689–694 (1997)
2. Bunke, H., Allermann, G.: Inexact graph matching for structural pattern recognition. Pattern Recognition Letters 1, 245–253 (1983)
3. Conte, D., Foggia, P., Sansone, C., Vento, M.: Thirty years of graph matching in pattern recognition. International Journal of Pattern Recognition and Artificial Intelligence 18(3), 265–298 (2004)
4. Geusebroek, J.M., Burghouts, G.J., Smeulders, A.W.M.: The Amsterdam library of object images. International Journal of Computer Vision 61(1), 103–112 (2005)
5. Gibert, J., Valveny, E., Bunke, H.: Graph embedding in vector spaces by node attribute statistics. Pattern Recognition 45(9), 3072–3083 (2012)
6. Ren, P., Wilson, R., Hancock, E.: Graph characterization via Ihara coefficients. IEEE Transactions on Neural Networks 22(2), 233–245 (2011)
7. Riesen, K., Bunke, H.: IAM Graph Database Repository for Graph Based Pattern Recognition and Machine Learning. In: da Vitoria Lobo, N., et al. (eds.) S+SSPR 2008. LNCS, vol. 5342, pp. 287–297. Springer, Heidelberg (2008)
8. Riesen, K., Bunke, H.: Approximate graph edit distance computation by means of bipartite graph matching. Image and Vision Computing 27, 950–959 (2009)
9. Riesen, K., Bunke, H.: Graph Classification and Clustering Based on Vector Space Embedding. World Scientific (2010)
10. Sanfeliu, A., Fu, K.S.: A distance measure between attributed relational graphs for pattern recognition. IEEE Transactions on Systems, Man, and Cybernetics - Part B 13(3), 353–363 (1983)
11. Tarr, M.J.: The object databank, http://www.cnbc.cmu.edu/tarrlab/stimuli/objects/index.html

Approximate Axial Symmetries from Continuous Time Quantum Walks

Luca Rossi[1], Andrea Torsello[1], and Edwin R. Hancock[2]

[1] Department of Environmental Science, Informatics and Statistics,
Ca' Foscari University of Venice, Italy
[2] Department of Computer science, University of York, UK

Abstract. The analysis of complex networks is usually based on key properties such as small-worldness and vertex degree distribution. The presence of symmetric motifs on the other hand has been related to redundancy and thus robustness of the networks. In this paper we propose a method for detecting approximate axial symmetries in networks. For each pair of nodes, we define a continuous-time quantum walk which is evolved through time. By measuring the probability that the quantum walker to visits each node of the network in this time frame, we are able to determine whether the two vertices are symmetrical with respect to any axis of the graph. Moreover, we show that we are able to successfully detect approximate axial symmetries too. We show the efficacy of our approach by analysing both synthetic and real-world data.

Keywords: Complex Network, Symmetry, Quantum Walk.

1 Introduction

The study of complex networks [1] has recently attracted considerable interest because of the large variety of complex systems that can be modeled and analysed using graphs. A non-exhaustive list of examples includes metabolic networks [2], protein interactions [3], brain networks [4], vascular systems [5], scientific collaboration networks [6] and road maps [7]. Properties such as small-worldness and the power-law distribution of vertex degrees [1] have been observed in several real-world networks, suggesting a marked difference with Erdös-Rényi random graphs [8].

More recently there has been some interest in characterizing the presence of symmetries in networks [9] [10]. Recall that, given a graph $G = (V, E)$, an automorphism is a permutation σ of the set of vertices V of the graph which preserve the adjacency relations, i.e. if $(u, v) \in E$ then $(\sigma(u), \sigma(v)) \in E$. Hence we can view the group of automorphisms $\mathrm{Aut}(G)$ of a graph as a representation of its symmetries. MacArthur et al. [9] observe that many real-world graphs possess a very large automorphism group, in contrast to classical random graph models. In particular the authors observe the presence of a certain number of small symmetric subgraphs, such as tree-like or clique-like structures, and relate this to the redundancy and thus robustness of real-world networks. Note however

G.L. Gimel' farb et al. (Eds.): SSPR & SPR 2012, LNCS 7626, pp. 144–152, 2012.

that the problem of finding the set of automorphisms of a graph is actually an instance of the graph isomorphism problem, and thus it belongs to the NP class. Xiao et al. [10] study the origin of symmetry in real-world networks. In common with [9], their work is based on the analysis of local symmetric motifs such as symmetric bicliques, i.e. an induced complete bipartite subgraph, denoted as K_{V_1,V_2}, in which every vertex of V_1 is connected to every vertex of V_2. Their analysis reveals that the symmetry of complex networks is a consequence of a particular linkage pattern, where vertices with similar degrees tend to share common neighbors. It is also worth mentioning the work of Mowshowitz [11], which links the complexity of a graph to the entropy of the distribution of symmetric orbits.

Recently there has been a considerable interest in quantum walks, as an alternative to the well studied classical random walks. Although similar in its definition, the quantum walk is remarkably different from its classical counterpart. Most notably, its evolution is governed by a unitary matrix instead of a stochastic one and the state vector is complex valued instead of real valued. This in turn produces interference effects which yield completely different probability distributions on the graph. Moreover, these interference effects seem to be enhanced by the presence of symmetrical motifs in the graph. Emms et al. [12] showed that quantum commute time embeddings are tightly related to the presence of symmetries. In particular the authors found that the embedding co-ordinates of nodes are degenerate in dimensions that correspond to global symmetries. In a related paper, Emms et al. [13] demonstrate how to lift the cospectrality of strongly regular graphs using the third power of the support matrix derived from a discrete time quantum walk. Thus it seems reasonable to investigate the use of quantum walks as a means of detecting symmetries in networks.

In real-world data, however, we have to deal with the presence of noise, which will eventually break the symmetries of the network. In this paper we propose a new method for detecting the approximate axial symmetries of a graph using continuous time quantum walks. The remainder of this paper is organised as follows. First we review the definition of the continuous time quantum walk on a graph, then we show how to exploit the interference patterns to detect both exact and approximate axial symmetries and then we briefly discuss the proposed algorithm. Finally our approach is evaluated on a set of synthetic graphs and real-world networks.

2   Continuous-Time Quantum Walks

Quantum walks are the quantum analogue of classical random walks [14]. In this paper we consider only continuous-time quantum walks, as first introduced by Farhi and Gutmann in [15].

As in the classical random walk, given a graph $G = (V, E)$, the state space of the continuous-time quantum walk defined on G is the set of the vertices V of the graph. Unlike the classical case, where the evolution of the walk is governed by a stochastic matrix (i.e. a matrix whose columns sum to unity), in the quantum case the dynamics of the walker is governed by a complex unitary matrix i.e., a matrix that multiplied by its conjugate transpose yields the identity matrix. Hence the evolution of the quantum walk is reversible, which implies that quantum walks are non-ergodic and do not possess a limiting distribution. Using Dirac notation, we denote the basis state corresponding to the walk being at vertex $u \in V$ as $|u\rangle$. A general state of the walk is a complex linear combination of the basis states, such that the state of the walk at time t is defined as

$$|\psi_t\rangle = \sum_{u \in V} \alpha_u(t) |u\rangle \tag{1}$$

where the amplitude $\alpha_u(t) \in \mathbb{C}$ and $|\psi_t\rangle \in \mathbb{C}^{|V|}$ are both complex.

At each point in time the probability of the walker being at a particular vertex of the graph is given by the square of the norm of the amplitude of the relative state. More formally, let X^t be a random variable giving the location of the walker at time t. Then the probability of the walker being at the vertex u at time t is given by

$$\Pr(X^t = u) = \alpha_u(t)\alpha_u^*(t) \tag{2}$$

where $\alpha_u^*(t)$ is the complex conjugate of $\alpha_u(t)$. Moreover $\sum_{u \in V} \alpha_u(t)\alpha_u^*(t) = 1$ and $\alpha_u(t)\alpha_u^*(t) \in [0,1]$, for all $u \in V$, $t \in \mathbb{R}^+$.

We now introduce the evolution operator of the quantum walk. First though, recall that the adjacency matrix of the graph G has elements

$$A_{uv} = \begin{cases} 1 \text{ if } (u,v) \in E \\ 0 \text{ otherwise} \end{cases} \tag{3}$$

Let D be the diagonal degree matrix with elements $d_u = \sum_{v=1}^n A(u,v)$, where $n = |V|$ is the number of vertices of the graph. The Laplacian of G is then defined as the degree matrix minus the adjacency matrix, i.e. $L = D - A$.

The evolution of the walk is then given by Schrödinger equation, where we take the Hamiltonian of the system to be the graph Laplacian, which yields

$$\frac{d}{dt} |\psi_t\rangle = -iL |\psi_t\rangle \tag{4}$$

Given an initial state $|\psi_0\rangle$, we can solve Equation 4 to determine the state vector at time t

$$|\psi_t\rangle = e^{-iLt} |\psi_0\rangle \tag{5}$$

Given the Laplacian matrix we can compute its spectral decomposition $L = \Phi\Lambda\Phi^T$, where Φ is the $n \times n$ matrix $\Phi = (\phi_1|\phi_2|...|\phi_n)$ with the ordered eigenvectors as columns and $\Lambda = \text{diag}(\lambda_1, \lambda_2, ..., \lambda_n)$ is the $n \times n$ diagonal matrix with the ordered eigenvalues as elements, such that $0 = \lambda_1 \leq \lambda_2 \leq ... \leq \lambda_n$.

Using the spectral decomposition of the graph Laplacian and the fact that $\exp[-iLt] = \Phi^T \exp[-i\Lambda t]\Phi$ we can finally write

$$|\psi_t\rangle = \Phi^T e^{-i\Lambda t}\Phi |\psi_0\rangle \tag{6}$$

3   Approximate Axial Symmetries Detection

In order to detect the axial symmetries of a graph, we exploit the interference properties exhibited by quantum walks. In particular, our analysis will rely on the destructive interference which arises when a symmetrical structure is present. Note, however, that we are interested in both exact and approximate axial symmetries. In fact, due to the presence of different noise sources, most real-world networks are not perfectly symmetric. In consequence, the search for exact axial symmetries would fail to discover those global symmetries which are more likely to be affected by noise. On the other hand, we argue that our algorithm is capable of detecting both exact and approximate axial symmetries, thus making it suitable for real-world network analysis.

3.1   Methodology

Given a pair of vertices u, v, we initialise the quantum walk as follows

$$\alpha_j(0) = \begin{cases} +\frac{1}{\sqrt{2}} & \text{if } j = u \\ -\frac{1}{\sqrt{2}} & \text{if } j = v \\ 0 & \text{otherwise} \end{cases} \tag{7}$$

If u and v are symmetrical with respect to a symmetry axis A, then it is easy to show that when the walk is initialised as above we have $\alpha_w(t) = 0$, $\forall w \in A$ and $\forall t$.

Theorem 1. *If u, v are symmetrical with respect to a symmetry axis A and $\alpha_u(0) = -\alpha_v(0)$, then $\alpha_w(t) = 0$, $\forall w \in A$ and $\forall t$.*

Sketch of Proof. Assume that the graph G has at least one symmetry axis A, and $w \in A$ is a vertex of G. Then, for each path from u to w, there will be a symmetrical path from v to w. As a consequence of this, both walkers starting from u and from v will arrive at w at the same time t. Moreover, since we initialised the amplitudes such that $\alpha_u(0) = -\alpha_v(0)$, the two walkers will be in antiphase and thus their contribution to the observation probability of w will cancel out, i.e. $\Pr(X^t = w) = 0, \forall t$.

Note that due to its oscillatory behaviour, the observation probability of the quantum walker on any node might temporarily collapse to zero. However, only if the vertex belongs to a symmetry axis its observation probability will remain constantly null.

Hence the procedure to detect the symmetry axes in a graph is as follows. First we define a quantum walk according to Equation 7. We then let the quantum walk evolve for a time interval T and we measure the total observation probability $\pi_w = \sum_{t \in T} \Pr(X^t = w)$ for each node $w \neq u, v$ during T. If $\pi_w = 0$, then we say that the node belongs to the symmetry axis. We repeat this procedure for each pair of nodes of the graph, and we detect all the exact symmetry axes of the network. Finally, we can estimate the symmetry axes sizes by counting, for a given pair of nodes, the number of nodes w where $\pi_w = 0$.

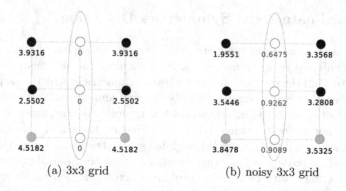

(a) 3x3 grid   (b) noisy 3x3 grid

Fig. 1. When applied to a 3x3 grid, our algorithm is able to detect all its 4 symmetry axes. Moreover, even if some noise is present, a suitable choice of τ will still allow us to detect the original axes

Figure 1(a) shows the result of running the proposed algorithm on a 3x3 grid. If we initialise the bottom corners of the graph with equal but opposite amplitudes, we see that the observation probabilities of the vertices belonging to the vertical axis remain constantly equal to zero. In particular here we let the quantum walk evolve from $t = 0$ to $t = 10$, sampling this time interval 100 times uniformly. For each axial node w we have $\pi_w = 0$, while the nodes with initial non-zero amplitudes have the highest total observation probability. The remainder of the vertices of the graph have $\pi_w > 0$, and symmetrically placed nodes with respect to the vertical axis share the same π_w. Note that we detect the same axis if we choose another pair of vertices which are symmetrical with respect to this axis, such as the top corners.

Moreover, we argue that we can detect approximate axial symmetries as well, i.e., axial symmetries which are affected by noise, by selecting those vertices in which $\pi < \tau$, where τ is a suitably chosen threshold. Figure 1(b) shows what happens if we add some noise the the 3x3 grid by removing one edge. Note that due to the small size of the graph the deletion of one single edge can actually deeply change the graph structure. Although the total observation probability has clearly changed and it is now non-zero everywhere, we still see that it is lower on the vertices corresponding to the vertical axis. Hence, by choosing a suitable value for τ, we are still able to detect this approximate symmetry.

4   Experimental Results

In this section, we validate the proposed approach by performing a series of experiments on both synthetic data and real-world data.

4.1   Synthetic Data

The synthetic data is composed of Erdös-Rényi random graphs [8], small-world graphs, scale-free graphs, stochastic Kronecker graphs [16] (which exhibit both

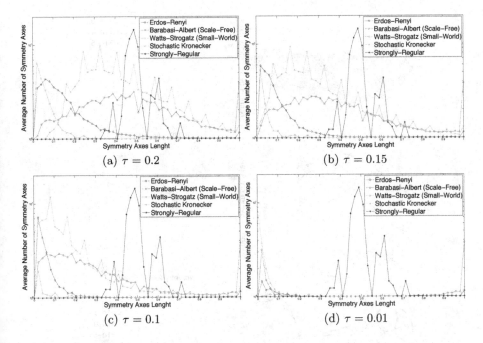

Fig. 2. Symmetry axes distribution. Note that as the threshold varies, the shape of the strongly-regular graphs distribution remains unaltered, as the symmetry present in this category are all exact.

small-world and scale-free properties), and strongly regular graphs. A regular graph with ν vertices and degree k is said to be strongly regular if there are two integers λ and μ such that every two adjacent vertices have λ common neighbours and every two non-adjacent vertices have μ common neighbors. We choose strongly regular graphs because they are known to be highly symmetric and this should be reflected in the experimental results.

For each graph in the dataset, we compute its symmetry axes together with their sizes, as explained in the previous section. Figure 2 shows the distribution of the symmetry axial length for each type of network, for different choices of the threshold τ. Note that local symmetries correspond to larger axes, since the axis size is equal to the number of nodes of the graph minus the size of the symmetric orbit, which in the case of a local symmetry is clearly small. On the other hand, a global symmetry will correspond to smaller symmetry axis. In other words, a left peaked distribution indicates the presence of global symmetries, while a right peaked distribution indicates the presence of local symmetries.

Note that the distribution for the strongly-regular graphs remains unaltered when we change τ. This is because the graphs in this category possess exact symmetries, due to their regular structure. Hence the probability of the walker being found at a node belonging to a symmetry axis is exactly zero and we recover the same axes independently of the threshold value. Note, moreover,

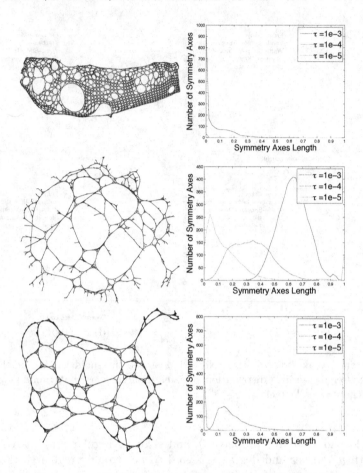

Fig. 3. Sample cities embeddings and their corresponding axes length distributions. Note how different layouts give rise to different symmetries.

that the high number of symmetry axes belonging to this class of graphs is exactly what we would expect given the high degree of symmetry displayed by strongly-regular graphs.

As for the other network models, Figure 2 shows that the number of exact symmetries is clearly lower. We observe the presence of a high number of exact local symmetries in the scale-free graphs, which are probably due to the presence of small trees rooted in a hub node. It is interesting to note that the behaviour of the stochastic Kronecker graphs, which possess both scale-free and small-world properties, seems to be dominated by their scale-free behaviour, although the number of local symmetries is clearly reduced. More generally, Figure 2 shows that we can easily separate graphs belonging to different network models on the basis of their symmetry axes distributions.

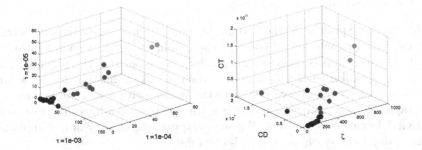

Fig. 4. City maps embeddings obtained using the mean of the symmetry axes plot distributions for three different values of τ as co-ordinates (left). The cities are coloured according to the labels induced by k-means. The labeling is consistent when the embedding is done using Communicability Distance [1], the classical Commute Time and the zeta function as co-ordinates (right), but there are still some differences which suggest that the information extracted with our algorithm is meaningful and novel.

4.2 Real-World Data

Road networks are a typical example of technological networks, i.e. man-made networks designed for the distribution of resources. Other examples include power grids, airline routes, river networks and the Internet. In this paper we apply our algorithm to a dataset of 33 city maps. For each city, we compute the approximate symmetry axes and their length. Figure 3 shows the embeddings of 3 different cities and the corresponding plots. We observe that different layouts of the cities give rise to different symmetries. As expected, the first city, which shows a very regular grid-like structure, seems to possess only approximate global symmetries, but no perfect symmetries. On the other hand, the second city displays a wide variety of approximate symmetry axes, and a few exact local symmetries, similarly to those displayed by the scale-free network model. A visual inspection of its embedding confirms the presence of several small hubs, as predicted. Finally, the third city shows a large number of local symmetries, which arise as a consequence of its very particular linkage pattern, where pairs of nodes are connected in an quasi-bipartite fashion.

In order to take the analysis on step further, we describe each city with a three-dimensional feature vector whose co-ordinates are respectively the means of the axes length distributions for decreasing thresholds. The resulting embedding is shown in Figure 4 (left). Here we can clearly see 3 well separated clusters, where the labels of the clusters have been assigned using k-means. Note that each city of Figure 3 actually belongs to a different cluster of Figure 4. Moreover, we compare our embedding with the one obtained using the Communicability Distance [1], the classical Commute Time and the zeta function $\zeta(s) = \sum_{\lambda_i \neq 0} \lambda_i^{-s}$ associated with the Laplacian eigenvalues as co-ordinates. Figure 4 (right) shows that the result is still quite consistent with our labeling, although we can clearly see some differences. This indicates that the our algorithm is indeed extracting some meaningful and novel information from the data.

5 Conclusions

In this paper we proposed a method for detecting approximate axial symmetries in networks. For each pair of nodes of the graph, we define a continuous-time quantum walk which is evolved through time. By measuring the probability of the quantum walker to visit each node of the network in this time frame, we are able to tell whether if two vertices are symmetrical with respect to any axis of the graph. Moreover, we showed that we are able to successfully detect approximate axial symmetries as well. We demonstrated the efficacy of our approach by analysing both synthetic and real-world data.

Acknowledgments. Edwin Hancock was supported by a Royal Society Wolfson Research Merit Award.

References

1. Estrada, E.: The Structure of Complex Networks. Oxford University Press (2011)
2. Jeong, H., Tombor, B., Albert, R., Oltvai, Z., Barabási, A.: The large-scale organization of metabolic networks. Nature 407, 651–654 (2000)
3. Ito, T., Chiba, T., Ozawa, R., Yoshida, M., Hattori, M., Sakaki, Y.: A comprehensive two-hybrid analysis to explore the yeast protein interactome. Proceedings of the National Academy of Sciences 98, 4569 (2001)
4. Sporns, O.: Network analysis, complexity, and brain function. Complexity 8, 56–60 (2002)
5. West, G., Brown, J., Enquist, B.: A general model for the structure, function, and allometry of plant vascular systems. Nature 400, 664–667 (1999)
6. Newman, M.: Scientific collaboration networks. i. network construction and fundamental results. Physical Review E 64, 016131 (2001)
7. Kalapala, V., Sanwalani, V., Moore, C.: The structure of the united states road network. University of New Mexico (2003) (preprint)
8. Erdös, P., Rényi, A.: On random graphs. Publ. Math. Debrecen 6, 290–297 (1959)
9. MacArthur, B., Sánchez-García, R., Anderson, J.: Symmetry in complex networks. Discrete Applied Mathematics 156, 3525–3531 (2008)
10. Xiao, Y., Xiong, M., Wang, W., Wang, H.: Emergence of symmetry in complex networks. Phys. Rev. E 77, 066108 (2008)
11. Mowshowitz, A.: Entropy and the complexity of graphs: I. an index of the relative complexity of a graph. Bulletin of Mathematical Biology 30, 175–204 (1968)
12. Emms, D., Wilson, R., Hancock, E.: Graph Embedding Using Quantum Commute Times. In: Escolano, F., Vento, M. (eds.) GbRPR. LNCS, vol. 4538, pp. 371–382. Springer, Heidelberg (2007)
13. Emms, D., Severini, S., Wilson, R., Hancock, E.: Coined Quantum Walks Lift the Cospectrality of Graphs and Trees. In: Rangarajan, A., Vemuri, B.C., Yuille, A.L. (eds.) EMMCVPR 2005. LNCS, vol. 3757, pp. 332–345. Springer, Heidelberg (2005)
14. Kempe, J.: Quantum random walks: an introductory overview. Contemporary Physics 44, 307–327 (2003)
15. Farhi, E., Gutmann, S.: Quantum computation and decision trees. Physical Review A 58, 915 (1998)
16. Mahdian, M., Xu, Y.: Stochastic Kronecker Graphs. In: Bonato, A., Chung, F.R.K. (eds.) WAW 2007. LNCS, vol. 4863, pp. 179–186. Springer, Heidelberg (2007)

A Clustering-Based Ensemble Technique for Shape Decomposition

Sergej Lewin, Xiaoyi Jiang, and Achim Clausing

Department of Mathematics and Computer Science
University of Münster
Einsteinstrasse 62, 49149 Münster, Germany
{slewin,xjiang,achim.clausing}@uni-muenster.de

Abstract. Ensemble techniques have been very successful in pattern recognition. In this work we investigate ensemble solution for shape decomposition. A clustering-based approach is proposed to determine a final decomposition from an ensemble of input decompositions. A recently published performance evaluation framework consisting of a benchmark database with manual ground truth together with evaluation measures is used to demonstrate the benefit of the proposed ensemble technique.

1 Introduction

Ensemble techniques have been very successful in pattern recognition. In addition to the classification problem [1], the fundamental fusion approach has been introduced to many other domains including strings [2], graphs [3], clusterings [4], and segmentations [5]. In this work we investigate ensemble solution of shape decomposition.

The shape decomposition methods from the literature have their respective strengths and weaknesses, and their performance varies for different shapes. There exists no dominating approach that outperforms the others for all shapes. The variety of decomposition results is exemplarily demonstrated in Fig. 1(a) with overlayed five decompositions (produced by five different algorithms). Some cuts cross each other while there are cuts, which are almost congruent.

Combination (fusion) is a useful technique for taking advantages of the strengths and compensating the weaknesses of different approaches. In this particular case, a simple union of all cuts from different decompositions is obviously not meaningful. Instead, it suggests itself that some cuts should be removed and almost congruent cuts should be unified. In this paper we present an ensemble technique for combining shape decompositions. Although there exist quite a number of shape decompositions algorithms (see [6–8] and the references therein), multiple decomposition combination has not been studied before.

The remainder of this paper is organized as follows. Section 2 describes our decomposition ensemble technique. A performance evaluation framework from [9] applied in this work is summarized in Section 3, followed by the evaluation results in Section 4. Finally, some discussions conclude this paper.

G.L. Gimel' farb et al. (Eds.): SSPR & SPR 2012, LNCS 7626, pp. 153–161, 2012.
© Springer-Verlag Berlin Heidelberg 2012

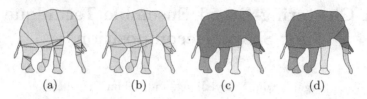

(a)   (b)   (c)   (d)

Fig. 1. (a) Overlayed five decompositions; (b) representative cuts of clusters; (c) combined decomposition with threshold 3 (majority voting); (d) combined decomposition with threshold 2

2   Combined Shape Decomposition Approach

Our combination approach consists of two steps. First, the set of all cuts from the input decompositions is clustered into groups of similar cuts and each cluster is represented by one single cut by averaging all cuts within it. Then, a selection procedure is performed to determine the cuts of the result decomposition. An outline of the decomposition ensemble algorithm is given in Fig. 2. It is invariant to transformations like scaling and rotation (although the baseline decomposition algorithms may be effected by such transformations).

2.1   Cut Clustering

The set of all cuts from the input decompositions can be partitioned into groups of similar ones by clustering. In case of agglomerative hierarchical clustering, for instance, each cut builds its own cluster at the very beginning. Then, the clusters are merged until the distance between any two clusters is larger than a threshold t_1. Depending on how the distance between clusters is computed, different types of hierarchical clustering can be distinguished. In our work the so-called unweighted pair group method with arithmetic mean is used. This distance between two clusters X and Y of size $|X|$ and $|Y|$, respectively, is defined by:

$$D(X,Y) = \frac{1}{|X| \cdot |Y|} \sum_{x \in X} \sum_{y \in Y} dist(x,y)$$

where $dist(x,y)$ is the distance between two cuts x and y, which can be computed based on distances between cut endpoints along shape boundary. The distance between two boundary points is defined by the minimal arc length of boundary curve between these points. Then, the distance between two cuts x (with endpoints a and b) and y (with endpoints c and d), respectively, can be simply defined as a sum of distances between corresponding endpoints:

$$dist(x,y) = \min\{pdist(a,c) + pdist(b,d), \ pdist(a,d) + pdist(b,c)\}$$

where $pdist$ is the distance along shape boundary.

Since after clustering each cluster contains similar cuts, we can determine one representative cut for a cluster by averaging all cuts in it. In order to guarantee

Input: n decompositions of a shape;
Output: combined decomposition;
Parameter: thresholds t_1 and t_2;

Cluster the cuts of all n input decompositions (t_1 needed);
Determine a representative cut for each cluster;
Compute the weight for each representative cut;
Select representative cuts based on the weights (t_2 needed);
Resolve intersecting selected cuts;

Fig. 2. Outline of ensemble decomposition algorithm

that the endpoints of the average cut lie on the shape boundary, the averaging is performed based on boundary parametrization. Assuming that $\gamma : [0, 1] \to \mathbb{R}^2$ is a boundary curve. The parameter $p \in [0, 1]$ describes the position of curve point $\gamma(p)$ relative to start point $\gamma(0)$. The endpoints of the average cut are not computed by averaging the positions of left (right) endpoints of the cuts in a cluster geometrically, but in the parameter space. For averaging the left endpoints $x_1, x_2, \ldots, x_k$ of k representative cuts, for instance, let's assume their corresponding parameter values be $p_1, p_2, \ldots, p_k$, then the average endpoint is $\gamma(\frac{1}{k} \sum_{i=1}^{k} p_i)$. Fig. 1(b) shows the representative cuts of the clusters computed from the cuts in Fig. 1(a).

2.2   Selection of Cuts

After clustering and computing representative cuts, the cuts for the combined decomposition should be selected. This selection step should reject perceptually unreasonable cuts and resolve intersections of cuts. This can be performed based on weights derived from clusters. Each representative cut represents one cluster consisting of one or more cuts. Since all involved shape decomposition methods try to mimic human being's behavior, it can be assumed that most of the cuts created are perceptually reasonable. Therefore, the more methods create one cut, the larger is the likelihood that this cut is perceptually reasonable and should thus be kept. At the same time, it can be surely assumed that the likelihood of a perceptually unreasonable cut, which is coevally produced by two different methods, is relatively small. According to these considerations, the representative cut of a cluster can be weighted by the number of input shape decompositions, whose cuts are contained in the cluster. Note that by this definition the weight is usually smaller than the number of cuts in a cluster, since two cuts from a decomposition can be clustered into the same cluster.

Given the weight of representative cuts, the selection of cuts can be performed by using thresholding. As a first thresholding scheme, the *majority voting* can be realized. In this case the threshold t_2 is equal to the half of the number of input shape decompositions to be combined. All representative cuts, which appear in at least 50% of shape decompositions, are selected. Alternatively, a smaller threshold t_2 can be used.

After the cut selection it can happen that some of the remaining cuts intersect each other. Such intersection can be resolved based on the same weights by simply removing the lower weighted cut of two crossing cuts.

The decomposition ensemble algorithm described above is summarized in Fig. 2. Fig. 1(c) and 1(d) show two combined decompositions created from the representative cuts in Fig. 1(b). The first one corresponds to the majority voting scheme, while the second one rejects all cuts created by one decomposition only. As expected, the majority voting rule results in a coarser shape decomposition' and a smaller threshold retains more representative cuts for a finer shape decomposition. In this particular case, the second combined decomposition contains all important cuts. In general, however, the lower the threshold value is, the higher likelihood is that some perceptually unreasonable cuts would be selected for the combined decomposition.

Our ensemble approach introduces two extra thresholding parameters t_1 and t_2. They can be optimized using training data, similarly as we have to deal with the parameters of the involved shape decomposition methods. This issue will be further discussed in Section 4.1.

3   Performance Evaluation Framework

In computer vision there exist today a lot of ground truth data and quantitative evaluations. However, shape decomposition is an exception. In this domain performance evaluation still remains qualitative. This is even true in recent publications (e.g. [10]). Typically, statements like "As we can see, ..., our method decomposes shapes into parts with high visual naturalness comparable to [x]" [10] are made[1].

Recently, a supervised quantitative performance evaluation framework was proposed [9], which consists of a benchmark database with manually specified ground truth (GT) and dissimilarity measures. A decomposition generated by an algorithm is compared with the GT decomposition of the same shape and quantitative dissimilarity measures are computed for performance comparison. We apply this framework to demonstrate the benefit of our ensemble technique. In the following its main components are briefly summarized.

3.1   Dissimilarity Measures

A dissimilarity measure is a distance function: $\gamma : D_S \times D_S \to \mathbb{R}$, where D_S denotes the set of all valid decompositions of a given shape S. It is reasonable to require that this distance function be semi-metric, i.e. it fulfils the following three conditions: $\forall x, y \in D_S$, 1) non-negativity: $\gamma(x, y) \geq 0$; 2) identity: $\gamma(x, y) = 0$ if and only if $x = y$; 3) symmetry: $\gamma(x, y) = \gamma(y, x)$. Several variants of distance

---

[1] Note that the quantitative comparison presented in [10] concerns the number of computed parts in the sense of data reduction, which is not related to the perceptual quality of decomposition.

measures have been discussed in [9]. In this work we apply the following two measures for experimental work.

Hamming Distance. A shape can be regarded as a non-rectangular image and a shape decomposition as a region-based segmentation accordingly. Thus, comparison of shape decompositions becomes that of comparing image segmentations. There exist a number of distance measures for this purpose [11]. One such measure is the Hamming distance (D_H). Considering two decompositions $A = \{A_1, \ldots, A_n\}$ and $B = \{B_1, \ldots, B_m\}$ of the same shape S, the directional Hamming distance between A and B is defined by: $D_H(A \to B) = \sum_i |A_i \setminus B_{k_i}|$, where B_{k_i}, $k_i = \arg\max_j |A_i \cap B_j|$, is the best fitting of A_i in B. The reversed distance $D_H(B \to A)$ can be similarly computed. Finally, the overall Hamming distance is given by:

$$D_H(A, B) = \frac{D_H(A \to B) + D_H(B \to A)}{2|S|}$$

Obviously, $H_D \in [0, 1]$ is semi-metric.

Jaccard Measure. If we interpret a shape as a set of pixels and a decomposition as a clustering, then the distance measures for comparing clusterings can be applied to quantify the difference between two shape decompositions. Several such measures have been proposed in the literature [11]. Particularly intuitive are those measures based on counting pairs. Given two decompositions A and B, we consider all pairs of pixels and count: a) n_{11}: #pairs of pixels, which are in the same part of A and in the same part of B; b) n_{10}: #pairs of pixels, which are in the same part of A and in different parts of B; c) n_{01}: #pairs of pixels, which are in different parts of B and in the same part of A; d) n_{00}: #pairs of pixels, which are in different parts of both A and of B. Several distance measures for comparing clusterings are based on these four counts. For instance, the Jaccard measure is defined by:

$$J(A, B) = \frac{n_{10} + n_{01}}{n_{10} + n_{01} + n_{11}}$$

Obviously, $J \in [0, 1]$ is semi-metric.

3.2 Benchmark Database

To perceptually evaluate shape decomposition methods, i.e. how well they mimic decompositions created by human beings, we need a database with manually specified GT. In [9] it is proposed to use the benchmark database from a large-scale psychological study [12], which fits the requirement perfectly and is publicly available. It consists of human-generated decompositions of 88 different shapes. Each shape is decomposed by 38 human subjects on average. Fig. 3 (upper row) shows overlayed decompositions (represented by cuts) of five shapes. As can be seen, a high degree of decomposition consistency was achieved among human subjects. Nevertheless, there are a lot of outliers among the cuts. Some shapes, e.g. the fourth shape in Fig. 3, are decomposed less consistently.

Fig. 3. Top: human–generated decompositions. Bottom: corresponding majority-voted decompositions.

When using this database, the question arises how to handle the one-to-many situation (one decomposition D generated by an algorithm vs. multiple GT decompositions). The standard practice is to average the dissimilarity measures computed between D and each GT decomposition. An alternative solution is as follows. In the ideal case, the automatic decomposition should mimic as well as possible the shape decomposition performed by the majority of the involved human beings. Thus, it makes sense to determine a "majority-voted" decomposition from the decompositions created by human beings. The ensemble technique presented in this paper is a suitable tool for this task and thus will be used in our experimental testing. Fig. 3 (bottom row) shows majority-voted decompositions computed by our method.

4   Experimental Validation

For validation purpose we have performed a study using five shape decomposition methods and the quantitative performance evaluation framework.

4.1   Shape Decomposition Algorithms

In order to involve as many shape decomposition approaches as possible, we contacted several authors for their code. Unfortunately, only very few authors responded and finally, only one working implementation [6] (with source code) was provided. Thus, we implemented four other shape decomposition methods. In total, we have the following five algorithms for our study: 1) **ACD**: approximate convexity-based approach [6]; 2) **DCE**: evolution-based approach [13]; 3) **IFD**: flow-based approach [14]; 4) **MD**: morphology-based approach [7]; 5) **SD**: skeleton-based approach [8].

In order to treat all approaches equally, their parameters were individually adapted to the benchmark database. All five algorithms depend on a single parameter, which essentially specifies the fineness of shape decomposition. Its optimal value was trained by computing the average Hamming distances and the average Jaccard distances for different parameter values and selecting the best one. Taking **ACD** as an example, Table 1 shows the Hamming and Jaccard

Table 1. Distances for different values of the single parameter θ of **ACD**

θ	0.04	0.05	0.06	0.07	0.08
Hamming	0.142	0.132	0.128	0.132	0.133
Jaccard	0.350	0.329	0.323	0.330	0.332

distances for different values of its single parameter θ. In both cases an optimal parameter value $\theta = 0.06$ is found and used for performance evaluation.

Our ensemble approach introduces two extra thresholding parameters t_1 and t_2. In our experiments the majority voting scheme for t_2 turns out to clearly outperform lower threshold values. For this reason this option is chosen for the experimental testing. For t_1 the same training procedure was done to find its optimal parameter value.

4.2 Validation Results

The first part of Table 2 shows the average dissimilarity measures over all shapes of the benchmark database for the five individual shape decomposition methods (sorted by their performance). For each shape the computed decomposition is either compared to all corresponding GT instances and the average value is reported (left half of the table). Alternatively, the "majority-voted" decomposition determined by our ensemble technique is used to compute the measures (right half of the table). It is important to mention that the performance listed in Table 2 and the resulting ranking of the involved shape decomposition algorithms must be treated with care. Only **ACD** was provided by the algorithm developer. The other four algorithms were implemented by ourselves and we may not reach the optimal performance, partly due to the missing details in the algorithm description. However, this fact is not harmful for our study because our goal is to demonstrate improved performance by means of ensemble solution.

The performance of combining four (five variants in total) and all five shape decomposition algorithms are listed in the second part of of Table 2. All combinations outperform the individual baseline algorithms. In the former case the combination **ACD/IFD/MD/SD** dominates because the worst-performer **DCE** (in our implementation) is not involved. It is remarkable that the four test configurations (all GT instances vs. majority-voted, Hamming vs. Jaccard) produced virtually the same ranking list.

Table 2 reveals that a comparison against the "majority-voted" decomposition consistently leads to higher performance measures than against all corresponding GT instances. This fact may be credited to the majority voting rule used in the evaluation.

Fig. 4 shows the decompositions of four selected shapes. The results of the quantitative comparison coincide well with visual comparison of decompositions. It is also confirmed that the ensemble technique outperforms the baseline algorithms. This superiority of combination is attributed to the fact that they can compensate the absence of some important cuts. For example, the cut separating the front wheel of tricycle in Fig. 4 (fourth shape) is not created by **IFD** and **SD**, but contained in the combined decomposition due to the support of other

Table 2. Average dissimilarity measures over the benchmark database

	all GT instances		majority-voted	
	Hamming	Jaccard	Hamming	Jaccard
ACD	0.128	0.323	0.092	0.251
IFD	0.145	0.350	0.112	0.267
MD	0.151	0.371	0.126	0.328
SD	0.163	0.402	0.131	0.335
DCE	0.208	0.497	0.188	0.466
ACD/IFD/MD/SD	0.114	0.302	0.069	0.190
ACD/IFD/MD/DCE	0.117	0.305	0.074	0.201
ACD/IFD/SD/DCE	0.118	0.311	0.069	0.188
ACD/MD/SD/DCE	0.117	0.305	0.076	0.206
IFD/MD/SD/DCE	0.121	0.317	0.076	0.206
ACD/IFD/MD/SD/DCE	0.111	0.288	0.069	0.186

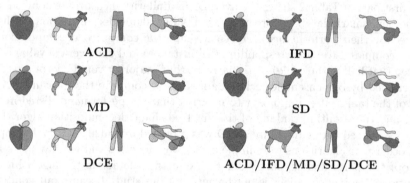

Fig. 4. Decompositions of four shapes generated by involved algorithms

decompositions. In addition, a lot of perceptually unreasonable cuts are rejected. For example, **DCE** produces a lot of unimportant cuts, which are not contained in the combined decomposition.

5   Conclusion

Although there exist quite a number of shape decompositions algorithms, multiple decomposition combination has not been studied before. In this paper we presented a clustering-based ensemble solution for shape decomposition. A recently published performance evaluation framework consisting of a benchmark database with manual ground truth together with evaluation measures was used to demonstrate the benefit of the proposed ensemble technique. We will make the source code for our decomposition ensemble method publicly available.

The proposed ensemble technique is useful in its right to improve the decomposition performance. In addition, it can also be adopted to solve the parameter selection problem as done in [15, 16]. The parameter space is explored *without* the need of ground truth and combined towards high-quality results.

Acknowledgement. The authors thank J.-M. Lien for providing his program [6] for our tests.

References

1. Tulyakov, S., Jaeger, S., Govindaraju, V., Doermann, D.S.: Review of Classifier Combination Methods. In: Marinai, S., Fujisawa, H. (eds.) Machine Learning in Document Analysis and Recognition. SCI, vol. 90, pp. 361–386. Springer, Heidelberg (2008)
2. Jiang, X., Wentker, J., Ferrer, M.: Generalized median string computation by means of string embedding in vector spaces. Pattern Recognition Letters 33, 842–852 (2012)
3. Jiang, X., Münger, A., Bunke, H.: On median graphs: Properties, algorithms, and applications. IEEE Trans. Pattern Anal. Mach. Intell. 23, 1144–1151 (2001)
4. Vega-Pons, S., Ruiz-Shulcloper, J.: A survey of clustering ensemble algorithms. Int. Journal of Pattern Recognition and Artificial Intelligence 25, 337–372 (2011)
5. Franek, L., Abdala, D.D., Vega-Pons, S., Jiang, X.: Image Segmentation Fusion Using General Ensemble Clustering Methods. In: Kimmel, R., Klette, R., Sugimoto, A. (eds.) ACCV 2010, Part IV. LNCS, vol. 6495, pp. 373–384. Springer, Heidelberg (2011)
6. Lien, J.M., Amato, N.M.: Approximate convex decomposition of polygons. Comput. Geom. Theory Appl. 35, 100–123 (2006)
7. Kim, D.H., Yun, I.D., Lee, S.U.: A new shape decomposition scheme for graph-based representation. Pattern Recognition 38, 673–689 (2005)
8. Zeng, J., Lakaemper, R., Yang, X., Li, X.: 2D Shape Decomposition Based on Combined Skeleton-Boundary Features. In: Bebis, G., Boyle, R., Parvin, B., Koracin, D., Remagnino, P., Porikli, F., Peters, J., Klosowski, J., Arns, L., Chun, Y.K., Rhyne, T.-M., Monroe, L. (eds.) ISVC 2008, Part II. LNCS, vol. 5359, pp. 682–691. Springer, Heidelberg (2008)
9. Lewin, S., Jiang, X., Clausing, A.: Framework for quantitative performance evaluation of shape decomposition algorithms. In: Proc. of ICPR (2012)
10. Ren, Z., Yuan, J., Li, C., Liu, W.: Minimum near-convex decomposition for robust shape representation. In: Proc. of ICCV, pp. 303–310 (2011)
11. Jiang, X., Marti, C., Irniger, C., Bunke, H.: Distance measures for image segmentation evaluation. EURASIP Journal on Adv. in Signal Processing, 1–10 (2006)
12. DeWinter, J., Wagemans, J.: Segmentation of object outlines into parts: A large-scale integrative study. Cognition 99, 275–325 (2006)
13. Latecki, L.J., Lakämper, R.: Convexity rule for shape decomposition based on discrete contour evolution. CVIU 73, 441–454 (1999)
14. Dey, T., Giesen, J., Goswami, S.: Shape Segmentation and Matching with Flow Discretization. In: Dehne, F., Sack, J.-R., Smid, M. (eds.) WADS 2003. LNCS, vol. 2748, pp. 25–36. Springer, Heidelberg (2003)
15. Wattuya, P., Jiang, X.: A Class of Generalized Median Contour Problem with Exact Solution. In: Yeung, D.-Y., Kwok, J.T., Fred, A., Roli, F., de Ridder, D. (eds.) SSPR 2006 and SPR 2006. LNCS, vol. 4109, pp. 109–117. Springer, Heidelberg (2006)
16. Wattuya, P., Jiang, X.: Ensemble Combination for Solving the Parameter Selection Problem in Image Segmentation. In: da Vitoria Lobo, N., Kasparis, T., Roli, F., Kwok, J.T., Georgiopoulos, M., Anagnostopoulos, G.C., Loog, M. (eds.) S+SSPR 2008. LNCS, vol. 5342, pp. 392–401. Springer, Heidelberg (2008)

Laplacian Eigenimages in Discrete Scale Space

Martin Tschirsich[1] and Arjan Kuijper[1,2]

[1] Technische Universität Darmstadt, Germany
[2] Fraunhofer IGD, Darmstadt, Germany

Abstract. Linear or Gaussian scale space is a well known multi-scale representation for continuous signals. However, implementational issues arise, caused by
discretization and quantization errors. In order to develop more robust scale space
based algorithms, the discrete nature of computer processed signals has to be
taken into account. Aiming at a computationally practicable implementation of
the discrete scale space framework we used suitable neighborhoods, boundary
conditions and sampling methods. In analogy to prevalent approaches, a discretized diffusion equation is derived from the continuous scale space axioms
adapted to discrete two-dimensional images or signals, including requirements
imposed by the chosen neighborhood and boundary condition. The resulting discrete scale space respects important topological invariants such as the Euler number, a key criterion for the successful implementation of algorithms operating on
its deep structure. In this paper, relevant and promising properties of the discrete
diffusion equation and the eigenvalue decomposition of its Laplacian kernel are
discussed and a fast and robust sampling method is proposed. One of the properties leads to Laplacian eigenimages in scale space: Taking a reduced set of images
can be considered as a way of applying a discrete Gaussian scale space.

1 Introduction

In the field of computer vision, deriving information from observed images or *signals*
is a central problem. Various strategies have been invented to do so in a performant
manner, usually by applying some kind of operator. These operators often detect or rely
on the presence of image structure or features such as edges or stationary points and are
of fixed size. Their performance then depends on the *inner scale*, the sampling density
or resolution of the image they operate on.

To overcome this dependence between operator and inner scale, various strategies
such as integral images used by the Viola-Jones Framework [1] or a whole range of
multi-scale representations have been proposed [2]. Almost all those strategies consist
of transforming the given images into a scale independent representation first before
applying an operator on this representation. A common requirement for such preliminary transformations is to mask as little information present in the original image as
possible, that is, not to rely on prior information not present in the image data itself and
therefore not unnecessarily limiting the range of tasks they can be applied to.

Various scale-invariant or multi-scale signal representations satisfying these requirements exist, such as the lowpass pyramid, wavelet and scale space representations, however there are qualitative differences. A pyramid representation consists of several fixed

G.L. Gimel' farb et al. (Eds.): SSPR & SPR 2012, LNCS 7626, pp. 162–170, 2012.

images or *scales* with preselected resolution, each computed by smoothing and sub-sampling a finer scale with the first and finest scale being the initial image. A sliding window approach is then used to evaluate a fixed-size operator at every position and on every scale of the pyramid. Because of its fixed number and decreasing size of scales, the pyramid representation can be efficiently computed and stored. The subsampling operator however, primarily implemented for performance reasons, introduces often undesired subsampling artifacts. It also complicates and sometimes prevents tracing of image features over multiple scales. The Gaussian scale space is used to overcome these drawbacks and can be thought of as the natural generalization of the lowpass pyramid. It is also popular for its theoretical foundation. A Gaussian scale space representation of a given signal is a family of derived signals, progressively smoothed with a Gaussian filter.

The focus of this paper is on the linear or Gaussian scale space representation of discrete images or signals and the various implementational issues that have to be addressed in order for this representation to be useful and applicable to real world data.

The discrete scale space proposed by Lindeberg [3] takes the discrete nature of computer processed signals into account. It is based on equivalent assumptions and axioms that have been used to derive the continuous Gaussian scale space adapted to discrete signals. It is our belief that porting scale space algorithms from a discretized continuous to the discrete scale space will eventually lead to more accurate, robust and possibly faster implementations. The discrete scale space formalized by Lindeberg however does not respect important topological invariants such as the Euler number [4]. Since most algorithms that operate on the deep structure of the Gaussian scale space require this topological invariant to hold [5, 6], we had to give a modified definition of the discrete scale space respecting the Euler number [7]. In this paper we investigate and discuss relevant and promising properties of the discrete diffusion equation and the eigenvalue decomposition of its Laplacian kernel. We propose a fast and robust sampling method. One of the properties leads to what we coined as Laplacian eigenimages in scale space, where taking a reduced set of images can be considered as a way of applying a discrete Gaussian scale space.

2   Continuous and Discrete Scale Space

Linear or *Gaussian scale space* was introduced in western literature by Witkin [8] and extended into two dimensions by Koenderink [9] and has become a useful framework for multi-scale signal representation. However, Gaussian scale space was first described in Japan as of 1959 [10]. In 1962, Taizo Iijima proposed convolution with the Gaussian kernel as the canonical way to construct the Gaussian scale space [11]. Other, more general scale spaces exist besides the Gaussian scale space. Hereafter, scale space will refer to the two-dimensional linear Gaussian scale space.

2.1   Continuous Scale Space

The *scale space representation* of a continuous two-dimensional signal $f : \mathbb{R}^2 \to \mathbb{R}$ is a one-parameter family of derived signals $L : \mathbb{R}^2 \times \mathbb{R}_+$ defined by one of the following equivalent definitions.

Convolution with a Scale Space Kernel: $L(\cdot,\cdot;t) = g(\cdot,\cdot;t) * f(\cdot,\cdot)$ where $g(x,y;t) = \frac{1}{4\pi t}e^{-\frac{x^2+y^2}{4t}}$. Commonly, the factor $4t$ is taken instead of $2\sigma^2$. The Gaussian g is also called the canonical scale space kernel.

Scale Space Axioms: A common definition of the scale space is by a limited set of scale space axioms derived from real world requirements. Koenderink [9] formulated the axioms of causality, homogeneity and isotropy. Another equivalent set of axioms states that $L(\cdot,\cdot,t) = \mathcal{L}_t f(\cdot,\cdot)$ where $\mathcal{L}_t$ is a linear shift invariant operator and thus representable as a convolution. Together with non-enhancement of local extrema stating that $\partial_t L < 0$ for maxima and $\partial_t L > 0$ for minima and some other additional requirements, these axioms all lead to the Gaussian kernel as a unique choice as the scale space kernel. An overview of a wide range of continuous Gaussian scale-space axiomatics used by different authors can be found in [10].

Diffusion Equation: The scale space representation L of a signal f is the solution of the diffusion equation

$$\partial_t L = \Delta L = \partial_{xx} L + \partial_{yy} L$$

with $L(\cdot;0) = f(\cdot)$ as initial condition. Δ denotes the Laplacian operator.

2.2 Discrete Scale Space

In this section, we will give a concise definition of the Gaussian scale space for two-dimensional discrete signals comprising a continuous scale parameter. The scale space representation of a discrete signal $f : \mathbb{Z}^2 \to \mathbb{R}$ is a one-parameter family of derived signals $L : \mathbb{Z}^2 \times \mathbb{R}^+$ defined by one of the following equivalent definitions:

Convolution with a Scale Space Kernel: $L(\cdot,\cdot;t) = k(\cdot,\cdot;t) * f(\cdot,\cdot)$ where $k : \mathbb{Z}^2 \times \mathbb{R}_+ \to \mathbb{R}$ is the discrete version of the Gaussian scale space kernel.

$$L(x,y;t) = \sum_{m=-\infty}^{\infty} k(m;t) \sum_{n=-\infty}^{\infty} k(n;t)f(x-m,y-n)$$

Scale Space Axioms: For a complete set of axioms describing the discrete scale space, we refer to those chosen by Lindeberg [3].

Diffusion Equation: The scale space representation L of a signal f is the solution of the diffusion equation

$$\partial_t L = \alpha \nabla_2^5 L + \beta \nabla_\times^2 L \tag{1}$$

with $L(\cdot,\cdot;0) = f(\cdot,\cdot)$ as initial condition and some $\alpha, \beta \geq 0$. The five-point operator ∇_2^5 and the cross operator $\nabla_\times^2$ approximate the continuous Laplacian operator. They correspond to convolution of $L(\cdot,\cdot;t)$ with kernel $\begin{bmatrix} & 1 & \\ 1 & -4 & 1 \\ & 1 & \end{bmatrix}$ respectively $\begin{bmatrix} \frac{1}{2} & 0 & \frac{1}{2} \\ 0 & -2 & 0 \\ \frac{1}{2} & 0 & \frac{1}{2} \end{bmatrix}$ for fixed scale t. For $\alpha = 1$ and $\beta = 0$ (1) simplifies to $\partial_t L = \nabla_2^5 L$ and results in k being separable.

3   Eigenbasis Decomposition of the Laplacian

The discrete scale space representation L of signal f holds various useful properties. Sampling of the scale space for example requires the ability to efficiently compute a scale $L(\cdot, \cdot; t_0)$ for fixed t_0 and relies on certain characteristics of the Laplacian operator.

The discrete scale space representation L of f is continuous in scale t. A computational investigation of L however must rely on a finite number of sampled scales. There are multiple approaches to sampling L differing in accuracy, runtime complexity and memory usage. One apparent approach is given by the definition of L via discrete convolution with a scale space kernel. The scale space kernel is of infinite domain and must be truncated in order to compute an individual scale, thus introducing truncation errors. A periodic boundary condition for f further complicates the computation. In this case, circular convolution with a Laplacian kernel provides for a more elegant but still computationally complex solution. Applied in its eigenspace however, the circular convolution operator reduces to a simple and much less complex scaling transformation. This section details how to efficiently decompose a scale of L and its derivative $\partial_t L$ into a sum of eigenimages of the Laplacian circular convolution operator. The next section then provides a simple solution of the discretized diffusion equation, enabling for fast and accurate sampling of L.

For periodic discrete signals f with discrete domain $D(f) = [1, M] \times [1, N]$, the diffusion equation $\partial_t L = \nabla_5^2 L$ can be written as a circular convolution with finite Laplacian kernel

$$\partial_t L = \nabla_5^2 L = \begin{bmatrix} & 1 & \\ 1 & -4 & 1 \\ & 1 & \end{bmatrix} \circledast L,$$

where $\circledast$ denotes the circular convolution operator.

The discrete circular convolution is a linear operator and can be expressed in matrix form if we consider $L(\cdot, \cdot; t)$ to designate a vector. Scale $L(\cdot, \cdot; t)$ of the scale space representation L can be represented as a vector $\mathbf{L}(t) \in \mathbb{R}^{MN}$ with $\mathbf{f} = \mathbf{L}(0)$.

$$\mathbf{L}(t) = \begin{bmatrix} L(1, 1; t) \\ L(1, 2; t) \\ \vdots \\ L(M, N; t) \end{bmatrix} \in \mathbb{R}^{MN}$$

For periodic f, the diffusion equation can be written in matrix form as $\partial_t L = \nabla_5^2 L = \Delta_{M,N} \mathbf{L}$ where $\Delta_{M,N} \in \mathbb{R}^{MN \times MN}$ denotes a circulant block matrix corresponding to the Laplacian operator ∇_5^2. For $M, N \geq 3$ it takes the form

$$\Delta_{M,N} = \begin{bmatrix} \mathbf{A}_N & \mathbf{I}_N & & \mathbf{I}_N \\ \mathbf{I}_N & \ddots & \ddots & \\ & \ddots & \ddots & \mathbf{I}_N \\ \mathbf{I}_N & & \mathbf{I}_N & \mathbf{A}_N \end{bmatrix} \in \mathbb{R}^{MN \times MN}, \mathbf{A}_N = \begin{bmatrix} -4 & 1 & & 1 \\ 1 & \ddots & \ddots & \\ & \ddots & \ddots & 1 \\ 1 & & 1 & -4 \end{bmatrix} \in \mathbb{R}^{N \times N},$$

where $\mathbf{I}_N$ is the identity of $\mathbb{R}^{N \times N}$. Let the normalized eigenvectors and eigenvalues of Δ_{MN} be $\mathbf{u}_{i,j}$ and $\lambda_{i,j}$. Since Δ_{MN} is a real, symmetric and therefore diagonalizable matrix, its $M \cdot N$ orthonormal eigenvectors $\mathbf{u}_{i,j}$ form an eigenbasis $\mathbf{U} = \left[\mathbf{u}_{1,1}, \mathbf{u}_{2,1}, \cdots, \mathbf{u}_{M,N} \right]$ and $\Delta_{MN} = \mathbf{U} \Lambda \mathbf{U}^T$ with $\Lambda = \mathrm{diag}(\lambda_{1,1}, \lambda_{2,1}, \cdots, \lambda_{M,N})$ and $\lambda_{i,j} \leq 0$ since Δ_{MN} is a negative-semidefinite matrix.

The scale $\mathbf{L}(t)$ of the scale space representation $\mathbf{L}$ can be written as a weighted sum of eigenimages of the Laplacian operator, i.e. as a scalar product of the orthonormal eigenvectors $\mathbf{u}_{i,j}$ of $\Delta_{M,N}$ and the scalar coefficients $c_{i,j}(t) = \langle \mathbf{L}(t), \mathbf{u}_{i,j} \rangle$ resulting from the projection of $\mathbf{L}(t)$ to $\mathbf{u}_{i,j}$:

$$\mathbf{L}(t) = \sum_{i,j} c_{i,j}(t) \mathbf{u}_{i,j}$$

Its partial derivative $\partial_t \mathbf{L}(t)$ can then be computed from scaling each projected component separately by the corresponding eigenvalue.

$$\partial_t \mathbf{L}(t) = \mathbf{U} \Lambda \mathbf{U}^T \mathbf{L}(t) = \sum_{i,j} c_{i,j}(t) \lambda_{i,j} \mathbf{u}_{i,j}.$$

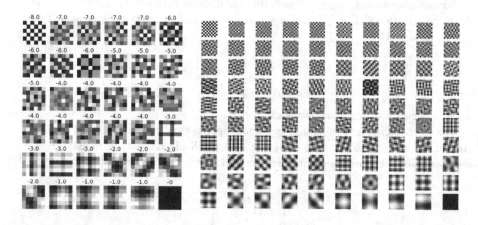

Fig. 1. Eigenimages and eigenvalues of the Laplacian $\Delta_{MN} \in \mathbb{R}^{6 \times 6}$ and $\Delta_{MN} \in \mathbb{R}^{10 \times 10}$

4   Efficient Computation of the Eigenimages

The size and values of the Laplacian matrix $\Delta_{M,N}$ depend uniquely on the dimension of f equal to the cardinality of $D(f)$. They are easily computed following the specification given in the previous section. Unfortunately, even for images of moderate size of e.g. 128×128 pixel, f would have $2^{7 \cdot 2}$ dimensions, resulting in a Laplacian matrix of 2^{28} entries taking up to $2GB$ memory. Even under consideration of less computationally and memory intensive sparse matrix formats, such matrices are too large to be

handled efficiently. Therefore, in order to reduce the overall complexity of e.g. computing the eigenbasis $\mathbf{U}$, we have to exploit the symmetric structure of the Laplacian. The
Laplacian kernel is of rank > 1, thus not separable. However, using the distributivity
of the circular convolution, it can be rewritten as a summation of two one dimensional
kernels.

$$
\partial_t L = \begin{bmatrix} & 1 & \\ 1 & -4 & 1 \\ & 1 & \end{bmatrix} \circledast L = \begin{bmatrix} 1 \\ -2 \\ 1 \end{bmatrix} \circledast L + \begin{bmatrix} 1 & -2 & 1 \end{bmatrix} \circledast L
$$

In matrix form, this translates to a direct summation of two substantially smaller matrices. The Laplacian matrix $\Delta_{M,N} \in \mathbb{R}^{MN \times MN}$ can be written as the direct sum of two
∇_3^2 operators $\Delta_M \in \mathbb{R}^{M \times M}$ and $\Delta_N \in \mathbb{R}^{N \times N}$

$$
\Delta_{M,N} = \Delta_M \oplus \Delta_N = (\Delta_M \otimes \mathbf{I}_N) + (\mathbf{I}_M \otimes \Delta_N)
$$

where Δ_M and Δ_N are the matrix representations of the row wise applied central difference operator of second order. They differ only in their dimensions. $\otimes$ denotes the
Kronecker product. For $M \geq 3$, Δ_M has the form of a Toeplitz matrix.

$$
\Delta_M = \begin{bmatrix} -2 & 1 & & & 1 \\ 1 & -2 & 1 & & \\ & \ddots & \ddots & \ddots & \\ 1 & & & 1 & -2 \end{bmatrix} \in \mathbb{R}^{M \times M}
$$

Each eigenvector $\mathbf{u}_{i,j}$ of $\Delta_{M,N}$ can be expressed as the outer product of two eigenvectors $\mathbf{v}_i$ and $\mathbf{w}_j$ of Δ_M and Δ_N. The corresponding eigenvalue $\lambda_{i,j}$ is then the sum of
the corresponding eigenvalues v_i and ω_j of Δ_M and Δ_N.

$$
\Delta_{M,N} \mathbf{u}_{i,j} = \lambda_{i,j} \mathbf{u}_{i,j}
$$
$$
\Leftrightarrow (\Delta_M \oplus \Delta_N)(\mathbf{v}_i \otimes \mathbf{w}_j) = (v_i + \omega_j)(\mathbf{v}_i \otimes \mathbf{w}_j)
$$

Omitting all details that can be found in [12], we finally have an analytic formula ex-
pressing the eigenvalues $\lambda_{i,j}$ and eigenvectors $\mathbf{u}_{i,j}$ of the Laplacian matrix $\Delta_{M,N}$ with
$i = 0 \dots M - 1, j = 0 \dots N - 1$.

$$
\lambda_{i,j} = v_i + \omega_j = \frac{(\lambda_{M,i}^F)^2}{(\lambda_{M,i}^F + 1)} + \frac{(\lambda_{N,j}^F)^2}{(\lambda_{N,j}^F + 1)} = \frac{\left(e^{\left(\frac{2\pi\iota}{M}i\right)} - 1\right)^2}{e^{\left(\frac{2\pi\iota}{M}i\right)}} + \frac{\left(e^{\left(\frac{2\pi\iota}{N}j\right)} - 1\right)^2}{e^{\left(\frac{2\pi\iota}{N}j\right)}}
$$

$$
\mathbf{u}_{i,j} = v_i \otimes w_j = d_{M,i}^F \otimes d_{N,j}^F = d_{M,i}^F =
$$
$$
\left[\exp\left(\frac{2\pi\iota}{M}i\right)^0, \dots, \exp\left(\frac{2\pi\iota}{M}i\right)^{M-1} \right] \otimes \left[\exp\left(\frac{2\pi\iota}{N}j\right)^0, \dots, \exp\left(\frac{2\pi\iota}{N}j\right)^{N-1} \right]
$$

These eigenvectors are not guaranteed to be real, although $\Delta_{M,N}$ always possesses a
real eigenbase. Exploiting further symmetries of the eigenvectors v_i and w_j, we can
derive a real eigenbase $\tilde{\mathbf{U}}$ given by $\mathbf{u}\tilde{}_{i,j} = \tilde{v}_i \otimes \tilde{w}_j$ with $\tilde{v}_i = \Re(v_i) + \Im(v_i)$ and
$\tilde{w}_j = \Re(w_j) + \Im(w_j)$.

5   Discrete Scale Space Eigenbasis Decomposition

As shown in a previous section, scale $\mathbf{L}(t)$ and its partial derivative $\partial_t \mathbf{L}(t)$ of the scale space representation $\mathbf{L}$ can be written written as a weighted sum of the eigenimages $\mathbf{u}_{1,1}, \ldots, \mathbf{u}_{M,N}$ of the Laplacian operator. This implicit change of base allows us to give a simple solution for the discretized diffusion equation.

$$\partial_t L(t) = \Delta_5^2 L(t)$$
$$\Leftrightarrow \sum_{i,j} \partial_t c_{i,j}(t)\, \mathbf{u}_{i,j} = \sum_{i,j} c_{i,j}(t)\, \lambda_{ij} \mathbf{u}_{i,j}$$

Multiplying both sides with $\mathbf{u}_{k,l}$ and exploiting the orthonormality $\langle \mathbf{u}_{i,j}, \mathbf{u}_{k,l} \rangle = \delta_{i,k}\delta_{j,l}$ where δ represents the Kronecker symbol gives us the partial derivate $\partial_t \mathbf{L}$ projected onto eigenvector $\mathbf{u}_{k,l}$. This differential equation can be easily solved for $c(t)$.

$$\left\langle \mathbf{u}_{k,l}, \sum_{i,j} \partial_t c_{i,j}(t)\, \mathbf{u}_{i,j} \right\rangle = \left\langle \mathbf{u}_{k,l}, \sum_{i,j} c_{i,j}(t)\, \lambda_{i,j} \mathbf{u}_{i,j} \right\rangle$$
$$\Leftrightarrow \partial_t c_{k,l}(t) = c_{k,l}(t)\, \lambda_{k,l}$$
$$\Leftrightarrow c_{k,l}(t) = \exp(\lambda_{k,l} t)\, c_{k,l}(0)$$

We finally get an explicit formula for $\mathbf{L}(t)$: The scale space representation $\mathbf{L}$ is the solution of the discretized diffusion equation and has the form

$$\mathbf{L}(t) = \sum_{i,j} c_{i,j}(t)\, \mathbf{u}_{i,j} = \sum_{i,j} \exp(\lambda_{i,j} t)\, c_{i,j}(0)\, \mathbf{u}_{i,j}$$

with scalar coefficients $c_{i,j}(t) = \langle \mathbf{L}(t), \mathbf{u}_{i,j} \rangle$. In matrix representation, the solution simplifies to $\mathbf{L}(t) = \mathbf{U} \exp(\Lambda t)\, \mathbf{U}^T \mathbf{L}(0)$. Partial derivatives of any order $\partial_{t^n} \mathbf{L}(t)$ can be easily computed using

$$\partial_{t^n} \mathbf{L}(t) = \sum_{i,j} \partial_{t^n} \exp(\lambda_{i,j} t)\, c_{i,j}(0)\, \mathbf{u}_{i,j} = \sum_{i,j} \lambda_{i,j}^n \exp(\lambda_{i,j} t)\, c_{i,j}(0)\, \mathbf{u}_{i,j}$$

which, in matrix representation, simplifies to $\mathbf{L}(t) = \mathbf{U} \Lambda^n \exp(\Lambda t)\, \mathbf{U}^T \mathbf{L}(0)$.

5.1   Reduced Eigenimages

Analyzing the time evolution $\mathbf{L}(t) = \sum_{i,j} \exp(\lambda_{i,j} t)\, c_{i,j}(0)\, \mathbf{u}_{i,j}$, which is merely a weighted sum of eigenimages, it becomes obvious that eigenimages with smaller eigenvalues have less influence on scales for $t > 0$ than those with eigenvalues near 0.

Omitting less influential eigenimages allows us to reduce memory and time complexity with only moderate changes in the scale space representation. The sum of squared differences SSD between the $\mathbf{L}(t)$ and $\tilde{\mathbf{L}}(t)$ with

$$\tilde{\mathbf{L}}(t) = \sum_{i,j,\lambda_{i,j} > \lambda_{min}} \exp(\lambda_{i,j} t)\, c_{i,j}(0)\, \mathbf{u}_{i,j}$$

converges rapidly to 0.

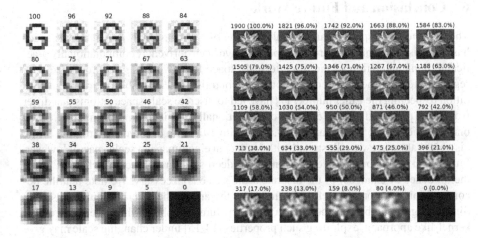

Fig. 2. Lossy compression of signal **f**. Each image is a reconstruction of **f** from a reduced set of eigenimages. The number of eigenimages is shown above each image.

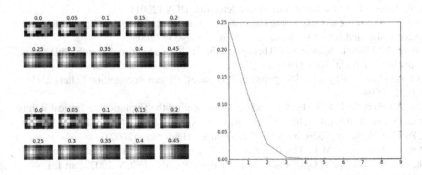

Fig. 3. Sampled scale space representation **L** (left top), sampled scale space representation $\tilde{\mathbf{L}}$ with reduced eigenimages (left bottom) and the sum of squared differences from low scale to high scale images (right)

This confirms the intuitive expectation that eigenimages with lower eigenvalues correspond to those high frequency components that vanishing very fast with increasing scale. The low frequency components or eigenimages with bigger eigenvalues are more robust against smoothing. Therefore, eliminating eigenimages with small eigenvalues roughly compares to smoothing.

The aforementioned method is comparable to a dimensionality reduction in eigenspace of the Laplacian operator. It is independent of prior information of the content of **L**. Taking the initial image **L** (0) and thus $c_{i,j}(0)$ into account would allow for further error reduction.

6   Conclusion and Future Work

The discrete scale space as an equivalent to the two-dimensional Gaussian scale space has been discussed and important properties have been derived. A computationally practicable implementation of the discrete scale space framework has been outlined. Our regular 6-neighborhood, a periodic boundary condition and a suitable critical point definition respecting the Euler number [7] have led to discrete scale space axioms that differ from those proposed by Lindeberg. A computationally efficient sampling method, based on properties of the Laplacian kernel. A first investigation of the deep structure of the discrete scale space was illustrated with Laplacian eigenimage in scale space. However, further and more formal investigation of the deep structure (structural changes under the influence of changes) of the discrete scale space is necessary. We note that the decomposition can borrow useful insights from the area of heat kernels on graphs [13]. Since our grid can be considered as a well-structured graph on which we apply a heat kernel-like approach, exploiting such properties [14, 15] under changing scale may give new insights.

References

1. Viola, P., Jones, M.: Robust real-time object detection. IJCV (2001)
2. Kuijper, A.: Exploring and exploiting the structure of saddle points in gaussian scale space. Computer Vision and Image Understanding 112(3), 337–349 (2008)
3. Lindeberg, T.: Discrete Scale-Space Theory and the Scale-Space Primal Sketch. PhD thesis, Royal Institute of Technology (1991)
4. Kuijper, A.: On detecting all saddle points in 2D images. Pattern Recognition Letters 25(15), 1665–1672 (2004)
5. Kuijper, A., Florack, L.M.J.: Understanding and Modeling the Evolution of Critical Points under Gaussian Blurring. In: Heyden, A., Sparr, G., Nielsen, M., Johansen, P. (eds.) ECCV 2002, Part I. LNCS, vol. 2350, pp. 143–157. Springer, Heidelberg (2002)
6. Kuijper, A., Florack, L.M.J.: The Relevance of Non-generic Events in Scale Space Models. In: Heyden, A., Sparr, G., Nielsen, M., Johansen, P. (eds.) ECCV 2002, Part I. LNCS, vol. 2350, pp. 190–204. Springer, Heidelberg (2002)
7. Tschirsich, M., Kuijper, A.: A Discrete Scale Space Neighborhood for Robust Deep Structure Extraction. In: S+SSPR 2012, vol. 7626, pp. 124–132. Springer, Heidelberg (2012)
8. Witkin, A.P.: Scale-space filtering. In: Proc. 8th Int. Joint Conf. Art. Intell., pp. 1019–1022. Karlsruhe, Germany (August 1983)
9. Koenderink, J.J.: The structure of images. Biological Cybernetics 50, 363–370 (1984)
10. Weickert, J., Ishikawa, S., Imiya, A.: Scale-space has been discovered in Japan. Technical report, Department of Computer Science, University of Copenhagen (August 1997)
11. Iijima, T.: Basic theory on normalization of a pattern (in case of typical one-dimensional pattern). Bulletin of Electrical Laboratory, 368–388 (1962)
12. Tschirsich, M.: The discrete scale space as a base for robust scale space algorithms. Technical report, Department of Computer Science, Technical University of Darmstadt (June 2012)
13. Hancock, E.R., Wilson, R.C.: Pattern analysis with graphs: Parallel work at Bern and York. Pattern Recognition Letters 33(7), 833–841 (2012)
14. Xiao, B., Hancock, E.R., Wilson, R.C.: Geometric characterization and clustering of graphs using heat kernel embeddings. Image Vision Comput. 28(6), 1003–1021 (2010)
15. Xiao, B., Hancock, E.R., Wilson, R.C.: Graph characteristics from the heat kernel trace. Pattern Recognition 42(11), 2589–2606 (2009)

A Relational Kernel-Based Framework for Hierarchical Image Understanding

Laura Antanas, Paolo Frasconi, Fabrizio Costa,
Tinne Tuytelaars, and Luc De Raedt

Katholieke Universiteit Leuven, Belgium

Abstract. While relational representations have been popular in early work on syntactic and structural pattern recognition, they are rarely used in contemporary approaches to computer vision due to their pure symbolic nature. The recent progress and successes in combining statistical learning principles with relational representations motivates us to reinvestigate the use of such representations. More specifically, we show that statistical relational learning can be successfully used for hierarchical image understanding. We employ kLog, a new logical and relational language for learning with kernels to detect objects at different levels in the hierarchy. The key advantage of kLog is that both appearance features and rich, contextual dependencies between parts in a scene can be integrated in a principled and interpretable way to obtain a qualitative representation of the problem. At each layer, qualitative spatial structures of parts in images are detected, classified and then employed one layer up the hierarchy to obtain higher-level semantic structures. We apply a four-layer hierarchy to street view images and successfully detect corners, windows, doors, and individual houses.

1 Introduction

Understanding images by recognizing its constituent objects is a challenging task and it could be solved, in principle, using computer vision techniques that employ low- to medium-level features, such as geometric primitives, patches, or invariant features [1]. Although helpful for the recognition process, these features do not suffice for higher-level tasks dealing with more complex patterns. In this case, it is more intuitive to describe visual scenes in terms of *structural hierarchical* (or *graph-like*) representations that build on visual image parts. They reflect the natural composition of scenes into objects and parts of objects. In particular, man-made (vs. natural) scenes exhibit considerable structure that can be captured using qualitative spatial relations. For example, a typical house consists of aligned elements such as: a roof, some windows, one or more doors and possibly a chimney. A hierarchical aspect is that a window itself is composed of rectangular-like corner configurations with a certain appearance.

This view on hierarchical image representation was embraced by early ideas that hierarchical structure and relations are key components of an image understanding system [2]. A key advantage of using relational representations [3] is their capability of exploiting contextual knowledge in images via symbolic relations. In addition, they abstract spatial information away from exact locations making it independent of metric details. Although popular in early work on syntactic or structural pattern recognition [4],

G.L. Gimel' farb et al. (Eds.): SSPR & SPR 2012, LNCS 7626, pp. 171–180, 2012.

relational approaches have been rarely used to solve computer vision problems (except [5, 6]). One reason is that low-and mid-level vision features were not always as mature as today to support such ambitious representations. Another reason is the limitation of pure relational approaches in handling noisy data. Yet, when combined with statistical techniques, they show robustness to noise [3, 7]. Motivated by our previous results on using distances between logical interpretations to hierarchically detect structures in images [5], we solve the same problem using kLog, a general purpose relational language for kernel-based learning. The resulting approach is more principled, as it is grounded in a statistical learning framework, is computationally more tractable and provides improved results. Our earlier approach relied on more expensive logical matching and generalization operations and was more tailored towards this particular application.

kLog [8] is a new statistical relational learning framework, which builds on ideas from statistics to address uncertainty, while incorporating a relational representation of the domain. Images are described in terms of automatically extracted semantic parts and relationships between them, thus as relational databases or (hyper)-graphs. Domain knowledge can easily be incorporated using logical rules. The novelty of kLog is that, starting from existing visual features, it can take relational contextual features into account in a principled and natural way. Furthermore, its declarative approach offers a flexible and interpretable way to consider both appearance and spatial information in an image. Finally, kLog transforms the relational databases into graph-based representations and uses graph kernels to extract the feature space. Thus, our contribution is a new approach to hierarchical image understanding, in which *spatial configurations* of scenes are combined with *kernel-based learning* for structured data to recognize objects throughout all layers of a hierarchy, in a unified way.

The goal of this paper is to understand images by recognizing objects at different layers of a hierarchy. The base layer relies on local interest points and their descriptors. A subsequent layer consists of objects, while higher layers consist of configurations of objects. We focus on the recognition of structures in street view images, yet, our approach can be used for other domains as well. We learn to recognize objects from a set of manually labeled examples of object categories, i.e., houses, windows and doors. Each house is annotated with the locations and shapes of its constituent windows and doors. The approach is evaluated on a dataset of 60 street view images.

2 Related Work

Thus far, most work in computer vision has focused on fixed compositional structures [9] or constellation models [10]. Recently, more attention was devoted to using high-level relational representations for image understanding or object recognition [11–13]. Yet, most of this work is restricted to a model-based approach and perform interpretation through image grammars. These have been well-studied [14], but need considerably more input from the user in terms of a set of grammar rules. This in contrast to our approach, which is based on learning from annotated examples and which uses domain knowledge to specify only basic qualitative spatial relations between image parts.

Several papers have addressed the problem of understanding images of house facades. In [15], structure models of meaningful facade concepts are learned from examples, while in [16], the authors tackle the house delineation problem by generating

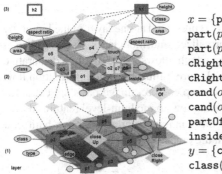

$x = \{\texttt{part}(p_1, botL, door), \texttt{part}(p_2, topL, door),$
$\texttt{part}(p_5, botL, win), \texttt{part}(p_6, botR, win),$
$\texttt{part}(p_7, topR, win), \texttt{cUp}(p_2, p_1, d3, edge),$
$\texttt{cRight}(p_3, p_1, d2, edge), \texttt{cRight}(p_6, p_5, d3, edge),$
$\texttt{cRight}(p_4, p_2, d5, noedge), \dots,$
$\texttt{cand}(o_1, thin, size3, h1), \texttt{cand}(o_5, thin, size2, h1),$
$\texttt{cand}(o_3, squared, size2, h2), \dots, \texttt{partOf}(p_5, o_2),$
$\texttt{partOf}(p_6, o_2), \texttt{partOf}(p_2, o_1) \dots$
$\texttt{inside}(o_7, o_2), \texttt{touch}(o_6, o_2), \dots\}.$
$y = \{\texttt{class}(o_1, door), \texttt{class}(o_5, window),$
$\texttt{class}(o_3, none), \dots\}.$

Fig. 1. A hierarchical description of a house image. Parts are squares (purple, yellow, red); relations are diamonds (green/blue – spatial/functional constraints, grey –memberships); properties are circles (pink). Parts not belonging to a class of interest are empty squares. A visual interpretation $i = (x, y)$ is on the right; x specifies the input features, while y is the learning target.

vertical separating lines on the facade and using a dissimilarity measure between these features. Finally, the works in [17, 18] assume having the structure or grammar of a building facade and estimate the parameters of the model. Closely related are graph matching and other kernel-based techniques for image understanding [19]. Different from these, our work combines the best of both worlds by using a kernel-based approach to learn from logical interpretations. The paper extends our recent results in [20] with more complex relationships and, thus, a richer feature space.

3 Hierarchical Image Understanding

In our hierarchical framework an image is described at several layers $(0), \dots, (k)$ in a hierarchy, with 0 the base layer and k the top layer. Figure 1 shows the hierarchical structure of a partial house facade. At each layer, the image consists of a set of parts, their properties and (spatial) relationships among them. The task then is to use this information at layer i to generate and classify candidate parts at the next higher layer $i + 1$ in the hierarchy. Thus, at each layer, parts belonging to classes of interest are detected and employed at the next layer to detect higher-level concepts. As training data, annotated images are available at all layers.

In the house facade problem, the *base layer* consists of the image itself, with the pixels as parts. In the *primitive layer* the parts are local patterns, e.g., a corner or an edge. The *object layer* is built from *spatial configurations* of such local patterns, forming higher-level parts that are *doors* and *windows*. These are then used at the next layer, i.e., the *house layer*, to find even higher-level parts representing *houses*. Each layer consists of parts and classes they belong to, and it is formed by making use of spatial configurations of parts from the previous lower-level layer. The hierarchical framework propagates the detected parts using a pipeline through each layer.

4   Object Detection at One Layer

Next, we describe how an image is relationally represented at one layer in the hierarchy and how our object detection problem is formalized and modeled with kLog. It is a domain specific language embedded in Prolog which allows to specify, in a declarative way, logical and relational learning problems. Figure 2 illustrates the information flow in kLog. We use the object layer as running example. Here, image parts are extracted from raw images via the primitive layer and using low-to medium-level features detectors, as described below. At the house layer the relational representation is built in a similar way using as parts the detections from the object layer.

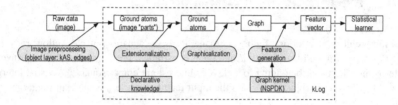

Fig. 2. From images to feature vectors in kLog

4.1   From Images to Primitive Parts

The primitive layer takes as input image pixels and groups them in corner-like features representing image parts at the object layer together with their properties. We employ the KAS detector [21] to detect corners formed by chains of 2 connected, roughly straight contour segments. Because we can get many detections we only keep square-like corners with an angle of $\approx 90°$. Also, we train a binary classifier to discard irrelevant corners found on other structures than buildings (e.g., car), using the HOG descriptor [22] on the corners. We use the training annotations of windows and doors.

Each corner-like part can be one of the types in the set $\{topR, topL, botR, botL\}$ representing top-right, top-left, bottom-right and bottom-left corners, respectively. The corner type is given by the orientation of the segments composing the 2AS. We use the HOG descriptor[1] to characterize the appearance of each corner. Yet, instead of the raw descriptor we train another classifier to map each HOG to a discrete attribute, either a *window* or a *door* label. A final characteristic of a part is its estimated bounding box.

4.2   Data and Problem Modeling

We represent this information at a higher level using the classic entity/relationship (E/R) data model, a paradigm frequently used in database theory [23]. The E/R model for our problem, with some further assumptions required by kLog, is shown in Figure 3(a). It provides an abstract representation of the examples, i.e. class of interest candidate instance in this case. The elements of an E/R model are *entity sets* (in Figure 3(a) depicted

---

[1] A variation of the HOG descriptor with 16 orientation bins, window size of 128x128 pixels and a block size of 8x8 cells showed improved results.

as rectangles), *relationships* linking entity sets (depicted as diamonds) and *attributes* that describe objects and relationships (depicted as ovals). In kLog, the database scheme is directly derived from the E/R model, and contains two kinds of relations: those introducing entity sets (*E-relations*) and those introducing relationships (*R-relations*). As in database theory, they correspond to tuples (or facts) in the database.

In our problem, E-relations are parts of the image and candidate objects of interest. Each entity has properties and a unique identifier (underlined ovals). They can be visualized as relational facts, in Figure 1 (right). The tuple $\text{part}(p_1, botL, door)$ specifies a part entity, where p_1 is the identifier and the other arguments are properties extracted by the previous layer in the hierarchy. As already indicated, they are the corner type and category. The tuple $\text{cand}(o_1, thin, size3)$ represents a possible object of interest. It has identifier o_1 and properties describing its discretized aspect ratio and size. These are estimated from the extracted bounding box of the candidate. R-relations are linked to the entities that participate in the relationships. In our problem, we have spatial relationships amongst parts and, respectively, amongst candidates, as well as membership relations between parts and candidates. Spatial R-relations are derived from the spatial localization of the entities, i.e., bounding boxes, and extension. An example is the relationship $\text{cRight}(p_3, p_1, d2, edge)$, which indicates that part entities p_3 and p_1 are spatially close to each other and aligned on the X axis with p_3 to the right of p_1. It has as properties the discretized Euclidian distance between the bounding boxes and a property indicating if the two part entities are linked by a detected contour segment.

A key advantage of kLog is that it supports *extensional* as well as *intensional* relations. Extensional relations are explicitly listed sets of given relations, whereas intensional relations are defined implicitly using logical rules. In other words, intensional relations are derived from other intensional or extensional relations given a set of rules and they represent domain-related feature construction.

Declarative Feature Construction. Intensional relations are $\text{cUp}/4$, $\text{cRight}/4$, $\text{inside}/2$ and $\text{touch}/2$, derived using notions of spatial theory, $\text{cand}/4$ and $\text{partOf}/2$. As an example, the spatial relation $\text{cRight}/4$ is defined as a logical rule in the following way:
$$\text{cRight}(A, B, D, Edge) \leftarrow \text{part}(A, \_, \_), \text{part}(B, \_, \_), \text{edge}(Edge, A),$$
$$\text{edge}(Edge, A), \text{right}(A, B), \text{close}(A, B, D).$$
where $\text{close}(A, B, D) \leftarrow \text{bb}(A, BB_1), \text{bb}(B, BB_2), \text{dist}(BB_1, BB_2, D), D < th$.
and $\text{right}(A, B)$ is similarly defined based the bounding boxes BB_i of the part entities. In words, A is to the right of B if the min and max X coordinates of BB_1 are smaller than the minimum and the maximum X coordinates of BB_2, respectively, and if A is not too much above or below (in a fuzzy way) of B. The R-relation $\text{cUp}/2$ is defined in a similar way. The atom $\text{edge}(Edge, A)$ is true if the entity A belongs to a contour segment $Edge$.

The intensional E-relation $\text{cand}/4$ defines possible objects of a class of interest at one layer, i.e., doors/windows at the object layer. It is defined using the rule:
$$\text{cand}(Id, Ar, A, H) \leftarrow \text{sprl}(A, B), \text{sprl}(B, C), \text{edge}(E_{ab}, A), \text{edge}(E_{ab}, B,),$$
$$\text{edge}(E_{bc}, B), \text{edge}(E_{bc}, C), \text{getid}([A, B, C], Id), \text{getprop}([A, B, C], Ar, A, H).$$
where $\text{sprl}/2$ brings the pairs of parts that satisfy any of the spatial relations $\{\text{cRight}, \text{cUp}, \text{cDown}, \text{cLeft}\}$; $\text{getid}/2$ associates a unique identifier to the newly generated

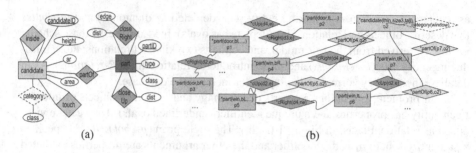

(a)   (b)

Fig. 3. a) E/R modeling of the object detection problem. Rectangles denote entity vertices, diamonds denote relationships, and circles denote properties. b) Part of the graphicalized interpretation of the image.

candidate (based on the combination of parts) and `getprop` calculates the discretized properties of the candidate relation, i.e., aspect ratio, area and height, based on the bounding box of the candidate, given the set of parts. Each candidate relation groups the three parts that satisfy a square-like spatial constraint. The membership relation `partOf/2` indicates that a part belongs to a candidate.

Other intensional relations are `touch/2`, indicating if two candidate entities are spatially touching and `inside/2`, which holds if one candidate is spatially inside the other. The grounding of intensional relations is computed using Prolog's deduction mechanism and represents the extensionalization step in kLog's information flow. In the setting established above, each image is an instance of a relational database or an *interpretation*. An interpretation of an image at the object layer is exemplified in Figure 1.

Problem Definition. kLog learns from interpretations, a well-established setting in relational learning [3]. We are given a training set of n independent interpretations $D = \{(x_1, y_1), (x_2, y_2), \ldots, (x_n, y_n)\}$ sampled identically from some unknown but fixed distribution; x_i is a set of input ground atoms and y_i a set of output target ground atoms. In our problem and in Figure 1 the target is the unary relation `category/1`. The goal is to learn a mapping $h : X \to Y$, from the inputs X to the outputs Y. During prediction, we are given a partial interpretation of an image consisting of ground atoms x, and are required to complete the interpretation using h to predict the output atoms y.

4.3   Graphicalization and Feature Generation

Next, each interpretation x is converted into a bipartite graph G that has a vertex for each ground relation. Vertices correspond to grounded atoms, either E-relations or R-relations, but identifiers are removed. Edges connect E-relations and R-relations: there is an undirected edge $\{e, r\}$ if the entity identifier in e appears as an argument in r (see Figure 3(b)). Thus, edges connect vertices that share identifiers in the tuples. Role information (i.e., the position of an entity in a relationship) is retained as an edge annotation. The graph can be seen as the result of unrolling (or grounding) the E/R diagram for a particular image. There is no loss of information associated with this step.

Once interpretations are represented as graphs, any graph kernel in conjunction with a statistical learner can be used to solve the classification problem in the supervised setting. The kLog implementation uses a variant of the fast neighborhood subgraph pairwise distance kernel (NSPDK) [24]. It has two advantages: i) it allows fast computations with respect to the graph size, as the graphicalization step can yield large graphs; ii) it is a general purpose kernel with a flexible bias, allowing us to integrate multiple heterogeneous features and context knowledge through the way it is defined.

NSPDK belongs to the large family of decomposition kernels [25] that count the number of common parts between two objects. Parts in this case are pairs of subgraphs defined as follows. Given a graph $G = (V, E)$ and a radius $r \in \mathbb{N}$, we denote by $N_r^v(G)$ the subgraph of G rooted in v and induced by the set of vertices $V_r^v \doteq \{x \in V : d^\star(x, v) \leq r\}$, where $d^\star(x, v)$ is the shortest-path distance between x and v. For a given distance $d \in \mathbb{N}$, the *neighborhood-pair* relation is then defined as $R_{r,d} = \{(N_r^v(G), N_r^u(G), G) : d^\star(u, v) = d\}$. The kernel between two graphs is then the decomposition kernel defined by relations $R_{r,d}$ for $r = 0, \ldots, R$ and $d = 0, \ldots, D$:,

$$K(G, G') = \sum_{r=0}^{R} \sum_{d=0}^{D} \sum_{\substack{A, B : R_{r,d}(A, B, G) \\ A', B' : R_{r,d}(A', B', G')}} \kappa((A, B), (A', B')). \tag{1}$$

Several choices are possible for κ. In our experiments we used an exact matching kernel where $\kappa((A, B), (A', B')) = 1$ iff (A, B) and (A', B') are pairs of isomorphic graphs, but also a soft matching kernel (see [8] for details). The maximum radius R and the maximum distance D are kernel hyperparameters. kLog provides a flexible architecture in which only the specification language is fixed. The actual features are determined by the choice of the graph kernel but also by the definition of intensional relations.

5   Summary of Experiments

We experimented on a dataset containing 60 street view images of rows of houses [5]. They commonly display a rich structure (and variety), yet, same row houses are quite consistent in terms of structure. All images show near-frontal views of the houses and no further rectification was performed. On these images, windows, doors and houses were manually annotated. We used three layers in the hierarchy: *primitive, object* and *house* layers. We experimented with kLog at the object and house layers, since these provide the most structure. The primitive layer serves as a preprocessing step. We measure performance in terms of precision P, recall R and F1 score and use the PASCAL VOC criterion[2] to compare the positive predicted candidate's bounding box to the ground-truth. If the overlap is larger than 50%, it is a true positive, otherwise a false positive.

Primitive Layer. To asses the accuracy of the parts categories the object layer builds on, we also report results at the primitive layer. For the first classification step establishing whether a corner is relevant or not we obtain $F1 = 0.85$. For the second classification steps distinguishing between window and door corners, $F1 = 0.64$.

---

[2] Available at http://pascallin.ecs.soton.ac.uk/challenges/VOC/

Method	R	P	F1
RD hierarchy [5]	0.61	0.65	0.63
Boosting60	0.54	0.49	0.51
Boosting120	0.57	0.48	0.52
kLog	0.74	0.64	0.68

Method	R	P	F1
RD window	0.61	0.35	0.44
RD door	0.42	0.47	0.44
kLog window	0.60	0.55	0.57
kLog door	0.51	0.42	0.50

Fig. 4. kLog performance compared to baselines; classes *house* (left), *door* and *window* (right). For the feature boosting detector we use a different number of weak classifiers (Boosting60/120).

Object Layer. The experiments at the object layer are performed starting from sparse, previously detected, 2AS at the primitive layer that belong to windows or doors. We used the following features: part entity relation `part`, spatial relationships between parts `cRight`, `cUp`, candidate entity relation `cand`, membership relationship `partOf` and other spatial/functional relationships between candidate entities (such as `inside` and `touch`). At this layer, similarly, we solve the problem in two steps. First, we establish whether a candidate is relevant or not and then we distinguish between windows and doors. We vary the parameters of the kernel r and d to assess the impact of contextual features on the performance of detecting windows and doors. We obtain the best result, $F1 = 0.57$ for class *window* and $F1 = 0.50$ for class *door*, when $r = 2, d = 4$.

House Layer. Candidates classified as *window* or *door* become parts at the house layer. We used a variation of the same relations (e.g., the absence of property *edge*). Again, we vary the parameters r and d to assess the impact of contextual features on the performance of detecting houses and obtain $P = 0.64, R = 0.74, F1 = 0.68$.

Many alternative statistical learners can be used on the feature vectors created by kLog. In our experiments, we used a standard implementation of support vector machines [26], which was integrated via a wrapper in kLog, together with a linear kernel. We performed 5-fold cross-validation on the dataset with fixed folds. The cost c of the SVM was chosen via internal 5-fold cross-validation on the training set, for each split.

Comparison to Baselines. Our aim is not to compete with strong detectors using dense features, but to evaluate how structure and contextual knowledge can be flexibly exploited in our problem. We show that even if we start from sparse cues, the detection problem is solvable with good results thanks to the use of relational representations and kLog's flexible language and kernel. One baseline is the feature boosting approach with template matching [27]. We train an ensemble of weak detectors for the class *house*. Individual houses can be more effectively detected using a template matching approach than a texture-based one, since houses in the same row have the same texture and street scenes greatly vary in texture across the dataset. A second baseline is our relational distance-based approach (RD) [5]. It uses the same sparse features and data splits. Figure 4 shows results for comparison. The baselines perform well for our detection problem, however, by incorporating more structural context, kLog improves results. Also, in [5] we employed an extra candidate selection step, which resulted in higher precision. This step is not performed in the experiments with kLog.

6   Conclusions

We presented a new statistical relational learning approach to hierarchically understand images of houses. To this end, we employ kLog, a framework for logical and relational learning with kernels. The declarative, relational representation used by kLog allows a flexible exploitation of the structural and contextual knowledge in visual scenes. We show that even if we start from sparse cues, our problem is solvable with good results thanks to the use of relational representations and kLog's flexible language and kernel. This work explores a new relational scheme for solving computer vision problems. This result can be improved using a collective classification setting, in which target predictions are also considered during training and testing. Additionally, hierarchical features could be used as top-down feedback. For example, a detected house can constraint the number of doors composing the house, and thus, improve door detection results.

Acknowledgements. Laura Antanas is supported by the grant agreement First-MM-248258.

References

1. Tuytelaars, T., Mikolajczyk, K.: Local invariant feature detectors: A survey. Foundations and Trends in Computer Graphics and Vision 3(3), 177–280 (2007)
2. Hanson, A., Riseman, E.: Visions: A computer system for interpreting scenes. In: CVS, pp. 303–333 (1978)
3. De Raedt, L.: Logical and Relational Learning. Springer (2008)
4. Fu, K.: Syntactic methods in pattern recognition, vol. 112. Elsevier Science (1974)
5. Antanas, L., van Otterlo, M., Tuytelaars, T., Raedt, L.D., Oramas Mogrovejo, J.: A relational distance-based framework for hierarchical image understanding. In: ICPRAM, vol. (2), pp. 206–218 (2012)
6. Pearce, A.R., Caelli, T., Bischof, W.F.: Learning relational structures: Applications in computer vision. Applied Intelligence 4, 257–268 (1994)
7. Getoor, L., Friedman, N., Koller, D., Taskar, B.: Learning probabilistic models of relational structure. In: ICML, pp. 170–177 (2001)
8. Frasconi, P., Costa, F., Raedt, L.D., Grave, K.D.: klog: A language for logical and relational learning with kernels. CoRR (2012)
9. Felzenszwalb, P., Girshick, R., McAllester, D., Ramanan, D.: Object detection with discriminatively trained part-based models. IEEE TPAMI 32(9), 1627–1645 (2010)
10. Fergus, R., Perona, P., Zisserman, A.: Weakly supervised scale-invariant learning of models for visual recognition. IJCV 71(3), 273–303 (2007)
11. Han, F., Zhu, S.: Bottom-up/top-down image parsing with attribute grammar. IEEE Transactions on Pattern Analysis and Machine Intelligence 31(1), 59–73 (2009)
12. Zhu, L., Chen, Y., Lin, Y., Lin, C., Yuille, A.: Recursive segmentation and recognition templates for image parsing. IEEE TPAMI 34(2), 359–371 (2012)
13. Girshick, R., Felzenszwalb, P., McAllester, D.: Object detection with grammar models. IEEE TPAMI 33(12) (2011)
14. Zhu, S.C., Mumford, D.: A stochastic grammar of images. Found. Trends. Comput. Graph. Vis. 2(4), 259–362 (2006)
15. Hartz, J.: Learning probabilistic structure graphs for classification and detection of object structures. In: ICMLA, pp. 5–11 (2009)

16. Zhao, P., Fang, T., Xiao, J., Zhang, H., Zhao, Q., Quan, L.: Rectilinear parsing of architecture in urban environment. In: CVPR, pp. 342–349 (2010)
17. Koutsourakis, P., Simon, L., Teboul, O., Tziritas, G., Paragios, N.: Single view reconstruction using shape grammars for urban environments. In: ICCV, pp. 1795–1802 (2009)
18. Terzic, K., Hotz, L., Sochman, J.: Interpreting structures in man-made scenes - combining low-level and high-level structure sources. In: ICAART, pp. 357–364 (2010)
19. Tuytelaars, T., Fritz, M., Saenko, K., Darrell, T.: The nbnn kernel. In: ICCV, pp. 1824–1831 (2011)
20. Antanas, L., Frasconi, P., Tuytelaars, T., De Raedt, L.: Employing relational languages for image understanding. In: IEEE Workshop on Kernels and Distances for Computer Vision, pp. 1–2 (2011)
21. Ferrari, V., Fevrier, L., Jurie, F., Schmid, C.: Groups of adjacent contour segments for object detection. TPAMI, 36–51 (2008)
22. Dalal, N., Triggs, B.: Histograms of oriented gradients for human detection. In: CVPR, pp. 886–893 (2005)
23. Garcia-Molina, H., Ullman, J.D., Widom, J.: Database Systems: The Complete Book, 2nd edn. Prentice Hall Press, Upper Saddle River (2008)
24. Costa, F., Grave, K.D.: Fast neighborhood subgraph pairwise distance kernel. In: ICML, pp. 255–262 (2010)
25. Haussler, D.: Convolution kernels on discrete structures. Technical Report UCSC-CRL-99-10, University of California at Santa Cruz (1999)
26. Fan, R.E., Chang, K.W., Hsieh, C.J., Wang, X.R., Lin, C.J.: Liblinear: A library for large linear classification. J. Mach. Learn. Res. 9, 1871–1874 (2008)
27. Torralba, A., Murphy, K.P., Freeman, W.T.: Sharing features: Efficient boosting procedures for multiclass object detection. In: CVPR, pp. 762–769 (2004)

A Jensen-Shannon Kernel for Hypergraphs

Lu Bai[1], Edwin R. Hancock[1,*], and Peng Ren[2,**]

[1] Department of Computer Science
University of York, UK
Deramore Lane, Heslington, York, YO10 5GH, UK
[2] College of Information and Control Engineering
China University of Petroleum (Huadong), China

Abstract. In this paper we explore how to construct a Jensen-Shannon kernel for hypergraphs. We commence by calculating probability distribution over the steady state random walk on a hypergraph. The Shannon entropies required to construct the Jensen-Shannon divergence for pairs of hypergraphs are obtained from steady state probability distributions of the random walk. The Jensen-Shannon divergence between a pair of hypergraphs is the difference between the Shannon entropies of the separate hypergraphs and a composite structure. Our proposed kernel is not restricted to hypergraphs. Experiments on (hyper)graph datasets extracted from bioinformatics and computer vision datasets demonstrate the effectiveness and efficiency of the Jensen-Shannon hypergraph kernel for classification and clustering.

1 Introduction

Hypergraph based strategies have recently been investigated for representing and processing structures where the relations present between objects are higher order. A hypergraph is a generalization of a graph. Unlike the pairwise edges in a graph, hypergraph representations allow a hyperedge to encompass an arbitrary number of vertices, and can hence capture multiple relationships among features. There have been several successful methods for characterizing hypergraphs, which include a) marginalizing higher order relationships to unary order [14], b) marginalizing the higher order relationships to pairwise order and then adopt pairwise graph matching methods [3], c) performed visual clustering by adopting tensors for representing uniform hypergraphs [10], and d) exploiting a set of coefficients from hypergraph Ihara zeta function to capture frequency of the cycle structures in a hypergraph [9]. One main limitation of the existing methods for hypergraph characterization is that they are usually limited to uniform structures, and do not fully capture hypergraph characteristics. On the other hand, existing hypergraph characterization methods also tend to require prohibitive computational overheads. In order to overcome these problems, an

* Edwin R. Hancock is supported by a Royal Society Wolfson Research Merit Award.
** Peng Ren is supported by the National Natural Science Foundation of China Grant 61105005.

attractive alternative is to use kernel methods. Kernel methods are popular in statistical learning theory and offer an elegant way to formulate efficient algorithms to deal with high dimensional data, without the need to construct an explicit high dimensional feature space. As one of the special case of ILP [13], a number of graph kernels have been developed and proven to be powerful in graph clustering and classification. These graph kernels can be generally categorized into three classes [11], i.e. graph kernels based on a) walks, b) paths and c) restricted subgraph and subtree structures. To generalize the graph kernels to construct hypergraph kernels, Wachman has summarized the existing graph kernels based on walks and then proposed a rooted kernel for hypergraphs [13]. However, the definitions of these kernels, no matter for graphs or hypergraphs, highly rely on the enumerations of topology features such that most of them cannot be efficiently computed in an algebraic manner.

Recently, information theory has been used to define a new family of kernels based on probability distributions over the elements of the objects being compared, and these have been applied to structured data [7]. These so-called nonextensive information theoretic kernels are derived from the mutual information between probability distributions on different structures, and are related to the Shannon entropy. An example is the Jensen-Shannon kernel [7]. Our aim in this paper is to explore whether the Jensen-Shannon kernel can be applied to hypergraphs. The kernel is computed using the Jensen-Shannon divergence between pairs of hpyergraphs. The Jensen-Shannon divergence between a pair of hypergraphs is defined as the difference in entropies between a composite hypergraph formed from the two hypergraphs, and the sum of the entropies for the two separate hypergraphs. The required entropies are computed using the Shannon entropy with the probability distributions associated with the steady state random walks on the separate hypergraphs and their composite hypergraph. Since the probability distribution of a hypergraph can be calculated directly from the incidence matrix of the hypergraph, and the adjacency matrix of a graph can be easily converted into an incidence matrix representation. Hence, our kernel can be applied to undirected graphs. We perform experiments on several bioinformatics and computer vision datasets. We empirically demonstrate that our kernel can not only readily accommodate nonuniform hypergraphs but also easily scale to large hypergraphs. The performance of our kernel is competitive to state of the art graph kernels and hypergraph based methods.

2   Definitions and Notations

2.1   Hypergraph Fundamentals

A hypergraph is a generalization of a undirected graph, it is usually denoted by a pair set $G(V, E)$ where V is a set of vertices and E is a set of non-empty subsets of V called hyperedges. A hypergraph can be represented in terms of a matrix. For a hypergraph $G(V, E)$ with I vertices and J hyperedges, its incidence matrix $\mathcal{H}$ is defined as a $I \times J$ matrix with element $\mathcal{H}(i, j)$ as follows:

$$\mathcal{H}(i,j) = \begin{cases} 1 \text{ if } v_i \in e_j \\ 0 \text{ otherwise.} \end{cases} \qquad (1)$$

An example hypergraph is shown in Fig.1(a). Here the vertex set is $V = \{v_1, v_2, v_3, v_4, v_5, v_6\}$ and the hyperedge set is $E = \{e_1 = \{v_1, v_2, v_3\}, e_2 = \{v_3, v_4, v_5\}, e_3 = \{v_5, v_6\}\}$. The incidence matrix is shown in Fig.1(b)

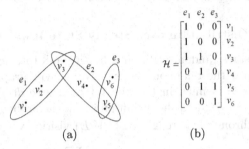

(a)   (b)

Fig. 1. (a) Hypergraph example. (b) Incidence matrix

2.2   Jensen-Shannon Kernel

The Jensen-Shannon kernel is a nonextensive mutual information kernel [7]. It is defined on probability distributions over structured data. The kernel for structures p and q is positive definite (**pd**) with the following kernel function

$$k_{JSK}(P_p, P_q) = \log 2 - JSD(P_p, P_q) \qquad (2)$$

where $JSD(P_p, P_q)$ is the Jensen-Shannon divergence between the probability distributions P_p and P_q defined as

$$JSD(P_p, P_q) = H_S\left(\frac{P_p + P_q}{2}\right) - \frac{1}{2}(H_S(P_p) + H_S(P_q)) \qquad (3)$$

where H_S is the Shannon entropy. We are interested in computing a hypergraph kernel between pairs of hypergraphs using the Jensen-Shannon divergence. For a pair of hypergraphs $G_p(V_p, E_p)$ and $G_q(V_q, E_q)$, the Jensen-Shannon divergence is given by

$$JSD(P\{G_p\}, P\{G_q\}) = H_S(P\{G_p \oplus G_q\}) - \frac{H_S(P\{G_p\}) + H_S(P\{G_q\})}{2} \qquad (4)$$

where $P\{G_p\}$ and $P\{G_q\}$ are the probability distributions on $G_p(V_p, E_p)$ and $G_q(V_q, E_q)$, and $P\{G_p \oplus G_q\}$ is the probability distribution on the composite hypergraph $G_p \oplus G_q$ of $G_p(V_p, E_p)$ and $G_q(V_q, E_q)$. Here $H_S(P\{G_p\})$ is the Shannon entropy for the probability distribution $P\{G_p\}$ given by

$$H_S(P\{G_p\}) = -\sum_{k=1}^{|V_p|} P\{G_p\}(k) \log P\{G_p\}(k) \qquad (5)$$

where $P\{G_p\}(k)$ is the k-th element of the probability distribution $P\{G_p\}$. We will use the disjoint union to construct the composite hypergraph $G_p \oplus G_q$.

3   Jensen-Shannon Hypergraph Kernel

In this section, we show how to establish a Jensen-Shannon kernel for hypergraphs. We commence by defining the probability distribution associated with the steady state random walk on a hypergraph. Then we show how the probability distributions of hypergraphs can be used to compute the required Shannon entropies for the Jensen-Shannon kernel between pairs of hypergraphs.

3.1   Probability Distribution over Steady State Random Walk

We use the steady state random walk on a hypergraph to calculate the probability distribution for the Shannon entropy. For a hypergraph $G(V, E)$ with the incidence matrix $\mathcal{H}$ defined in (1), the vertex degree $d(v_i)$ for $v_i \in V$ is $d(v_i) = \sum_{e_j \in E} \mathcal{H}(i, j)$. Based on the definition in [4], the probability of a steady state random walk through hyperedges on $G(V, E)$ visiting vertex v_i is

$$P\{G\}(i) = P_G(v_i) = d(v_i)/\sum_{u \in V} d(u) \tag{6}$$

3.2   Composite Structure: Disjoint Union Hypergraph

We use the disjoint union of a pair of hypergraphs as the composite structure to compute the Jensen-Shannon kernel of hypergraphs. Based on the definition in [1], the disjoint union of a pair of hypergraphs is a binary operation that combines all distinct elements of the pair of hypergraphs, while retaining the original set of memberships as a distinguishing characteristic. For a pair of hypergraphs $G_p(V_p, E_p)$ and $G_q(V_q, E_q)$, the disjoint union hypergraph $G_U(V_U, E_U)$ of $G_p(V_p, E_p)$ and $G_q(V_q, E_q)$ is denoted as

$$G_U(V_U, E_U) = G_p(V_p, E_p) \cup G_q(V_q, E_q) = \{V_p \cup V_q, E_p \cup E_q\} \tag{7}$$

where $G_p(V_p, E_p)$ and $G_q(V_q, E_q)$ is the connected components of the disjoint union hypergraph $G_U(V_U, E_U)$. The probabilities of the steady state random walks visiting vertices v_p and v_q through hyperedges in the individual components $G_p(G_p, G_p)$ and $G_q(G_q, G_q)$ of the union are $P\{G_p\}(i_p) = d(v_{p_{i_p}})/\sum_{u_p \in V_p} d(u_p)$ and $P\{G_q\}(i_q) = d(v_{q_{i_q}})/\sum_{u_q \in V_q} d(u_q)$ respectively. A steady state random walk which departs from a vertex in one of the components is unable to visit any vertices in the other component. In the disjoint union, the probabilities of a steady state random walk departing from $G_p(V_p, E_p)$ and $G_q(V_q, E_q)$ are $\alpha_p = |V_p|/(|V_p| + |V_q|)$ and $\alpha_q = |V_q|/(|V_p| + |V_q|)$ respectively. Then the probabilities of such a steady state random walk departing from a random vertex in $G_U(V_U, E_U)$ and visiting vertices $v_{p_{i_p}}$ and $v_{q_{i_q}}$ in components $G_p(V_p, E_p)$ and $G_q(V_q, E_q)$ are $\alpha_p P\{G_p\}(i_p)$ and $\alpha_q P\{G_q\}(i_q)$ respectively. In this context, we obtain the probability distribution of a steady state random walk visiting vertices through hyperedges in $G_U(V_U, E_U)$ as

$$P\{G_U\} = P_{G_U} = \alpha_p P\{G_p\} + \alpha_q P\{G_q\} \tag{8}$$

where $P\{G_p\}$ and $P\{G_q\}$ are the probability distributions of individual components of $G_p(V_p, E_p)$ and $G_q(V_q, E_q)$ associated with their own steady state random walks respectively. The Shannon entropy of the disjoint union hypergraph $G_U(V_U, E_U)$ is then defined as

$$H_S(G_U) = H_S(P\{G_U\}) = H_S(\alpha_p P\{G_p\} + \alpha_q P\{G_q\}) \tag{9}$$

3.3   Jensen-Shannon Kernel on Hypergraphs

We define a Jensen-Shannon kernel on hypergraphs. Suppose the hypergraphs under consideration are represented by the set $\{G_1, \cdots, G_p, \cdots, G_q, \cdots, G_N\}$. For a pair of hypergraphs $G_p(V_p, E_p)$ and $G_q(V_q, E_q)$, we construct the disjoint union hypergraph $G_U(V_U, E_U)$ as the composite structure of $G_p(V_p, E_p)$ and $G_q(V_q, E_q)$. Associated with the function defined in (4) and (9), the Jensen-Shannon hypergraph kernel based on $G_U(V_U, E_U)$ is defined as

$$
\begin{aligned}
k_{JSHK}(P\{G_p\}, P\{G_q\}) &= \log 2 - (\alpha_p - \frac{1}{2})H_S(P\{G_p\}) - (\alpha_q - \frac{1}{2})H_S(P\{G_q\}) \\
&= \log 2 - \frac{2|V_p| - (|V_p| + |V_q|)}{2(|V_p| + |V_q|)}H_S(P\{G_p\}) - \frac{2|V_q| - (|V_p| + |V_q|)}{2(|V_p| + |V_q|)}H_S(P\{G_q\}) \\
&= \log 2 - \frac{|V_p| - |V_q|}{2(|V_p| + |V_q|)}H_S(P\{G_p\}) - \frac{|V_q| - |V_p|}{2(|V_p| + |V_q|)}H_S(P\{G_q\}) \tag{10}
\end{aligned}
$$

Since the probability distributions associated with the steady state random walks and the disjoint union hypergraphs of pairs of hypergraphs can be established through the incidence matrices directly, our proposed hypergraph kernel can hence accommodate both uniform and nonuniform hypergraphs.

3.4   Algorithmic Complexity

The computational complexity of the proposed Jensen-Shannon hypergraph kernel depends on three factors, these include 1) the construction of the disjoint union of hypergraphs, 2) the computation of probability distributions for pairs of hypergraphs and their disjoint union hupergraphs, and 3) the construction of the kernel matrices. Consider a hypergraph dataset with size N and two sample hypergraphs $G_p(V_p, E_p)$ and $G_q(V_q, E_q)$ with number of vertices m and n respectively. The construction of the disjoint union hypergraph $G_U(V_U, E_U)$ of $G_p(V_p, E_p)$ and $G_q(V_q, E_q)$ requires $O((m+n)^2)$ operations. Then the computation of the probability distribution for the disjoint union hypergraph $G_U(V_U, E_U)$ requires $O(m+n)$ operations. The computations of the probability distributions from $G_p(V_p, E_p)$ and $G_q(V_q, E_q)$ require $O(m)$ and $O(n)$ operations respectively. The construction of the kernel matrix requires $O(N^2/2)$ operations.

4   Experimental Results

4.1   Stability Evaluation

We commence by evaluating the stability of our Jensen-Shannon hypergraph kernel based on the disjoint unions on pairs of hypergraphs. The evaluation employs

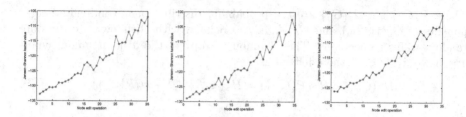

Fig. 2. Stability evaluation under hyperedge edit operation

three seed hypergraphs which have 350 vertices, 300 vertices and 250 vertices respectively. For each seed hypergraph, we perform random edit operations to simulate the effects of noise. The edit operations are based on hyperedge deletions. Since our proposed kernel can accommodate both uniform and nonuniform hypergraphs, these edit operated hypergraphs are easily accommodated by our proposed kernel. Fig.2 shows the effects of hyperedge deletions. In each plot the x-axis represents the fraction of hyperedges deleted, and the y-axis represents the value of the kernel $k_{JSHK}(G_o, G_e)$ between the original hypergraph G_o and its noise corrupted counterpart G_e. The plots show that there is an approximately linear relationship between the Jensen-Shannon hypergraph kernel and the number of the deleted hyperedges, i.e. the hypergraph edit distance. This implies that our method possess the ability to distinguish hypergraphs under controlled structural-errors.

4.2   Real-World Datasets

We compare our proposed Jensen-Shannon hypergraph kernel (JSHK) with several alternative state of the art structural characterization methods, these include 1) hypergraph characterizations using the Ihara zeta function (HCIZF) [9], 2) the truncated Laplacian spectra (TLS) and truncated normalized Laplacian spectra (TNLS) [8], 3) the Weisfeiler-Lehman subtree kernel [12], 4) the Ramon & Gaertner graph kernel [5], 5) the p-random walk graph kernel [6] 6) the random walk graph kernel [5], 7) the shortest path graph kernel [2], and 8) graphlet count graph kernel. We use three stantard graph based datasets extracted from bioinformatics datasets [12] and a hypergraph based dataset extracted from the COIL image dataset for experimental evaluation.

MUTAG: The MUTAG benchmark is based on graphs representing 188 chemical compounds, and aims to predict whether each compound possesses mutagenicity. The maximum and average number of vertices are 28 and 17.93 respectively. As the vertices and edges of each compound are labeled with a real number, we transform these graphs into unweighted graphs.

ENZYMES: The ENZYMES dataset is a dataset based on graphs representing protein tertiary structures consisting of 600 enzymes from the BRENDA enzyme database. In this case the task is to correctly assign each enzyme to one of the 6

EC top-level classes. The maximum and average number of vertices are 126 and 32.63 respectively.

D&D: The D&D dataset contains 1178 protein structures. Each protein is represented by a graph, in which the nodes are amino acids and two nodes are connected by an edge if they are less than 6 Angstroms apart. The prediction task is to classify the protein structures into enzymes and non-enzymes. The maximum and average number of vertices are 5748 and 284.32 respectively.

COIL: The COIL dataset consists of images of 100 objects. In our experiments, we use selected images for three similar cups, three similar bottles and three pieces of similar vegetable. For each object we employ 18 images captured from different viewpoints. The hypergraph are extracted using feature hypergraph method [9] The maximum and minimum vertices of COIL dataset are 549 and 213 respectively.

4.3   Experiments on Graphs Extracted from Bioinformatics Datasets

We evaluate the performance of our kernel (JSHK) on the graphs extracted from the bioinformatics datasets. We then perform 10-fold cross-validation associated with SMO-Support Vector Machine Classification to evaluate the performance of our kernel and the alternative methods, using nine samples for training and one for testing. All parameters of the SVMs were optimized. We report the average prediction accuracies of each method in Table 1, the runtime were measured under Matlab R2011a running on a ThinkPad T61P with 2.2GHz Intel 2-Core processor with 2GB RAM. We also compare our kernel with several state of the art graph kernels. Shervashidze et al.[12] have reported the accuracies of the graph kernels based on the same classification method and datasets as ours. The runtime of these methods were measured under Matlab R2008a running on an ApplePro with 3.0GHz Intel 8-Core processor with 16GB RAM. We report these accuracies and runtimes in Table.1.

The graphs in the D&D dataset are on average more than 284 nodes and at maximum 5748 nodes. The accuracy and runtime of our JSHK is competitive to that of the Weisfeiler-Lehman subtree kernel, graphlet count graph kernel and the shortest path kernel. The other alternative kernels did not finish in two days.

The graphs of the MUTAG dataset are of similar sizes, but correspond to very different structures. On this dataset, the accuracy of our JSHK outperforms all the other alternatives. The runtime of JSHK is competitive to that of the Weisfeiler-Lehman subtree kernel,graphlet count graph kernel and the shortest path kernel, and outperforms the other alternatives.

The graphs in the ENZYMES are of variable sizes. On this dataset, the accuracy of the JSHK is 27.05%. It is lower than the accuracies of the Weisfeiler-Lehman subtree kernel, graphlet count kernel and the shortest path kernel, but higher than that of the other alternatives. The runtime of JSHK outperforms the alternatives.

On the whole, the accuracy and runtime of our JSHK outperforms or is competitive to that of all the alternative kernels. Only the Weisfeiler-Lehman subtree kernel and the shortest path kernel are competitive to our JSHK.

Table 1. Performance and CPU Runtime Comparison
on Bioninformatics Datasets

Datasets	MUTAG	ENZYMES	D&D
JSHK	87.76%	27.05%	78.00%
Weisfeiler-Lehman	82.05%	46.42%	79.78%
Ramon& Gaertner	83.78%	13.35	57.27
p-random walk	79.19%	27.67%	66.64
random walk	80.72%	21.68%	71.70
shortest path	87.28%	41.68%	78.45
graphlet count	75.61%	32.70%	78.59
Datasets	MUTAG	ENZYMES	D&D
JSHK	2"	19"	14'59"
Weisfeiler-Lehman	6"	20"	11'
Ramon& Gaertner	40'60"	38$days$	103$days$
p-random walk	4'42"	10'	4$days$
random walk	12"	12'19"	48$days$
shortest path	2"	5"	23h17'2"
graphlet count	3"	25"	30'21"

4.4 Hypergraph Clustering Comparisons

In this subsection, we illustrate the clustering performance of proposed JSHK on
the hypergraph dataset extracted from the COIL image dataset. We also compare
our methods with several alternative state of the art hypergraph based learning
methods which include HCIZF, TLS and TLNS. We compute the kernel matrices or embedding vectors using our methods and the alternatives respectively.
Then we apply the K-means clustering method to compute the classification
accuracies for the three groups of testing hypergraphs. We report the highest
prediction accuracies of each method in Table.1(-:over computing, i.e.infinite
value). Table.2 indicates that high accuracies for our methods are achievable.
Our JSHK outperforms all the alternatives. For the cup object images based
hypergraphs which the maximum and minimum vertices are 310 and 213 respectively, the accuracy of HCIZF is competitive to that of our methods. But for the
bottle and vegetable object images based hypergraphs, HCIZF generates over
computing (i.e.infinite value), since the maximum and minimum vertices are 549
and 305 respectively. The experiments reveal that our proposed JSHK can easily
scale up even to large size hypergraph data.

Table 2. Accuracy of Classification Comparison
on Hypergraphs

Datasets	**JSHK**	TLS	TNLS	HCIZF
Cups	97.55%	86.60%	52.78%	96.29%
Bottles	100%	77.41%	83.39%	–
Vegetable	94.44%	77.20%	69.65%	–

5 Conclusion

In this paper, we have shown how to construct a Jensen-Shannon kernel for hypergraphs using the Jensen-Shannon divergence. The method is based on the probability distribution over the steady state random walk on a hypergraph. We ues the Shannon entropy to measure the mutual information between a pair of hypergraphs and establish hypergraph kernel. Experimental results reveal that our kernel is competitive to the state of the art graph kernels and hypergraph based learning methods.

References

1. Berge, C.: Hypergraphs: combinatorics of finite sets. North-Holland (1989)
2. Borgwardt, K.M., Kriegel, H.P.: Shortest-path kernels on graphs. In: Proceedings of the IEEE International Conference on Data Mining, pp. 74–81 (2005)
3. Chertok, M., Keller, Y.: Efficient high order matching. IEEE Transactions on Pattern Analysis and Machine Intelligence 32, 2205–2215 (2010)
4. Cooper, C., Frieze, A.M., Radzik, T.: The Cover Times of Random Walks on Hypergraphs. In: Kosowski, A., Yamashita, M. (eds.) SIROCCO 2011. LNCS, vol. 6796, pp. 210–221. Springer, Heidelberg (2011)
5. Gärtner, T., Flach, P.A., Wrobel, S.: On graph kernels: Hardness results and efficient alternatives. In: Proceedings of the Conference on Computational Learning Theory, pp. 129–143 (2003)
6. Kashima, H., Tsuda, K., Inokuchi, A.: Marginalized kernels between labeled graphs. In: Proceedings of the International Conference on Machine Learning, pp. 321–328 (2003)
7. Martins, A.F.T., Smith, N.A., Xing, E.P., Aguiar, P.M.Q., Figueiredo, M.A.T.: Nonextensive information theoretic kernels on measures. Journal of Machine Learning Research 10, 935–975 (2009)
8. Oliva, A., Torralba, A.: Modeling the shape of the scene: A holistic representation of the spatial envelope. International Journal of Computer Vision 42, 145–175 (2001)
9. Ren, P., Aleksic, T., Wilson, R.C., Hancock, E.R.: A polynomial characterization of hypergraphs using the ihara zeta function. Pattern Recognition 44, 1941–1957 (2011)
10. Shashua, A., Levin, A.: Linear image coding for regression and classification using the tensor-rank principle. In: Proceedings of the IEEE Conference on Computer Vision and Pattern Recognition, pp. 42–49 (2001)
11. Shervashidze, N., Borgwardt, K.M.: Fast subtree kernels on graphs. In: Proceedings of the Neural Information Processing Systems, pp. 1660–1668 (2009)
12. Shervashidze, N., Schweitzer, P., van Leeuwen, E., Mehlhorn, K., Borgwardt, K.: Weisfeiler-lehman graph kernels. Journal of Machine Learning Research 1, 1–48 (2010)
13. Wachman, G., Khardon, R.: Learning from interpretations: a rooted kernel for ordered hypergraphs. In: Proceedings of the International Conference on Machine Learning, pp. 943–950 (2007)
14. Zass, R., Shashua, A.: Probabilistic graph and hypergraph matching. In: Proceedings of the IEEE Conference on Computer Vision and Pattern Recognition (2008)

Heat Flow-Thermodynamic Depth Complexity in Directed Networks

Francisco Escolano[1], Boyan Bonev[1], and Edwin R. Hancock[2]

[1] University of Alicante
{sco,boyan}@dccia.ua.es
[2] University of York
erh@cs.york.ac.uk

Abstract. In this paper we extend the heat diffusion-thermodynamic depth approach for undirected networks/graphs to directed graphs. This extension is motivated by the need to measure the complexity of structural patterns encoded by directed graphs. It consists of: a) analyzing and characterizing heat diffusion traces in directed graphs, b) extending the thermodynamic depth framework to capture the second-order variability of the diffusion traces to measure the complexity of directed networks. In our experiments we characterize several directed networks derived from different natural languages. We show that our proposed extension finds differences between languages that are blind to the classical analysis of degree distributions.

1 Introduction

The quantification of the complexity of patterns plays a fundamental role in pattern recognition and machine learning. Information theory [1] provides principled approaches to the analysis complexity that include minimum description length (MDL) and minimum message length (MML) which allow us to find the model that parsimoniously describes vectorial data. However, the latter principles have not been incorporated to the graph domain until recently (see [2] for trees and [3] for edge-weighted undirected graphs). In fact, the intersection between structural pattern recognition and complex networks has proved to be fruitful and has inspired several interesting measures of graph complexity. Most of these measures rely on quantifying the degree of randomness of the structural representation. For instance, Körner entropy was motivated by the need to measure how much information can flow through a graph, when pairs of symbols can be confused [4]. This implies admitting a probability that a memoryless source emits a symbol. For each node in the graph there is a symbol, and two nodes are adjacent if their symbols are distinguishable. In this setting entropy is defined as the minimal cross entropy between the probability distribution and the vertex packing polytope of the graph. Since the vertex packing polytope is the convex hull of all characteristic vectors of stable sets of the graph, the task of measuring Körner entropy relies on solving an NP problem. More recently, Passerini and Severini have applied the quantum (von Neumann) entropy to graphs [5]. The state of a quantum mechanical system of a finite dimension is defined by a

G.L. Gimel' farb et al. (Eds.): SSPR & SPR 2012, LNCS 7626, pp. 190–198, 2012.
© Springer-Verlag Berlin Heidelberg 2012

density matrix for networks that can be modeled the combinatorial Laplacian. As a result the von Neumann entropy is given by the Shannon entropy of the Laplacian eigenvalues (normalized by the volume of the graph). This definition of graph entropy is maximal for random graphs, minimal for complete ones and intermediate for star graphs. Although the computation of the von Neumann entropy is cubic with the size of the network, it has recently been shown how this can be approximated using quadratic degree statistics and successfully applied to structural discrimination [6].

The Körner and von Neumann entropies are two examples of *randomness* complexity applied to graphs. An alternative is to use the so called *statistical* complexity and to quantify the regularities of the structure beyond its randomness [7]. The general underlying principle of statistical complexity is that it is zero for both random and regular/complete (completely ordered) graphs. A very recent example of statistical complexity is the Estrada heterogeneity index [8]. This index is defined as the Dirichlet sum of root squared degree differences. The obtained index is related to the Randić index [9]. Estrada's heterogeneity index is obviously zero for both random graphs and regular or complete ones. Another method which goes beyond randomness complexity is our recent application of thermodynamic depth [10] to the domain of graphs [11]. This involves defining both the macro-states (the graph) and the micro-states (the nodes). Complexity is quantified in terms of the amount of structural variability as a node evolves through a subgraph containing adjacent nodes (first order expansion) and eventually encompasses the full graph (if possible). Each expansion step is characterized by the temporal trace of heat flowing through the network. The sequence of expansions for a given node is referred to as a *history* [12], which contains the heat flow traces of each node. The variability of a given history quantifies how uniformly the full graph emerges from the corresponding node. The average heat flow traces of all the nodes can be combined to yield a second-order variability measure, the so called thermodynamic depth of the graph. Shallow (low-variability) graphs are characterized by similar histories with low variance and this means that heat flows satisfies similar topological constraints at each node. In contrast, deep graphs emerge from histories with large variance. Both random and complete graphs have zero depth, whereas grids and linear graphs have larger depths. A nice property of our thermodynamic depth approach is that it can be applied not only to heat flow traces but also to the heterogeneity index and the von Neumann entropy. It has been successfully used to correlate PPI networks with the phyla of bacteria.

All of the above approaches are confined to undirected graphs since many networks or graphs in the real world can be modeled with them: (e.g. protein-protein interaction (PPI) networks, shapes as Delaunay triangulations, adjacency graphs in images). However, considering the orientation of edges (e.g. directed trees and causal graphs in Bayesian networks) adds meaningful information which allows us to model networks such as metabolic pathways (cascades of chemical reactions) as well as natural languages (where the relative order of words matters) together with social networks (e.g. citation networks). The Internet is a clear example of a directed network and Pagerank is an example of well known algorithm [13] which exemplifies the formal difficulty of analyzing directed graphs and the study of how the information flows through them in the context of the Internet. When a sink node (a node with zero outdegree) is reached by a random walk, there is a given (small) probability of making a transition to any other node in the network (this is called *teleporting*). The Laplacian of a directed graph can be defined through a symmetrization process provided that the transition matrix for the random walks allows for teleporting. Given a symmetric semi-definite operator such as the directed Laplacian, it is straightforward to compute heat kernels and thus to evaluate diffusion flow traces. A natural way of quantifying the complexity of directed graphs is to extend our thermodynamic depth approach to deal with oriented edges.

In this paper we will address that challenging point as follows. In Section 2 we will describe the directed Laplacian. Section 3 is devoted to the analysis of the fundamental formal differences between undirected and directed graphs in terms of heat flow diffusion using heat kernels. In Section 4 we redefine the thermodynamic depth for digraphs. Experiments and discussion (Section 5) are focused on the analysis of directed networks derived from natural languages and the quantification of their complexities. For instance, we show that our proposed extension can identify differences between languages that are blind to the classical analysis of degree distributions. We conclude this paper with a summary of our contributions and suggestions for future work.

2   The Laplacian of a Directed Graph

A directed graph (digraph) $G = (V, E)$ with $n = |V|$ vertices and edges $E \subseteq E \times E$ is encoded by and adjacency matrix $\mathbf{A}$ where $A_{ij} > 0$ if $i \to j \in E$ and $A_{ij} = 0$ otherwise (this definition includes weigthed adjacency matrices). The outdegree matrix $\mathbf{D}$ is a diagonal matrix where $D_{ii} = \sum_{j \in V} A_{ij}$. The transition matrix $\mathbf{P}$ is defined by $P_{ij} = \frac{A_{ij}}{D_{ii}}$ if $(i, j) \in E$ and $P_{ij} = 0$ otherwise. The transition matrix is key to defining random walks on the digraph and P_{ij} is the probability of reaching node j from node i. Given these definitions we have that $\sum_{j \in V} P_{ij} \neq 1$ in general. In addition, $\mathbf{P}$ is irreducible iff G is strongly connected (there is path from each vertex to every other vertex). If $\mathbf{P}$ is irreducible, the Perron-Frobenius theorem ensures that there exists a left eigenvector ϕ satisfying $\phi^T \mathbf{P} = \lambda \phi^T$ and $\phi(i) > 0 \, \forall i$. If $\mathbf{P}$ is aperiodic (spectral radius $\rho = 1$) we have $\phi^T \mathbf{P} = \rho \phi^T$ and all the other eigenvalues have an absolute value smaller that $\rho = 1$. By ensuring strong connection and aperiodicity we also ensure that any random walk in a directed graph satisfying these two properties converges to a unique stationary distribution.

Normalizing ϕ so that $\sum_{i \in V} \phi(i) = 1$, we encode the eigenvector elements as a probability distribution. This normalized row vector ϕ corresponds to the stationary distribution of the random walks defined by $\mathbf{P}$ since $\phi \mathbf{P} = \phi$. Therefore, $\phi(i) = \sum_{j, j \to i} \phi(j) P_{ji}$, that is, the probability of that the random walk is at node i is the sum of all incoming probabilities from all nodes j satisfying $j \to j$.

If we define $\Phi = diag(\phi(1) \ldots \phi(n))$, we have that the j-th column of $\Phi\mathbf{P}$ has the form $(\Phi\mathbf{P})_j = [\phi(1)P_{j1}\, \phi(2)P_{j2}\, \ldots\, \phi(n)P_{jn}]^T$, that is, $\sum_{i=1}^{n}(\Phi\mathbf{P})_j = \phi(i)$. Since $(\Phi\mathbf{P})^T = \mathbf{P}^T\Phi$ the i-th row of $\mathbf{P}^T\Phi$ is identical to the j-th column of $\Phi\mathbf{P}$ and thus $\sum_{j=1}^{n}(\mathbf{P}^T\Phi)^i = \phi(i)$. Consequently, the matrix $\Phi\mathbf{P} + \mathbf{P}^T\Phi$ is also symmetric and the sum of the elements in the columns in i-th row (or the sum of the elements in the rows in the same column) is like

$$\sum_{j=1}^{n}(\Phi\mathbf{P} + \mathbf{P}^T\Phi)^i = \sum_{j=1}^{n}(\Phi\mathbf{P} + \mathbf{P}^T\Phi)_i = \underbrace{\sum_{i,i\to j}\phi(i)P_{ij} + \underbrace{\sum_{j,j\to i}\phi(j)P_{ji}}_{\phi(i)}}, \quad (1)$$

i.e. the sum of both incoming and outcoming probabilities. Since ϕ corresponds to the stationary distribution we have that $\sum_{i,i\to j}\phi(i)P_{ij} = \phi(i)$ for $(\Phi\mathbf{P})^T = \mathbf{P}^T\Phi = \phi^T$. Consequently, $\sum_{j=1}^{n}(\Phi\mathbf{P} + \mathbf{P}^T\Phi)^i = 2\phi(i)\ \forall i$. This leads to the definition of the following matrices:

$$\mathbf{L} = \Phi - \frac{\Phi\mathbf{P} + \mathbf{P}^T\Phi}{2} \quad \text{and} \quad \mathcal{L} = I - \frac{\Phi^{1/2}\mathbf{P}\Phi^{-1/2} + \Phi^{-1/2}\mathbf{P}^T\Phi^{1/2}}{2}, \quad (2)$$

where $\Phi = diag(\phi(1)\ldots\phi(n))$, $\mathbf{L}$ is the *combinatorial directed Laplacian* and $\mathcal{L}$ is the *normalized directed Laplacian* [14]. Focusing on $\mathbf{L}$ we have

$$L_{ij} = \begin{cases} \phi(i) & \text{if } i = j \\ -\left(\frac{\phi(i)P_{ij}+\phi(j)P_{ji}}{2}\right) & \text{otherwise .} \end{cases}, \quad (3)$$

where it is assumed that $P_{ii} = 0\ \forall i$. Otherwise $L_{ii} = \phi(i)(1 - P_{ii})$.

Symmetrizing $\mathbf{P}$ leads to real valued eigenvalues and eigenvectors. In addition Φ plays the role of a degree matrix and off-diagonal entries are designed so that the all-ones vector $\mathbf{1}$ is the eigenvector f_1 of the combinatorial Laplacian (the vector with eigenvalue $\lambda_1 = 0$). This is due to the fact that the sum of the i-th row of $\mathbf{L}$ is $\sum_{j=1}^{n}(\mathbf{L})^i = \phi(i) - 2\phi(i)/2 = 0$. In any case, satisfying irreducibility is difficult in practice since sink vertices may arise frequently. For instance, a circular graph C_n given by $1 \to 2 \to 3 \to \ldots \to n \to 1$ is clearly irreducible. However, the linear graph L_n obtained by removing $n \to 1$ from the cycle is reducible since we have a sink at n and the graph is no longer strongly connected. Sink vertices introduce rows of zeros in $\mathbf{A}$ and consequently in $\mathbf{P}$. The consequence is the non-existence of a left Perron eigenvector and this renders computing the Laplacians is impossible. A formal trick consists of replacing $\mathbf{P}$ by $\mathbf{P}'$ so that $P'_{ij} = \frac{1}{n}$ if $A_{ij} = 0$ and $D_{ii} = 0$. This strategy is adopted in Pagerank [13] and allows for *teleporting* acting on the random walk to any other node in the graph. Teleporting is modeled by redefining $\mathbf{P}$ in the following way: $\mathbf{P} = \eta\mathbf{P}' + (1-\eta)\frac{\mathbf{1}\mathbf{1}^T}{n}$ with $0 < \eta < 1$. The new $\mathbf{P}$ ensures both irreducibility and aperiodicity and this allows us to both apply $\mathbf{P}'$ with probability η and to teleport from any node with $A_{ij} = 0$ with probability $1 - \eta$. In [15] a trade-off between large values η (preserving more the structure of $\mathbf{P}'$) and small ones (potentially increasing the spectral gap) is recommended. For instance, in [16], where the task is to learn classifiers on directed graphs, the setting is $\eta = 0.99$. When using the new $\mathbf{P}$ we always have that $P_{ii} \neq 0$ due to the Pagerank masking. Such masking may introduce significant interferences in heat diffusion when the Laplacian is used to derive the heat kernel.

3 Directed Heat Kernels and Heat Flow

The definition of $\mathbf{P}$ is critical for finding both the directed combinatorial Laplacian $\mathbf{L}$ and the directed normalized Laplacian $\mathcal{L}$. Consequently it is also critical in determining the behavior of the heat kernel derived from the latter matrices. If the graph is strongly connected and aperiodic the original $\mathbf{P}$ has a unique equilibrium distribution and the components of the combinatorial Laplacian are given by Eq. 3. Otherwise the above conditions are enforced by exploiting the Pagerank transformation. In any case, the $n \times n$ *heat/diffusion kernel* $\mathbf{K}_\beta(G)$ of the graph is the solution to the heat/diffusion equation: $\frac{\partial \mathbf{K}_\beta}{\partial \beta} = -\mathbf{L}\mathbf{K}_\beta$, and is given by the matrix exponentiation $\mathbf{K}_\beta(G) = exp(-\beta\mathbf{L})$, for $\beta \geq 0$. Using the Taylor series (which may be useful for large n) we have: $\mathbf{K}_\beta(G) = e^{-\beta\mathbf{L}} = \mathbf{I}_n - \beta\mathbf{L} + \frac{\beta^2}{2!}\mathbf{L}^2 - \frac{\beta^3}{3!}\mathbf{L}^3 + \ldots$, where $\mathbf{I}_n$ is the $n \times n$ identity matrix. In this regard, the matrix $\mathbf{W} = \frac{\Phi\mathbf{P} + \mathbf{P}^T\Phi}{2}$ can be seen as the *weight matrix* of the undirected graph G_u associated with G (which may be also weighted) through $\mathbf{P}$ and Φ. Therefore, the analysis of how the heat flows through G is equivalent to the analysis of how it flows through G_u.

We commence by reviewing the concept of *heat flow* [11]. Firstly, the spectral decomposition of the diffusion kernel is $\mathbf{K}_\beta(G) = exp(-\beta\mathcal{L}) \equiv \Psi\Lambda\Psi^T$, where $\Lambda = diag(e^{-\beta\lambda_1}, e^{-\beta\lambda_2}, \ldots, e^{-\beta\lambda_n})$, $\Psi = [\psi_1, \psi_2, \ldots, \psi_n]$, and $\{(\lambda_i, \psi_i)\}_{i=1}^n$ are the eigenvalue-eigenvector pairs of $\Phi - \mathbf{W}$. Hence $K_{\beta_{ij}} = \sum_{k=1}^n \psi_k(i)\psi_k(j)e^{-\lambda_k\beta}$, and $K_{\beta_{ij}} \in [0, 1]$ is the (i, j) entry of a doubly stochastic matrix. Doubly stochasticity for all β implies *heat conservation* in the system as a whole. That is, not only in the nodes and edges of the graph but also in the *transitivity links* eventually established between non-adjacent nodes (if i is not adjacent to j, eventually will appear an entry $K_{\beta_{ij}} > 0$ for β large enough). The total *directed heat* flowing through the graph at a given β *(instantaneous directed flow)* is given by

$$F_\beta(G) = \sum_{i \to j} A_{ij} \left(\sum_{k=1}^n \psi_k(i)\psi_k(j)e^{-\lambda_k\beta} \right), \tag{4}$$

A more compact definition of the flow is $F_\beta(G) = \mathbf{A} : \mathbf{K}_\beta$, where $\mathbf{X} : \mathbf{Z} = \sum_{ij} X_{ij}Z_{ij} = trace(\mathbf{X}\mathbf{Z}^T)$ is the Frobenius inner product. While instantaneous flow for the heat flowing through the edges of the graph, it accounts neither for the heat remaining in the nodes nor for that in the transitivity links. The limiting cases are $F_0 = 0$ and $F_{\beta_{max}} = \frac{1}{n}\sum_{i \to j} A_{ij}$ which is reduced to $\frac{|E|}{n}$ if G is unattributed ($A_{ij} \in \{0, 1\} \forall ij$). Defining F_β in terms of $\mathbf{A}$ instead of $\mathbf{W}$, we retain the *directed* nature of the original graph G. The function derived from computing $F_\beta(G)$ from $\beta = 0$ to β_{max} is the so called *directed heat flow trace.* These traces exhibit the following differences with respect to those of unattributed undirected graphs:

1. They satisfy the *phase transition principle* [11] (although the formal proof is out of the scope of this paper).
2. In general heat flow diffuses more slowly than in the undirected case and phase transition points (PTPs) appear later. This is due to the constraints imposed by **A**.
3. PTPs may coincide with equilibrium points even when the directed graph is not the complete one. This happens in strongly connected graphs with many cycles (where connectivity constraints are relaxed) but the traces of single cycles do not have this behavior.
4. The sum of all walks connecting every pair of nodes is maximal (if we exclude the sum of all cycles for each node) for all components corresponding to non-zero values in **A**. This is straightforward to prove by expressing the kernel in terms of sums of walks.
5. Graphs with at least one sink require the Pagerank mastering strategy which introduces noise in the diffusion process. This noise has no practical effect even for moderate/small values of η (e.g. $\eta = 0.15$).
6. The heat diffusion process does not only allow increasing heat values for setting transitivity links but it may also happen at directed edges. The main reason is that $\mathbf{K}_\beta$ is expressed in terms of an undirected attributed graph given by **W** even for non-attributed strongly connected and aperiodic digraphs.

4 Heat Flow - Thermodynamic Depth Complexity

The application of thermodynamic depth (TD) to characterize the complexity of directed graphs demands the formal specification of the micro-states whose history leads to the macro-state (of the network). Here we define such micro-states in terms of *expansion subgraphs*.

Let $G = (V, E)$ with $|V| = n$. Then the *directed history of a node* $i \in V$ is $h_i(G) = \{e(i), e^2(i)), \ldots, e^p(i)\}$ where: $e(i) \subseteq G$ is the *first-order expansion subgraph* given by i and all $j : i \to j$. If there are nodes j also satisfying $j \to i$ then these edges are included. If node i is a sink then $e(i) = i$. Similarly $e^2(i) = e(e(i)) \subseteq G$ is the *second-order expansion* consisting on $j \to z : j \in V_{e(i)}, z \notin V_{e(i)}$, including also $z \to j$ if these edges exists and $j \to z$. This process continues until p cannot be increased. If G is strongly connected $e^p(i) = G$, otherwise $e^p(i)$ is the strongly connected component to which i belongs.

Every $h_i(G)$ defines a different causal trajectory which may lead to G itself, if it is strongly connected, or to one of its strongly connected components otherwise. Thus, in terms of TD the full graph G or the union of its strongly connected components is the macro-state (macroscopic state). The *depth* of such macro-states relies on the variability of the causal trajectories leading to them. The higher the variability, the more complex it is to explain how the macro-state is reached and the deeper is this state. Therefore, in order to characterize each trajectory we combine the heat flow complexities of its expansion subgraphs by means of defining *minimal enclosing Bregman balls* (MEBB) [18]. Here we use the I-Kullback-Leibler (I-KL) Bregman divergence between traces f and g: $D_F(\boldsymbol{f}\|\boldsymbol{g}) = \sum_{i=1}^{d} f_i \log \frac{f_i}{g_i} - \sum_{i=1}^{d} f_i + \sum_{i=1}^{d} g_i$ with convex generator $F(\boldsymbol{f}) = \sum_{i=1}^{d} (f_i \log f_i - f_i)$.

Given $h_i(G)$, the heat flow complexity $\boldsymbol{f}_t = f(e^t(i))$ for the $t-th$ expansion of i, a generator F and a Bregman divergence D_F, the *causal trajectory* leading to G (or one of its strongly connected components) from i is characterized by the center $\boldsymbol{c}_i \in R^d$ and radius $r_i \in R$ of the MEBB $\mathcal{B}^{c_i, r_i} = \{\boldsymbol{f}_t \in \mathcal{X} : D_F(\boldsymbol{c}_i\|\boldsymbol{f}_t) \leq r_i\}$. Solving for the center and radius implies finding $\boldsymbol{c}^*$ and r^* minimizing r subject to $D_F(\boldsymbol{c}_i\|\boldsymbol{f}_t) \leq r \; \forall t \in \mathcal{X}$ with $|\mathcal{X}| = T$. Considering the Lagrange multipliers α_t we have that $\boldsymbol{c}^* = \nabla^{-1} F(\sum_{t=1}^{T} \alpha_t \boldsymbol{f}_t \nabla F(\boldsymbol{f}_t))$. The efficient algorithm in [18] estimates both the center and multipliers. This idea is closely related to Core Vector Machines [19], and it is interesting to focus on the non-zero multipliers (and their support vectors) used to compute the optimal radius. More precisely, the multipliers define a convex combination and we have $\alpha_t \propto D_F(\boldsymbol{c}^*\|\boldsymbol{f}_t)$, and the radius is simply chosen as: $r^* = \max_{\alpha_t > 0} D_F(\boldsymbol{c}^*\|\boldsymbol{f}_t)$.

Given the directed graph $G = (V, E)$, with $|V| = n$ and all the n pairs $(\boldsymbol{c}_i, r_i)$, the *heat flow-thermodynamic depth complexity* of G is characterized by the MEBB $\mathcal{B}^{c,r} = \{\boldsymbol{c}_t \in \mathcal{X}_i : D_F(\boldsymbol{c}\|\boldsymbol{c}_i) \leq r\}$. As a result, the *TD depth of the directed graph* is given by $\mathcal{D}(G) = r$. This definition of depth is highly consistent with summarizing node histories with second-order variability operators to find a global causal trajectory which is as tightly bounded as possible.

5   Experiments: Analysis of Language Complexity

We analyse networks extracted from the adjacency of words for different languages. We used a subset of the parallel corpora published in the Official Journal of the European Union. We used 100,000 lines of text from each language, all of them corresponding to the same text (human translation). The languages included in this study are: Bulgarian (BG), Czech (CS), Danish (DA), German (DE), Greek (EL), English (EN), Spanish (ES), Estonian (ET), Finnish (FI), French (FR), Hungarian (HU), Italian (IT), Lithuanian (LT), Latvian (LV), Maltese (MT), Dutch (NL), Polish (PL), Portuguese (PT), Slovak (SK), Slovene (SL) and Swedish (SV).

The directed adjacency graph represent words which appear consecutively in a text. We take the words as they appear in the text (surface form) and not only their lemmas. In this way we retain morphology, which imposes different restrictions in each language. In the graphs we construct the edges commencing from each node (word) V_i connect to the words which follow V_i. Thus a language with no restrictions is represented by a fully connected graph. We also take into account the frequency of each connection occurring in the corpus, and we store this information as attributes for the directed edges. This means that we give a greater importance to those adjacencies between words which are used more frequently (in the corpus). Although we do not store the frequency of each word

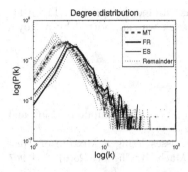 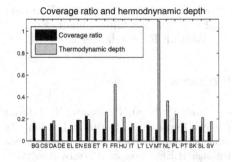

Fig. 1. Left: Log-log plot of the degree distribution of three languages (MT, FR, ES). The remaining languages are represented as well. All distributions behave in a similar way. We found no significant difference between the degree distributions of different languages. Right: The coverage measures the amount of text covered by the n most frequent surface forms.

in the graph representation we do use it for selecting the n most frequent words which constitute the nodes V_i.

In Fig. 1-left we show that the classical analysis based on the degree distribution is blind to differences of complexities between languages (all languages follow a similar degree distribution). In Fig. 1-right we compare the thermodynamic depth for different languages and show the amount of text that was covered by the graph of each language. Each of the graphs have $n = 500$ nodes which correspond to the n most frequent surface forms. These n surfaces cover part of the corpus of the language and the remainder of the surfaces in the corpus are not represented by the graph because of their lower frequency. We may take as a baseline for the complexity of a language the coverage ratio of n surface forms. An intuitive explanation is that if all the languages had a similar number of different lemmas in the parallel corpora, then the number of different surface forms would depend on the morphology of each language. A simpler morphology would enable the n surfaces to cover a larger amount of text than that covered by a rich morphology. This baseline does not capture all the subtle complexities of the network formed by the adjacency relation between words. The bar plot shows that there are some languages which do have the same tendencies both for thermodynamic depth and coverage. This is not the case of FR and MT.

6 Conclusions and Future Work

In this work we extend the heat-diffusion TD initially designed for unattributed undirected graphs to digraphs. We analyze the Laplacian operator used to that end and the consequences of using it to compute the heat flow. We enunciate several properties of heat diffusion traces in digraphs. In our experiments we compute the complexity of several languages and find differences that are blind to degree distribution analysis. Our future work includes the formal proof of the properties and the exploration of other graph-based representations of languages.

Acknowledgements. Francisco Escolano was funded by project TIN2011-27043 of the Spanish Government. Edwin Hancock was supported by a Royal Society Wolfson Research Merit Award.

References

1. Escolano, F., Suau, P., Bonev, B.: Information Theory in Computer Vision and Pattern Recognition. Springer, London (2009)
2. Torsello, A., Hancock, E.R.: Learning Shape-Classes Using a Mixture of Tree-Unions. IEEE Tran. on Pattern Analysis and Mach. Intelligence 28(6), 954–967 (2006)
3. Torsello, A., Lowe, D.L.: Supervised Learning of a Generative Model for Edge-Weighted Graphs. In: Proc. of ICPR (2008)
4. Körner, J.: Coding of an Information Source Having Ambiguous Alphabet and the Entropy of Graphs. In: Transactions of of the 6th Prague Conference on Information Theory, pp. 411–425 (1973)
5. Passerini, F., Severini, S.: The von Neumann Entropy of Networks. arXiv:0812.2597v1 (December 2008)
6. Han, L., Escolano, F., Hancock, E.R., Wilson, R.: Graph Characterizations From Von Neumann Entropy. Pattern Recognition Letters (in press, 2012)
7. Feldman, D.P., Crutchfield, J.P.: Measures of Statistical Complexity: Why? Physics Letters A 238(4-5), 244–252 (1998)
8. Estrada, E.: Quantifying Network Heterogeneity. Phys. Rev. E 82, 066102 (2010)
9. Randić, M.: Characterization of Molecular Branching. Journal of the American Chemical Society 97(23), 6609–6615 (1975)
10. Lloyd, S., Pagels, H.: Complexity as Thermodynamic Depth Ann. Phys. 188, 186 (1988)
11. Escolano, F., Hancock, E.R., Lozano, M.A.: Heat Diffusion: Thermodynamic Depth Complexity of Networks. Phys. Rev. E 85, 036206 (2012)
12. Escolano, F., Lozano, M.A., Hancock, E.R., Giorgi, D.: What Is the Complexity of a Network? The Heat Flow-Thermodynamic Depth Approach. In: Hancock, E.R., Wilson, R.C., Windeatt, T., Ulusoy, I., Escolano, F. (eds.) SSPR&SPR 2010. LNCS, vol. 6218, pp. 286–295. Springer, Heidelberg (2010)
13. Page, L., Brin, S., Motwani, R., Winograd, T.: The PageRank Citation Ranking: Bring Order to the Web (Technical Report). Stanford University (1998)
14. Chung, F.: Laplacians and the Cheeger Inequailty for Directed Graphs. Annals of Combinatorics 9, 1–19 (2005)
15. Johns, J., Mahadevan, S.: Constructing Basic Functions from Directed Graphs for Value Functions Approximation. In: Proc. of ICML (2007)
16. Zhou, D., Huang, J., Schölkopf, B.: Learning from Labeled and Unlabeled Data on a Directed Graph. In: Proc. of ICML (2005)
17. Brasseur, C.E., Grady, R.E., Prassidis, S.: Coverings, Laplacians and Heat Kernels of Directed Graphs. Electr. J. Comb 01/2009 16(1) (2009)
18. Nock, R., Nielsen, F.: Fitting Smallest Enclosing Bregman Ball. In: Gama, J., Camacho, R., Brazdil, P.B., Jorge, A.M., Torgo, L. (eds.) ECML 2005. LNCS (LNAI), vol. 3720, pp. 649–656. Springer, Heidelberg (2005)
19. Tsang, I.W., Kocsor, A., Kwok, J.T.: Simple Core Vector Machines with Enclosing Balls. In: Proc. of ICLM (2007)

Shape Similarity Based on a Treelet Kernel with Edition

Sébastien Bougleux[1], François-Xavier Dupé[2], Luc Brun[3],
and Myriam Mokhtari[3]

[1] Université de Caen Basse-Normandie
GREYC CNRS UMR 6072
`bougleux@unicaen.fr`
[2] Aix-Marseille Université
CNRS UMR 7279 - LIF
`francois-xavier.dupe@lif.univ-mrs.fr`
[3] ENSICAEN
GREYC CNRS UMR 6072
`{luc.brun,myriam.brun}@ensicaen.fr`

Abstract. Several shape similarity measures, based on shape skeletons, are designed in the context of graph kernels. State-of-the-art kernels act on bags of walks, paths or trails which decompose the skeleton graph, and take into account structural noise through edition mechanisms. However, these approaches fail to capture the complexity of junctions inside skeleton graphs due to the linearity of the patterns. To overcome this drawback, tree patterns embedded in the plane have been proposed to decompose the skeleton graphs. In this paper, we reinforce the behaviour of kernel based on tree patterns by explictly incorporating an edition mechanism adapted to tree patterns.

Keywords: shape similarity, kernel methods, tree patterns, edition.

1 Introduction

Several 2D shape representations and signatures have been proposed as a basis of shape recognition and classification, in particular the medial axis (or skeleton) and the associated medial axis transform. Indeed, the medial axis is a geometric graph homotopic to the shape and the medial axis transform allows to reconstruct the shape. However, the medial axis does not highlight enough the local shape properties needed for shape comparison, especially for the design of similarity measures. To overcome this drawback, suitable local shape properties are attached to the elements of the graph encoding the skeleton, leading to graph-based similarity measures.

Graph comparison can be performed by various methods, for example, graph edit distance and graph matching algorithms [1] form a first family. However, they are defined in graph space which almost contains no mathematical structure, thus prohibiting the use of many common tools. One solution is to project

G.L. Gimel' farb et al. (Eds.): SSPR & SPR 2012, LNCS 7626, pp. 199–207, 2012.
© Springer-Verlag Berlin Heidelberg 2012

graphs into a richer (or more flexible) space. Such a projection can be done through graph kernels. With appropriately defined kernels, graphs can be implicitly (sometimes explicitly) mapped into a vector space whose dot product corresponds to the kernel. Most of graph kernels rely on graph decomposition into walks, paths or trails [2–5]. However, these patterns fail to capture the complexity of junctions inside graphs, and so the branching points of the skeletons. One solution has been proposed in the chemioinformatics framework, where several graph kernels based on nonlinear patterns have been proposed. These patterns include unlabeled subgraphs [6], tree patterns [7], i.e. trees where a node can appear more than once, and subtrees of limited size [8]. Following [8], we have recently proposed a kernel based on a decomposition of skeletons into bags of subtrees embedded in the plane [9]. While this kernel provides good classification results compared to more sophisticated ones, it does not include any mechanism that would allow to be robust to spurious branches inside skeletons.

As the skeleton is very sensitive to small variations of the shape boundary (noise or small elongations), spurious nodes and edges (structural noise) are present inside its graph structure. In order to tackle such problems, an edition mechanism has been proposed for kernels based on bags of paths [4, 5]. Given a pertinence measure of each egde and node, the idea is to compute for each path, a sequence of reduced paths by successively removing their less pertinent part. Then, the resulting graph kernels are based on hierarchical comparisons between features attached to the elements of the rewritten paths.

This paper presents an extension of the treelet kernel proposed in [9] by incorporating an edition mechanism inspired by [4, 5]. First we recall our shape representation, which is based on a combinatorial map encoding of the skeleton, allowing to explicitly take into account its embedding in the plane (Section 2). Based on this encoding, we describe our extension of the treelet kernel which improves its robustness against structural noise (Section 3). Finally, several experiments are proposed in order to evaluate the performance of the resulting kernel and to measure the performances of our edition mechanism (Section 4).

2   Shape Representation

Usual graph-based encoding of the skeleton of a 2D shape do not take into account its planar properties, and thus remain invariant for any permutation of adjacent branches. To overcome this drawback, the skeleton can be encoded by a 2D combinatorial map [9]. Such a model may be understood as an encoding of a planar graph taking explicitly into account the orientation of the plane.

Combinatorial Map Encoding. As illustrated by Fig.1(b), a 2D combinatorial map (e.g. [10]) is defined by the triplet $M = (D, \sigma, \alpha)$, where D corresponds to the set of darts (or half-edges) obtained by decomposing each edge into two darts, $\sigma : D \to D$ is a permutation whose cycles correspond to the sequence of darts encountered when turning counter-clockwise around each node. Note that permutation σ explicitly encodes the orientation of edges around each node. Finally, $\alpha : D \to D$ is a fixed point free involution whose cycles correspond to pairs

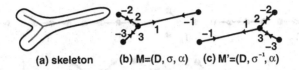

(a) skeleton   **(b) M=(D, σ, α)**   **(c) M'=(D, σ⁻¹, α)**

Fig. 1. Skeleton encoding: $\sigma = (-1)(1\,2\,3)(-2)(-3)$, $\alpha = (1 - 1)(2 - 2)(3 - 3)$

of darts, each pair corresponding to an edge. The encoding of a skeleton by such a map is performed by representing each branch by two darts defining one edge (a cycle of α). The orientation of branches, around branching points, is explictly encoded by the cycles of permutation σ.

The kernel between two shapes, described in Section 3, is based on the decomposition of their associated combinatorial maps into submaps having a tree structure. The identification of similar submaps relies on the computation of the symmetry group between submaps. The symmetry group $\mathrm{Sym}(M_1, M_2)$ from a map $M_1 = (D_1, \sigma_1, \alpha_1)$ to a map $M_2 = (D_2, \sigma_2, \alpha_2)$ defines the set of bijections $\psi : M_1 \to M_2$ that align the edges of M_1 onto the edges of M_2, while preserving or reversing their orientation around the nodes. Such bijections describe both rotational and mirror symmetries needed to align the two maps. They satisfy the following relations: (i) $\psi \circ \alpha_1 = \alpha_2 \circ \psi$, and (ii) $\psi \circ \sigma_1 = \sigma_2 \circ \psi$ or (iii) $\psi \circ \sigma_1 = \sigma_2^{-1} \circ \psi$. Relations (i) and (ii) correspond to a rotational symmetry, in which case ψ is a map isomorphism [11]. Relations (i) and (iii) correspond to a mirror symmetry and ψ is considered as a reflection [9, 12]. If $M_1 = M_2 = M$, then the symmetry group $\mathrm{Sym}(M_1, M_2)$ is equal to the set of permutations $\psi : M \to M$ which satisfy (i), and (ii) or (iii). This set is respectively composed of the automorphism group of M, noted $\mathrm{Aut}(M)$, and the automorphism group of the trivial mirror symmetric of M[1], noted $\mathrm{Aut}_R(M)$ [9]. These two groups can be computed by Cori's algorithm (see [11, 12] for more details).

Shape Features. In order to attach features to a combinatorial map encoding a skeleton, we define a set of node and edge labels (V and E), each node and edge label being respectively associated to a single cycle σ and α of the map [10]. We use mainly the same shape features as [9]. Let $f_E = (f_{E,i}(e))_i$ be the features attached to each edge of E, and $f_V = (f_{V,i}(v))_i$ the ones attached to each node.

Following [3, 5], a first edge feature corrresponds to the 4 polynomial coefficients of a regression polynomial of order 4 that modelize the evolution of the radius (of the inscribed disk) along the branch. The second edge feature associates the length of the shape boundary which contributes to the creation of the branch, normalized by the total length of the shape boundary in order to be invariant to scaling (see [5] for more details). This measure, defined as a function $w : E \to \mathbb{R}_+$, may thus be understood both as a relevant feature of an edge and as a measure of its relevance according to the shape.

---

[1] The trivial mirror symmetric of M is the map $M' = (D, \sigma^{-1}, \alpha)$ constructed by reversing the orientation of the darts around nodes (see Fig.1(c)).

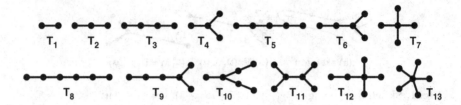

Fig. 2. The set $\mathbb{T}$ of tree patterns

Regarding nodes of V, the first feature corresponds to the sum of the relevances of its incident edges. The second feature associates to each node its Euclidean distance to the gravity center of the shape, normalized by the square root of the shape area in order to be invariant to scaling.

3   Shape Similarity

Based on the previous combinatorial map representation, similarity between two shapes relies on a decomposition of each map into a bag of subtrees [9], and on a hierarchical kernel between these subtrees in order to be robust to structural noise. This kernel replaces the edition mechanisms proposed in [5] in the context of subtrees.

3.1   Bag of Treelets

Given a shape represented by a combinatorial map M, and features attached to its nodes and edges, M is transformed into a bag of submaps having a tree structure. Each submap, together with its corresponding features, represent a part of the shape. Following [8, 9], the enumeration of the submaps is restricted to unlabeled and unrooted trees having between 2 and 6 nodes. As illustrated by Fig. 2, these trees of limited size form a dictionary of 13 *tree patterns*, denoted by $\mathbb{T} = \{T_p\}_{p=2,\ldots,13}$. The choice of the bounds on the number of nodes corresponds to a compromise between the expressiveness of the resulting bag and the time required to enumerate predefined subtrees.

An instance t of a tree pattern of $\mathbb{T}$ in M is called a *treelet*. It is represented as a 5-uplet (V, E, f_V, f_E, w), where f_V and f_E denote the features associated to the part of the shape described by t, and $w(t) = \sum_{e \in E(t)} w(e)$ represents its relevance according to the shape (the normalized boundary length induced by the edges of t). In practice, a treelet can be encoded by the index p of the corresponding tree pattern $T_p \in \mathbb{T}$, and an injection from edges of T_p to edges of M. The extraction of all the treelets from M can be performed by an enumeration process similar to the one proposed by [8]. The only difference is the preservation of the orientation of edges around each node.

3.2   Kernel between Bags of Treelets

Let $\mathcal{B}$ and $\mathcal{B}'$ denote two bags of treelets extracted from combinatorial maps M and M' respectively. Inspired by marginalized kernels [2], we have proposed in [9] a kernel defined as a weighted sum of minor kernels between all pairs of treelets of $(\mathcal{B} \times \mathcal{B}') \setminus \{(t, t') \in \mathcal{B} \times \mathcal{B}' : |V(t)| = |V(t')| = 2\}$ (we do not consider treelets isomorphic to tree pattern T_1):

$$K_{\mathbb{T}}(M, M') = \frac{1}{|\mathcal{B}||\mathcal{B}'|} \sum_{t \in \mathcal{B}} \sum_{t' \in \mathcal{B}'} \lambda_{\mathcal{B}}(t) \lambda_{\mathcal{B}'}(t') K(t, t'). \tag{1}$$

Kernel K corresponds to a minor kernel between treelets (see the following sections). The function $\lambda_{\mathcal{B}} : \mathcal{B} \to \mathbb{R}_+$ represents the relevance of each treelet relatively to its bag, which is defined by $\lambda_{\mathcal{B}}(t) = w(t) / \max_{t' \in \mathcal{B}} w(t')$. This weight allows to reduce the influence of treelets encoding non relevant parts of a shape.

3.3   Treelet Kernel

Let t and t' be two treelets representing parts of shapes. When they correspond to the same tree pattern $T_p \in \mathbb{T}$ (t and t' are structurally isomorphic to T_p), they can differ according to the features attached to their nodes and edges. Also, depending on the tree pattern, several matches between the two treelets are possible. In order to take into account both rotational and mirror symmetries of the shapes, the set of mappings between t and t' must preserve their orientations, but also reverse their orientations. This set corresponds to the symmetry group $\mathrm{Sym}(t, t')$, which is equivalent to $\mathrm{Aut}(T_p) \cup \mathrm{Aut}_R(T_p)$, and which can thus be easily pre-computed for each tree pattern [12].

   In order to measure the similarity between the treelets, we have proposed in [9] a positive-definite kernel defined as the average of similarities between their different matches derived from $\mathrm{Sym}(t, t')$:

$$K_{\text{treelet}}(t, t') = \begin{cases} \frac{1}{|\mathrm{Sym}(t, t')|} \sum_{\psi \in \mathrm{Sym}(t, t')} K_{\psi}(t, t') & \text{if } \mathrm{Sym}(t, t') \neq \emptyset, \\ 0 & \text{else.} \end{cases} \tag{2}$$

Kernel K_{ψ} is defined as the product of the similarities between each pair of nodes and each pair of edges provided by the mapping $\psi : t \to t'$:

$$K_{\psi}(t, t') = \prod_{v \in V(t)} K_V(v, \psi(v)) \prod_{e \in E(t)} K_E(e, \psi(e)),$$

where kernel K_V (resp. K_E) encodes the similarity between node's features (resp. edge's feature). It is defined as a tensor product of Gaussian kernels between each feature:

$$K_A(a, a') = \prod_{k=1}^{n_A} \exp \left(-\frac{\|f_{A,k}(a) - f_{A,k}(a')\|^2}{2\sigma_k^2} \right),$$

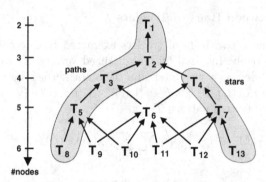

Fig. 3. Edition rules for the tree patterns of $\mathbb{T}$. Arcs represent transitions between treelets by either node suppression or edge contraction.

where A corresponds to V or E and a corresponds to a node v or an edge e. Kernel K_{Sym} can be seen as an extension of kernels on paths, trails or walks [2, 5] to trees embedded in the plane.

Experiments in [9] show the efficiency of treelet kernel K_{treelet}. In particular, the results are close to the one obtained by [5], which includes an edition process to reduce structural noise and to enhance similarity between closely related treelets. In the sequel, we extend this process to treelets in order to improve the robustness of kernel $K_{\mathbb{T}}$.

3.4   Hierarchical Treelet Kernel

Since shape skeletons are sensitive to small deformations of the shape, the similarity measure K_{treelet} between two treelets can be affected by structural noise. Also, two treelets not corresponding to the same tree pattern $(\text{Sym}(t, t') = \emptyset)$ may be similar up to some node suppressions or edge contractions. Following [5] in the case of paths, each treelet of a bag is transformed into a sequence of smaller ones through an edition process. Since deformations of the shape can be formalized by additions of nodes and edges, the two operations used to construct the sequence of treelets are *node suppression* and *edge contraction*. Node suppression corresponds to cut the parts of the shape connected to the treelet by the node. Edge contraction corresponds to a contraction of the shape. Each edge of the treelet is candidate to this operation. Node suppression is restricted to nodes of degree 2. This operation is topologically equivalent to the contraction of one of the two edges incident to the node. The set of possible rewritings of treelets defines an acyclic graph on the set $\mathbb{T}$ of tree patterns (see Fig. 3).

Let t be a treelet with k nodes, structurally equivalent to a tree pattern $T_p \in \mathbb{T}$. Depending on T_p, several nodes or edges can be suppressed or contracted in order to obtain a treelet with $k - 1$ nodes. The operation which induces a minimal distortion of the shape is retained. In order to encode this notion of distortion, a cost is assigned to each operation. This cost corresponds to the boundary of the part of the shape which is deleted: the relevance of the edge in the case

Table 1. Matching on Kimia25 dataset

	Method	$k=1$	$k=2$	$k=3$
1	Edit distance [13]	23	19	18
2	SID [14]	23	21	20
3	K_T with K_treelet only restricted to paths [9]	24	22	21
4	Syntactic matching [15]	25	21	19
5	Shape Context [16]	25	24	22
6	K_T with K_treelet only [9]	25	24	22
7	ID-Shape Context [17]	25	24	25
8	K_T with K_edit	25	25	24

of contraction, and the sum of relevances of the deleted edges in the case of suppression (see [5] for more details). The retained operation is the one having a minimal cost. Let κ be the application of this cheapest treelet edition, and let κ^k be the application of k successive editions. Then, the similarity between two treelets is measured by the kernel:

$$K_\mathrm{edit}(t,t') = \sum_{k=0}^{m_t} \sum_{l=0}^{m_{t'}} \exp\left(-\frac{w_k(t) + w_l(t')}{2\sigma_\mathrm{edit}^2}\right) K_\mathrm{treelet}(\kappa^k(t), \kappa^l(t'))$$

where m_t is the number of editions needed to transform t into a treelet equivalent to the tree pattern T_1 (an edge), and $w_k(t)$ is the cost associated to each reduced treelet operation $\kappa^k(t)$, defined as the sum of the costs of the k editions. Each feature associated to a reduced treelet is defined as a modification of the initial features associated to t according to the deformation of the shape (see [5] for more details).

Contrary to K_treelet, K_edit allows to compare two treelets which are not equivalent to a same tree pattern. Also, one can note that K_edit relies upon reacher structures than its counterpart based on paths [5], and thus more candidate operations need to be tested during the construction of the sequence of reduced treelets. But the sequences can be easily pre-computed for each treelet during the construction of the bags, as well as the associated features which have been modified by the edition process. So the proposed extension does not affect the computation of the kernel, and since the maximal number of editions is always 4, it is less time consuming than [5] (as long as the number of editions used in [5] is more than 4).

4   Experiments

In order to illustrate the behaviour of the proposed kernel K_T with the treelet kernel K_edit, we have considered the same experiments as in [9], that is k-NN matching and classification of the shapes of Kimia25 and Kimia99 datasets [14]. They contain respectively 25 and 99 discrete shapes, which are organized into 6 and 11 classes.

Table 2. Classification accuracy

Method	Accuracy			
	Kimia25		Kimia99	
	k-NN	Maha.	k-NN	Maha.
Edit distance [13]	0.89	0.84	0.927	0.907
Trails [5]	0.96	0.952	0.921	0.92
K_T with K_{treelet} only [9]	**0.953**	**0.946**	**0.936**	**0.933**
K_T with K_{edit}	**0.981**	**0.975**	**0.962**	**0.958**

k-NN Matching. For each shape of Kimia25 dataset, its $k = 1, 2, 3$ closest shapes are computed according to a similarity measure, ours being defined by kernel K_T. Values displayed in Table 1 represent, for each value of k, the number of closest shapes belonging to the same class than the input one. The parameters of K_T (the σ_k associated to each feature as well as σ_{edit}) have been optimized through experiments in order to obtain the best global match. As shown by lines 3 and 6 of Table 1, the use of nonlinear patterns over linear ones improves the efficiency of kernel K_T. Line 8 shows the improvement obtained by incorporating K_{edit} into K_T. One can also remark that the proposed kernel provides a result very close to the optimum, and slightly improves the one obtained by [17]. A similar behaviour has been observed on Kimia99 dataset. Note that [17] proposed a matching method which does not induce a definite positive similarity measure. Such a drawback prevents [17] to readily combine its similarity measure with complex numerical tools such as PCA or SVM.

Classification. The second experiment compares the proposed kernel K_T with two state-of-the-art kernels. For each method, the best kernel parameters have been estimated with a cross-validation on a reduced training set of Kimia25 or Kimia99 datasets. Then, a k-fold cross-validation, based on a Mahalanobis distance to each class and a k-NN, is computed to evaluate the efficiency of the kernels ($k = 4$ for Kimia25, and $k = 5$ for Kimia99). The resulting accuracies (number of true positive divided by the total number of shapes) are reported in Table 2. Again, our kernel with edition outperforms our previous kernel based on treelets, as well as the the one based on a Gaussian edit distance [13] and the one provided by trail kernels [5]. Note that this last kernel also use convering mechanisms to reduce the size of the bags.

5   Conclusion

To measure the similarity between 2D shapes, we have presented an extension of the kernel based on a decomposition of skeleton graphs into treelets embedded in the plane [9]. The extension, designed to take explicitly into account structural noise, relies on a hierarchical comparison of the treelets through an edition mechanism. Experiments show that the proposed kernel improves the results obtained with our previous kernel without edition mechanisms, as well as the ones obtained by several state-of-the-art methods.

Acknowledgments. This work is supported by ANR-09-SECU-02-02 CARTES funded by *Agence Nationale de la Recherche* within its CSOSG program.

References

1. Pelillo, M., Siddiqi, K., Zucker, S.: Matching hierarchical structures using association graphs. IEEE Trans. on PAMI 21(11), 1105–1120 (1999)
2. Kashima, H., Tsuda, K., Inokuchi, A.: Marginalized kernels between labeled graphs. In: Proc. of the 20st Int. Conf. on Machine Learning, pp. 321–328 (2003)
3. Goh, W.B.: Strategies for shape matching using skeletons. Computer Vision and Image Understanding 110(3), 326–345 (2008)
4. Dupé, F.X., Brun, L.: Edition within a Graph Kernel Framework for Shape Recognition. In: Torsello, A., Escolano, F., Brun, L. (eds.) GbRPR 2009. LNCS, vol. 5534, pp. 11–20. Springer, Heidelberg (2009)
5. Dupé, F.X., Brun, L.: Tree Covering within a Graph Kernel Framework for Shape Classification. In: Foggia, P., Sansone, C., Vento, M. (eds.) ICIAP 2009. LNCS, vol. 5716, pp. 278–287. Springer, Heidelberg (2009)
6. Shervashidze, N., Vishwanathan, S.V.N., Petri, T.H., Mehlhorn, K., Borgwardt, K.M.: Efficient graphlet kernels for large graph comparison. In: Proc. of the 12th Int. Conf. on Artificial Intelligence and Statistics, pp. 488–495 (2009)
7. Mahé, P., Vert, J.P.: Graph kernels based on tree patterns for molecules. Machine Learning 75, 3–35 (2009)
8. Gaüzère, B., Brun, L., Villemin, D.: Two New Graph Kernels and Applications to Chemoinformatics. In: Jiang, X., Ferrer, M., Torsello, A. (eds.) GbRPR 2011. LNCS, vol. 6658, pp. 112–121. Springer, Heidelberg (2011)
9. Bougleux, S., Dupé, F.X., Brun, L., Gaüzère, B., Mokhtari, M.: Shape similarity based on combinatorial maps and a tree pattern kernel. In: 21st Int. Conf. on Pattern Recognition (2012)
10. Brun, L., Mokhtari, M., Domenger, J.P.: Incremental modifications on segmented image defined by discrete maps. Journal of Visual Communication and Image Representation 14, 251–290 (2003)
11. Cori, R.: Computation of the automorphism group of a topological graph embedding. Technical Report I-8612, Université Bordeaux 1, France (1985)
12. Bougleux, S., Brun, L.: Symmetry group of 2d combinatorial maps. Technical report, GREYC, France (2012)
13. Neuhaus, M., Bunke, H.: Edit-distance based kernel for structural pattern classification. Pattern Recognition 39, 1852–1863 (2006)
14. Sharvit, D., Chan, J., Tek, H., Kimia, B.B.: Symmetry-based indexing of image databases. Journal of Visual Communication and Image Representation 9(4), 366–380 (1998)
15. Gdalyahu, Y., Weinshall, D.: Flexible syntactic matching of curves and its application to automatic hierarchical classification of silhouettes. IEEE Trans. on PAMI 21(2), 1312–1328 (1999)
16. Belongie, S., Malik, J., Puzicha, J.: Shape matching and object recognition using shape contexts. IEEE Trans. on PAMI 24(4), 509–522 (2002)
17. Ling, H., Jacobs, D.W.: Shape classification using the inner-distance. IEEE Trans. on PAMI 29(2), 286–299 (2007)

3D Shape Classification Using Commute Time

Muhammad Haseeb and Edwin R. Hancock*

Department of Computer Science, The University of York, UK

Abstract. This paper describes a commute-time based 3D shape descriptor that is robust with respect to changes in pose and topology. A new and completely unsupervised mesh segmentation algorithm is proposed, which is based on the commute time embedding of the mesh and the k-means clustering using the embedded mesh vertices. We use the discrete Laplace-Beltrami operator to construct the graph Laplacian.

Keywords: 3D Mesh Clustering, Commute Time Embedding, Shape Descriptor.

1 Introduction

Despite significant efforts in the past 10 to 15 years, graph clustering and classification remain an open challenge in the machine learning community. One of the most promising approaches is to use spectral clustering methods which exploits graph representations of the data and locate clusters by partitioning the graph that optimize an edge cut criterion. Early spectral approaches recursively compute the normalized cut [1] over the graph using the first non-zero Laplacian eigenvector (also known as the Fiedler vector [2] and are referred to as spectral bi-partitioning (SB) methods. Unfortunately, this does not guarantee good clusters as the normalized cut is computed recursively irrespective of the global structure of the data [3]. Qiu and Hancock [4] have used commute time for the purpose of image segmentation and show that the commute time method outperforms the normalized cut.

Recently, the graph spectral methods defined in the context of clustering have been applied to 3D shape processing. Here the discrete representation of 3D shape in the computer is a mesh, or sometimes a point set. In this context, spectral invariants such as the eigenfunctions of the Laplacian operator can be used for near-isometric shape matching. For instance, Mateus et al. [5] used eigenmaps obtained by the first k eigenfunctions of the Laplace operator as low-dimensional Euclidean representations of non-rigid shapes for the purpose of 3D point registration. Cuzzolin et al. [6] and Yamasaki et al. [7] have performed segmentation for mesh sequences. However, the former method computes only protrusions, while the latter uses an additional skeleton. In [6], the authors use locally linear embedding (LLE) to represent a cloud of points and perform segmentation in the LLE space. The segments obtained are then propagated across time to obtain a temporally coherent segmentation of a voxel-sequence into protrusions of the shape. The method works well for rigid body parts (such as head, hands and legs etc), but cannot be used directly for identifying rigid body-parts (for example, separating the upper-arm from the lower-arm).

* Edwin R. Hancock is supported by a Royal Society Wolfson Research Merit Award.

G.L. Gimel' farb et al. (Eds.): SSPR & SPR 2012, LNCS 7626, pp. 208–215, 2012.

Spectral methods can also be used to measure the similarity of 3D shapes. For instance, diffusion geometry methods were used to define low dimensional representations for manifolds. Rustamov [8] has suggested using the eigendecomposition of the Laplace-Beltrami operator to construct an isometric invariant surface representation, aiming to measure similarity between non-rigid shapes, rather than for correspondence detection. The Global Point Signature (GPS) suggested by Rustamov [8] for shape comparison employs the discrete Laplace-Beltrami operator, which globally captures the shapes geometry. The Laplace-Beltrami operator was later employed by many other researchers. For instance, Sun et al. [9] defined a point signature based on the properties of the heat diffusion process on a shape, referred to as the Heat Kernel Signature (HKS). HKS is obtained by restricting the well-known heat kernel to the temporal domain. Ovsjanikov et al. [10] employed a heat diffusion process to construct the Heat Kernel Maps for the shape matching. Castellani et al. [11] have extended Heat Kernel Signature (HKS). The local heat kernel values observed at each point are accumulated into a histogram for a fixed number of scales leading to the so-called Global Heat Kernel Signature (GHKS).

In this paper we construct a novel 3D shape distribution for the purpose of 3D object classification. The method commence from a modification of the 3D shape distribution reported in [12]. Firstly, instead of using Euclidean distances between pair of points on the shape, we use commute time distance computed from the eigenvalues and the eigenfunctions of the Laplace-Beltrami operator. Secondly we put a restriction on the pair of points being selected more than once. The empirical results show that the distribution computed using our method gives a better shape signature than [12] and [8].

2   Commmute Time

In this section, we briefly review how to compute the commute time and describe the relationships to the graph Laplacian. Commute time is the time taken by a random walker on a graph walking from a node u to node v and then back to node u. The commute time can be computed from the Laplacian spectrum as it has a close relationship with the graph Laplacian and heat kernel.

Consider a weighted graph by the triple $\Gamma = (V, E, \Omega)$, where V is the set of nodes, $E \subseteq V \times V$ is the set of edges, and Ω is the weighted adjacency matrix.

$$\Omega(u, v) = \begin{cases} w(u, v) & \text{if } (u, v) \in E \\ 0 & \text{otherwise} \end{cases}$$

where $w(u, v)$ is the weight on the edge $(u, v) \in E$. Furthermore, let $T = diag(d_u; u \in V)$ be the diagonal weighted degree matrix with elements given by the degrees of the nodes, $d_u = \sum_{v=1}^{|V|} w(u, v)$. The *unnormalized* weighted Laplacian matrix is given by $L = T - \Omega$ and the *normalized* weighted Laplacian matrix is defined to be $\mathcal{L} = T^{-1/2} L T^{-1/2}$ and has elements

$$\mathcal{L}(u, v) = \begin{cases} 1 & \text{if } u = v \\ -\frac{w(u,v)}{\sqrt{d_u d_v}} & \text{if } u \neq v \text{ and } (u, v) \in E \\ 0 & \text{otherwise} \end{cases}$$

The spectral decomposition of the normalized Laplacian is $\mathcal{L} = \Phi\Lambda\Phi^T$ where $\Lambda = diag(\lambda_1, \lambda_2, ..., \lambda_{|V|})$ is the diagonal matrix with the ordered eigenvalues as the elements satisfying the condition $0 = \lambda_1 \leq \lambda_2 \leq ..., \leq \lambda_{|V|}$ and $\Phi = (\phi_1|\phi_2|...|\phi_{|V|})$ is the matrix with the ordered eigenvectors as columns.

The *hitting time* $O(u, v)$ of a random walk on a graph is defined as the expected number of steps before node v is visited, commencing from node u. The *commute time* $CT(u, v)$, on the other hand, is the expected time for the random walk to travel from node u to reach node v and then return. As a result $CT(u, v) = O(u, v) + O(v, u)$. In terms of the eigenvectors of the *normalized* Laplacian the commute time matrix is given by

$$CT(u, v) = vol \sum_{i=2}^{|V|} \frac{1}{\lambda_i} \left(\frac{\phi_i(u)}{\sqrt{d_u}} - \frac{\phi_i(v)}{\sqrt{d_v}} \right)^2 \tag{1}$$

where $vol = \sum_{v \in V} d_v$ is the volume of the graph.

The commute time embedding is a mapping from the data space into a Hilbert subspace, which preserves the original commutes times. It has some properties similar to existing embedding methods including principal component analysis [13] (PCA), the Laplacian eigenmap [3] and the diffusion map [14]. The embedding of the nodes of the graph into a vector space that preserves commute time has the co-ordinate matrix

$$\Theta = \sqrt{vol}\Lambda^{-1/2}\Phi^T T^{-1/2} \tag{2}$$

The columns of the matrix are vectors of embedding co-ordinates for the nodes of the graph.

3   Laplace-Beltrami Operator

Let f be a real valued funtion defined on a differentiable manifold $\mathcal{M}$ with Riemannian metric. The LaplaceBeltrami operator, like the Laplacian, is the divergence of the gradient of f i.e.

$$\Delta f = \text{div}(\text{grad}(f)) \tag{3}$$

where grad and div are the gradient and divergence on the manifold respectively. The Laplace-Beltrami operator is a semi-positive definite operator. Most of the techniques [8] [15] for characterizing points on non rigid 3D shapes use the eingenpairs of the Laplace-Beltrami operator. The combinatorial Laplacian is suitable for the meshes and its does not contain much information about the shape. The discrete Laplacian or Laplace-Beltrami operator captures the geometric and topological properties of the surface. There are many schemes proposed to construct the discrete Laplacian that estimates the Laplace-Beltrami operator. Majority of them use the method of cotagents. However, the method described by Destrun et al. [16] and Meyer at al. [15] are more stable than the others. Xu [17] modified the method proposed by Meyer et al. This modification gives better convergence properties. In this paper we will follow Xu's method to construct the discrete Laplacian (Laplace-Beltrami operator).

3.1   The Generalized Eigenvalue Problem

For a function f defined on the surface, the Laplacian Δf is approximated as

$$\Delta f \approx \frac{1}{s_i} \sum_{j \in N(i)} w_{ij}[f(p_j) - f(p_i)]$$

where $N(i)$ are the neighbors for the vertex p_i and w_{ij} is the weight assigned to the edge between point p_i and p_j . The above formula can be written as $\Delta f \approx Lf$. Here L is the discrete Laplacian matrix. The weight w_{ij} of the edge is given by

$$w_{i,j} = \frac{\cot \alpha_{ij} + \cot \beta_{ij}}{2} \tag{4}$$

The angles appearing in this formula i.e. α_{ij} and β_{ij} are shown in the figure 1. The area s_i is also shown as the shaded region in the same figure. We compute the Laplacian, which has the entries as follows

$$L(i,j) = \begin{cases} \sum_k w(i,k)/s_i & \text{if } i = j \\ -w(i,j)/s_i & \text{if } i \text{ and } j \text{ are adjacent} \\ 0 & \text{otherwise} \end{cases}$$

The standard eigenvalue problem for L is $L\phi = \lambda\phi$, where λ is the eigenvalue of L and ϕ is the corresponding eigenvector. The area s_i at each vertex is computed as

$$s_i = \frac{\cot \alpha_{ij} + \cot \beta_{ij}}{8} ||p_i - p_j||^2 \tag{5}$$

Since the areas s_i computed at the vertices of the mesh are different, hence, the discrete Laplacian matrix L computed is not symmetric. This may cause the eigenvalues and eigenfunction to be complex. Therefore, we solve the generalized eigenvalue problem. Let S be the diagonal matrix with entries $S_{ii} = s_i$ and $W_{ij} = w_{ij}$ be the symmetric weight matrix. Since $L = S^{-1}W$, therefore, we can rewrite the equation $L\phi = \lambda\phi$ as $S^{-1}W\phi = \lambda\phi$ or

$$W\phi = \lambda S\phi \tag{6}$$

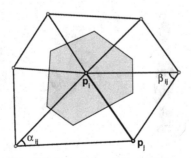

Fig. 1. Definitions of the angles and the area appearing in the discrete Laplace-Beltrami operator

Once we have the eigenvalues and eigenfunction of L to hand, we can compute the commute time matrix using the equation 1. Where we replace the degree of the nodes (i.e. d_u and d_v) by the area associated with the vertices (i.e. s_i. and s_j respectively). We replace the *vol* in the original equation by $\sum_i s_i$.

4   Shape Classification Using Commute Time

The commute time embedding gives a deformation independent embedding of a 3D shape into a high dimensional space. In this paper, we compute a shape descriptor from the commute time embedding. We use Laplace-Beltrami operator detailed in Section 3 to estimate the Laplacian of the shape. From the eigenvalues and eigenvectors of the Laplacian obtained, we compute the commute time matrix using the procedure given in Section 2. We use a modification of *D2* distributions introduced in [12]. *D2* distribution is essentially, the histogram of pairwise Euclidean distance between the points uniformly sampled from the surface. To compute our new shape descriptor, we use the commute time distance instead of the Euclidean distance. We also restrict a pair of point from being sampled more than once.

5   Experimental Results

In this section, we provide some experimental investigations. We focus on the use of commute time embedding of 3D shapes in two different settings. The first is an investigation of using the the commute time embedding for the purpose of partitioning the 3D shape into its parts. The second investigation is about using the modified shape distribution of Osada et al [12] computed by employing the commute time distance instead of the Euclidean distance.

In our first experiment we use the commute time embedding coordinates computed using equation 2 to partition six deformations of a human body selected from the Nonrigid world 3D database [18] shown in figure 2. The database contains a total of 148 objects, including 9 cats, 11 dogs, 3 wolves, 17 horses, 15 lions, 21 gorillas, 1 shark, 24 female figures, and two different male figures, containing 15 and 20 poses. The database also contains 6 centaurs, and 6 seahorses for partial similarity experiments.

Fig. 2. The k-means clustering on the Commute Time coordinates results in segmentation of six deformations of a 3D shape

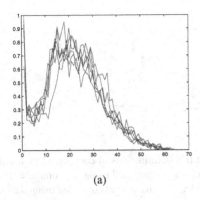 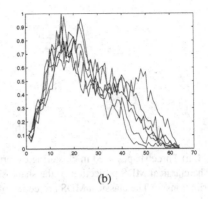

(a)   (b)

Fig. 3. The histogram for the six 3D shapes shown in figure 2. a) The commute time histogram b) The Euclidean histogram.

Each object contains approximately 3500 vertices. Figure 2 shows the result of the 3D shape, pose invariant segmentation using the k-means clustering on the commute time coordinates.

In the second experiment, we construct the shape distribution for different six different deformations of each of the three 3D shapes shown in figure 4(a). Figure 3(a) shows the shape descriptors for the six deformations using commute times. The shape descriptors for the same six deformations using Euclidean distances are shown in figure 3(b). This shows that the shape descriptor computed using commute time is more robust to shape deformations. We find the distance between each pair of the distributions using Bhattacharyya distance [19]. We project the distance matrix into vector space using classical multi-dimensional scaling (MDS). Figure 4 shows that the commute time shape distribution clusters similar shapes better than the Euclidean shape distribution.

For the final experiment we use the Watertight Benchmark which contains 400 closed surface shapes, grouped into 20 classes with 20 shapes each. We query every shape in the benchmark against all the other shapes. We compute three retrieval statics i.e. nearest first tier (FT), second tier (ST) and nearest neighbor (NN) for Osada's D2, Rastamov's GPS, Rastamov's Volumetric Shape Descripor (VSD) and the commute time. The shape retrieval results of the experiment are summarized in the table 1 which suggest that the commute time gives a better shape signature.

Table 1. Shape retrieval statistics

Descriptor	FT	ST	NN
Osada's D2	49.2%	67.8%	71.5%
Rustamov's GPS	46.3%	58.0%	79.1%
Rustamov's VSD [20]	48.7%	62.0%	81.3%
Commute Time	49.7%	69.2%	86.7%

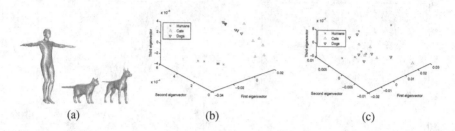

Fig. 4. a) Three shapes used in clustering experiment (six deformations of each shape are used). b) The classical MDS projection of the shape similarities as computed using the commute time distributions. c) The classical MDS projection of the shape similarities as computed using the D2 distributions.

6   Conclusions

In this paper we have investigated how the commute time between the vertices on mesh can be used to partition the 3D shape. We also used commute time distance to construct the 3D shape distribution for the purpose of 3D shape clustering and 3D shape classification. The empirical results show that commute time is a better choice for shape classification problem. In future we would like to extend our shape descriptor by employing the geodesic distances between each pair of vertices.

Acknowledgement. Edwin R. Hancock was supported by a Royal Society Wolfson Research Merit Award.

References

1. Shi, J., Malik, J.: Normalized cuts and image segmentation. IEEE Transactions on Pattern Analysis and Machine Intelligence 22, 888–905 (2000)
2. Chung, F.R.K.: Spectral Graph Theory. American Mathematical Society (1997)
3. Belkin, M., Niyogi, P.: Laplacian eigenmaps for dimensionality reduction and data representation. Neural Computation 15, 1373–1396 (2003)
4. Qiu, H., Hancock, E.R.: Clustering and embedding using commute times. IEEE Trans. Pattern Anal. Mach. Intell. 29, 1873–1890 (2007)
5. Mateus, D., Horaud, R., Knossow, D., Cuzzolin, F., Boyer, E.: Articulated shape matching using laplacian eigenfunctions and unsupervised point registration. In: CVPR (2008)
6. Cuzzolin, F., Mateus, D., Knossow, D., Boyer, E., Horaud, R.: Coherent laplacian 3-D protrusion segmentation. In: CVPR (2008)
7. Lee, N.S., Yamasaki, T., Aizawa, K.: Hierarchical mesh decomposition and motion tracking for time-varying-meshes. In: ICME, pp. 1565–1568. IEEE (2008)
8. Rustamov, R.M.: Laplace-beltrami eigenfunctions for deformation invariant shape representation. In: Symposium on Geometry Processing, pp. 225–233 (2007)
9. Sun, J., Ovsjanikov, M., Guibas, L.: A concise and provably informative multi-scale signature based on heat diffusion. In: Proceedings of the Symposium on Geometry Processing, SGP 2009, pp. 1383–1392. Eurographics Association, Aire-la-Ville (2009)

3D Shape Classification Using Commute Time   215

10. Ovsjanikov, M., Mérigot, Q., Mémoli, F., Guibas, L.J.: One point isometric matching with the heat kernel. Comput. Graph. Forum 29, 1555–1564 (2010)
11. Castellani, U., Mirtuono, P., Murino, V., Bellani, M., Rambaldelli, G., Tansella, M., Brambilla, P.: A New Shape Diffusion Descriptor for Brain Classification. In: Fichtinger, G., Martel, A., Peters, T. (eds.) MICCAI 2011, Part II. LNCS, vol. 6892, pp. 426–433. Springer, Heidelberg (2011)
12. Osada, R., Funkhouser, T., Chazelle, B., Dobkin, D.: Matching 3d models with shape distributions. In: Proceedings of the International Conference on Shape Modeling & Applications, SMI 2001, pp. 154–167. IEEE Computer Society, Washington, DC (2001)
13. Jolliffe, I.: Principal Component Analysis, 2nd edn. American Mathematical Society. Springer-Verlag New York, Inc., New York (2002)
14. Lafon, S., Lee, A.: Diffusion maps: a unified framework for dimension reduction, data partitioning and graph subsampling. IEEE Transactions on Pattern Analysis and Machine Intelligence (2005)
15. Meyer, M., Desbrun, M., Schröder, P., Barr, A.H.: Discrete Differential-Geometry Operators for Triangulated 2-Manifolds. In: Hege, H.C., Polthier, K. (eds.) Visualization and Mathematics III, pp. 35–57. Springer, Heidelberg (2003)
16. Desbrun, M., Meyer, M., Schröder, P., Barr, A.H.: Implicit fairing of irregular meshes using diffusion and curvature flow. In: Proceedings of the 26th Annual Conference on Computer Graphics and Interactive Techniques, SIGGRAPH 1999, pp. 317–324. ACM Press/Addison-Wesley Publishing Co., New York (1999)
17. Xu, G.: Discrete laplace-beltrami operator on sphere and optimal spherical triangulations. Int. J. Comput. Geometry Appl. 16, 75–93 (2006)
18. Alexander, Bronstein, M.: Nonrigid world 3d database v 1.0 @ONLINE (2009)
19. Bhattacharyya, A.: On a measure of divergence between two statistical populations defined by their probability distributions. Bulletin of the Calcutta Mathematics 35, 99–110 (1943)
20. Rustamov, R.M.: Robust volumetric shape descriptor. In: Eurographics Workshop on 3D Object Retrieval, pp. 1–5 (2010)

Conditional Random Fields for Land Use/Land Cover Classification and Complex Region Detection

Gulcan Can[1,*], Orhan Firat[1,*], and Fatos Tunay Yarman Vural[1]

[1] Department of Computer Science, Middle East Technical University, Ankara, Turkey
{gulcan,orhan.firat,vural}@ceng.metu.edu.tr

Abstract. Developing a complex region detection algorithm that is aware of its contextual relations with several classes necessitates statistical frameworks that can encode contextual relations rather than simple rule-based applications or heuristics. In this study, we present a conditional random field (CRF) model that is generated over the results of a robust local discriminative classifier in order to reveal contextual relations of complex objects and land use/land cover (LULC) classes. The proposed CRF model encodes the contextual relation between the LULC classes and complex regions (airfields) as well as updates labels of the discriminative classifier and labels the complex region in a unified framework. The significance of the developed model is that it does not need any explicit parameters and/or thresholds along with heuristics or expert rules.

Keywords: conditional random fields, land use/land cover, complex region de-tection, satellite imagery.

1 Introduction

Visual patterns and object occurrences in remote sensing images exhibit high intra-class variance, meaning that two or more instances of the same object or object groups may look coercively different. For example, two airfields may have entirely different color structures, composing roads, shapes, sizes and configurations of their sub-parts (e.g. one may have just one, crossing, parallel runway(s) having hammer shaped, circular, polygonal dispersal areas and located in sandy, snowy, coastal or urban terrain). Occasionally these objects and object groups may even look more similar to instances within other classes that to instances within their own class, e.g. circular oil tanks of a refinery and circular dispersal area of a military airfield.

Contextual models are significantly useful in order to handle the huge variability within classes in the image because of their expressive representations. By forming a contextual framework, any object can be accurately classified not only by considering its low-level vision features but also its local context (spatial relations) over a probabilistic graphical model. Figure 1 illustrates some real world examples that exhibit high intra-class variance.

---

* Corresponding author.

G.L. Gimel' farb et al. (Eds.): SSPR & SPR 2012, LNCS 7626, pp. 216–224, 2012.

Fig. 1. Examples of airfields from satellite images. Despite the high variance in color, texture, deployment terrain and composition, we can still recognize them as remotely sensed views of airfields.

The ability to recognize such complex objects comes from both appearance cues of the object itself and contextual relations of the complex objects with their surroundings. These contextual relations can be defined as co-occurrence frequency of other classes in a predefined neighborhood of the complex object. For instance, if the complex object in consideration is an airfield, we would expect urban, vegetation, water existence nearby to be less than a certain ratio. This information comes either from domain knowledge or by explicit observations. However, deciding this ratio by a static threshold is not desirable, since less likely configurations are not allowed at all. As in the case of urban areas in the surrounding of an airfield, we may set a 20% urban co-occurrence threshold in a 300-meters neighborhood by domain knowledge. Yet, Figure 1 demonstrates cases contradicting with such a threshold. Hence rather than determining crisp thresholds for the recognition task of a complex object, constructing a probabilistic model is much more flexible and suitable.

Probabilistic graphical models are the state-of-art approach for modeling contextual relations between semantic classes [1] and have many applications in remote sensing [2-3]. Since labels in spatial data are not independent as well as observations, assumptions on data being "independent and identically distributed" (i.i.d.) is violated by using traditional classifiers. Therefore such classifiers may produce undesirable results when applied to such data.

This problem motivates the use of Markov Random Fields (MRFs) and more recently Conditional Random Fields (CRFs) for spatial data. In the proposed approach, contextual relations between a complex object and its surroundings, which is characterized by LULC classes, are modeled within a CRF framework. The major contribution of the proposed model is that a random field is constructed over semantic classes rather than pixels or super-pixels as in the literature. Our model aims to correctly identify the complex object by recognizing the co-occurrence pattern of all other classes in its surrounding as well as updating previously assigned class labels which can be obtained by any kind of classification method.

The rest of this paper is organized as follows: In section 2, recent motivating studies in the field of probabilistic graphical models are stated, followed by section 3, the adopted methods are explained. In section 4, proposed algorithm is introduced and in section 5 dataset and experiments are described. Finally in section 6 some conclusion remarks and future work are given.

2   Recent Studies

Studies modeling spatial structures vary both in their representations used to encode the spatial information and their approaches for learning. Inference on the generated graphical models depends on the model selection in the studies and may be thought as a representation dependent step.

One of the pioneering studies employs both contextual and hierarchical representations with a *relationship learning* process and Bayesian inference algorithm is proposed by Porway et. al [4]. Their approach presents a grammar-based hierarchical and contextual model for object recognition. This grammar-based model combines a stochastic context free grammar (SCFG) [5] with a Markov Random Field (MRF) to capture both local and global context and combines bottom-up information with top-down knowledge. They represent the frequency of occurrence and type of object parts with a SCFG and model the spatial and appearance relationships between them using MRFs, thus create a constrained grammar that can represent a huge number of instances for a single category. Another contribution of this study is that, this contextual and hierarchical model learns statistical constraints on the appearances and relationships between different parts of the image classes with a minimax entropy framework [6]. This framework selects the set of contextual relationships necessary for modeling the object class; begins with a large set of relationships that could potentially exist between parts, then iteratively selects only those relationships that help the model best match true statistics for that image class. They separated hierarchy into two sets for objects and scene which enables to plug-in any object detection algorithm for bottom-up detection procedure. They employed compositional boosting [7] for some specific bottom-up proposals.

In [8], a region and object based model for object-detection is proposed through a hierarchy of CRFs. In the bottom level, a CRF is comprised of pixels as probabilistic graphical model nodes and features are extracted in pixel level accordingly, a unified energy function made it possible to incorporate bottom-middle and top level random fields. In the middle level, segments are formed as the model nodes and contextual relations between segments are revealed with region statistics. Finally as the top-most level of the proposed hierarchical graphical model, segments and objects are connected to each other and contextual relations between objects are tried to be extracted from positional relations of the objects both considering segment level interactions at once. The model employed for this graphical model is a conditional MRF (CRF) that is trained by labeled images from both levels with logistic regression and inference is conducted by use of hill-climbing.

Jiang et.al. propose a *context based concept fusion* model for semantic concept detection [9]. In this study, posterior probabilities for several classifiers are fed to a CRF model for generating updated posterior probabilities through a fully-connected CRF where each node represents a concept. This corresponds to class labels in our case.

Lee et.al. propose a model, namely support vector random fields (SVRF), which combines the ability of CRFs to model different types of spatial dependencies and the appealing generalization properties of support vector machines (SVMs) [10]. Their approach employs an observation-matching potential by changing the association potential in CRF model. Therefore they combined the discriminative classification power of SVMs with spatial context encoding power of CRFs.

3   Methods Adopted

SVM is a supervised classification approach which makes use of kernels (mapping functions) and sparsity [11]. By the help of kernels, samples in n-dimensional space are carried to a higher dimensional Hilbert space, therefore samples become linearly separable by a hyper plane. In this model a hyper plane that separates nearest samples from different classes with maximum margin is selected and named as support vectors. Tolerance and cost parameters can be used to allow or penalize outliers. SVM is also known as max-margin classifier.

As stated previously, spatial relations between neighboring pairs can be modeled by MRFs and CRFs. More specifically, the class labels can be assigned by maximum a posterior (MAP) estimation in image classification task as $y_{MAP} = argmax_y P(y|x)$. This can be interpreted as CRF framework that models directly the posterior probability of labels given the observed data. Consequently, besides the contextual information in labels, the CRF framework has ability to capture the contextual information in observed data.

The discriminative CRF framework considers Markovian property of y conditioned on x and directly models the posterior as a Gibbs distribution with the following form:

$$P(y|x, \theta) = \frac{1}{Z} \exp \left\{ -\sum_{c \in C} \varphi_c (y_c, x, \theta) \right\} \tag{1}$$

where $Z = \sum_y \exp \{ -\sum_{c \in C} \varphi_c(y_c, x, \theta) \}$ is partition function (normalization constant), φ_c is potential defined on clique c with parameters θ, C is set of cliques, and y_c is set of labels over clique c. Then the pair-wise CRF models can be written as

$$P(y|x, \theta) = \frac{1}{Z} \exp \left\{ -\sum_{i \in S} \varphi_i (y_i, x, w) - \sum_{i \in S} \sum_{j \in \eta_i} \varphi_{ij}(y_i, y_j, x, v) \right\} \tag{2}$$

where η_i is the set of neighbors of site i, φ_i and φ_{ij} are the unary and pair-wise clique potentials with parameters w and v, respectively, then θ denotes the parameter set $\theta = \{w, v\}$. The unary potential φ_i represents the association of a single site to semantic labels, whereas pair-wise potentials φ_{ij} can be seen as a measure of how the labels at neighboring sites at i and j should interact given image x. In fact, the unary and pair-wise potentials in CRF should be designed as discriminative as possible according to the domain it is applied to.

In our approach, we construct a fully connected random field over classes similar to concepts proposed by Jiang et.al. SVM is an intermediate step to assign class labels to segments. We could have applied other methods here, but preferred SVM due to its high performance. With this step, initial class map for LULC classes is obtained and they are then updated in the proposed CRF model according to context of the complex object.

4   Proposed Algorithm

In this study, *airports* are chosen as the complex regions and *water, forest, green-land, urban, concrete, soil* as LULC classes. The most significant cue of an airfield is the existence of runway(s), which consists of basically long straight parallel lines. We propose an algorithm for categorization of Parallel Line Bounded Regions (PLBR) which is stated as a strong indicator and invariant of airfields in highly variant contextual environments [12]. In the proposed algorithm, context information of airfields is formulated over LULC classes. Proposed model is also for updating labels assigned to LULC classes in the means of airport context.

Proposed model is a fully-connected conditional random field which can be seen in Figure 2. Fully-connected graphical model is selected in order to reveal contextual relations between all classes and their mutual influence.

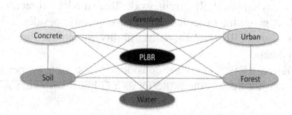

Fig. 2. Proposed CRF model, which is a fully connected graph aiming to capture all possible pairwise relations between semantic categories

Figure 3 illustrates workflow of the proposed algorithm. In the proposed algorithm, the first step is the preprocessing step in which segmentation of the input image via the mean-shift algorithm [13] and PLBRs extraction. The PLBRs are extracted by finding the line segments [14] on the steerable-filter [15] response of the image first and then extracting the parallel ones. Note that PLBRs are treated as regular seg-ments.

Pre-processing step is followed by a feature extraction step. For each segment, fea-tures are extracted as illustrated in Figure 4. For this purpose, fundamental maps are obtained first. These are spectral values (red, green, blue, near-infrared), DTED map, Gabor filter response, normalized difference water index (NDWI) map [16] and nor-malized differ-ence vegetation index (NDVI) map calculated using (3) and (4) respectively.

$$NDWI = \frac{Green - NIR}{Green + NIR} \qquad (3)$$

$$NDVI = \frac{NIR - Red}{NIR + Red} \qquad (4)$$

Gabor response image is obtained by taking maximum response of a pixel at eight directions and one scale.

An initial labeling for LULC classes over segments is obtained with an SVM classifier by using features extracted. SVM is trained with a labeled dataset and the parameters are determined using cross-validation in validation set. Note that, the initial labeling in this step is conducted only for segments, rather than PLBRs, with considering only LULC classes (6 classes).

For each PLBR, a fully-connected CRF model as depicted in Figure 2 is constructed. No further operation is conducted over a PLBR before embedding them into this CRF model as a node. For each PLBR, the 300-meter neighborhood is analyzed in the ini-tial class label map obtained by SVM classifier. All segments with the same class label are treated as one single node in the CRF model of the corresponding PLBR. As an example, in Figure 5, there are several segments initially labeled as urban class (cyan) by SVM. For the Urban node in the CRF model, we extract unary potentials considering all these urban segments.

After obtaining the CRF model for a PLBR, node and edge features are extracted for all seven nodes. Node features for a node in the CRF model are extracted as in Figure 4, but this time; extraction is conducted over all segments of the corresponding 7 class label not separately but as a whole (e.g. mean and standard deviation of urban class in Figure 5, is computed over all cyan area and used as the node features of the Urban node).

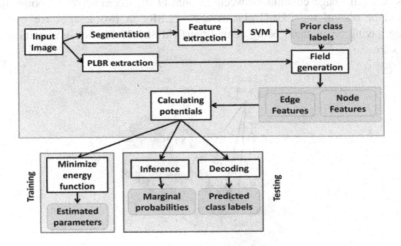

Fig. 3. Workflow of the algorithm

For the edge features, overlapping, adjacent and in-neighborhood class frequencies are used (see Figure 5). These are calculated based on pixel counts of each class in corresponding area.

For the CRF framework to be function reasonably, node features and edge features are converted into unary and pair-wise potentials using multi-class logistic function.

Then potentials are fed to energy function and parameters are learnt for the model by minimizing this energy function (2). In test phase, CRF model for each PLBR is decoded according to trained parameters.

During parameter estimation in training phase, L-BFGS is used. It is a limited-memory quasi-Newton method for unconstrained optimization. Two distinct loss functions are applied during training, namely loopy belief propagation (LBP) and pseudo negative log-likelihood (NLL). As the decoding method, Iterated Conditional Modes (ICM) is used [17].

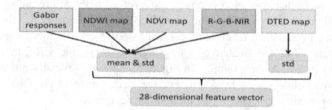

Fig. 4. Feature extraction

5   Dataset and Experiments

In this study, 4 GEOEYE multispectral images with size of ~3800x3800 pixels are used. Each image contains between 23 and 89 PLBRs either corresponds to an air field or one of the LULC classes. 112 PLBRs in two images are used for training having 53 of them being actually airfield and 77 PLBRs in remaining two images are

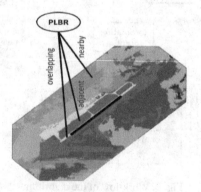

Fig. 5. Proposed edge features over SVM labels

used for testing having 49 of them actually airfield. Ground truths are prepared by labeling segments around each PLBR region.

There are seven classes in this study, namely *water, urban, forest, green land, soil, concrete*, and *airfield*. In our experiments, we used an eighth state, an extra state to enable the model to reject. Since we combine segments according to their prior labels, mixing cases may confuse the CRF model. Eighth state corresponds to "other" or "mixing" class.

As performance measures, recall and precision are used for airfield detection.

Table 1. Performance results of the proposed algorithm using ICM decoding

	Pseudo Negative Log-likelihood	92	46.94
Loss Function	Negative Log-likelihood with LBP	93.33	57.14
		Precision	Recall

6   Conclusion and Future Work

The proposed model categorizes PLBRs and corrects/updates SVM results in the context of an airport-PLBR. The actual goal of this paper is focusing on detecting actual complex objects, airfields in our case, accurately. Recall values in Table 1 demonstrate that it is a hard task. PLBRs could be decoded as urban or forest-green land which may semantically correspond to roads and edges between green lands respectively. Other studies about airfield detection in the literature have specific datasets and employ generally heuristic or threshold-based approaches. This makes comparison with our approach inapplicable.

Instead of using SVM output of training images for training the CRF model, ground truth of training images can be directly employed. This would probably be better for CRF to learn true relationships among classes. However, in this case, train and test phases would have different steps, since SVM would not be involved in training phase to obtain areas under each class node.

The proposed CRF model is able to embed spatial information around complex objects in terms of LULC classes; however it is open to improvements. The model can be designed as a star shape to make pair-wise relations more representative. For the CRF model corrects SVM results in segment level, not as a whole semantic class, segments layer can be added to the model which is connected to class nodes in the originally proposed model. This hierarchical model would make our model more relaxed and capable. However, computational tractability would be lost in this dense graph structure and only approximate methods could be applied.

References

1. Galleguillos, C., Belongie, S.: Context based object categorization: A critical survey. Computer Vision and Image Understanding 114, 712–722 (2010)

2. Benedek, C., Sziranyi, T.: Change Detection in Optical Aerial Images by a Multilayer Conditional Mixed Markov Model. IEEE Transactions on Geoscience and Remote Sensing 47, 3416–3430 (2009)
3. Zhong, P., Wang, R.: Learning conditional random fields for classification of hyperspectral images. IEEE Transactions on Image Processing: a Publication of the IEEE Signal Processing Society 19, 1890–1907 (2010)
4. Porway, J., Wang, Q., Zhu, S.C.: A Hierarchical and Contextual Model for Aerial Image Parsing. International Journal of Computer Vision 88, 254–283 (2009)
5. Lin, L., Wu, T., Porway, J., Xu, Z.: A stochastic graph grammar for compositional object representation and recognition. Pattern Recognition 42, 1297–1307 (2009)
6. Zhu, S.C., Wu, Y.N., Mumford, D.: Minimax Entropy Principle and Its Application to Texture Modeling (1997)
7. Wu, T.-F., Xia, G.-S., Angeles, L.: Compositional Boosting for Computing Hierarchical Image Structures. Learning
8. Gould, S.: Probabilistic models for region-based scene understanding (2010)
9. Jiang, W., Chang, S.-F., Loui, A.C.: Context-Based Concept Fusion with Boosted Conditional Random Fields. In: 2007 IEEE International Conference on Acoustics, Speech and Signal Processing, ICASSP 2007, vol. 1, pp. I-949–I-952 (2007)
10. Lee, C.-H., Greiner, R., Schmidt, M.: Support Vector Random Fields for Spatial Classification. In: Jorge, A.M., Torgo, L., Brazdil, P.B., Camacho, R., Gama, J. (eds.) PKDD 2005. LNCS (LNAI), vol. 3721, pp. 121–132. Springer, Heidelberg (2005)
11. Cortes, C., Vapnik, V.: Support-vector networks. Machine Learning 20, 273–297 (1995)
12. Firat, O., Tursun, O.T., Yarman Vural, F.T.: Application of Context Invariants in Airport Region of Interest Detection for Multi-spectral Satellite Imagery. In: IEEE 20th Conference on Signal Processing and Communications Applications, SIU (2012)
13. Comaniciu, D., Meer, P., Member, S.: Mean shift: A robust approach toward feature space analysis. IEEE Transactions on Pattern Analysis and Machine Intelligence 24, 603–619 (2002)
14. Grompone von Gioi, R., Jakubowicz, J., Morel, J.-M., Randall, G.: LSD: a fast line segment detector with a false detection control. IEEE Transactions on Pattern Analysis and Machine Intelligence 32, 722–732 (2010)
15. Freeman, W.T., Adelson, E.H.: The design and use of steerable filters. IEEE Transactions on Pattern Analysis and Machine Intelligence 13, 891–906 (1991)
16. Mcfeeters, S.K.: The use of the Normalized Difference Water Index (NDWI) in the delineation of open water features. International Journal of Remote Sensing 17, 1425–1432 (1996)
17. Besag, J.: Spatial Interaction and the Statistical Analysis of Lattice Systems. Journal of the Royal Statistical Society. Series B (Methodological) 36, 192–236 (1974)

Recognition of Long-Term Behaviors by Parsing Sequences of Short-Term Actions with a Stochastic Regular Grammar

Gerard Sanromà, Gertjan Burghouts, and Klamer Schutte

TNO, Oude Waalsdorperweg 63, 2597 AK The Hague, The Netherlands
gerard.sanromaguell@tno.nl

Abstract. Human behavior understanding from visual data has applications such as threat recognition. A lot of approaches are restricted to limited time actions, which we call *short-term actions*. Long-term behaviors are sequences of short-term actions that are more extended in time. Our hypothesis is that they usually present some structure that can be exploited to improve recognition of short-term actions. We present an approach to model long-term behaviors using a syntactic approach. Behaviors to be recognized are hand-crafted into the model in the form of grammar rules. This is useful for cases when few (or no) training data is available such as in threat recognition. We use a stochastic parser so we handle noisy inputs. The proposed method succeeds in recognizing a set of predefined long-term interactions in the CAVIAR dataset. Additionally, we show how imposing prior knowledge about the structure of the long-term behavior improves the recognition of short-term actions with respect to standard statistical approaches.

Keywords: long-term behavior, stochastic context-free grammars, human activity analysis, visual surveillance.

1 Introduction

Automated recognition of long-term behavior is relevant for many applications, where particular events have to be signaled. As examples: theft of truck cargo; dwelling of people in elderly homes; shopping behavior inside a mall.

Where short-term action recognition has received much attention [5,2,3], automated recognition of long-term behavior has been studied less. Long-term behavior is an interesting research topic, as it requires temporal modeling of sequences of short term actions. In this paper, we consider long-term behavior as a sequence of short-term actions. As contribution, we will provide a method to improve the recognition of such sequences by a parsing mechanism.

A complicating factor for recognition of behaviors, is that the potential number of temporally ordered combinations of actions is very high. One way to deal with this is to learn the limited set of likely combinations. The learning of temporal sequences has been studied intensively in the past, for instance, by a HMM [3] or by the related discriminative CRF [4]. They have both shown

G.L. Gimel' farb et al. (Eds.): SSPR & SPR 2012, LNCS 7626, pp. 225–233, 2012.

their merit for solving various problems, for a comparison see [9]. A problem with these methods is that they are known to require a large training set to learn the sequences. For the applications that we envisage in this paper, like the prevention of an unwanted or even hazardous situation, typically only very few positive examples are available. This makes the HMM and CRF intractable. Often world-knowledge is available on how situations evolve. The goal of this paper is to exploit such prior knowledge explicitly for the recognition of long-term behaviors. We consider an alternative modeling of sequences that requires few learning examples by including world-knowledge by means of a hand-crafted rule set which is enforced using a stochastic grammar.

Grammars enable the encoding of sequences by simple expressions that limit the possibilities and capture the world-knowledge about how long-term behaviors evolve [8,6]. An example of such a sequence is *browse*, consisting of the actions: *walk standing still look around walk* etc. Such sequences are perfectly suited to be specified by a grammar. For the recognition of behaviors grammars have been studied previously. Our starting point is that we have only few learning examples, so the learning of grammar rules is out of this papers scope [10]. Grammars have mostly been used as a second-stage recognizer of situations or long-term behaviors, based on first-stage detectors of the constituent short-term actions [10,8]. We will follow the same strategy in this paper. In [8], context-free grammars were considered. A disadvantage of context-free grammars is that they cannot be directly interpreted as finite-state machines (FSM). Often, a FSM is the means by which expert knowledge is encoded, because it is an easy tool to model and understand. Instead, we will use a regular grammar which is built on top of an FSM, thus allowing to model the expert knowledge as a state diagram.

Our contributions in this paper are two-fold. First, we propose a regular grammar that exploits world-knowledge that is encoded by a FSM. To demonstrate the power of this grammar, we show it on the publicly available CAVIAR dataset that includes videos of realistic long-term behaviors. CAVIAR defines the included behaviors in terms of FSMs, which we will integrate into the grammar. Second, we show experimentally that long-term interactions can be recognized by the proposed grammar, and thereby the recognition of the constituent short-term actions is improved.

In section 2 we introduce the method used for recognition of short-term actions which is based on [2]. In section 3 our method for behavior recognition based on stochastic parsing is presented. In section 4 experimental validation is presented and some discussion is given. Finally, we conclude in section 5.

2   Short-Term Action Detection

As the basic observations, we are interested in recognizing a vocabulary of short-term interactions between two people from a set $\mathcal{A} = \{action_1, \ldots, action_L\}$ We use the non-parametric approach by [2]. This approach uses trajectory information from a set of previously extracted tracks by some standard method. So, for each clip we have a set of N tracks X^i, $i = 1, \ldots, N$ each one corresponding to the trajectory followed by one person. This is, for i-th track, $X^i = \{\mathbf{x}^i_t, t \in T^i\}$, where $T^i = [t^i_1, \ldots, t^i_m]$ is the index-set of the frame interval that track X^i is present on the scene.

For each pair of persons (i, j) at each time t we compute the following feature vector.

$$\mathbf{f}^{(ij)}_t = \left[s^t_i, s^t_j, a^t_{ij}, d^t_{ij}, d^{dif}_{ij}, s^{dif}_{ij}\right] \tag{1}$$

where s^t_i is the distance covered by person i in between frames $t - w$ and t, a^t_{ij} is the alignment between persons, d^t_{ij} is the distance between two persons, d^{dif}_{ij} is the difference of distances w frames apart, and s^{dif}_{ij} is the difference in velocities (check [2] for more details).

We segment the clips into windows of ws frames with a certain overlap controlled by the offset wo. Therefore, for each window k we obtain the following set of feature vectors

$$F^{(ij)}_k = \left\{\mathbf{f}^{(ij)}_t, t \in \mathcal{T}^k\right\} \tag{2}$$

where $\mathcal{T}^k = [\tau^k_a, \ldots, \tau^k_b]$ contains the indices of the frames belonging to the k-th window.

The goal of this section is to compute the probability $P\left(F^{test} | action_p\right)$ of a test window given the action $action_p \in \mathcal{A}$. Ground truth action labels $l^{(ij)}_t \in \mathcal{A}$ are attached to each feature vector $\mathbf{f}^{ij}_t$ in the training set. We define a label indicator function that returns the prevalence of an action inside a window (normalized to sum up to one). This is,

$$L\left(F^{(ij)}_k ; action_p\right) = \#\left\{l^{(ij)}_t \mid l^{(ij)}_t = action_p \wedge t \in \mathcal{T}^k\right\} \Big/ |F^{(ij)}_k| \tag{3}$$

where $\#\{\bullet\}$ corresponds to the cardinality of a set and $|F^{(ij)}_k|$ is the amount of vectors in the window.

In order to remove noise, we compute the PCA projections of the feature vectors $\tilde{\mathbf{f}}^{(ij)}_t, \forall i, j, t$ in the training set so as to retain an 80% of the total variance, obtaining also the projected windows $\tilde{F}^{(ij)}_k$. Given a test window of projected features $\tilde{F}^{test}$, we define the probability of being produced by a certain $action_p$ in the following way:

$$P\left(\tilde{F}^{test} | action_p\right) = \text{K-nn}\left(\tilde{F}^{test}, action_p\right) \Big/ K , \tag{4}$$

where the function K-nn($\tilde{F}, action_p$) accumulates the prevalence of the action $action_p$ over the K nearest windows of $\tilde{F}$ in the training set. As distance measures to do the sort we use two variants, namely, the originally used Hausdorff distance [2] and, as an alternative, the Earth Mover's Distance (EMD) [11].

3   Behavior Recognition by Stochastic Parsing

Consider a sequence of observations $F_1, \ldots, F_T$ generated by the interaction between two people as explained in the previous section. With the model developed in the previous section each short-term action observable F_t in the sequence is classified regardless their relationships with past or future observables. Short-term actions usually follow some activity patterns which, on the other hand, depend on the context in which they are found. Often, such patterns, or long-term behaviors, can only be characterized at long time extents comprising tens or even hundreds of short-term action observations. Therefore, we claim that more robust detection of short-term actions is achieved when shifting up to the level of long-term behavior analysis.

Inspired by the work in [8], we model long-term behaviors as grammar production rules. Recognition of long-term behaviors transforms then to finding the sequence compatible with the rules that best fits to the observables, which is essentially a parsing problem.

Stochastic grammars provide a proper framework to do so since they allow for probabilistic measurements both in the observations and the production rules. Stolcke [12] proposed an efficient parsing algorithm for stochastic grammars. A stochastic grammar is a tuple $G = (\mathcal{N}, \Sigma, \mathcal{R}, S, \mathcal{P})$, where $\mathcal{N}$ are the non-terminals, Σ the terminals, $\mathcal{R}$ the rules, S the starting non-terminal and $\mathcal{P}$ the rule probabilities. We mainly restrict to the sub-type of regular grammars. Regular grammars have the same expressive power as finite-state machines (FSM) [7], the latter ones traditionally used for representing human activity. Moreover, it is possible a direct interpretation of rule probabilities as transition probabilities, which facilitates the task of estimating them. In the FSM formalism, which we use to illustrate our method, observations and states correspond to terminal and non-terminal symbols in the grammar. The rules of a regular grammar have the following forms.

$C \rightarrow s$, where C is in $\mathcal{N}$ and s is in Σ
$C \rightarrow sD$, where C, D are in $\mathcal{N}$ and s is in Σ

As a more appropriate abstraction, we transform the sequence of observables $F_1 \ldots F_T$ into a sequence $S_1 \ldots S_T$, where each position $S_t^{action_p} = P(F_t | action_p)$ accounts for the probability of observation of each short-term action at each time step. Given any sub-sequence $\mathcal{S}_a \ldots \mathcal{S}_b$, the stochastic parser delivers:

- The Viterbi parse. This is, the most likely sequence of (unambiguous) short-term actions that we would observe if they were produced following the rules of behavior $C \in \mathcal{N}$,

$$s_a \ldots s_b = Viterbi\_parse\left(\mathcal{S}_a \ldots \mathcal{S}_b | C\right) , \qquad (5)$$

where $s_t \in \mathcal{A}$.
- The Viterbi probability of such a sequence, which in our case is a product of observation probabilities and transition probabilities as defined by the Viterbi parse. This is,

$$P\left(\mathcal{S}_a \ldots \mathcal{S}_b | C\right) = P\left(F_a | s_a\right) \prod_{t=a+1}^{b} P\left(s_t | s_{t-1}\right) P\left(F_t | s_t\right) \qquad (6)$$

where $s_a \ldots s_b$ is the Viterbi parse of $\mathcal{S}_a \ldots \mathcal{S}_b$, and $P\left(s_i | s_j\right)$ is the probability of transition from action $s_i \in \mathcal{A}$ to action $s_j \in \mathcal{A}$.

A novel contribution of our method is that we divide our grammar into two parts: the constrained and the unconstrained part. The constrained part is responsible for interpreting the sequence of incoming short-term actions S according to the specified rules of the behaviors. The unconstrained part provides a straightforward interpretation that does not impose any structure at all. Separation of the grammar in constrained and unconstrained part has two advantages. On one hand it allows to parse any input sequence without interruptions (since the unconstrained part accepts any sequence). On the other hand it also provides a reference to validate candidate recognitions (as we will see later).

To illustrate this idea suppose that we want to recognize a set of M predefined long-term behaviors. Suppose that the i-th long-term behavior is composed by the sequence of short-term actions *join - interact - split*, and that the terminals of our grammar are $\Sigma = \{join, interact, split\}$. Figure 1 shows a representation.

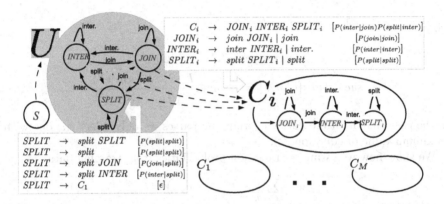

Fig. 1. Terminals and non-terminals are in non-captials and capitals, respectively. Non-terminal S is the starting symbol. The unconstrained part is encapsulated in non-terminal U and represented in grey. As an example of this part we show the production rules of the $SPLIT$ non-terminal. The unconstrained part is encapsulated in non-terminals C_i which contain the rules of the i-th pre-specified behavior. Solid arrows represent transitions associated with observations. Dashed arrows represent transitions not associated with any observation that have a fixed probability ϵ. They give the capability of detecting behaviors starting at any moment in time.

The procedure for recognizing long-term behaviors is the following. At each time step t, the stochastic parser processes the whole set of detections from that time, $S_t^{action_p}, \forall action_p$. Operation consists of a series of *prediction*, *scanning* and *completion* steps. Each time step that a non-terminal C_i is *completed* means that the parser has found a sub-string that is compatible with the rules of C_i. This traduces to a candidate detection of the long-term behavior C_i from which the time interval $[a, b]$ can be easily retrieved (check [12] for details). Final decision is based upon comparison of the constrained and unconstrained interpretations. This is, behavior is recognized if the probability of the sequence $S_a \ldots S_b$ being generated by the constrained rule C_i is not too low with respect to the probability of being generated by the unconstrained rule U. More precisely, behavior is recognized if

$$\frac{P\left(\mathcal{S}_a,\ldots,\mathcal{S}_b|C_i\right)}{P\left(\mathcal{S}_a,\ldots,\mathcal{S}_b|U\right)} \geq \rho^{(b-a+1)} \tag{7}$$

where $0 \leq \rho \leq 1$ controls the tolerance to false positives / false negatives, and the exponent makes this measure invariant to the length of the sequence.

As previously stated, the aim of our method is to deliver an unambiguous sequence of short-term action detections $s_1 \ldots s_T$ from an input sequence of probabilistic observations $\mathcal{S}_1 \ldots \mathcal{S}_T$. We define the *null* action *ignore* for the cases when no decision can be made. Final action detection is decided as

$$s_t = \begin{cases} ignore & if \ \nexists a,b,C_i \ s.t. \ a \leq t \leq b \land \dfrac{P\left(\mathcal{S}_a,\ldots,\mathcal{S}_b|C_i\right)}{P\left(\mathcal{S}_a,\ldots,\mathcal{S}_b|U\right)} \geq \rho^{(b-a+1)} \\ s_t' & otherwise \end{cases}, \tag{8}$$

where

$$s_a' \ldots s_b' = Viterbi\_parse\left(\mathcal{S}_a \ldots \mathcal{S}_b|C\right)$$

$$such \ that \ \{C,a,b\} = \underset{C',a',b'}{\arg\max} \frac{P\left(\mathcal{S}_{a'},\ldots,\mathcal{S}_{b'}|C'\right)}{P\left(\mathcal{S}_{a'},\ldots,\mathcal{S}_{b'}|U\right)} \tag{9}$$

In the case that multiple partly overlapping parses we only select the one with maximum value of equation (7).

We show how we estimate transition probabilities from training data:

$$P\left(action_q|action_p\right) = \frac{\displaystyle\sum_{(i,j)}\sum_k L\left(F_k^{(ij)} ; action_p\right) \cdot L\left(F_{next(k)}^{(ij)} ; action_q\right)}{\displaystyle\sum_{action_{q'}}\sum_{(i,j)}\sum_k L\left(F_k^{(ij)} ; action_p\right) \cdot L\left(F_{next(k)}^{(ij)} ; action_{q'}\right)} \tag{10}$$

where F_{ij}^k is a particular window, $next(k)$ is a function that returns the next window in time to k, and $L\left(\bullet\right)$ is the label indicator function of equation (3).

4   Experiments and Results

We have performed experiments on the CAVIAR dataset [1]. The CAVIAR database consists of a set of clips showing long-term behaviors. There are annotations of the bounding boxes as well as labels of short-term interactions between pairs of people. Such interactions are: *join, fight, interact, move, leave victim, leave object* and *split*. We have created an additional label *ignore* corresponding to the *null* action for the cases when two people are close to each other without interacting. Due to the extremely low prevalence of the labels *leave victim* and *leave object* as well as to some arbitrariness of the human annotator in the case of the *leave object* label discard them by assigning to the *ignore* label. In terms of positional features, the actions *fight* and *interact* are equivalent (i.e., both consist on two people interacting close to each other). Therefore, we have decided to merge both labels into one called *fighteract*.

According to the structure of the behaviors defined in the CAVIAR documentation [1] and the modifications that we have made to the labels, we have identified the two long-term behaviors of figure 2 as the ones represented in the clips.

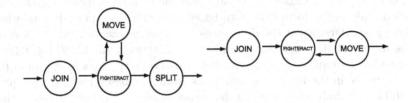

Fig. 2. Long-term behaviors shown by the CAVIAR clips.

From a total of 27 clips 7 of them are discarded because they contain no pairwise interactions of people at all (neither from the *ignore* class). From the 20 videos used in the experiments, 8 of them contain exclusively interactions of the type *ignore* (i.e., they do not show any interaction between actors but some of them get close to each other at some moment). The remaining 12 clips contain relevant interactions and eventually also *ignore*-type interactions.

We segment videos into windows of size ws with an overlap defined by offset wo, identical to our learning framework for the observations of short-term actions. We use ground truth annotations of bounding boxes to get the trajectories of each person by projecting the position of the feet with the homography relating the image plane with the ground plane. Because of the low prevalence of certain classes in the dataset (e.g., *split*), we use the data from all videos except the current test one as training set for the short-term action detectors.

In order to see the benefits of imposing the structure of the behavior through a grammar, we compare the accuracy of short-term action detection obtained using either the K-nn classifier of equation (4) or the output of the stochastic parsing as defined in equation (8).

We show both mean Matthew's Correlation Coefficient (MCC) between classes and confusion matrices. MCC is a measure of quality of two-class classification defined as

$$\text{MCC} = \frac{\text{TP} \times \text{TN} - \text{FP} \times \text{FN}}{\sqrt{(\text{TP}+\text{FP})(\text{TP}+\text{FN})(\text{TN}+\text{FP})(\text{TN}+\text{FN})}}, \tag{11}$$

where $\text{TP}, \text{TN}, \text{FP}, \text{FN}$ account for true positives, true negatives, false positives and false negatives, respectively. An MCC value of 1 means perfect prediction. A value of 0 means not better than random prediction. A value of -1 indicates total disagreement.

In the table below we show both the mean MCC among the classes and the MCC between the *ignore* and the rest of the classes obtained by each method.

Per-class (MCC)	
K-nn (Hausdorff)	0.33
K-nn (EMD)	0.37
Grammar + K-nn (Hausdorff)	0.46
Grammar + K-nn (EMD)	0.48

ignore vs. all (MCC)	
Grammar + K-nn (Hausdorff)	0.77
Grammar + K-nn (EMD)	0.75

As we see in the per-class results, grammar-based methods obtain better classification accuracies than the others, specially the EMD-based one. EMD-based variants usually outperform Hausdorff-based ones. This is especially true when not using grammars, when the differences are more noticeable. From these results we deduce that recognition of short-term actions in the CAVIAR dataset is improved when imposing their expected long-term structures as shown in figure 2. As we see in the *ignore* vs. all results, grammar-based methods are quite successful in discriminating between the *ignore* class and the rest. It means that they succeed in detecting when some predefined behavior happens.

Confusion matrices are shown in figure 3. Rows represent actual detections while columns represent ground truth classes. Perfect detections would show a matrix with ones in the diagonal and zeros elsewhere.

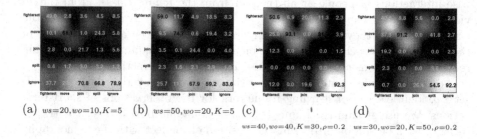

(a) $ws=20, wo=10, K=5$ (b) $ws=50, wo=20, K=5$ (c) $ws=40, wo=40, K=30, \rho=0.2$ (d) $ws=30, wo=20, K=50, \rho=0.2$

Fig. 3. Values for the parameters corresponding to the best results are shown under each confusion matrix. The methods are: (a) K-nn (Hausdorff), (b) K-nn (EMD), (c) Grammar + K-nn (Hausdorff) and (d) Grammar + K-nn (EMD).

As we see in the confusion matrices (and from the tables above) the EMD-based variant of the grammar method outperforms the rest. Short-term detectors tend to overestimate the *ignore* class. The *split* action has a significantly low prevalence in the training set. Due to this, short-term detectors tend to confuse it with the most prevalent classes *ignore* and *move*. Grammar-based methods correctly deduce that they are part of some long-term behavior but miss-classify them as *move* because action detectors tend to do so and also due to the structural compatibility between *split* and *move* states in the rules of figure 2.

5 Conclusions

We have presented a method to improve the recognition of short-term actions as well as to recognize long-term behaviors by imposing a behavior structure. It is useful for the cases when few (or no) training data about long-term behaviors is available. It uses non-parametric detection of short-term actions in the bottom layer which are input to a stochastic parser in the top layer.

We propose a new methodology for estimating probabilities of the grammar rules as well as we introduce a novel criterion for recognizing long-term behaviors based on the allowed deviation from the straightforward interpretation. We propose a new variant of the short-term action detector based on the EMD.

We perform experiments of recognition of long-term interactions between people in the CAVIAR dataset [1]. Results show that the EMD variants usually outperform the Hausdorff-based ones. Moreover, grammars are quite successful in recognizing pre-defined behaviors in the CAVIAR dataset as we see in the *ignore* vs. *the rest* classification results. Regarding per-label classification, they present an average improvement of $\sim$ 25%. This demonstrates that imposing long-term behavior structure improves short-term action detection.

Acknowledgments. This work has been carried out as part of the EU FP SEC project ARENA.

References

1. (2004), http://homepages.inf.ed.ac.uk/rbf/caviar/
2. Blunsden, S., Andrade, E.L., Fisher, R.B.: Non Parametric Classification of Human Interaction. In: Martí, J., Benedí, J.M., Mendonça, A.M., Serrat, J. (eds.) IbPRIA 2007. LNCS, vol. 4478, pp. 347–354. Springer, Heidelberg (2007)
3. Brdiczka, O., Yuen, P.C., Zaidenberg, S., Reignier, P., Crowley, J.L.: Automatic acquisition of context models and its application to video surveillance. In: ICPR, pp. 1175–1178 (2006)
4. Burghouts, G.J., Marck, J.W.: Reasoning about threats: From observables to situation assessment. IEEE Transactions on Systems, Man, and Cybernetics, Part C 41(5), 608–616 (2011)
5. Burghouts, G., Schutte, K.: Correlations between 48 human actions improve their detection. In: ICPR (2012)
6. Fernández-Caballero, A., Castillo, J.C., Rodríguez-Sánchez, J.M.: Human activity monitoring by local and global finite state machines. Expert Syst. Appl. 39(8), 6982–6993 (2012)
7. Hays, D.G.: Chomsky hierarchy. In: Encyclopedia of Computer Science, pp. 210–211. John Wiley and Sons Ltd., Chichester
8. Ivanov, Y.A., Bobick, A.F.: Recognition of visual activities and interactions by stochastic parsing. Pattern Anal. Mach. Intell. 22(8), 852–872 (2000)
9. Kasteren, T.L., Englebienne, G., Kröse, B.J.: An activity monitoring system for elderly care using generative and discriminative models. Personal Ubiquitous Comput. 14(6), 489–498 (2010)
10. Kitani, K.M., Sato, Y., Sugimoto, A.: Recovering the basic structure of human activities from a video-based symbol string. In: WMVC, p. 9 (2007)
11. Rubner, Y., Tomasi, C., Guibas, L.J.: The earth mover's distance as a metric for image retrieval. Int. J. Comput. Vision 40(2), 99–121 (2000)
12. Stolcke, A.: An efficient probabilistic context-free parsing algorithm that computes prefix probabilities. Comput. Linguist. 21(2), 165–201 (1995)

A Comparison between Structural and Embedding Methods for Graph Classification*

Albert Solé-Ribalta, Xavier Cortés, and Francesc Serratosa

Universitat Rovira i Virgili - Spain
{albert.sole,xavier.cortes,francesc.serratosa}@urv.cat

Abstract. Structural pattern recognition is a well-know research field that has its birth in the early 80s. Throughout 30 years, structures such as graphs have been compared through optimization of functions that directly use attribute values on nodes and arcs. Nevertheless, in the last decade, kernel and embedding methods appeared. These new methods deduct a similarity value and a final labelling between nodes through representing graphs into a multi-dimensional space. It seems that lately kernel and embedding methods are preferred with respect to classical structural methods. However, both approaches have advantages and drawbacks. In this work, we compare structural methods to embedding and kernel methods. Results show that, with the evaluated datasets, some structural methods give slightly better performance and therefore, it is still early to discard classical structural methods for graph pattern recognition.

1 Introduction and Literature Review

Classical graph approaches for pattern recognition applications rely on computing distances between graphs on the graph domain. That is, the distance between two graphs is obtained by directly optimizing some objective function which consider node and edge attributes. However, in the graph domain these distances cannot be computed in polynomial time. To overcome this problem, a large set of approximation algorithms have been developed since the 80 [1]. Undoubtedly, pattern recognition applications which work on the graph domain need to rely on graph class prototypes to represent sets of data. The synthesis of these representatives is usually done using either a sequential or hierarchical synthesis. In this type of synthesis, the graph prototype is constructed in an iterative fashion. These iterative approaches start from a model computed using two graphs and they iteratively refine the current model by sequentially considering all the graphs in the training set. This process of synthesis, similar to generative models, relies on the fact that the model is able to capture the global information also in the initial steps. If this is not the case, usually the process derives in a bad model and so performance of the application at hand decreases. To overcome this drawback of classical models one can rely on

---

* We acknowledge Consolider Ingenio (CSD2007-00018) and CICYT (DPI2010-17112).

G.L. Gimel' farb et al. (Eds.): SSPR & SPR 2012, LNCS 7626, pp. 234–242, 2012.

new procedures to synthesize the prototype such as a Common Labelling solutions [2–4]. Traditional graph prototypes seems that lately have been move
apart in favour of graph embeddings and kernels. These methods allow exploiting existing classical pattern recognition algorithms such as Bayes Classifiers,
SVM or K-Means. Both methods, graph embeddings and kernels, rely on the
same intuition, in both cases the graph is somehow encoded in a vector which
correspond to a point in a multi-dimensional vector space. The main difference
between kernels and embeddings rely on how the embeddings are performed.
In graph embeddings, one knows the destination space and the transformation
function. In this way, one explicitly does the embedding by transforming the
graph into a vector. Since in graph embeddings the destination space is known,
the similarity/distance/dissimilarity function is directly applied to vectors in the
destination vector space. On the contrary, graph kernels go one step forward by
embedding this initial vectorial representation in a larger vectorial space. This
destination space is usually unknown. However, by using the so-called kernel
trick, one is able to compute distances on the destination space by using the
dot product between the initial vectorial representations of the graphs. Graph
embeddings and graph kernels share the basic intuition. That is, they rely in
encoding the graph in a some-dimensional vector space and performing operations on that space. Consequently, they share some drawbacks. Two of the main
drawbacks are the following. The first one is related to the fact that usually,
both approaches rely on a non-complete representation of the graphs such as
[5–7], the Laplacian matrix [8] or the walk kernel [9], sacrificing, in some cases,
performance for speed. The second one is related to the unfolding of the embedding in case graph information needs to be used or represented. This process
might be of crucial importance if learned or classified data must be shown to
the user, to either perform supervised learning or just validation. This unfolding
process might be of great difficulty [10] if not impossible. The objective of this
paper is manifold. On the one hand, we introduce the graph prototype synthesis from the Common Labelling point of view. This new formulation defines in
a clear and uniform way the synthesis of graph under a global framework. On
the other hand, we present a comparative study of state-of-the-art methods for
graph classification based on graph embeddings and common labelling synthesis
for graph prototyping.

2   Graph Prototypes

In this section, the graph prototypes that are evaluated are introduced, and a
synthesis based on the common labelling is defined. To this aim, we start by
giving some notation and basic definitions.

Attributed Graphs. Given vertex attribute domain Δ_v and edge attribute do-
main Δ_e, we define an attributed graph with a four-tuple $G^p = (\Sigma_v^p, \Sigma_e^p, \gamma_v^p, \gamma_e^p) \subseteq$
$\mathcal{H}$, where $\Sigma_v^p = \{v_k^p | k = 1, , N\}$ is the set of vertices, $\Sigma_e^p = \{e_{ij}^p | i, j \in 1, , N, i \neq j\}$
is the set of arcs and $\gamma_v^p : \Sigma_v^p \rightarrow \Delta_v$ and $\gamma_e^p : \Sigma_e^p \rightarrow \Delta_e$ assign attribute values to
vertices and arcs respectively. Since the article is focussed on graph prototyping

for pattern recognition, usually, a training set of elements is provided for learning this prototype. In this case, we denote the training set of prototypes by symbol $S = \{G^1, ..., G^P\}$, we consider that all graphs in S have the same order. If this is not the case, graphs can be extended with null nodes and arcs. These null nodes and arcs are labelled with a special attribute $\emptyset \in \Delta_v$ for nodes and $\emptyset \in \Delta_e$ for edges. This is the usual mechanism to deal with graphs of different cardinality, the reader is referred to [11].

2.1   Generalized Median Graph

Given a set of graph $S = \{G^1, ..., G^P\}$ and a dissimilarity/distance function $d(\cdot, \cdot)$, the Median Graph $\overline{M}$ [12] is the attributed graph that minimizes the sum of distances between it and all the graphs in the set of Attributed Graphs. That is,

$$\overline{M} = arg \min_{M \in \mathcal{H}} \sum_{G^p \in S} d(G^p, M) \qquad (1)$$

Notice that the Generalized Median Graph is usually not a member of the set, and in general, more than one Generalized Median Graph may exist for a given set of graphs. The computation of a Median Graph is at least NP-complete since the general graph matching problem it is. Nevertheless, several suboptimal methods to obtain approximate solutions for the Median Graph, in reasonable time, have been presented [2, 3, 5, 10, 12–14]. These methods apply some heuristic functions in order to reduce the complexity of the graph distance computation and the size of the search space. Most of the existing methods to compute the Generalized Median Graph use a classical synthesis, that is, either sequential or hierarchical. Here, we present a synthesis based on the Common Labelling framework [4]. In this way, given a Common Labelling $H = \{h^1, ..., h^P\}$ where $h^p : \Sigma_v - > L$, a Median Graph $\overline{M} = (\overline{\Sigma}_v, \overline{\Sigma}_e, \overline{\gamma}_v, \overline{\gamma}_e)$ is defined as another Attributed Graph where attributes on nodes and arcs are computed as:

$$\overline{\gamma}_v(\overline{v}_i) = \frac{1}{\zeta_v} \sum_{G^p \in S} (1 - \delta(h^{p^{-1}}(\overline{v}_i), \emptyset)) \gamma_v(h^{p^{-1}}(\overline{v}_i)) \qquad (2)$$

$$\overline{\gamma}_e(\overline{e}_{i,j}) = \frac{1}{\zeta_e} \sum_{G^p \in S} (1 - \delta(\epsilon_{i,j}^{G^p}, \emptyset)) \epsilon_{i,j}^{G^p} \qquad (3)$$

where,

$$\epsilon_{i,j}^{G^p} = \gamma(e_{h^{p-1}(\overline{v}_i), h^{p-1}(\overline{v}_j)}) \qquad (4)$$

$$\zeta_v = \sum_{G^p \in S} (1 - \delta(h^{p^{-1}}(\overline{v}_i), \emptyset)), \zeta_e = \sum_{G^p \in S} (1 - \delta(\epsilon_{i,j}^{G^p}, \emptyset)) \qquad (5)$$

and δ represents the Kronecker Delta function and $h^{p^{-1}}$ the inverse function of h^p. The main idea of (2) and (3) is that the attribute values of the Median Graph is the mean of the values of all the nodes or arcs of the Attributed Graphs that their nodes or arcs have been matched to a concrete node of the virtual structure L. Moreover, in the case that there does not exist a node or arc in the Attributed Graph, its value is not considered to compute the mean.

2.2   Set Median Graph

The Set Median Graph [12] is an alternative to Generalized Median Graphs. The difference between the two models consists in the search space where the Median Graph is searched. The search space for the Median Graph is the domain of the Attributed Graphs $\mathcal{H}$. In contrast, the search space for the Set Median Graph is restricted the set of graphs that represents S. The computation of Set Median Graph is, in general, exponential with respect to the order of the graphs, due to the complexity of graph isomorphism problem, but quadratic with respect to the number of graphs in S. In some applications, Set Median Graphs are preferred to Median Graphs due to two main reasons. Firstly, practical evaluations show that the capacity of Set Median Graphs to represent a set is almost similar to the capacity of Median Graphs [10]. Secondly, the synthesis (using the whole set of graphs or incrementally) is less computationally demanding. The Set Median Graph is the Attributed Graph in the set such that it has the minimum distance between it and the rest of the Attributed Graphs. Note the computation of the set median does not need any labelling only just graph-to-graph distances.

2.3   Closure Graphs

The closure graph [15] is a graph prototype originally applied to graph databases. In the closure graph, attributes of nodes or edges in the training set are represented with a set of attributes. Closure graphs are restricted to discrete attributes, if this is not the case, an extra discretization step must be performed in order to discretize them. Closure graphs need more physical space than median graphs. Formally, a closure graph $\overline{M} = (\overline{\Sigma}_v, \overline{\Sigma}_e, \overline{\gamma}_v, \overline{\gamma}_e)$ is a graph where node and arc attributes are a set of domains of the nodes and arcs in the training set, $\overline{\Delta}_v = \{\Delta_v, \Delta_v, ..., \Delta_v\}$ and $\overline{\Delta}_e = \{\Delta_e, \Delta_e, ..., \Delta_e\}$. We synthetize a Closure Graph from a set of Attributed Graphs S and a Common Labelling φ as follows:

$$\gamma_v(\overline{v}_i) = \{a | a = \gamma(h^{p^{-1}}(\overline{v}_i)), 1 \leq p \leq P, \gamma(h^{p^{-1}}(\overline{v}_i)) \neq \emptyset\} \tag{6}$$

$$\gamma_e(\overline{e}_{i,j}) = \{b_k | b_k = \epsilon_{i,j}^{G_p}, \epsilon_{i,j}^{G_p} \neq \emptyset\} \tag{7}$$

The basic intuition of (6) and (7) is that nodes and edges of the Closure graph can take all values that nodes and edges of the training set have taken. In the case that there does not exist a node or arc in the Attributed Graph, its value is not considered. Practical evaluations show that Median Graphs and Closure Graphs tend to generalize too much the set that they represent; allowing graphs that are distant from the ones that have not been used to synthesize them. To alleviate this weakness, the following probabilistic models have been defined.

2.4   First Order Random Graphs

A First-Order Random Graph (FORG) [11] is a model graph that contains first-order probabilities on nodes and arcs to describe a set of Attributed Graphs. To deal with the first-order probabilities, there is a random variable associated with each vertex or arc, which represents the attribute information of the corresponding graph nodes and arcs in the set of Attributed Graphs. This random variable has a one-dimensional probability density function defined over the same attribute domain of the Attributed Graphs, including a null value, $\emptyset$, that denotes the non-instantiation of a FORG graph node or arc in an Attributed Graphs. This was the first probabilistic model that appeared in the literature to represent a set of Attributed Graphs. It assumes that the Attributed Graphs in a set or cluster have similar local parts. Nevertheless, in practical applications, some graphs can be quite different despite of belonging to the same class. For this reason, in several applications representing a set of attributed graphs with only first order probabilities seems to be too restrictive. A First Order Random Graph $\overline{M} = (\overline{\Sigma}_v, \overline{\Sigma}_e, P_v, P_e)$ is a graph where the node and arc attribute domains are random variables with values in Δ_v and Δ_e. Probabilities at the nodes and arcs of the FORG are related to the number of times the values have appeared at the nodes or arcs of the Attributed Graphs related to this node or arc. In the case of the nodes,

$$p_i(a) = \frac{1}{P} \sum_{\forall p \in 1..P} \delta(h^{p^{-1}}(\overline{v}_i), a) \tag{8}$$

where P corresponds to the number of graphs given to construct the prototype. Arc probabilities are computed in an equivalent form.

2.5   Function Described Graphs

A Function Described Graph (FDG) [16, 17] is a model graph that appeared with the aim of overcoming the representational power of FORGs. It contains first-order probabilities of attributes and second-order structural information to describe a set of Attributed Graphs. The first order information was represented in the same way than FORGs trough probability density functions. The second-order structural information represents qualitative information of the second-order joint probability of each pair of vertices or arcs. This information is represented by binary relations called Antagonisms, Occurrences and Existences between nodes and arcs. FDGs increased the representational power at the cost of increasing also the required physical space. Two nodes or arcs are antagonistic if they have never taken place together in any graph used to synthesise the FDG although these two nodes or arcs are included in the FDG as different elementary parts. There is an occurrence relation between two nodes or arcs of the FDG if always that one of related nodes or arcs in the graph has appeared; also the other node or arc of the same graph has appeared. Finally, there is an existence relation between two nodes or arcs if all the graphs in the class described by the FDG have at least one of the two nodes or arcs. A Function-Described Graph $\overline{M} = (\overline{\Sigma}_v, \overline{\Sigma}_e, P_v, P_e, R_v)$ is a graph where $\overline{\Sigma}_v, \overline{\Sigma}_e, P_v$ and P_e are defined exactly in the same way than FORGs; including, in addition, a set of binary relations R. See [16] for the construction of these binary relations.

2.6   Second Order Random Graphs

A Second-Order Random Graph (SORG) [18] is a probabilistic model closely related to FDGs. The main difference lies in the fact that the second-order structural information is not defined as binary relations but with the specific information of the second-order joint probability. Thus, the physical space needed to represent SORGs is much higher than FDGs but also its ability to represent the set of Attributed Graphs increases. In [17], it is shown how to convert a SORG to an FDGs simply by analysing the second-order probabilities and deciding if the binary relations hold. A Second-Order Random Graph $\overline{M} = (\overline{\Sigma}_v, \overline{\Sigma}_e, P_v, P_e)$ is a graph where $\overline{\Sigma}_v$ and $\overline{\Sigma}_e$ are defined similarly to FORGs but in P there are first order and also second order probability densities. See [18] for the construction of these second-order probability densities given a Common Labelling.

3   Performance Evaluation

3.1   Datasets

To evaluate the structural graph prototypes for graph classification and compare its results with embedding methods, we have selected three datasets form the repository presented in [19]. The datasets are Letter HIGH and LOW and GREC. Table 1 summarizes the characteristics of the datasets. The **Letter HIGH / LOW** database contains examples of the 15 capital letters (classes) of the Roman alphabet which are composed of straight lines. Each class contains 150 graph examples. The straight lines are represented by edges and the terminal points of the lines are represented by the nodes. The **GREC** database is composed of a set of 150 symbols from architecture, electronics and other technical fields. We have used a subset of 22 different symbols (classes) which are composed only of straight lines. These images are converted into graphs by assigning a node to each junction or terminal point and an edge to each line.

Table 1. Summary of graph data set characteristics, viz. the size of the training (tr), the validation (va) and the test set (te), the number of classes (Ω), the label alphabet of both nodes and edges, the average number of nodes and edges (mean nodes/edges)

Dataset	Size (tr, va, te)	$\|\Omega\|$	V labels	E labels	mean nodes	mean edges
Let. L-H	750,750,750	15	(x,y) coord.	None	5	9
GREC	286,286,528	22	(x,y) coord.	None	11	23

3.2   Results

Table 2 presents classification accuracies achieved with different structural [11, 12, 15, 16, 18] and embedding [7, 10, 20] methods. The embedding methods in [7, 20] are based on representing each graph with a vector of distances which related each graph to a set of prototypes extracted from the training set. Different prototype selection methods are presented: sps-c, bps-c and k-cps-c.

Once selected the prototypes and the embedding is performed, training is done using a SVM. In the original article, these methods are compared with the K-NN classifier under the Graph Edit Distance. The method in [10] is addressed to compute the Median Graph to later apply a K-NN classifier. The Median Graph is computed in an iterative form using an embedding space, the selected embedding is inspired in [7, 20]. For the structural methods on the Letter dataset, in addition to the standard procedure of constructing a single prototype to represent the class, we also analyzed the effect of representing the class using several prototypes. Specifically, we also test the system using 5 prototypes to represent each class of the Letters datasets (elements have been chosen randomly). Intuitively, representing the class using several prototypes may increase performance since the representer should reduce overgeneralization of the elements it represents. It is important to highlight that, results extracted from [7, 20] may not be evaluated with the same version of the dataset presented here, since, even the dataset we used is downloaded from the same website cited in [7, 20], the number of training, validation and test elements do not seem to correspond. Even though, we assume that results can be compared since the results we obtained with the reference K-NN (Table 2 (T4)) method seem to correlate with

Table 2. Comparative study between embedding and structural methods for graph classification. Numbers indicate: (T1) results extracted from [7], (T2) results extracted from [20], (T3) distance computed with the Graduated Assigment [21], (T4) results extracted from [10], (T5) with 5 prototypes to model the class and (T6) with 1 prototype to model the class.

Alg. type	Method	Let. LOW	Let. HIGH	GREC
	K-NN (T1)	91.1	61.6	86.1
Ref. System	K-NN (T2)	89.1	-	-
	K-NN (T3)	94.3	82.2	95.0
	K-NN (T4)	-	-	97.9
	Embed. Kernel(T1)	91.8	74.3	89.2
Embed. Mthds	sps-c (T2)	92.3	-	-
	bps-c (T2)	92.9	-	-
	k-cps-c (T2)	92.0	-	-
	Set Median (T4)	-	-	76.7
	Generalized Median (T4)	-	-	78.5
	Generalized Median	90.3 (T5)	70.0 (T5)	90.9 (T5)
		89.2 (T6)	70.9 (T6)	-
	Set Median	96.4 (T5)	74.5 (T5)	80.5 (T5)
		96.3 (T6)	68.9 (T6)	-
	Closure Graph	93.6 (T5)	49.1 (T5)	57.0 (T5)
Struct. Mthds		69.2 (T6)	22.1 (T6)	-
	FORG	93.9 (T5)	80.1 (T5)	85.8 (T5)
		92.3 (T6)	79.2 (T6)	-
	FDG	93.9 (T5)	81.6 (T5)	85.8 (T5)
		92.3 (T6)	81.7 (T6)	-
	SORG	94.0 (T5)	80.9 (T5)	91.2 (T5)
		92.8 (T6)	79.9 (T6)	-

the results obtained in [7, 20]. Under these considerations, obtained results show that structural/classical methods achieve recognition ration on the same range as new methodologies based on graph embeddings. Some structural methods, such as the Generalized Median Graph, obtain less classification ration than the embedding methodologies. However, more advanced structural methods, such as First Order Random Graphs, Function Described Graphs, Second Order Graphs and Closure Graphs, obtain greater recognition ration, even if the classification algorithm, i.e. the K-NN, is much simpler than the SVM used in [7, 20]. With respect to the embedding method in [10], which also uses a K-NN classifier, we see that the improvement of structural methods is notable. More advanced classification algorithms may further increase the recognition ratio. With respect to the approach with different representers per class on structural algorithms, it is possible to note a general tendency on the decrease of the recognition ration. This tendency increases in Closure Graphs. Thus, we could conclude that, in these datasets, it is better to use all the information available to generate a single prototype. One could think that some clustering schema to group elements, instead of random selection, should improve results. However, since, in this particular dataset, noise on elements is uniformly distributed, a random selection algorithm is as valid as any other grouping scheme.

4   Conclusions and Discussion

In this article, we presented a comparison between the two main approaches to perform graph classifications. The comparison is focussed in structural and embedding methods for graph data. The main objective of embedding methods is to encode/embed each graph as a point in some-dimensional vector space. In this way, they take the advantage of classical classification mechanisms for vector spaces to work with graph data in the embedding space. Once vectors are embedded in the vectorial space, operations between them are usually done in polynomial time. However, in most of the cases the process embedding requires non-polynomic computations. The main drawback of embedding methods is that usually vectorial representations of graphs are not complete, so they do not contain all the information available. On the other hand, most of the structural methods required either non-polynomic or approximated algorithms to compute the distance between two graphs. However, good approximation methods exist. Results presented in this article, show that classical graph prototype with combination with advanced synthesis mechanisms, such as the common labelling, improve or at least give the same classification accuracy than embedding methods. In addition, if embedding is done in non-polynomic time the classification phase is faster with structural methods since the queried graph do not need to be embedded and fewer graph-to-graph comparisons are required.

References

1. Conte, D., Foggia, P., Sansone, C., Vento, M.: Thirty years of graph matching in pattern recognition. IJPRAI 18(3), 265–298 (2004)

2. Mukherjee, L., Singh, V., Peng, J., Xu, J., Zeitz, M., Berezney, R.: Generalized median graphs and applications. Journal of Combinatorial Optimization 17, 21–44 (2009)
3. Hlaoui, A., Wang, S.: Median graph computation for graph clustering. Soft Computing - A Fusion of Foundations, Methodologies and Applications 10, 47–53 (2006)
4. Solé-Ribalta, A., Serratosa, F.: Graduated Assignment Algorithm for Finding the Common Labelling of a Set of Graphs. In: Hancock, E.R., Wilson, R.C., Windeatt, T., Ulusoy, I., Escolano, F. (eds.) SSPR&SPR 2010. LNCS, vol. 6218, pp. 180–190. Springer, Heidelberg (2010)
5. Ferrer, M., Serratosa, F., Sanfeliu, A.: Synthesis of Median Spectral Graph. In: Marques, J.S., Pérez de la Blanca, N., Pina, P. (eds.) IbPRIA 2005. LNCS, vol. 3523, pp. 139–146. Springer, Heidelberg (2005)
6. Gibert, J., Valveny, E., Bunke, H.: Graph embedding in vector spaces by node attribute statistics. Pattern Recognition 45(9), 3072–3083 (2012)
7. Riesen, K., Bunke, H.: Graph classification based on vector space embedding. IJPRAI 23(6), 1053–1081 (2009)
8. Kondor, R.I., Lafferty, J.D.: Diffusion kernels on graphs and other discrete input spaces. In: ICML, pp. 315–322 (2002)
9. Gärtner, T., Flach, P.A., Wrobel, S.: On graph kernels: Hardness results and efficient alternatives. In: COLT, pp. 129–143 (2003)
10. Ferrer, M., Valveny, E., Serratosa, F., Riesen, K., Bunke, H.: Generalized median graph computation by means of graph embedding in vector spaces. Pattern Recognition 43(4), 1642–1655 (2010)
11. Wong, A.K.C., You, M.: Entropy and distance of random graphs with application to structural pattern recognition. IEEE Transactions on Pattern Analysis and Machine Intelligence PAMI-7(5), 599–609 (1985)
12. Jiang, X., Müunger, A., Bunke, H.: On median graphs: Properties, algorithms, and applications. IEEE Trans. Pattern Anal. Mach. Intell. 23(10), 1144–1151 (2001)
13. Ferrer, M., Valveny, E., Serratosa, F.: Median graphs: A genetic approach based on new theoretical properties. Pattern Recognition 42(9), 2003–2012 (2009)
14. Ferrer, M., Serratosa, F., Valveny, E.: Evaluation of Spectral-Based Methods for Median Graph Computation. In: Martí, J., Benedí, J.M., Mendonça, A.M., Serrat, J. (eds.) IbPRIA 2007, Part II, LNCS, vol. 4478, pp. 580–587. Springer, Heidelberg (2007)
15. He, H., Singh, A.K.: Closure-tree: An index structure for graph queries. In: ICDE, p. 38 (2006)
16. Serratosa, F., Alquézar, R., Sanfeliu, A.: Synthesis of function-described graphs and clustering of attributed graphs. IJPRAI 16(6), 621–656 (2002)
17. Serratosa, F., Alquézar, R., Sanfeliu, A.: Function-described graphs for modelling objects represented by sets of attributed graphs. Pattern Recognition 36(3), 781–798 (2003)
18. Sanfeliu, A., Serratosa, F., Alquézar, R.: Second-order random graphs for modeling sets of attributed graphs and their application to object learning and recognition. IJPRAI 18(3), 375–396 (2004)
19. Riesen, K., Bunke, H.: IAM Graph Database Repository for Graph Based Pattern Recognition and Machine Learning. In: da Vitoria Lobo, N., Kasparis, T., Roli, F., Kwok, J.T., Georgiopoulos, M., Anagnostopoulos, G.C., Loog, M. (eds.) S+SSPR 2008. LNCS, vol. 5342, pp. 287–297. Springer, Heidelberg (2008)
20. Bunke, H., Riesen, K.: Towards the unification of structural and statistical pattern recognition. Pattern Recognition Letters 33(7), 811–825 (2012)
21. Gold, S., Rangarajan, A.: A graduated assignment algorithm for graph matching. IEEE Trans. Pattern Anal. Mach. Intell. 18(4), 377–388 (1996)

Improving Fuzzy Multilevel Graph Embedding through Feature Selection Technique

Muhammad Muzzamil Luqman[1,2], Jean-Yves Ramel[1], and Josep Lladós[2]

[1] Laboratoire d'Informatique, Université François Rabelais de Tours, 37200 France
[2] Computer Vision Center, Universitat Autònoma de Barcelona, 08193 Spain
{luqman,ramel}@univ-tours.fr, josep@cvc.uab.es

Abstract. Graphs are the most powerful, expressive and convenient data structures but there is a lack of efficient computational tools and algorithms for processing them. The embedding of graphs into numeric vector spaces permits them to access the state-of-the-art computational efficient statistical models and tools. In this paper we take forward our work on explicit graph embedding and present an improvement to our earlier proposed method, named "fuzzy multilevel graph embedding - FMGE", through feature selection technique. FMGE achieves the embedding of attributed graphs into low dimensional vector spaces by performing a multilevel analysis of graphs and extracting a set of global, structural and elementary level features. Feature selection permits FMGE to select the subset of most discriminating features and to discard the confusing ones for underlying graph dataset. Experimental results for graph classification experimentation on IAM letter, GREC and fingerprint graph databases, show improvement in the performance of FMGE.

Keywords: graphics recognition, graph classification, explicit graph embedding, feature selection.

1 Introduction and Related Works

Over decades of research in pattern recognition, the research community has developed a range of expressive and powerful approaches for diverse problem domains. Graph based structural representations are widely employed for extracting the structure, topology and geometry, in addition to the statistical details of underlying data [1]. During next step in the processing chain, generally these representations could not be exploited to their full strength because of limited availability of computational tools for them. On the other hand, the efficient and mature computational models offered by statistical approaches, work only on vector data and cannot be directly applied to these high-dimensional representations. The emerging domain of graph embedding in pattern recognition, addresses this problem of the lack of efficient computational tools for graph based representations.

Graph embedding is a methodology aimed at representing a whole graph, along-with the attributes attached to its nodes and edges, as a point in a suitable

G.L. Gimel' farb et al. (Eds.): SSPR & SPR 2012, LNCS 7626, pp. 243–253, 2012.

vector space. Graph embedding is a natural outcome of parallel advancements in structural and statistical pattern recognition. It offers a straightforward solution, by employing the representational power of symbolic data structures and the computational superiority of feature vectors [2]. It acts as a bridge between structural and statistical approaches [3][4], and allows a pattern recognition method to benefit from computational efficiency of state-of-the-art statistical models and tools along-with the convenience and representational power of classical symbolic representations [5]. This permits the last three decades of research on graph based structural representations in various domains [1], to benefit from the state-of-the-art machine learning models and tools. Graph embedding has its application to the whole variety of domains that are entertained by pattern recognition and where the use of a relational data structure is mandatory for performing high level semantic tasks. Apart from reusing the computational efficient methods for vector spaces, another important motivation behind graph embedding methods is to solve the computationally hard problems geometrically [6]. We refer the interested reader to [7][8] for further reading on graph embedding.

The graph embedding methods are formally categorized as *implicit* graph embedding or *explicit* graph embedding. The implicit graph embedding methods are based on graph kernels. A graph kernel is a function that can be thought of as a dot product in some implicitly existing vector space. Instead of mapping graphs from graph space to vector space and then computing their dot product, the value of the kernel function is evaluated in graph space. Such an embedding satisfies the main mathematical properties of dot product. However, since it does not explicitly map a graph to a point in vector space, a strict limitation of implicit graph embedding is that it does not permit all the operations that could be defined on vector spaces. We refer the interested reader to [7][9][10] for further reading on graph kernels and implicit graph embedding.

On the other hand, the more useful, explicit graph embedding methods explicitly embed an input graph into a feature vector and thus enable the use of all the methodologies and techniques devised for vector spaces.

Definition 1. Attributed graph (AG). *Let A_V and A_E denote the domains of possible values for attributed vertices and edges respectively. These domains are assumed to include a special value that represents a null value of a vertex or an edge. An attributed graph AG over (A_V, A_E) is defined to be a four-tuple:*

$$AG = (V, E, \mu^V, \mu^E)$$

where,

V *is a set of vertices,*
$E \subseteq V \times V$ *is a set of edges,*
$\mu^V : V \longrightarrow A_V^k$ *is function assigning k attributes to vertices and*
$\mu^E : E \longrightarrow A_E^l$ *is a function assigning l attributes to edges.*

Definition 2. Explicit graph embedding. *Explicit graph embedding maps a graph to a point in suitable vector space. It encodes the graphs* *by equal size vectors and produces one vector per graph. Mathematically, for a graph $AG = (V, E, \mu^V, \mu^E)$, explicit graph embedding is a function ϕ, which maps graph AG from graph space G to a point $(f_1, f_2, ..., f_n)$ in n dimensional vector space $\mathbb{R}^n$:*

$$\phi : G \longrightarrow \mathbb{R}^n$$
$$AG \longmapsto \phi(AG) = (f_1, f_2, ..., f_n)$$

The vectors obtained by an explicit graph embedding method can also be employed in a standard dot product for defining an implicit graph embedding function between two graphs [11]. An interesting property of explicit graph embedding is that the graphs are embedded in pattern spaces in a manner that similar structures come close to each other and different structures goes far away i.e. an implicit clustering is achieved [12]. Another important property of explicit graph embedding is that the graphs of different size and order need to be embedded into a fixed size feature vector. This means that for constructing the feature vector, an important step is to mark the important details that are available in all the graphs and are applicable to a broad range of graph types. We refer the interested reader to [7] for further reading on explicit graph embedding.

1.1 Related Works

Recently, two interesting series of works on explicit graph embedding for pattern recognition, with an application to graphics recognition, have been proposed in literature.

The first method is from Bunke et al. [11] and is based on dissimilarity of a graph from a set of prototypes. The main idea of this work is to construct a vector of graph edit distances from the graph to be embedded and a set of k prototypes selected in the graph database. The embedding of the graph is thus a vector of k distances. Formally, let $\Gamma = g_1, ..., g_n$ be a set of graphs and $p = p_1, ..., p_k \subset \Gamma$ be a subset of selected prototypes from Γ. The graph embedding is defined as the function $\Phi : \Gamma \longmapsto (\mathbb{R})^k$, such that $\Phi(g) = [d(g, p_1), ..., d(g, p_k)]$ where $d(g, p_i)$ is the graph edit distance between graph g and the ith prototype graph in p. In [13], the authors propose an improvement of the graph embedding method by using feature selection methods. This type of projection is very interesting as it offers computational advantages over the traditional graph based algorithms. However, the limitation of setting the edit distance is found in this method. In addition, the choice of prototype graphs is also a significant parameter as it determines the size of the vector and its capacity to effectively represent the graph in the vector space. Also, it remains highly dependent on the application and its learning set.

The second method is from Gibert et al. [14] and is based on the frequencies of appearance of specific knowledge-dependent substructures in graph. The main idea of this work is to construct vector representation of graphs by counting the frequency of appearance of specific set of representatives of node labels and their corresponding edges. In [15] the authors propose an improvement of their graph embedding technique by dimensionality reduction of the obtained feature vector. In [16] the authors have applied multiple classifiers to their graph embedding method. In [14] the authors have studied the application of feature selection algorithms for their graph embedding method. This type of graph embedding algorithms provides an embedding of graph into feature vector, in linear time complexity. Their simplicity of implementation is an important advantage. However, the features that have been used, are very localized to nodes and arcs. The graph embedding contains little information on the topology, which can have a negative impact on the classification results.

1.2   Main Contribution of This Paper

This paper is a continuation of our work on explicit graph embedding. The method is originally proposed in [17][18] and is named as "fuzzy multilevel graph embedding - FMGE". FMGE embeds a graph into feature vector space by extracting a large number of features from graph. The use of high dimensional feature vector permits FMGE to achieve generalization to diverse graphs in an unsupervised fashion. However this also results into high dimensionality and sparsity of feature vector. In [19] we studied the application of dimensionality reduction techniques on FMGE extracted features. Motivated from the similar works in [13] and [14] where the authors have applied feature selection algorithms on explicit graph embedding methods, in this paper we take forward our work on graph embedding and study the application of feature selection algorithms on FMGE extracted features.

The rest of this paper is organized as follows. In Section 2, we briefly outline the Fuzzy Multilevel Graph Embedding (FMGE). In Section 3 we describe the application of feature selection algorithms on FMGE. Experimentation and discussion is presented in Section 4. In Section 5 we conclude this paper with future lines of research.

2   Fuzzy Multilevel Graph Embedding (FMGE)

Most of the existing works on graph embedding deal only the graphs that are comprised of edges with a single attribute and vertices with either no or only symbolic attributes. These methods are only useful for specific application domains for which they are designed. FMGE does not require any dissimilarity measure between graphs and to the best of our knowledge, FMGE extends the methods in literature by offering the embedding of attributed graphs with many numeric as well as symbolic attributes on both nodes and edges. It is applicable to directed as well as undirected attributed graphs. The time complexity of FMGE is linear to number of attributes and size of the graphs [18]. Many existing solutions for graph embedding offer to utilize the statistical significant details in graphs for embedding them into feature vectors. FMGE exploits the topological, structural and attribute information of the graphs along-with the statistical significant information, for constructing feature vectors of adapted and optimal size. It employs fuzzy overlapping trapezoidal intervals for minimizing the information loss while mapping from continuous graph space to discrete feature vector space. The proposed feature vector is very significant for application domains where the use of graphs is mandatory for representing rich structural and topological information, and an approximate but computational efficient solution is needed. The unsupervised learning abilities of FMGE and the fact that it does not require a labeled graph dataset for learning allows its inexpensive deployment to various application domains [18]. FMGE performs multilevel analysis of graph to extract discriminatory information of three different levels. These include the graph level information, structural level information and the elementary level information (see Fig. 1). The three levels of information represent three different views of graph for extracting global details, details on topology of graph and details on elementary building units of graph. The feature vector of FMGE is named Fuzzy Structural Multilevel Feature Vector - FSMFV (see Fig. 2).

The features for graph level information represent a coarse view of graph and give general information about the graph. These features include graph order and graph size.

The features for structural level information represent a deeper view of graph and are extracted from the node degrees and subgraph homogeneity in graph. Subgraph homogeneity is represented by computing resemblance attributes for the nodes and edges of graph. The resemblance attributes for an edge is computed from the attributes on its neighboring nodes. The resemblance for a numeric attribute (a) is computed as a ratio of this attribute's values on neighboring nodes of an edge $(a_1$ and $a_2)$ (see Eq. 1). Whereas the resemblance for a symbolic attribute (b) is computed as a ratio of this attribute's values on neighboring nodes of an edge $(b_1$ and $b_2)$ (see Eq. 2).

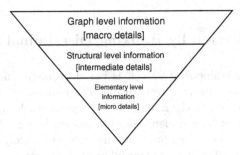

Fig. 1. Multi-facet view of discriminatory information in graph

Graph order	Graph size	Embedding of node degree	Embedding(s) of subgraph(s) homogenity	Embedding(s) of node attribute(s)	Embedding(s) of edge attribute(s)

Fig. 2. Feature vector of FMGE

$$resemblance(a_1, a_2) = min(|a_1|, |a_2|)/max(|a_1|, |a_2|) \qquad (1)$$

$$resemblance(b_1, b_2) = \begin{vmatrix} 1 & b_1 = b_2 \\ 0 & otherwise \end{vmatrix} \qquad (2)$$

The third level of information is extracted by penetrating into further depth and more granular view of graph and employing details of the elementary building blocks of graph. These features represent the information extracted from the node and edge attributes.

The node degree, numeric resemblance attributes, numeric node attributes and numeric edge attributes are embedded by fuzzy histograms whereas the symbolic resemblance attributes, symbolic node attributes and the symbolic edge attributes are embedded by crisp histograms. FMGE learns the intervals for constructing these histograms, during an unsupervised learning phase and employs these parameters during graph embedding phase [17][18][19].

Length of Feature Vector (FSMFV): The length of the feature vector is strictly dependent on the size of histograms used for encoding the three levels of information. The length of feature vector is uniform for all graphs in an input collection and is given by Eq.3.

$$\text{Length of FSMFV} = 2 + \sum s_i + \sum c_j \qquad (3)$$

where,
- 2 refers to the features for graph order and graph size.
- s_i refers to the number of bins in fuzzy interval encoded histogram for each numeric attribute i, in graph.
- c_j refers to the number of bins in crisp interval encoded histogram for each symbolic attribute j, in graph.

3   Feature Selection by Ranking Discriminatory Features

The feature vector obtained by FMGE is based on histogram encoding of the multilevel information extracted from graph. The number of features in the vector is directly dependent on the number of bins employed for constructing these histograms. The use of high dimensional histograms is explicitly built into the method as it enables FMGE to provide a more robust encoding of information and enables it to generalize to unseen graphs. However, this results into a serious drawback as well, that the feature vector becomes sparse and confuses between classes of graphs. Previously we have tried to reduce the dimensionality of the feature vector by using principal component analysis (PCA) [19]. PCA is based on linear transformation of data to a low dimensional space that describes most of variance in data. But we feel that instead of an unsupervised dimensionality reduction technique like PCA, the use of a supervised dimensionality reduction technique (a.k.a. feature selection) will result into a more meaningful ranking of the FMGE extracted features. This ranking will permit to select the high discriminatory features and ignore the confusing features in the FMGE embedded vector space. Thus producing more compact feature vector representation of graphs and adding strength to the discriminatory power of the feature vector.

We have employed the Relief algorithm based feature selection [20]. The Relief algorithm is a classical ranking method that is based on the ability of features to discriminate between different classes. For each instance of a given feature, the near-hit (closest value among elements of same class) and the near-miss (closest value among element of other classes) are computed. A weight is calculated for every feature in terms of the distances of each sample to its near-hit and near-miss. Mathematically, for a set S of m samples of feature f_i, the rank value ω_{f_i} is computed as:

$$\omega_{f_i} = \frac{1}{m} \sum_{x \in S} |x - Z_x^-| - |x - Z_x^+| \tag{4}$$

where,
- Z_x^- is near-miss of sample x.
- Z_x^+ is near-hit of sample x.

A high ranking value of ω_{f_i} is desirable as it indicates that the feature is important and has high discriminatory capabilities. In order to reduce the size of FMGE feature vector and to remove the unimportant features from it, we select the subset of top-ranked features, on the basis of ranks obtained through the Relief algorithm.

4 Experimentation

The experimentation has been performed on *'IAM Graph Database Repository for Graph Based Pattern Recognition and Machine Learning'*. The IAM graph database repository is publicly available from the website of IAPR technical committee on graph based representations (TC-15)[1], and contains graph datasets from the field of document image analysis and graphics recognition, describing both synthetic and real data [21].

4.1 Datasets

The summary of the letter, GREC and fingerprint datasets, together with some characteristic properties, is given in Table 1. The letter graph dataset is comprised of graphs extracted from drawings of 15 capital letters of Roman alphabet that consists of straight lines only. The prototype drawing of letters are converted into prototype graphs by representing lines by undirected edges and ending points of lines by nodes. Each node is labeled with a two-dimensional attribute giving its position relative to a reference coordinate system. The GREC graph dataset is comprised of graphs representing 22 symbols from architectural and electronic drawings. Graphs are extracted from the denoised images

---

[1] http://www.greyc.ensicaen.fr/iapr-tc15/index.php

Table 1. IAM graph database

		Letter LOW	GREC	Fingerprint
Size	Train	750	836	500
	Valid	750	836	300
	Test	750	1628	2000
Classes		15	22	4
Average	$|V|$	4.7	11.5	5.4
	$|E|$	3.1	12.2	4.4
Maximum	$|V|$	8	25	26
	$|E|$	6	30	25
Numeric attribute	$|V|$	2	2	2
	$|E|$	0	1	1
Symbolic attribute	$|V|$	0	1	0
	$|E|$	0	1	0

by representing ending points, corners, intersections and circles by nodes and labeled with a two-dimensional attribute giving their position. The nodes are connected by undirected edges that are labeled as line or arc and have the angle with respect to the horizontal direction as attribute. Fingerprint images are converted into graphs by representing the ending points and bifurcation points of the skeletonized regions as nodes. Each node is labeled with a two-dimensional attribute giving its position. The edges are attributed with an angle denoting the orientation of the edge with respect to the horizontal direction.

4.2   Experimental Setup and Results

We have evaluated the application of Relief feature selection algorithm on FMGE, by classification rate obtained by a nearest neighbor classifier. The experiments are performed by first tuning the parameters on the validation set and then using the best configuration on the test set.

The first validation parameter is the number of fuzzy intervals for embedding numeric information in graph i.e. the node degree, numeric resemblance attributes, numeric node attributes and numeric edge attributes. Starting from 2 intervals, the number of fuzzy intervals for embedding the numeric information is increased until 25 (in steps of 1).

The second validation parameter is selection of top-ranked features. For each of the 25 configurations of FMGE, we applied the Relief feature selection algorithm (with a neighborhood size of 10 for calculating near-hit and near-miss), to obtain rankings of features. We used this ranking information to generate all subsets of high to low ranked features (for each of the 25 configuration of FMGE), i.e. subset containing top-1 feature, subset containing top-2 feature, subset containing top-3 feature and so on. We validated the classification rate for all of these subsets of features and selected the subset of features that produced the best classification rate on validation set (for each of the 25 configurations of FMGE). Fig. 3 shows the validation results for letter LOW, GREC and Fingerprint datasets. The colored curves in the plot represent the first validation parameter i.e. number of the fuzzy intervals for encoding numeric information. Whereas, each point on a curve gives the classification rate obtained on the n top-ranked features. The plots clearly demonstrate that the method can obtain its maximum classification rates on only a small subset of top-ranked features.

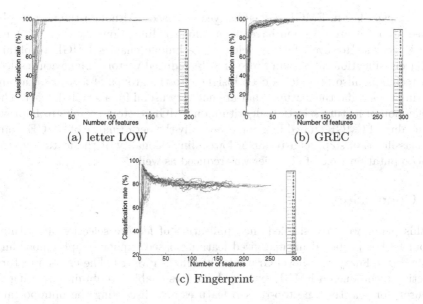

(a) letter LOW   (b) GREC

(c) Fingerprint

Fig. 3. Validation results showing the classification rates obtained for different subsets of top-ranked features for the 25 configurations of FMGE. The horizontal axis contains sorted (high to low ranked by Relief algorithm) features extracted by FMGE.

Table 2. Results on test sets - IAM graph database. CR is the classification rate (%) obtained by k-nn classifier and DIM is the dimensionality of feature vector.

	Letter LOW		GREC		Fingerprint	
	CR	DIM	CR	DIM	CR	DIM
Reference system [21]	99.6		95.5		76.6	
Gibert et al. [14] (best CR)	100		98.7		80.5	
Gibert et al. [14] (Relief feature selection)	99.1		96.4		77.6	
Full FMGE vector	96.5	58	97.2	79	76.6	127
FMGE vector (PCA reduced [19])	96.3	10	96.8	14	77.2	16
FMGE vector (Relief feature selection)	99.2	37	**99.8**	47	**85.5**	61

The configuration of FMGE with optimized subset of top-ranked features (w.r.t. classification rate on validation set), was used to embed the test set and to compute the final classification rates on test set.

Table 2 presents the classification rates of a k-nn classifier in FMGE embedded vector space on test set, before and after the application of feature selection technique. The PCA dimensionality reduction results are reported from [19]. For comparison of results, the k-nn classification rates are reported for the graph edit distance plus k-nn classifier based reference system [21] and k-nn classifier based system of Gibert et al. [14].

The results show that the application of feature selection technique successfully improves the performance of FMGE representation of graphs. The Relief feature selection technique selects much lower number of features (as compared to full FMGE vector) and in all cases improves the classification rates obtained by k-nn classifier in FMGE embedded vector spaces. Although the vectors obtained after feature selection have slightly higher dimensionality than PCA reduced vectors. However, feature selection technique enables FMGE to obtain better classification rates than those of PCA reduced vectors. The feature selection technique also provides a deep insight into the groups of features that are discriminatory and the features that are not very useful for a graph dataset. The application of feature selection algorithm on FMGE vectors permits to improve the quality of FMGE embedding and eventually the performance of FMGE. During classification step, as a result of lower dimensionality of the feature vector the computation time of classifier was reduced as well.

5   Conclusion

In this paper we have studied the application of feature selection algorithms to our earlier proposed unsupervised learning based explicit graph embedding method (the Fuzzy Multilevel Graph Embedding - FMGE). The use of a feature selection technique on FMGE extracted features enables to obtain a ranking of features, for selecting the top-ranked features and discarding the unimportant features. This permits to increase the quality of embedding and improves the performance of FMGE. Feature selection step also improves the unsupervised learning capabilities of FMGE by adapting its feature-set to underlying graph dataset. The initial experimental results are very encouraging and in future we plan to study the application of sophisticated feature selection techniques for further improving the quality of Fuzzy Multilevel Graph Embedding.

References

1. Conte, D., Foggia, P., Sansone, C., Vento, M.: Thirty years of graph matching in pattern recognition. International Journal of Pattern Recognition and Artificial Intelligence 18(3), 265–298 (2004)
2. Bunke, H., Irniger, C., Neuhaus, M.: Graph Matching – Challenges and Potential Solutions. In: Roli, F., Vitulano, S. (eds.) ICIAP 2005. LNCS, vol. 3617, pp. 1–10. Springer, Heidelberg (2005)
3. Bunke, H., Gunter, S., Jiang, X.: Towards Bridging the Gap between Statistical and Structural Pattern Recognition: Two New Concepts in Graph Matching. In: Singh, S., Murshed, N., Kropatsch, W.G. (eds.) ICAPR 2001. LNCS, vol. 2013, pp. 1–11. Springer, Heidelberg (2001)
4. Roth, V., Laub, J., Kawanabe, M., Buhmann, J.: Optimal cluster preserving embedding of nonmetric proximity data. IEEE Transactions on Pattern Analysis and Machine Intelligence 25(12), 1540–1551 (2003)
5. Chen, T., Yang, Q., Tang, X.: Directed graph embedding. In: International Joint Conference on Artificial Intelligence, pp. 2707–2712 (2007)
6. Shaw, B., Jebara, T.: Structure preserving embedding. In: International Conference on Machine Learning, pp. 1–8 (2009)
7. Foggia, P., Vento, M.: Graph Embedding for Pattern Recognition. In: Ünay, D., Çataltepe, Z., Aksoy, S. (eds.) ICPR 2010. LNCS, vol. 6388, pp. 75–82. Springer, Heidelberg (2010)

8. Lee, G., Madabhushi, A.: Semi-Supervised Graph Embedding Scheme with Active Learning (SSGEAL): Classifying High Dimensional Biomedical Data. In: Dijkstra, T.M.H., Tsivtsivadze, E., Marchiori, E., Heskes, T. (eds.) PRIB 2010. LNCS, vol. 6282, pp. 207–218. Springer, Heidelberg (2010)

9. Riesen, K., Bunke, H.: Graph Classification and Clustering Based on Vector Space Embedding. World Scientific (2010)

10. Riesen, K., Bunke, H.: Graph Classification And Clustering Based On Vector Space Embedding. World Scientific Publishing Co., Inc. (2010)

11. Bunke, H., Riesen, K.: Recent advances in graph-based pattern recognition with applications in document analysis. Pattern Recognition 44(5), 1057–1067 (2011)

12. Wilson, R.C., Hancock, E.R., Luo, B.: Pattern vectors from algebraic graph theory. IEEE Transactions on Pattern Analysis and Machine Intelligence 27(7), 1112–1124 (2005)

13. Bunke, H., Riesen, K.: Improving vector space embedding of graphs through feature selection algorithms. Pattern Recognition 44(9), 1928–1940 (2011)

14. Gibert, J., Valveny, E., Bunke, H.: Feature selection on node statistics based embedding of graphs. Pattern Recognition Letters (April 2012)

15. Gibert, J., Valveny, E., Bunke, H.: Dimensionality Reduction for Graph of Words Embedding. In: Jiang, X., Ferrer, M., Torsello, A. (eds.) GbRPR 2011. LNCS, vol. 6658, pp. 22–31. Springer, Heidelberg (2011)

16. Gibert, J., Valveny, E., Terrades, O.: Multiple classifiers for graph of words embedding. Multiple Classifier Systems, 1–10 (2011)

17. Luqman, M.M., Lladós, J., Ramel, J.Y., Brouard, T.: A Fuzzy-Interval Based Approach for Explicit Graph Embedding. In: Ünay, D., Çataltepe, Z., Aksoy, S. (eds.) ICPR 2010. LNCS, vol. 6388, pp. 93–98. Springer, Heidelberg (2010)

18. Luqman, M.M., Ramel, J.Y., Lladós, J., Brouard, T.: Fuzzy Multilevel Graph Embedding. Pattern Recognition (accepted, 2012), http://dx.doi.org/10.1016/j.patcog.2012.07.029

19. Luqman, M.M., Lladós, J., Ramel, J.Y., Brouard, T.: Dimensionality Reduction for Fuzzy-Interval Based Explicit Graph Embedding. In: GREC, pp. 117–120 (2011)

20. Kira, K., Rendell, L.A.: The feature selection problem: traditional methods and a new algorithm. In: Proceedings of the Tenth National Conference on Artificial Intelligence, AAAI 1992, pp. 129–134. AAAI Press (1992)

21. Riesen, K., Bunke, H.: IAM Graph Database Repository for Graph Based Pattern Recognition and Machine Learning. In: da Vitoria Lobo, N., Kasparis, T., Roli, F., Kwok, J.T., Georgiopoulos, M., Anagnostopoulos, G.C., Loog, M. (eds.) S+SSPR 2008. LNCS, vol. 5342, pp. 287–297. Springer, Heidelberg (2008)

Dynamic Learning of SCRF for Feature Selection and Classification of Hyperspectral Imagery

Ping Zhong*, Zhiming Qian, and Runsheng Wang

ATR National Laboratory, School of Electronic Science and Engineering,
National University of Defense Technology, 410073, Changsha, Hunan, China
{zhongping,qianzhiming,rswang}@nudt.edu.cn

Abstract. This paper investigates the feature selection and contextual classification of hyperspectral images through the sparse conditional random field (SCRF) model. To relieve the heavy degeneration of classification performance caused by the characteristics of the hyperspectral data and the oversparsity when SCRF selects a small feature subset, we develop a dynamic learning framework to train the SCRF. Under the piecewise training framework, the proposed dynamic learning method of SCRF can be implemented efficiently through separated dynamic sparse trainings of simple classifiers defined by corresponding potentials. Experiments on the real-world hyperspectral images attest to the effectiveness of the proposed method.

Keywords: Conditional random field, classification, feature selection.

1 Introduction

Hyperspectral image analysis is attracting a growing interest in real world applications, such as urban planning, mapping, agriculture, forestry, and disaster prevention and monitoring. Many these applications can be finally transformed into some classification tasks. In the literature, many techniques have been developed for the classification purpose, including support vector machines [1, 2], neural networks[3], graph method[4–6], and others. Many algorithms take into consideration only spectral variations, ignore spatial correlations, and treat each site independently. However, hyperspectral images show strong correlations across spatial and spectral neighbors[6], which have been proved to be very useful for image analysis in both the remote sensing and computer vision communities.

Markov random fields (MRFs) are the classical probabilistic approaches for modeling the contextual information in label images. However, for computational tractability, the observed data are assumed to be conditional independent, which neglects the contextual information in the observed data of a given class. Conditional random fields (CRFs) have recently gained popularity since they have the ability to incorporate contextual information in the labels as well as the

---

* This research was conducted with support of the NSF of China (Grant No. 60902088 and 61271439) and NDTF Project of ATR Lab. (Grant No. 9140C8004011005).

G.L. Gimel' farb et al. (Eds.): SSPR & SPR 2012, LNCS 7626, pp. 254–263, 2012.
© Springer-Verlag Berlin Heidelberg 2012

observations[7]. But as for other supervised classifiers, excessive large number of spectral features may bring on the well-known overfitting problem for CRFs[8, 9]. Moreover, it is inefficient to use many irrelevant features due to the increased computational complexity.

Reduction in the number of features thus can be a direct way to overcome the overfitting and save the computational cost. Recently, there have several approaches to select the relevant features for the classical log-linear CRFs with potentials defined as simple linear combinations of features. But for the extended CRF with potentials defined as discriminative classifiers, the log-likelihood cannot guarantees to be an additive function of features. Thus it may be difficult to use the methods directly to select features for the extended CRFs. In contrast, we addressed the feature selection problem during training by adding a sparsity-promoting regularizer to the log-likelihood in the form of a log Laplacian prior on the model parameters[9]. The trained sparse model is named sparse CRF (SCRF) model.

In this work, we go one step further to demonstrate that as the generalized linear models (GLMs), SCRFs may suffer from the heavy degeneration of classification performance when they select small feature subset. This work develops a dynamic learning method of the SCRF (D-SCRF, for short) to relieve the negative effects of the problem on the classification performance. Moreover, we will show that under the piecewise training framework, the dynamic learning of SCRF can be efficiently implemented through two separated dynamic trainings of Sparse Multinomial Logistic Regression (SMLR) models.

2 SCRF for Feature Selection and Classification

In hyperspectral image classification, the observed data y is considered to be a set of spectral vectors $\{y_1, y_2, ..., y_I\}$, where $y_i = [y_{i1}, y_{i2}, ..., y_{iD}]^T$ denotes a spectral vector associated with an image site $i \in S$. D is the number of spectral bands and $S = \{1, 2, ..., I\}$ is the set of image sites. The label set is given by $x = \{x_1, x_2, ..., x_I\}$, where $x_i \in \{1, 2, ..., L\}$ and L is the number of classes.

The CRF for hyperspectral image classification directly models the posterior as

$$P(x|y, \theta) = \frac{1}{Z} \exp\left\{ \sum_{i \in S} \phi_i(x_i, y, w) + \sum_{i \in S} \sum_{j \in \eta_i} \xi_{ij}(x_i, x_j, y, v) \right\} \quad (1)$$

where Z is a normalizing constant known as the partition function. The unary clique potential $\phi_i(.)$ is defined as multinomial logistic regression (MLR) model:

$$\phi_i(x_i, y, w) = \sum_{l=1}^{L} \delta(x_i = l) \log P(x_i = l | y, w) \quad (2)$$

where

$$P(x_i = l | y, w) = \begin{cases} \frac{\exp(w_l^T y_i)}{1 + \sum_{k=1}^{L-1} \exp(w_k^T y_i)} & if \ l < L \\ \frac{1}{1 + \sum_{k=1}^{L-1} \exp(w_k^T y_i)} & if \ l = L \end{cases} \quad (3)$$

w_k is the parameter vector $[w_{k1}, ..., w_{kD}]^T$ for kth class. The pairwise clique potential $\xi_{ij}(.)$ is defined as a generalization of the Ising model[10]:

$$\xi_{ij}(x_i, x_j, y, v) = \sum_{k,l \in \{1,...,L\}} v_{kl}^{\mathrm{T}} \mu_{ij}(y) \delta(x_i = k) \delta(x_j = l) \qquad (4)$$

where v_{kl} is the parameter vector and $\mu_{ij}(y)$ is a spectral feature vector obtained by concatenating all elements of two vectors y_i and y_j.

The parameters $\theta = \{v, w\}$ is said to be sparse if and only if many of its entries are exactly zero. The sparsity is associated with the definition of feature selection. So the feature selection can be implemented by the sparse trainings of the model parameters. Let $\{\tilde{x}, \tilde{y}\} = \{\tilde{x}_c, \tilde{y}_c\}_{c \in \tilde{C}}$ be the selected training samples. The sparse training is implemented as a maximum a posteriori (MAP) estimate

$$\tilde{\theta} = \arg\max_{\theta} Q(\theta) = \arg\max_{\theta} (L(\theta) - \lambda_\theta \|\theta\|_1) \qquad (5)$$

where $\|\theta\|_1 = \sum_n |\theta_n|$ denotes the l_1 norm of the parameters θ in the sparsity-promoting Laplacian distribution and $L(\theta)$ is the log-likelihood.

3   Dynamic Learning of SCRF

The sparsity of the parameter set θ is controlled by the regularization parameter λ_θ. The larger is λ_θ, the greater is sparsity. Excessively large values of λ_θ will result in under-fitting, while excessively small values of λ_θ could result in over-fitting. In the literature of l_1 regularization, the cross-validation method is usually used to select the optimum λ_θ from predefined values [11], which are fixed through the whole training procedures. However, as for the generalized linear model (GLM), the fixed-value-based method may bring two problems for the SCRF. Firstly, to select relative small feature subset, SCRF should be trained with large values of λ_θ. But the fixed excessively large parameter can result in the over-sparsity. Secondly, each band of hyperspectral data contains some information but only some of the bands have significant effects on output. Such characteristics also prevent the optimal λ_θ derived from fixed-value-based methods from obtaining high level of performances[12].

Both the problems are derived essentially from negative effects of the too many irrelevant or weakly relevant features on the classifier. So a direct method dealing with the problems is to get rid of the obvious irrelevant features on the basis of their relevance or discriminant powers with regard to the targeted classes before training. But the primary feature selection procedure is not correlated to the SCRF model. In contrast, we develop a dynamic learning method to incorporate the primary feature selection procedure into the training of SCRF. As mentioned earlier, the larger is λ_θ, the greater is the sparsity, which means more features are discarded. Based on this conclusion, the dynamic learning makes the λ_θ vary during iterative training: the large values of the λ_θ are utilized to get rid of the obvious irrelevant or weakly relevant features at the earlier iterations; then the later iterations arrives the convergence and obtains the superior classifier. For the remainder of this work, the variable parameter is denoted as λ_θ^α, and then we get the objective function of dynamic learning framework as

$$Q^\alpha(\theta) = L(\theta) - \lambda_\theta^\alpha \|\theta\|_1 \qquad (6)$$

3.1 Piecewise Implementation of Dynamic Learning

Because $\|\theta\|_1 = \sum_n |\theta_n|$ is a nondifferentiable term at the origin, the usual gradient-based methods cannot be directly utilized to maximize the objective function. In this work, we develop an efficient sparse training method under the piecewise training framework. Firstly, $L(\theta)$ is divided according to the types of the cliques. Let $\tilde{C}_m$ be the set of the type of cliques with m sites selected for model training. Then the divided graph factor a is a clique c in the set $A = \left\{\tilde{C}_m\right\}_{m=1,2,\dots} \triangleq \tilde{C}$, and consequently, the divided factor $f_a(\tilde{x}_a, y)$ of $L(\theta)$ is exactly the potential $\psi_c(\tilde{x}_c, \tilde{y}, \theta)$. Finally, the piecewise dynamic training of SCRF with the special division is to maximize the objective function

$$Q_{PW}^\alpha(\theta) = \sum_{c \in \tilde{C}} \log \frac{\psi_c(\tilde{x}_c, \tilde{y}, \theta)}{\sum_{x_c} \psi_c(x_c, \tilde{y}, \theta)} - \lambda_\theta^\alpha \|\theta\|_1 \qquad (7)$$

Consider only up to pairwise clique potentials, then Eq. (7) can be rewritten as

$$Q_{PW}^\alpha(w, v) = \underbrace{\left(\sum_{i \in \tilde{C}_1} \log \frac{\exp\{\phi_i(\tilde{x}_i, \tilde{y}, w)\}}{\sum_{x_i} \exp\{\phi_i(x_i, \tilde{y}, w)\}} - \lambda_w^\alpha \|w\|_1 \right)}_{Q_w^\alpha}$$
$$+ \underbrace{\left(\sum_{(i,j) \in \tilde{C}_2} \log \frac{\exp\{\xi_{ij}(\tilde{x}_i, \tilde{x}_j, \tilde{y}, v)\}}{\sum_{x_i, x_j} \exp\{\xi_{ij}(x_i, x_j, \tilde{y}, v)\}} - \lambda_v^\alpha \|v\|_1 \right)}_{Q_v^\alpha} \qquad (8)$$

Eq. (8) shows that under piecewise training framework with the special division, D-SCRF can be trained by independently dynamic training the local sparse classifiers over each kind of cliques.

In the first term in Eq.(8), the unary potential modeled as MLR in Eq. (3) has the normalization condition as $\sum_{l=1}^L P(x_i = l | y, w) = 1$. So the denominator of the first term in Eq. (8) is just the constant one. We then immediately have

$$Q_w^\alpha = \sum_{i \in \tilde{C}_1} \log P(\tilde{x}_i | \tilde{y}, w) - \lambda_w^\alpha \|w\|_1 \triangleq L_{MLR}(w) - \lambda_w^\alpha \|w\|_1 \qquad (9)$$

Since $P(\tilde{x}_i | \tilde{y}, w)$ is defined as MLR (Eq. (4)), $L_{MLR}(w)$ is log-likelihood of MLR and then Eq. (9) is exactly the objective function of D-SMLR [13].

In the second term in Eq.(8), Q_v^α can be written as

$$Q_v^\alpha = \sum_{i,j \in \tilde{C}_2} \log P(\tilde{x}_i, \tilde{x}_j | \mu_{ij}(\tilde{y}), v) - \lambda_v^\alpha \|v\|_1 \triangleq L_{MLR}(v) - \lambda_v^\alpha \|v\|_1 \quad (10)$$

where

$$P(\tilde{x}_i = k, \tilde{x}_j = l | \mu_{ij}(\tilde{y}), v) = \frac{\exp(v_{kl}^T \mu_{ij}(\tilde{y}))}{\sum_{m=1}^L \sum_{n=1}^L \exp(v_{mn}^T \mu_{ij}(\tilde{y}))} \quad (11)$$

Eq. (11) shows that $P(\tilde{x}_i, \tilde{x}_j | \mu_{ij}(\tilde{y}), v)$ acts as a MLR model with L^2 classes, and then $L_{MLR}(v)$ is also the log-likelihood of MLR and Eq. (10) is exactly the objective function of D-SMLR.

Therefore, we can draw the conclusion that with the potentials defined as Eq. (3) and (4), the dynamic training of the SCRF can be implemented as exactly two kinds of dynamic sparse MLR (D-SMLR) models under the piecewise training framework. The D-SMLR is implemented through changing the hyperparameter λ_θ^α under the iterative training framework.Then the varied hyperparameter is relevant to the iterations and λ_θ^α can be further denoted as $\lambda_\theta^{(t)}$. In this work, we use the following function of varied hyperparameter with the variable t

$$\lambda_\theta^{(t)} = \rho_{\theta,1} * \beta^t + \rho_{\theta,2} \quad (12)$$

where $0 \leq \beta < 1$, $\rho_{\theta,1}$ and $\rho_{\theta,2}$ are positive constants. More details of derivation of the D-SMLR algorithm can be found in [13].

3.2   Model Combination in Inference

We noted that the D-SCRF training through independent D-SMLR trainings may leads to problems with over-counting during inference[14].We introduce scalar powers for each term, and then combine the independently trained models during inference as

$$P(x|y) \propto \exp\left\{\gamma_1\left[\sum_{i \in S} \phi_i(x_i, y, \tilde{w})\right] + \left[\sum_{i \in S} \sum_{j \in \eta_i} \xi_{ij}(x_i, x_j, \mu_{ij}(y), \tilde{v})\right]\right\} \quad (13)$$

where $\tilde{w}$ and $\tilde{v}$ are the optimal D-SMLR parameters learned independently, and γ_1 is the fixed power for the unary potential. The inference of the form (13) can be efficiently implemented by loopy belief propagation (LBP).

4   Experimental Results

4.1   Data Set for Experiments

The proposed algorithm was tested on real world hyperspectral image. The data consist of a 145x145 pixels portion of an AVIRIS image acquired over NW Indian

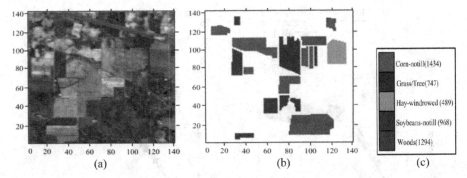

Fig. 1. Indian Pine data set. (a) is original image produced by the mixture of three bands. (b) is ground truth with five classes. (c) is map colour and number of samples.

Table 1. Number of total, training, and test Samples in Indian Pine data set

class Name	total	training	test
corn-notill	1434	500	934
grass/Tree	747	260	487
hay-windrowed	489	172	317
soybeans-notill	968	340	628
woods	1294	452	842
total	4932	1724	3208

Pine in June 1992[15]. In our experiments, all of the 220 original spectral channels were employed and five classes were selected from the efficiency point of view only (see Fig. 1). We randomly select the spatially joint pairwise pixels to create the training dataset. The details of training and testing pixels for each class are listed in table 1.

4.2 Convergence

At first, we evaluate performances of the dynamic training method. The convergence property of the training method is illustrated in Fig. 2 through the plots of gradients with change of iteration times. Since there are total 880 w_{ij} $(i = 1, ..., 220, j = 1, ..., 4)$ and 2200 v_{ij} $(i = 1, ..., 440, j = 1, ..., 5)$ in this experiment setup, it is impossible to demonstrate the gradients of all parameters. Without losing generality, we present only the gradients of the parameters corresponding to the first dimension in the feature vectors. As shown in Fig. 2, both the training processes show convergences with more than 100 iterations.

4.3 Classification Behavior with Different Number of Selected Features

Then, we present the classification performances of SCRF and D-SCRF with the different number of selected features. The SCRF is also trained by the piecewise

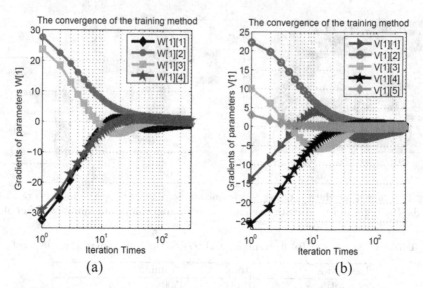

Fig. 2. Convergence of the training method. (a) is the plots of the gradients of $\{w_{1j}, j = 1, ..., 4\}$. (b) is the plots of the gradients of $\{v_{1j}, j = 1, ..., 5\}$.

training method presented in Section 3.1. At first, we demonstrate the classification behaviors of the two models with only unary (setting v as 0) or pairwise (setting w as 0) clique potentials respectively. Then we combine the unary and pairwise clique potentials to get the full SCRF and full D-SCRF through Eq. (13). Similar to that in work[14], the power parameter γ_1 in Eq. (13)was learned as 0.1 through cross validation.

In all the figures, SCRFs and D-SCRFs show similar classification accuracies when relatively large numbers of features are selected. However, with the decreasing number of selected features, the plot of the SCRFs drops sharply for the undersparsity and the characteristics of the hyperspectral data, while the D-SCRFs show more stable classification performance. This means that the D-SCRFs relieves the heavy degeneration of classification performance caused by the undersparsity in the SCRFs and can be more fit for the feature selection in the classification of hyperspectral data. Fig. 3(a) also demonstrates that the full CRFs show better results than the corresponding CRF models with only unary (Fig. 3(b)) or pairwise (Fig. 3(c)) clique potentials since the full CRFs combine their strengths.

4.4   Quantitative Evaluation

Table 2 presents the performances of SCRF and D-SCRF with the selected 5% of total features over the Indian Pine. The SCRF obtained 90.85% overall classification accuracies, in contrast the D-SCRF achieved higher 93.56% accuracies. The inspection of the accuracy for each class confirms that except the grass/tree, D-SCRF obtained higher accuracies than SCRF for other classes. The higher

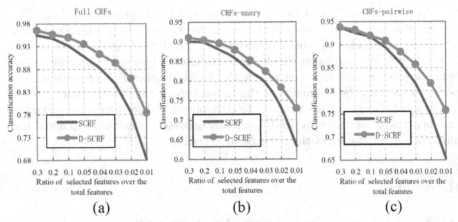

Fig. 3. Classification accuracies of different SCRFs and D-SCRF against number of selected features. (a) is results of full SCRF and full D-SCRF. (b) and (c) show the results of SCRFs and D-SCRFs with only unary and pairwise clique potentials respectively.

accuracy of SCRF for the grass/tree class may derive from the fact that the D-SCRF used the same varied hyperparameter for all the classes. The D-SCRF model can further improve the classification accuracy of each class by setting different varied hyperparameters for different classes. We also give the performance of D-SMLR, which uses only single site spectral data to predict the corresponding label, with the selected 5% of total features and compare it with the SCRF and D-SCRF. It can be noted from table 2 that the classification accuracies of both the SCRF and D-SCRF are much higher than the 87.87% accuracy of D-SMLR. This comparison demonstrates the importance of contextual information for the hyperspectral image classification.

Table 2. Classification Accuracies of D-SMLR, SCRF and D-SCRF

class	D-SMLR	SCRF	D-SCRF
corn-no till	81.91	85.97	90.26
grass/trees	97.74	98.77	98.36
hay-windrowed	98.06	99.05	99.68
soybeans-no till	71.34	77.39	84.55
woods	97.28	98.57	98.81
overall accuracy	87.87	90.85	93.56

5 Conclusion

In this work, we investigated the D-SCRF on the feature selection in the contextual classification of hyperspectral data and developed a dynamic learning framework to relieve heavy degeneration of classification performance caused by over-sparsity in SCRF and the characteristics of the hyperspectral data. The results on real-world hyperspectral data validate the efficiency and effectiveness of the D-SCRF. The experimental results of current form also indicate several future works. We developed the dynamic training framework to use the varied hyperparameters and thus can relieve the heavy degeneration of classification performance. But the optimality of the varied hyperparameters is difficult to be investigated. In the future, we hope to develop the methods to select the optimal hyperparameters, or to use the adaptive sparseness methods to avoid the adjusting or estimating of the hyperparameters[16].

References

1. Chi, M., Bruzzone, L.: Semisupervised Classification of Hyperspectral Images by SVMs Optimized in the Primal. IEEE Trans. Geosci. Remote Sens. 45, 1870–1880 (2007)
2. Muñoz-Marí, Bruzzone, L., Camps-Valls, G.: A Support Vector Domain Description Approach to Supervised Classification of Remote Sensing Images. IEEE Trans. Geosci. Remote Sens. 45, 2683–2692 (2007)
3. Ashish, D., McClendon, R.W., Hoogenboom, G.: Land-use classification of multispectral aerial images using artificial neural networks. Int. Jour. Remote Sens. 30, 1989–2004 (2009)
4. Camps-Valls, G., Marsheva, T.V.B., Zhou, D.: Semi-Supervised Graph-Based Hyperspectral Image Classification. IEEE Trans. Geosci. Remote Sens. 45, 3044–3054 (2007)
5. Zhong, P., Wang, R.: Modeling and Classifying Hyperspectral Imagery by CRFs with Sparse Higher Order Potentials. IEEE Trans. Geosci. Remote Sens. 49, 688–705 (2011)
6. Zhong, P., Wang, R.: Learning conditional random fields for classification of hyperspectral images. IEEE Trans. Image Process. 19, 1890–1907 (2010)
7. Lafferty, J., McCallum, A., Pereira, F.: Conditional random fields: probabilistic models for segmenting and labeling sequence data. In: International Conference on Machine Learning, pp. 282–289 (2001)
8. Zhong, P., Wang, R.: A multiple Conditional random fields ensemble model for urban area detection in remote sensing optical images. IEEE Trans. Geosci. Remote Sens. 45, 3978–3988 (2007)
9. Zhong, P., Wang, R.: Learning Sparse CRFs for Feature Selection and Classification of Hyperspectral Imagery. IEEE Trans. Geosci. Remote Sens. 46, 4186–4197 (2008)
10. Kumar, S.: Models for learning spatial interactions in natural images for context-based classification. PhD thesis. Carnegie Mellon University (2005)
11. Krishnapuram, B., Carin, L., Figueiredo, M.A.T., Hartemink, A.J.: Sparse multinomial logistic regression: fast algorithms and generalization bounds. IEEE Trans. Pattern Anal. Machine Intell. 27, 957–968 (2005)
12. Ng, A.Y.: Feature selection, L1 vs. L2 regularization, and rotational invariance. In: International Conference on Machine Learning (2004)
13. Zhong, P., Zhang, P., Wang, R.: Dynamic learning of sparse multinomial logistic regression for feature selection and classification of hyperspectral data. IEEE Geosci. Remote Sens. Lett. 5, 280–284 (2008)

14. Shotton, J., Winn, J., Rother, C., Criminisi, A.: TextonBoost for image understanding: multi-class object recognition and segmentation by jointly modeling texture, layout, and context. Int. Jour. Comp. Vision. 81, 2–23 (2009)
15. Landgrebe, D.A.: Signal Theory Methods in Multispectral Remote Sensing. Wiley, Hoboken (2003)
16. Figueiredo, M.A.T.: Adaptive sparseness for supervised learning. IEEE Trans. Pattern Anal. Machine Intell. 25, 1150–1159 (2003)

Entropic Selection of Histogram Features for Efficient Classification

Ákos Utasi

Computer Automation Research Institute, Hungarian Academy of Sciences
Kende u. 13-17, H-1111 Budapest, Hungary
akos.utasi@sztaki.mta.hu
http://web.eee.sztaki.hu/~ucu

Abstract. This paper addresses the problem of local histogram-based image feature selection for learning binary classifiers. We show a novel technique which efficiently combines histogram feature projection with the conditional mutual information (CMI) based classifier selection scheme. Moreover, we investigate cost-sensitive modifications of the CMI-based selection procedure, which further improves the classification performance. Extensive evaluations show that the proposed methods are suitable for object detection and recognition tasks.

Keywords: classifier selection, mutual information, histogram feature.

1 Introduction

Histogram-based local image features are widely used in many pattern recognition and computer vision applications. In object detection, categorization, or recognition algorithms such features are usually combined with an efficient classification technique. Among the vast variety of histogram features, local binary patters (LBP) [1] or histogram of oriented gradients (HOG) [2] are widely adopted in many applications, because they can be calculated easily, and they are robust against small deformations and varying illumination. In monolithic classification approaches a single feature vector is constructed by concatenating the local features extracted over a dense predefined 2D grid. Finally, the feature vector is combined with a classifier, e.g. linear support vector machine (SVM) [3]. However, such monolithic approaches suffer from high computation costs since either (a) features are extracted at a large number of locations, or (b) the combined feature vector has a high dimension, resulting in slow classification.

One possible solution to overcome the above drawbacks is to limit the feature extraction step to grid locations where the extracted feature has a high discriminative power for classification. In each location we can train a weak learner, which usually has moderate classification accuracy. However, from the combination of several weak learners we can construct a strong classifier, which achieves a high classification performance. AdaBoost [4] is one of the most widely used techniques for boosting weak learners, and has been successfully applied e.g. for face detection using local Haar-like image features [5].

G.L. Gimel' farb et al. (Eds.): SSPR & SPR 2012, LNCS 7626, pp. 264–272, 2012.
© Springer-Verlag Berlin Heidelberg 2012

The binary feature selection technique proposed in [6] is based on the fundamental concepts of information theory to quantify the uncertainty of random variables and to measure the information shared between them. The Conditional Mutual Information (CMI) estimates the information shared between the training data and a classifier, given another classifier. This can be utilized to select the feature which best describes the training data, and is the most independent from other features selected previously. One main advantage of this technique is that it is able to cope with overfitting, while AdaBoost is known to be sensitive to this phenomenon. The CMI-based feature selection technique has been successfully applied for facial expression recognition using LBP features [7].

Fisher Linear Discriminant (FLD) [8] analysis is frequently used to find the projection of histogram features which best separates two object classes, *e.g.* [9] embedded the projected features into the AdaBoost learning framework to detect faces. [10] proposed the Weighted Fisher Linear Discriminant (WFLD) as the weak learner in the AdaBoost framework. Thereby the WFLD minimizes the weighted classification error computed from the sample weights, which are updated by the AdaBoost procedure. The main advantage of this technique is that it eliminates the need of re-sampling the training data, and it leads to a more efficient use of the training samples.

In the proposed method we adopt the CMI-based feature selection technique, but we employ weak learner parameter optimization during the feature selection process to further improve classification accuracy, as opposed to previous methods [6] where these parameters are assumed to be already set. In AdaBoost the sample weights are used for WFLD, which are updated in each iteration using the weights of misclassified samples. However, sample misclassification is not defined in the CMI-based feature selection. Therefore, by using the concepts of information theory we introduce a novel method for updating the sample weights. Finally, we introduce cost-sensitive modifications of the CMI feature selection, which improves the classification accuracy of imbalanced datasets, where learning methods usually end up preferring the larger class.

The rest of the paper is organized as follows. In Sec. 2 we briefly overview the AdaBoost classifier learning method [4] and the WFLD weak learner technique [10]. In Sec. 2.1 we discuss the CMI-based feature selection [6] in more detail. The proposed method is presented in Sec. 3. Finally, in Sec. 4 we show our experimental results using two public image databases.

2 Classifier Learning with WFLD

We denote by $\mathbf{X} = \{x_1, \ldots, x_N\}$ a set of N training images, where each image x_i has a binary class label $y_i \in \{1, 0\}$ and $\mathbf{Y} = \{y_1, \ldots, y_N\}$. We extract $\mathbf{F} = \{f_1, \ldots, f_K\}$ a set of K features from an image x, where $f_k(x) \in \mathbb{R}^m$ is a histogram feature extracted at a given position. Finally, each feature is projected by the $g_k : \mathbb{R}^m \to \mathbb{R}$ function. In our case g is the WFLD [10], which guarantees optimal classification of the two classes, and is defined as $g = w^{\mathsf{T}} f$, such that

$$w = (\Sigma_1 + \Sigma_0)^{-1} (\mu_1 - \mu_0) , \qquad (1)$$

where μ denotes the weighted mean and Σ is the weighted covariance matrix of the training set of a given class, *i.e.*

$$\mu = \frac{1}{n\sum_i d_i} \sum_i d_i f(x_i), \quad \Sigma = \frac{1}{(n-1)\sum_i d_i^2} \sum_i d_i^2 (f(x_i)-\mu)(f(x_i)-\mu)^{\mathsf{T}}, \quad (2)$$

where n denotes the number of samples in the given class, and d_i denotes the weight of a particular sample. Hereafter we use $g_k(\cdot) = g_k(f_k(\cdot))$ as a shorthand. Similarly to [5] our weak classifier h_k at a given position is defined in the form

$$h_k(x) = \begin{cases} 1 & \text{if } p_k g_k(x) < p_k \theta_k \\ 0 & \text{otherwise} \end{cases}, \quad (3)$$

where p_k is the parity and θ_k is a threshold. Having a subset of weak learners the strong classifier $H(x)$ is defined as

$$H(x) = \text{sgn}\left(\sum_{t=1}^{T} \alpha_t h_{\nu(t)}(x) - b \right), \quad (4)$$

where T denotes the number of selected weak learners, $\nu(t)$ returns the index of the tth weak learner, $\{\alpha_t\}$ are the weights and b is the bias. In an iterative boosting scheme the weak learner is selected in each step , which minimizes an error function $\epsilon_k = \epsilon(h_k)$ describing the fitness of the weak learner on the labeled training data $(\mathbf{X}, \mathbf{Y})$. In AdaBoost ϵ_k is the classification error, which is expressed as the sum of the $\mathbf{D} = \{d_1, \ldots, d_N\}$ weights of the misclassified samples, α_t is estimated from ϵ_k, and the bias is expressed as $b = \frac{1}{2}\sum_t \alpha_t$.

Thus in iteration t first the optimal WFLD projection w_k is determined from $\mathbf{D}$ using Eqs. 1–2, then the optimal $p_k^\star$ and $\theta_k^\star$ parameters are determined in a brute force manner, and the $h_{\nu(t)}$ classifier with minimal ϵ_k is selected, *i.e.*

$$(p_k^\star, \theta_k^\star) = \underset{p_k, \theta_k}{\text{argmin}} \{\epsilon_k\}, \quad (5)$$

$$\nu(t) = \underset{k}{\text{argmin}} \{\epsilon_k\}. \quad (6)$$

2.1   CMI-Based Classifier Selection

[6] proposed an iterative *binary* feature selection method based on CMI. In each iteration the feature is selected, which maximizes the mutual information on training samples $(\mathbf{X}, \mathbf{Y})$, depending on the output of any feature selected in previous iterations. This procedure can be formalized as follows. Let $\hat{x}_i^k \in \{1, 0\}$ denote the response of the kth classifier on the ith sample, *i.e.* $\hat{x}_i^k$ is a *binary* feature, and $\hat{\mathbf{X}}^k = \{\hat{x}_i^k\}$. In the first step the feature which maximizes the $I(\mathbf{Y}; \mathbf{X})$ mutual information (MI) on the samples is selected, *i.e.*

$$\nu(1) = \underset{k}{\text{argmax}} \left\{ I\left(\mathbf{Y}; \hat{\mathbf{X}}^k\right) \right\}. \quad (7)$$

Note that the mutual information $I(\mathbf{Y}; \mathbf{X})$ of two random variables $\mathbf{Y}$ and $\mathbf{X}$ can be expressed in terms of entropy as $I(\mathbf{Y}; \mathbf{X}) = H(\mathbf{Y}) - H(\mathbf{Y}|\mathbf{X})$, where the conditional entropy $H(\mathbf{Y}|\mathbf{X})$ quantifies the uncertainty of $\mathbf{Y}$ when $\mathbf{X}$ is selected. By minimizing this uncertainty in Eq. 7 we obtain the classifier which best describes the training data. By similar considerations in subsequent iterations the $I(\mathbf{Y}; \mathbf{X}|\mathbf{Z})$ CMI is utilized for feature selection, thus for $t = 2, \ldots, T$

$$\nu(t) = \operatorname*{argmax}_{k} \left\{ \min_{s<t} I\left(\mathbf{Y}; \hat{\mathbf{X}}^k | \hat{\mathbf{X}}^{\nu(s)}\right) \right\} . \tag{8}$$

3   Proposed Method

We introduce the generalization of the CMI-based technique of Sec. 2.1 for boosting arbitrary features, where the optimal $p_k^\star$ and $\theta_k^\star$ parameters of the weak learners are determined using the

$$\epsilon = 1 - \min I(\mathbf{Y}; \mathbf{X}|\mathbf{Z}) \tag{9}$$

error function in Eq. 5, and the weak learner with minimal ϵ_k is selected as in Eq. 6. Finally, the classifier weights $\{\alpha_t\}$, and the bias b of the strong classifier are estimated by a linear SVM [3]. During the SVM learning we also utilize cost factors c_1 and c_0 for the two classes $\mathcal{C}_1$ and $\mathcal{C}_0$, which are chosen to satisfy $c_0/c_1 = n_1/n_0$ [11], where n_1 and n_0 denote the cardinality of the two classes. This re-balancing technique is necessary when the training data is imbalanced, *i.e.* when the size of one of the two classes is significantly larger then the other's. Without re-balancing the resulting classifier will tend to favor the larger class, and the samples of the smaller class will be misclassified with a higher probability.

3.1   Sample Weights for CMI-Based Feature Selection

The generalized CMI feature selection technique uses the error function Eq. 9 to determine the optimal parameters of the weak learners in Eq. 5, and to select the optimal weak learner using Eq. 6. However, in the original CMI procedure no sample weights and no update procedure are available for computing the WLFD projection vector w of Eq. 1. Therefore, we extend this method with sample weights together with an update procedure, and we use information theory concepts to define sample misclassification for the update.

Recall that in AdaBoost the sample weights are updated using the weights of the misclassified samples, *i.e.* by defining the

$$e_i = \mathbb{1}\{h(x_i) \neq y_i\} \in \{0, 1\} \tag{10}$$

indicator function the classifier error ϵ_i is calculated as $\epsilon_i = \sum d_i \cdot e_i$, and the sample weights are updated as $d_{t+1,i} = d_{t,i}\beta^{1-e_i}$, such that $\beta = \frac{\epsilon}{1-\epsilon}$ for $\epsilon < 0.5$. In the proposed CMI-based method we assume that a particular sample x_i is misclassified by the weak learner h_k when changing its response $\hat{x}_i^k$ would imply an increase of the MI ($t = 1$) or of the CMI ($t = 2, \ldots, T$). We use the following notations to formally define our technique. Let $p(\varphi, v) = p_{X,Y}(\varphi, v)$ denote the joint distribution of the two random variables, similarly $p(\varphi) = p_X(\varphi)$, and $p(v) = p_Y(v)$, where φ and v are boolean variables. By definition MI is

$$I\left(\mathbf{Y}; \hat{\mathbf{X}}\right) := \sum_{\varphi, v} p(\varphi, v) \log \frac{p(\varphi, v)}{p(\varphi)\, p(v)} \ . \tag{11}$$

However, in our case the above probabilities are determined from a limited training set having N elements. Therefore, we can re-write it using a fast look-up-table (LUT) solution as follows. We express the above distributions in terms of frequencies of the random variables' occurrences as $p(\varphi, v) = \frac{1}{N}n(\varphi, v)$, $p(\varphi) = \frac{1}{N}n(\varphi)$, and $p(v) = \frac{1}{N}n(v)$, where $n(\cdot)$ denotes the cardinality. Note that $n(v{=}0) = n_0$, $n(v{=}1) = n_1$ denote the cardinality of the two classes $\mathcal{C}_0$ and $\mathcal{C}_1$ of the training set. We create a LUT $\mathcal{L}$ on the $0 \leq n \leq N$ integer range as $\mathcal{L}[n] = n \log n$, and we rewrite Eq. 11 as

$$I\left(\mathbf{Y}; \hat{\mathbf{X}}\right) = \frac{1}{N} \left(\sum_{\varphi, v} \mathcal{L}[n(\varphi, v)] - \sum_{\varphi} \mathcal{L}[n(\varphi)] - \sum_{v} \mathcal{L}[n(v)] + \mathcal{L}[N] \right) . \tag{12}$$

We can see that the terms $\sum \mathcal{L}[n(v)]$ and $\mathcal{L}[N]$ in the above equation are constants during feature selection. Furthermore, the $1/N$ normalizing constant can be neglected, and we refer to this unnormalized MI as $\tilde{I}(\mathbf{Y}; \hat{\mathbf{X}})$. According to our original assumption, changing the value of response $\hat{x}_i$ will affect $n(\varphi)$ and $n(\varphi, v)$ only, since $n(v)$ depends solely on the training set. For example assuming class label $y_i = 0$ and changing the response value $\hat{x}_i = 0$ to 1 will increase $n(0, 1)$ and $n(1)$ but will decrease $n(0, 0)$ and $n(0)$. Using this property we can express the change of the unnormalized MI denoted by $\tilde{I}_\Delta(y_i = v; \hat{x}_i = \varphi)$ as

$$\begin{aligned}\tilde{I}_\Delta(v; \varphi) = \ &\mathcal{L}[n(v, \varphi){-}1] + \mathcal{L}[n(v, 1{-}\varphi){+}1] - \mathcal{L}[n(v, \varphi)] - \mathcal{L}[n(v, 1{-}\varphi)] \\ &+ \mathcal{L}[n(\varphi)] + \mathcal{L}[n(1{-}\varphi)] - \mathcal{L}[n(\varphi) - 1] - \mathcal{L}[n(1{-}\varphi) + 1] \ . \end{aligned} \tag{13}$$

Similarly to [5] in our method the sample weights are initialized to $d_i = \frac{1}{2n_0}, \frac{1}{2n_1}$ for $y_i = 0, 1$ respectively, but for updating their value we utilize Eq. 13. First, we define the indicator function

$$e_i = \mathbb{1}\{\tilde{I}_\Delta(y_i; \hat{x}_i) > 0\} \in \{0, 1\} \tag{14}$$

to indicate whether sample x_i is misclassified or not. Then we define the classification error γ of the selected weak learner as the sum of the weights of the misclassified samples, i.e. $\gamma = \sum d_i \cdot e_i$. Finally, sample weights are updated as

$$d_{t+1,i} = d_{t,i} \gamma^{1-e_i} \ . \tag{15}$$

Thus the above update rule decreases the weights of the samples which were classified correctly by the selected weak learner. Note that in the case of CMI we can define the rules similarly to Eqs. 12–13 in a straightforward way, but these were omitted in the present paper due to space limitations. In the following we refer to this method as CMISVM.

3.2 Balanced Feature Selection and Weight Update

The method presented in Sec. 3 uses cost factors in the final step of constructing the strong classifier. In our second method first we incorporate re-balancing into the weak learner selection by utilizing the weighted mutual information (wMI), which is defined as

$$I_w\left(\mathbf{Y};\hat{\mathbf{X}}\right) = \sum_{\varphi,v} w(\varphi,v)\, p(\varphi,v) \log \frac{p(\varphi,v)}{p(\varphi)\,p(v)} , \tag{16}$$

where we use the cost factors of Sec. 3 to define the weights as $w(\varphi,1) = 1$ and $w(\varphi,0) = n_1/n_0$. The weighted conditional mutual information (wCMI) is defined similarly. Finally, we incorporate a re-balancing technique into the weight update rule defined in Eq. 15 by taking into account the distribution of $\tilde{I}_\Delta(y_i;\hat{x}_i)$. Our goal is to achieve a more aggressive change in the weight of the correctly classified samples (where $\tilde{I}_\Delta(\cdot;\cdot) \leq 0$), which do not change the MI significantly, i.e. if $|\tilde{I}_\Delta(y_i;\hat{x}_i)| < |\tilde{I}_\Delta(y_j;\hat{x}_j)|$ then d_i can be decreased more aggressively. Therefore, we modify Eq. 15 as

$$d_{t+1,i} = d_{t,i}\gamma^{1-e_i} \cdot \frac{I(\mathbf{Y};\hat{\mathbf{X}}) + \min_j\{\tilde{I}_\Delta(y_j;\hat{x}_j)\}}{I(\mathbf{Y};\hat{\mathbf{X}}) + \tilde{I}_\Delta(y_i;\hat{x}_i)} . \tag{17}$$

In the rest of the paper this method will be referred as *wCMISVM*.

4 Experiments

In our experiments we used two public datasets. From the FERET face database [12,13] we used the annotations to align the heads into the same eye positions. For detection the smaller class contains faces cropped from the aligned images and are resized to 112×128 pixels. Moreover, the other class contains randomly cropped parts from background images. For recognition we used a slightly larger part of the head and a 128×128 pixels size. From the available annotations we defined three classification problems for recognition: a) *race*: Asian or White, b) *glasses*: wearing or not, and c) *gender*: female or male. The second dataset we used is the MIT CBCL Car [14] database, which contains front and rear view of cars, and the size of the images were 128×128 pixels. Again, the samples of the other class are random background images not containing any cars. We extended the datasets by adding the mirrored version of each sample in the set. Note that there is a significant difference between the two experiments. In case of recognition the samples of a class are similar, while in the detection experiment the larger class contains very different samples as they are random parts of backgrounds.

Our features are HOG blocks[2] which are computed in a single cell of 8×8 pixels, and a 9-bin histogram ($0° - 180°$) is calculated using linear gradient voting and L2-Hys normalization. In our evaluation we selected *AdaBoost* with WFLD weak learners [10] (see Sec. 2) as baseline, and the number of weak learners T was limited to the $\{2, 4, \ldots, 20\}$ range.

4.1   Recognition

For the race recognition experiment the C_1 class contains faces of Asian people, the size of *training* data is $n_1 = 512$ and for *testing* 98 samples were used. The C_0 class contains 3080 faces of White people, from which we used $n_0 = 2628$ for *training* and 452 for *testing*. For recognizing people wearing glasses we *trained* the classifiers with $n_1 = 194$ faces with glasses, and $n_0 = 1698$ without glasses. For *testing* 68 and 446 samples were used. Finally, in the gender recognition experiment the C_1 class contains 1828 female faces, from which we used $n_1 = 1532$ samples for *training* and 296 for *testing*. C_0 contains $n_0 = 2574$ male faces for *training* and 428 for *testing*. All these datasets are imbalanced, in order to present the advantages of the proposed approach, and we can also see that the *glasses* dataset is the most imbalanced (approx. 1:9 ratio). Fig. 1 shows sample images from recognition experiment.

Fig. 1. Example images from the recognition experiment. Left: Asian vs White; Center: glasses vs no glasses; Right: female vs male.

4.2   Detection

In the face detection experiment the C_1 class contained 4818 faces, from which we used $n_1 = 3170$ samples for *training* and 1648 for *testing*. C_0 contained 18880 non-faces, from which $n_0 = 12208$ samples were used for *training* and 6672 for *testing*. For the car detection the *training* set of C_1 contained $n_1 = 828$ car images, and the size of the *testing* set was 204. The C_0 class contained $n_0 = 4990$ *training* samples, and 1282 *test* samples. Examples from the datasets are shown in Fig. 2.

4.3   Evaluation

For evaluation we selected the *G-mean* from the available metrics [15], which is accepted as a good metric for imbalanced classification problems. After obtaining the *G-mean* values for the three classifiers containing $T \in \{2, 4, \ldots, 20\}$ weak learners we selected the classifier with maximal *G-mean* then we compared the other classifiers to this value and computed the difference which was considered as the error score of the classifier. Summing these differences for all T configurations we obtained a total error score for each method in a particular classification problem. Table 1 shows the error scores of the three methods both

Fig. 2. Example images from the detection experiment. Top: face vs non-face; Bottom: car vs non-car.

for recognition and detection tasks. We can see that the re-balancing techniques of Sec. 3.2 are beneficial for the feature selection, as the *wCMISVM* classifier clearly outperformed the other methods. However, this method is slightly less effective for detection tasks. This may be due to the nature of the data since in this experiment the samples of the larger class contain very different images, and a single cost-factor may not suitable to represent such a large variation.

Table 1. Error scores of the three methods in different classification tasks

	AdaBoost[10]	*CMISVM*	*wCMISVM*
Race	0.1822	0.1001	**0.0527**
Glasses	0.2049	0.2129	**0.0251**
Gender	**0.0889**	0.2389	0.1240
Recognition	0.4760	0.5519	**0.2018**
Car	0.0388	**0.0112**	0.0453
Face	**0.0334**	0.0606	0.0593
Detection	0.0722	**0.0718**	0.1046

5   Conclusions

In this paper we investigated the difficulties of CMI-based classifier selection using WFLD as weak learners. We proposed a novel technique for updating the sample weights of the training data. To improve the efficiency of the CMI-based method on imbalanced datasets we proposed re-balancing techniques for both the feature selection and the weight update procedures. We performed extensive evaluations on two public datasets. The experiments confirmed that the proposed methods improve the efficiency of CMI-based boosting in case of imbalanced datasets. As a part of our future work we plan to extend our experiments with additional datasets and with more tests with various degree of data imbalance.

Acknowledgement. Portions of the research in this paper use the FERET database of facial images collected under the FERET program, sponsored by the DOD Counterdrug Technology Development Program Office. This work was supported by the Hungarian Scientific Research Fund under grant number 80352.

References

1. Ojala, T., Pietikäinen, M., Harwood, D.: A comparative study of texture measures with classification based on feature distributions. Pattern Recognition 29(1), 51–59 (1996)
2. Dalal, N., Triggs, B.: Histograms of oriented gradients for human detection. In: International Conference on Computer Vision and Pattern Recognition (2005)
3. Vapnik, V.N.: The nature of statistical learning theory. Springer-Verlag New York Inc. (1995)
4. Freund, Y., Schapire, R.E.: A decision-theoretic generalization of on-line learning and an application to boosting. In: European Conference on Computational Learning Theory (1995)
5. Viola, P., Jones, M.: Robust real-time face detection. International Journal of Computer Vision 57(2), 137–154 (2004)
6. Fleuret, F.: Fast binary feature selection with conditional mutual information. Journal of Machine Learning Research 5, 1531–1555 (2004)
7. Shan, C., Gong, S., McOwan, P.W.: Conditional mutual information based boosting for facial expression recognition. In: British Machine Vision Conference (2005)
8. Fisher, R.A.: The use of multiple measurements in taxonomic problems. Annals of Eugenics 7(2), 179–188 (1936)
9. Wang, H., Li, P., Zhang, T.: Histogram feature-based Fisher linear discriminant for face detection. Neural Computing and Applications 17(1), 49–58 (2008)
10. Laptev, I.: Improving object detection with boosted histograms. Image and Vision Computing 27(5), 535–544 (2009)
11. Morik, K., Brockhausen, P., Joachims, T.: Combining statistical learning with a knowledge-based approach – A case study in intensive care monitoring. In: International Conference on Machine Learning (1999)
12. Phillips, P.J., Moon, H., Rizvi, S.A., Rauss, P.J.: The FERET evaluation methodology for face recognition algorithms. IEEE Transactions on Pattern Analysis and Machine Intelligence 22(10), 1090–1104 (2000)
13. Phillips, P.J., Wechsler, H., Huang, J., Rauss, P.J.: The FERET database and evaluation procedure for face recognition algorithms. Image and Vision Computing 16(5), 295–306 (1998)
14. Papageorgiou, C., Poggio, T.: A trainable system for object detection. International Journal of Computer Vision 38(1), 15–33 (2000)
15. García, V., Mollineda, R.A., Sánchez, J.: Theoretical analysis of a performance measure for imbalanced data. In: International Conference on Pattern Recognition (2010)

2D Shapes Classification Using BLAST

Pietro Lovato and Manuele Bicego

Computer Science Department - University of Verona, Italy

Abstract. This paper presents a novel 2D shape classification approach, which exploits in this context the huge amount of work carried out by bioinformaticians in the biological sequence analysis research field. In particular, in the approach presented here, we propose to encode shapes as biological sequences, employing the widely known sequence alignment tool called BLAST (Basic Local Alignment Search Tool) to devise a similarity score, used in a nearest neighbour scenario. Obtained results on standard datasets show the feasibility of the proposed approach.

Keywords: 2D shape classification, sequence alignment, biological sequences.

1 Introduction

The classification of 2D shapes represent an old and widely investigated research field in computer vision and pattern recognition. Many approaches have been proposed in the past (see e.g. the reviews [1–3]), many of them based on the analysis of the boundary: actually, object contours have shown to be very effective in many applications, with several different approaches presented over the past years, exhibiting different characteristics: robustness to noise and occlusions, invariance to translation, rotation, and scale, computational requirements, and accuracy.

In this paper, a novel approach for contour-based 2D shape classification is proposed, which exploits techniques and solutions coming from the biological sequence alignment context [4]. From a very general point of view, the proposed approach starts from the observation that, in the past, the huge and profitable interaction between pattern recognition and biology/bioinformatics was mainly unidirectional, namely devoted at studying and applying PR tools and ideas to the analysi of biological data [5][1]. In this paper a somehow unexplored alternative way of interaction is investigated: the idea is to employ advanced bioinformatics solutions to solve pattern recognition problems. Actually, there are application scenarios in the bioinformatics field – like sequence modelling, phylogeny, database searches – which have been deeply and successfully investigated

---

[1] In some other cases, biological/bioinformatics problems have led to the definition of novel methodological pattern recognition issues – a clear example is the biclustering problem (simultaneous clustering of features and patterns), which was initially introduced to analyse expression microarray data in order to discover subsets of genes with a coherent behaviour in subsets of samples [6].

G.L. Gimel' farb et al. (Eds.): SSPR & SPR 2012, LNCS 7626, pp. 273–281, 2012.
© Springer-Verlag Berlin Heidelberg 2012

for many years by bioinformaticians. We are convinced that such fields can offer interesting solutions to pattern recognition problems, if we are able to encode our problem in biological terms. A very recent and interesting example of such an alternative way of thinking is the Video Genome Project[2], where internet videos were encoded as "video DNA sequences" and analysed with phylogenetic related tools [7].

In this paper we follow this line of investigation by exploiting the huge amount of work carried out in the field of biological sequence analysis [4] to face the 2D shape classification problem. In particular, we propose to transform a sequence contour into an aminoacid sequence, employing the most famous biological sequence alignment tool – the BLAST (Basic Local Alignment Search Tool [8]), – to devise a similarity measure between sequences. Such similarity is then used in a standard nearest neighbour classification scenario. The proposed approach has been tested with two standard datasets, the Chicken Pieces Database [9] and the Vehicle Shape dataset [10]; even if we applied a very simple "shape to biological sequence" mapping, obtained results were very promising, also in comparison with the state of the art.

2   Background: Sequence Alignment with BLAST

Research in biology is very often based on the analysis of biological sequences, both nucleotide sequences – i.e. strings made with the 4 symbols of DNA, namely $ATCG$ – and aminoacid sequences – i.e. strings with symbols coming from a 22 letters alphabet. Many different kinds of biological analyses are based on a preliminary sequence alignment step. As can be intuitively understood, the alignment of two sequences is aimed at finding the best registration between them (namely the best way of superimposing one sequence on the other); the registration is done by taking into account the biological nature of the input sequence, so that biological (usually evolutionary) events, such as mutations and rearrangements, can be clearly expressed [4].

From a practical point of view, alignment is obtained by inserting spaces inside the sequences (the so called gaps) in order to maximize the point-wise similarity between them – see Fig. 1.

In the past, a huge amount of approaches have been proposed to deal with this task (see [11–13] for recent reviews and perspectives on the topic), with already effective methods aged in the seventies or early eighties [14, 15]. A thorough treatment of this topic is of course out of the scope of this paper. Two distinctions are important from our perspective: the former distinguishes between pairwise and multiple alignment approaches, with the former devoted at finding the best registration of two sequences and the latter aimed ad finding a simultaneous alignment of more than two sequences. The latter subdivides the approaches in global and local alignment methods: the global ones try to find the best overall alignment between sequences, whereas the local ones aim at finding short regions of high similarity.

---

[2] See http://v-nome.org/about.html

Sequence 1   **TACTAGGCATGAC**
Sequence 2   **ACAGGTCAGTC**

Aligned Sequence 1   **TACTAGG–CATGAC**
Aligned Sequence 2   **–AC–AGGTCA–GTC**

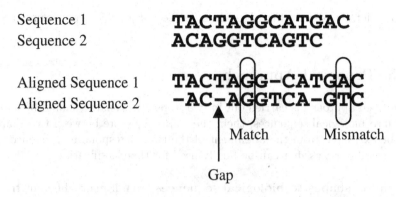

Match   Mismatch

Gap

Fig. 1. Alignment of two sequences

The BLAST (Basic Local Alignment Search Tool) algorithm is for sure the most widely known alignment tool (the Scopus database indicates more than 30 thousands citations to the orignial paper, whereas for GoogleScholar they are more than 40 thousands), introduced by Altschul and colleagues in the 1990. Many different versions have been lately introduced, some of them being now very popular (e.g. psiBLAST [16]). In few words, the BLAST algorithm permits to find the sub-optimal alignment of a query sequence with respect to a dataset of other sequences, providing also a score to every pairwise alignment. BLAST is an approximate algorithm (only giving a sub-optimal yet accurate result), whose success is devoted to the simple but effective heuristics implemented inside which permit a really fast implementation (dynamic programming solutions to the same problem are nowadays absolutely not employable).

Briefly, given in input a sequence (query) to be aligned to a dataset, the algorithm performs the following steps:

1. remove low complexity regions from the query sequence
2. extract from the query sequence all the K-mers (i.e. all the possible subsequences, with overlap, of length K). These subsequences are called "words"
3. search, in the whole database, all the words having a reasonably good match with the words of the query sequence – these words are called "hits"
4. use these words as seeds, attempting to extend both forward and backward from the match to produce an alignment. The algorithm will continue this extension as long as the alignment score continues to increase or until it drops by a critical amount owing to the negative scores given by mismatches. These extended segments are called HSP (High Scoring segment Pairs), and represent the aligned part of the two sequences. In other words, the the alignment is *local*, namely is based on the alignment of a small part of the two sequences.
5. To the alignments found by BLAST during a search a statistical value is assigned, called the "Expect Value" (E-value). This number represents the number of times that an alignment as good as or better than that found by BLAST would be expected to occur by chance.

For more details about this algorithm, interested readers can refer to the book [17][3].

3   The Proposed Approach

The proposed approach is carried out in two steps: first, shapes should be transformed into biological sequences; then, the similarity score between two shapes should be extracted from the alignment of the two corresponding sequences. A nearest neighbour classifier can be finally used for the classification.

1. **From 2D shapes to biological sequences** Even if many different transformations can be adopted, involving complicate shape descriptors as well complicated mappings from them to aminoacids[4], here we adopted a rather simple scheme, in order to analyse the basic potentialities of our approach. In particular, every shape is described by encoding the contour with the 8 directional chain code [18], representing one of the simplest shape coding strategy; then, each chaincode value is directly mapped into one of eight aminoacids: A, R, N, D, C, Q, E, and G – which are the first 8 as given in Matlab ordering.

2. **From alignment to similarity** Given two shapes encoded as biological sequences, it is natural to link similarity between two shapes to the alignment similarity score: such quantity, which is a by-product of the alignment process, measures how "well aligned" the two shapes are, and is the objective function which is maximed during the alignment process. The computation of this quantity is based on the so called "scoring matrix", representing a matrix which, in a position i, j, gives a measure of the "price" we have to pay in a given alignment when substituing the aminoacid i with the aminoacid j. Different scoring matrices have been presented in the biological literature, each one starting from different biological assumptions and observations[5].

Given a testing sequence, we use the BLAST algorithm to align it to all the sequences in the training set, assigning it to the class of the most aligned training sequence. Clearly, since BLAST is a local alignment technique, multiple hits can be found of the same sequence. Nevertheless, similarly to what done in biology, we retain and consider only the first (and thus best) match. A further note: the BLAST algorithm returns a matching score (of the HSP) and the E-value. It is widely accepted in the biology to rank the aligments on the base of the E-value (the smaller the better) rather than on the alignment scores. Actually, after some

---

[3] Available from http://www.ncbi.nlm.nih.gov/books/NBK1734/

[4] Reasonably, we decided to encode shapes into aminoacid sequences, these allowing more sophisticated description if compared with nucleotide sequences (alphabet of 22 symbols rather than 4).

[5] The possibility of defining a scoring matrix which is specific for the shape problem is currently under investigation.

preliminary experiments, we noticed that results obtained with the E-value are substantially better than those obtained with the matching score, therefore we chose to use such value for our classification scheme.

As a final comment, we can observe that this scheme is rather simple and in some cases approximated: for example the closeness of the boundary in 2D shapes does not have a clear biological counterpart in biological sequences; moreover, many enhancements can be derived – as learning the mapping from a dataset, using quantized continuous shape descriptors to cover all the 22 aminoacids, defining a proper shape specific scoring matrix and so on. In any case, the obtained results are already very promising, encouraging us in going ahead along this research direction.

4   Results

The proposed idea has been tested on two different benchmarks, the *Chicken Pieces* dataset[6] [9] and the Vehicle Shape dataset[7] [10]. The first set is composed by 446 silhouettes of chicken pieces, each belonging to one of five classes representing specific chicken parts: wing (117 samples), back (76), drumstick (96), thigh and back (61), and breast (96). This represents a really challenging classification task, with the baseline classification accuracy of about 67% [19]. The second dataset contains 120 vehicle shapes extracted from traffic videos using motion information – as described in [10] –, classified in four classes: sedan, pickup, minivan or SUV. Some examples of shapes belonging to the two datasets are shown in Fig. 2 and 3. The classification accuracies have been computed in two different ways, in order to compare the proposed approach with the state of the art. In particular, for the chicken dataset we used Leave One Out accuracy (as in many nearest neighbour approaches dealing with the chicken dataset), whereas in the vehicle shape dataset we used 10-fold cross validation (as specified in [10]). As specified in the previous section, the classification, in both cases, has been carried out with the nearest neighbour rule.

In the alignment process of two sequences there are two crucial parameters that should be defined: the scoring matrix and the gap opening/extending penalty. As explained in the previous Sections, the former defines the price we have to pay in the alignment score for every substitution, whereas the latter defines the penalty in the similarity introduced by opening (or extending) a gap region. It is important to note that in biology these two parameters have a clear meaning, and can change drastically the final result. In this preliminary evaluation, we performed two sets of experiments: in the former (first row of Table 1) we tried to keep the easiest possible scheme, leaving such parameters as set by default in the BLAST implementation[8]. The only change we did was to remove the filter, applied within BLAST, which removes zones of low complexity (such

---

[6] http://algoval.essex.ac.uk:8080/data/sequence/chicken/

[7] http://visionlab.uta.edu/shape_data.htm

[8] Downloadable from
   ftp://ftp.ncbi.nlm.nih.gov/blast/executables/blast+/LATEST/

Wing

Back

Drumstick

Thigh and back

Breast

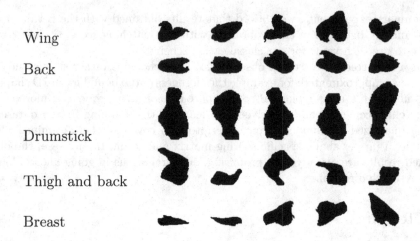

Fig. 2. Some examples from the Chicken Dataset

Seda

Pickup

Minivan

SUV

Fig. 3. Some examples from the Vehicle Dataset

as repetitions of the same symbol). Of course this has a clear meaning in biology, whereas in shapes such parts are indeed very informative (representing straight parts of the shape, for example) and should not be removed.

Table 1. Accuracies for the proposed methods

Method	Chichen	Vehicle
BLAST - Default Settings	0.7892	0.8208
BLAST - Reduced gap penalty	0.8206	0.8437
BLAST - Reduced gap penalty and BLOSUM90	0.8341	0.8542

In the second set of experiments we tried to exploit the fact that we are working with 2D shapes, using this information to properly set the two parameters. As a first trial, we relax one biological assumption which does not hold in the

2D shape classification case: in biology the gap penalty is typically high: it's not really desirable to break a biological sequence. In the shape case, nevertheless, such a strong constraint does not hold: actually, gaps can really help in dealing with occlusions and – mainly – scale changes. The second row of the Table 1 report results obtained by setting the gap opening penalty to 6 and the gap extending penalty to 2 (default values are 11 and 1, respectively[9]). It seems evident the beneficial effect of such operation.

As a second trial, we chose a substitution matrix which highly penalizes changes in the sequences (namely the algorithm is forced to try to align the sequences in the best possible way). The idea here is that whereas in biology there are somehow "equivalent" aminoacids (which can likely exchanged), in the 2D shapes context an exact matching can preferred. Results obtained by using a BLOSUM90 matrix (default is BLOSUM62, the higher the number after the word "BLOSUM" the more "conservative" the substitution matrix is) are reported in the third line of Table 1 (the gap opening/gap extending penalties were set as in previous experiment). Also in this case it can be noted the beneficial impact of such choice, even if not so evident as in the gap penalty case. We are currently continuing with further analysis of the impact of the substituion matrix on the performances and on the alignments.

Table 2. Comparative results: (a) Chicken dataset; (b) Vehicle dataset

Methodology	Accuracy
1-NN + Levenshtein edit distance	≈ 0.67
1-NN + approximated cyclic distance	≈ 0.78
K-NN + cyclic string edit distance	0.743
1-NN + mBm-based features	0.765
1-NN + HMM-based distance	0.738
1-NN + IT kernels on n-grams	0.814
Our best	0.834

(a)

Methodology	Accuracy
SVM + curvature	0.6250
SVM + Fourier Descriptors	0.8250
SVM + Zernike moments	0.7917
Ergodic HMM + Max Lik.	0.6250
Circular HMM + Max Lik.	0.7333
Left Right HMM + Max Lik.	0.7083
HMM + Weighted likelihood	0.8417
Our best	0.8542

(b)

---

[9] Unfortunately, in the BLAST implementation the choice should be made among a pre-fixed set of pair gap opening-gap extending penalties.

As a final comment, in Table 2 we reported some other recent results from the state of the art on the same datasets. Many different approaches have been tested on the Chicken dataset, using simple as well complicated classifiers (see for example comparisons reported in [20, 21]): in Table 2(a) we reported only those based on nearest neighbour rules – taken from [20]. Even if in some cases different experimental protocols have been employed, it seems evident that the proposed approach represents a promising alternative to classic as well as to advanced schemes. It is interesting to observe that the proposed approach, based on approximated matching, also outperforms exact matching techniques, as those based on edit distance. Moreover, as can be seen from Table 2(b), our approach also comparably compares with other techniques employing more sophisticated classifiers (as SVM) – here the results, all taken from [10], are fully comparable (the same validation protocol was employed).

5   Conclusions

In this paper we preliminary investigated the idea of exploiting bioinformatics tools to solve Pattern Recognition problems. In particular we cast the 2D shape analysis problem into the biological sequence aligment problem, for which a huge amount of approaches have been proposed in the bioinformatics community. Obtained results encourage us to go ahead along this research line.

References

1. Loncaric, S.: A survey of shape analysis techniques. Pattern Recognition 31(8), 983–1001 (1998)
2. Zhang, D., Lu, G.: Review of shape representation and description techniques. Pattern Recognition 37, 1–19 (2004)
3. Mingqiang, Y., Kidiyo, K., Joseph, R.: A survey of shape feature extraction techniques. In: Yin, P.Y. (ed.) Pattern Recognition Techniques, Technology and Applications (2008)
4. Durbin, R., Eddy, S., Krogh, A., Mitchison, G.: Biological sequence analysis: probabilistic models of proteins and nucleic acids. Cambridge Univ. (1998)
5. Baldi, P., Brunak, S.: Bioinformatics: the Machine Learning Approach, 2nd edn. MIT Press (2001)
6. Madeira, S., Oliveira, A.: Biclustering algorithms for biological data analysis: a survey. IEEE Trans. on Computational Biology and Bioinformatics 1, 24–44 (2004)
7. Bronstein, A., Bronstein, M., Kimmel, R.: The video genome. arXiv:1003.5320v1 (2010)
8. Altschul, S., Gish, W., Miller, W., Myers, E., Lipman, D.: Basic local alignment search tool. Journal of Molecular Biology 215, 403–410 (1990)
9. Andreu, G., Crespo, A., Valiente, J.: Selecting the toroidal self-organizing feature maps (TSOFM) best organized to object recognition. In: Proc. of IEEE ICNN 1997, vol. 2, pp. 1341–1346 (1997)
10. Thakoor, N., Gao, J., Jung, S.: Hidden markov model-based weighted likelihood discriminant for 2-d shape classification. IEEE Transactions on Image Processing 16(11), 2707–2719 (2007)

11. Li, H., Homer, N.: A survey of sequence alignment algorithms for next-generation sequencing. Briefings in Bioinformatics 11(5), 473–483 (2010)
12. Kemena, C., Notredame, C.: Upcoming challenges for multiple sequence alignment methods in the high-throughput era. Bioinformatics 25(19) (2009)
13. Notredame, C.: Recent evolutions of multiple sequence alignment algorithms. PLoS Computational Biology 3(8) (2007)
14. Needleman, S., Wunsch, C.: A general method applicable to the search for similarities in the amino acid sequence of two proteins. Journal of Modelecular Biology 48(3), 443–453 (1970)
15. Smith, T., Waterman, M.: Identification of common molecular subsequences. Journal of Molecular Biology 147, 195–197 (1981)
16. Altschul, S., Madden, T., Schaffer, A., Zhang, J., Zhang, Z., Miller, W., Lipman, D.: Gapped blast and psi-blast: a new generation of protein database search programs. Nucleic Acids Research 25, 3389–3402 (1997)
17. Bergman, N.: Comparative Genomics, vol. 1 and 2. Humana Press (2007)
18. Gonzalez, R., Woods, R.: Digital Image Processing, 2nd edn. Prentice Hall (2002)
19. Mollineda, R., Vidal, E., Casacuberta, F.: Cyclic sequence alignments: Approximate versus optimal techniques. Int. Journal of Pattern Recognition and Artificial Intelligence 16(3), 291–299 (2002)
20. Bicego, M., Martins, A., Murino, V., Aguiar, P., Figueiredo, M.: 2d shape recognition using information theoretic kernels. In: Proc. Int. Conf on Pattern Recognition, pp. 25–28 (2010)
21. Daliri, M., Torre, V.: Shape recognition based on kernel-edit distance. Computer Vision and Image Understanding 114(10), 1097–1103 (2010)

A New Random Forest Method for One-Class Classification

Chesner Désir[1], Simon Bernard[2], Caroline Petitjean[1], and Laurent Heutte[1]

[1] Université de Rouen, LITIS EA 4108, BP 12, 76801 Saint-Etienne-du-Rouvray, France
Laurent.Heutte@univ-rouen.fr
[2] Université de Liège, Department of EECS et GIGA-Research, Grande Traverse, 10 - B-4000
Liège - Belgium

Abstract. We propose a new one-class classification method, called One Class Random Forest, that is able to learn from one class of samples only. This method, based on a random forest algorithm and an original outlier generation procedure, makes use of the ensemble learning mechanisms offered by random forest algorithms to reduce both the number of artificial outliers to generate and the size of the feature space in which they are generated. We show that One Class Random Forests perform well on various UCI public datasets in comparison to few other state-of-the-art one class classification methods (gaussian density models, Parzen estimators, gaussian mixture models and one-class SVMs).

Keywords: One-class classification, decision trees, ensemble methods, random forests, outlier generation.

1 Introduction

One-class classification (OCC) is a binary classification task for which only one class of objects (the target class) is available for learning. OCC paradigm mainly deals with applications for which collecting counter-example samples (outliers) is impossible, like authorship verification, handwritten character or typist recognition [1,2], mobile-masquerader detection, machine or structure health monitoring [3], etc. As for traditional supervised learning, OCC literature usually opposes density-based methods to discriminative (or frontier-based) methods [4]. Density-based methods aim at estimating the probability density function of the target data and are thus straightforwardly applicable to OCC. The most used techniques among these methods are Parzen windowing and Mixtures of Gaussians (MoG) [5]. However, density-based methods are rarely effective for high dimensional data and usually require a large number of training samples to provide a reasonably good estimate of the distribution [5]. Discriminative approaches, based on the construction of a decision frontier between classes to discriminate, have also been introduced for OCC [2]. Their main difficulty is to synthesize the class of outliers in order to model the decision frontier. This is usually done by either using kernels, as in SVM-based methods [2], or by artificially generating outliers during training as in [1]. In this latter case, artificially generated outliers are often assumed to be uniformly distributed, so as to cover the whole domain of variation of the feature space. This implies to generate an exponential and thus expensive amount of outliers

G.L. Gimel' farb et al. (Eds.): SSPR & SPR 2012, LNCS 7626, pp. 282–290, 2012.

with respect to the dimension of the feature space, and as a consequence, this way of generating outliers is often inaccurate or unusable, especially with high dimensional data.

Ensemble methods are not so used to tackle OCC [4,5], though these methods are known to be powerful for traditional learning tasks [6]. As we will show, they offer some interesting randomization mechanisms that may be used to reduce both the number of outliers to generate and the dimension of the feature space in which outliers are generated. We investigate in this paper the use of ensembles of decision trees, such as random forests [7], that embed the interesting randomization mechanisms evoked above and that have proved their efficiency over single classifiers on various standard classification tasks [6]. We thus propose a new ensemble approach for OCC, called One-Class Random Forest (OCRF), based on a random forest algorithm and designed to tackle issues relative to the generation of outliers. The remainder of the paper is organized as follows. In Section 2, our method is detailed. Section 3 is devoted to the experimental protocol and results, and Section 4 gives conclusions and future works.

2   One-Class Random Forests

The new discriminative approach for OCC proposed in this paper, and named One-Class Random Forests (OCRF), is an ensemble approach based on a random forest algorithm. Let us recall that the random forest (RF) principle is one of the most successful and general purpose ensemble techniques, and has shown to be competitive with state-of-the-art classifiers like SVM and Adaboost [7,8]. It uses randomization to produce a diverse pool of individual tree-based classifiers. In the reference RF learning algorithm, two powerful randomization processes are used: bagging and Random Feature Selection (RFS). The first principle, bagging, consists in training each individual tree on a bootstrap replica of the training set. It is typically used to create the expected diversity among the individual classifiers and is particularly effective with unstable classifiers, like tree-based classifiers, in which small changes in the training set result in large changes in predictions. The second principle, RFS, is a randomization principle specifically used in tree induction algorithms. It consists, when growing the tree, in randomly selecting at each node of the tree a subset of features from which the splitting test is chosen. RFS contributes to the reduction of the dimensionality and has been shown to significantly improve RF accuracy over bagging alone [9,10].

Our OCRF algorithm includes these two randomization principles (bagging and RFS), combined with an original outlier generation process. This latter technique is usually difficult to implement since the number of outliers to generate for having reasonably good performance is exponential with respect to the size of the feature space, and may also increase as the number of available training samples increases. This issue may be addressed by sub-sampling the training set for each component classifier of the ensemble, as it is done in RF with bagging and RFS. Another popular randomization principle, the Random Subspace Method (RSM) [11], may also contribute to solve the dimensionality issue for outlier generation. It consists in randomly selecting a different subset of features for the training of each individual classifier. These two latter principles, RFS and RSM, are thus used in our method to generate outliers in smaller feature spaces.

Now let us describe our outlier generation process. The first naive approach would be to generate outliers uniformly, before the induction of the RF. But, as mentioned above, such a process is difficult to use because of computational costs with quite large datasets, and in addition would not allow to take full advantage from ensemble methods. We thus propose to generate outliers in each bootstrap sample before the induction of each individual tree, as shown in Figure 1. It allows to reduce the number of outliers to generate, thanks to RFS and RSM that reduces the dimensionality upstream. Then, regarding the distribution of outliers, our idea is to identify areas where the target data are sparsely located in the original feature space, and to generate a lot of outliers in these areas. Conversely, fewer outliers are generated in areas containing a lot of target samples. The distribution of outliers is designed to be complementary to the distribution of targets.

The OCRF algorithm is thus made of two main steps: (i) extraction of prior information from the target data in the original feature space, in order to guide the learning process, and (ii) induction of a random forest using both RSM, that notably reduces the dimension of the feature space, and so the number of outliers to generate, and, bagging and RFS that create diversity in the pool of tree classifiers (see Figure 1). Algorithm 1 presents the detailed training algorithm of OCRF.

In summary, the OCRF method takes advantage of: (i) combining a diverse ensemble of weak and unstable classifiers, which is known to be accurate and to increase the generalization performance over single classifiers, and (ii) sub-sampling the training dataset, in terms of training samples and features, in order to efficiently generate outliers by controlling their location and their number.

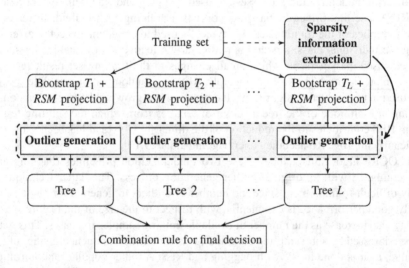

Fig. 1. Overview of the OCRF induction. Additional procedures, in comparison to a traditional RF, are highlighted (in green and boldface).

Algorithm 1. OCRF training algorithm

Require: a training set T, the number of outliers to be generated $N_{outlier}$, the domain of definition for the generation of outliers $\Omega_{outlier}$, the number of trees in the forest L, the parameter of RSM K_{RSM}

Ensure: a one-class random forest classifier

1: (A) ***Prior information extraction***
2: Compute H_{target} the normalized histogram of the target data
3: Compute $H_{outlier}$ the normalized histogram of the outlier data, so that $H_{outlier}$ is the complementary of H_{target}, i.e. $H_{outlier} = 1 - H_{target}$

4: (B) ***Outlier generation and forest induction***
5: **for** $l = 1$ to L **do**
6:    (*i*) Draw a bootstrap sample T_l from the training set
7:    (*ii*) Project this bootstrap sample onto a random subspace of dimension K_{RSM}
8:    (*iii*) Generate $N_{outlier}$ outliers according to the complementary histogram $H_{outlier}$ in the domain $\Omega_{outlier}$, so that the probability that a generated outlier falls in a bin of the histogram $H_{outlier}$ is proportional to the value associated to that bin
9:    (*iv*) Train a random tree on the augmented dataset composed of the target data and the newly generated outlier data
10: **end for**
11: **return** one-class random forest model

---

3 Experiments

In this section, we propose to experimentally assess the performance of OCRF on several public datasets and compare our approach with a few state-of-the-art one-class methods. In the following, we present the public datasets, the evaluation metrics, the one-class methods used in our comparison, and the parameters fixed for this experiment.

Datasets. Genuine one-class datasets are rare as outliers may be difficult or even impossible to sample. For testing OCC methods, authors generally transform multi-class problems into several binary classification tasks "target versus outlier", for each class of the dataset. Some authors select one class as target and label the remaining classes as outliers [1,2]; other authors do exactly the opposite, i.e. select one class as outlier and consider the remaining classes as the target class [12]. Thus, elaborating fair comparisons with other works based on such datasets is difficult as no clear consensus exists. We will use in our experiment the first approach that is more frequent in the literature, with one class as target and the others as outliers. We tackled in this experiment several problems of the literature, taken from 14 datasets of the recognized UC Irvine Machine Learning public repository (see Table 1). We have selected these datasets as they are often used for OCC comparison.

Evaluation Criteria. In our experiments, results are presented in terms of accuracy, but also in terms of target and outlier recognition rates, in order to allow for an analysis

Table 1. Description of the datasets taken from the UC Irvine repository [13]

Dataset	Number of		
	attributes	classes	instances
Sonar	60	2	208
Ionosphere	34	2	351
OptDigits	64	10	5620
Iris	4	3	150
Breast Cancer W. (bcw)	9	2	699
PenDigits	16	10	10994
Diabetes	8	2	768
Mfeat-factors	216	10	2000
Total number of one-class datasets		**41**	

of the "target vs outlier performance" trade-off. However, these evaluation measures, widely used in the binary classification literature, do not take into account the imbalanced nature of OCC datasets [14]. Since, there is still no consensus for the performance assessment of OCC algorithms we have also used in our experiments the Matthews correlation coefficient (MCC) or "phi coefficient". It is particularly well-adapted to imbalanced problems since it takes into account the disparities in the data [14]. The MCC is based on the contingency table from the confusion matrix and is given by:

$$\text{MCC} = \frac{TP \times TN - FP \times FN}{\sqrt{(TP+FP)(TP+FN)(TN+FP)(TN+FN)}}$$

where TP, TN, FN and FP respectively stands for true positive, true negative, false negative (or non-detection) and false positive (or false alarm). MCC values range from -1 if all predictions are wrong to $+1$ for perfect classification, zero values indicate that one of the two classes has not been correctly classified at all (the classifier predicts only one of the two classes).

A standard 10-fold stratified cross-validation has been repeated 5 times. The classifier performance are then averaged over the different runs.

State-of-the-Art OCC Methods and Parameterization. The OCRF algorithm is compared to four state-of-the-art OCC algorithms: the one-class SVM (OCSVM) [2] taken from the LibSVM toolbox and three density estimators, Gaussian estimator (Gauss), Parzen windows (Parzen) and Mixture of Gaussians models (MoG) taken from the Pattern Recognition Toolbox (PRTools) [15]. Each algorithm is run with the default parameterization of its toolbox[1]. Note that the definition of the threshold on the density estimator output is defined thanks to the parameter $fracrej = 0.05$ of PRTools. This parameter corresponds to the fraction of legitimate target cases that will be considered as outliers during training. OCRF is also run with standard values for the parameters [10,16]:

- the number of trees in the random forest is $L = 200$, a value commonly considered as sufficient in practice to ensure statistical convergence of the algorithm;

---

[1] Except for the v coefficient of the OCSVM, a lower bound on the fraction of support vectors, which is set to a more frequently cited value $v = 0.1$, instead of $v = 0.5$.

Table 2. OCC results with MCC (accuracy is indicated in % in brackets)

Dataset	OCRF	OCSVM	Gauss	Parzen	MoG
iris_versicolour	0,579 (81,5)	0,897 (95,3)	**0,903** (95,6)	0,685 (85,6)	0,607 (82,9)
iris_virginica	0,614 (82,7)	**0,900** (95,5)	0,813 (90,9)	0,716 (87,3)	0,604 (82,5)
iris_setosa	0,722 (87,1)	0,903 (95,6)	**0,921** (96,4)	0,799 (90,9)	0,643 (83,3)
bcw_benign	**0,919** (96,2)	0,848 (92,1)	0,902 (95,3)	0,709 (83,2)	0,867 (93,3)
bcw_malignant	**0,629** (81,3)	0,208 (68,2)	0,179 (46,3)	0,273 (69,1)	0,084 (49,6)
ionosphere_good	0,683 (83,3)	**0,785** (89,5)	0,781 (89,3)	0,180 (40,8)	0,584 (75,4)
ionosphere_bad	**0,169** (56,7)	-0,348 (28,2)	-0,410 (26,0)	0,106 (64,7)	-0,346 (33,2)
sonar_mines	0,048 (53,3)	**0,882** (93,6)	0,342 (65,9)	0 (46,2)	0,222 (47,8)
sonar_rocks	0,179 (59,0)	**0,889** (94,0)	0,120 (56,3)	0 (53,8)	0,274 (56,1)
diabetes_positive	0,139 (46,4)	0 (65,2)	0,147 (35,2)	0,188 (55,3)	**0,219** (39,2)
diabetes_negative	**0,241** (68,7)	0 (34,8)	-0,046 (66,5)	0,064 (53,9)	0,020 (68,3)
pendigits_0	**0,976** (99,6)	0 (89,6)	0,970 (99,4)	0,100 (89,7)	0,961 (99,3)
pendigits_1	0,585 (85,8)	0 (89,6)	0,652 (90,0)	0,212 (90,1)	**0,835** (96,6)
pendigits_2	0,835 (96,3)	0 (89,6)	**0,957** (99,2)	0 (89,6)	0,956 (99,2)
pendigits_3	0,918 (98,5)	0 (90,4)	**0,969** (99,5)	0,092 (90,4)	0,949 (99,1)
pendigits_4	0,961 (99,3)	0 (89,6)	**0,969** (99,4)	0 (89,6)	0,953 (99,1)
pendigits_5	0,756 (94,1)	0 (90,4)	0,880 (97,8)	0,092 (90,4)	**0,942** (99,0)
pendigits_6	**0,985** (99,7)	0 (90,4)	0,970 (99,5)	0 (90,4)	0,954 (99,2)
pendigits_7	0,887 (97,6)	0 (89,6)	0,887 (97,7)	0 (89,6)	**0,937** (98,8)
pendigits_8	0,634 (89,3)	0 (90,4)	0,716 (93,2)	0 (90,4)	**0,951** (99,2)
pendigits_9	0,577 (85,9)	0 (90,4)	0,577 (86,9)	0,093 (90,4)	**0,936** (98,9)
optdigits_0	0,776 (94,2)	0,165 (90,5)	**0,954** (99,2)	0 (90,1)	0,745 (95,9)
optdigits_1	0,147 (26,2)	0,054 (89,9)	**0,937** (98,9)	0 (89,8)	0,803 (96,7)
optdigits_2	0,143 (25,8)	0 (90,1)	**0,953** (99,2)	0 (90,1)	0,755 (96,0)
optdigits_3	0,121 (21,7)	0 (89,8)	**0,914** (98,4)	0 (89,8)	0,727 (95,5)
optdigits_4	0,077 (15,6)	0 (89,9)	**0,905** (98,3)	0 (89,9)	0,766 (96,1)
optdigits_5	0,041 (11,5)	0 (90,1)	**0,954** (99,2)	0 (90,1)	0,738 (95,8)
optdigits_6	0,410 (70,3)	0,026 (90,1)	**0,956** (99,2)	0 (90,1)	0,778 (96,3)
optdigits_7	0,264 (48,2)	0 (89,9)	**0,933** (98,8)	0 (89,9)	0,777 (96,3)
optdigits_8	0,043 (11,7)	0 (90,1)	**0,719** (93,6)	0 (90,1)	0,696 (95,2)
optdigits_9	0,077 (15,2)	0 (90,0)	**0,860** (97,4)	0 (90,0)	0,739 (95,7)
mfeat_factors_0	**0,844** (97,2)	0 (90,0)	0,737 (95,8)	0 (10,0)	0 (10,0)
mfeat_factors_1	**0,873** (97,8)	0 (90,0)	0,712 (95,4)	0 (10,0)	0 (10,0)
mfeat_factors_2	**0,879** (97,9)	0 (90,0)	0,740 (95,8)	0 (10,0)	0 (10,0)
mfeat_factors_3	**0,887** (98,0)	0,017 (90,0)	0,695 (95,1)	0 (10,0)	0 (10,0)
mfeat_factors_4	**0,884** (98,0)	0 (90,0)	0,743 (95,8)	0 (10,0)	0 (10,0)
mfeat_factors_5	**0,843** (97,3)	0,013 (90,0)	0,738 (95,8)	0 (10,0)	0 (10,0)
mfeat_factors_6	**0,910** (98,5)	0,068 (90,2)	0,770 (96,2)	0 (10,0)	0 (10,0)
mfeat_factors_7	**0,879** (97,9)	0,017 (90,0)	0,841 (97,3)	0 (10,0)	0 (10,0)
mfeat_factors_8	0,613 (90,6)	0 (90,0)	**0,647** (94,5)	0 (10,0)	0 (10,0)
mfeat_factors_9	**0,866** (97,6)	0,026 (90,1)	0,751 (96,0)	0 (10,0)	0 (10,0)

- the number of attributes for the Random Subspace Method is empirically set to $K_{RSM} = 10$ or $K = M$ if $M < 10$, where M is the dimension of the feature space;
- the number of attributes for the Random Feature Selection is $K_{RFS} = \sqrt{K_{RSM}}$.

Table 3. Case study for results of OCRF on (a) Optical Digit and (b) MFeat-factors datasets

(a)

		OCRF	OCSVM	Gauss
optdigits_0	MCC	**0,776**	0,165	0,954
	Acc	0,94	0,90	0,99
	T	0,99	0,04	0,92
	O	0,94	1,00	1,00
optdigits_1	MCC	0,147	0,054	**0,937**
	Acc	0,26	0,90	0,99
	T	1,00	0,01	0,90
	O	0,18	1,00	1,00
optdigits_2	MCC	0,143	0,000	**0,953**
	Acc	0,26	0,90	0,99
	T	1,00	0,00	0,92
	O	0,18	1,00	1,00
optdigits_3	MCC	0,121	0,000	**0,914**
	Acc	0,22	0,90	0,98
	T	1,00	0,00	0,92
	O	0,13	1,00	0,99
optdigits_4	MCC	0,077	0,000	**0,905**
	Acc	0,16	0,90	0,98
	T	1,00	0,00	0,92
	O	0,06	1,00	0,99
optdigits_5	MCC	0,041	0,000	**0,954**
	Acc	0,12	0,90	0,99
	T	1,00	0,00	0,92
	O	0,02	1,00	1,00
optdigits_6	MCC	0,410	0,026	**0,956**
	Acc	0,70	0,90	0,99
	T	1,00	0,00	0,92
	O	0,67	1,00	1,00
optdigits_7	MCC	0,264	0,000	**0,933**
	Acc	0,48	0,90	0,99
	T	1,00	0,00	0,91
	O	0,42	1,00	1,00
optdigits_8	MCC	0,043	0,000	**0,719**
	Acc	0,12	0,90	0,94
	T	1,00	0,00	0,91
	O	0,02	1,00	0,94
optdigits_9	MCC	0,077	0,000	**0,860**
	Acc	0,15	0,90	0,97
	T	1,00	0,00	0,90
	O	0,06	1,00	0,98

(b)

		OCRF	OCSVM	Gauss
mfeat-factors_0	MCC	**0,844**	0,000	0,737
	Acc	0,97	0,90	0,96
	T	0,86	0,00	0,58
	O	0,98	1,00	1,00
mfeat-factors_1	MCC	**0,873**	0,000	0,712
	Acc	0,98	0,90	0,95
	T	0,79	0,00	0,54
	O	1,00	1,00	1,00
mfeat-factors_2	MCC	**0,879**	0,000	0,740
	Acc	0,98	0,90	0,96
	T	0,85	0,00	0,58
	O	0,99	1,00	1,00
mfeat-factors_3	MCC	**0,887**	0,017	0,695
	Acc	0,98	0,90	0,95
	T	0,87	0,00	0,51
	O	0,99	1,00	1,00
mfeat-factors_4	MCC	**0,884**	0,000	0,743
	Acc	0,98	0,90	0,96
	T	0,82	0,00	0,58
	O	1,00	1,00	1,00
mfeat-factors_5	MCC	**0,843**	0,013	0,738
	Acc	0,97	0,90	0,96
	T	0,82	0,00	0,58
	O	0,99	1,00	1,00
mfeat-factors_6	MCC	**0,910**	0,068	0,770
	Acc	0,98	0,90	0,96
	T	0,85	0,02	0,62
	O	1,00	1,00	1,00
mfeat-factors_7	MCC	**0,879**	0,017	0,841
	Acc	0,98	0,90	0,97
	T	0,83	0,00	0,73
	O	1,00	1,00	1,00
mfeat-factors_8	MCC	0,613	0,000	**0,647**
	Acc	0,91	0,90	0,94
	T	0,83	0,00	0,45
	O	0,91	1,00	1,00
mfeat-factors_9	MCC	**0,866**	0,026	0,751
	Acc	0,98	0,90	0,96
	T	0,85	0,01	0,60
	O	0,99	1,00	1,00

Regarding the generation of outliers during training, one must define their number and the range of their values. We have chosen the generation domain of outliers to be 1.2 times greater than the target domain estimated through the training set, assuming that the outlier domain needs to cover the whole target domain. The number of outliers to generate is empirically set to $N_{outlier} = 10 \cdot N_{target}$ where N_{target} is the sample size of the available training data.

Experimental Results. The results of these experiments, in terms of accuracies and Matthews correlation coefficient (MCC) values, are presented in Table 2. In this table, the compared algorithms have MCC values of zero mostly when they always predict the outlier class except for MoG, for which MCC values of zero occur when it always predict the target class. The OCSVM classifier performs very well on *Iris*, *bcw*, *Ionosphere* datasets and even the best on *Sonar* dataset but it fails to identify outlier data on the remaining datasets.

We can observe that our method (OCRF) has no negative MCC values nor MCC values of zero. OCRF has the highest MCC values for 16 datasets among 41 while being competitive on all datasets except *OptDigits* datasets for which it fails to identify correctly the outlier data. The results show that all methods, except OCRF, have several negative MCC values or MCC values of zero. These methods seem to be rather unstable when dealing with some datasets: 33 datasets for OCSVM, 27 for Parzen, 11 for MoG and 2 datasets over 41 for Gauss. Gauss has an important negative value for *Ionosphere*$_{bad}$ dataset while it has good results in turn with *Ionosphere*$_{good}$ dataset. This latter observation often appears in Table 2: in a two-class problem, when the classifier performs well on the target class, it often fails if the second class is considered in turn as the target class.

We present in Table 3 detailed results for two handwritten digit datasets: *OptDigits* and *MFeat-factors*. For these datasets, OCRF has different results: the method performs poorly on *OptDigits* as it fails to identify correctly the outlier data whereas it performs the best on *MFeat-factors*. We can observe that, for all digit datasets, Gauss seems to be well adapted to describe each target digit cluster while OCSVM, for the same datasets, always predicts the outlier class. If we compare the MCC values to the accuracy (*Acc*), we can observe that MCC is more reliable to assess the performance of the method. For instance for *optdigits*$_0$ dataset, the accuracy of OCSVM is 90% while the target recognition rate is as low as 4%.

4   Conclusion and Future Works

In this paper, we have proposed a new OCC method that is general purpose and has proved its efficiency on various public datasets. The proposed method, called One-Class Random Forest, is based on the reference random forest algorithm combined with an original procedure for generating artificial outliers. This kind of process is often used with discriminative learning methods but is difficult to implement since the number of outliers to generate for having reasonably good performance is exponential with respect to the dimension of the feature space, and may also increase as the number of available training samples increases. We have shown that the random principles used in traditional RF can be powerful tools to overcome this issue: by sub-sampling the training set for each component classifier of the ensemble, through the selection of both the training samples (with bagging) and the features (with Random Feature Selection and Random Subspace method), and by then combining all of them, we reduce the minimum number of outliers to generate and increase the generalization accuracy of the ensemble.

To assess the efficiency of our method, experiments have been conducted on several public datasets from the UCI repository and OCRF has been compared to the four most used OCC algorithms. On most of these datasets and using the default parameterization of each method, results have shown that OCRF performs equally well or better than these state-of-the-art OCC algorithms. Besides, OCRF appears to be rather stable on these various applications.

References

1. Hempstalk, K., Frank, E., Witten, I.: One-Class Classification by Combining Density and Class Probability Estimation. In: Daelemans, W., Goethals, B., Morik, K. (eds.) ECML PKDD 2008, Part I. LNCS (LNAI), vol. 5211, pp. 505–519. Springer, Heidelberg (2008)
2. Scholkopf, B., Platt, J., Shawe-Taylor, J., Smola, A., Williamson, R.: Estimating the support of a high-dimensional distribution. Neural Computation 13(7), 1443–1471 (2001)
3. Tarassenko, L., Clifton, D., Bannister, P., King, S., King, D.: Novelty detection. Encyclopedia of Structural Health Monitoring (2009)
4. Khan, S., Madden, M.: A survey of recent trends in one class classification. Artificial Intelligence and Cognitive Science, 188–197 (2010)
5. Tax, D., Duin, R.: Combining One-Class Classifiers. In: Kittler, J., Roli, F. (eds.) MCS 2001. LNCS, vol. 2096, pp. 299–308. Springer, Heidelberg (2001)
6. Dietterich, T.: Ensemble Methods in Machine Learning. In: Kittler, J., Roli, F. (eds.) MCS 2000. LNCS, vol. 1857, pp. 1–15. Springer, Heidelberg (2000)
7. Breiman, L.: Random forests. Machine Learning 45(1), 5–32 (2001)
8. Robnik-Sikonja, M.: Improving Random Forests. In: Boulicaut, J.-F., Esposito, F., Giannotti, F., Pedreschi, D. (eds.) ECML 2004. LNCS (LNAI), vol. 3201, pp. 359–370. Springer, Heidelberg (2004)
9. Bernard, S., Heutte, L., Adam, S.: Forest-rk: A new random forest induction method. In: Advanced Intelligent Computing Theories and Applications. With Aspects of Artificial Intelligence, pp. 430–437 (2008)
10. Geurts, P., Ernst, D., Wehenkel, L.: Extremely randomized trees. Machine Learning 63(1), 3–42 (2006)
11. Ho, T.: The random subspace method for constructing decision forests. IEEE Transactions on Pattern Analysis and Machine Intelligence 20(8), 832–844 (1998)
12. Tax, D., Ypma, A., Duin, R.: Support vector data description applied to machine vibration analysis. In: Proc. 5th Annual Conference of the Advanced School for Computing and Imaging, Heijen, NL, Citeseer (1999)
13. Blake, C., Merz, C.: Uci repository of machine learning databases. Department of Information and Computer Science, vol. 55. University of California, Irvine (1998), http://www.ics.uci.edu/~mlearn/mlrepository.html
14. Baldi, P., Brunak, S., Chauvin, Y., Andersen, C., Nielsen, H.: Assessing the accuracy of prediction algorithms for classification: an overview. Bioinformatics 16(5), 412–424 (2000)
15. Duin, R.: PRTools version 3.0: A matlab toolbox for pattern recognition. In: Proc. of SPIE, Citeseer (2000)
16. Bernard, S., Heutte, L., Adam, S.: Influence of Hyperparameters on Random Forest Accuracy. In: Benediktsson, J.A., Kittler, J., Roli, F. (eds.) MCS 2009. LNCS, vol. 5519, pp. 171–180. Springer, Heidelberg (2009)

A New Index Based on Sparsity Measures for Comparing Fuzzy Partitions

Romain Quéré and Carl Frélicot

Mathématiques, Image et Applications – Université de La Rochelle
{romain.quere,carl.frelicot}@univ-lr.fr

Abstract. This article adresses the problem of assessing how close two strict and/or fuzzy partitions are. A new index based on a measurement of the sparsity of the contingency matrix crossing the partitions is proposed that satisfies the required properties formulated within the paper and presents a low complexity. It is compared to well-known existing indices of the literature, such as the Rand and the Jaccard indices, the transfert distance and some of their recent fuzzy counterparts.

Keywords: Cluster analysis, Rand index, Jaccard Index, Transfert distance, Sparsity measure, Fuzzy residual implications.

1 Introduction

A *partition* of a set $X = \{\mathbf{x_1}, ..., \mathbf{x_n}\}$ of n objects is a set of c non-empty subsets of X, called *clusters*, that group objects along common attributes they share. Partitions are usually characterized by a $(c \times n)$ *partition matrix* $U = (u_{ik})_{i=1,c;k=1,n}$, identified with a *c-partition* of X for the sake of simplicity. Each u_{ik} represents the degree with which the kth datum is associated to the ith cluster, each column $\mathbf{u_k}$ gathers the degrees for the kth object and each row U^i defines the ith cluster. In this paper, we focus on *fuzzy/probabilistic* partitions such that $u_{ik} \in [0, 1]$ and $\sum_{i=1}^{c} u_{ik} = 1$, and on *strict* partitions such that u_{ik} are binary and sum up to unity, *e.g.*:

$$U_h = \begin{pmatrix} 1 & 1 & 0 & 1 \\ 0 & 0 & 1 & 0 \end{pmatrix} \begin{matrix} U_h^1 \\ U_h^2 \end{matrix} \quad \text{and} \quad U_f = \begin{pmatrix} 0.6 & 0.8 & 0.3 & 0.9 \\ 0.4 & 0.2 & 0.7 & 0.1 \end{pmatrix} \begin{matrix} U_f^1 \\ U_f^2 \end{matrix}.$$

with column labels $\mathbf{u_1}\ \mathbf{u_2}\ \mathbf{u_3}\ \mathbf{u_4}$ for both.

Since clustering algorithms always produce a partition U even if there is no cluster structure in the data, assessing the quality of U is a problem of great interest. It can be tackled using the data itself, by mean of an *internal index* as in *cluster validity* [1, 2] or by assessing how close U is to a ground truth/ expert assessed (mostly strict) partition or a set of ordinary partitions, respectively by mean of an *external* and a *relative* index, both refered as *comparison indices* [3, 4]. This approach have been largely explored for both strict and fuzzy domains, see section 2. We propose to use sparsity measures [5] and fuzzy residual implications [6] to define a new fuzzy index in section 3. In section 4, numerical experiments show its good properties as compared to other indices.

G.L. Gimel' farb et al. (Eds.): SSPR & SPR 2012, LNCS 7626, pp. 291–300, 2012.
© Springer-Verlag Berlin Heidelberg 2012

2 Necessary Tools for the Index Construction

2.1 Comparing Partitions

Plenty of indices have been proposed in the literature for comparing partitions. Depending on the nature of the latter, *i.e.* strict or fuzzy, these indices rely on multiple and different techniques or theoretical frameworks. For the strict case, let us cite the well-known *Rand* and *Jaccard* indices [7] based on a set-theoretical approach and repectively denoted RI and JI hereafter, and the *transfert distance* TD based on graph theory [8]. Among fuzzy comparison indices, let us cite the recent Anderson et al. [9] and the Quéré and Frélicot's [10] extensions, both relying on fuzzy logics and denoted respectively RI_A, JI_A, RI_{QF}, JI_{QF}, the Huellermeier and Rifqi extension of the Rand index HR based on a geometrical approach [11], and the Campello's fuzzy extension of the transfert distance FTD [12]. The ideal index $I^\star$, no matter the nature of the partitions U and V it is meant to compare, must satisfy the following properties (see Table 1 for mentioned indices): $(I1)$ $I^\star(U,V) = 1 \Leftrightarrow U \equiv V$ *(identity)*, $(I2)$ $I^\star(U,V) \geq 0$ *(non-negativity)*, $(I3)$ $I^\star(U,V) = I^\star(V,U)$ *(symmetry)*. Moreover, we consider that such an index should satisfy an additional informal property: $(I4)$ $I^\star(U,V) >> I^\star(U,W)$ if V is known to be much more closer to U than W *(dynamics)*. If a practitioner decides whether or not two partitions are compatible by thresholding the index value, such an informal property ensures him that the index is known to present very different values while comparing close and distant partitions. Because of lack of space, we do not go further into the details of each index. It is not the purpose of this paper and we invite the interested reader to refer to surveys of quality [7, 9]. Yet, let us describe a well-known construction to go one step further in our proposition.

Table 1. Properties satisfied by some indices of the literature

Property	RI	JI	TD	RI_A	JI_A	RI_{QF}	JI_{QF}	HR	FTD
$(I1)$ Identity	•	•	•					•	•
$(I2)$ Non-negativity	•	•	•	•	•	•	•	•	•
$(I3)$ Symmetry	•	•	•	•	•	•	•	•	•

Crossing a c-partition U and a r-partition V results in a $(c \times r)$ contingency matrix $N(U,V) = (n_{ij})_{i=1,c;j=1,r}$ whose general term n_{ij} represents the number of data being in the ith cluster of U and in the jth cluster of V. If both U and V are strict, the cardinal of the intersection between each pair (U^i, V^j) of clusters is given by [13]:

$$N(U,V) = U \, {}^tV. \tag{1}$$

where ${}^tV$ stands for the transpose of V. This is the basis of some set-theoretical indices such as the strict Rand and Jaccard indices, or the fuzzy Anderson et al. extension [9] where a new computation of N is proposed, whose elements n_{ij} are replaced by:

$$n_{ij}^\top(U,V) = \sum_{k=1}^{n} \top(u_{ik}, v_{jk}) \tag{2}$$

where $\top$ is t-norm[1], see [14]. Basic t-norms are the minimum $\top_M(a, b) = min(a, b)$ and the product $\top_P(a, b) = a\,b$. There also exist parametrized families of t-norms, e.g. the Hamacher's one $\top_{H_\gamma}(a, b) = \frac{a\,b}{\gamma + (1-\gamma)\,(a+b-a\,b)}$, $\gamma \in [0, +\infty)$. Note that n_{ij} induced by (1) is strictly equivalent to (2) computed with $\top_P$.

2.2 Sparsity Measures

A fundamental problem in many data analysis problems is to find a suitable representation of the data, say $\mathbf{y} = \{y_1, ..., y_c\} \in \mathbb{R}^c$. A sparsity (or spareness) measure aims at assessing to which extend most values in $\mathbf{y}$ are close to zero while only few ones are non-zero so that they can be used to represent the data. Many sparsity measures are found in the literature, mainly coming from fields such as signal analysis, e.g. in [15]. For a comparison of fifteen well-known sparsity measures, the reader should refer to [5]. With no loss of generality, we restrict ourselves to sparsity measures $S : \mathbb{R}^c \to [0, 1]$. Among the properties such measures may have, let us cite the two that are required for the comparison index we propose: $(S1)$ adding a constant to each value decreases sparsity, $(S2)$ as one value becomes infinite, as sparse as possible is the distribution. Two of the sparsity measures reviewed in [5] have these properties. The first one is the *Hoyer's sparsity measure* which is based on the relationship between the L_1 norm and the L_2 norms. It is defined as:

$$H(\mathbf{y}) = \frac{\sqrt{c} - \frac{\sum_{j=1}^{c} y_j}{\sqrt{\sum_{j=1}^{c} y_j^2}}}{\sqrt{c} - 1}. \tag{3}$$

It varies from 0, *i.e.* $\mathbf{y}$ is not sparse, if all components are equal (up to signs) to unity if $\mathbf{y}$ contains a single non-zero element. The second one is called *kurtosis sparsity measure* by analogy to the well-known measure of peakedness of a probability distribution. It is defined as:

$$\kappa_4(\mathbf{y}) = \frac{\sum_{j=1}^{c} y_j^4}{\left(\sum_{j=1}^{c} y_j^2\right)^2}. \tag{4}$$

In order to show how these two measures behave, we have driven a short experiment inspired by the work in [5]. Consider a vector $\mathbf{y}$ of 500 values in $\{0, 1\}$ drawn from a Bernoulli distribution, so that 1 and 0 have a respective probability of occurence of p and $q = 1 - p$. When q is barely null, $\mathbf{y}$ is then composed of very few zeros while only a small number of values are 1 when q is close to 1. Thus as q increase so should the sparsity measure as exhibited in Fig. 1. One can observe that H presents more granularity than κ_4 which only gives a strong response for $q > 0.9$.

---

[1] A t-norm is binary operation on the unit interval $\top : [0, 1]^2 \to [0, 1]$ which is commutative, associative, non decreasing and has 1 for neutral element.

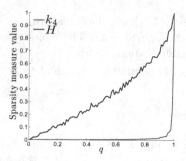

Fig. 1. Sparsity measures of 500 data drawn from a Bernoulli distribution

2.3   Fuzzy Implications

A fuzzy residual implication is an application $\mathcal{I} : [0,1]^2 \to [0,1]$, $(a,b) \mapsto \mathcal{I}(a,b)$, such that:

$$\mathcal{I}(a,b) = \sup_{t}\{t \in [0,1] : \top(a,t) \leq b\} \tag{5}$$

where $\top$ is t-norm. We speak about an implication function if $\mathcal{I}$ is non-increasing in the first variable, non-decreasing in the second variable and $I(0,0) = I(1,1) = 1$, and $I(1,0) = 0$, see [6] for a large survey on fuzzy implication functions. Within theses implications, the well-known *Gödel* is obtained with $\top_M$ and given by:

$$\mathcal{I}_M(a,b) = \begin{cases} 1 \text{ if } b \geq a \\ b \text{ if } b < a \end{cases} \tag{6}$$

As well, parametrical fuzzy implications are defined, *e.g.* the Hamacher's ones, defined by [16]:

$$\mathcal{I}_{H_\gamma}(a,b) = \begin{cases} 1 & \text{if } b \geq a \\ \frac{b(\gamma+a-\gamma a)}{b(\gamma+a-\gamma a)+a-b} & \text{if } b \leq a \end{cases}. \tag{7}$$

3   The New Index

Let us consider, for pedagogical purpose, the following two strict partitions :

$$U_h = \begin{array}{c} \mathbf{u_1}\ \mathbf{u_2}\ \mathbf{u_3}\ \mathbf{u_4} \\ \begin{pmatrix} 1 & 1 & 0 & 1 \\ 0 & 0 & 1 & 0 \end{pmatrix} \begin{array}{c} U_h^1 \\ U_h^2 \end{array} \end{array} \text{ and } V_h = \begin{array}{c} \mathbf{v_1}\ \mathbf{v_2}\ \mathbf{v_3}\ \mathbf{v_4} \\ \begin{pmatrix} 0 & 1 & 0 & 1 \\ 1 & 0 & 0 & 0 \\ 0 & 0 & 1 & 0 \end{pmatrix} \begin{array}{c} V_h^1 \\ V_h^2 \\ V_h^3 \end{array} \end{array}.$$

Both share in common the information that elements $\mathbf{x_2}$ and $\mathbf{x_4}$ belong to the same cluster, while $\mathbf{x_3}$ belongs to another one. Actually, the only difference between U_h and V_h is about $\mathbf{x_1}$, which is grouped with $\mathbf{x_2}$ and $\mathbf{x_4}$ in U_h while it is put aside in its own cluster in V_h because it probably differs in some subtle ways of $\mathbf{x_2}$ and $\mathbf{x_4}$. The idea here is that V_h can be seen as a refinement of U_h, so that V_h^1 is included in U_h^1, as pointed out by the contingency matrix crossing the two partitions $N(U_h, V_h) = \begin{pmatrix} 2 & 1 & 0 \\ 0 & 0 & 1 \end{pmatrix}$ by (1). Indeed, the 1st row $\mathbf{n_1} = (\mathbf{2}\ \mathbf{1}\ \mathbf{0})$ and the 1st column $^t\mathbf{n_1} = (\mathbf{2}\ \mathbf{0})$ of $N(U_h, V_h)$ show that the only two elements of V_h^1

also belong to U_h^1, leading thus to conclude that $V_h^1 \subset U_h^1$. Moreover, let us have a look at the matrices $N(U_h, U_h) = \begin{pmatrix} 3 & 0 \\ 0 & 1 \end{pmatrix}$ and $N(V_h, V_h) = \begin{pmatrix} 2 & 0 & 0 \\ 0 & 1 & 0 \\ 0 & 0 & 1 \end{pmatrix}$ crossing each partition with itself. Both logically present only one non-zero element on each of their rows and columns, so that it can be directly connected with the concept of sparsity. This is the main idea of our proposition : the closest the partitions, the more sparse the rows and columns of their contingency matrix. This idea is also valid when crossing two fuzzy partitions U_f and V_f. The inner product between U_f^i and V_f^j induced by (1) will be high if and only if both clusters are similar and are not too fuzzy, $i.e.$ they have a certain amount $n_s < n$ of same components close to 1, so that $U_f^i \, {}^t V_f^j$ tends to n_s. Follows that considering the whole set of clusters of U_f and V_f, the elements of N will be large for crossed fuzzy clusters having a lot of values in common, and will be small for frankly different ones, so that the idea of sparsity as defined in section 2.2 is preserved. The same reasoning holds while N is computed with (2), whatever the t-norm $\top$. The t-norm only slightly emphasizes or reduces the gaps between high and low values, in the same manner as exhibited in [10].

Concretely, the new index is constructed as follows. Given a sparsity measure $\mathcal{S}$, it is easy to compute $\mathcal{R}_\mathcal{S} = \{\mathcal{S}(\mathbf{n_r}), ..., \mathcal{S}(\mathbf{n_r})\}$ and $\mathcal{C}_\mathcal{S} = \{\mathcal{S}({}^t\mathbf{n_1}), ..., \mathcal{S}({}^t\mathbf{n_c})\}$ from the contingency matrix N crossing two partitions U and V. For each set, we propose to combine the sparsities using a suitable aggregation function $\mathcal{A}$, $e.g.$ the arithmetic means $\overline{\mathcal{R}}_\mathcal{S}$ and $\overline{\mathcal{C}}_\mathcal{S}$, to get two representative values. Many families of aggregation functions exist, see [17] for a recent monograph. In our proposition, we restrict to functions $\mathcal{A}$ taking values in $[0, 1]$ while computed for $\mathcal{R}_\mathcal{S}$ and $\mathcal{C}_\mathcal{S}$, so that the resulting two representatives of the sparsities in row/column of N can be inputs of a fuzzy residual implication to assess wether partitions U and V are compatible or not. Therefore, we propose a new comparison index of strict/fuzzy partitions as follows:

$$QF_{(\mathcal{S},\mathcal{A},\mathcal{I})}(U, V) = min\bigg(\mathcal{I}\Big(\mathcal{A}(\mathcal{R}_\mathcal{S}), \mathcal{A}(\mathcal{C}_\mathcal{S})\Big), \mathcal{I}\Big(\mathcal{A}(\mathcal{C}_\mathcal{S}), \mathcal{A}(\mathcal{R}_\mathcal{S})\Big)\bigg). \qquad (8)$$

This index is in $[0, 1]$ by construction, and it is required that $\mathcal{I}$ satisfies the so-called ordering property[2], so that $QF_{(\mathcal{S},\mathcal{A},\mathcal{I})}(U, V) = 1$ whenever $U \equiv V$. Finally, it is worthy on note that the asymptotic complexity of this new index is $\mathcal{O}(n)$, as Anderson et al. and Campello's ones, while the other considered indices are in $\mathcal{O}(n^2)$. To sum up, the proposed comparison index satisfies properties ($I1$), ($I2$), ($I3$) and ($I4$), and has a triple of user-defined parameters $(\mathcal{S}, \mathcal{A}, \mathcal{I})$:

- a sparsity measure $\mathcal{S} : \mathbb{R} \times ... \times \mathbb{R} \mapsto [0, 1]$, $e.g.$ the Hoyer H and the kurtosis κ_4 ones given by (3) and (4),
- an aggregation function $\mathcal{A} : [0, 1] \times ... \times [0, 1] \mapsto [0, 1]$, $e.g.$ the arithmetic mean $M(\mathcal{R}) = \frac{1}{card(\mathcal{R})} \sum_{s \in \mathcal{R}} s$,
- a fuzzy residual implication $\mathcal{I} : [0, 1] \mapsto [0, 1]$, $e.g.$ the Gödel $\mathcal{I}_M$ and the Hamacher $\mathcal{I}_{H_\gamma}$ ones respectively given by (6) and (7).

---

[2] $\forall a, b \in [0, 1], \ a \leq b$ iff $\mathcal{I}(a, b) = 1$.

4 Numerical Experiments

Some of the tested comparison indices require to choose a t-norm for their computation, some others use the product. For sake of simplicity and fairness, we choose $\top_P$ whenever a t-norm is required.

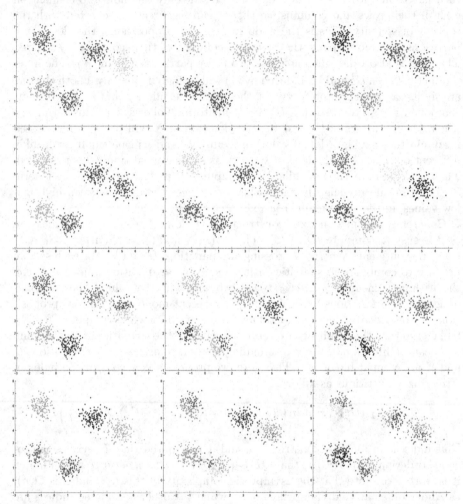

Fig. 2. From top to bottom and from left to right : ground truth clustering $U_{k^\star}$ and clusterings V_k ($k = 2, ..., 12$) obtained with the standard $k - means$ algorithm

4.1 Strict Partitions

First, we compare strict partitions of a 2-dimensional synthetic dataset composed of $k^\star = 5$ Gaussian clusters centered at $(-1, 6)$, $(1, 1)$, $(-1, 2)$, $(5, 4)$ and $(3, 5)$ with the same standard deviation $\Sigma = \frac{1}{2}Id$, so that two pairs of them present a slight overlap, see Fig. 2 (*top-left*). Remaining subfigures present the clusterings V_k obtained with the standard $k - means$ algorithm for $k = 2, ..., 12$. Each partition V_k is compared to the ground truth partition U_{k^*} using different strict indices, resulting in the curves plotted in Fig. 3 (*left*). The tested indices are the Rand Index RI, the Jaccard Index JI, the Transfert Distance TD, and the new sparsity based index $QF_{(\mathcal{S},\mathcal{A},\mathcal{I})}$ with different triples of parameters: $(\kappa_4, M, \mathcal{I}_M)$, $(H, M, \mathcal{I}_M)$, $(\kappa_4, M, \mathcal{I}_{H_5})$ and $(H, M, \mathcal{I}_{H_5})$. As expected, all the indices exhibit their maximum value when $k = k^* = 5$, but they do not have the same dynamics. In particular, the Rand index presents the smallest dynamics, so that we will stop using it and prefer the Jaccard indices in the remaining experiments. For $QF_{(\mathcal{S},\mathcal{A},\mathcal{I})}$, this property clearly depends on S. The difference of sparsity between the contingency matrices crossing U with each V_k is less marked with H because the Hoyer's sparsity measure presents more granularity than the kurtosis κ_4 in its measurement of the sparsity, as previously exhibited in Fig. 1. Another interesting point is that the proposed index better exhibit compatibility of partitions with respect to cluster refinements. Indeed, one can see that $QF_{(\mathcal{S},\mathcal{A},\mathcal{I})}$ considers that V_4 is closer to U_{k^*} more than the other indices do. We think it represents a slight improvement since V_4 differs with U_{k^*} about only one cluster, so that the clusters should be considered as being quite close, see cluster refinements in Fig. 2. Moreover, this behaviour is clearly reinforced with the Hamacher fuzzy implication $\mathcal{I}_{H_5}$, as shown for instance for V_6 and V_7. This is because $\mathcal{I}_{H_\gamma} > \mathcal{I}_M$ by construction. However, since $QF_{(\mathcal{S},M,\mathcal{I}_M)}$ and $QF_{(\mathcal{S},M,\mathcal{I}_{H_\gamma})}$ mostly give analogous results for both sparsities measures H and κ_4, the influence of the chosen fuzzy residual implication is no more studied.

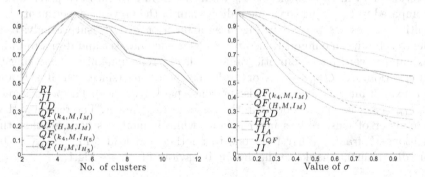

Fig. 3. Comparison of the ground truth strict partition U_{k^*} to the 11 strict partitions V_k shown in Fig. 2 (*left*). Comparison of the ground truth strict partition W_{k^*} to the 10 fuzzy 3-partitions V_σ with increasing overlap (*right*).

4.2 Strict vs. Fuzzy Partitions

Inspired by the work in [18], this second experiment aims at comparing a strict reference partition to a collection of fuzzy ones. Ten 3-dimensional datasets composed of $k^* = 3$ isotropic Gaussian clusters centered at $(1,0,0)$, $(0,0,0)$ and $(-1,0,0)$ are generated for increasing standard deviations $\sigma = \{\frac{1}{k}\}_{k=10,9,...,1}$, so that they evolve from no overlap to a strong one. The *Fuzzy C-Means (FCM)* algorithm is run for each dataset to produce 10 fuzzy 3-partitions V_σ, with a fuzzifier exponent and a termination parameter respectively set to 2 and 10^{-3}.

Fig. 3 (*right*) shows the resulting curves of comparison of those 10 partitions to the strict ground truth partition W_{k^*} for different indices. As expected, all indices achieve their maximum value for the smaller σ and decrease as σ increases. However, four of them present a higher dynamics so that their value for the most overlapping dataset is significantly lower: our fuzzy $QF_{(H,M,\mathcal{I}_M)}$, the Anderson et al. JI_A, the strict[3] JI, and the Quéré and Frélicot's JI_{QF}. Hoyer's H sparsity measure gives a better result than κ_4 because increasing σ can be seen as increasing a kind of amount of noise within the data, and Hoyer's measure is known to overperform κ_4 in such cases, see [15].

4.3   Fuzzy Partitions : Real Datasets

The last experiment is driven for several datasets from the UCI Machine Learning Repository [19], presenting various characteristics in terms of: number n of observations, number p of attributes, true number c^* of classes/clusters and degree of overlap between clusters. Since it has lead to convergent outcomes, we only give the results obtained on the following three well-known ones:

- Fisher iris ($n = 150, p = 4, c^* = 3$, slight overlap between two classes),
- Pima diabetes ($n = 768, p = 8, c^* = 2$, strong overlap between both classes),
- Italian wine ($n = 178, p = 13, c^* = 3$, (slight overlap between 3 classes).

For each dataset, the *FCM* algorithm is run under the same parametrization as in the previous experiment to produce a reference fuzzy c^*-partition U_{c^*} and a collection of 14 fuzzy c-partitions U_c, c varying from 2 to 15. Each partition U_c is compared to U_{c^*} using the same indices than in the previous experiment. The resulting curves are plotted in Fig. 4 as a function of c. Unsurprisingly, each index reaches its maximum value at $c = c^*$ for each dataset and drops from its maximum toward an asymptotic value, with different dynamics. According to this criterion, our $QF_{(\kappa_4,M,\mathcal{I}_M)}$ outperforms the others (since even if HR and FTD exhibit higher values, they also present a poor dynamics). The performance of $QF_{(H,M,\mathcal{I}_M)}$ is not as good as in the previous experiment. The reason is that the number of clusters was not changing while it increases here, so that the partitions V_c frankly differ from one to another and lead $QF_{(\kappa_4,M,\mathcal{I}_M)}$ to be more discriminant thanks to the drastic behaviour of κ_4, see Fig.1.

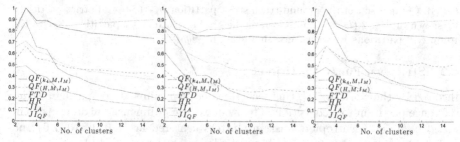

Fig. 4. Indices values obtained for the Iris, Pima and Wine datasets (from left to right)

---

[3] For the strict Jaccard Index JI, hardened partitions obtained from fuzzy partitions V_σ are considered.

5 Conclusion

In this article, we propose a new index for comparing strict and fuzzy partitions, lying on the original idea of measuring the sparsity of the contingency matrix crossing two partitions. Its construction, involving sparsity measures and fuzzy residual implications, is simple but efficient, so that as shown by numerous experimental results, this index outperforms the existing ones, in particular with respect to the dynamics property. Moreover, by its low computational complexity, the proposed index could become a privilegied tool for many practitioners.

References

[1] Pal, N., Bezdek, J.: On cluster validity for the fuzzy c-means model. IEEE Trans. on Fuzzy Systems 3(3) (1995)

[2] Wang, W., Zhang, Y.: On fuzzy cluster validity indices. Fuzzy Sets and Systems 158(19) (2007)

[3] Borgelt, C.: Prototype-based classification and clustering Habilitation Thesis. Habilitation Thesis, Fakultat fur Informatik der Otto von Guericke, Universitat Magdeburg (2005)

[4] Patrikainen, A.: Methods for comparing subspace clusterings. PhD thesis, Helsinki University of Technology (2005)

[5] Hurley, N., Rickard, S.: Comparing measures of sparsity. IEEE Trans. on Information Theory 55(10) (2009)

[6] Mas, M., Monserrat, M., Torrens, J., Trillas, E.: A survey on fuzzy implication functions. IEEE Trans. on Fuzzy Systems 15(6) (2007)

[7] Albatineh, A., Niewiadomska-Bugaj, M., Mihalko, D.: On similarity indices and correction for chance agreement. J. of Classification 23 (2006)

[8] Charon, I., Denoeud, L., Guenoche, A., Hudry, O.: Maximum transfer distance between partitions. J. of Classification 23(1) (2006)

[9] Anderson, D., Bezdek, J., Popescu, M., Keller, J.: Comparing fuzzy, probabilistic, and possibilistic partitions. IEEE Trans. on Fuzzy Systems 18(5) (2010)

[10] Quéré, R., Capitaine, H.L., Fraisseix, N., Frélicot, C.: On normalizing fuzzy coincidence matrices to compare fuzzy and/or possibilistic partitions with the rand index. In: 10th IEEE International Conference on Data Mining, pp. 977–982 (2010)

[11] Hüllermeier, E., Rifqi, M.: A fuzzy variant of the rand index for comparing clustering structures. In: 13th IFSA World Congress (2009)

[12] Campello, R.: Generalized external indexes for comparing data partitions with overlapping categories. Pattern Recognition Letters 31(9) (2010)

[13] Brouwer, R.: Extending the rand, adjusted rand and jaccard indices to fuzzy partitions. J. of Intelligent Information Systems 32(3) (2009)

[14] Klement, E., Mesiar, R.: Logical, Algebraic, Analytic, and Probabilistic Aspects of Triangular Norms. Elsevier (2005)

[15] Karvanen, J., Cichocki, A.: Measuring sparseness of noisy signals. In: 4th Int. Symp. on Independent Component Analysis and Blind Signal Separation (2003)

[16] Le Capitaine, H., Frélicot, C.: Classification with reject options in a logical framework: a fuzzy residual implication approach. In: 13th IFSA World Congress (2009)

[17] Grabisch, M., Marichal, J., Mesiar, R., Pap, E.: Aggregation Functions. Encyclopedia of Mathematics and its Applications. Cambridge University Press (2009)

[18] Ceccarelli, M., Maratea, A.: A Fuzzy Extension of Some Classical Concordance Measures and an Efficient Algorithm for their Computation. In: Lovrek, I., Howlett, R.J., Jain, L.C. (eds.) KES 2008, Part III. LNCS (LNAI), vol. 5179, pp. 755–763. Springer, Heidelberg (2008)

[19] Frank, A., Asuncion, A.: UCI machine learning repository (2010)

Polichotomies on Imbalanced Domains by One-per-Class Compensated Reconstruction Rule

Roberto D'Ambrosio and Paolo Soda

Integrated Research Centre, Universitá Campus Bio-Medico of Rome, Rome, Italy
{r.dambrosio,p.soda}@unicampus.it

Abstract. A key issue in machine learning is the ability to cope with recognition problems where one or more classes are under-represented with respect to the others. Indeed, traditional algorithms fail under class imbalanced distribution resulting in low predictive accuracy over the minority classes. While large literature exists on binary imbalanced tasks, few researches exist for multiclass learning. In this respect, we present here a new method for imbalanced multiclass learning within the One-per-Class decomposition framework. Once the multiclass task is divided into several binary tasks, the proposed reconstruction rule discriminates between safe and dangerous classifications. Then, it sets the multiclass label using information on both data distributions and classification reliabilities provided by each binary classifier, lowering the effects of class skew and improving the performance. We favorably compare the proposed reconstruction rule with the standard One-per-Class method on ten datasets using four classifiers.

1 Introduction

In data mining and machine learning, we deal with imbalanced (or skewed) recognition problem when one of the classes is largely under-represented in comparison to the others. Most traditional learning algorithms cannot cope with this case since they are biased towards the majority classes, resulting in poor predictive accuracy over the minority ones. This happens because they are designed to minimize errors over training samples, ignoring classes composed of few instances.

In real world applications, such as text classification, currency validation, and medical diagnosis, very often the a priori distributions of samples are different among the classes, thus resulting in an imbalanced classification task. Due to the relevance of the topic and its potential impact on the development of learning algorithms, in the recent years there have been several works proposing learning methods coping with skewed training set (TS). In particular, most of the existing literature focus on class imbalance learning methods for binary problems (also referred to as dichotomies), proposing solutions both at data and algorithmic levels, e.g. [1, 2]. Approaches working at data level provide different forms of resampling, e.g. random or direct oversampling and undersampling, whereas those working at algorithmic level introduce a bias to compensate the skewness of the classes, e.g. adjusting the costs of classes, adjusting decision thresholds and recognition-based learning.

Otherwise, learning under a skewed multiclass TS has received little attention despite the fact that such problems can be found in a large number of domains. The few recent

G.L. Gimel' farb et al. (Eds.): SSPR & SPR 2012, LNCS 7626, pp. 301–309, 2012.

works focused on multiclass learning (also named as polichotomy) can be roughly divided into two categories [3–7].

The first compensates class imbalance directly on the polychotomy [3, 5, 7]. In [5] the authors present a two-stage evolutionary neural network algorithm with the entropy and area fitness functions, under the assumption that a good classifier should combine a high classification rate level in the global dataset with an acceptable level for each class. In [7] the authors propose a small sphere and large margin approach for novelty detection problems, where the majority of training data are normal examples, and the training data also contain a small number of abnormal examples or outliers. Their basic idea is to construct a hypersphere solving a convex optimization problem containing most of the normal examples, such that the volume of this sphere is as small as possible, while at the same time the margin between the surface of this sphere and the outlier training data is as large as possible. In [3] the authors introducing several cost functions in the learning algorithm of a neural network in order to improve its generalization ability and speed up the convergence process.

The second category faces with polychotomy using a decomposition approach [4, 6], which consists in reducing the multiclass problem complexity in less complex binary subtasks, each one addressed by a classifier usually referred to as dichotomizer [8–11]. To provide the final classification, dichotomizers' outputs are combined according to a *reconstruction rule*. We found only two attempts proposing reconstruction rule suited for imbalance datasets [4, 6]. In particular, in [4] the authors applied the SMOTE algorithm [1], i.e. an oversampling method for binary skewed task, in each dichotomizer and then combine their outputs via a fuzzy model. In [6] the authors applied in the learning phase of each dichotomizer several methods compensating binary class imbalance.

The short analysis of the literature reported so far shows that multiclass learning in class imbalance circumstances is an issue deserving more research efforts. In this respect, we propose here a new reconstruction rule combining dichotomizers' outputs within the One-per-Class decomposition framework, thus falling into the second of the aforementioned branch. Once distinguished between safe and dangerous dichotomizers classifications on the basis of sample classification reliability, it applies different reconstruction rules for each of these two cases thus permitting to reduce effects due to the skewness between classes. We test our approach on ten databases using four different classification architectures, and we compare its results with those provided both by a well known One-per-Class reconstruction rule and by a multiclass classifier.

2 Background

In this section we first introduce decomposition methods with particular reference to the One-per-Class approach, and then we discuss performance metrics suited for class imbalance classification tasks.

2.1 One-per-Class Decomposition Method

Decomposition methods reduce multiclass problem complexity in less complex dichotomies, each one addressed by a classifier. Their rationale lies in observing that, on the one hand, discriminating between two classes is much easier than simultaneously distinguishing among many classes [12] and, on the other hand, most of the available classification algorithms are best suited to learn binary functions [13, 14].

Decomposition methods can be unified in a common framework wherein the output space is represented by a binary code matrix named as *decomposition matrix*. On its basis, it is then possible to distinguish between the different approaches as follows. Let $\Omega = \{\omega_1, \omega_2, \ldots, \omega_K\}$ represents the label set of the K-classes problem, with $K > 2$.

The decomposition of the polychotomy generates a pool of L dichotomizers, with the value of L depending upon the decomposition approach adopted. The dichotomizer M_i is a discriminating function that classifies each input sample in two separate superclasses, represented by the label set $\Omega_i = \{-1, 1\}$, each label identifying a subset of polychotomy classes. Therefore, the overall decomposition scheme can be set by a *decomposition matrix* $D \in \Re^L \times \Re^c$, whose elements are defined as:

$$
d_{ij} = \begin{cases} 1 & \text{if class } j \text{ is in the subgroup associated to label 1 of } M_i \\ -1 & \text{if class } j \text{ is in the subgroup associated to label -1 of } M_i \\ 0 & \text{if class } j \text{ is in neither groups associated to label -1 or 1 of } M_i \end{cases} \tag{1}
$$

Hence, the dichotomizer M_i is trained to associate patterns belonging to class ω_j with values d_{ij}. Within this framework, the main decomposition approaches are named as *One-per-Class* (OpC) [8, 11], *Error-Correcting Output Code* [8, 9], and *PairWise Coupling* [10]. In the following we further present the OpC approach for four reasons. First, it is applied in the reconstruction rule proposed in this paper; second, it introduces a strong degree of imbalance for each dichotomizer since it collapses samples of all classes except one into a superclass; third, it is a very popular method with the lowest computational complexity and, fourth, it is very often used to derive multiclass classifier from learning algorithms which are intrinsically binary.

OpC reduces the multiclass problems into K binary problems (i.e. $L = K$), each one addressed by one dichotomizer, thus achieving a squared decomposition matrix. We say that the jth dichotomizer is specialized in the jth class when it aims at recognizing if the input sample belongs either to the jth class or, alternatively, to any other class. Without loss of generality and to simplify the notation we use label 0 instead of -1. Therefore, the binary label of the jth dichotomizer is 1 if the sample x belongs to the jth class, and 0 otherwise.

It is worth noting that dichotomizers, besides labelling each pattern, may supply other information typically related to the degree that the sample belongs (or does not belong) to the corresponding class. Indeed, it has been proved that exploiting information derived from classifiers working at the measurement level permits to define reconstruction rules that are potentially more effective [15]. Since measurement classifiers can provide more information than other classifiers, we assume that only measurement experts are used in OpC scheme. This assumption is not a limitation since it is always possible to obtain a measurement for each classification act of any kind of classifiers [16].

2.2 Performance Metrics

The confusion matrix is usually used to assess the performance of a recognition system permitting to compute several indexes. The most used one is the global recognition accuracy (acc) defined as $acc = \frac{\sum_{j=1}^{K} n_{jj}}{N}$, where n_{jj} is the number of elements of class j correctly labelled and N is the total number of samples.

However, in case of imbalanced TS the recognition performance cannot be measured in terms of classification accuracy only, since this measure is strongly biased to favor the majority class. Hence, it would be more interesting to use a performance measure dissociating the hits (or the errors) that occur in each class. To this aim, we can compute the *accuracy by class* as $acc_j = \frac{n_{jj}}{N_j}$, where N_j is the number of samples in the j-th class. Notice that acc_j is independent of prior probabilities and, thus, it is robust when class distribution might be different in training and test sets or change over time. From such metrics we can compute an overall index by extending the geometric mean of accuracies (g) typically used to assess the performance in binary skewed problems. It is given by $g = (\prod_{j=1}^{K} acc_j)^{\frac{1}{K}}$, being a non-linear measure since a change in one of its arguments has a different effect on g depending on its magnitude. For instance, if a classifier misses the labels of all samples in the jth class, it results $acc_j = 0$, and $g = 0$.

3   Compensated Reconstruction Rule

In this section we propose a reconstruction rule suited for OpC scheme that, combining dichotomizers' outputs, reduces drawbacks given by learning under data skewness and improves system performance. Indeed, the typical error rate minimisation performed by most of learning algorithms over the TS introduces a bias in favour of the the majority class, resulting in low accuracies on classes composed of few instances. Hereinafter, the proposed rule is referred to as *Compensated Reconstruction Rule* (CRR).

On the basis of the outputs provided by the pool of dichotomizers we distinguish between *safe* and *dangerous* classifications. Safe classifications are those in which each dichotomizer is strongly confident about its output. Dangerous classifications are those in which two or more dichotomizers are weakly confident about their predictions. Considering the skewed nature of the dataset, the low confidence of two or more dichotomizers with their outputs increases the likelihood of providing wrong classification. The confidence of a classifier on its output should be measured by using classification reliability, i.e. a measure lying in $[0, 1]$ computed by any measurement classifier [16, 17]. Now, let us introduce the following notation:

- $\Omega = \{\omega_1, \omega_2, \ldots, \omega_K\}$ is the set of class labels, as reported in section 2.1;
- N is the total number of samples as reported in section 2.2;
- N_j is the number of samples belonging to the class ω_j, as reported in section 2.2;
- $x \in \Re^n$ is a sample;
- the binary profile $\mathbf{M}(x)$ is the K-bit codeword of x collecting dichotomizers' outputs as $\mathbf{M}(x) = [M_1(x), M_2(x), \ldots, M_K(x)]$. $M_j(x)$ is 1 if $x \in \omega_j$, 0 otherwise;
- the reliability profile $\boldsymbol{\Psi}(x)$ is a K elements vector collecting dichotomizers' reliabilities; each entry measures the reliability of the jth dichotomizer's output and represents the degree that x belong or not to predicted class. It is given by $\boldsymbol{\Psi}(x) = [\psi_1(x), \psi_2(x), \ldots, \psi_K(x)]$;

- the reverse a-priori probability profile **R** contains the knowledge on the a-priori classes distribution; it is a K elements vector $\mathbf{R} = [r_1, r_2, \ldots, r_K]$, where $r_j = 1 - N_j/N$;
- τ_0 is a threshold for detecting classifications more likely to be corrected in case of dichotomizer suggesting that x does not belong to its class ($M_j(x) = 0$). Its value is estimated on a validation set maximizing acc_j;
- τ_1 is a threshold for detecting classifications more likely to be corrected in case of dichotomizer suggesting that x belongs to its class ($M_j(x) = 1$). Its value is set as for τ_0.

On this basis, the classification of x is said to be a safe classification if:

$$(\alpha_1(x) > \tau_1 \wedge \alpha_0(x) > \tau_0) \vee (\alpha_1(x) > \tau_1 \wedge \alpha_0(x) < \tau_0 \wedge \overline{\alpha_0}(x) > \tau_0) \quad (2)$$

where $\alpha_1(x) = max_j(\psi_j(x)|M_j(x) = 1)$ is the largest value of reliability among those provided by dichotomizers whose outputs is 1, i.e. the dichotomizers suggesting that x belongs to their corresponding class. $\alpha_0(x) = min_j(\psi_j(x)|M_j(x) = 0)$ is the lowest reliability value among those provided by dichotomizers whose outputs are 0, i.e. the dichotomizers suggesting that x does not belong to their corresponding class. Furthermore, $\overline{\alpha_0}(x) = E(\psi_j(x)|M_j(x) = 0)$ is the average value of reliabilities associated with dichotomizers whose outputs are 0. In the following, for brevity we omit to indicate that α_0, α_1 and $\overline{\alpha_0}$ depend on (x) when this does not introduce any ambiguity.

The classification of x is defined as dangerous when:

$$(\alpha_1 < \tau_1 \wedge \alpha_0 < \tau_0) \vee (\alpha_1 > \tau_1 \wedge \alpha_0 < \tau_0 \wedge \overline{\alpha_0} < \tau_0) \quad (3)$$

To set the final classification, CRR applies different criteria for safe and dangerous classifications. In the former case, the index s of the dichotomizer setting the final class $\omega_s \in \Omega$ is given by:

$$s = \begin{cases} argmax_j(M_j(x) \cdot \psi_j(x)) \; if \; m \in [1, K] \\ argmim_j(\overline{M_j(x)} \cdot \psi_j(x)) \; if \quad m = 0 \end{cases} \quad (4)$$

where $\overline{M_j(x)}$ is the negate output of the dichotomizer and $m = \sum_{j=1}^{K} M_j(x)$. Notice that in such a case the final decision depends on both $\mathbf{M}(x)$ and $\mathbf{\Psi}(x)$ without considering data related to the degree of imbalance presented in the dataset.

In case of dangerous classification, when an error due to class skew is more likely to occur, we advantage the minority class in order to compensate classifier bias. To this aim, we exploit information contained in **R**, whereas information provided by $\mathbf{M}(x)$ and $\mathbf{\Psi}(x)$ has been used to detected the dangerous situation itself. To set the final decision we consider only two dichotomizers. The first is the dichotomizer which more likely suggests that x belongs to its class, since its output is one and it has the largest reliability among the others providing the same output. The second is the dichotomizer which more likely suggests that x does not belong to class j where the jth dichotomizer is specialized on, but x should belong to class i, with $j \neq i$. Indeed, the output $M_j(x)$ of this dichotomizer is zero and it has the lowest reliability among the others providing the same output (i.e. α_0). The CRR sets the index s as follows:

$$s = \begin{cases} argmax_j(r_j^0, r_j^1) \; if \; \alpha_1 \geq \alpha_0 \\ argmim_j(r_j^0, r_j^1) \; if \; \alpha_1 < \alpha_0 \end{cases} \quad (5)$$

where
$$r_j^i = (r_j | M_j(x) = i \wedge \psi_j = \alpha_i), \text{ with } i = \{0, 1\} \tag{6}$$

To explain the rationale of eq. 5, recall that we are considering dangerous classification where an error due to class skew is more likely to occur. Indeed, when $\alpha_1 \geq \alpha_0$ and taking into account eq. 3, two cases may happen. In the first one, values of both α_0 and α_1 should be below the thresholds τ_0 and τ_1 respectively. In the second one, α_1 is large but the average reliability $\overline{\alpha_0}$ of dichotomizers suggesting that the x does not belong to their class ($M_j(x) = 0$) is below a threshold τ_0. In both cases, it is reasonable to assume that wrong classification is given by the bias in favor of majority class; hence, CRR sets the final label as the class where the dichotomizer with the largest value of the reverse a-priori probability is specialized, i.e. the minority class among those considered.

Ortherwise, when $\alpha_1 < \alpha_0$, CRR sets the final class as the one where the dichotomizer with the lowest reverse a-priori probability is specialized. In this case, considering again that this is a dangerous classification, α_0 is smaller than threshold τ_0 (eq. 3), and α_1 is smaller than τ_1 but it is smaller than α_0 (eq. 5) . This observation implies that reliabilities values are small and noisy, suggesting that there isn't a clear trend among the dichotomizer outputs to set the final decision: hence, the class with the largest a-priori probability should be preferred to set the final classification.

Table 1. Summary of the used datasets. Symbols $*$ and $+$ marks public and private datasets, respectively

					Dataset					
	FER*	GLASS*	ISOLET*	LETTER*	OPTDIGIT*	IIFI*	IIFH2+	SAT*	WFRN*	WINE*
N. of samples	876	205	2000	2561	5620	600	573	6425	5456	178
N. of classes	6	5	26	26	10	3	5	6	4	3
N. of features	50	9	30	16	60	57	159	36	24	13
Majority class	28.1%	37.0%	4.7%	4.5%	10.2 %	36.0%	37.0%	23.9%	40.4%	39.9%
Minority class	7.5%	6.3 %	3.1%	2.9%	9.8%	31.5%	8.2%	9.7%	6.0%	27.0%

4 Experimental Evaluation

The method proposed in this paper has been compared to both a multiclass classifier and a well established OpC reconstruction rule, namely the *Hamming decoding* (HAMDEC) [8]. According to its original definition and assuming that each dichotomizer provides 1 or -1 labels, the index s of the dicothomizer setting the final class $\omega_s \in \Omega$ is given by:
$$s = argmin_i d_H(\mathbf{D}(\omega_i), \mathbf{M}(x)) \tag{7}$$

where
$$d_H(\mathbf{D}(\omega_i), \mathbf{M}(x)) = \sum_{j=1}^{K} \left(\frac{1 - sign(D(\omega_i, j)M_j(x))}{2} \right) \tag{8}$$

The experiments have been carried out by employing three different paradigms for the base dichotomizers: a k-Nearest Neighbor (kNN) as a statistical classifier, a Multi-Layer Perceptron (MLP) as a neural network, and a Support Vector Machine (SVM) as a kernel machine. With reference to multiclass classifier, we have considered the kNN and MLP network. Note that we have not considered multiclass SVM since it is usually implemented using a decomposition approach.

In particular, we used a kNN classifier choosing the k value within the range $\{1, 3, 5, 7\}$ that optimize performance on five validation set. To evaluate the reliability of kNN decisions we adopted a method that estimates the test patterns credibility on the basis of their quality in the feature space [17].

For the MLP, the neural network has a number of hidden layers equal to half of the sum of features number plus class number. The number of neurons is given in the input layer by the number of the features, whereas in the output layer is equal to the number of the classes. As with the kNN classifier, to evaluate the reliability of MLP decisions we adopted the method reported in [17].

In case of SVM we used both a SVMs with a Gaussian Radial Basis Function (RBF) and linear kernels, denoted as SVM_{rbf} and SVM_l, respectively. For SVM_{rbf}, the values of regularization parameter C and scaling factor σ have been selected within $\{1, 10, \ldots, 10^4\}$ and $\{10^{-4}, 10^{-3}, \ldots, 10\}$, respectively. For SVM_l, the values of cost C is selected within $\{2^{\pm 4}, 2^{\pm 3}, 2^{\pm 2}, 2^{\pm 1}, 2^0\}$. Each parameter value is selected according to average performance of classifier on five validation set. The reliability of a SVM classification is estimated as proposed in [18].

In our experiments, we used nine public and one private datasets. These datasets are characterized by a large variability in the number of features, classes and samples, allowing the assessment of the performance in different conditions (Table 1). They are: Facial Expression recognition (FER) [19, 20], Indirect Immunofluorescence Intensity (IIFI) [21], Indirect Immunofluorescence HEp-2 cells staining pattern (IIFH2) [22], Glass Identification (GLASS) [23], ISOLET [23], Letter Recognition (LETTER) [23], Optical Recognition of Handwritten Digits Data Set (OPTDIGIT) [23], Statlog (SAT) [23], Wall Following Robot Navigation Data (WFRN) [23], and Wine (WINE) [23]. For every dataset classification accuracy acc and geometric mean of accuracies g are computed averaging out values obtained performing ten fold cross validation. Each fold is computed maintaining the a-priori distribution of data showed by by original TS.

Results reported in Table 2 permit us to compare performance of CRR-based classification with those achieved both by a multiclass approach and a popular OpC scheme, employing also different classification paradigms.

With regard to the comparison between the CRR approach and the multiclass classification we observe that, generally, the first performs better than the latter both in term of acc and g. Indeed, the kNN with the CRR approach achieves better results than the corresponding multiclass scheme on all the tested datasets. In case of MLP neural network, the CRR scheme outperforms the multiclass classifier in eight cases out of ten datasets. In both cases acc and g values improvements are up to 88%.

Let us now focusing our attention to the comparisons between CRR and HAMDEC reconstruction schemes. With reference to the kNN classification paradigm, we notice that CRR outperforms the HAMDEC scheme in nine out of ten datasets both in terms of acc and g,showing improvements up to 18.0% and 14.0%, respectively. Exploring MLP results in terms of acc, we observe that CRR scheme outperforms HAMDEC in eight case out of ten, with the improvements ranging between from 1.1% up to 8.7%. In case of

Table 2. MLP, kNN and SVM performance (*acc*% and *g*% values) on ten public datasets

		Metrics	FER	GLASS	ISOLET	LETTER	OPTDIGIT	IIFI	IIFH2	SAT	WFRN	WINE
KNN	Multiclass	acc	40.3	51.00	4.3	4.1	10.0	67.2	48.7	86.1	87.5	78.4
		g	38.8	49.11	3.6	3.5	10.0	67.0	25.6	84.2	84.6	78.1
	CRR	acc	81.8	71.1	41.1	83.5	98.0	69.2	63.7	90.3	87.8	94.5
		g	79.1	14.1	3.9	82.4	98.0	69.9	55.7	87.9	87.3	98.0
	HAMDEC	acc	76.4	70.9	33.1	77.2	97.9	67.5	55.0	72.5	84.5	95.9
		g	70.8	0	0	75.2	97.8	65.2	55.7	88.1	83.3	96.5
MLP	Multiclass	acc	86.6	58.4	4.5	4.3	10.0	65.7	52.4	87.9	89.6	71.9
		g	85.8	55.9	0	0	10.1	61.8	28.9	83.9	86.6	71.8
	CRR	acc	73.6	69.0	54.8	77.8	98.3	71.5	62.1	90.8	88.3	97.6
		g	69.9	0	23.8	61.0	98.3	69.4	59.0	88.4	83.5	94.6
	HAMDEC	acc	85.8	66.7	46.1	76.0	97.2	67.2	62.6	89.1	87.5	96.5
		g	79.0	0	7.8	58.8	97.2	63.2	48.7	84.7	84.1	96.1
SVM$_l$	CRR	acc	72.3	44.7	47.0	52.6	88.9	51.6	53.7	61.4	49.1	97.6
		g	45.3	13.8	0	0	88.4	49.4	23.3	44.4	47.7	97.8
	HAMDEC	acc	41.0	42.5	7.7	6.7	43.7	65.8	53.6	49.8	48.2	94.6
		g	0	0	0	0	33.3	62.7	24.1	0	38.8	93.9
SVM$_{rbf}$	CRR	acc	88.1	68.5	24.5	72.9	11.8	55.5	59.5	90.9	65.6	98.8
		g	84.7	7.0	0	56.7	0	67.2	41.8	88.0	88.3	97.2
	HAMDEC	acc	85.2	58.9	9.9	55.7	10.4	56.5	46.1	88.6	59.8	95.9
		g	33.3	80.7	0	40.3	0	61.2	30.4	84.4	87.2	93.3

g values, we observe that CRR has performance improvement up to 16.0% in seven cases out of ten. SVM architecture employing CRR scheme shows larger accuracy than the one adopting HAMDEC scheme, independently of the used kernel. In particular, in nine out of ten datasets SVM$_l$ and SVM$_{rbf}$ improvements are up to 46% and 17%, respectively. In term of *g* and using SVM$_{rbf}$ CRR scheme outperforms HAMDEC in eight out of ten cases with improvements up to 55%. Using SVM$_{rbf}$ improvements concern nine out of ten datasets, with the *g*value raising up to 16%.

Before concluding the paper, let us introduce two further considerations on the results measured in terms of *g*. First, it is worth noting that *g* improvements are large (up to 55%), especially when the base classifier is the SVM$_l$. Second, in several cases where the application of the HAMDEC reconstruction rules provides values of *g* equal to zero, which correspond to misclassifying all samples of one or more minority classes, the CRR rule attains larger and non zero *g* values.

These last observations together with the results previously discussed point out that CRR scheme provides larger relative accuracies among all classes, thus achieving better classifications on classes with few samples and reducing class imbalance effects.

5   Conclusion

In this paper we presented a reconstruction rule for the One-per-Class decomposition framework overcoming issues related to learning under class skew in a multiclass scenario. It applies different criteria to set the final decision on the basis of both the outputs of the dichotomizers and the a-priori probabilities of the classes. The approach has been tested on ten datasets and successfully compared against a multiclass classifier and a well established OpC reconstruction rule.

References

1. Chawla, N.V., et al.: SMOTE: Synthetic minority over-sampling technique. J. of Artificial Intelligence Research 16(3), 321–357 (2002)
2. Soda, P.: A multi-objective optimisation approach for class-imbalance learning. Pattern Recognition 44, 1801–1810 (2011)
3. Alejo, R., et al.: An empirical study for the multi-class imbalance problem with neural networks. Progress in Pattern Recog., Image Analysis and Applications, 479–486 (2008)
4. Fernández, et al.: Multi-class imbalanced data-sets with linguistic fuzzy rule based classification systems based on pairwise learning. Comp. Intel. for Knowledge-Based Systems Design, 89–98 (2010)
5. Martínez-Estudillo, F.J., et al.: Evolutionary learning by a sensitivity-accuracy approach for multi-class problems. In: IEEE WCCI, pp. 1581–1588 (2008)
6. Soda, P., et al.: Decomposition methods and learning approaches for imbalanced dataset: An experimental integration. In: 20th Int. Conf. on Pattern Recognition, pp. 3117–3120 (2010)
7. Wu, M., et al.: A small sphere and large margin approach for novelty detection using training data with outliers. IEEE Tran. on Pattern Anal. and Machine Intel. 31(11), 2088–2092 (2009)
8. Allwein, E.L., et al.: Reducing multiclass to binary: a unifying approach for margin classifiers. J. of Machine Learning Research 1, 113–141 (2001)
9. Dietterich, T.G., et al.: Solving multiclass learning problems via error-correcting output codes. J. of Artificial Intelligence Research 2, 263 (1995)
10. Fürnkranz, J.: Round robin classification. JMLR 2, 721–747 (2002)
11. Masulli, F., et al.: Comparing decomposition methods for classication. In: 4th Intl. Conf. on Knowledge-Based Intel. Engineering Systems & Allied Technologies, pp. 788–791 (2000)
12. Rajan, S., et al.: An Empirical Comparison of Hierarchical vs. Two-Level Approaches to Multiclass Problems. In: Roli, F., Kittler, J., Windeatt, T. (eds.) MCS 2004. LNCS, vol. 3077, pp. 283–292. Springer, Heidelberg (2004)
13. Dietterich, T., et al.: Solving multiclass learning problem via error-correcting output codes. J. of Artificial Intel. Research 2, 263–286 (1995)
14. Mayoraz, E., et al.: On the decomposition of polychotomies into dichotomies. In: Proc. of the 14th Int. Conf. on Machine Learning, pp. 219–226 (1997)
15. Iannello, G., et al.: On the use of classification reliability for improving performance of the one-per-class decomposition method. Data & Knowledge Eng. 68, 1398–1410 (2009)
16. Foggia, P., et al.: On Rejecting Unreliably Classified Patterns. In: Haindl, M., Kittler, J., Roli, F. (eds.) MCS 2007. LNCS, vol. 4472, pp. 282–291. Springer, Heidelberg (2007)
17. Cordella, L.P., et al.: Reliability parameters to improve combination strategies in multi-expert systems. Pattern Analysis & Applications 2(3), 205–214 (1999)
18. Platt, J.: Probabilistic output for support vector machines and comparisons to regularize likelihood methods. Advanced in Large Margin Classifiers. MIT Press (2000)
19. Kanade, T., et al.: Comprehensive database for facial expression analysis. In: 4h IEEE Int. Conf. on Automatic Face and Gesture Recognition, pp. 46–53 (2000)
20. D'Ambrosio, R., et al.: Automatic Facial Expression Recognition Using Statistical-Like Moments. In: Maino, G., Foresti, G.L. (eds.) ICIAP 2011, Part I. LNCS, vol. 6978, pp. 585–594. Springer, Heidelberg (2011)
21. Soda, P., et al.: A multiple experts system for classifying fluorescence intensity in antinuclear autoantibodies analysis. Pattern Analysis & Applications 12(3), 215–226 (2009)
22. Soda, P., et al.: Aggregation of classifiers for staining pattern recognition in antinuclear autoantibodies analysis. IEEE T Inf. Technol. B 13(3), 322–329 (2009)
23. Frank, A., et al.: UCI machine learning repository (2010)

The Dipping Phenomenon

Marco Loog and Robert P.W. Duin

Pattern Recognition Laboratory
Delft University of Technology
Delft, The Netherlands
prlab.tudelft.nl

Abstract. One typically expects classifiers to demonstrate improved
performance with increasing training set sizes or at least to obtain their
best performance in case one has an infinite number of training samples at ones's disposal. We demonstrate, however, that there are classification problems on which particular classifiers attain their optimum
performance at a training set size which is finite. Whether or not this
phenomenon, which we term dipping, can be observed depends on the
choice of classifier in relation to the underlying class distributions. We
give some simple examples, for a few classifiers, that illustrate how the
dipping phenomenon can occur. Additionally, we speculate about what
generally is needed for dipping to emerge. What is clear is that this kind
of learning curve behavior does not emerge due to mere chance and that
the pattern recognition practitioner ought to take note of it.

1 On Learning Curves and Peaking

The analysis of learning curves, which describe how a classifier's error rate
behaves under different training set sizes, is an integral part of almost any
proper investigation into novel classification techniques or unexplored classification problems [7]. Though sometimes interest goes only to its asymptotics [9],
the learning curve is especially informative in the comparison of two or more
classifiers when considering the whole range of training set sizes. It indicates
at what samples sizes the one classifier may be preferable over the other for a
particular type of problem. Also, by means of extrapolation, the curve may give
us some clue on how many additional samples may be needed in a real-world
problem to reach a particular error rate. Such analyses are readily impossible
on the basis of a point estimate as, for example, obtained by means of leave one
out cross-validation on the whole data set at hand.

The learning curve one typically expects to observe falls off monotonically with
increasing training set size (see Figure 1). The rate of decrease depends on the
particular problem considered and the complexity of the classifier employed. Such
behavior can indeed be demonstrated in certain settings in which the classifier
selected typically fits the underlying data assumptions well, see for instance
[1,10]. In a similar spirit, various bounds on learning curves also show monotonic
decrease for the expected true error rate with increasing training set sizes [5,16].

G.L. Gimel' farb et al. (Eds.): SSPR & SPR 2012, LNCS 7626, pp. 310–317, 2012.
© Springer-Verlag Berlin Heidelberg 2012

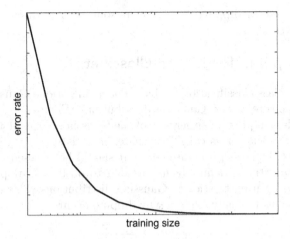

Fig. 1. An idealized learning curve in which the error rate drops monotonically with an increasing training set size

That such monotonic behavior can, however, not always be guaranteed has already been known at least since the mid nineties. Both Opper and Kinzel [12] and Duin [4] describe what is nowadays referred to in pattern recognition as the peaking phenomenon for learning curves: the error rate attains a local maximum that does not coincide with the smallest training sample size considered. This phenomenon has been described and investigated, for instance, for the Fisher discriminant classifier [4,13,14], for particular perceptron rules [12,11], and for lasso regression [8]. The naming of this phenomenon alludes to the peaking phenomenon for increasing feature sizes (as opposed to increasing training set sizes, which this paper is concerned with) as originally identified by Hughes [6] in the 1960s. Hughes' phenomenon for such feature curves shows that, for a fixed training sample size, the error initially drops but beyond a certain dimensionality typically starts to rise again.

On the basis of what we know about peaking, we may adjust our expectation about learning curves and speculate classifiers to at least obtain their best performance when an infinite number of training samples is used. But also this turns out to be false hope as this work demonstrates. It appears there are classification problems on which particular classifiers attain their optimal performance at a training set size which is finite. In contrast with peaking, we term this phenomenon dipping as it concerns a minimum in the learning curve, in fact, a non-asymptotic, global minimum.

The next three section of the paper, Sections 2, 3, and 4, give some simple examples, for three artificial classification problems in combination with specific classifiers, which demonstrate how the dipping phenomenon emerges. Though artificial, the examples clearly illustrate that this kind of learning curve behavior does not merely emerge due to chance, e.g. due some unfortunate draw of training data, but that it is an issue structurally present in particular problem-classifier combinations. The final section, Section 5, speculates on what generally is needed for dipping. It also offers some further discussions and concludes this contribution.

2   Basic Dipping for Linear Classifiers

Consider a two-class classification problem consisting of one Gaussian distribution and one mixture of two Gaussian distributions (Figure 2). The Gaussians of the second class appear on either side of the Gaussian of the first class. A perfectly symmetric situation is considered here: there is symmetry in the overall distribution and the class priors are equal. It should be stressed, however, that this perfect symmetry is definitely not needed to observe a dipping behavior, just like there is no need to stick to Gaussian distributions. This configuration, however, enables us to easily explain why dipping occurs.

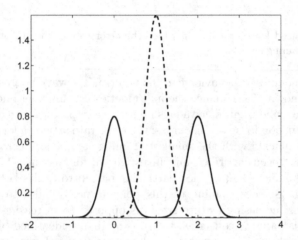

Fig. 2. Distribution of two-class data used to illustrate basic dipping

Let us consider what happens when we would make an expected learning curve for the nearest mean classifier (NMC, [3]). In the case of large total training set sizes, both means will be virtually on top of each other and the expected classification error will reach a worst case performance of 0.5. If, however, we go to smaller and smaller sample sizes, these means will in expectation be further and further apart due to their difference in variance. In the extreme case in which we have one observation from both classes, the one mean will be around the mode of class one and the other will be near one of the two modes of class two. Though one will still have means that lead to an error rate of about 0.5, chances are very slim. There will, however, be many configurations that both classify the first class and one lobe of the second class more or less correctly, which gives an expected error of around 0.25 as only the second lobe of the second class gets misclassified.

In conclusion, the smaller the sample size is the higher the probability is that the NMC delivers a performance considerably better than chance. Figure 3 gives

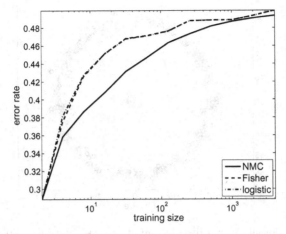

Fig. 3. Dipping learning curves for three linear classifiers, viz., NMC, Fisher discriminant, and logistic regression, based on the one-dimensional distribution presented in Figure 2

the expected learning curve (an average over 1000 repetitions) for training set sizes ranging from 2^1 to 2^{12} (compare to Figure 1). The same figure displays learning curves for the Fisher discriminant and logistic regression as well. Both linear classifiers also suffer from dipping and an explanation for this goes along the same lines as for the NMC.

3   Delayed Dipping

The following example demonstrates that the occurrence of the dip can be at any point along the learning curve. Let us again consider the NMC but now the classification problem changes to the one illustrated in Figure 4. The first class is a Gaussian distribution and the second class is a noisy ring positioned around the first class with a variable radius. Again the priors are taken equal.

When the radius of the ring is small, we are basically back in a situation similar to the one in Section 2 and one would observe dipping as in Figure 3. The more training samples one would have, the closer the two means would get. Though this is bad in case the ring is near the center class, when the ring grows larger and larger, while the noise level stays the same, more observations in fact lead to improved performance up to a certain level. Having one observation per class would mean that the larger part of the ring is going to be misclassified to the center class. Increasing the total training set size, however, moves the mean of the second class closer and closer to the mean of the first class. As long as the second class mean does not move into the region where the first class becomes dense, moving closer to the center will lead to a better classification of class two and therefore a better overall performance. When the ring grows infinitely large, the two means can be virtually the same (relative to the size of the ring) while the first class is classified nearly perfect and as good as half of the ring is correctly classified. This happens when the training set grows infinite as well.

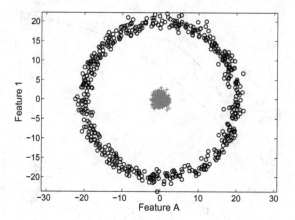

Fig. 4. Single instantiation of a distribution of two-class data used to illustrate early and late dipping for the NMC. The outer ring can vary in diameter based on which the time of dipping can be controlled.

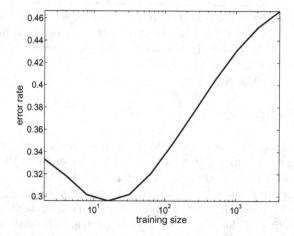

Fig. 5. Learning curve for the NMC based on the two-class distribution in Figure 4. It shows the dipping phenomenon to occur away from the smallest sample sizes.

In conclusion, by means of the variable ring diameter, one can tune the occurrence of the dip for the NMC to an arbitrary position along the learning curve. Figure 5 gives a learning curve that dips at a training sample of 16, which is obtained for a radius of 20 with a Gaussian standard deviation and a ring noise level standard deviation of 1.

4   Dipping of QDA

Our final example shows that dipping is not limited to linear discriminants but may also be encountered when employing more flexible classifiers. Here we consider classical quadratic discriminant analysis (QDA, [10]). Figure 6 shows the class configurations used—a variation to the one from Section 2. Figure 7 displays the learning curve obtained by QDA[1]. A reason rather similar to the one given in Section 2 can be given for the observed dipping, though it is slightly more involved because of the more complex classifier considered. Here we merely note that in case of large sample sizes the decision boundary is close to the middle and the error rate gets close to the worst case solution, which is slightly less than 0.5. For smaller sample sizes the decision boundary shifts away from the middle, which on average leads to an improvement in classification error as can be observed in the learning curve from Figure 7.

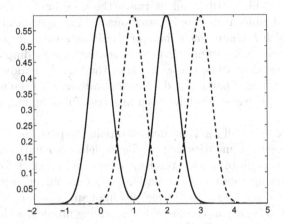

Fig. 6. Distribution of two class data used to let QDA dip

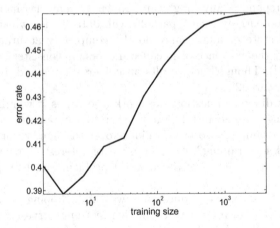

Fig. 7. Learning curve for QDA related to the two-class distribution in Figure 6, illustrating that dipping is not limited to the simplest of classifiers.

---

[1] As the per class sample sizes sometimes equals one, the covariance matrices in QDA were moderately regularized in this experiment.

5   Discussion and Conclusion

For four different classifiers we have demonstrated that the dipping phenomenon can be observed. We explained why it emerges in a basic setting using linear classifiers, sketched how the dipping point can attain an arbitrary location along the learning curve, and illustrated the possibility that also discriminants more complicated than linear can show dipping behavior. What seems to be a essential requirement is that the model underlying the classifier does not suit the classification problem considered very well. Curves similar to those on Figure 2 can be generated for linear discriminant analysis (LDA), the perceptron, or the linear support vector machine. All in all, it raises the question at what complexity classifiers will not suffer from dipping any longer. More specifically: can we find problems for which k nearest neighbors or the Parzen classifier show this type of behavior? Or are nonparametric techniques immune to dipping? Certainly for the Parzen classifier, when one would keep the kernel's bandwidth fixed, we would not be surprised if particular data configurations will even make this classifier dip. To date, however, we have been unsuccessful in finding an illustration of such behavior.

It may even be that still less is needed for dipping to potentially happen. Even if the type of decision boundaries that can be modeled by a particular classifier is in principle rich enough to include the Bayes decision boundary for the problem at hand, the learning routine or estimation procedure might be unable to find the correct fit. An example is the Fisher discriminant, which is not always able to separate linearly separable classes. The underlying problem is that we want to minimize the expected classification error but in reality we always have to settle for a surrogate loss that is all but a bad approximation to the 0-1 loss. Maybe due to this discrepancy, "anything" can happen: for any classifier one might be able to find a, potentially rather pathological, data set for which the classifier dips. That this state of affairs may not be completely accurate is, however, demonstrated by the existence of so-called universally consistent classifiers (see, for instance, [15]). Though such results on universality should, in turn, also be interpreted with care [2].

A completely different question this work also raises is whether one should treat the training set size just like any other free parameter a classifier has. Should one, for example, also cross-validate over the number of training samples to be used for training? Another issue of interest is whether the phenomenon can be observed in any real-world problem and how it affects such setting.

Irrespective of the previous questions, we think dipping is a phenomenon that one should keep in mind when studying learning curves. When observed, it may not be ascribed blindly to chance or a bad training sample. It might just be inherent in the combination of problem at hand and classifier employed.

References

1. Amari, S., Fujita, N., Shinomoto, S.: Four types of learning curves. Neural Computation 4(4), 605–618 (1992)
2. Ben-David, S., Srebro, N., Urner, R.: Universal learning vs. no free lunch results. In: Philosophy and Machine Learning Workshop NIPS 2011 (December 2011), http://www.dsi.unive.it/PhiMaLe2011/
3. Duda, R., Hart, P.: Pattern classification and scene analysis. John Wiley & Sons (1973)
4. Duin, R.: Small sample size generalization. In: Proceedings of the Scandinavian Conference on Image Analysis, vol. 2, pp. 957–964 (1995)
5. Haussler, D., Kearns, M., Seung, H., Tishby, N.: Rigorous learning curve bounds from statistical mechanics. Machine Learning 25(2), 195–236 (1996)
6. Hughes, G.: On the mean accuracy of statistical pattern recognizers. IEEE Transactions on Information Theory 14(1), 55–63 (1968)
7. Jain, A., Duin, R., Mao, J.: Statistical pattern recognition: A review. IEEE Transactions on Pattern Analysis and Machine Intelligence 22(1), 4–37 (2000)
8. Krämer, N.: On the peaking phenomenon of the lasso in model selection. Arxiv preprint arXiv:0904.4416 (2009)
9. Langley, P.: Machine learning as an experimental science. Machine Learning 3(1), 5–8 (1988)
10. McLachlan, G.: Discriminant Analysis and Statistical Pattern Recognition. John Wiley & Sons (1992)
11. Opper, M.: Learning to generalize. In: Frontiers of Life, vol. 3(part 2), pp. 763–775. Academic Press (2001)
12. Opper, M., Kinzel, W.: Statistical mechanics of generalization. In: Models of Neural Networks III, ch. 5. Springer (1995)
13. Raudys, S., Duin, R.: Expected classification error of the fisher linear classifier with pseudo-inverse covariance matrix. Pattern Recognition Letters 19(5), 385–392 (1998)
14. Skurichina, M., Duin, R.: Stabilizing classifiers for very small sample sizes. In: Proceedings of the 13th International Conference on Pattern Recognition, vol. 2, pp. 891–896. IEEE (1996)
15. Steinwart, I.: Consistency of support vector machines and other regularized kernel classifiers. IEEE Transactions on Information Theory 51(1), 128–142 (2005)
16. Vapnik, V.: Estimation of dependences based on empirical data. Springer (1982)

Colour Matching Function Learning

Luis Romero-Ortega[1] and Antonio Robles-Kelly[1,2,3]

[1] School of Eng. and Inf. Tech., UNSW@ADFA, Canberra ACT 2600, Australia
[2] NICTA *, Locked Bag 8001, Canberra ACT 2601, Australia
[3] Research School of Eng., ANU, Canberra ACT 0200, Australia

Abstract. In this paper, we aim at learning the colour matching functions making use of hyperspectral and trichromatic imagery. The method presented here is quite general in nature, being data driven and devoid of constrained setups. Here, we adopt a probabilistic formulation so as to recover the colour matching functions directly from trichromatic and hyperspectral pixel pairs. To do this, we derive a log-likelihood function which is governed by both, the spectra-to-colour equivalence and a generative model for the colour matching functions. Cast into a probabilistic setting, we employ the EM algorithm for purposes of *maximum a posteriori* inference, where the M-step is effected making use of Levenberg-Marquardt optimisation. We present results on real-world data and provide a quantitative analysis based upon a colour calibration chart.

1 Introduction

The accurate capture and reproduction of colours as acquired by digital camera sensors is an active area of research. This is not a straightforward task since digital cameras are comprised by three spectral broad-band color sensors which are not colorimetric. This implies that the RGB values yielded by the camera are not a linear combination of the device-independent CIE color matching functions [1]. Further, colours, as acquired by digital cameras, are, in general, device dependent.

Whereas colorimetry focuses on the accuracy of the colours acquired by the camera, spectroscopy has as object of study the spectrum of light absorbed, transmitted, reflected or emitted by objects and illuminants in the scene. In contrast with trichromatic sensors, multispectral and hyperspectral sensing devices can acquire wavelength-indexed reflectance and radiance data in tens of hundreds of bands across a broad spectral range. Recently, there has been renewed interest in multispectral imaging as related to color constancy [2], the analysis of the spectral properties of objects in the scene [3] and the optimal multiplexing of bandpass filtered illumination [4,5].

Moreover, making use of photogrammetry and spectroscopy techniques based upon monochromatic narrow-band illuminants, it is possible to recover the spectral response of the camera under study [6]. Methods which employ calibration

---

* NICTA is funded by the Australian Government as represented by the Department of Broadband, Communications and the Digital Economy and the Australian Research Council through the ICT Centre of Excellence program.

G.L. Gimel' farb et al. (Eds.): SSPR & SPR 2012, LNCS 7626, pp. 318–326, 2012.

targets and charts have also been proposed. These methods employ quadratic programming [7], monochromators [8] or spectrophotometers [9].

The methods above require calibration charts and, in many cases, complicated setups and constrained environments. Here we present a method which employs two sets of images, one trichromatic and another one hyperspectral so as to learn the colour matching functions. The method presented here hinges in an inference process based upon a maximum-likelihood formulation which leads to the application of the EM (Expectation-Maximisation) algorithm. In the following section, we provide some background on the relationship between the colour output of the camera, the colour matching functions and the spectral image radiance.

2   Background

To better understand the relation between the spectra in hyperspectral imagery and the color output of trichromatic cameras, we commence by providing some background on the expression of the image radiance at pixel v as given in [10]. Let the image radiance be given by

$$I(\lambda, v) = L(\lambda)P(\varphi_i, \phi_i, \varphi_s, \phi_s)S(v, \lambda) \tag{1}$$

where $P(\cdot)$ is the mean scattered power in the direction (φ_s, ϕ_s), $L(\lambda)$ is the power spectrum of the light impinging on the object surface in the direction (φ_i, ϕ_i) and $S(v, \lambda)$ is the surface reflectance at wavelength λ.

The expression above is important since it allows the use of the shorthand $R(\lambda, v) = P(\varphi_i, \phi_i, \varphi_s, \phi_s)S(v, \lambda)$ so as to write the image radiance as follows

$$I(\lambda, v) = L(\lambda)R(\lambda, v) \tag{2}$$

This expression has been used widely in the literature [11] and is consistent with reflectance models in the computer vision literature, such as that in [12].

Recall that in a trichromatic camera, a fraction of the light incident on the surface of the object being observed is reflected towards the camera. The light then pases through the camera lens, which focuses the incoming light beam onto the image plane of the camera. After reaching the image plane, the colour channel values for each pixel in the image are determined by the responses to the incoming light of the R, G, B receptors of the camera.

This is important since it permits us to consider two sample-sets. We denote the first of these, which corresponds to trichromatic pixels, as $\mathcal{A}^{RGB}$. The second sample-set, $\mathcal{A}^{HS}$, corresponds to the spectra at hyperspectral pixels. If the trichromatic pixel u is a match to the hyperspectral pixel v, i.e. $u \sim v$, then, using the notation above we can write

$$I_k(u) = \kappa_k \sum_{\lambda \in \Lambda} Q_k(\lambda)L(\lambda)R(\lambda, v) \tag{3}$$

where κ_k is a constant that depends on the sensor geometry, $I_k(u)$ is the colour value for the channel $k = \{R, G, B\}$, Λ is the visible spectral range, i.e. $\Lambda = [400nm, 700nm]$, and the colour matching functions for the three colour channels are denoted by $Q_k(\lambda)$, $k = \{R, G, B\}$.

Here, we note that the illuminant power spectrum may be recovered making use of methods elsewhere in the literature such as that in [13]. We can assume it to be in hand. Also, in the following section, we assume that the matches between pixels are available. We will ellaborate further on this in Section 4.

3 Maximum Likelihood Formulation

Note that the treatment above permits us to view each pixel value as a product of a per-pixel, per-wavelength factor which applies equally to all the pixels in the sample sets $\mathcal{A}^{RGB}$ and $\mathcal{A}^{HS}$. This in turn allows a statistical treatment of the problem. In this section, we cast the problem into an Expectation-Maximization setting.

3.1 Log-Likelihood Function

The idea underpinning the EM algorithm is to recover maximum likelihood solutions to problems involving missing or hidden data. To do this, we view the colour matching functions as a set of hidden variables to be estimated. Thus, we cast the problem as a *maximum a posteriori* (MAP) one which aims at maximizing the probability of the colour matching function given the input trichromatic and hyperspectral image pixels. This can be expressed as follows

$$P(Q_k(\lambda) \mid \Omega, \Theta_k) = P(Q_k(\lambda) \mid \Omega)P(Q_k(\lambda) \mid \Theta_k)$$

where Ω is the set of matching spectra-colour pixel tuples and Θ_k is the set of hyperparameters for the colour matching function $Q_k(\lambda)$.

Note that, in the expression above, the first term of the right-hand side is the conditional probability governed by the hyperspectral image radiance value and the corresponding colour value. Thus, the maximization of the probability $P(Q_k(\lambda) \mid \Omega)$ implies that the colour matching function $Q_k(\lambda)$ should satisfy the relationship between the spectral radiance and the trichromatic colour values. The second term accounts for the dependency of the colour matching function $Q_k(\lambda)$ upon the hyperparameters in Θ_k. Note that these hyperparameters can be viewed as a means to enforcing a cumulative distribution function in a manner akin to histogram equalization methods [14].

We can take our analysis further by considering a probability distribution function for $P(Q_k(\lambda) \mid \Omega)$ of the form

$$P(Q_k(\cdot) \mid \Omega) = \frac{1}{\gamma_k \sqrt{2\pi}} \prod_{\substack{u \in \mathcal{A}^{RGB} \\ v \in \mathcal{A}^{HS} \\ u \sim v}} \exp\left\{ -\frac{1}{2\gamma_k^2} \left| I_k(u) - \kappa_k \sum_{\lambda \in \Lambda} Q_k(\lambda)I(\lambda, v) \right|^2 \right\} \quad (4)$$

where γ_k is the variance variable and the second term in the argument of the exponential function arises from Equations 2 and 3.

In a similar fashion, we can consider the colour matching function values to be distributed in accordance to a mixture of N Gaussians. This is a good approximation for the colour matching functions used in practice by camera manufacturers [15]. As a result, we can write

$$P(Q_k(\cdot) \mid \Theta_k) = \prod_{\lambda \in \Lambda} \sum_{q=1}^{N} \alpha_{q,k} \frac{1}{\sigma_{q,k}\sqrt{2\pi}} \exp\left\{ -\frac{(\mu_{q,k} - \lambda)^2}{2\sigma_{q,k}^2} \right\}$$

where, $\Theta_k = \{\theta_{q,k}\}_{q=1}^{N}$ and, as usual, $\theta_{q,k} = \{\mu_{q,k}, \sigma_{q,k}\}$ are the mean and the covariance for the kth colour response and the qth Gaussian in the mixture and $\alpha_{q,k}$ is the mixture weight.

With these ingredients, the log-likelihood function becomes

$$\mathcal{L}(Q_k(\cdot) \mid \Omega, \Theta_k) = -\frac{1}{2\gamma_k^2} \sum_{\substack{u \in \mathcal{A}^{RGB} \\ v \in \mathcal{A}^{HS} \\ u \sim v}} \left| I_k(u) - \kappa_k \sum_{\lambda \in \Lambda} Q_k(\lambda) I(\lambda, v) \right|^2 +$$

$$\sum_{\lambda \in \Lambda} \log\left\{ \sum_{q=1}^{N} \alpha_{q,k} \frac{1}{\sigma_{q,k}\sqrt{2\pi}} \exp\left\{ -\frac{(\mu_{q,k} - \lambda)^2}{2\sigma_{q,k}^2} \right\} \right\} \tag{5}$$

where we have removed the term $\log\left\{ \frac{1}{\gamma_k\sqrt{2\pi}} \right\}$ from further consideration since it does not depend on the colour matching function $Q_k(\cdot)$ or the hyperparameter set and, hence, does not affect the inference process.

3.2 Expectation-Maximization

Note that, in the equation above, only the last term in the right-hand side depends on the hyperparameter-set Θ_k. This is an important observation, since it suggests an iterative update scheme in which the hyperparameters and the colour matching functions $Q_k(\lambda)$ can be recovered using the EM algorithm [16], which we describe in the following.

Expected Log-Likelihood Function. In Equation 5, the two terms on the right-hand side are log-likelihoods in their own right. That is, the first of these is the log-likelihood of $Q_k(\lambda)$ given the colour responses and spectral values. The second term corresponds to the likelihood of $Q_k(\lambda)$ given the hyperparameters Θ_k.

Thus, we can index the expected log-likelihood to iteration number n and write

$$\mathcal{Q}(Q_k^{n+1}(\cdot) \mid \Omega, \Theta_k^n) = -\tau \sum_{\substack{u \in \mathcal{A}^{RGB} \\ v \in \mathcal{A}^{HS} \\ u \sim v}} \left| I_k(u) - \kappa_k \sum_{\lambda \in \Lambda} Q_k^{n+1}(\cdot) I(\lambda, v) \right|^2 +$$

$$\sum_{\lambda \in \Lambda} \log\left\{ \sum_{q=1}^{N} \alpha_{q,k}^n \frac{1}{\sigma_{q,k}^n \sqrt{2\pi}} \exp\left\{ -\frac{(\mu_{q,k}^n - \lambda)^2}{2(\sigma_{q,k}^n)^2} \right\} \right\} \tag{6}$$

where we have used the shorthand $\tau = \frac{1}{2\gamma_k^2}$.

M-Step. In the M-step, we aim at maximising the expected log-likelihood with respect to the colour matching functions. Note that the maximisation of the log-likelihood can also be cast as a minimisation of the form

$$
\operatorname*{argmin}_{Q_k^{n+1}(\cdot)} \left\{ \tau \sum_{\substack{u \in \mathcal{A}^{RGB} \\ v \in \mathcal{A}^{HS} \\ u \sim v}} \left| I_k(u) - \kappa_k \sum_{\lambda \in \Lambda} Q_k^{n+1}(\cdot) I(\lambda, v) \right|^2 - \right.
$$
$$
\left. \sum_{\lambda \in \Lambda} \log \left\{ \sum_{q=1}^{N} \alpha_{q,k}^n \frac{1}{\sigma_{q,k}^n \sqrt{2\pi}} \exp \left\{ -\frac{\left(\mu_{q,k}^n - \lambda\right)^2}{2(\sigma_{q,k}^n)^2} \right\} \right\} \right\}
\tag{7}
$$

This observation is important since it allows the M-step to be viewed as a regularised nonlinear least-squares minimsation which can be tackled using the Levenberg-Marquardt algorithm (LMA), which is an iterative trust region procedure [17] aimed at recovering a numerical solution to the problem of minimising a function over a space of parameters.

E-Step. To estimate the hyperparameter set $\theta_{q,k}^{n+1}$, we introduce the posterior probability

$$
P(\theta_{q,k}^n \mid Q_k^{n+1}(\lambda)) = \frac{P(\theta_{q,k}^n \mid Q_k^{n+1}(\lambda)) P(Q_k^{n+1}(\lambda))}{P(\theta_{q,k}^n)}
\tag{8}
$$

so as to write the gradient of the log-likelihood function as follows

$$
\nabla_{\Theta_k^n} \mathcal{L}(Q_k^{n+1}(\cdot) \mid \Theta_k^n) = \sum_{\lambda \in \Lambda} P(\Theta_k^n \mid Q_k^{n+1}(\lambda)) \nabla_{\Theta_k^n} \log \left\{ P(\Theta_k^n \mid Q_k^{n+1}(\lambda)) \right\}
\tag{9}
$$

Recall that the maximum likelihood corresponds to the values of Θ_k^{n+1} for which $\nabla_{\Theta_k^{n+1}} \mathcal{L}(Q_k^{n+1}(\cdot) \mid \Theta_k^n) = 0$. Since we have assumed $P(Q_k^{n+1}(\lambda) \mid \Theta_k^n)$ to be a mixture of Gaussians, we can recover the maximum likelihood estimates of Θ_k^n by differentiating Equation 5 with respect to Θ_k^n, substitute the results into Equation 9 and solve $\nabla_{\Theta_k^n} \mathcal{L}(Q_k^{n+1}(\lambda) \mid \Theta_k^n) = 0$.

This is a well-known estimation problem [18], which after some algebraic manipulation, yields the following update rules

$$
\mu_{q,k}^{n+1} = \frac{\sum_{\lambda \in \Lambda} \lambda \, h_{q,k}^n(\lambda)}{\sum_{\lambda \in \Lambda} h_{q,k}^n(\lambda)}
$$

$$
\sigma_{q,k}^{n+1} = \frac{\sum_{\lambda \in \Lambda} \left(\mu_{q,k}^{n+1} - \lambda\right)^2 h_{q,k}^n(\lambda)}{\sum_{\lambda \in \Lambda} h_{q,k}^n(\lambda)}
\tag{10}
$$

$$
\alpha_{q,k}^{n+1} = \frac{1}{|\Lambda|} \sum_{\lambda \in \Lambda} h_{q,k}^n(\lambda)
\tag{11}
$$

where $P(Q_k^{n+1}(\lambda) \mid \theta_{q,k}^n)$ can be computed, in a straightforward manner, making use of Equation 5 and

$$h_{q,k}^n(\lambda) = \frac{P(Q_k^{n+1}(\lambda) \mid \theta_{q,k}^n)}{\sum_{r=1}^N P(Q_k^{n+1}(\lambda) \mid \theta_{r,k}^n)} \tag{12}$$

where we have followed [19] and set

$$P(Q_k^{n+1}(\lambda) \mid \theta_{q,k}^n) = \frac{1}{\Psi} \sum_{q=1}^N \alpha_{q,k}^n \frac{Q_k^{n+1}(\lambda)}{\sigma_{q,k}^n \sqrt{2\pi}} \exp\left\{ -\frac{(\mu_{q,k}^n - \lambda)^2}{2(\sigma_{q,k}^n)^2} \right\}$$

and Ψ is a normalisation constant.

4   Implementation Issues

Having presented the theoretical background of our approach, we now turn to the implementation of the method. Note that our method is not limited to a particular number of bands and applies equally to each of the three colour channels. Moreover, so far, we have assumed that the colour-spectra matches, i.e. $I_k(u)$ and $I(\cdot, v)$ with $u \sim v$ are available. In practice, this is not the case. Moreover, acquiring spectro-colourimetric image pairs is impractical in many cases due the error that may be introduced by registering the two views, i.e. that captured by a trichromatic camera and that acquired using the hyperspectral imager.

Thus, we opt for a discriminative approach based upon two code books [20] with sufficient amount of samples so as to have statistical relevance. We commence by building two pixel sets. The first of these from images taken using the colour camera for which we aim at learning the colour matching functions and the other one from imagery captured using the hyperspectral imager. Then, we build the two codebooks, one for each of these pixel-sets. We do this making use of k-means clustering [18].

The codebook for the hyperspectral sample is then converted into RGB values making use of the current estimate of the colour matching functions. This permits matches between the two codebooks to be recovered making use of nearest-neighbours in the RGB chromaticity colour space. These codebooks are, hence, used as an alternative to $\mathcal{A}^{RGB}$ and $\mathcal{A}^{HS}$, where the elements indexed u and v in the respective sets are a match to each other if they are the nearest neighbour to one another.

5   Experiments

Here we show results on real-world multispectral and trichromatic imagery. To this end, we have used two trichromatic commercial cameras, i.e. Nikon D80 and Nikon D5100, so as to acquire 210 pictures, 105 images with each camera. For each of the color cameras, we have used approximately 5.2 Mega pixels sampled

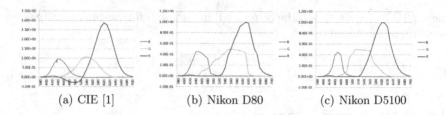

| (a) CIE [1] | (b) Nikon D80 | (c) Nikon D5100 |

Fig. 1. Colour matching functions for the CIE 1955 standard and those recovered by our algorithm for the Nikon D80 and Nikon D5100 camera models

Fig. 2. From left-to-right: imagery rendered with the CIE colour matching functions and those recovered by our algorithm for the Nikon D80 and Nikon D5100 camera models

over the 105 images in a grid like fashion. All our trichromatic imagery has been acquired in raw output mode with manual exposure calibration.

For our hyperspectral data, we have used 1 Mpixel taken from 131 images. Similarly to our trichromatic data, these have been sampled in a grid like fashion with 7.6 Kpixels selected from each image using 16×16 pixel tiles. All our hyperspectral imagery was acquired with a Liquid Crystal Tunable Filter (LCTF) at 2 Mpixel resolution and 33 spectral bands in the visible range in $10nm$ steps. In all our experiments, the number of mixtures N is set to two and initialised the colour matching functions and hyperparameters making use of the CIE1955 standard [1]. Here we have used $\tau = 1$ and iterated until the L_2-norm between the hyperparameter set Θ^n and Θ^{n+1} is below a user-provided threshold, which we set to $1e^{-15}$. In average, the algorithm converged in 10 iterations.

In Figure 1, we show the colour matching functions for the CIE 1955 standard [1] and those recovered using our approach. Note that the CIE standard does contain negative values, whereas the colour matching functions for commercial cameras, by definition, should be positive. In Figure 2, we show example results for one of our input images. In the figure, we show the image rendered with the CIE colour matching functions and those recovered by our algorithm. Note the differences between the images. In particular, with respect to the CIE colour matching functions.

To provide a more quantitative result, we have acquired imagery, hyperspectral and trichromatic, for the XRite Color Checker chart. This is a tiled

Table 1. Mean and standard deviation for the error when using the CIE color matching functions and those learnt by our method

	Nikon D80	Nikon D5100		Nikon D80	Nikon D5100
CIE Matching Functions [1]	0.048 ± 0.025	0.040 ± 0.019	Our Method	0.032 ± 0.017	0.036 ± 0.0215

colorimetric calibration board containing 12 colours and 6 shades of gray. We have rendered the hyperspectral image of the colour chequer with the CIE colour matching functions and those learnt by our method. Once the RGB images are generated, we compare the colours on the chart with those on the imagery acquired with the trichromatic cameras. To do this, we have performed white balancing using the shades of gray and computed the mean-squared differences between the trichromatic imagery and that yielded by the colour matching functions applied to the hyperspectral image.

In Table 1, we show the mean and standard deviation for the CIE colour matching functions and those learnt by our method. In the table, the mean and standard deviation have been normalised to be between zero and unity. This is so as to allow scale variations between the two sets of colour matching functions. This also permits comparison with colour difference measures often used in colorimetry. Note that our learning method outperforms the CIE color matching functions for both cameras.

6   Conclusions

In this paper, we have introduced an approach aimed at learning the colour matching functions from hyperspectral and trichromatic imagery. We do this based upon a probabilistic formulation where the EM algorithm is employed so as to recover the colour matching functions directly from trichromatic and hyperspectral pixel pairs. We have derived a log-likelihood function which is governed by both, the accordance of the spectra-to-colour equivalence and a generative model for the colour matching functions. The method is quite general in nature, being data driven and devoid of constrained setups. Our results on real-world data show that our method is capable of learning colour matching functions which deliver colours in close accordance to those acquired by sample trichromatic cameras.

References

1. Stiles, W.S., Burch, J.M.: Interim report to the Commission Internationale de l'Éclairage Zurich, 1955, on the National Physical Laboratory's investigation of colour-matching. Optica Acta 2, 168–181 (1955)
2. Wandell, B.A.: The synthesis and analysis of color images. IEEE Transactions on Pattern Analysis and Machine Intelligence 9(1), 2–13 (1987)

3. Schechner, Y., Nayar, S.: Uncontrolled modulation imaging. In: IEEE Conference on Computer Vision and Pattern Recognition, pp. II:97–II:204 (2004)
4. Schechner, Y., Nayar, S., Belhumeur, P.N.: A theory of multiplexed illumination. In: Int. Conference on Computer Vision, pp. II:808–II:816 (2003)
5. Chi, C., Yoo, H., Ben-Ezra, M.: Multi-spectral imaging by optimized wide band illumination. Int. Journal of Computer Vision 86, 140–151 (2010)
6. Vora, P.L., Farrell, J.E., Tietz, J.D., Brainard, D.H.: Linear models for digital cameras. In: Proceedings of the IS&T's 50th Annual Conference, pp. 377–382 (1997)
7. Finlayson, G., Hordley, S., Hubel, P.: Recovering device sensitivities with quadratic programming. In: Proceedings of the IS&T/SID Color Imaging Conference, pp. 90–95 (1998)
8. Urban, P., Desch, M., Happel, K., Spiehl, D.: Recovering camera sensitivities using target-based reflectances captured under multiple led-illuminations. In: Proceedings of the Workshop on Color Image Processing, pp. 9–16 (2010)
9. Martin, R., Arno, Z., Reinhard, K.: Practical spectral characterization of trichromatic cameras. In: SIGGRAPH Asia (2011)
10. Horn, B.K.P., Brooks, M.J.: The variational approach to shape from shading. CVGIP 33(2), 174–208 (1986)
11. Kimmel, R., Elad, M., Shaked, D., Keshet, R., Sobel, I.: A variational framework for retinex. International Journal of Computer Vision 52(1), 7–23 (2003)
12. Finlayson, G.D., Schaefer, G.: Solving for colour constancy using a constrained dichromatic reflection model. International Journal of Computer Vision 42(3), 127–144 (2001)
13. Huynh, C.P., Robles-Kelly, A.: A solution of the dichromatic model for multispectral photometric invariance. International Journal of Computer Vision 90(1), 1–27 (2010)
14. Gonzalez, R.C., Wintz, P.: Digital Image Processing. Addison-Wesley Publishing Company Limited (1986)
15. Vora, P., Farrell, J., Tietz, J., Brainard, D.: Image capture: Simulation of sensor responses from hyperspectral images. IEEE Transactions on Image Processing 10(2), 307–316 (2001)
16. Dempster, A., Laird, N., Rubin, D.: Maximum-likelihood from incomplete data via the EM algorithm. J. Royal Statistical Soc. Ser. B Methodological 39, 1–38 (1977)
17. Nocedal, J., Wright, S.: Numerical Optimization. Springer (2000)
18. Bishop, C.M.: Pattern Recognition and Machine Learning. Springer (2006)
19. Barber, D.: Bayesian Reasoning and Machine Learning. Cambridge University Press (2012)
20. Dalai, N., Triggs, B.: Histogram of oriented gradients for human detection. In: Computer Vision and Pattern Recognition, pp. I:886–I:893 (2005)

Constrained Log-Likelihood-Based
Semi-supervised Linear Discriminant Analysis

Marco Loog[1] and Are C. Jensen[2]

[1] Pattern Recognition Laboratory, Delft University of Technology
Delft, The Netherlands
prlab.tudelft.nl
[2] Department of Informatics, University of Oslo
Oslo, Norway

Abstract. A novel approach to semi-supervised learning for classical Fisher linear discriminant analysis is presented. It formulates the problem in terms of a constrained log-likelihood approach, where the semi-supervision comes in through the constraints. These constraints encode that the parameters in linear discriminant analysis fulfill particular relations involving label-dependent and label-independent quantities. In this way, the latter type of parameters, which can be estimated based on unlabeled data, impose constraints on the former. The former parameters are the class-conditional means and the average within-class covariance matrix, which are the parameters of interest in linear discriminant analysis. The constraints lead to a reduction in variability of the label-dependent estimates, resulting in a potential improvement of the semi-supervised linear discriminant over that of its regular supervised counterpart. We state upfront that some of the key insights in this contribution have been published previously in a workshop paper by the first author. The major contribution in this work is the basic observation that a semi-supervised linear discriminant analysis can be formulated in terms of a principled log-likelihood approach, where the previous solution employed an ad hoc procedure. With the current contribution, we move yet another step closer to a proper formulation of a semi-supervised version of this classical technique.

1 Introduction

Supervised learning aims to learn from examples. That is, given a limited number of instances of a particular input-output relation, its goal is to generalize this relationship to new and unseen data in order to enable the prediction of the associated output given new input. Specifically, supervised classification aims to infer an unknown feature vector-class label relation from a finite, potentially small, number of input feature vectors and their associated, desired output class labels. Now, an elementary question is whether and, if so, how the availability of additional unlabeled data can significantly improve the training of such classifier. This is what constitutes the problem of semi-supervised classification or, generally, semi-supervised learning [3,22].

G.L. Gimel' farb et al. (Eds.): SSPR & SPR 2012, LNCS 7626, pp. 327–335, 2012.

The hope or, rather, belief is that semi-supervision can bring enormous progress to many scientific and application areas in which classification problems play a key role, simply by exploiting the often enormous amounts of unlabeled data available (think computer vision, text mining, retrieval, medical diagnostics, but also social sciences, psychometrics, econometrics, etc.). The matter of the fact, however, is that up to now semi-supervised methods have not been widely accepted outside of the realms of computer science, being little used in other domains. Part of the reason for this may be that current methods offer no performance guarantees [2,20] and often deteriorate in the light of large amounts of unlabeled samples [4,5,18].

References [10] and [11] identify as main reason for the frequent failure of semi-supervision that current semi-supervised approaches typically rely on assumptions extraneous to the classifier being considered. A similar point has been raised in [13]. Indeed, the main current approaches to semi-supervised learning stress the need for presuppositions such as the cluster assumption: points from the same class cluster, the smoothness assumption: neighboring point have the same label, the assumption of low density separation: the decision boundary is located in low density areas, and the like [3,22]. Given a particular assumption holds, one is able to extract relevant information not only from the labeled, but especially from the unlabeled examples. While it is undeniably true that having more precise knowledge on the distribution of data could, or even should, help in training a better classifier, in many real-world settings it may be questionable if one can at all check if such conditions are indeed met. Moreover, as soon as these additional model assumptions do not fit the data, there obviously is the real risk that adding unlabeled data actually leads to a severe deterioration of classification performance [4,5,10,11,18]. Note that this is in contrast with the supervised setting, where most classifiers, generative or not, are capable of handling mismatched data assumptions rather well, in the sense that adding more training data generally improves the performance of the classifier.

This work continues in the spirit of the earlier research presented in [10] and [11]. Reference [10] introduces a semi-supervised version of the simple, at times still topical [9,19], nearest mean classifier (NMC, [16]). It suggests to exploit known relationships between the class means and the label-independent overall mean. Enforcing these constraints during semi-supervision, yields label-dependent estimates that have smaller expected deviation from the true parameter value, which, in turn, leads to reduced classification errors. In fact, despite its simplicity, semi-supervised NMC in some cases provides error rates that are competitive with state-of-the-art methods (compare [10] and [3]). Where [10] presents a straightforward way to enforce labeled-unlabeled constraints merely involving class means and overall means, [11] shows how to deal with a known constraint on the average within-class covariance matrix as well. The constraint is relevant to linear discriminant analysis (LDA) but more difficult to deal with. Results in [11] show the overall good performance of semi-supervised LDA, not only when compared to standard supervised LDA setting but also in the light of earlier approaches to semi-supervised LDA, which often show detrimental performance with large amounts of unlabeled samples indeed.

Where the earlier approach provide an ad hoc ways to enforce the constraints, this paper casts the problem into a principled log-likelihood framework, basically proposing to optimize the regular likelihood underlying LDA under the constraint proposed in [11].

After the next section, which presents a brief overview of related work, Section 3 briefly recapitulates the relevant details of the approach presented in [10]. The main focus in that section will, however, be on semi-supervised LDA as presented in [11] and the better founded approach to semi-supervision through the log-likelihood. Section 4 offers an experiment by means of which we try to get an initial impression about how different the new principled approach is from the earlier ad hoc technique. In addition, it reports on the results obtained. Section 5 wraps up the work with a discussion and conclusion.

2   Related Work

There are few works that focus on semi-supervised LDA. Most relevant contributions come from statistics and have been published mainly in the 1960s and 1970s. Reference [8] suggests to maximize the likelihood over all permutations of possible labelings of unlabeled objects. A computationally more feasible approach is proposed by McLachlan [14,15], which follow an iterative procedure. Firstly, the linear discriminant is trained on the labeled data only and used to label all unlabeled instances. Using the now-labeled data, the classifier is retrained and employed to relabel the initially unlabeled data. This process of relabeling originally unlabeled data is repeated until none of the samples changes label.

The above approach to semi-supervised learning is basically a form of so-called self-training or self-learning, which has been suggested in different guises [14,17,21]. This iterative method also relates directly to the well-known approach to semi-supervision based on expectation maximization (see [18] and the discussion papers related to [6]). We note that employing expectation maximization to infer the missing labels will in many cases also lead to worsened error rates, particularly if too many unlabeled examples are included.

Finally, we remark that there are also semi-supervised approach to LDA as a dimensionality reduction technique (as opposed to LDA as a classifier) but we refrain from reviewing these works here.

3   Constrained Log-Likelihood-Based LDA

3.1   Semi-supervised NMC Basics

The semi-supervised version of the (NMC) proposed in [10] is simple but has been proven to be effective notwithstanding. To start with, note that when employing a regular supervised NMC, the K class means, m_i with $i \in \{1, \ldots, K\}$, and the overall mean of the data, m, fulfill the linear constraint [7]

$$m = \sum_{i=1}^{K} p_i m_i,\tag{1}$$

where p_i is the prior of class i. Having additional unlabeled data, one can improve the estimate of m because it does not depend on any labels. In this case, however, the constraint in Equation (1) will typically be violated. The core idea in [10] is that one can get improved estimates of the class means by adapting them such that the constraint is satisfied again. The solution chosen is to simply alter the K sample class means m_i by the same shift such that the new total sample mean $m' = \sum_{i=1}^{K} p_i m'_i$ of the shifted class means m'_i coincides with the total sample mean μ. The total mean m' has been obtained using all data available. All in all, the following update of the class means is suggested

$$m'_i = m_i - \sum_{i=1}^{K} p_i m_i + \mu.\tag{2}$$

3.2   Ad Hoc Semi-supervised LDA

For LDA, next to Equation (1), an additional known constraint equates the sum of the estimates of the between-class covariance matrix $\mathbf{B}$ and within-class covariance $\mathbf{W}$ to the total covariance over all data $\mathbf{T}$ (cf. [7]). That is

$$\mathbf{T} = \mathbf{W} + \mathbf{B},\tag{3}$$

where

$$\mathbf{T} := \frac{1}{N} \sum_{i=1}^{K} \sum_{j=1}^{N_i} (x_{i,j} - m)(x_{i,j} - m)^{\mathrm{t}},\tag{4}$$

in which $x_{i,j}$ is the jth feature vector from class i, m is the estimated overall mean, N_i is the number of samples from class i, and N is the total number of samples. The remaining variables in the equation have the following definitions:

$$\mathbf{W} := \sum_{i=1}^{K} p_i \mathbf{C}_i,\tag{5}$$

where $\mathbf{C}_i$ is the sample covariance matrix for class i, and

$$\mathbf{B} := \sum_{i=1}^{K} p_i (m_i - m)(m_i - m)^{\mathrm{t}}.\tag{6}$$

The parameters of interest are the class means m_i, the within-class covariance matrix $\mathbf{W}$, and the priors p_i. These parameters should be estimated from both labeled and unlabeled data under the constraints provided in Equations (1) and (3), in which the left hand side is fixed and determined by all data available.

Now, denote the estimated total mean based on all the data by μ, as in Subsection 3.1, and let the corresponding total covariance matrix be denoted by Θ. The corresponding mean m and covariance matrix $\mathbf{T}$ are based merely on the labeled data. Reference [11] now suggests the following easy and effective solution in order to meet the constraints. To start with, transform every labeled datum x as follows:

$$x \leftarrow \Theta^{\frac{1}{2}} \mathbf{T}^{-\frac{1}{2}} (x - m) + \mu. \tag{7}$$

The transformation sees to it that the overall mean and covariance statistics of the labeled data match the respective statistics as measured on all data. That is, on the transformed data, the corresponding m and $\mathbf{T}$ equal μ and Θ, respectively. The next step is to simply train a regular LDA on this transformed training data, providing the semi-supervised estimates for m_i and $\mathbf{W}$. By means of Equation (6), the corresponding $\mathbf{B}$ in the transformed space can be determined. By construction, one has

$$\mu = \sum_{i=1}^{K} p_i m_i \tag{8}$$

and

$$\Theta = \mathbf{W} + \mathbf{B}. \tag{9}$$

As the transformation applied is affine, we can actually directly estimate the m_is and the $\mathbf{W}$ in the original space. Given the class means m_i' and the within-class covariance matrix $\mathbf{W}'$ determined on the *untransformed* labeled data only, the following holds:

$$m_i = \Theta^{\frac{1}{2}} \mathbf{T}^{-\frac{1}{2}} (m_i' - m) + \mu \tag{10}$$

$$\mathbf{W} = \mathbf{T}^{-\frac{1}{2}} \Theta^{\frac{1}{2}} \mathbf{W}' \Theta^{\frac{1}{2}} \mathbf{T}^{-\frac{1}{2}}. \tag{11}$$

This expresses the m_is and $\mathbf{W}$ in terms of first and second order moment statistics in the original space.

Experiments have demonstrated that this approach outperforms standard LDA in most cases. What might be even more important, however, is that it is better behaved than the self-learning and EM-type of approaches, not showing the extreme detrimental behavior the latter methods can display at times.

3.3 Constrained Log-Likelihood-Based LDA

A basic problem with the foregoing solution is that it is unclear in which way it can be considered optimal. It delivers the necessary means and covariance matrices such that the constraints are satisfied but there are infinitely many solutions fulfilling the same constraints. Some of these can be easily constructed by assigning arbitrary labels to the unlabeled data and training a standard LDA. As can be checked easily, the parameters estimated in this way will necessarily satisfy the constraints given in Equations (1) and (3). The simplicity of the solution from [11] may be appealing but there really seems to be little reason to prefer it over any of the others[1]. There is, however, a solution to this matter that, by now, should not come as a surprise (see [11]).

The maximum likelihood solution typical employed for LDA, finds the parameters p_i, m_i, and $\mathbf{W}$ by optimizing the log-likelihood of the complete model (see, for instance, [16]). In the paper's notation, for a given labeled data set, this log-likelihood L can be expressed as follows:

$$L(p_i, m_i \, \mathbf{W}) =$$
$$\sum_{i=1}^{K} \sum_{j=1}^{N_i} \left(\log p_i - \tfrac{d}{2} \log 2\pi - \tfrac{1}{2} \log |\mathbf{W}| - \tfrac{1}{2}(x_{i,j} - m_i)\mathbf{W}^{-1}(x_{i,j} - m_i)^{\mathrm{t}} \right). \tag{12}$$

Maximizing this expression directly leads to the standard sample estimates for the class priors, the class means, and the average within-class covariance matrix, provided all p_is are positive and add up to 1. Now, next to the constraint on the priors, as an elegant solution to the semi-supervised estimation problem, the constraints in Equations (1) and (3), which rely on unlabeled data as well, could be imposed in addition.

This choice is much more attractive than the ad hoc solution from [11], in the sense that there is a clear optimality criterion at the basis of this semi-supervised solution to LDA. In addition, it is based on the log-loss that is at the basis of LDA in the first place. The price we pay is that we run into a more complicated optimization problem as the constrained log-likelihood formulation does not allow for a closed-form solution nor is it convex.

4   Experiment and Results

A small experiment was conducted, basically to get an impression of how our new approach to semi-supervision compares to the earlier suggestion from [11]. Experiments with standard supervised LDA are included as well, for comparison but also to remind us that we are still not there: occasionally supervised LDA will still outperform its semi-supervised variants.

The constrained log-likelihood is optimized by means of a Hessian-corrected gradient ascent on the constrained means m_i (Newton's method). In every iteration, given the updated means, the new within-class covariance matrix $\mathbf{W}$ can be determined and we can check if it fails to be positive definite. If this happens, we decrease the step size of the ascend and reevaluate $\mathbf{W}$. A similar action is taken in case the log-likelihood decreases. In the experiments, two different starting points satisfying the various constraints are employed. The first one, which is referred to as α, initializes all class means by the total data mean μ and the

---

[1] Maybe, though, one might be able to demonstrate that the data transformation as suggested in Subsection 3.2 is the one that deforms the original data in some sort of minimal way. Up to now, however, we have been unsuccessful in showing this.

within covariance matrix by the total covariance $\mathbf{T}$. The second starting point is provided by the ad hoc solution from Subsection 3.2 as presented in [11], which is referred to as Ω.

Table 1. Some basic properties of the fourteen real-world data sets

data set	number of objects	dimensionality (original)	dimensionality (after PCA)	number of class	smallest class size	largest class size
glass	214	9	6	6	9	76
haberman	306	3	3	2	81	225
ionosphere	351	33	30	2	126	225
iris	150	4	3	3	50	50
parkinsons	195	22	3	2	48	147
pendigits	10992	16	13	10	1055	1144
pima	768	8	5	2	268	500
sat	6435	36	16	6	626	1533
segmentation	210	19	5	7	30	30
sonar	208	60	29	2	97	111
spambase	4601	57	2	2	1813	2788
transfusion	748	3	3	2	178	570
vowel	990	10	9	11	90	90
wdbc	569	30	2	2	212	357

For the experiments, fourteen real-world data sets are taken from the UCI Machine Learning Repository [1]. The data sets used together with some specifications can be found in Table 1. To avoid any problems with singular total covariance matrices, the dimensionality of all data sets is initially reduced using PCA so to retain a fraction of 0.99 of the total variance. The data dimensionalities after PCA can be found in column four of the table.

The largest effect of semi-supervision may be expected when the labeled training set is small. Training set sizes are therefore set to equal the dimensionality plus the number of classes, which makes sure that the within-class covariance is still invertible. The remainder of the data set is both used as unlabeled data for the semi-supervised learners and as test set, meaning that we are in a transductive setting. This random split of data is repeated 10 times for every experiment from which averaged error rates are calculated. Table 2 reports on these results and, in addition, compares the four different LDAs by means of a paired t-test.

5   Discussion and Conclusion

A more sound and appealing approach to semi-supervised learning for classical LDA has been suggested. It is based on a direct optimization of the log-likelihood subject to the constraints that have been studied before in [11]. It is through these constraints that unlabeled data has its influence on the final solution.

By construction, the approach using the Ω initialization will give a higher log-likelihood than the ad hoc solution. As it turns out, experimentally, the α initialization shows the same: it gives a log-likelihood higher than the ad hoc procedure. The resulting error rates show, however, that this does not necessarily lead to a significant performance improvement in terms of classification error

Table 2. Results for the supervised and the three semi-supervised approaches to LDA are displayed. Underlining indicates the best performing method. Bold faced fonts indicate that these results do not differ significantly from the best result (based on a t-test).

data set	supervised	constrained (ad hoc, [11])	constrained (α)	constrained (Ω)
glass	**.604**	.607	.607	.621
haberman	.426	**.385**	**.376**	.396
ionosphere	.367	.239	.244	**.231**
iris	**.090**	.169	.191	.201
parkinsons	.392	.383	**.348**	.357
pendigits	.559	.558	.580	**.454**
pima	.385	.346	**.345**	**.341**
sat	.538	.501	.505	**.444**
segmentation	.503	.553	**.431**	.570
sonar	.440	.358	.350	**.344**
spambase	**.414**	.489	.494	.496
transfusion	.444	**.388**	.406	.402
vowel	.730	.744	.768	**.715**
wdbc	**.181**	.219	.322	.261

(cf. [12]). Therefore, even though our approach is another step in the right direction, we did not arrive at the point yet where we can guarantee reductions in expected error rates. Overall, however, the results in Table 2 at least show that the constrained approach is to be preferred over the original suggestion made in [11] when compared on the basis of classification performance.

We suspect that one of the key issues that should be studied in more depth in the future is the optimization of the constrained log-likelihood. At this point, we have little insight in how close we come to a global optimum or, generally, what would be the most effective way of reaching a satisfactory solution.

References

1. Asuncion, A., Newman, D.: UCI machine learning repository (2007), http://www.ics.uci.edu/~mlearn/MLRepository.html
2. Ben-David, S., Lu, T., Pál, D.: Does unlabeled data provably help? Worst-case analysis of the sample complexity of semi-supervised learning. In: COLT 2008, pp. 33–44 (2008)
3. Chapelle, O., Schölkopf, B., Zien, A.: Semi-Supervised Learning. MIT Press, Cambridge (2006)
4. Cohen, I., Cozman, F., Sebe, N., Cirelo, M., Huang, T.: Semisupervised learning of classifiers: Theory, algorithms, and their application to human-computer interaction. IEEE Transactions on Pattern Analysis and Machine Intelligence, 1553–1567 (2004)
5. Cozman, F., Cohen, I.: Risks of semi-supervised learning. In: Semi-Supervised Learning, ch. 4. MIT Press (2006)

6. Dempster, A.P., Laird, N.M., Rubin, D.B.: Maximum likelihood from incomplete data via the em algorithm. Journal of the Royal Statistical Society. Series B (Methodological) 39(1), 1–38 (1977)

7. Fukunaga, K.: Introduction to Statistical Pattern Recognition. Academic Press (1990)

8. Hartley, H.O., Rao, J.N.K.: Classification and estimation in analysis of variance problems. Review of the International Statistical Institute 36(2), 141–147 (1968)

9. Liu, Q., Sung, A.H., Chen, Z., Liu, J., Huang, X., Deng, Y.: Feature selection and classification of MAQC-II breast cancer and multiple myeloma microarray gene expression data. PLoS ONE 4(12), e8250 (2009)

10. Loog, M.: Constrained Parameter Estimation for Semi-supervised Learning: The Case of the Nearest Mean Classifier. In: Balcázar, J.L., Bonchi, F., Gionis, A., Sebag, M. (eds.) ECML PKDD 2010, Part II. LNCS, vol. 6322, pp. 291–304. Springer, Heidelberg (2010)

11. Loog, M.: Semi-supervised Linear Discriminant Analysis Using Moment Constraints. In: Schwenker, F., Trentin, E. (eds.) PSL 2011. LNCS (LNAI), vol. 7081, pp. 32–41. Springer, Heidelberg (2012)

12. Loog, M., Duin, R.P.W.: The Dipping Phenomenon. In: Gimel' farb, G.L., Hancock, E., Imiya, A., Kudo, M., Kuijper, A., Omachi, S., Windeatt, T., Yamada, K. (eds.) SSPR & SPR 2012. LNCS, vol. 7626, pp. 310–317. Springer, Heidelberg (2012)

13. Mann, G.S., McCallum, A.: Generalized expectation criteria for semi-supervised learning with weakly labeled data. The Journal of Machine Learning Research 11, 955–984 (2010)

14. McLachlan, G.J.: Iterative reclassification procedure for constructing an asymptotically optimal rule of allocation in discriminant analysis. Journal of the American Statistical Association 70(350), 365–369 (1975)

15. McLachlan, G.: Estimating the linear discriminant function from initial samples containing a small number of unclassified observations. Journal of the American Statistical Association 72(358), 403–406 (1977)

16. McLachlan, G.: Discriminant Analysis and Statistical Pattern Recognition. John Wiley & Sons (1992)

17. McLachlan, G., Ganesalingam, S.: Updating a discriminant function on the basis of unclassified data. Communications in Statistics - Simulation and Computation 11(6), 753–767 (1982)

18. Nigam, K., McCallum, A., Thrun, S., Mitchell, T.: Learning to classify text from labeled and unlabeled documents. In: Proceedings of the Fifteenth National Conference on Artificial Intelligence, pp. 792–799 (1998)

19. Salazar, R., Roepman, P., Capella, G., Moreno, V., Simon, I., Dreezen, C., Lopez-Doriga, A., Santos, C., Marijnen, C., Westerga, J., et al.: Gene expression signature to improve prognosis prediction of stage II and III colorectal cancer. Journal of Clinical Oncology 29(1), 17–24 (2011)

20. Singh, A., Nowak, R., Zhu, X.: Unlabeled data: Now it helps, now it doesn't. In: Advances in Neural Information Processing Systems, vol. 21 (2008)

21. Yarowsky, D.: Unsupervised word sense disambiguation rivaling supervised methods. In: Proceedings of the 33rd Annual Meeting on Association for Computational Linguistics, pp. 189–196 (1995)

22. Zhu, X., Goldberg, A.: Introduction to Semi-Supervised Learning. Morgan & Claypool Publishers (2009)

Out-of-Sample Embedding by Sparse Representation

Bogdan Raducanu[1] and Fadi Dornaika[2,3]

[1] Computer Vision Center, 08193 Bellaterra, Spain
[2] University of the Basque Country (UPV/EHU), San Sebastian, Spain
[3] IKERBASQUE, Basque Foundation for Science, Bilbao, Spain

Abstract. A critical aspect of non-linear dimensionality reduction techniques is represented by the construction of the adjacency graph. The difficulty resides in finding the optimal parameters, a process which, in general, is heuristically driven. Recently, sparse representation has been proposed as a non-parametric solution to overcome this problem. In this paper, we demonstrate that this approach not only serves for the graph construction, but also represents an efficient and accurate alternative for out-of-sample embedding. Considering for a case study the Laplacian Eigenmaps, we applied our method to the face recognition problem. Experimental results conducted on some challenging datasets confirmed the robustness of our approach and its superiority when compared to existing techniques.

1 Introduction

In recent years, a new family of non-linear dimensionality reduction techniques for manifold learning has emerged. The most known ones are: Kernel Principal Component Analysis (KPCA) [1], Locally Linear Embedding (LLE) [2,3], Isomap [4], Supervised Isomap [5], Laplacian Eigenmaps (LE)[6,7]. This family of non-linear embedding techniques appeared as an alternative to their linear counterparts which suffer of severe limitation when dealing with real-world data: i) they assume the data lie in an Euclidean space and ii) they may fail to get a faithful representation of data distribution when the number of samples is too small. On the other hand, the non-linear dimensionality techniques are able to discover the intrinsic data structure by exploiting the local topology. In general, they attempt to optimally preserve the local geometry around each data sample while using the rest of the samples to preserve the global structure of the data.

The non-linear embedding approaches model the structure of data by preserving some geometrical property of the underlying manifold. For instance, while the Isomap method attempts to maintain global properties, LE and LLE aim at preserving local geometry which implicitly tends to keep the global layout of the data manifold.

An inherent limitation of these approaches is that they do not provide an explicit mapping function between low and high dimensional spaces. Such function is essential for ensuring continuity of low dimensional representation and projecting data between spaces. This issue has been addressed quite satisfactorily by applying Radial Basis Function network to approximate the optimal mapping function [8]. However, the quality of RBFN relies on the careful selection of a few parameters which are chosen empirically. In [9], the authors cast MDS, ISOMAP, LLE, and LE in a common

G.L. Gimel' farb et al. (Eds.): SSPR & SPR 2012, LNCS 7626, pp. 336–344, 2012.
© Springer-Verlag Berlin Heidelberg 2012

framework, in which these methods are seen as learning eigenfunctions of a kernel. The authors try to generalize the dimensionality reduction results for the unseen data samples.

Due to this limitation, the 'out-of-sample' problem (projection of unseen samples on the embedded space) is not a straightforward process and it is less intuitive than in the case of linear manifolds. For this reason, it hasn't received too much attention so far. In this paper, we adopt the sparse representation approach as an optimal solution to the 'out-of-sample' problem. In the past, it was used as an efficient alternative [10] to the parametric construction of the adjacency graph. Without any loss of generality, we chose the Laplacian Eigenmaps as one of the non-linear dimensionality reduction techniques to test our method.

The paper is structured as follows. In section 2, we briefly review the Laplacian Eigenmaps. In section 3, we introduce our proposed approach for the out-of-sample problem based on sparse representation. Section 4 contains the experimental results. We evaluate the performance of proposed out-of-sample method for the face recognition problem. Finally, in section 5 we present our conclusions and provide the guidelines for future work.

2   Review of Laplacian Eigenmaps

Laplacian Eigenmaps is a recent non-linear dimensionality reduction technique that aims to preserve the local structure of data [6]. Using the notion of the Laplacian of the graph, this non-supervised algorithm computes a low-dimensional representation of the data set by optimally preserving local neighborhood information in a certain sense. We assume that we have a set of N samples $\{\mathbf{x}_i\}_{i=1}^{N} \subset \mathbb{R}^D$. Let's define a neighborhood graph on these samples, such as a K-nearest-neighbor or ϵ-ball graph, or a full mesh, and weigh each edge $\mathbf{x}_i \sim \mathbf{x}_j$ by a symmetric affinity function $W_{ij} = K(\mathbf{x}_i; \mathbf{x}_j)$, typically Gaussian:

$$W_{ij} = \exp(-\frac{\|\mathbf{x}_i - \mathbf{x}_j\|^2}{\beta}) \tag{1}$$

where β is suitable positive scalar. It is usually set to the average of squared distances between all pairs.

LE seeks latent points $\{\mathbf{y}_i\}_{i=1}^{N} \subset \mathbb{R}^L$ that minimize $\frac{1}{2}\sum_{i,j} \|\mathbf{y}_i - \mathbf{y}_j\|^2 W_{ij}$, which discourages placing far apart latent points that correspond to similar observed points. If $\mathbf{W} \equiv W_{ij}$ denotes the symmetric affinity matrix and $\mathbf{D}$ is the diagonal weight matrix, whose entries are column (or row, since $\mathbf{W}$ is symmetric) sums of $\mathbf{W}$, then the Laplacian matrix is given $\mathbf{L} = \mathbf{D} - \mathbf{W}$. The objective function can also be written as:

$$\frac{1}{2}\sum_{i,j} \|\mathbf{y}_i - \mathbf{y}_j\|^2 W_{ij} = tr(\mathbf{Z}^T \mathbf{L} \mathbf{Z}) \tag{2}$$

where $\mathbf{Z}^T = \mathbf{Y} = [\mathbf{y}_1, \ldots, \mathbf{y}_N]$ is the $L \times N$ embedding matrix and $tr(.)$ denotes the trace of a matrix. The ith row of the matrix $\mathbf{Z}$ provides the vector $\mathbf{y}_i$—the embedding coordinates of the sample $\mathbf{x}_i$.

The embedding matrix $\mathbf{Z}$ is the solution of the optimization problem:

$$\min_{\mathbf{Z}} tr(\mathbf{Z}^T \mathbf{L} \mathbf{Z}) \quad s.t. \quad \mathbf{Z}^T \mathbf{D} \mathbf{Z} = \mathbf{I}, \quad \mathbf{Z}^T \mathbf{L} \mathbf{1} = \mathbf{0} \tag{3}$$

where $\mathbf{I}$ is the identity matrix and $\mathbf{1} = (1, \ldots, 1)^T$. The first constraint eliminates the trivial solution $\mathbf{Z} = \mathbf{0}$ (by setting an arbitrary scale) and the second constraint eliminates the trivial solution $\mathbf{1}$ (all samples are mapped to the same point). Standard methods show that the embedding matrix is provided by the matrix of eigenvectors corresponding to the smallest eigenvalues of the generalized eigenvector problem,

$$\mathbf{L} \mathbf{z} = \lambda \mathbf{D} \mathbf{z} \tag{4}$$

Let the column vectors $\mathbf{z}_0, \ldots, \mathbf{z}_{N-1}$ be the solutions of (4), ordered according to their eigenvalues, $\lambda_0 = 0 \leq \lambda_1 \leq \ldots \leq \lambda_{N-1}$. The eigenvector corresponding to eigenvalue 0 is left out and only the next eigenvectors for embedding are used. The embedding of the original samples is given by the row vectors of the matrix $\mathbf{Z}$, that is, $\mathbf{Y} = [\mathbf{y}_1, \mathbf{y}_2, \ldots, \mathbf{y}_N] = \mathbf{Z}^T$.

$$\mathbf{x}_i \longrightarrow \mathbf{y}_i = (z_1(i), \ldots, z_L(i))^T \tag{5}$$

where $L < N$ is the dimension of the new space.

From equation (4), we can observe that the dimensionality of the subspace obtained by LE is limited by the number of samples N.

3   Proposed Out-of-Sample Embedding

3.1   Projection of New Samples

Assume we have obtained a LE embedding $\mathbf{Y}_s = (\mathbf{y}_1, \ldots, \mathbf{y}_N)$ of seen samples $\mathbf{X}_s = (\mathbf{x}_1, \ldots, \mathbf{x}_N)$ and consider unseen (out-of-sample) sample in observed space $\mathbf{x}_{N+1}$. The natural way to embed the new sample would be to recompute the whole embedding $(\mathbf{Y}_s, \mathbf{y}_{N+1})$ for $(\mathbf{X}_s, \mathbf{x}_{N+1})$ from Eq. (3). This is computationally costly and does not lead to defining a mapping for new samples; we seek a way of keeping the old embedding fixed and embed new sample based on that. Then, the next most natural way is to recompute the embedding but keeping the old embedded samples fixed and imposing that the embedding of the new sample (vector $\mathbf{y}_{N+1}$) should minimize the following target function:

$$\sum_{i=1}^{N} \|\mathbf{y}_{N+1} - \mathbf{y}_i\|^2 W_{(N+1)i} \tag{6}$$

$$\sum_{i=1}^{N} (\mathbf{y}_{N+1} - \mathbf{y}_i)^T (\mathbf{y}_{N+1} - \mathbf{y}_i) W_{(N+1)i} \tag{7}$$

The above should correspond to a minimum, and thus the derivative with respect to $\mathbf{y}_{N+1}$ of the target function should vanish:

$$2 \sum_{i=1}^{N} (\mathbf{y}_{N+1} - \mathbf{y}_i) W_{(N+1)i} = 0 \tag{8}$$

From the above, we can conclude that the embedding $\mathbf{y}_{N+1}$ is given by:

$$\mathbf{y}_{N+1} = \frac{\sum_{i=1}^{N} W_{(N+1)i}\, \mathbf{y}_i}{\sum_{i=1}^{N} W_{(N+1)i}} \tag{9}$$

The above formula stipulates that the embedding of an unseen sample is simply the linear combination of all fixed embedded samples where the linear coefficients are set to the similarity between the unseen sample and the existing sample.

Whenever $W_{(N+1)i}$ is set to a Kernel function (i.e., $W_{(N+1)i} = K(\mathbf{x}_{N+1}, \mathbf{x}_i)$, Eq. (9) is equivalent to the Laplacian Eigenmaps Latent Variable Model (LELVM) introduced in [11].

3.2 Computation of the Similarity Coefficients via Sparse Representation

The problem of out-of-sample embedding boils down to the estimation of the similarities $W_{(N+1)i}, i = 1, \ldots, N$. In [11], these $W_{(N+1)i}$ were computed using a K nearest neighbor and a Heat Kernel. However, it is well known that the neighborhood size as well as the Kernel parameter may affect the embedding process. We will bypass this limitation by using the coding provided by sparse representation.

In traditional graph construction process, the graph adjacency structure and the graph weights are derived separately. It was argued that the graph adjacency structure and the graph weights are interrelated and should not be separated. Thus it is desired to develop a procedure which can simultaneously completes these two tasks within one step. In [10], the authors proposed to simultaneously build the adjacency graph and its weights. To this end, they used the sparse representation of each training sample as a linear superposition of basis functions (rest of the training samples) plus the noise.

We apply the sparse coding/representation principle for computing the set of coefficients $W_{(N+1)i}$. Let the vector $\mathbf{a} = (W_{(N+1)1}, W_{(N+1)2}, \ldots, W_{(N+1)N})^T$. Thus, the objective is to compute the vector $\mathbf{a}$ given the unseen sample and the training data. Based on sparse coding, the unseen sample $\mathbf{x}_{N+1}$ can be written as

$$\mathbf{x}_{N+1} = \sum_{i=1}^{N} a_i \mathbf{x}_i + \mathbf{e} = \mathbf{X}\mathbf{a} + \mathbf{e} \tag{10}$$

The goal is to minimize both the reconstruction error and the L_1 norm of the vector $\mathbf{a}$:

$$\min_{\mathbf{a},\mathbf{e}} (\|\mathbf{a}\|_{L_1} + \|\mathbf{e}\|_{L_1}) \quad s.t. \quad \mathbf{x}_{N+1} = \mathbf{X}\mathbf{a} + \mathbf{e} \tag{11}$$

Let $\mathbf{a}'$ denote the vector $\mathbf{a}' = (\mathbf{a}^T, \mathbf{e}^T)^T$ and $\mathbf{I}$ denote the $D \times D$ identity matrix, then the objective function (11) can be written as:

$$\min \|\mathbf{a}'\|_{L_1} \quad s.t. \quad [\mathbf{X}\ \mathbf{I}]\, \mathbf{a}' = \mathbf{x}_{N+1} \tag{12}$$

Although no sparse priors are imposed, the sparse property of the coefficient vector $\mathbf{a}$ is generated naturally by the L_1 optimization. Once the vector $(\mathbf{a}^T, \mathbf{e}^T)^T$ is computed, the similarity coefficients $W_{(N+1)i}$ are set to:

$$W_{(N+1)i} = |a_i|, i = 1, \ldots, N$$

3.3 Advantages of the Proposed Out-of-Sample Embedding Scheme

Although our proposed out-of-sample formula (Eq. (9)) is similar to that of the Latent Variable Model [11], it has the two following interesting differences and advantages:

1. For the LVM scheme, the neighborhood size must be set manually, and the optimal setting may be different for different data sets. In our scheme, the computation of similarity coefficients adapts to the dataset through the use of sparse coding. No parameter is required.
2. There have been many ways to compute the similarity coefficients and the most popular one among them is the typical Heat Kernel (Gaussian weighting function) described in Eq.(1). However, the Gaussian aperture may affect the final classification results significantly, and how to optimally determine this parameter is still an open problem. Our scheme get rid of this since we exploit the sparseness property of the deduced coefficients in order to express both adjacency structure and the associated weights without any predefined parameter.

4 Performance Evaluation

To validate the effectiveness of our proposed approach, we applied it to the face recognition problem.

4.1 Data Sets

We considered in our experiments four public face data sets. All these databases are characterized by a large variation in face appearance.

1. **Yale**[1]**:** The YALE face data set contains 165 images of 15 persons. Each individual has 11 images. The images demonstrate variations in lighting condition, facial expression. Each image is resized to 32×32 pixels.
2. **ORL**[2]**:.** There are 10 images for each of the 40 human subjects, which were taken at different times, varying the lighting, facial expressions (open/closed eyes, smiling/not smiling) and facial details (glasses/no glasses). The images were taken with a tolerance for some tilting and rotation of the face up to $20°$.
3. **UMIST**[3]**:.** The UMIST data set contains 575 gray images of 20 different people. The images depict variations in head pose.
4. **Extended Yale - part B**[4]**:.** It contains 16128 images of 28 human subjects under 9 poses and 64 illumination conditions. In our study, a subset of 1800 images has been used. Figure 1 shows some face samples in the extended Yale Face Database B.

---

[1] http://see.xidian.edu.cn/vipsl/database_Face.html
[2] http://www.cl.cam.ac.uk/research/dtg/attarchive/facedatabase.html
[3] http://www.shef.ac.uk/eee/research/vie/research/face.html
[4] http://vision.ucsd.edu/ ∼ leekc/ExtYaleDatabase/ExtYaleB.html

Fig. 1. Some samples in Extended Yale data set

4.2 Experimental Results

To make the computation of the embedding process more efficient, the dimensionality of the original face samples was reduced by applying random projections [12]. It has a similar role to that of PCA yet with the obvious advantage that random projections do not need any training data.

We have compared our method with other two approaches. One of them is the Latent Variable Model (LVM), proposed in [11]. The other one, is a linearization of the existing mapping $\mathbf{X}_s \rightarrow \mathbf{Y}_s$. To this end, we use simple linear regression in order to infer a linear matrix transform $\mathbf{A}$ that best approximates the existing mapping through the linear equation $\mathbf{Y}_s = \mathbf{A}^T \mathbf{X}_s$. We stress the fact the linearization has not been thoroughly tested as an out-of-sample method. Instead, this linearization was used for spectral regression (e.g., [13]).

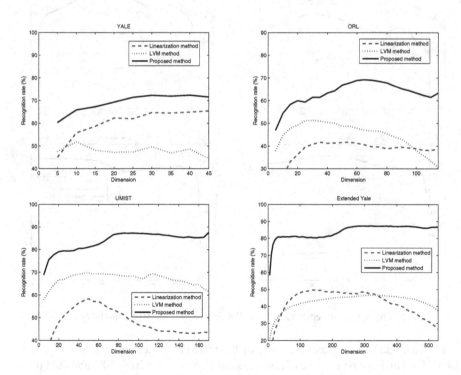

Fig. 2. Experimental results on all 4 datasets for the 30-70 modality

Table 1. Maximum average recognition rate

Dataset \ Method	Sparse Rep.	LVM			Linearization
30%-70%		$\epsilon = 3$	$\epsilon = 5$	$\epsilon = 7$	
YALE	**72.36%**	51.84%	41.66%	33.15%	65.43%
ORL	**69.25%**	51.35%	37.71%	30.25%	41.71%
UMIST	**87.56%**	69.72%	60.49%	52.65%	58.31%
Ext. Yale	**87.29%**	46.66%	31.33%	24.25%	49.90%
50%-50%		$\epsilon = 3$	$\epsilon = 5$	$\epsilon = 7$	
YALE	**81.85%**	70.12%	61.60%	52.09%	68.14%
ORL	**82.50%**	72.05%	60.35%	49.25%	46.38%
UMIST	**95.03%**	85.90%	76.04%	70.03%	76.25%
Ext. Yale	**91.46%**	61.09%	46.85%	39.03%	53.14%
70%-30%		$\epsilon = 3$	$\epsilon = 5$	$\epsilon = 7$	
YALE	**86.73%**	77.15%	73.87%	67.95%	75.51%
ORL	**88.75%**	82.16%	73.66%	65.41%	53.25%
UMIST	**97.74%**	93.06%	85.20%	79.94%	80.52%
Ext. Yale	**92.12%**	70.97%	58.36%	48.74%	57.14%

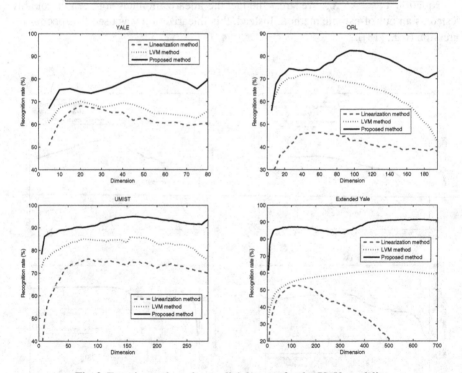

Fig. 3. Experimental results on all 4 datasets for the 50-50 modality

For each face data set and for every method, we conducted three groups of experiments for which the percentage of training samples was set to 30%, 50% and 70% of the whole data set. The remaining data was used for testing. Here, the testing implies: (i) the

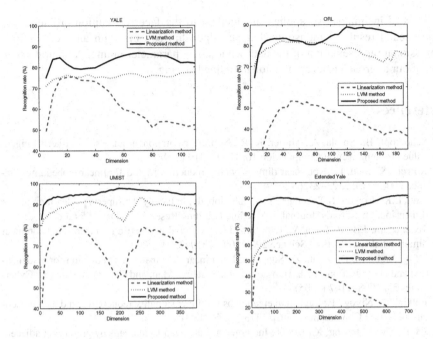

Fig. 4. Experimental results on all 4 datasets for the 70-30 modality

out-of-sample embedding of the unseen sample (face), and (ii) recognizing it through the use of the Nearest Neighbor classifier in the embedded space. The partition of the data set was done randomly. For a given embedding method, the recognition rate was computed for several dimensions belonging to $[5, L_{max}]$, where L_{max} is a parameter directly related with the number of training samples.

In figures 2, 3, and 4 we show the average recognition rates for all 4 datasets, based on the average of 10 random splits.

In table 1, we present the best (average) performance obtained by each 'out-of-sample' method, based on 10 random splits. For the case of LVM method, the ϵ parameter corresponds to the number of neighbors used to approximate the unseen sample. We could appreciate that the smaller this number is, the better the result.

The above results confirm the superiority of our approach when compared with existing ones. We can observe that this superiority was obtained for all data sets and for all dimensions tested for the obtained embedding space. We can also observe that the linearization method provided the poorest results, which can be explained by the fact that the linear method is global and does not take into account the local adjacency information.

5   Conclusion

In this paper, we demonstrated that sparse representation can serve as an efficient and accurate alternative for out-of-sample embedding. Considering for a case study the

Laplacian Eigenmaps, we applied our method to the face recognition problem. The experimental results demonstrate that our algorithm can maintain an accurate low-dimensional representation of the data without any parameter tuning. A natural extension of our approach is its application to online learning.

References

1. Schölkopf, B., Smola, A., Müller, K.-R.: Nonlinear component analysis as a kernel eigenvalue problem. Neural Computation 10, 1299–1319 (1998)
2. Roweis, S., Saul, L.: Nonlinear dimensionality reduction by locally linear embedding. Science 290(5500), 2323–2326 (2000)
3. Saul, L.K., Roweis, S.T., Singer, Y.: Think globally, fit locally: Unsupervised learning of low dimensional manifolds. Journal of Machine Learning Research 4, 119–155 (2003)
4. Tenenbaum, J.B., de Silva, V., Langford, J.C.: A global geometric framework for nonlinear dimensionality reduction. Science 290(5500), 2319–2323 (2000)
5. Geng, X., Zhan, D., Zhou, Z.: Supervised nonlinear dimensionality reduction for visualization and classification. IEEE Transactions on Systems, Man, and Cybernetics-Part B: Cybernetics 35, 1098–1107 (2005)
6. Belkin, M., Niyogi, P.: Laplacian eigenmaps for dimensionality reduction and data representation. Neural Computation 15(6), 1373–1396 (2003)
7. Jia, P., Yin, J., Huang, X., Hu, D.: Incremental Laplacian Eigenmaps by preserving adjacent information between data points. Pattern Recognition Letters 30(16), 1457–1463 (2009)
8. Elgammal, A., Lee, C.: Non-linear manifold learning for dynamic shape and dynamic appearance. Computer Vision and Image Understanding 106(1), 31–46 (2007)
9. Bengio, Y., Paiement, J., Vincent, P.: Out-of-sample extensions for LLE, Isomap, MDS, eigenmaps and spectral clustering. In: Advances in Neural Information Processing (2004)
10. Yan, S., Wang, H.: Semi-supervised learning by sparse representation. In: SIAM International Conference on Data Mining (2009)
11. Carreira-Perpinan, M.A., Lu, Z.: The Laplacian Eigenmaps latent variable model. Journal of Machine Learning Research 2, 59–66 (2007)
12. Goel, N., Bebis, G., Nefian, A.: Face recognition experiments with random projections. In: SPIE Conference on Biometric Technology for Human Identification (2005)
13. Cai, D., He, X., Han, J.: Spectral regression for efficient regularized subspace learning. In: Proc. Int. Conf. Computer Vision, ICCV 2007 (2007)

Extended Analyses for an Optimal Kernel in a Class of Kernels with an Invariant Metric

Akira Tanaka[1], Ichigaku Takigawa[2], Hideyuki Imai[1], and Mineichi Kudo[1]

[1] Division of Computer Science, Hokkaido University,
N14W9, Kita-ku, Sapporo, 060-0814 Japan
{takira,imai,mine}@main.ist.hokudai.ac.jp
[2] Creative Research Institution, Hokkaido University,
N21W10, Kita-ku, Sapporo, 001-0021 Japan
takigawa@cris.hokudai.ac.jp

Abstract. Learning based on kernel machines is widely known as a powerful tool for various fields of information science such as pattern recognition and regression estimation. An appropriate model selection is required in order to obtain desirable learning results. In our previous work, we discussed a class of kernels forming a nested class of reproducing kernel Hilbert spaces with an invariant metric and proved that the kernel corresponding to the smallest reproducing kernel Hilbert space, including an unknown true function, gives the best model. In this paper, we relax the invariant metric condition and show that a similar result is obtained when a subspace with an invariant metric exists.

Keywords: kernel regressor, reproducing kernel Hilbert space, orthogonal projection, invariant metric.

1 Introduction

Learning based on kernel machines [1], represented by the support vector machine (SVM) [2] and the kernel ridge regression [3,4], is widely known as a powerful tool for various fields of information science such as pattern recognition, regression estimation, and density estimation. In general, an appropriate model selection is required in order to obtain a desirable learning result by kernel machines. Although the model selection in a fixed model space such as selection of a regularization parameter is sufficiently investigated in terms of theoretical and practical senses (See [5,6] for instance), the selection of a model space itself is not sufficiently investigated in terms of a theoretical sense, while practical algorithms for selection of a kernel (or its parameters) such as the cross-validation are revealed. The difficulty of the theoretical analyses for selection of a kernel (or its parameters) lies on the fact that the metrics of two reproducing kernel Hilbert spaces (RKHS)[7,8] corresponding to two different kernels may differ in general, which means that we do not have a unified framework to evaluate learning results obtained by different kernels. In order to avoid this difficulty, we considered a class of kernels whose corresponding RKHS's have an invariant

G.L. Gimel' farb et al. (Eds.): SSPR & SPR 2012, LNCS 7626, pp. 345–353, 2012.

metric and proved that the kernel corresponding to the smallest reproducing kernel Hilbert space, including an unknown true function, gives the best model in [9].

In this paper, we relax the invariant metric condition for the whole space and prove that a similar result is obtained when a subspace with an invariant metric exists in the smallest reproducing kernel Hilbert space.

2 Mathematical Preliminaries for the Theory of Reproducing Kernel Hilbert Spaces

In this section, we prepare some mathematical tools concerned with the theory of reproducing kernel Hilbert spaces [7,8].

Definition 1. *[7] Let $\mathbf{R}^n$ be an n-dimensional real vector space and let $\mathcal{H}$ be a class of functions defined on $\mathcal{D} \subset \mathbf{R}^n$, forming a Hilbert space of real-valued functions. The function $K(\boldsymbol{x}, \tilde{\boldsymbol{x}})$, $(\boldsymbol{x}, \tilde{\boldsymbol{x}} \in \mathcal{D})$ is called a reproducing kernel of $\mathcal{H}$, if*

1. *For every $\tilde{\boldsymbol{x}} \in \mathcal{D}$, $K(\cdot, \tilde{\boldsymbol{x}})$ is a function belonging to $\mathcal{H}$.*
2. *For every $\tilde{\boldsymbol{x}} \in \mathcal{D}$ and every $f \in \mathcal{H}$,*

$$f(\tilde{\boldsymbol{x}}) = \langle f(\cdot), K(\cdot, \tilde{\boldsymbol{x}}) \rangle_{\mathcal{H}}, \tag{1}$$

where $\langle \cdot, \cdot \rangle_{\mathcal{H}}$ denotes the inner product of the Hilbert space $\mathcal{H}$.

The Hilbert space $\mathcal{H}$ that has a reproducing kernel is called a reproducing kernel Hilbert space (RKHS). The reproducing property Eq.(1) enables us to treat a value of a function at a point in $\mathcal{D}$. Note that reproducing kernels are positive definite [7]:

$$\sum_{i,j=1}^{N} c_i c_j K(\boldsymbol{x}_i, \boldsymbol{x}_j) \geq 0, \tag{2}$$

for any N, $c_1, \ldots, c_N \in \mathbf{R}$, and $\boldsymbol{x}_1, \ldots, \boldsymbol{x}_N \in \mathcal{D}$. In addition, $K(\boldsymbol{x}, \tilde{\boldsymbol{x}}) = K(\tilde{\boldsymbol{x}}, \boldsymbol{x})$ holds for any $\boldsymbol{x}, \tilde{\boldsymbol{x}} \in \mathcal{D}$ [7]. If a reproducing kernel $K(\boldsymbol{x}, \tilde{\boldsymbol{x}})$ exists, it is unique [7]. Conversely, every positive definite function $K(\boldsymbol{x}, \tilde{\boldsymbol{x}})$ has the unique corresponding RKHS [7]. Hereafter, the RKHS corresponding to a reproducing kernel $K(\boldsymbol{x}, \tilde{\boldsymbol{x}})$ is denoted by $\mathcal{H}_K$.

Next, we introduce the Schatten product [10] that is a convenient tool to reveal the reproducing property of kernels.

Definition 2. *[10] Let $\mathcal{H}_1$ and $\mathcal{H}_2$ be Hilbert spaces. The Schatten product of $g \in \mathcal{H}_2$ and $h \in \mathcal{H}_1$ is defined by*

$$(g \otimes h)f = \langle f, h \rangle_{\mathcal{H}_1} g, \quad f \in \mathcal{H}_1. \tag{3}$$

Note that $(g \otimes h)$ is a linear operator from $\mathcal{H}_1$ onto $\mathcal{H}_2$. It is easy to show that the following relations hold for $h, v \in \mathcal{H}_1$, $g, u \in \mathcal{H}_2$.

$$(h \otimes g)^* = (g \otimes h), \quad (h \otimes g)(u \otimes v) = \langle u, g \rangle_{\mathcal{H}_2}(h \otimes v), \tag{4}$$

where the superscript * denotes the adjoint operator.

We give some theorems concerned with sum and difference of reproducing kernels used in the following contents.

Theorem 1. *[7] If K_i is the reproducing kernel of the class F_i with the norm $\|\cdot\|_i$, then $K = K_1 + K_2$ is the reproducing kernel of the class F of all functions $f = f_1 + f_2$ with $f_i \in F_i$, and with the norm defined by*

$$\|f\|^2 = \min\left[\|f_1\|_1^2 + \|f_2\|_2^2\right], \tag{5}$$

the minimum taken for all the decompositions $f = f_1 + f_2$ with $f_i \in F_i$.

Theorem 2. *[7] If K is the reproducing kernel of the class F with the norm $\|\cdot\|$, and if the linear class $F_1 \subset F$ forms a Hilbert space with the norm $\|\cdot\|_1$, such that $\|f\|_1 \geq \|f\|$ for any $f \in F_1$, then the class F_1 possesses a reproducing kernel K_1 such that $K^c = K - K_1$ is also a reproducing kernel.*

Theorem 3. *[7] If K and K_1 are the reproducing kernels of the classes of F and F_1 with the norms $\|\cdot\|$, $\|\cdot\|_1$, and if $K - K_1$ is a reproducing kernel, then $F_1 \subset F$ and $\|f_1\|_1 \geq \|f_1\|$ for every $f_1 \in F_1$.*

3   Formulation of Regression Problems

Let $\{(y_i, \boldsymbol{x}_i) | i = 1, \ldots, \ell\}$ be a given training data set with $y_i \in \mathbf{R}$, $\boldsymbol{x}_i \in \mathbf{R}^n$, satisfying

$$y_i = f(\boldsymbol{x}_i) + n_i, \tag{6}$$

where f denotes the unknown true function and n_i denotes a zero-mean additive noise. The aim of regression problems is to estimate the unknown function f by using the given training data set and statistical properties of the noise.

In this paper, we assume that the unknown function f belongs to the RKHS $\mathcal{H}_K$ corresponding to a certain kernel function K. If $f \in \mathcal{H}_K$, then Eq.(6) is rewritten as

$$y_i = \langle f(\cdot), K(\cdot, \boldsymbol{x}_i) \rangle_{\mathcal{H}_K} + n_i, \tag{7}$$

on the basis of the reproducing property of kernels. Let $\boldsymbol{y} = [y_1, \ldots, y_\ell]'$ and $\boldsymbol{n} = [n_1, \ldots, n_\ell]'$ with the superscript $'$ denoting the transposition operator, then applying the Schatten product to Eq.(7) yields

$$\boldsymbol{y} = \left(\sum_{k=1}^{\ell} [e_k^{(\ell)} \otimes K(\cdot, \boldsymbol{x}_k)]\right) f(\cdot) + \boldsymbol{n}, \tag{8}$$

where $e_k^{(\ell)}$ denotes the k-th vector of the canonical basis of $\mathbf{R}^\ell$. For a convenience of description, we write

$$A_{K,X} = \left(\sum_{k=1}^{\ell} [e_k^{(\ell)} \otimes K(\cdot, \boldsymbol{x}_k)] \right), \tag{9}$$

where $X = \{\boldsymbol{x}_1, \ldots, \boldsymbol{x}_\ell\}$. Note that $A_{K,X}$ is a linear operator that maps an element in $\mathcal{H}_K$ onto $\mathbf{R}^\ell$ and Eq.(8) can be written by

$$\boldsymbol{y} = A_{K,X} f(\cdot) + \boldsymbol{n}, \tag{10}$$

which represents the relation between the unknown true function f and an output vector $\boldsymbol{y}$. Therefore, a regression problem can be interpreted as an inversion problem of the linear equation Eq.(10) [11]. In general, an estimated function $\hat{f}(\boldsymbol{x})$ is represented as

$$\hat{f}(\cdot) = L\boldsymbol{y}, \tag{11}$$

where L denotes a learning operator such as the support vector machine and the kernel ridge regressor.

4   Kernel Specific Generalization Ability and Some Known Results

In general, a learning result by kernel machines is represented by a linear combination of $K(\boldsymbol{x}, \boldsymbol{x}_i)$, which means that the learning result is an element in $\mathcal{R}(A_{K,X}^*)$ (the range space of the linear operator $A_{K,X}^*$) since

$$\hat{f}(\cdot) = A_{K,X}^* \boldsymbol{\alpha} = \left(\sum_{k=1}^{\ell} [K(\cdot, \boldsymbol{x}_k) \otimes e_k^{(\ell)}] \right) \boldsymbol{\alpha} = \sum_{k=1}^{\ell} \alpha_k K(\cdot, \boldsymbol{x}_k) \tag{12}$$

holds, where $\boldsymbol{\alpha} = [\alpha_1, \ldots, \alpha_\ell]'$ denotes an arbitrary vector in $\mathbf{R}^\ell$. The point at issue of this paper is to discuss goodness of a model space, that is, the generalization ability of $\mathcal{R}(A_{K,X}^*)$ which is independent from criteria of learning machines. Therefore, we define the generalization ability of kernel machines specified by a kernel K and a set of input vectors X as the distance between the unknown true function f and $\mathcal{R}(A_{K,X}^*)$ written as

$$J(f; K, X) = \|f - P_{K,X} f\|_{\mathcal{H}_K}^2, \tag{13}$$

where $P_{K,X}$ denotes the orthogonal projector onto $\mathcal{R}(A_{K,X}^*)$ and $\|\cdot\|_{\mathcal{H}_K}$ denotes the induced norm of $\mathcal{H}_K$. Note that the orthogonality of $P_{K,X}$ is also defined by the metric of $\mathcal{H}_K$. Selection of an element in $\mathcal{R}(A_{K,X}^*)$ as a learning result is out of the scope of this paper since the selection depends on learning criteria. We also ignore the observation noise in the following contents since the noise does not affect Eq.(13).

Here, we give some propositions as preparations to evaluate Eq.(13).

Lemma 1. *[9]*

$$P_{K,X} = \sum_{i=1}^{\ell} \sum_{j=1}^{\ell} (G_{K,X}^+)_{i,j} \left[K(\cdot, \boldsymbol{x}_i) \otimes K(\cdot, \boldsymbol{x}_j) \right], \tag{14}$$

where $G_{K,X}$ denotes the Gramian matrix of K with X and the superscript $^+$ denotes the Moore-Penrose generalized inverse[12].

Lemma 2. *[9] For any $f \in \mathcal{H}_K$ and X,*

$$\| P_{K,X} f \|_{\mathcal{H}_K}^2 = \boldsymbol{f}' G_{K,X}^+ \boldsymbol{f} \tag{15}$$

holds, where $\boldsymbol{f} = [f(\boldsymbol{x}_1), \dots, f(\boldsymbol{x}_\ell)]'$.

Since $P_{K,X} = A_{K,X}^+ A_{K,X}$, the learning result

$$\hat{f}(\cdot) = A_{K,X}^+ \boldsymbol{y} \tag{16}$$

gives the minimum norm least-squares solution of f and it gives the orthogonal projection of f onto $\mathcal{R}(A_{K,X}^*)$ in noise free cases.

In [9], we discussed a class of nested RKHS's with an invariant metric. We review important results given in [9].

Let K_1 and K_2 be reproducing kernels satisfying

$$\mathcal{H}_{K_1} \subset \mathcal{H}_{K_2} \tag{17}$$

and

$$\| f \|_{\mathcal{H}_{K_1}}^2 = \| f \|_{\mathcal{H}_{K_2}}^2 \quad \text{for any } f \text{ in } \mathcal{H}_{K_1}. \tag{18}$$

Then we have the following theorem.

Theorem 4. *[9] Let K_1 and K_2 be kernels. If Eqs.(17) and (18) hold, then for any $f \in \mathcal{H}_{K_1}$ and X,*

$$\| f - P_{K_1,X} f \|_{\mathcal{H}_{K_2}}^2 \leq \| f - P_{K_2,X} f \|_{\mathcal{H}_{K_2}}^2 \tag{19}$$

holds.

This theorem claims that the kernel corresponding to the smallest RKHS in the class of RKHS's with an invariant metric gives the best model space if an unknown function f belongs to the smallest one. Note that Theorem 4 does not hold in general without the invariant metric condition. In fact, we gave an example that Theorem 4 does not hold without the invariant metric condition in [13].

From Eq.(18) and Theorem 2, there exists the reproducing kernel K^c satisfying

$$K_2 = K_1 + K^c. \tag{20}$$

In [14], we discussed the relationship between these kernels (or corresponding RKHS's) and the condition of the invariant metric; and obtained the following theorem.

Theorem 5. *Let K_1 and $K_2 = K_1 + K^c$ be kernels whose corresponding RKHS's satisfy $\mathcal{H}_{K_1} \subset \mathcal{H}_{K_2}$. The following three statements are equivalent each other.*

1) For any $f \in \mathcal{H}_{K_1}$, $\|f\|^2_{\mathcal{H}_{K_1}} = \|f\|^2_{\mathcal{H}_{K_2}}$,

2) $\mathcal{H}_{K_1} \cap \mathcal{H}_{K^c} = \{0\}$,

3) For any $f_1 \in \mathcal{H}_{K_1}$ and $f_2 \in \mathcal{H}_{K^c}$, $\langle f_1, f_2 \rangle_{\mathcal{H}_{K_2}} = 0$.

5   Analyses under Relaxed Invariant Metric Condition

Let K_1 and $K_2 = K_1 + K^c$ be kernels where K^c is also a kernel, then corresponding RKHS's satisfy

$$\mathcal{H}_{K_1} \subset \mathcal{H}_{K_2} \tag{21}$$

from Theorem 1; and we have

$$\|f\|_{\mathcal{H}_{K_1}} \geq \|f\|_{\mathcal{H}_{K_2}} \tag{22}$$

for any $f \in \mathcal{H}_{K_1}$ from Theorem 3.

We assume that a linear class $F \subset \mathcal{H}_{K_1}$ forms a Hilbert space with the norm $\|\cdot\|_F$ and assume that

$$\|f\|_F = \|f\|_{\mathcal{H}_{K_i}} \quad \text{for any } f \in F, \ (i \in \{1,2\}). \tag{23}$$

Then on the basis of Theorems 2, there exists a kernel K_F such that

$$K_i^c = K_i - K_F, \ (i \in \{1,2\}) \tag{24}$$

is also a kernel and

$$\mathcal{H}_{K_F} \cap \mathcal{H}_{K_i^c} = \{0\} \tag{25}$$

holds from Theorem 5. Note that it is trivial that Eq.(24) can be rewritten as

$$K_i = K_i^c + K_F, \ (i \in \{1,2\}). \tag{26}$$

Since K_F is guaranteed to be a kernel, we use $\mathcal{H}_{K_F}$ instead of F, hereafter. Note that we can also represent K_2 as

$$K_2 = K_F + K_1^c + K^c. \tag{27}$$

If we have an explicit form of K_F, then $\mathcal{R}(A^*_{K_F,X})$ gives a better model than $\mathcal{R}(A^*_{K_1,X})$ and $\mathcal{R}(A^*_{K_2,X})$ for any $f \in \mathcal{H}_{K_F}$ according to Theorem 4. However in general, we can not always obtain K_F from K_1 and K_2 (or K^c). When $\mathcal{H}_{K_1^c} \cap \mathcal{H}_{K^c} = \{0\}$, $\mathcal{R}(A^*_{K_1,X})$ gives a better model than $\mathcal{R}(A^*_{K_2,X})$ for any $f \in \mathcal{H}_{K_1}$ according to Theorems 4 and 5. However, when $\mathcal{H}_{K_1^c} \cap \mathcal{H}_{K^c} \neq \{0\}$, $\mathcal{R}(A^*_{K_2,X})$ may be a better model than $\mathcal{R}(A^*_{K_1,X})$ for some $f \in \mathcal{H}_{K_1}$ since the metrics of $\mathcal{H}_{K_1}$ and $\mathcal{H}_{K_2}$ may differ.

The aim of this paper is to show that for any $f \in \mathcal{H}_{K_F}$, $\mathcal{R}(A^*_{K_1,X})$ gives a better model than $\mathcal{R}(A^*_{K_2,X})$ even if $\mathcal{H}_{K_1^c} \cap \mathcal{H}_{K^c} \neq \{0\}$.

Lemma 3. *[9] Let H_1 and H_2 be n.n.d. Hermitian matrices and let $\boldsymbol{y} \in \mathcal{R}(H_1)$, then*

$$J = \boldsymbol{y}^*(H_1^+ - (H_1 + H_2)^+)\boldsymbol{y} \geq 0 \tag{28}$$

holds.

The following theorem is the main result of this paper.

Theorem 6. *Let K_1 and $K_2 = K_1 + K^c$ be kernels where K^c is also a kernel. If Eq.(23) holds, then for any $f \in \mathcal{H}_{K_F}$ and X,*

$$\|f - P_{K_1,X}f\|_{\mathcal{H}_{K_2}}^2 \leq \|f - P_{K_2,X}f\|_{\mathcal{H}_{K_2}}^2 \tag{29}$$

holds.

Proof. From Eq.(22), we have

$$Z = \|f - P_{K_2,X}f\|_{\mathcal{H}_{K_2,X}}^2 - \|f - P_{K_1,X}f\|_{\mathcal{H}_{K_2,X}}^2$$
$$\geq \|f - P_{K_2,X}f\|_{\mathcal{H}_{K_2,X}}^2 - \|f - P_{K_1,X}f\|_{\mathcal{H}_{K_1,X}}^2 = Z_1$$

and from Lemma 2, the Pythagorean theorem, the invariant metric condition,

$$Z_1 = \|f\|_{\mathcal{H}_{K_2}}^2 - \boldsymbol{f}'G_{K_2,X}^+\boldsymbol{f} - (\|f\|_{\mathcal{H}_{K_1}}^2 - \boldsymbol{f}'G_{K_1,X}^+\boldsymbol{f})$$
$$= \|f\|_{\mathcal{H}_{K_2}}^2 - \boldsymbol{f}'G_{K_2,X}^+\boldsymbol{f} - (\|f\|_{\mathcal{H}_{K_2}}^2 - \boldsymbol{f}'G_{K_1,X}^+\boldsymbol{f})$$
$$= \boldsymbol{f}'G_{K_1,X}^{-1}\boldsymbol{f} - \boldsymbol{f}'G_{K_2,X}^{-1}\boldsymbol{f}$$
$$= \boldsymbol{f}'\left(G_{K_1,X}^{-1} - G_{K_2,X}^{-1}\right)\boldsymbol{f} = Z_2$$

is obtained. Since

$$G_{K_1,X} = G_{K_F,X} + G_{K_1^c,X} \tag{30}$$

and

$$G_{K_2,X} = G_{K_F,X} + G_{K_1^c,X} + G_{K^c,X} \tag{31}$$

hold, the fact that

$$\boldsymbol{f} \in \mathcal{R}(A_{K_F,X}) = \mathcal{R}(A_{K_F,X}A_{K_F,X}^*) = \mathcal{R}(G_{K_F,X}) \subset \mathcal{R}(G_{K_F,X} + G_{K_1^c,X})$$

and Lemma 3 yield $Z_2 \geq 0$ which concludes the proof.   □

According to Theorem 6, it is concluded that if the smaller RKHS include a subspace with an invariant metric, the kernel corresponding to the smaller RKHS gives a better model than larger one for any function in the subspace, which is an extension of the results obtained in [9].

6 Conclusion

In this paper, we discussed a class of kernels forming a nested class of RKHS's; and proved that if the smallest RKHS in the class has a subspace with an invariant metric, the kernel corresponding to the smallest RKHS gives the best model for any function in the subspace, which is a direct extension of our previous result obtained in [9]. Drastic relaxation of the invariant metric condition and extending the obtained results to practical learning machines such as the SVM and the kernel ridge regressor are ones of our future works.

Acknowledgement. This work was partially supported by the Ministry of Education, Science, Sports and Culture, Grant-in-Aid for Scientific Research (C), 24500001.

References

1. Muller, K., Mika, S., Ratsch, G., Tsuda, K., Scholkopf, B.: An Introduction to Kernel-Based Learning Algorithms. IEEE Transactions on Neural Networks 12, 181–201 (2001)
2. Vapnik, V.N.: The Nature of Statistical Learning Theory. Springer, New York (1999)
3. Shawe-Taylor, J., Cristianini, N.: Kernel Methods for Pattern Recognition. Cambridge University Press, Cambridge (2004)
4. Cristianini, N., Shawe-Taylor, J.: An Introduction to Support Vector Machines and other kernel-based learning methods. Cambridge University Press, Cambridge (2000)
5. Sugiyama, M., Ogawa, H.: Subspace Information Criterion for Model Selection. Neural Computation 13, 1863–1889 (2001)
6. Sugiyama, M., Kawanabe, M., Muller, K.: Trading Variance Reduction with Unbiasedness: The Regularized Subspace Information Criterion for Robust Model Selection in Kernel Regression. Neural Computation 16, 1077–1104 (2004)
7. Aronszajn, N.: Theory of Reproducing Kernels. Transactions of the American Mathematical Society 68, 337–404 (1950)
8. Mercer, J.: Functions of Positive and Negative Type and Their Connection with The Theory of Integral Equations. Transactions of the London Philosophical Society A, 415–446 (1909)
9. Tanaka, A., Imai, H., Kudo, M., Miyakoshi, M.: Optimal Kernel in a Class of Kernels with an Invariant Metric. In: da Vitoria Lobo, N., Kasparis, T., Roli, F., Kwok, J.T., Georgiopoulos, M., Anagnostopoulos, G.C., Loog, M. (eds.) S+SSPR 2008. LNCS, vol. 5342, pp. 530–539. Springer, Heidelberg (2008)
10. Schatten, R.: Norm Ideals of Completely Continuous Operators. Springer, Berlin (1960)
11. Ogawa, H.: Neural Networks and Generalization Ability. IEICE Technical Report NC95-8, 57–64 (1995)

12. Rao, C.R., Mitra, S.K.: Generalized Inverse of Matrices and its Applications. John Wiley & Sons (1971)
13. Tanaka, A., Imai, H., Kudo, M., Miyakoshi, M.: Theoretical Analyses on a Class of Nested RKHS's. In: 2011 IEEE International Conference on Acoustics, Speech, and Signal Processing (ICASSP 2011), pp. 2072–2075 (2011)
14. Tanaka, A., Miyakoshi, M.: Theoretical Analyses for a Class of Kernels with an Invariant Metric. In: 2010 IEEE International Conference on Acoustics, Speech, and Signal Processing, pp. 2074–2077 (2010)

Simultaneous Learning of Localized Multiple Kernels and Classifier with Weighted Regularization

Naoya Inoue[1] and Yukihiko Yamashita[2]

[1] Planning & Development Division, Planex Communications Inc. 2-11-9,
Ebisu-Nishi, Shibuya-ku, Tokyo 150-0021 Japan
n708i@yy.ide.titech.ac.jp
[2] Graduate School of Science and Engineering, Tokyo Institute of Technology,
2-12-1-S6-19, O-okayama, Meguro-ku, Tokyo 152-8553, Japan
yamasita@ide.titech.ac.jp

Abstract. Kernel classifiers have demonstrated their high performance for many classification problems. For the proper selection of kernel functions, multiple kernel learning (MKL) has been researched. Furthermore, the localized MKL (LMKL) enables to set the weights for the kernel functions at each point. However, the training of the weight functions for kernel functions is a complex nonlinear problem and a classifier can be trained separately after the weights are fixed. The iteration of the two processes are often necessary. In this paper we propose a new framework for MKL/LMKL. In the framework, not kernel functions but mappings to the feature space are combined with weights. We also propose a new learning scheme to train simultaneously weights for kernel functions and a classifier. We realize a classifier by our framework with the Gaussian kernel function and the support vector machine. Finally, we show its advantages by experimental results.

1 Introduction

Kernel classifiers such as the support vector machine (SVM) [8], the kernel Fisher discriminant (KFD) [5], etc. have demonstrated their high performance for many classification problems. By mapping an input vector into a high dimensional feature space, nonlinear classification functions in the input space are provided even if they are linear in the feature space. However, to realize such excellent performances, the proper selection of kernel functions is very important. For the purpose the multiple kernel learning (MKL) has been intensively researched [6,1,4,7,3]. Their kernel function is given as a weighed combination of simple kernel functions. In MKL, the weights for kernel functions are spatially uniform. Because the spacial distribution of samples may not be uniform, the best kernel function should depend on its position. To solve this problem, Gönen and Alpaydin proposed the localized multiple kernel learning (LMKL) that can change the weights for kernel functions at each point [2]. However, because the weights are necessary for an unknown input pattern, we have to train the weight

G.L. Gimel' farb et al. (Eds.): SSPR & SPR 2012, LNCS 7626, pp. 354–362, 2012.
© Springer-Verlag Berlin Heidelberg 2012

functions for kernel functions before training SVM. It is a complex nonlinear problem. Therefore, their proposed learning process consists of the iteration of 2 processes, training of weight functions for kernel functions and training of SVM.

In this paper, we propose a new framework for MKL or LMKL. A kernel function is given by a weighted combination of kernel functions, and weights for kernel functions and coefficients of classifiers are separated in the original framework, whereas a nonlinear mapping to the feature space is given by a weighted combination of feature mappings, and combined weights for nonlinear mappings and classifiers are introduced in the proposed framework. We also propose a learning process to train the combined weights. We call this learning framework the simultaneous LMKL (SLMKL).

We apply the framework to SVM with the Gaussian kernel function (GKF). This classifier is called SLMKL-SVM. The problem to train the combined weights for SLMKL-SVM is given as a standard convex quadratic problem that can be calculated by using a optimization package such as CPLEX. We also show the advantages of SLMKL-SVM over SVM by experimental results.

We explain the framework of SLMKL in Section 2, and SLMKL-SVM in Section 3. We show experimental results in Section 4, and we conclude this paper and provide future works in Section 5.

2   Framework for SLMKL

Let $\{(\boldsymbol{x}_n, y_n)\}_{n=1}^{N}$ be a set of samples, where $\boldsymbol{x}_n$ is a sample pattern and y_n is its label (± 1). The inner product of $\boldsymbol{x}$ and $\boldsymbol{z}$ is denoted by $\langle \boldsymbol{x}, \boldsymbol{z} \rangle$. In this paper, we consider a linear discriminant model for a binary problem.

In case of the standard kernel method, we fix a Mercer kernel function $k(\boldsymbol{x}, \boldsymbol{z})$ and the discriminant function is given by

$$d(\boldsymbol{x}) = \sum_{n=1}^{N} \alpha_n k(\boldsymbol{x}_n, \boldsymbol{x}) + \theta, \tag{1}$$

where α_n and θ are coefficients and a threshold, respectively. From Mercer's theorem, we can consider there exists a feature mapping $\boldsymbol{\Phi}(\boldsymbol{x})$ such that $k(\boldsymbol{x}, \boldsymbol{z}) = \langle \boldsymbol{\Phi}(\boldsymbol{x}), \boldsymbol{\Phi}(\boldsymbol{z}) \rangle$. Then, the eq.(1) can be written as

$$d(\boldsymbol{x}) = \langle \boldsymbol{w}, \boldsymbol{\Phi}(\boldsymbol{x}) \rangle + \theta, \qquad \boldsymbol{w} = \sum_{n=1}^{N} \alpha_n \boldsymbol{\Phi}(\boldsymbol{x}_n). \tag{2}$$

In MKL, we prepare several kernel functions $k_j(\boldsymbol{x}, \boldsymbol{z})$ $(j = 1, 2, \ldots, J)$ and construct a kernel function as

$$k(\boldsymbol{x}, \boldsymbol{z}) = \sum_{j=1}^{J} \eta_j k_j(\boldsymbol{x}, \boldsymbol{z}), \tag{3}$$

where η_i $(j = 1, 2, \ldots, J)$ are weights for kernel functions. The discriminant function is also given by eq.(1). In this framework, α_i and η_i are separately trained and the problem to obtain η_i is very complex and nonlinear.

In LMKL, the position dependent weight functions $\eta_j(\boldsymbol{x})$ $(j = 1, 2, \ldots, J)$ are used. The kernel function is given by

$$k(\boldsymbol{x}, \boldsymbol{z}) = \sum_{j=1}^{J} \eta_j(\boldsymbol{x}) k_j(\boldsymbol{x}, \boldsymbol{z}) \eta_j(\boldsymbol{z}), \tag{4}$$

The discriminant function is given by

$$d(\boldsymbol{x}) = \sum_{n=1}^{N} \alpha_n \sum_{j=1}^{J} \eta_j(\boldsymbol{x}_n) k_j(\boldsymbol{x}_n, \boldsymbol{x}) \eta_j(\boldsymbol{x}) + \theta, \tag{5}$$

The weights for kernel functions can be chosen at every point. However, it seems to be very difficult to obtain $\eta_j(\boldsymbol{x})$ from training data because $\boldsymbol{x}$ is an unknown data in the classification stage so that another training process is necessary to obtain $\eta_j(\boldsymbol{x})$. Therefore, its learning process consists of the iteration of 2 processes, training of $\eta_j(\boldsymbol{x})$ and training of α_n and θ.

In the proposing SLMKL, we assume that we can prepare $\boldsymbol{\Phi}_j(\boldsymbol{x})$ $(j = 0, 1, 2, \ldots, J)$ of which inner product can be calculated analytically. Their kernel functions are denoted by

$$k_{i,j}(\boldsymbol{x}, \boldsymbol{z}) \equiv \langle \boldsymbol{\Phi}_i(\boldsymbol{x}), \boldsymbol{\Phi}_j(\boldsymbol{z}) \rangle. \tag{6}$$

Although the existence of inner product seems to be a strong condition, in Section 4, we show the inner product of GKFs of which kernel parameters are different. Because GKF is widely used in many recognition problems, the proposing framework can be used widely.

The model of $\boldsymbol{w}$ in eq.(2) is defined by J feature mappings as

$$\boldsymbol{w} = \sum_{n=1}^{N} \sum_{j=1}^{J} \alpha_{j,n} \boldsymbol{\Phi}_j(\boldsymbol{x}_n). \tag{7}$$

The parameters $\alpha_{j,n}$ express both weights for kernel functions and coefficients for a classifier. Since it is difficult to fix weights for feature mappings for an unknown input pattern, we used only $\boldsymbol{\Phi}_0$ for the feature mapping into which an unknown input pattern is substituted. Then, the discriminant function is given by

$$d(\boldsymbol{x}) = \langle \boldsymbol{w}, \boldsymbol{\Phi}_0(\boldsymbol{x}) \rangle + \theta = \sum_{n=1}^{N} \sum_{j=1}^{J} \alpha_{j,n} k_{j,0}(\boldsymbol{x}_n, \boldsymbol{x}) + \theta. \tag{8}$$

The advantages of this frame work are as follows. (1) The weights for feature mappings can be changed at each sample point individually. (2) It does not need a continuous weight functions for feature mappings whereas LMKL has $\eta_j(\boldsymbol{x})$. (3) The coefficients $\alpha_{j,n}$ can express simultaneously weights for kernel functions and coefficients for a classifier and can be trained similarly to the original classifier.

3 SLMKL-SVM

The criterion of SVM is given as minimizing

$$\|w\|^2 + C\sum_{n=1}^{N}\xi_i \tag{9}$$

under the condition that for $n = 1, 2, \ldots, N$,

$$y_n d(x_n) - 1 + \xi_n \geq 0, \quad \xi_n \geq 0, \tag{10}$$

where C is a constant. The first term of eq.(9) comes from minimizing the margin $1/\|w\|$. It can be considered as a regularization term. The second term is the hinge loss.

In this section, we consider the simplest case of SLMKL in order to investigate basic features of SLMKL. Let $J = 2$ and we use two feature mappings Φ_1 and Φ_2. We use Φ_2 for Φ_0 that is the feature mapping into which an unknown input pattern is substituted. Then, w can be expressed by $w = w_1 + w_2$, where

$$w_j = \sum_{n=1}^{N}\alpha_{j,n}\Phi_j(x_n) \quad (j = 1, 2). \tag{11}$$

However, even if we substitute eqs.(8) and (11) to eqs.(9) and (10), respectively, the result of the training is reduced to a standard SVM. The representer theorem [9] ensures that w has to be spanned only by $\{\Phi_2(x_n)\}_{n=1}^{N}$. That implies $\alpha_{1,n} = 0$ for all n and the classifier is a standard SVM.

We have another problem in the framework of SLMKL with respect to the ratio of weights between for $\alpha_{1,n}$ and for $\alpha_{2,n}$. The main term in the discriminant function or the hinge loss is

$$\langle w, x \rangle = \sum_{n=1}^{N}\sum_{i=1}^{2}\alpha_{i,n}k_{i,2}(x_n, x). \tag{12}$$

In the training stage, sample points are substituted into x. The dominant weights for $\alpha_{1,n}$ and $\alpha_{2,n}$ are given by $k_{1,2}(x_n, x_n)$ and $k_{2,2}(x_n, x_n)$, respectively. For example, we assume that GKF in eq.(16) is used and we let $M = 20$ and $\sigma_1 = 3\sigma_2$. Then, the former is approximately 10^{-8} and the latter is 1. On the other hand, in the regularization term

$$\|w\|^2 = \sum_{m=1}^{N}\sum_{n=1}^{N}\sum_{i=1}^{2}\sum_{j=1}^{2}k_{i,j}(x_m, x_n)\alpha_{i,m}\alpha_{j,n}, \tag{13}$$

their dominant weights for them are $k_{1,1}(x_n, x_n)$ and $k_{2,2}(x_n, x_n)$, respectively, and both are 1. Accordingly, the regularization for $\alpha_{1,n}$ is much stronger than that for $\alpha_{2,n}$.

We explain this problem in the point of view of linear algebra. Let S_j be the subspace spanned by $\{\boldsymbol{\Phi}_j(\boldsymbol{x}_n)\}_{n=1}^{N}$ $(j = 1, 2)$. When $k_{1,2}(\boldsymbol{x}_n, \boldsymbol{x}_n)$ is small, S_1 and S_2 are nearly orthogonal. The vector $\boldsymbol{w}$ is included in $S_1 + S_2$, and $\boldsymbol{w}$ is evaluated with vectors in $S_1 + S_2$ in the regularization term. However, it is evaluated only with vectors in S_2 in the hinge loss term. Then, in the term, the weight for $\alpha_{1,n}$ becomes smaller than that for $\alpha_{2,n}$.

In order to solve this problem, we have to weaken the regularization for $\alpha_{1,n}$ and we introduce the following five new weighted regularization terms.

(1) Projected regularization (PrR): Let P_2 be the orthogonal projection operator onto S_2. The regularization term is defined by $\|P_2\boldsymbol{w}\|^2$. By projecting $\boldsymbol{w}$ onto S_2 in the regularization, the norm of projection of a vector in S_1 becomes be comparable with the values of inner products with a vector in S_2.

(2) Training sample regularization (TSR): The regularization term is given by $\sum_{n=0}^{N} |\langle \boldsymbol{w}, \boldsymbol{\Phi}_{\sigma_2}(x_n)\rangle|^2$. Because the sum of squared norms of $\langle \boldsymbol{w}, \boldsymbol{\Phi}_2(x_n)\rangle$, which are the same terms in the hinge loss, is used for the regularization, the strength of regularization is balanced.

(3) Peak Regularization (PkR): $\alpha_{j,n}$ in the regularization term is replaced by $k_{j,2}(\boldsymbol{x}_n, \boldsymbol{x}_n)\alpha_{j,n}$. It weaken the regularization of $\alpha_{1,n}$ by multiplying $k_{1,2}(\boldsymbol{x}_n, \boldsymbol{x}_n)$ whereas that of $\alpha_{2,n}$ does not change it because $k_{2,2}(\boldsymbol{x}_n, \boldsymbol{x}_n) = 1$. The regularization term is given by

$$\sum_{m=1}^{N}\sum_{n=1}^{N}\sum_{i=1}^{2}\sum_{j=1}^{2} k_{i,2}(\boldsymbol{x}_m, \boldsymbol{x}_m)k_{j,2}(\boldsymbol{x}_n, \boldsymbol{x}_n)k_{i,j}(\boldsymbol{x}_m, \boldsymbol{x}_n)\alpha_{i,m}\alpha_{j,n}. \qquad (14)$$

(4) Sum of kernel function regularization (SKFR): $\alpha_{j,n}$ in the regularization term is replaced by $\sum_{m=1}^{N} k_{j,2}(\boldsymbol{x}_n, \boldsymbol{x}_m)\alpha_{j,n}$. The strengths of the regularizations for $\alpha_{1,n}$ and $\alpha_{2,n}$ are changed by $\sum_{m=1}^{N} k_{1,2}(\boldsymbol{x}_n, \boldsymbol{x}_m)$ and $\sum_{m=1}^{N} k_{2,2}(\boldsymbol{x}_n, \boldsymbol{x}_m)$, respectively. PkR uses the value of a kernel function with $\boldsymbol{x}_n$ itself for $\alpha_{j,n}$ but SKFR uses sum with all samples.

(5) Square root of sum of kernel function regularization (SSKFR): $\alpha_{j,n}$ in the regularization term is replaced by $\sqrt{\sum_{m=1}^{N} k_{j,2}(\boldsymbol{x}_n, \boldsymbol{x}_m)}\alpha_{j,n}$. Since the change of weights in SKFR seems to be large, we use the square root of SKFR although it seems to be heuristic.

With each regularization term, the criterion can be transformed to a constrained quadratic optimization problem.

4   Experiments

We show experimental results of a toy problem and 13 types of UCI datasets used in [5]. We use the Gaussian kernel function (GKF) for a feature mapping. We prepare two kernel parameters σ_1 and σ_2. Then, the feature mapping with σ_j from a point $\boldsymbol{x}$ to a function is defined by

$$(\boldsymbol{\Phi}_j(\boldsymbol{x}))(\boldsymbol{z}) = \left(\frac{2}{\pi\sigma_j^2}\right)^{\frac{M}{2}} \exp(-\|\boldsymbol{z} - \boldsymbol{x}\|^2/\sigma_j^2). \qquad (15)$$

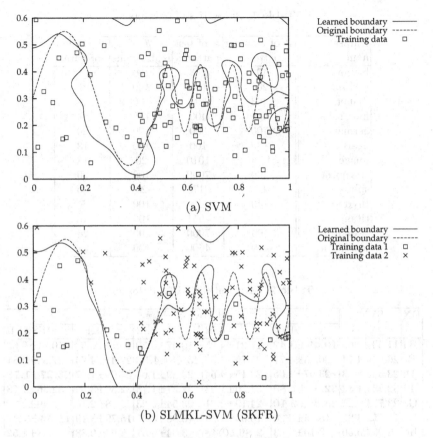

(a) SVM

(b) SLMKL-SVM (SKFR)

Fig. 1. Results for artificial data

for $j = 1, 2$. From eq.(6), the kernel function for SLMKL is given by

$$k_{i,j}(\boldsymbol{x}, \boldsymbol{z}) = \left(\frac{2\sigma_i \sigma_j}{\sigma_i^2 + \sigma_j^2}\right)^M \exp\left(-\frac{\|\boldsymbol{x} - \boldsymbol{z}\|^2}{\sigma_i^2 + \sigma_j^2}\right). \tag{16}$$

We let $\sigma_1^2 = r\sigma^2$ and $\sigma_2^2 = (2 - r)\sigma^2$. If $r = 1$, it reduces to the standard SVM. We set $r = 1.5$ and $r = 1.8$ for SLMKL-SVM.

In order to see the basic performance of SLMKL-SVM, we conducted an experiment with artificial data. As shown in Fig. 1, the domain is a two-dimensional region $D = \{(x, y)|x \in [0, 1], y \in [0, 0.6]\}$. The original boundary (dashed line) consists of two kinds of sinusoidal functions. It is more complex in the right section of D. Furthermore, samples (small square or cross points) are denser in the region $[0.4, 1] \times [0.1, 0.5]$. The error rate is evaluated by 201×201 uniform samples in D. The parameters (σ^2 and C) are set to provide the best recognition rate for each classifier. The error rates for SVM, and SLMKL-SVM with PrR, TSR, PkR, SKFR, and SSKFR are 12.95%, 11.18%, 14.07%, 11.42%, 10.79%,

Table 1. Properties of datasets

Dataset name	# of training patterns	# of test patterns	# of realizations	dimension of data
banana	400	4900	100	2
breast cancer	200	77	100	9
diabetis	468	300	100	8
flare solar	666	400	100	9
german	700	300	100	20
heart	170	100	100	13
image	1300	1010	20	18
ringnorm	400	7000	100	20
splice	1000	2175	20	60
thyroid	140	75	100	5
titanic	150	2051	100	3
twonorm	400	7000	100	20
waveform	400	4600	100	21

Table 2. Error rates for UCI datasets

DN	SVM	SLMKL-SVM $r = 1.5$				
		PrR	TSR	PkR	SKFR	SSKFR
Ba	11.53±0.66	**10.43±0.42**	11.21±4.25	10.47±0.50	10.71±0.53	10.91±0.63
Br	26.04±4.74	26.43±4.77	25.70±4.32	**25.48±4.53**	27.49±4.74	27.60±4.81
Di	23.53±1.73	23.97±1.65	24.15±4.54	23.49±1.64	24.09±1.77	**23.27±1.70**
Fl	32.43±1.82	**32.33±1.80**	33.70±1.98	32.37±1.78	32.86±3.50	32.36±1.78
Ge	23.71±2.20	**23.47±2.10**	24.19±2.21	23.54±2.23	23.68±2.16	23.49±2.37
He	15.95±3.26	15.65±3.28	16.63±7.29	15.96±3.31	15.79±3.19	**15.53±3.37**
Im	2.96±0.60	3.10±0.53	**2.89±0.48**	3.04±0.51	3.63±0.88	3.68±1.52
Ri	1.66±0.12	**1.47±0.11**	30.31±21.77	1.48±0.11	1.72±0.12	1.58±0.10
Sp	10.88±0.66	10.80±0.67	**10.79±0.68**	11.00±0.66	11.06±0.70	11.09±0.75
Th	4.80±2.19	**4.31±2.21**	7.47±3.20	4.97±2.25	4.72±2.25	4.60±2.40
Ti	22.42±1.02	22.77±1.11	23.06±5.34	22.75±1.12	22.77±1.18	**22.36±1.00**
Tw	2.96±0.23	**2.39±0.12**	2.95±0.34	2.41±0.13	2.42±0.13	2.49±0.16
Wa	**9.88±0.43**	9.98±0.49	10.52±4.94	9.99±0.52	10.07±0.40	10.19±0.47

(Values in bold font indicates the best result in the row.)

and 12.23%, respectively. The calculated boundaries (solid lines) of SVM and SLMKL-SVM with SKFR are shown for their best parameters.

In Fig. 1 (b), a small square point expresses training data x_n where $\langle w_1, \Phi_2(x_n) \rangle > \langle w_2, \Phi_2(x_n) \rangle$ and a small cross point expresses the opposite. We can see large and small kernel parameters are selected in smooth and complicated region, respectively, by the training.

We also conducted experiments using the 13 UCI datasets listed in Table 1. For example, the 'banana' dataset has 100 realizations and each realization has 400 training and 4900 test patterns of which dimension is two.

Table 3. Error rates for UCI datasets

DN	SVM	SLMKL-SVM $r = 1.8$				
		PrR	TSR	PkR	SKFR	SSKFR
Ba	11.53±0.66	**10.41±0.44**	10.80±0.50	10.69±0.55	10.69±0.53	10.70±0.54
Br	26.04±4.74	**25.56±4.71**	26.23±5.67	25.74±4.71	27.58±4.78	27.38±4.76
Di	23.53±1.73	23.38±1.67	**23.37±1.81**	23.85±1.59	23.60±1.76	23.67±1.79
Fl	32.43±1.82	32.37±1.78	33.42±1.51	**32.36±1.78**	32.59±2.85	**32.36±1.79**
Ge	23.71±2.20	23.66±2.19	24.15±2.13	23.73±2.17	**23.55±2.33**	23.70±2.14
He	15.95±3.26	**15.26±3.16**	15.62±3.09	16.00±3.23	15.57±3.22	16.10±3.28
Im	2.96±0.60	3.08±0.56	**2.85±0.45**	3.05±0.54	3.14±0.62	3.25±0.58
Ri	1.66±0.12	**1.45±0.10**	28.45±22.68	**1.45±0.10**	1.91±0.20	1.63±0.12
Sp	10.88±0.66	**10.77±0.71**	10.84±0.61	**10.77±0.71**	11.14±0.71	11.11±0.71
Th	4.80±2.19	4.44±2.22	6.55±3.12	4.83±2.26	4.64±2.22	**4.36±2.06**
Ti	**22.42±1.02**	22.43±1.02	22.46±1.07	24.91±8.47	23.00±4.81	22.69±1.56
Tw	2.96±0.23	2.45±0.14	3.00±0.23	**2.40±0.12**	2.54±0.16	2.45±0.13
Wa	9.88±0.43	10.60±2.30	**9.73±0.45**	10.38±0.47	10.49±0.43	10.55±0.35

(Values in bold font indicates the best result in the row.)

Table 4. p-value of t-test against SVM

DN	$r = 1.5$					$r = 1.8$				
	PrP	TSR	PkR	SKFR	SSKFR	PrP	TSR	PkR	SKFR	SSKFR
Ba	**0.000**	0.230	**0.000**	**0.000**	**0.000**	**0.000**	**0.000**	**0.000**	**0.000**	**0.000**
Br	0.718	0.299	0.198	0.984	0.989	0.238	0.601	0.328	0.988	0.976
Di	0.966	0.897	0.434	0.987	0.144	0.268	0.263	0.912	0.611	0.712
Fl	0.349	1.000	0.407	0.860	0.392	0.407	1.000	0.392	0.681	0.393
Ge	0.319	0.971	0.410	0.592	0.352	0.565	0.964	0.655	0.424	0.618
He	0.260	0.801	0.509	0.364	0.187	0.066	0.233	0.543	0.205	0.626
Im	0.775	0.347	0.670	0.995	0.969	0.736	0.263	0.685	0.816	0.931
Ri	**0.000**	1.000	**0.000**	1.000	**0.000**	**0.000**	1.000	**0.000**	1.000	**0.040**
Sp	0.356	0.341	0.711	0.790	0.817	0.312	0.424	0.312	0.875	0.846
Th	0.059	1.000	0.705	0.400	0.270	0.126	1.000	0.538	0.305	0.073
Ti	0.989	0.879	0.984	0.987	0.338	0.527	0.606	0.998	0.879	0.924
Tw	**0.000**	0.404	**0.000**	**0.000**	**0.000**	**0.000**	0.889	**0.000**	**0.000**	**0.000**
Wa	0.936	0.900	0.947	0.999	1.000	0.999	**0.009**	1.000	1.000	1.000

The parameters σ and C were selected by 5-fold cross validation as explained
in [5]. The averaged error rate and its standard deviation over all realizations
are shown in Tables 2 and 3. Their p-values of t-test against SVM are shown in
Table 4. The results of SVM are referred from [5].

We can see the advantage of the proposed method especially for 'banana',
'ringnorm', 'twonorm' and 'waveform'. However, because the result for 'ringnorm' with TSR is very bad, we have to research on the combination of mappings
to the feature space and regularization in the criterion.

5 Conclusion

In this paper, we proposed a new framework for MKL or LMKL. In the framework, not kernel functions but feature mappings are linearly combined. We also proposed the SLMKL-SVM with weighted regularizations. We conducted experiments and showed its advantages.

For future work, we have to research on feature mappings of which inner product can be obtained analytically and investigate regularization terms to improve its performance.

References

1. Bach, F.R., Lanckriet, G.R.G., Jordan, M.I.: Multiple kernel learning, conic duality, and the smo algorithm. In: ICML (July 2004)
2. Gönen, M., Alpaydin, E.: Localized multiple kernel learning. In: Proceedings of the 25th International Conference on Machine Learning, ICML 2008, pp. 352–359. ACM, New York (2008), http://doi.acm.org/10.1145/1390156.1390201
3. Gönen, M., Alpaydin, E.: Multiple kernel learning algorithm. Journal of Machine Learning Research 12, 2211–2268 (2011)
4. Lanckriet, G.R.G., Bie, T.D., Cristianini, N., Jordan, M.I., Noble, W.S.: A statistical framework for genomic data fusion. Some Fine Journal 20(16), 2626–2635 (2004)
5. Mika, S., Rätsch, G., Weston, J., Schölkopf, B., Müller, K.R.: Fisher discriminant analysis with kernels. In: Neural Networks for Signal Processing IX, pp. 41–48. IEEE (1999)
6. Pavlidis, P., Weston, J., Cai, J., Grundy, W.N.: Gene functional classification from heterogeneous data. In: Proceedings of the Fifth Annual International Conference on Computational Biology, RECOMB 2001, pp. 249–255. ACM, New York (2001), http://doi.acm.org/10.1145/369133.369228
7. Sonnenburg, S., Rätsch, G., Schäfer, C., Schölkopf, B.: Large scale multiple kernel learning. J. Mach. Learn. Res. 7, 1531–1565 (2006), http://dl.acm.org/citation.cfm?id=1248547.1248604
8. Vapnik, V.N.: Statistical Learning Theory. Wiley, New-York (1998)
9. Wahba, G.: Spline Models for Observational Data (CBMS-NSF Regional Conference Series in Applied Mathematics). Society for Industrial and Applied Mathematics, Philadelphia (1990)

Change-Point Detection in Time-Series Data by Relative Density-Ratio Estimation

Song Liu[1], Makoto Yamada[2], Nigel Collier[3], and Masashi Sugiyama[1]

[1] Tokyo Institute of Technology
2-12-1 O-okayama, Meguro-ku, Tokyo 152-8552, Japan
{song@sg.,sugi@}cs.titech.ac.jp
[2] NTT Communication Science Laboratories
2-4, Hikaridai, Seika-cho, Kyoto, Japan 619-0237
yamada.makoto@lab.ntt.co.jp
[3] National Institute of Informatics
2-1-2 Hitotsubashi, Chiyoda-ku, Tokyo 101-8430, Japan
collier@nii.ac.jp

Abstract. The objective of change-point detection is to discover abrupt property changes lying behind time-series data. In this paper, we present a novel statistical change-point detection algorithm that is based on non-parametric divergence estimation between two retrospective segments. Our method uses the relative Pearson divergence as a divergence measure, and it is accurately and efficiently estimated by a method of direct density-ratio estimation. Through experiments on real-world human-activity sensing, speech, and Twitter datasets, we demonstrate the usefulness of the proposed method.

Keywords: change-point detection, distribution comparison, relative density-ratio estimation, kernel methods, time-series data.

1 Introduction

Detecting abrupt changes in time-series data, called *change-point detection*, has attracted researchers in the statistics and data mining communities for decades [1–6].

Some pioneer works demonstrated good change-point detection performance by comparing the probability distributions of time-series samples over past and present intervals [1]. As both the intervals move forward, a typical strategy is to issue an alarm for a change point when the two distributions are becoming significantly different. Various change-point detection methods follow this strategy, for example, the *cumulative sum* [1], the *generalized likelihood-ratio method* [2], and the *change finder* [3].

Another group of methods that have attracted high popularity in recent years is the *subspace methods* [4, 5]. By using a pre-designed time-series model, a subspace is discovered by principle component analysis from trajectories in past

G.L. Gimel' farb et al. (Eds.): SSPR & SPR 2012, LNCS 7626, pp. 363–372, 2012.

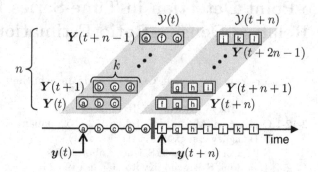

Fig. 1. Notation

and present intervals, and their dissimilarity is measured by the distance between the subspaces. One of the major approaches is called *subspace identification* [5], which compares the subspaces spanned by the columns of an *extended observability matrix* generated by a state-space model with system noise.

However, the methods explained above rely on pre-designed parametric models such as underlying probability distributions [1, 2], auto-regressive models [3], and state-space models [4, 5], for tracking some specific statistics such as the mean, the variance, and the spectrum. Thus, they are not robust against different types of changes, which significantly limits the range of applications in practice. To cope with this problem, non-parametric estimation methods such as *kernel density estimation* may be used. However, non-parametric methods tend to be less accurate in high-dimensional problems because of the so-called *curse of dimensionality*.

To overcome this difficulty, a new strategy was introduced recently which estimates the *ratio* of probability densities directly without going through density estimation [7]. In the context of change-point detection, a direct density-ratio estimation method called the *Kullback-Leibler importance estimation procedure* (KLIEP) [8] was reported to outperform competitive approaches [6] such as the *one-class support vector machine* [9] and *singular-spectrum analysis* [4].

The goal of this paper is to further advance this line of research. More specifically, our contributions in this paper are two folds.

• We apply a recently-proposed density-ratio estimation method called the *unconstrained least-squares importance fitting* (uLSIF) [10] to change-point detection. Notable advantages of uLSIF are that an analytical solution can be obtained, it achieves the optimal non-parametric convergence rate, it has optimal numerical stability, and it has higher robustness [7].
• We further improve the uLSIF-based change-point detection method by employing a state-of-the-art extension of uLSIF called *relative uLSIF* (RuLSIF) [11], which was proved to have an even better non-parametric convergence property than plain uLSIF [11], with other advantages of uLSIF maintained.

2 Problem Formulation

In this section, we formulate our change-point detection problem (see Figure 1).
Let $\boldsymbol{y}(t) \in \mathbb{R}^d$ be a d-dimensional time-series sample at time t. Let

$$\boldsymbol{Y}(t) := [\boldsymbol{y}(t)^\top, \boldsymbol{y}(t+1)^\top, \ldots, \boldsymbol{y}(t+k-1)^\top]^\top \in \mathbb{R}^{dk}$$

be a subsequence of time series at time t with length k, where $^\top$ represents the
transpose. Following the previous work [6], we treat the subsequence $\boldsymbol{Y}(t)$ as
a sample, instead of a single point $\boldsymbol{y}(t)$, by which time-dependent information
can be incorporated naturally. Let $\mathcal{Y}(t)$ be a set of n retrospective subsequence
samples starting at time t:

$$\mathcal{Y}(t) := \{\boldsymbol{Y}(t), \boldsymbol{Y}(t+1), \ldots, \boldsymbol{Y}(t+n-1)\}.$$

For change-point detection, let us consider two consecutive segments $\mathcal{Y}(t)$ and
$\mathcal{Y}(t+n)$. Our strategy is to compute a certain dissimilarity measure between $\mathcal{Y}(t)$
and $\mathcal{Y}(t+n)$, and use it as the plausibility of change points. More specifically,
the larger the dissimilarity is, the more likely the point is a change point.

   Now the problems that need to be addressed are what kind of dissimilarity
measure we should use and how we estimate it from data. We will discuss these
issues in the next section.

3 Change-Point Detection via Density-Ratio Estimation

In this section, we first define our dissimilarity measure, and then show methods
for estimating the dissimilarity measure.

3.1 Divergence-Based Dissimilarity Measure

We use a dissimilarity measure of the following form:

$$D(P_t \| P_{t+n}) + D(P_{t+n} \| P_t), \tag{1}$$

where P_t and P_{t+n} are probability distributions of samples in $\mathcal{Y}(t)$ and $\mathcal{Y}(t+n)$,
respectively. $D(P \| P')$ denotes the f-divergence [12, 13]:

$$D(P \| P') := \int p'(\boldsymbol{Y}) f\left(\frac{p(\boldsymbol{Y})}{p'(\boldsymbol{Y})}\right) \mathrm{d}\boldsymbol{Y},$$

where f is a convex function such that $f(1) = 0$, and $p(\boldsymbol{Y})$ and $p'(\boldsymbol{Y})$ are
probability density functions of P and P', respectively. Because the f-divergence
is not symmetric, we use a symmetrized divergence in Eq.(1).

   The f-divergence includes various popular divergences such as the *Kullback-
Leibler (KL) divergence* by $f(t) = t \log t$ and the *Pearson (PE) divergence* by
$f(t) = \frac{1}{2}(t-1)^2$:

$$\mathrm{KL}(P \| P') := \int p(\boldsymbol{Y}) \log \frac{p(\boldsymbol{Y})}{p'(\boldsymbol{Y})} \mathrm{d}\boldsymbol{Y} \quad \text{and} \quad \mathrm{PE}(P \| P') := \frac{1}{2} \int p'(\boldsymbol{Y}) \left(\frac{p(\boldsymbol{Y})}{p'(\boldsymbol{Y})} - 1\right)^2 \mathrm{d}\boldsymbol{Y}.$$

In the rest of this section, we explain three methods of directly estimating the density ratio $\frac{p(\boldsymbol{Y})}{p'(\boldsymbol{Y})}$ from samples $\{\boldsymbol{Y}_i\}_{i=1}^n$ and $\{\boldsymbol{Y}'_j\}_{j=1}^n$ drawn from $p(\boldsymbol{Y})$ and $p'(\boldsymbol{Y})$: the *KL importance estimation procedure* (KLIEP) [8] in Section 3.2, *unconstrained least-squares importance fitting* (uLSIF) [10] in Section 3.3, and *relative uLSIF (RuLSIF)* [11] in Section 3.4.

3.2 Kullback-Leibler Importance Estimation Procedure (KLIEP)

KLIEP [8] is a direct density-ratio estimation algorithm that is suitable for estimating the KL divergence.

Density-Ratio Model: Let us model the density ratio $\frac{p(\boldsymbol{Y})}{p'(\boldsymbol{Y})}$ by the following kernel model:

$$g(\boldsymbol{Y};\boldsymbol{\theta}) := \sum_{\ell=1}^n \theta_\ell K(\boldsymbol{Y}, \boldsymbol{Y}_\ell), \tag{2}$$

where $\boldsymbol{\theta} := (\theta_1, \ldots, \theta_n)^\top$ are parameters to be learned from data samples, and $K(\boldsymbol{Y}, \boldsymbol{Y}')$ is a kernel basis function. In practice, we use the Gaussian kernel and the kernel width is chosen by cross-validation (see [8] for details).

Learning Algorithm: The parameters $\boldsymbol{\theta}$ in the model $g(\boldsymbol{Y};\boldsymbol{\theta})$ are determined so that the empirical KL divergence from $p(\boldsymbol{Y})$ to $g(\boldsymbol{Y};\boldsymbol{\theta})p'(\boldsymbol{Y})$ is minimized:

$$\max_{\boldsymbol{\theta}} \frac{1}{n}\sum_{i=1}^n \log\left(\sum_{\ell=1}^n \theta_\ell K(\boldsymbol{Y}_i, \boldsymbol{Y}_\ell)\right) \text{ s.t. } \frac{1}{n}\sum_{j=1}^n\sum_{\ell=1}^n \theta_\ell K(\boldsymbol{Y}'_j, \boldsymbol{Y}_\ell)=1, \quad \theta_1, \ldots, \theta_n \geq 0.$$

The equality constraint is for normalization purposes because $g(\boldsymbol{Y};\boldsymbol{\theta})p'(\boldsymbol{Y})$ should be a probability density function. The inequality constraint comes from the non-negativity of the density-ratio function. Since this is a convex optimization problem, the unique global optimal solution $\widehat{\boldsymbol{\theta}}$ can be simply obtained, for example, by a gradient-projection iteration. Finally, a density-ratio estimator is given as

$$\widehat{g}(\boldsymbol{Y}) = \sum_{\ell=1}^n \widehat{\theta}_\ell K(\boldsymbol{Y}, \boldsymbol{Y}_\ell). \tag{3}$$

KLIEP was shown to achieve the optimal non-parametric convergence rate [8].

Change-Point Detection by KLIEP: Given a density-ratio estimator $\widehat{g}(\boldsymbol{Y})$, an approximator of the KL divergence is given as

$$\widehat{\mathrm{KL}} := \frac{1}{n}\sum_{i=1}^n \log \widehat{g}(\boldsymbol{Y}_i).$$

In the previous work [6], this KLIEP-based KL-divergence estimator was applied to change-point detection and demonstrated to be promising in experiments.

3.3   Unconstrained Least-Squares Importance Fitting (uLSIF)

Recently, another direct density-ratio estimator called uLSIF was proposed [10], which is suitable for estimating the PE divergence.

Learning Algorithm: In uLSIF, the same density-ratio model $g(\boldsymbol{Y};\boldsymbol{\theta})$ as KLIEP (see Eq.(2)) is used. However, its training criterion is different; the density-ratio model is fitted to the true density ratio under the squared loss. More specifically, the parameter $\boldsymbol{\theta}$ in the model $g(\boldsymbol{Y};\boldsymbol{\theta})$ is determined so that the following squared loss $J(\boldsymbol{Y})$ is minimized:

$$
\begin{aligned}
J(\boldsymbol{Y}) &:= \frac{1}{2}\int\left(\frac{p(\boldsymbol{Y})}{p'(\boldsymbol{Y})}-g(\boldsymbol{Y};\boldsymbol{\theta})\right)^2 p'(\boldsymbol{Y})\,\mathrm{d}\boldsymbol{Y} \\
&= \frac{1}{2}\int\frac{p(\boldsymbol{Y})}{p'(\boldsymbol{Y})}^2 p'(\boldsymbol{Y})\,\mathrm{d}\boldsymbol{Y} - \int p(\boldsymbol{Y})g(\boldsymbol{Y};\boldsymbol{\theta})\,\mathrm{d}\boldsymbol{Y} + \frac{1}{2}\int g(\boldsymbol{Y};\boldsymbol{\theta})^2 p'(\boldsymbol{Y})\,\mathrm{d}\boldsymbol{Y}.
\end{aligned}
$$

Since the first term is a constant, we focus on the last two terms. By approximating the expectations by the empirical averages, the uLSIF optimization problem is given as follows:

$$
\min_{\boldsymbol{\theta}\in\mathbb{R}^n}\left[\frac{1}{2}\boldsymbol{\theta}^\top\widehat{\boldsymbol{H}}\boldsymbol{\theta}-\widehat{\boldsymbol{h}}^\top\boldsymbol{\theta}+\frac{\lambda}{2}\boldsymbol{\theta}^\top\boldsymbol{\theta}\right], \tag{4}
$$

where the penalty term $\frac{\lambda}{2}\boldsymbol{\theta}^\top\boldsymbol{\theta}$ is included for regularization purposes, and λ (≥ 0) denotes the regularization parameter, which is chosen by cross validation. (see [10] for details). $\widehat{\boldsymbol{H}}$ is the $n\times n$ matrix and $\widehat{\boldsymbol{h}}$ is the n-dimensional vector defined as

$$
\widehat{H}_{\ell,\ell'} := \frac{1}{n}\sum_{j=1}^n K(\boldsymbol{Y}'_j,\boldsymbol{Y}_\ell)K(\boldsymbol{Y}'_j,\boldsymbol{Y}_{\ell'}) \quad\text{and}\quad \widehat{h}_\ell := \frac{1}{n}\sum_{i=1}^n K(\boldsymbol{Y}_i,\boldsymbol{Y}_\ell).
$$

It is easy to confirm that the solution $\widehat{\boldsymbol{\theta}}$ of (4) can be analytically obtained as

$$
\widehat{\boldsymbol{\theta}} = (\widehat{\boldsymbol{H}}+\lambda\boldsymbol{I}_n)^{-1}\widehat{\boldsymbol{h}}, \tag{5}
$$

where $\boldsymbol{I}_n$ denotes the n-dimensional identity matrix. Finally, a density-ratio estimator is given by Eq.(3) with Eq.(5).

Change-Point Detection by uLSIF: Given a density-ratio estimator $\widehat{g}(\boldsymbol{Y})$, an approximator of the PE divergence can be constructed as

$$
\widehat{\mathrm{PE}} := -\frac{1}{2n}\sum_{j=1}^n \widehat{g}(\boldsymbol{Y}'_j)^2 + \frac{1}{n}\sum_{i=1}^n \widehat{g}(\boldsymbol{Y}_i) - \frac{1}{2}.
$$

This approximator is derived from the following expression of the PE divergence:

$$
\mathrm{PE}(P\|P') = -\frac{1}{2}\int\left(\frac{p(\boldsymbol{Y})}{p'(\boldsymbol{Y})}\right)^2 p'(\boldsymbol{Y})\mathrm{d}\boldsymbol{Y} + \int\left(\frac{p(\boldsymbol{Y})}{p'(\boldsymbol{Y})}\right)p(\boldsymbol{Y})\mathrm{d}\boldsymbol{Y} - \frac{1}{2}. \tag{6}
$$

Notable advantages of uLSIF are that its solution can be computed analytically, it possesses the optimal non-parametric convergence rate, it has the optimal numerical stability, and it has higher robustness [7]. As experimentally demonstrated in our supplementary technical report [14], uLSIF-based change-point detection compares favorably with the KLIEP-based method.

3.4   Relative uLSIF (RuLSIF)

Depending on the condition of the denominator density $p'(Y)$, the density-ratio value $\frac{p(Y)}{p'(Y)}$ can be unbounded (i.e., they can be infinity). This is actually problematic because the non-parametric convergence rate of uLSIF is governed by the "sup"-norm of the true density-ratio function: $\max_Y \frac{p(Y)}{p'(Y)}$. To overcome this problem, *relative density-ratio estimation* was introduced [11].

Relative PE Divergence: Let us consider the α-*relative PE-divergence* for $0 \le \alpha < 1$:

$$\mathrm{PE}_\alpha(P\|P') := \mathrm{PE}(P\|\alpha P + (1-\alpha)P') = \int p'_\alpha(Y)\,(r_\alpha(Y) - 1)^2\, dY,$$

where $p'_\alpha(Y) = \alpha p(Y) + (1-\alpha)p'(Y)$ and $r_\alpha(Y) = \frac{p(Y)}{p'_\alpha(Y)}$. We refer to $r_\alpha(Y)$ as the α-*relative density ratio*. The α-relative density ratio is reduced to the plain density ratio if $\alpha = 0$, and it tends to be "smoother" as α gets larger. Indeed, the α-relative density ratio is bounded above by $1/\alpha$ for $\alpha > 0$, even when the plain density ratio $\frac{p(Y)}{p'(Y)}$ is unbounded. This was proved to contribute to improving the estimation accuracy [11].

Learning Algorithm: In the same way as the uLSIF method, the parameter θ of the model $g(Y;\theta)$ is learned by minimizing the squared difference between true and estimated ratios:

$$\begin{aligned}
J(Y) &= \frac{1}{2}\int p'_\alpha(Y)(r_\alpha(Y) - g(Y;\theta))^2\, dY \\
&= \frac{1}{2}\int p'_\alpha(Y)r_\alpha^2(Y)dY - \int p(Y)r_\alpha(Y)g(Y;\theta)\, dY \\
&\quad + \frac{\alpha}{2}\int p(Y)g(Y;\theta)^2\, dY - \frac{1-\alpha}{2}\int p'(Y)g(Y;\theta)^2\, dY,
\end{aligned}$$

where the first term is a constant term. Note that we still use the same kernel model (2) as $g(Y;\theta)$ for approximating the α-relative density ratio.

Again, by ignoring the constant and approximating the expectations by empirical averages, the α-relative density ratio can be learned in the same way as the plain density ratio. Indeed, the optimization problem of a relative variant of uLSIF, called RuLSIF, is given as the same form as uLSIF; the only difference is the definition of the matrix $\widehat{H}$, which is now given by

$$\widehat{H}_{\ell,\ell'} := \frac{\alpha}{n} \sum_{i=1}^{n} K(\boldsymbol{Y}_i, \boldsymbol{Y}_\ell) K(\boldsymbol{Y}_i, \boldsymbol{Y}_{\ell'}) + \frac{(1-\alpha)}{n} \sum_{j=1}^{n} K(\boldsymbol{Y}'_j, \boldsymbol{Y}_\ell) K(\boldsymbol{Y}'_j, \boldsymbol{Y}_{\ell'}).$$

RuLSIF inherits the advantages of uLSIF, i.e., its solution can be computed analytically, it has the superior numerical stability, and it has higher robustness; furthermore, RuLSIF possesses an even better non-parametric convergence property than uLSIF [11].

Change-Point Detection by RuLSIF: By using an estimator $\widehat{g}(\boldsymbol{Y})$ of the α-relative density ratio, the α-relative PE divergence can be approximated as

$$\widehat{\mathrm{PE}}_\alpha := -\frac{\alpha}{2n} \sum_{i=1}^{n} \widehat{g}(\boldsymbol{Y}_i)^2 - \frac{1-\alpha}{2n} \sum_{j=1}^{n} \widehat{g}(\boldsymbol{Y}'_j)^2 + \frac{1}{n} \sum_{i=1}^{n} \widehat{g}(\boldsymbol{Y}_i) - \frac{1}{2}.$$

As experimentally demonstrated in our supplementary technical report [14], the RuLSIF-based change-point detection performs even better than the plain uLSIF-based method. Thus, we focus on RuLSIF in the experiments in Section 4.

4   Experiments

In this section, we experimentally investigate the performance of the proposed and existing change-point detection methods.

First, we use a human activity dataset and a speech dataset. The human activity dataset is a subset of the *Human Activity Sensing Consortium (HASC) challenge 2011*, which provides human activity information collected by portable three-axis accelerometers. The speech dataset is the *IPSJ SIG-SLP Corpora and Environments for Noisy Speech Recognition* (CENSREC) dataset provided by National Institute of Informatics (NII), which records human voice in a noisy

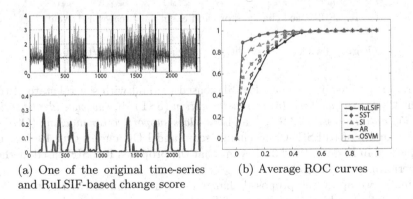

(a) One of the original time-series and RuLSIF-based change score

(b) Average ROC curves

Fig. 2. HASC human-activity dataset (http://hasc.jp/hc2011/)

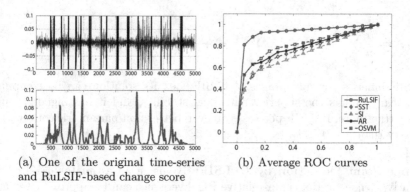

(a) One of the original time-series and RuLSIF-based change score

(b) Average ROC curves

Fig. 3. NII speech dataset (http://research.nii.ac.jp/src/eng/list/index.html)

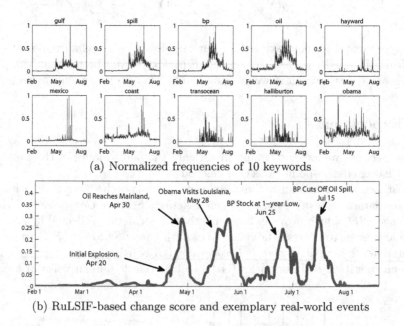

(a) Normalized frequencies of 10 keywords

(b) RuLSIF-based change score and exemplary real-world events

Fig. 4. Twitter dataset (http://www.ark.cs.cmu.edu/tweets/)

environment. We compare our RuLSIF-based method with several state-of-the-art methods: *Singular spectrum transformation* (SST) [4], *subspace identification* (SI) [5], *auto regressive* (AR) [3], and *one-class support vector machine* (OSVM) [9]. Examples of RuLSIF-based change score and ROC curves over 10 datasets are plotted in Figures 2 and 3, showing that the proposed RuLSIF-based method outperforms other methods.

Finally, we apply the proposed change-point detection method to the *CMU Twitter dataset*, which is an archive of Twitter messages that have been collected from April 2010 to October 2010 via the Twitter API. Here we track the degree of popularity of a given topic by monitoring the frequency of selected keywords. More specifically, we focus on events related to *"Deepwater Horizon oil spill in the Gulf of Mexico"* which occurred on April 20, 2010, and was widely broadcast among the Twitter community. We use the frequency of 10 keywords: *"gulf"*, *"spill"*, *"bp"*, *"oil"*, *"hayward"*, *"mexico"*, *"coast"*, *"transocean"*, *"halliburton"*, and *"obama"* (see Figure 4(a)). For quantitative evaluation, we referred to the Wikipedia entry "Timeline of the Deepwater Horizon oil spill" as a real-world event source. The change-point score obtained by the proposed RuLSIF-based method is plotted in Figure 4(b), where four occurrences of important real-world events show the development of this news story.

As we can see from Figure 4(b), the change-point score increases immediately after the initial explosion of the deepwater horizon oil platform and soon reaches the first peak when oil was found on the sea shore of Louisiana on April 30. Shortly after BP announced its preliminary estimation on the amount of leaking oil, the change-point score rises quickly again and reaches its second peak at the end of May, at which time President Obama visited Louisiana to assure local residents of the federal government's support. On June 25, the BP stock was at its one year's lowest price, while the change-point score spikes at the third time. Finally, BP cuts off the spill on July 15, as the score reaches its last peak.

5   Conclusion

We extended the existing KLIEP-based change detection method and proposed to use uLSIF or RuLSIF as a building block. Through experiments, we demonstrated that the RuLSIF-based change detection method is promising.

SL was supported by NII internship fund and the JST PRESTO program. MY and MS were supported by the JST PRESTO program. NC was supported by NII Grand Challenge project fund.

References

1. Basseville, M., Nikiforov, I.V.: Detection of Abrupt Changes: Theory and Application. Prentice-Hall, Inc., Upper Saddle River (1993)
2. Gustafsson, F.: The marginalized likelihood ratio test for detecting abrupt changes. IEEE Transactions on Automatic Control 41(1), 66–78 (1996)
3. Takeuchi, Y., Yamanishi, K.: A unifying framework for detecting outliers and change points from non-stationary time series data. IEEE Transactions on Knowledge and Data Engineering 18(4), 482–489 (2006)
4. Moskvina, V., Zhigljavsky, A.: Change-point detection algorithm based on the singular-spectrum analysis. Communications in Statistics: Simulation and Computation 32, 319–352 (2003)
5. Kawahara, Y., Yairi, T., Machida, K.: Change-point detection in time-series data based on subspace identification. In: Proceedings of the 7th IEEE International Conference on Data Mining, pp. 559–564 (2007)
6. Kawahara, Y., Sugiyama, M.: Sequential change-point detection based on direct density-ratio estimation. Statistical Analysis and Data Mining 5(2), 114–127 (2012)

7. Sugiyama, M., Suzuki, T., Kanamori, T.: Density Ratio Estimation in Machine Learning. Cambridge University Press, Cambridge (2012)
8. Sugiyama, M., Suzuki, T., Nakajima, S., Kashima, H., von Buenau, P., Kawanabe, M.: Direct importance estimation for covariate shift adaptation. Annals of the Institute of Statistical Mathematics 60(4), 699–746 (2008)
9. Desobry, F., Davy, M., Doncarli, C.: An online kernel change detection algorithm. IEEE Transactions on Signal Processing 53(8), 2961–2974 (2005)
10. Kanamori, T., Hido, S., Sugiyama, M.: A least-squares approach to direct importance estimation. Journal of Machine Learning Research 10, 1391–1445 (2009)
11. Yamada, M., Suzuki, T., Kanamori, T., Hachiya, H., Sugiyama, M.: Relative density-ratio estimation for robust distribution comparison. Advances in Neural Information Processing Systems 24, 594–602 (2011)
12. Ali, S.M., Silvey, S.D.: A general class of coefficients of divergence of one distribution from another. Journal of the Royal Statistical Society, Series B 28(1), 131–142 (1966)
13. Csiszár, I.: Information-type measures of difference of probability distributions and indirect observation. Studia Scientiarum Mathematicarum Hungarica 2, 229–318 (1967)
14. Liu, S., Yamada, M., Collier, N., Sugiyama, M.: Change-point detection in time-series data by relative density-ratio estimation. arXiv 1203.0453 (2012)

Online Metric Learning Methods Using Soft Margins and Least Squares Formulations*

Adrian Perez-Suay and Francesc J. Ferri

Dept. Informàtica, Universitat de València, Burjassot 46100, Spain
{Adrian.Perez,Francesc.Ferri}@uv.es

Abstract. Online metric learning using margin maximization has been introduced as a way to learn appropriate dissimilarity measures in an efficient way when information as pairs of examples is given to the learning system in a progressive way. These schemes have several practical advantages with regard to global ones in which a training set needs to be processed. On the other hand, they may suffer from a poor performance depending on the quality of the examples and the particular tuning or other implementation details. This paper formulates several online metric learning alternatives using a passive-aggressive schema. A new formulation of the online problem using least squares is also introduced. The relative behavior of the different alternatives is studied and comparative experimentation is carried out to put forward the benefits and weaknesses of each alternative.

1 Introduction

Organizing, classifying and/or representing sets of data is of key importance in many different application domains from fields like image analysis, pattern recognition or data mining. Distance-based methods form a well established group of approaches to tackle classification, regression, estimation and clustering problems. The performance of such methods, depends on the metric that relates input instances which is intimately tied to the way objects are represented.

In recent years, Distance Metric Learning (DML) has been an active area of research. In most cases, DML aims at learning an appropriate Mahalanobis-like distance matrix. Although there are many approaches, the most common is to define a (usually convex) criterion function which expresses the desired goal [1,2] which basically consists of keeping similar objects close and dissimilar ones far away at the same time. Determining the solution of such global optimization problems can be computationally expensive specially when dealing with large-scale problems [3]. Consequently, the need of effective and efficient learning methods has led to the emergence of sequential methods [4,5], mainly based on optimizing a convenient criterion over only one instance (pair of objects) that is made available for learning at every time step.

* Work partially funded by FEDER and Spanish Government through projects TIN2009-14205-C04-03, TIN2012-38604-C05-02 and Consolider Ingenio 2010 CSD07-00018.

G.L. Gimel' farb et al. (Eds.): SSPR & SPR 2012, LNCS 7626, pp. 373–381, 2012.
© Springer-Verlag Berlin Heidelberg 2012

Unfortunately, many practical and theoretical problems arise. On one hand, different ways of sequentially enforcing additional constraints may lead to different solutions requiring different amount of computation. On the other hand, the performance of the final solution may deviate significantly from the ideal (global) goal depending of the particular instances used in the last iterations.

The present work jointly introduces a family of online metric learning algorithms which use margin maximization [4]. A novel formulation of the same online optimization problem is proposed using a least square formulation instead of the passive-aggressive schema [6]. Exhaustive comparative experimentation is carried out in order to fully characterize the advantages and drawbacks of each online algorithm with regard to other state of the art alternatives.

2   Online Metric Learning

Assume $\mathcal{R}^d$ is a real d−dimensional feature space and consider a set of points $\{x_i\}_{i=1}^N \in \mathcal{R}^d$, and a labeling function y_{ij} which indicates when a pair of points x_i, x_j is similar ($y_{ij} = 1$), or dissimilar ($y_{ij} = -1$). This labeling can come from a user or an appropriate oracle according to the practical problem at hand. A distance function (or simply distance) is a real function defined on $\mathcal{R}^d \times \mathcal{R}^d$ satisfying nonnegativity, identity and triangle equality. This function is a pseudo-metric if identity (it can be zero even for different objects) is not enforced. It is possible to represent a vast family of pseudo-metrics (including the Euclidean distance) by using a Positive Semi Definite (PSD) matrix M as:[1]

$$d_{ij}^M = d^M(x_i, x_j) = (x_i - x_j)^\top M (x_i - x_j). \tag{1}$$

The main goal of metric learning is to obtain a matrix M that reflects the (dis)similarity between pairs of points, x_i, x_j, leading to appropriately different distance values depending on whether these points are really (dis)similar or not. An idealized situation can be visualized as having a convenient threshold value, b, in such a way that all similar distance values are under b and all dissimilar distance values are above b.

2.1   The Separable Case

A pseudo-metric function is better if the corresponding separation between (dis)similar distance values is bigger. This condition can be expressed as maximizing the margin around the threshold value, b. In the separable case and following a similar approach as with support vector machines [7], maximizing the margin can be turned into fixing a margin value and minimizing the (Frobenius) norm of the matrix M. Setting a fixed value of 2 between both kind of distance values can be compactly expressed as

$$y_{ij}(b - d_{ij}^M) \geq 1. \tag{2}$$

---

[1] By convenience, we consider in this work squared versions of the distances.

Instead of considering constrained optimization using all information (examples) available, the above problem can be solved in a more convenient way both from the point of view of computation and robustness by using an online learning approach [4,8]. Under this sequential scheme, at each step k, a particular model formed by the pair (M^k, b^k) is available to make a prediction over the labeled pair $t_k = (x_i, x_j, y_{ij})$ which is revealed to the system at this step. First, a prediction with the previous model (M^k, b^k) is made and a loss corresponding to the violation of this particular constraint is measured. In particular, the hinge loss is used

$$\ell_H(M, b, t_k) = \max\left\{0, 1 - y_{ij}(b - d_{ij}^M)\right\}. \tag{3}$$

Only in the case when the predictor (M^k, b^k) fails, i.e. the hinge-loss is greater than zero, $\ell_H(M^k, b^k, t_k) > 0$, the system is forced to retrain their current model. The aim is to find the nearest model $(\hat{M}^{k+1}, \hat{b}^{k+1})$ to the previous one that attains zero loss in the received pair (provided that it exists). This can be written as the following (online) optimization problem

$$(\hat{M}^{k+1}, \hat{b}^{k+1}) = \underset{M,b}{\arg\min} \frac{1}{2}\|M - M^k\|_{\mathsf{Fro}}^2 + \frac{1}{2}(b - b^k)^2, \tag{4}$$

$$s.t. \quad \ell_H(M, b, t_k) = 0, \tag{5}$$

where $\|\cdot\|_{\mathsf{Fro}}$ is the Frobenius norm. As was shown in [4] the corresponding update becomes:

$$\hat{M}^{k+1} = M^k - \tau y_{ij}(x_i - x_j)(x_i - x_j)^\top, \tag{6}$$

$$\hat{b}^{k+1} = b^k + \tau y_{ij}, \tag{7}$$

where

$$\tau = \frac{\ell_H(M^k, b^k, t_k)}{1 + \|(x_i - x_j)(x_i - x_j)^\top\|_{\mathsf{Fro}}^2}. \tag{8}$$

Two additional constraints are needed in the definition of the problem [4]. The first one is that the matrix M must be Positive Semi Definite (PSD), that is, $M \succeq 0$ and consequently the threshold b must be above 1, $b \geq 1$. Since τ is nonnegative, the constraint $M \succeq 0$ is straightforwardly taken taken into account if $y_{ij} = -1$ when using the rank-one update in Eq. (6). In the case $y_{ij} = 1$, this update may introduce one negative eigenvalue in M at most. The closest PSD matrix can be found by setting this negative eigenvalue to zero after eigendecomposing M. Alternatively and equivalently, both the eigenvalue along with its corresponding eigenvector can be added to M after computing them using the Lanczos method [4].

2.2   Soft Margin Formulation

The previous formulation only makes sense if feasible solutions exist. Which means that similar/dissimilar pairs can be strictly separated. As in [6], different alternatives for the inseparable case can be considered. In particular, the inseparable case can be solved by adding the slack variable $\xi_{ij} \in \mathcal{R}$ to the original formulation and enforcing its positiveness by adding a new constraint to the problem. This slack variable ξ_{ij} is weighted by a hyper-parameter $C \in [0, +\infty[$, that preserves the trade off between closeness to the previous model and loss minimization. Following the passive-aggressive approach [6] the corresponding online optimization problem can be stated as:

$$(M^{k+1}, b^{k+1}) = \arg\min_{M,b,\xi_{ij}} \frac{1}{2}\|M - M^k\|_{\text{Fro}}^2 + \frac{1}{2}(b - b^k)^2 + C\xi_{ij}, \qquad (9)$$

$$s.t. \quad \ell_H(M, b, t_k) \le \xi_{ij}, \quad \xi_{ij} \ge 0 \qquad (10)$$

Alternatively, the objective function can be scaled quadratically with respect to ξ_{ij}. This fact avoids the need to restrict ξ_{ij} to be nonnegative. That is,

$$(M^{k+1}, b^{k+1}) = \arg\min_{M,b,\xi_{ij}} \frac{1}{2}\|M - M^k\|_{\text{Fro}}^2 + \frac{1}{2}(b - b^k)^2 + C\xi_{ij}^2, \qquad (11)$$

$$s.t. \quad \ell_H(M, b, t_k) \le \xi_{ij}. \qquad (12)$$

These two formulations lead to different update rules that will be referred to as PA-I and PA-II, respectively as in [6]. These update rules are the ones in Equations (6) and (7) but changing the rate τ by τ_1 and τ_2, respectively.

$$\tau_1 = \min\left\{C, \frac{\ell_H(M^k, b^k, t_k)}{1 + \|(x_i - x_j)(x_i - x_j)^\top\|_{\text{Fro}}^2}\right\}, \tau_2 = \frac{\ell_H(M^k, b^k, t_k)}{1 + \frac{1}{2C} + \|(x_i - x_j)(x_i - x_j)^\top\|_{\text{Fro}}^2}. \qquad (13)$$

2.3   Least Squares Formulation

An alternative formulation inspired on a Least-Squares approach [9] is also possible. Instead of forcing a soft margin and minimize the amount of violation on this condition, it is possible to force similar and dissimilar distance values to fall close to the "representative" values $b - 1$ and $b + 1$, respectively. To this end, one can sequentially minimize the corresponding squared error. This can be formulated as:

$$(M^{k+1}, b^{k+1}) = \arg\min_{M,b,\xi_{ij}} \frac{1}{2}\|M - M^k\|_{\text{Fro}}^2 + \frac{1}{2}(b - b^k)^2 + C\xi_{ij}^2, \qquad (14)$$

$$s.t. \quad 1 - y_{ij}(b - d_{ij}^M) = \xi_{ij}. \qquad (15)$$

The main change in this formulation is that the inequality in the restriction (12) has been changed by an equality at restriction (15) and, the loss function is now a function which measures how far is the distance value from its corresponding idealized one ($b-1$ or $b+1$). This fact brings more aggressiveness to the problem formulation and the final number of updates will consequently expected to be greater. In fact, the update rule can be derived in a very similar way as in the previous cases leading to a different rate given by

$$\tau_3 = \frac{1 - y_{ij}(b^k - d_{ij}^{M^k})}{1 + \frac{1}{2C} + \|(x_i - x_j)(x_i - x_j)^\top\|_{\mathsf{Fro}}^2}.$$ (16)

Note that τ_3 can take negative values and it holds that $\tau_2 = \max\{0, \tau_3\}$.

The corresponding algorithm will be referred to as PA-LS here, although it cannot be properly considered as passive-aggressive because only in the particular case when it holds $1 - y_{ij}(b - d_{ij}^M) = 0$, the system does a passive step (do nothing) and this only occurs when the distance value takes exactly its desired value. This is in contrast with the above PA-I and PA-II approaches which perform a passive step in the case when $\ell_H = 0$ that corresponds to $1 - y_{ij}(b - d_{ij}^M) \le 0$.

2.4 Tuning and Implementation Details

The performance of the different online learning algorithms using the above update rules strongly depend on the value of the parameter C (which needs to be adapted for each database) and also on the initial model given by M and b (that have been set to the zero matrix and 0 following considerations in [4]).

On the other hand, the PSD constraint needs to be enforced at each learning step but it is possible to relax this by allowing negative models during a fixed amount of learning steps after enforcing positiveness [3]. In fact, in our experiments we have obtained better results in general if the PSD constraint is not enforced until the end of the online learning process. Consequently, for each one of the above algorithms we consider a positive (+) version in which the PSD constraint (and $b \ge 1$) are enforced at each iteration, and a negative (-) one in which these constraints are only enforced at the end of the process.

3 Experiments and Results

In order to compare all the above online methods, an exhaustive experimentation has been designed. It has been mainly focused on classification, time execution and convergence-optimality trade off. Several different publicly available databases from [10,11] are taken into account. Moreover, a more realistic database previously used in CBIR tasks [12] has also been selected. This database is extracted from a commercial collection called "Art Explosion", distributed by the company Nova Development (http://www.novadevelopment.com). To perform more meaningful classification experiments, only classes with more than 100 elements have been selected. In all cases, objects are considered similar only if they share the same class label. Datasets used in the comparative study and their particular characteristics are shown in Table 1.

Table 1. Characteristics of Databases

	wine	ionosphere	balance	soybean	BDG100	nist16
Size	178	351	625	266	1710	2000
Dimension	13	34	4	35	104	256
no. of classes	3	2	3	15	10	10

Experimentation setup has been fixed as suggested in the work [5], where one of the most competitive metric learning algorithms has been introduced and studied. Precisely this algorithm, the Information Theoretic Metric Learning (ITML), has been adopted in the present study as a baseline. The ITML algorithm has been used as suggested in [5] using the software made available by the authors that has its own tuning mechanism which assigns appropriate parameters to the algorithm.

To study the behavior of all the online methods, initial values have been fixed as $(M^0, b^0) = 0$. In all cases, C is tuned using a validation set taking from the available training set. In particular, exponentially spaced values between $[10^{-4}, 10^2]$, have been considered. In order to feed all the methods considered with the same amount of information, a subset of $40c(c-1)$ pairs has been randomly selected for each run on each of the databases. The online algorithms are feed with random pairs from this subset. The subset of pairs is presented several times with different random order until the total number of time steps exceeds 20% of total number of pairs in each training set. As will be shown in the following, this amount of iterations has been proved as a good trade off between computational cost and performance. All results presented are the average of 10 independent runs with different random initializations but with exactly the same data for each one of the alternative algorithms.

To illustrate the behavior of the different approaches throughout the learning process, several loss and performance measures have been taken during the learning process. First, Figure 1 shows the predictive 0-1 loss defined as $\ell = \frac{1}{2T} \sum_{k=1}^{T} |y_k - \hat{y}_k|$, where T is the learning sequence length, k represents the corresponding pair supplied at $(k+1)$-th step and y_k and $\hat{y}_k$ are the true and predicted labels using the model (M^k, b^k). This measure illustrates the behavior of the different online algorithms throughout time when discriminating between similar and dissimilar objects.

All online learning algorithms have lead to reasonably good behavior in the experiments carried out according to this loss measure. In 5 out of 6 databases, negative methods have led to better results compared to positive versions of the same methods. The case of the two biggest databases is specially remarkable. On the other hand, the LS methods exhibit significantly worse behavior in wine, balance, soybean and BDG10 databases.

To measure the quality of the final outcome of the different algorithms, the corresponding matrices, M, have been used to construct a k-NN classifier. The classification error using up to the first 25 neighbors has been computed and the best results for each database and method are shown in Table 2.

The classification errors obtained with all algorithms including ITML, are all good classification results in the context of the experimentation carried out in this work. It must be noted that the differences among different classification results are not significant in most of the cases. Nevertheless, it can be concluded that all algorithms lead to very competitive results. Also worth noting is the fact that the combination of LS approach with negative matrices lead to the best results in the three last databases.

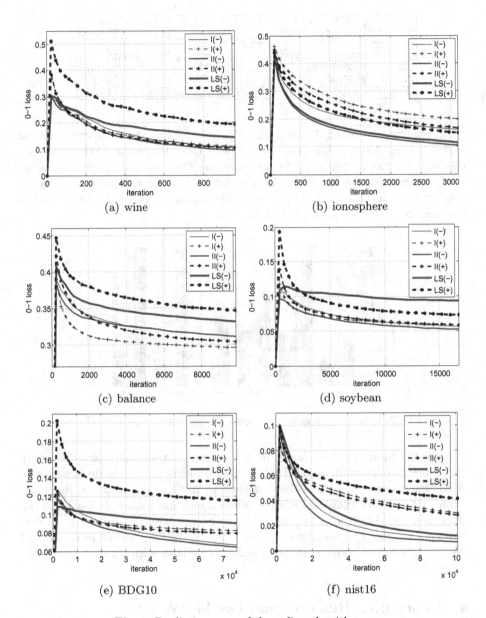

Fig. 1. Predictive error of the online algorithms

Table 2. Average classification errors and best number of neighbors (in brackets). Best result for each database is shown in bold.

	ITML	PA-I$^+$	PA-I$^-$	PA-II$^+$	PA-II$^-$	PA-LS$^+$	PA-LS$^-$
wine	3.33(3)	1.76(5)	1.68(6)	**1.59(3)**	1.68(9)	2.42(7)	1.87(3)
ionosphere	15.46(2)	**12.87(2)**	13.59(2)	14.00(2)	13.87(2)	14.26(2)	13.84(2)
balance	26.27(3)	**21.88(3)**	21.92(2)	25.81(3)	23.40(2)	31.23(6)	25.79(6)
soybean	8.63(1)	10.73(1)	10.00(1)	9.87(1)	9.70(1)	9.77(1)	**8.27(1)**
BDG10	20.27(6)	20.93(7)	21.74(6)	20.72(7)	21.40(7)	20.25(9)	**20.20(9)**
nist16	5.65(1)	5.67(1)	5.86(1)	5.67(1)	6.03(1)	6.64(1)	**5.62(1)**

All experiments on all databases have been run using the same computer. In particular, an AMD Athlon(tm) 64 X2 Dual Core Processor 4200+ has been used but restricting the code to use only one CPU to obtain more accurate measurements (except for the two biggest databases). Figure 2 shows the relative averaged running time spent by each online learning algorithm with regard to the time used by the ITML algorithm.

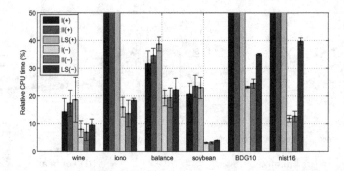

Fig. 2. Relative averaged execution CPU time with regard to ITML

From the running times shown, it can be seen that all online algorithms are very efficient compared to ITML. The case of positive versions on ionosphere, BDG10 and nist16 databases that get close to 200% is an exception. But more importantly, we see that the negative online algorithms reduce dramatically the time spent by their corresponding positive versions while preserving good performance results. The running times are kepts relatively low even in the case of PA-LS(-) that need about twice the number of updates with regard to the other negative online algorithms.

4   Concluding Remarks and Further Work

A family of online metric learning algorithms has been considered. The formulation using a soft margin and a passive-aggressive scheme has been extended by considering a least squares formulation. The algorithms have been implemented as positive versions in which the PSD constraint is enforced at each iteration, and negative ones in which the constraint is enforced only at the end. Performance results show that under an appropriate implementation and tuning, all online methods are able to arrive at good results, but the negative versions are appealing due to their low running times. All online methods presented in this work still have room for improvement. In particular, the number of iterations can be adapted by introducing convergence criteria that are now under study. Also, with regard to LS methods, we are currently improving its running time by adding an updating tolerance when the value of τ_3 is close enough to zero.

References

1. Globerson, A., Roweis, S.T.: Metric learning by collapsing classes. [13]
2. Weinberger, K.Q., Blitzer, J., Saul, L.K.: Distance metric learning for large margin nearest neighbor classification. [13]
3. Chechik, G., Sharma, V., Shalit, U., Bengio, S.: Large scale online learning of image similarity through ranking. Journal of Machine Learning Research 11, 1109–1135 (2010)
4. Shalev-Shwartz, S., Singer, Y., Ng, A.Y.: Online and batch learning of pseudo-metrics. In: Brodley, C.E. (ed.) ACM International Conference Proceeding Series, ICML, vol. 69, ACM (2004)
5. Davis, J.V., Kulis, B., Jain, P., Sra, S., Dhillon, I.S.: Information-theoretic metric learning. In: Ghahramani, Z. (ed.) ICML. ACM International Conference Proceeding Series, vol. 227, pp. 209–216. ACM (2007), http://www.cs.utexas.edu/users/pjain/itml/
6. Shalev-Shwartz, S., Crammer, K., Dekel, O., Singer, Y.: Online passive-aggressive algorithms. In: Thrun, S., Saul, L.K., Schölkopf, B. (eds.) NIPS. MIT Press (2003)
7. Scholkopf, B., Smola, A.J.: Learning with Kernels: Support Vector Machines, Regularization, Optimization, and Beyond. MIT Press, Cambridge (2001)
8. Perez-Suay, A., Ferri, F.J., Albert, J.V.: An Online Metric Learning Approach through Margin Maximization. In: Vitrià, J., Sanches, J.M., Hernández, M. (eds.) IbPRIA 2011. LNCS, vol. 6669, pp. 500–507. Springer, Heidelberg (2011)
9. Suykens, J.A.K., Vandewalle, J.: Least squares support vector machine classifiers. Neural Process. Lett. 9, 293–300 (1999)
10. Duin, R.P.W.: Prtools version 3.0: A matlab toolbox for pattern recognition. In: Proc. of SPIE, p. 1331 (2000)
11. Asuncion, A., Newman, D.J.: UCI machine learning repository (2007)
12. Arevalillo-Herrez, M., Ferri, F.J., Domingo, J.: A naive relevance feedback model for content-based image retrieval using multiple similarity measures. Pattern Recognition 43(3), 619–629 (2010)
13. Advances in Neural Information Processing Systems 18 [Neural Information Processing Systems, NIPS 2005, December 5-8, 2005, Vancouver, British Columbia, Canada]. In: NIPS (2005)

Shape Analysis Using the Edge-Based Laplacian

Furqan Aziz, Richard C. Wilson, and Edwin R. Hancock

Department of Computer Science, University of York, YO10 5GH, UK
{furqan,wilson,erh}@york.ac.uk

Abstract. This paper presents a novel analysis and application of the eigensystem of the edge-based Laplacian of a graph. The advantage of using the edge-based Laplacian over its vertex-based counterpart is that it significantly expands the set of differential operators that can be implemented in the graph domain. We commence by presenting a new mesh characterization based on the adjacency matrix of the mesh that captures both the geometric and topological properties of the shape. We use the edge-based eigenvalues to develop a novel method for defining pose-invariant signatures for non-rigid three-dimensional shapes based on the edge-based heat kernel. To illustrate the utility of our method, we perform numerous experiments applying the method to correspondence matching and classifying non-rigid three-dimensional shapes represented in terms of meshes.

1   Introduction

The key idea in the analysis of three-dimensional deformable shapes is to define an informative and discriminative feature descriptor that characterizes each point on the surface of the shape. Generally these techniques use a feature vector in $\mathbb{R}^n$ [1,2], which contains both local and global information for that point. These feature descriptors can be used in many ways for analyzing three-dimensional shapes. For correspondence matching, the descriptors are used to find potential correspondence among pairs of points on two different shapes [1,2]. For clustering the parts of a shape, the signatures can be used to identify semantically coherent parts of an object[3,4]. Local descriptors can be combined in different ways to define a global shape signature and this can be used for shape classification or recognition[5,6].

Recently, there is an increasing interest in descriptors obtained from the spectral decomposition of the Laplace-Beltrami operator associated with a shape. For example, Rustamov[3] has defined the Global Point Signature (GPS) that uses the spectrum of the discrete Laplace-Beltrami operator to represent three-dimensional non-rigid shapes. Sun et al[2] have used a heat diffusion process to define signatures and this is referred to as the Heat Kernel Signature (HKS). Castellani et al[5] have used HKS to define Global Heat Kernel Signature (GHKS) and have used this for brain classification. Aubry et al[1] have proposed the Wave Kernel Signature (WKS) which represents the average probability of measuring a quantum mechanical particle at a specific location. All of these techniques use

G.L. Gimel' farb et al. (Eds.): SSPR & SPR 2012, LNCS 7626, pp. 382–390, 2012.
© Springer-Verlag Berlin Heidelberg 2012

the spectrum of the discrete Laplacian, Δ, an operator which is defined only on the vertices of a graph.

The discrete Laplacian defined over the vertices of a graph has found applications in many areas including computer vision and complex networks[7]. However one of the limitation of discrete Laplacian is that it cannot link most results in analysis to a graph theoretic analogue. For example the wave equation $u_{tt} = \Delta u$, defined with the discrete Laplacian, does not have finite speed of propagation. In [8,9], Friedman and Tillich develop a calculus on graphs which provides a strong connection between graph theory and analysis. Their work is based on the fact that graph theory involves two different volume measures. i.e., a "vertex-based" measure and an "edge-based" measure. This approach has many advantages. Moreover it allows the direct application of many results from analysis to the graph domain.

Recently we have presented a new approach to characterizing points on a non-rigid three-dimensional shape[10]. This is based on the eigenvalues and eigenfunctions of the edge-based Laplacian, constructed over a mesh that approximates the shape. This leads to a new shape descriptor signature, called the Edge-based Heat Kernel Signature (EHKS). The EHKS was defined using the heat equation, which is based on the edge-based Laplacian. As first step we explored the application of the EHKS to shape segmentation. In this paper we take this study one step further. We use the EHKS for correspondence matching and show its robustness under noise. We also define a global signature (the GEHKS), which is based on the EHKS for shape classification. We perform numerous experiments and demonstrate the performance of the proposed methods on non-rigid three dimensional shapes and compare it to WKS.

2 Edge-Based Eigensystem

In this section we briefly review the eigenvalues and eigenfunction of the edge-based Laplacian[9]. Let $G = (\mathcal{V}, \mathcal{E})$ be a graph with a boundary ∂G. Let $\mathcal{G}$ be the geometric realization of G. The geometric realization is the metric space consisting of vertices $\mathcal{V}$ with a closed interval of length l_e associated with each edge $e \in \mathcal{E}$. We associate an edge variable x_e with each edge that represents the standard coordinate on the edge with $x_e(u) = 0$ and $x_e(v) = 1$. For our work, it will suffice to assume that the graph is finite with empty boundary (i.e., $\partial G = 0$) and $l_e = 1$.

The eigenpairs of the edge-based Laplacian can be expressed in terms of the eigenpairs of the normalized adjacency matrix of the graph. Let A be the adjacency matrix of the graph G, and $\tilde{A}$ be the row normalized adjacency matrix. i.e., the $(i,j)th$ entry of $\tilde{A}$ is given as $\tilde{A}(i,j) = A(i,j)/\sum_{(k,j)\in\mathcal{E}} A(k,j)$. Let $(\phi(v), \lambda)$ be an eigenvector-eigenvalue pair for this matrix. Note $\phi(.)$ is defined on vertices and may be extended along each edge to an edge-based eigenfunction. Let ω^2 and $\phi(e, x_e)$ denote the edge-based eigenvalue and eigenfunction. Here $e = (u, v)$ represents an edge and x_e is the standard coordinate on the edge (i.e., $x_e = 0$ at v and $x_e = 1$ at u). Then the eigenpairs of the edge-based Laplacian are given as follows:

1. For each $(\phi(v), \lambda)$ with $\lambda \neq \pm 1$, we have a pair of eigenvalues ω^2 with $\omega = \cos^{-1} \lambda$ and $\omega = 2\pi - \cos^{-1} \lambda$. Since there are multiple solutions to $\omega = \cos^{-1} \lambda$, we obtain an infinite sequence of eigenfunctions; if $\omega_0 \in [0, \pi]$ is the principal solution, the eigenvalues are $\omega = \omega_0 + 2\pi n$ and $\omega = 2\pi - \omega_0 + 2\pi n, n \geq 0$. The eigenfunctions are $\phi(e, x_e) = C(e) \cos(B(e) + \omega x_e)$ where

$$C(e)^2 = \frac{\phi(v)^2 + \phi(u)^2 - 2\phi(v)\phi(u)\cos(\omega)}{\sin^2(\omega)}$$

$$\tan(B(e)) = \frac{\phi(v)\cos(\omega) - \phi(u)}{\phi(v)\sin(\omega)}$$

   There are two solutions here, $\{C, B_0\}$ or $\{-C, B_0 + \pi\}$ but both give the same eigenfunction. The sign of $C(e)$ must be chosen correctly to match the phase.

2. $\lambda = 1$ is always an eigenvalue of $\tilde{A}$. We obtain a principle frequency $\omega = 0$, and therefore since $\phi(e, x_e) = C \cos(B)$ and so $\phi(v) = \phi(u) = C \cos(B)$, which is constant on the vertices.

3. If the graph is bipartite, then $\omega = -1$ is an eigenvalue of $\tilde{A}$. We obtain a principle frequency $\omega = \pi$, and then since $\phi(e, x_e) = C \cos(B + \pi x_e)$, so $\phi(v) = C \cos(B) = \phi(u)$ implying an alternating sign eigenfunction.

4. The sets $\{\pi + 2n\pi : n \geq 0\}$ and $\{2\pi + 2n\pi : n \geq 0\}$ occur with multiplicity $|E| - |V|$. Note that although the eigenfunctions corresponding to these eigenvalues are zero on vertices, they are not zero on edge interiors.

This comprises all the principal eigenpairs which are supported on the vertices.

Note that although these eigenfunctions are orthogonal, they are not normalized. To normalize these eigenfunctions we need to find the normalization factor corresponding to each eigenvalue. Let $\rho(\omega)$ denotes the normalization factor corresponding to eigenvalue ω. Then

$$\rho^2(\omega) = \sum_{e \in \mathcal{E}} \int_0^1 \phi^2(e, x_e) \, dx_e$$

Evaluating the integral, we get

$$\rho(\omega) = \sqrt{\sum_{e \in \mathcal{E}} C(e)^2 \left[\frac{1}{2} + \frac{\sin(2\omega + 2B(e))}{4\omega} - \frac{sin(2B(e))}{4\omega} \right]}$$

Once we have the normalization factor to hand, we can compute a complete set of orthonormal bases by dividing each eigenfunction with the corresponding normalization factor. Therefore the orthonormalized eigenfunctions corresponding to eigenvalues ω^2 are $\phi(e, x_e) = \frac{C(e)}{\rho(\omega)} \cos(B(e) + \omega x_e)$. Once normalized, these eigenfunctions form a complete set of orthonormal bases for $L^2(\mathcal{G}, \mathcal{E})$.

3   Shape Descriptors

Once the edge-based eigenpairs are known, we can use them to link most of the results in analysis to the graph domain. Our goal is to use the solution of partial differential equations based on the edge-based Laplacian over a mesh for characterizing points on non-rigid 3D shapes. The signatures we propose here are based on the heat diffusion process governed by the equation

$$\frac{\partial H_t}{\partial t} = -\Delta_E H_t \tag{1}$$

where Δ_E is the edge-based Laplacian and H_t is the heat kernel. The solution to above equation is called the heat kernel. The heat kernel has the following eigen-decomposition:

$$H_t(x,y) = \sum_{i=0}^{\infty} e^{-\omega^2 t}\phi(x)\phi(y) \tag{2}$$

where (ϕ, ω^2) are the edge-based eigenpairs.

3.1   Local Descriptor

Given a point x on the surface of a three-dimensional shape, its Edge-based Heat Kernel Signature (EHKS) is given as[10]:

$$EHKS(x) = [H_{t_0}(x,x), H_{t_1}(x,x), ..., H_{t_k}(x,x)] \tag{3}$$

In [10], we have experimentally shown the applications of EHKS for shape segmentation. In this paper we show the applications of EHKS for correspondence matching.

3.2   Global Descriptor

To extend our method to the problem of shape classification we define a global signature (GEHKS) for the whole shape which is based on the EHKS. Our approach of defining a global signature for the shape is closely related to the approach of [5]. Given a shape S, we define its global edge-based heat kernel signature as

$$GEHKS(S) = hist\left(EHKS(x_1), EHKS(x_2), ..., EHKS(x_n)\right) \tag{4}$$

where $hist(.)$ is the histogram operator. Since the GEHKS is defined on small and large values of t, it encodes both local and global information about the shape.

3.3   Discrete Settings

A three dimensional shape can be conveniently represented by a mesh which approximates the shape. Therefore to find the corresponding edge-based Laplacian we need to find the adjacency matrix of the mesh. The simplest way of defining the adjacency matrix of the mesh using the un-weighted (0-1) or the weighted (distance or proximity) matrix is sensitive to the regularity of the particular triangulation and give little information about the shape itself. In [10] we have proposed a new method for for constructing the adjacency matrix of the mesh that uses the angle information between the edges and the area around each vertex (see Figure 1).

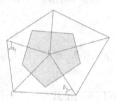

Fig. 1. Angles and the area appearing in the adjacency matrix

Let M is a matrix whose $(i, j)th$ entry is defined as

$$M(i,j) = \begin{cases} \frac{\cot \alpha_{ij} + \cot \beta ij}{2} & \text{if } (i,j) \in E \\ 0 & \text{otherwise} \end{cases} \qquad (5)$$

where α_{ij} and β_{ij} are the angles opposite to the edge (i, j), as shown in Figure 1. Let S be a diagonal matrix whose ith diagonal entry is the area associated with the triangles abetting the vertex i. We define the symmetric adjacency matrix as $A = S^{1/2}MS^{1/2}$. The $(i, j)th$ entry of the adjacency matrix, in terms of the elements of the matrices M and S, is given as follows[10]:

$$A(i,j) = \begin{cases} \sqrt{S(i,i)S(j,j)}M(i,j) & \text{if } (i,j) \in E \\ 0 & \text{otherwise} \end{cases} \qquad (6)$$

The matrix defined above not only captures more information about the geometric and topological properties of the shape itself but also minimizes the dependence of the adjacency matrix on the mesh.

4   Experiments

In this section we will present both the qualitative and the quantitative analysis of the proposed edge-based heat kernel signature. We perform our experiments on the SHREC 2010 dataset, which contains 10 different shapes each with 20 different non-rigid deformations. Figure 2 shows some of these shapes. To find the edge-based eigenpairs, we first construct the adjacency matrix, as described in the previous section. We compute the area associated with each vertex using the method proposed in [11]. We then find the eigenpairs of the normalized adjacency matrix. To find signatures for shape we compute first 300 smallest eigenvalues and corresponding eigenvectors using the *eigs* routine in Matlab which is used to solve the sparse eigenvalue problem. We compute the scaled EHKS by uniformly sampling 100 points for different values of t over the time interval $[t_{min}, t_{max}]$ where $t_{min} = 4\ln 10/\lambda_{300}$ and $t_{max} = 4\ln 10/\lambda_2[2]$. In Figure 3(a) we have

Fig. 2. The SHREC 2010 database of shapes

represented the values of EHKS with different colors on 6 different shapes of a human body, which shows the stability of the EHKS under different deformations of shapes. To prove the stability of our method, we illustrate the method on the problem of segmenting and classifying parts of a human body using EHKS. We select points on hands, feet, and head of 15 different poses a human body and compute their EHKS. To visualize the results, We apply PCA on these signatures and embed them in a three dimensional space. Figure 3(b) shows that EHKS can not only distinguish between different classes of features, it can also distinguish classes of features of different shapes.

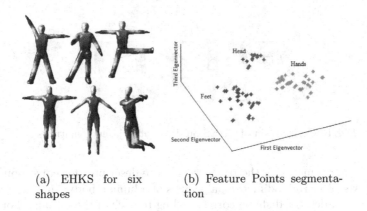

(a) EHKS for six shapes

(b) Feature Points segmentation

Fig. 3. Performance of EHKS

In our next experiment, we show the stability of EHKS under controlled noise. For this purpose we take three-dimensional shapes of a human body and a bear and their deformed shapes. We add gaussian noise to the deformed shapes with mean $\mu = 0$, and standard deviation $\sigma = 0.3$. We select three different points on each of the given shape and compute their EHKS. Next we compute EHKS for each point on the deformed shapes corrupted by gaussian noise. We compute the Euclidean distance of the feature descriptor of the selected points with feature descriptor of each point on the deformed shape. In Figure4, the lines between shapes show the first 50 best matches of each of the three points on the given shapes with the points on deformed shapes. Results show that the proposed method is robust under controlled gaussian noise. To demonstrate the usefulness

Fig. 4. Robustness under noise

of the proposed adjacency matrix, we compare the performance of the EHKS using different adjacency matrices. We select two different three-dimensional shapes of a human body. Next we randomly select a points on three different parts of the the shape and, for each point, find the first 50 best matches on the deformed shape. Figure 5 shows the results EHKS when computed from the proposed matrix(left), the matrix $A = \left(\frac{P+P^T}{2} \right)$ where $P = S^{-1}M$(middle), and the symmetric matrix M that uses the angle information only(right). Results shows that EHKS constructed using the proposed adjacency matrix is more stable. To

Fig. 5. Comparison of EHKS on different adjacency matrices

evaluate the performance of the proposed feature descriptor for correspondence matching, we select three-dimensional shapes of a human body, ant and glasses. We also select a deformed shape corresponding to each of these shape. For each three-dimensional shape we select a random point on five different parts of the shape and compute EHKS for each of these points. Next We compute the Euclidean distance of the feature descriptor of selected points on each shape with the feature descriptors of each point on the corresponding deformed shape. Figure 6(a) shows the first 50 best matches of each of the point on the shape with the points on deformed shapes. We perform a similar experiment for WKS(Figure 6(b)). Results show that EHKS is more robust and stable as compared to WKS. To compare the performance of EHKS with WKS, we select a three dimensional

(a) EHKS on three different shapes

(b) WKS on three different shapes

Fig. 6. Comparing EHKS and WKS

shape of a human body and its deformed shape. we randomly select a point on 10 different parts of the shape and find the best match for each point on the deformed shape using both the EHKS and WKS. We repeat this experiment for five times. The number of successful matches for both methods are given in table 1.

Table 1. Number of best matches

	1	2	3	4	5
EHKS	8	8	8	9	8
WKS	6	8	7	7	8

In our final experiment, we show the applications of GEHKS for shape classification. For this purpose, we select three-dimensional shapes of ant, plier and octopus from SHREC 2010 dataset with all of their deformations and compute GEHKS for each of these shapes. To visualize the results, We apply PCA on these GEHKS and embed them in a three dimensional space. Figure 7 shows that the proposed method can be useful for clustering different shapes. To compare the accuracy of the proposed method we perform a similar experiment with WKS and compute the rand indices for both methods. The accuracy of the proposed method was 0.8514 while the that of WKS was 0.7893. These results show that the EHKS is more informative than the WKS, and gives higher performance for correspondence matching and shape classification.

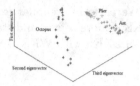

Fig. 7. Clustering of different shapes

5   Conclusion and Future Work

We have presented a method for analyzing three-dimensional non-rigid shapes which is based on heat equation defined over the edge-based Laplacian. Experimental results show that our method can be used for clustering, correspondence matching and classifying 3D shapes. In future, we would like to use the solutions of other partial differential equations over graph defined using the edge-based Laplacian, which have close relation to equation in analysis.

Acknowledgements. Edwin Hancock was supported by a Royal Society Wolfson Research Merit Award.

References

1. Aubry, M., Schlickewei, U., Cremers, D.: The wave kernel signature: A quantum mechanical approach to shape analysis. Tech. rep., TU München, Germany (2011)
2. Sun, J., Ovsjanikov, M., Guibas, L.: A concise and provably informative multi-scale signature based on heat diffusion. Comp. Grarph Forum, 1383–1392 (2010)
3. Rustamov, R.: Laplace-beltrami eigenfunctions for deformation invariant shape representation. In: Eurographics Symp. on Geom. Processing, pp. 225–233 (2007)
4. Aubry, M., Schlickewei, U., Cremers, D.: Pose-consistent 3d shape segmentation based on a quantum mechanical feature descriptor. In: Proc. 33rd DAGM Symposium, Frankfurt, Germany (2007)
5. Castellani, U., Mirtuono, P., Murino, V., Bellani, M., Rambaldelli, G., Tansella, M., Brambilla, P.: A New Shape Diffusion Descriptor for Brain Classification. In: Fichtinger, G., Martel, A., Peters, T. (eds.) MICCAI 2011, Part II. LNCS, vol. 6892, pp. 426–433. Springer, Heidelberg (2011)
6. Osada, R., Funkhouser, T., Chazelle, B., Dobkin, D.: Shape distributions. ACM Transactions on Graphics, 807–832 (2002)
7. Estrada, E.: Characterization of 3d molecular structure. Chemical Physics Letters 319, 713–718 (2000)
8. Friedman, J., Tillich, J.: Calculus on graphs. CoRR (2004)
9. Friedman, J., Tillich, J.: Wave equations for graphs and the edge based laplacian. Pacific Journal of Mathematics, 229–266 (2004)
10. Aziz, F., Wilson, R., Hancock, E.: Shape signature using the edge-based laplacian. In: International Conference on Pattern Recognition (2012)
11. Xu, G.: Discrete laplace-beltrami operator on sphere and optimal spherical triangulations. International Journal of Computational Geometry, 75–93 (2006)

One-Sided Prototype Selection
on Class Imbalanced Dissimilarity Matrices

Mónica Millán-Giraldo[1,2], Vicente García[2], and J. Salvador Sánchez[2]

[1] Intelligent Data Analysis Laboratory, University of Valencia
Av. Universitat s/n, 46100 Burjassot, Valencia, Spain
[2] Institute of New Imaging Technologies
Department of Computer Languages and Systems, University Jaume I
Av. Sos Baynat s/n, 12071 Castelló de la Plana, Spain

Abstract. In the dissimilarity representation paradigm, several prototype selection methods have been used to cope with the topic of how to select a small representation set for generating a low-dimensional dissimilarity space. In addition, these methods have also been used to reduce the size of the dissimilarity matrix. However, these approaches assume a relatively balanced class distribution, which is grossly violated in many real-life problems. Often, the ratios of prior probabilities between classes are extremely skewed. In this paper, we study the use of renowned prototype selection methods adapted to the case of learning from an imbalanced dissimilarity matrix. More specifically, we propose the use of these methods to under-sample the majority class in the dissimilarity space. The experimental results demonstrate that the one-sided selection strategy performs better than the classical prototype selection methods applied over all classes.

1 Introduction

In the traditional approach to Statistical Pattern Recognition, each object is represented in terms of n observable features or attributes, which can be regarded as a vector in an n-dimensional feature space. An alternative is the *dissimilarity space* proposed by Duin and Pekalska [1, 2]. To build the dissimilarity space, a representation set of r objects (or prototypes), $R = \{p_1, \ldots, p_r\}$, is needed. The dissimilarity representation allows to symbolize individual feature-patterns by pairwise dissimilarities computed between examples from the training set T and objects from the representation set R. Thus the dissimilarity vectors can be interpreted as numerical features and describe the relation between each object with the rest of objects [3].

Given a training set of m objects in the feature space, $T = \{x_1, \ldots, x_m\}$, the classifier is built using a dissimilarity matrix $D(T, R)$ that describes the proximities between the m training set objects and the r prototypes. The representation set can be chosen as the complete training set T, a set of constructed prototypes, a subset of T that covers all classes, or even an arbitrary set of labeled or unlabeled objects.

The dimensionality in the dissimilarity space is determined by the amount of prototypes in the set R. When $R = T$, the dissimilarity matrix $D(T, T)$ might impose high computational requirements on the classifier [4] and adversely affect the performance [5]. To face this drawback, several works have proposed to reduce the dimensionality of the dissimilarity space by selecting a small representation set from the training

G.L. Gimel' farb et al. (Eds.): SSPR & SPR 2012, LNCS 7626, pp. 391–399, 2012.

data [6]. Obviously, a pruned representation set will lead to reduce the distance matrix $D(T,T)$ to $D(T,R)$. In this context, prototype selection constitutes one of the most active research lines, which has primarily been addressed in two ways: (i) finding a small representation set capable of generating a low-dimensional dissimilarity space [4, 6, 7], and (ii) reducing the original dissimilarity matrix [8, 9].

Prototype selection methods have demonstrated to perform well in dissimilarity space classification when the classes are balanced. However, in many real-life problems the ratios of prior probabilities between classes can be extremely skewed. This situation is known as the class imbalance problem [10, 11]. A data set is said to be imbalanced when the examples from one class (the majority class) heavily outnumber the examples from the other (minority) class. This topic is particularly important in practical applications where it is costly to misclassify examples from the minority (or positive) class, such as medical diagnosis and monitoring, fraud/intrusion detection, credit risk and bankruptcy prediction, information retrieval and filtering tasks.

In this work, we explore the use of well-known prototype selection procedures (originally designed to be applied in the feature space) on the dissimilarity matrix $D(T,T)$ when this is imbalanced. Here, we propose to exploit these methods in a biased fashion, where only the majority class is pruned. In fact, this can be viewed as an under-sampling strategy, which is one of the common solutions to the class imbalance problem in feature spaces [12]. The experimental results show that this one-sided strategy performs significantly better than the standard application of prototype selection on both classes.

2   Prototype Selection Methods

Several prototype selection algorithms have been adapted and/or developed in order to select a small representation set R or to reduce the dissimilarity matrix $D(T,R)$. For example, Lozano et al. [13] employed prototype optimization methods often applied in vector spaces, such as editing and condensing, for constructing more general dissimilarity-based classifiers. Kim and Oommen [8] used the well-known condensed nearest neighbor rule [14] to reduce the original training set before computing the dissimilarity-based classifiers on the entire data. Other new methods have been evolved to be applied in the dissimilarity space, such as Kcentres, Edicon, ModeSeek, Featsel and a genetic algorithm [6, 9].

However, all these proposals do not consider the skewness in the class distribution. In this work, we concentrate on using four prototype selection methods, commonly applied to feature-based classification models, for the reduction of the Euclidean distance representation $D(T,T)$ (here called the original dissimilarity matrix) in domains with class imbalance. Two different families of prototype selection methods exist in the literature: editing and condensing. Editing removes erroneously labeled and atypical examples from the original set and "cleans" possible overlapping between classes, which usually leads to significant improvements in performance. Condensing, on the other hand, aims at selecting a sufficiently small subset of examples that yields approximately the same performance as using the whole training set.

The simplest procedure to pick up a small subset corresponds to random selection (RS). However, this may throw out potentially useful data. Paradoxically, it has empirically been shown to be an effective prototype selection method. Unlike the random approach, many other proposals are based upon a more intelligent selection strategy. For example, Wilson [15] introduced a popular editing algorithm (WE) that tries to remove noisy instances and/or border points. This algorithm discards training examples whose label does not agree with that of their majority k neighbors. Another early prototype selection method is the condensed nearest neighbor (CNN) proposed by Hart [14], which is focused on selecting a consistent subset from the training set but keeping or even improving the classification accuracy. Nevertheless, as this approach could retain noisy objects, the joint use of editing and condensing algorithms (e.g., WE+CNN) is commonly employed to select an appropriate reduced subset.

3   Performance Evaluation in Imbalanced Domains

Traditionally, standard performance metrics have been classification accuracy and/or error rates. For a two-class problem, these can be easily derived from a 2×2 confusion matrix as that given in Table 1.

Table 1. Confusion matrix for a two-class problem

	Predicted as positive	Predicted as negative
Positive class	True Positive (TP)	False Negative (FN)
Negative class	False Positive (FP)	True Negative (TN)

However, as pointed out by many authors [16,17], the performance of a classification process over imbalanced data sets should not be expressed in terms of the plain accuracy and/or error rates because these measures are strongly biased towards the majority class. This has motivated to search for new performance evaluation metrics based upon simple indices, such as the true positive rate (TPr) and the true negative rate (TNr). The TPr (or TNr) is the percentage of positive (or negative) examples correctly classified.

One of the most widely-used evaluation methods in the context of imbalanced class distributions is the ROC curve. Here, we will utilize the area under the ROC curve (AUC), which is a quantitative representation of a ROC curve. For a binary problem, the AUC criterion defined by a single point on the ROC curve is also referred to as balanced accuracy [18]:

$$AUC_b = \frac{TPr + TNr}{2} \qquad (1)$$

where $TPr = \frac{TP}{TP+FN}$ measures the percentage of positive examples that have been classified correctly, whereas $TNr = \frac{TN}{TN+FP}$ corresponds to the percentage of negative cases predicted as negative.

4   Experimental Setup

Eight real data sets were employed in the experiments. In order to force the class imbalance, all data sets were transformed into two-class problems by keeping one original

Table 2. Data sets used in the experiments

Data Set	#Positive	#Negative	#Classes	Majority Class	Source
Breast	81	196	2	1	UCI[1]
Ecoli	35	301	8	1,2,3,5,6,7,8	UCI
German	300	700	2	1	UCI
Haberman	81	225	2	1	UCI
Laryngeal$_2$	53	639	2	1	Library[2]
Pima	268	500	2	1	UCI
Vehicle	212	634	4	2,3,4	UCI
Yeast	429	1055	10	1,3,4,5,6,7,8,9,10	UCI

[1] UCI Machine Learning Database Repository http://archive.ics.uci.edu/ml/
[2] Library http://www.vision.uji.es/~sanchez/Databases/

class (the minority one) and joining the objects of the remaining classes. The fifth column in Table 2 indicates the original classes that were joined to shape the majority class.

A stratified five-fold cross-validation method was adopted for the present experiments. For each fold, four parts were pooled as the training data T, and the remaining block was employed as an independent test set S. Ten repetitions were run for each trail. The results from classifying the test samples were averaged across the 50 runs. For each database, the whole training set ($R = T$) was used to compute the original dissimilarity matrix $D(T, T)$, with the Euclidean distance as a dissimilarity measure. This procedure was also applied to the test set, $D(S, T)$, to be represented in the dissimilarity space.

The four prototype selection methods described in Sect. 2 were utilized for the experiments: random selection (RS), condensed nearest neighbor (CNN), Wilson's editing (WE), and the combination of this with Hart's condensing (WE+CNN). All these methods were implemented following two different strategies: (i) *hard selection* over both existing classes, and (ii) *one-sided selection* only over the majority (negative) class. In this latter case, like occurs in typical under-sampling processes, we did not remove minority (positive) examples because they are too limited and important to be discarded. The Fisher and the nearest neighbor (1-NN) learning algorithms were used to each original dissimilarity matrix and also to matrices that were previously pruned by the different prototype selection methods.

Note that the RS procedure allows to control the number of prototypes to be chosen. Here, we extracted 50% out of each class for the hard selection strategy, and a number of negative examples equal to the size of the positive class $|P|$ for the one-sided selection strategy.

5   Results

In order to analyze the effect of the class imbalance on the performance of the prediction models, we generated different dissimilarity matrices, each one with an amount of positive examples, by randomly increasing the minority class size until reaching its original size. The number of objects in the majority class keeps constant for all dissimilarity matrices. Figure 1 shows the TPr and TNr for two illustrative examples of these data sets when using the Fisher classifier, where the x-axis represents the number of positive samples in the dissimilarity matrix. Note that both TPr and TNr have been plotted in a different scale in order to make these graphics clearer.

As expected, when the dissimilarity matrices are strongly imbalanced, the Fisher performance on the minority class is significantly worse than that on the majority class: the TNr is close to 0.90, but the TPr is below 0.40. As the size of the minority class increases, the TPr improves and the TNr lessens. It is worth noting, however, that the poor results of TPr remain even when all the positive examples are put into the dissimilarity matrix. In such an imbalance scenario, this effect demonstrates the need of using some strategy to generate more appropriate (balanced) dissimilarity matrices.

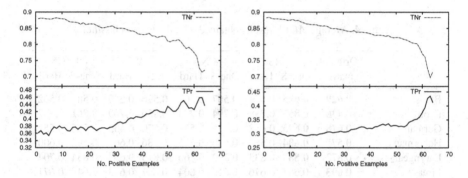

Fig. 1. Effect of the class imbalance on Fisher classifier performance for the Breast (left) and Haberman (right) databases

Tables 3 and 4 report the average AUC_b with the 1-NN and the Fisher classifiers respectively, when using the original dissimilarity matrix $D(T, T)$ and after pruning this by means of the prototype selection methods. The column "*One-S*" contains the results from applying the prototype selection procedures only over the majority class, whereas the column "*Hard*" refers to the results obtained when pruning both classes. For each data set, the best case has been highlighted in bold type. Average rankings of the Friedman statistic (distributed according to chi-square with 8 degrees of freedom) have also been included.

From the results in Tables 3 and 4, one can observe that both classifiers are affected by the class imbalance problem when they are trained with the original dissimilarity matrix, yielding relatively low AUC_b values. On the other hand, when employing the prototype selection methods over both classes (hard selection), the behavior varies from one data set to another: the AUC_b values are even worse than those achieved with the original dissimilarity matrix for some databases and better for others.

The results obtained with the application of prototype selection over the majority class (one-sided selection) show that all these techniques perform better, in terms of AUC_b, than the original dissimilarity matrix. It is also interesting to remark that this bi-

Table 3. Average AUC_b results obtained with the 1-NN classifier

	Original matrix	RS One-S	Hard	CNN One-S	Hard	WE One-S	Hard	WE+CNN One-S	Hard
Breast	0.575	0.574	0.562	0.620	0.598	**0.646**	0.587	0.626	0.610
Ecoli	0.791	**0.866**	0.774	0.706	0.668	0.838	0.776	0.815	0.676
German	0.535	0.551	0.539	**0.699**	0.693	0.681	0.587	0.692	0.623
Haberman	0.575	0.575	0.578	0.575	0.565	0.600	0.585	0.602	0.589
Laryngeal$_2$	0.775	0.846	0.746	0.830	0.793	**0.887**	0.741	0.849	0.738
Pima	0.624	0.632	0.625	0.687	0.674	**0.707**	0.686	0.694	0.690
Vehicle	0.579	0.606	0.580	0.699	0.673	0.679	0.572	**0.728**	0.588
Yeast	0.660	0.668	0.641	0.692	0.676	**0.719**	0.662	0.710	0.653
Average rankings	7.125	5.375	7.375	3.875	5.500	1.875	6.125	2.000	5.750

Table 4. Average AUC_b results obtained with the Fisher classifier

	Original matrix	RS One-S	Hard	CNN One-S	Hard	WE One-S	Hard	WE+CNN One-S	Hard
Breast	**0.629**	0.625	0.609	0.567	0.560	0.596	0.530	0.583	0.552
Ecoli	0.736	0.857	0.738	0.794	0.773	0.860	0.760	**0.861**	0.737
German	0.678	**0.693**	0.658	0.535	0.530	0.570	0.553	0.566	0.552
Haberman	0.575	**0.604**	0.580	0.586	0.575	0.586	0.601	0.592	0.600
Laryngeal$_2$	0.872	**0.883**	0.833	0.792	0.766	0.815	0.694	0.834	0.704
Pima	**0.693**	0.687	0.676	0.612	0.604	0.657	0.673	0.649	0.671
Vehicle	0.660	**0.742**	0.647	0.584	0.580	0.606	0.573	0.611	0.574
Yeast	0.690	**0.710**	0.672	0.659	0.643	0.688	0.663	0.680	0.654
Average rankings	3.437	1.500	4.375	6.312	7.687	4.312	6.375	4.125	6.875

ased selection is significantly better than the classical approaches to prototype selection over both classes.

As a further confirmation of the findings with the AUC_b values, we have run a Wilcoxon signed-ranks test [19] between each pair of techniques. The upper diagonal half of Tables 5 and 6 summarizes this statistic for a significance level of 0.10 (10% or less chance), whereas the lower diagonal half corresponds to a significance level of 0.05. The symbol "•" indicates that the method in the row significantly improves the method of the column, and the symbol "○" means that the method in the column performs significantly better than the method of the row. The two bottom rows show how many times the algorithm of the column has been significantly better than the rest of procedures for $\alpha = 0.10$ and $\alpha = 0.05$.

It is worth pointing out that, as can be observed in Tables 5 and 6, the one-sided selection has been significantly better than the hard selection strategy for all the prototype selection algorithms (for $\alpha = 0.10$ and $\alpha = 0.05$), both with the 1-NN classifier and the Fisher classifier. This allows to assert that such a biased selection of prototypes for

Table 5. Summary of the Wilcoxon statistic for the prototype selection methods with the 1-NN classifier

	(1)	(2)	(3)	(4)	(5)	(6)	(7)	(8)
(1) One-sided RS	-	•			○		○	
(2) Hard RS	○	-			○		○	
(3) One-sided CNN			-	•	○		○	•
(4) Hard CNN			○	-	○		○	
(5) One-sided WE	•	•	•	•	-	•		•
(6) Hard WE					○	-	○	
(7) One-sided WE+CNN	•	•	•		•		-	•
(8) Hard WE+CNN					○		○	-
$\alpha = 0.10$	1	0	2	0	6	0	6	0
$\alpha = 0.05$	1	0	1	0	6	0	5	0

Table 6. Summary of the Wilcoxon statistic for the prototype selection methods with the Fisher classifier

	(1)	(2)	(3)	(4)	(5)	(6)	(7)	(8)
(1) One-sided RS	-	•	•	•	•	•	•	•
(2) Hard RS	○	-		•				•
(3) One-sided CNN	○		-	•	○		○	
(4) Hard CNN	○	○	○	-	○		○	
(5) One-sided WE	○		•	•	-	•		•
(6) Hard WE	○				○	-	○	
(7) One-sided WE+CNN	○		•	•			-	•
(8) Hard WE+CNN	○			○				-
$\alpha = 0.10$	7	2	1	0	4	0	4	0
$\alpha = 0.05$	7	1	1	0	4	0	2	0

the construction of a more balanced dissimilarity matrix (with all the positive examples and only a subset of negative examples) can be deemed as an appropriate solution to the class imbalance problem in dissimilarity spaces.

In the case of the 1-NN classifier, it seems that the best prototype selection method corresponds to Wilson's editing, whose one-sided version has performed significantly better than other six algorithms at both significance levels. The WE+CNN procedure presents a very similar behavior, being significantly better than other five algorithms at a significance level of 0.05. Clearly, the random selection and Hart's condensing methods have achieved the worst results when statistically compared in terms of AUC_b.

Paradoxically, for the Fisher classifier, Table 6 shows that the one-sided random selection constitutes the best procedure, with a performance significantly better than any other algorithm. Not too far from the best alternative, one can see that the Wilson's editing with one-sided selection has been significantly better than other four strategies.

6 Conclusions

Prototype selection methods have been widely used in the dissimilarity-based approach for the selection of a small representation set (from the whole training set) and/or the reduction of the original dissimilarity matrix. When the data set and/or the dissimilarity matrix are imbalanced, however, the selection process could produce reduced data sets and/or dissimilarity matrics that do not accurately represent the true class distribution, what may lead to an increase in the class skewness.

In this paper, we have carried out some experiments using four renowned prototype selection algorithms for under-sampling the original dissimilarity matrix in domains with class imbalance. The empirical results suggest that the application of these techniques to both classes produces poor performance on the minority class. On the contrary, the strategy based upon the biased selection on the majority class significantly increases the prediction rate on the positive class and the value of average AUC_b, being statistically demonstrated by means of a Wilcoxon signed-ranks test.

Acknowledgment. This work has partially been supported by the Spanish Ministry of Education and Science under grants CSD2007–00018, AYA2008–05965–0596 and TIN2009–14205, the Fundació Caixa Castelló-Bancaixa under grant P1–1B2009–04, the Generalitat Valenciana under grant PROMETEO/2010/028, the Spanish Ministry of Economy and Competitiveness and FEDER through the EFIS project IPT-2011-0962-920000.

References

1. Duin, R.P.W., Pękalska, E.: The dissimilarity space: Bridging structural and statistical pattern recognition. Pattern Recognition Letters 33(7), 826–832 (2012)
2. Pekalska, E., Duin, R.P.W.: The Dissimilarity Representation for Pattern Recognition: Foundations and Applications. World Scientific (2005)
3. Pekalska, E., Duin, R.P.W.: Dissimilarity representations allow for building good classifiers. Pattern Recognition Letters 23(8), 943–956 (2002)
4. Kim, S.W.: An empirical evaluation on dimensionality reduction schemes for dissimilarity-based classifications. Pattern Recognition Letters 32(6), 816–823 (2011)
5. Duin, R.P.W., Pękalska, E.: The Dissimilarity Representation for Structural Pattern Recognition. In: San Martin, C., Kim, S.-W. (eds.) CIARP 2011. LNCS, vol. 7042, pp. 1–24. Springer, Heidelberg (2011)
6. Pekalska, E., Duin, R.P.W., Paclík, P.: Prototype selection for dissimilarity-based classifiers. Pattern Recognition 39(2), 189–208 (2006)
7. Plasencia-Calaña, Y., García-Reyes, E., Duin, R.P.W.: Prototype selection methods for dissimilarity space classification. Technical report, Advanced Technologies Application Center CENATAV
8. Kim, S.W., Oommen, B.J.: On using prototype reduction schemes to optimize dissimilarity-based classification. Pattern Recognition 40(11), 2946–2957 (2007)
9. Plasencia-Calaña, Y., García-Reyes, E., Orozco-Alzate, M., Duin, R.P.W.: Prototype selection for dissimilarity representation by a genetic algorithm. In: Proc. 20th International Conference on Pattern Recognition, pp. 177–180 (2010)

10. Chawla, N., Japkowicz, N., Kotcz, A.: Editorial: Special issue on learning from imbalanced data sets. SIGKDD Explorations 6(1), 1–6 (2004)
11. Sun, Y., Wong, A., Kamel, M.S.: Classification of imbalanced data: A review. International Journal of Pattern Recognition and Artificial Intelligence 23(4), 687–719 (2009)
12. Batista, G.E.A.P.A., Prati, R.C., Monard, M.C.: A study of the behavior of several methods for balancing machine learning training data. SIGKDD Explorations 6(1), 20–29 (2004)
13. Lozano, M., Sotoca, J.M., Sánchez, J.S., Pla, F., Pekalska, E., Duin, R.P.W.: Experimental study on prototype optimisation algorithms for prototype-based classification in vector spaces. Pattern Recognition 39, 1827–1838 (2006)
14. Hart, P.E.: The condensed nearest neighbor rule. IEEE Trans. on Information Theory 14, 515–516 (1968)
15. Wilson, D.L.: Asymptotic properties of nearest neighbor rules using edited data. IEEE Trans. on Systems, Man and Cybernetics 2(3), 408–421 (1972)
16. Daskalaki, S., Kopanas, I., Avouris, N.: Evaluation of classifiers for an uneven class distribution problem. Applied Artificial Intelligence 20(5), 381–417 (2006)
17. Provost, F., Fawcett, T.: Analysis and visualization of classifier performance: Comparison under imprecise class and cost distributions. In: Proc. 3rd International Conference on Knowledge Discovery and Data Mining, pp. 43–48 (1997)
18. Sokolova, M.V., Japkowicz, N., Szpakowicz, S.: Beyond Accuracy, F-Score and ROC: A Family of Discriminant Measures for Performance Evaluation. In: Sattar, A., Kang, B.-H. (eds.) AI 2006. LNCS (LNAI), vol. 4304, pp. 1015–1021. Springer, Heidelberg (2006)
19. Demšar, J.: Statistical comparisons of classifiers over multiple data sets. Journal of Machine Learning Research 7(1), 1–30 (2006)

Estimating Surface Characteristics and Extracting Features from Polarisation

Lichi Zhang, Edwin R. Hancock*, and Jing Wu

Department of Computer Science, University of York, UK
School of Computer Science and Informatics, Cardiff University, UK

Abstract. In this paper we develop a practical method for estimating shape, color and reflectance using only three images taken under polarised light. We develop a novel and practical framework to optimise the estimates and eliminate the redundant information, then investigate three different methods to compare their class discriminating capacities. We present experiment to demonstrate the validity of the proposed method for a database of fruit objects from 5 different classes, and we show that the proposed method is capable of accurately extracting the features of the input examples. The framework can further be applied in a variety fields of computer vision and pattern recognition domains including object recognition and classification.

1 Introduction

Accurately estimating and reproducing surface appearance is a task of pivotal importance in computer vision and graphics. Applications include object recognition and classification, and image rendering. The appearance of surfaces is determined by shape, color and reflectance [1]. These intrinsic surface properties are independent of each other and affect the observed image intensity in a complicated way. Therefore a robust way of simultaneously estimating these surface properties is required for successful object recognition.

There have been a number of attempts in the literature aimed at accurately measuring surface characteristics. However existing methods are limited by their requirement of high cost measurement systems, and a large number of input images. BRDF was firstly introduced in [2]. The direct measurement of the reflectance function requires a gonioreflectometer [3], which is both expensive and cumbersome to use. Other available methods use complicated devices such as light stages and geometric domes to build reflectance functions from image intensity variations under different light source directions [4]. Recently Ma et al.[5] presented a method to estimate surface normal maps of an object using four spherical gradient illumination patterns from either diffuse or specular reflectance components. The technique relies on structured light, and hence adds scanning time and system complexity to the overloads.

* Edwin Hancock was supported by a Royal Society Wolfson Research Merit Award.

G.L. Gimel' farb et al. (Eds.): SSPR & SPR 2012, LNCS 7626, pp. 400–408, 2012.
© Springer-Verlag Berlin Heidelberg 2012

Polarisation has proven to be an effective method in the analysis of light reflection in computer vision. Its applications are in reflectance component separation [6] and surface normal estimation. There have been a number of attempts in the literature aimed at surface orientation estimation of objects using polarisation, where the incident light is unpolarised. The specular and diffuse reflections from the objects become partially polarised, and their values analysed by placing a linear polariser in front of the camera and rotating its orientation. Such effects can be exploited for shape recovery using the Fresnel theory, which was used by Wolff and Boult to describe the direct reflection of electromagnetic waves with the given polarisation state of the incident light [7]. This leads to a means of surface normal estimation since the zenith angle of the reflected or re-emitted light is constrained by the degree of polarisation, and the azimuth angle is constrained by the phase angle.

In this paper we introduce a novel statistical framework for simultaneously obtaining shape, texture and reflectance properties from a single view using the theory of polarisation. We commence by acquiring the polarised images under retro-reflection settings, and separate the reflectance components by applying the method of blind source separation (BSS) following the work of Zhang et al.[8]. Then we optimise the estimates and eliminate redundant information. The estimates are converted into long vectors which is used for statistical feature extraction.

We also apply three statistical methods to the data for feature extraction. These are the tradiation method of principal component analysis (PCA), and the improved approaches which includes weight map, that are weighted PCA (WPCA) and supervised weighted PCA (SWPCA). The weight map indicates the importance of different locations in discriminating objects, thus the accuracy of results produced by feature extraction methods can be improved. The methods are developed based on the works of Wu et al. [9]. In summary, the novel contributions of this paper are:

1. We provide a novel framework which estimates shape, color and reflectance information using polarisation measurements, that only requires three input images for each object and low-cost devices.
2. We develop optimization methods which eliminate redundant information from the reflectance estimates, so that the feature extraction approaches can produce results accurately and efficiently.
3. We use three statistical methods for feature extraction, which are PCA, WPCA and SWPCA. All the approaches are also optimised to be applied in multi-class recognition.

2   Modeling Surface Characteristics

In this section we present the methods used for estimating reflectance, shape and color properties from polarised images. When light arrives at a surface, part of it undergoes isotropic subsurface scattering before being re-emitted which is denoted as diffuse reflection. The remainder is reflected in a specular manner. According to the dichromatic model [10], for every pixel in the images its intensity I is decided by two reflectance components which are diffuse I_d and specular I_s by $I = I_d + I_s$. The process of detailed reflectance estimation can be simplified if the specular and diffuse components are separated beforehand. Using the blind source separation (BSS) method introduced in [8], we obtain the separated reflectance components from the polarised images.

Reflectance and Shape Estimation: We apply two reflectance models to reflectance measurements. They are Lambertian model used for diffuse component, and Torrance-Sparrow (T-S)[11] for specular component estimation. The two models are simplified under retro-reflection and isotropy. Let the specular and diffuse reflectance models be $R_s(\theta_s, E_s)$ and $R_d(\theta_d, E_d)$ respectively, and let E_s and E_d be their parameter value sets. The estimated surface normal zenith angles for the two models are θ_s and θ_d. Associated with each model is a scaling coefficient, denoted by scaler k_s and k_d. We numerically invert the reflectance function to recover the surface zenith angle using

$$\theta_s = R_s^{-1}(I_s/k_s, E_s) = R_s^{-1}(I'_s, E_s) , \tag{1}$$
$$\theta_d = R_d^{-1}(I_d/k_d, E_d) = R_d^{-1}(I'_d, E_d) . \tag{2}$$

We compute k_s and k_d so that the two components $I'_s = I_s/k_s$ and $I'_d = I_d/k_d$ are normalized. Since they correspond to the same image location, θ_s and θ_d should be identical. As I_s and I_d are known, the parameter values can be found when the distributions of θ_s and θ_d are closest to each other. Because of its rapid (quadratic) convergence we use Newton's method to estimate the parameters E_s and E_d, using a mutual information criterion (M) for the distributions of θ_d and θ_s. Details of mutual information computation can be found in [12]. The Newton method for updating the parameter sets is $W^{(t+1)} = W^{(t)} - \gamma Q[R^{(t)}]^{-1}\nabla R^{(t)}$, where $W^{(m)} = [E_s^{(m)}, E_d^{(m)}]^T$, $Q[R^{(t)}]$ is the Hessian of the error-function and ∇R its gradient. Here we use $E_s^{(0)} = 0.5$ and $E_d^{(0)} = 0.5$ as it is valid for the parameter coefficient values of the chosen reflectance models.

The two zenith angle estimates θ_s and θ_d are ideally identical as they represent the same object. However, they differ from each other due to shadows and texture in the input images and the limited capacities of the chosen reflectance models. Here for simplification we follow the constraint that the actual surface normal θ is the mean value of the two estimates, which is $\theta = (\theta_s + \theta_d)/2$.

Color Estimation: From the dichromatic reflection model [10], each color vector $(I_R, I_G, I_B)^T$ is determined by a linear combination of specular reflection $(I_{Rs}, I_{Gs}, I_{Bs})^T$ and diffuse reflection $(I_{Rd}, I_{Gd}, I_{Bd})^T$, which is written as

$$\begin{pmatrix} I_R \\ I_G \\ I_B \end{pmatrix} = k_s \begin{pmatrix} I_{Rs} \\ I_{Gs} \\ I_{Bs} \end{pmatrix} + k_d \begin{pmatrix} I_{Rd} \\ I_{Gd} \\ I_{Bd} \end{pmatrix} . \tag{3}$$

The weights k_s and k_d depend only on the geometry of the objects in the input images, and we only focus on the color vector of diffuse reflection as it represent the hue properties of objects. We also assume that the estimated color vector is normalised for simplification. Using the diffuse reflectance component estimate I_d we can obtain the corresponding values of three color channels as I'_{Rd}, I'_{Gd} and I'_{Bd}, thus we can have I_{Rd}, I_{Gd} and I_{Bd} using the equations $I'_{Rd} = w_d I_{Rd}$, $I'_{Gd} = w_d I_{Gd}$, $I'_{Bd} = w_d I_{Bd}$ and $\sqrt{I_{Rd}^2 + I_{Gd}^2 + I_{Bd}^2} = 1$.

3   Estimates Analysis

The surface properties of shape, color and reflectance are stored in the form of matrices which correspond to every pixel of input images. In this section we show how to eliminate the redundant information in the estimates and convert them into long vectors.

Reflectance: For simplification we assume the illumination and viewing direction are identical, and the reflectance properties are independent of the surface azimuth angle. The reflectance functions for each experimental objects are created by using the method described in [13]. For every pixel in the image, its intensity has the corresponding value of the surface radiance function $g(\theta(x, y)) = I(x, y)$. By tabulating these two values against each other, we have a dense but noisy sampling of the function g. Then we bin the values of $g(\theta(x, y))$ into η bins whose width is τ. Let $\Gamma_i = (x, y)|(i - 1)\tau \le \theta(x, y) < i\tau$ be the set of pixels (x, y) for which $\theta(x, y)$ falls into the ith bin. For each bin we find the median value of g : $h(i) = \underset{(x,y)\in\Gamma_i}{\text{median}} I(x, y)$. Then the reflectance function is stored as a long vector $(h(1), ..., h(\eta))$.

Color: The hue value in HSI color representation ranges from 0 to 360 degrees, and each degree stands for a specific color which can be converted from RGB triplet. The aim here is to compute hue values from every pixel in the image, and create hue distribution vector which is the histogram of all hue values. We convert RGB values to hue by using the equations described in [14], which are

$$\begin{bmatrix} Y \\ C_1 \\ C_2 \end{bmatrix} = \begin{bmatrix} 1/3 & 1/3 & 1/3 \\ 1 & -1/2 & -1/2 \\ 0 & -\sqrt{3}/2 & \sqrt{3}/2 \end{bmatrix} \begin{bmatrix} I_{Rd} \\ I_{Gd} \\ I_{Bd} \end{bmatrix}, \tag{4}$$

$$\text{Hue} = \begin{cases} \arccos(C_2/\sqrt{C_1^2 + C_2^2}), & C_1 \ge 0 ; \\ 2\pi - \arccos(C_2/\sqrt{C_1^2 + C_2^2}), & C_1 < 0 . \end{cases} \tag{5}$$

Shape: Wu et al. [9] developed a framework which uses methods based on principal geodesic analysis (PGA) to extract surface shape features of facial needlemap recovered by shape from shading (SFS), and implement gender classification based on the estimates. PGA is a generalisation of PCA which can be applied to feature extraction for 3D shape analysis. However, there are two drawbacks. Firstly, as SFS is proved to be an ill-posed problem the method can not be used for general objects. Secondly, there is a requirement that the input images should be fully-aligned and subjects have no boundaries. Here we present a novel method to solve these problems.

Denote θ as the data matrix containing zenith angle information obtained in the previous section. We recover a height map B using shapelets method [15] without the need of azimuth angle estimates. The values of height are represented in Cartesian coordinate, which can lead to a loss of information and poor performance in feature extraction. To overcome this problem we embed the points on a spherical manifold system, which represents the size of object and shape variations in a convenient way. Suppose the center point of the object locates at (x_0, y_0, z_0), for any point in the surface whose location is (x, y, z) its spherical coordinate triplet (r, ϑ, φ) is computed by the following equations

$$r = \sqrt{(x - x_0)^2 + (y - y_0)^2 + (z - z_0)^2} \, , \tag{6}$$

$$\vartheta = \cos^{-1}[\sqrt{(x - x_0)^2 + (y - y_0)^2}/r] \, , \tag{7}$$

$$\varphi = \begin{cases} \cos^{-1}[(y - y_0)/\sqrt{(x - x_0)^2 + (y - y_0)^2}] \, , & x \geq x_0 \, ; \\ 2\pi - \cos^{-1}[(y - y_0)/\sqrt{(x - x_0)^2 + (y - y_0)^2}] \, , & x < x_0 \, . \end{cases} \tag{8}$$

where r is the radial distance between the surface point and the center point of object, the zenith angle ϑ and the azimuth angle φ are the directions from (x_0, y_0, z_0) to (x, y, z). Suppose a $3 \times xy$ matrix $B' = (b_1', b_2', b_3')$ as the height map presented in spherical coordinate, in which the three rows represent the values of r, ϑ and φ respectively. We create a 2D radius distribution matrix T in which the row is for azimuth angle ranges in $[0, 2\pi]$ while as the column is for zenith angle lies in the closed interval $[0, \pi/2]$. We bin the values of B' into 30×120 sqaure-shaped windows of size 3×3. Let $w(x, y)$ in B' be the set of rows that follows $w(x, y) = \{(x, y) | 3(x - 1) \leq h_1' < 3x, 3(y - 1) \leq h_2' < 3y\}$, for each set we find the median value of $b_3'(w(x, y))$ and stored in T as $T(x, y) = \underset{(x,y) \in w(i,j)}{\text{median }} B'(x, y)$. The values in T which the corresponding set is empty with B' in their range are set to be 0. The 2D matrix T is then converted to be long vectors for the feature extraction.

4   Feature Extraction

From a set of sample data, Principal Component Analysis (PCA) aims to find a linear subspace which maximises the variance of the projected data. It is widely applied in the fields of dimensionality reduction and feature extraction. However, the projections calculated by PCA usually are not those that best separate the data into distinct classes. Wu et. al [9] proposed several weighting schemes to improve the discriminating capacity of the leading PCA eigenvectors for gender classification. Here we extend their idea to multiple class recognition.

Firstly we incorporate a pre-computed weight map into PCA, namely weighted PCA (WPCA). The weight map is a representation of the discriminating capacity for each location in the long vectors. The locations that better identify objects are assigned higher weights than the rest part. Suppose there are m classes from the n input data $X = [x_1, x_2, ..., x_n]$, and the mean vectors for each class are denoted $[\bar{x}_1, \bar{x}_2, ..., \bar{x}_m]$. The weight at location l is computed as

$$w_l = 1 - \exp\left\{ -[\sum_{i=1}^{m}(\bar{x}_{i,l} - \bar{x}_l)]^2/m^2 \right\}, \tag{9}$$

where $\bar{x}$ is the mean vector of all the data in X. By making use of the mean vectors, the constructed weight map is less influenced by the difference in the data of the same class. Next, the weight map $W = [w_1, w_2, ..., w_n]^T$ is multiplied component-wise with each long vector in X, and we have the set of weighted data $X' = [W.*x_1, W.*x_2, ..., W.*x_n]$, where $.*$ denotes componentwise matrix multiplication. We apply Singular Vector Decomposition (SVD) to X' which gives $X' = CSV^T$, where C is the left eigenvector matrix represented as feature components, S the diagonal matrix of singular values, and V the right eigenvector matrix which consists of d dimensional feature vectors.

We also extend the above approach by learning the weight map in a supervised way, which is termed supervised weighted PCA (SWPCA). Suppose we apply PCA to a data set $X = [x_1, x_2, ..., x_n]$, and obtain the leading k eigenvectors $\Phi = [e_1, ..., e_k]$ and the corresponding eigenvalues $\Lambda = [\lambda_1, ..., \lambda_k]$. The PCA feature vector for data $x \in X$ is $v = \Phi^T x$, which can be expressed component-wise as

$$v_i = \sum_{l=1}^{d} \Phi_{il}^T x_l, \tag{10}$$

where d is the dimension of data x, Φ_i denotes the ith eigenvector, and Φ_{il} is its value at the location l. SWPCA extends the above component-wise feature extraction by incorporating a weight map as $v_i^{SW} = \sum_{l=1}^{d} \Phi_{il}^T w_l x_l$, where w_l is the weight at the location l. Because the weight map has a large absolute value in class-discriminating regions, SWPCA increases the influence of class-discriminating regions over the extracted features and decreases that of the non-discriminating regions. Suppose there are m classes in the data set $X = [x_1, x_2, ..., x_n]$, which are denoted $class_i$, $i = 1, ..., m$. The weight map is initialized as the one used in WPCA, and is optimised by minimising an error function,

$$\xi = \sum_{i=1}^{m} \sum_{j \in class_i} \frac{D(v_j, \bar{v}_i)^2}{D(v_j, \bar{\bar{v}}_i)^2}, \tag{11}$$

where v_j is the WPCA feature vector (normalized by eigenvalues S) of data $x_j \in X$, $\bar{v}_i$ is the mean feature vector for $class_i$, and $\bar{\bar{v}}_i$ is mean of the data not belonging to $class_i$. Function D calculates the Euclidean distance between the feature vectors. Substituting Equation (10) into Equation (11), we have ξ to update the weight map W using $W^{(t+1)} = W^{(t)} - \nabla\xi(W^{(t)})$. We use gradient descent method to optimise each w_l in the weight map.

5   Experiment Results

In this section we present experimental evaluations of our framework for surface characteristics estimation. During the acquisition we placed a vertical polarisation filter in front of a collimated light source, so that the object are illuminated by polarised light in the direction of the camera (frontal illumination). There is also a polariser in front of the camera, which can be rotated and change the intensities of input images following the equation of Transmitted Radiance Sinusoid (TRS)[16].

There are 35 fruits in 7 different categories for the experiments, and there are 5 objects in each class. The experimental objects include red and green apples, oranges, pears, tomatoes, lemons and apricots. The results of reflectance functions and hue distributions for all inputs are shown in Fig.1. The reflectance information of fruits in different classes is hard to distinguish, as it is easily

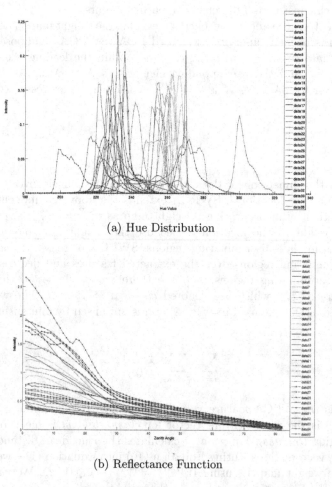

(a) Hue Distribution

(b) Reflectance Function

Fig. 1. The estimated hue distribution and reflectance functions for 35 fruit objects

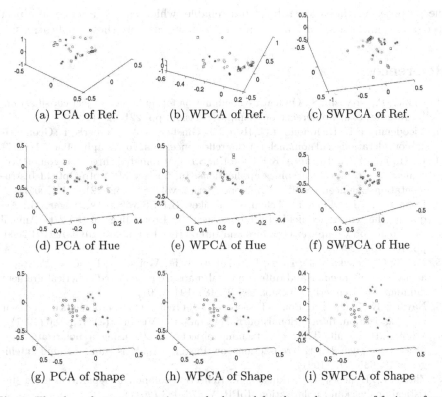

(a) PCA of Ref.   (b) WPCA of Ref.   (c) SWPCA of Ref.

(d) PCA of Hue   (e) WPCA of Hue   (f) SWPCA of Hue

(g) PCA of Shape   (h) WPCA of Shape   (i) SWPCA of Shape

Fig. 2. The three feature extraction methods used for three characters of fruit surface appearance

influenced by dirts such as dusts and oil. The hue distributions are easier to recognise as the light source conditions are identical for all inputs, however some objects in different classes have similar or even identical colors. Therefore using only color information cannot discriminate fruits accurately, but the properties of shape and reflectance should also be considered.

In Fig.2 we show the feature extraction results using the three methods, i.e. PCA, WPCA and SWPCA. From the figure, the features extracted using SW-PCA are better separated by different classes, and are more concentrated within the same class than those extracted using the other two methods. It is also clear that using techniques such as nearest neighbour or SVM the fruit classification results can be much improved when considering the three properties simultaneously. This is one of the topics for future research.

6   Conclusion

In this paper we provide a novel framework for obtaining shape, color and reflectance information using the polarisation techniques and then uses three feature extractions methods on the estimates. We demonstrate experimentally that the proposed methods are robust and reliable, which can be applied in object recognition and classification. Future research will explore these applications.

References

1. Gehler, P., Nowozin, S.: On feature combination for multiclass object classification. In: International Conference on Computer Vision, pp. 221–228 (2009)
2. Nicodemus, F.E., Richmond, J.C., Hsia, J.J., Ginsberg, I.W., Limperis, T.: Geometrical considerations and nomenclature for reflectance. Nbs Monograph 160, 1–52 (1977)
3. Li, H., Foo, S.C., Torrance, K.E., Westin, S.H.: Automated three-axis goniorefctometer for computer graphics applications. In: Society of Photo-Optical Instrumentation Engineers (SPIE) Conference Series, vol. 5878, pp. 221–231 (2005)
4. Debevec, P., Hawkins, T., Tchou, C., Duiker, H.P., Sarokin, W., Sagar, M.: Acquiring the reflectance field of a human face. In: Proceedings of the 27th Annual Conference on Computer Graphics and Interactive Techniques, SIGGRAPH 2000, pp. 145–156 (2000)
5. Ma, W.C., Hawkins, T., Peers, P., Chabert, C.F., Weiss, M., Debevec, P.: Rapid acquisition of specular and diffuse normal maps from polarized spherical gradient illumination. Rendering Techniques (9), 183–194 (2007)
6. Nayar, S.K., Fang, X.S., Boult, T.: Separation of Reflection Components using Color and Polarization. International Journal of Computer Vision 21(3), 163–186 (1997)
7. Wolff, L.B., Boult, T.E.: Constraining object features using a polarization reflectance model. IEEE Transactions on Pattern Analysis and Machine Intelligence 13, 635–657 (1991)
8. Zhang, L., Hancock, E., Atkinson, G.: Reflection component separation using statistical analysis and polarisation. IbPRIA, 476–483 (2011)
9. Wu, J., Smith, W.A.P., Hancock, E.R.: Gender discriminating models from facial surface normals. Pattern Recognition 44(12), 2871–2886 (2011)
10. Shafer, S.A.: Using color to separate reflection components. Color Research & Application 10(4), 210–218 (1985)
11. Torrance, K.E., Sparrow, E.M.: Theory for off-specular reflection from roughened surfaces. Journal of the Optical Society of America 57(9), 1105–1112 (1967)
12. Zhang, L., Hancock, E. R.: Robust Shape and Polarisation Estimation Using Blind Source Separation. In: Real, P., Diaz-Pernil, D., Molina-Abril, H., Berciano, A., Kropatsch, W. (eds.) CAIP 2011, Part I. LNCS, vol. 6854, pp. 178–185. Springer, Heidelberg (2011)
13. Smith, W., Hancock, E.R.: Recovering face shape and reflectance properties from single images. In: International Conference on Automatic Face & Gesture Recognition, pp. 1–8 (2008)
14. Zhang, C., Wang, P.: A new method of color image segmentation based on intensity and hue clustering. In: International Conference on Pattern Recognition, vol. 3, pp. 613–616 (2000)
15. Kovesi, P.: Shapelets correlated with surface normals produce surfaces. In: International Conference on Computer Vision, vol. 2, pp. 994–1001 (2005)
16. Atkinson, G.A., Hancock, E.R.: Recovery of surface orientation from diffuse polarization. IEEE Transactions on Image Processing 15(6), 1653–1664 (2006)

Extended Fisher Criterion Based on Auto-correlation Matrix Information

Hitoshi Sakano[1], Tsukasa Ohashi[2], Akisato Kimura[1],
Hiroshi Sawada[1], and Katsuhiko Ishiguro[1]

[1] NTT Communication Science Laboratories, NTT Corporation
2-4 Hikaridai, Seika-cho, Soraku-gun, Kyoto 619-0237, Japan
sakano.hitoshi@lab.ntt.co.jp
[2] Graduate School of Engineering, Doshisha University 1-3 Tatara Miyakodani,
Kyotanabe-shi, Kyoto 610-0394, Japan

Abstract. Fisher's linear discriminant analysis (FLDA) has been attracting many researchers and practitioners for several decades thanks to its ease of use and low computational cost. However, FLDA implicitly assumes that all the classes share the same covariance: which implies that FLDA might fail when this assumption is not necessarily satisfied. To overcome this problem, we propose a simple extension of FLDA that exploits a detailed covariance structure of every class by utilizing revealed by the class-wise auto-correlation matrices. The proposed method achieves remarkable improvements classification accuracy against FLDA while preserving two major strengths of FLDA: the ease of use and low computational costs. Experimental results with MNIST and other several data sets in UCI machine learning repository demonstrate the effectiveness of our method.

1 Introduction

This paper proposes a simple extension of Fisher's linear discriminant analysis (FLDA) that exploits a detailed covariance structure of every class. The major advantage of our method lies on its ease of use and low computational cost. This makes our method more useful for practitioners.

FLDA is widely used as a discriminative feature extractor, especially in the field of pattern recognition, computer vision and machine learning. Its application areas have a wide variety, which include character recognition and face recognition [3,4]. FLDA has been attracting a lot of researchers and practitioners for a long time thanks to its simple formulation and low computational costs. However, FLDA implicitly assumes that a distribution of each class should be Gaussian and all the classes share the same covariance matrix. When facing a classification problem with other circumstances, its classification performance might degrade drastically.

Many extensions of FLDA have been proposed to overcome this problem. They are roughly classified into two categories. The first category is (1) non-linear or

G.L. Gimel' farb et al. (Eds.): SSPR & SPR 2012, LNCS 7626, pp. 409–416, 2012.
© Springer-Verlag Berlin Heidelberg 2012

piecewise linear extensions of FLDA. Hastie, et al.[8],Zhu, et al.[9],Gkalelis, et al.[10] employed cluster analysis to fit multi-peak feature distributions. Baudat[6] and Sierra[7] studied non-linear transformation extension to represent complex feature distributions. This approach is very popular, however, it requires high computational cost that may eliminate one of the strengths of FLDA. Further, this approach may incur model selection difficulties such as the number of peaks and the type of transformations. Another approach is: (2) incorporating between-distribution metrics such as Kullback-Leibler divergence or Chernoff distance into the computation of between-class scatter matrices [13,12], instead of simple Euclidean norms. One major problem of this approach lies on the asymmetric structure of metrics, which leads to inconsistent formulations of the entire method. In other words, the second approach is attractive if we can avoid this problem.

Based on the above observations, this paper proposes yet another extension of FLDA along with the second approach. The main problem is how to inject covariance information of every class into between-class scatter matrix. Inspired by the class description of Class Featuring Information Compression (CLAFIC) [14,11], we describe this covariance information as a subspace spanned by eigenvectors of a class-specific auto-correlation matrix. Thus, we can acquire rich information of class-wise feature distributions by simply concatenating the subspace induced from auto-correlation matrix to the subspace obtained from the original FLDA. Our proposed formulation consists of simple matrix operations only, and the algorithm is still easy to use and enjoy low-computational cost. Further it is easy to extend the formulation to multi-class categorization problems.

The rest of the paper is organized as follows. Section 2 reviews the classical FLDA and clarify its fundamental problems. Section 3 describes our new criterion function for FLDA based on the description of class-wise feature distribution. Section 4 demonstrates the effectiveness of the proposed method through some experimental evaluations with standard benchmark datasets. Finally Section 5 concludes this paper and poses some future work.

2   Fisher's Discriminant Analysis and Its Problems

This section reviews the classical FLDA and clarifies its fundamental problems.

Let $X_c = \{x_1, \ldots, x_{n_c}\}$ be a set of D-dimensional samples in class c, where n_c is the number of samples assigned to the class c. To find the most discriminative basis for C-class classification problem, FLDA maximizes between-class distances represented as the following between-class scatter matrix:

$$\Sigma_B = \frac{1}{C} \sum_{c=1}^{C} (\boldsymbol{\mu} - \boldsymbol{\mu}_c)(\boldsymbol{\mu} - \boldsymbol{\mu}_c)^\top, \tag{1}$$

and minimizes within-class distances represented as the following within-class scatter matrix:

$$\Sigma_W = \frac{1}{C} \sum_{c=1}^{C} \Sigma_c, \tag{2}$$

where μ is the mean vector of all samples, μ_c is the mean vector of samples assigned in the class c, and Σ_c is the scatter matrix of the class c:

$$\Sigma_c = \frac{1}{n_c} \sum_{i=1}^{n_c} (\mu_c - x_{ci})(\mu_c - x_{ci})^\top. \tag{3}$$

The problem is easily solvable as the following generalized eigenvalue problem:

$$\Sigma_B a = \Sigma_W a \lambda, \tag{4}$$

where a is an eigenvector and λ is an eigenvalue obtained from the above generalized eigenvalue problem. Eigenvectors correspond to the most distinctive axes (projections) to the given dataset.

The number of valid eigenvectors of the above generalized eigenvalue problem should be less than C, since the number of class means is C and therefore the maximum rank of the between-class scatter matrix is $C-1$ if $D > C - 1$. When dealing with 2-class classification problems, only one dimensional subspace is available. If common covariance assumption is violated in high dimensional feature space, since most of samples are distributed out of discriminant axis true discriminant plane may be close to discriminant axis. This dimensionality limitation is the fundamental problem of FLDA.

3 DFDA: Describing Covariance Structure of Classes in FLDA

In this section we propose a new FLDA criterion that reflects unique covariance structures of each class. Our observation is that one reason of the dimensionality limitation of FLDA is that FLDA only focuses on separating class mean vectors. In other words, FLDA does not consider the difference of covariance matrices of classes, or information about sample distributions of classes, which might have certain discriminative power to the classification problem. From this point of view, a straightforward extension of FLDA has been proposed in [12] that is based on Kullback-Leibler divergence. However, its optimization procedure is much complex than the original FLDA and it weakens the usefulness of FLDA. A more simpler extension, which is based on the Chernoff criterion, has been proposed in [13]. However, this method does not scale to large class problems such as Chinese character classification because this model requires pairwise classification procedure for multiclass problems.

Our proposed method is inspired CLAFIC[14]: representing sample distribution information of classes as subspaces. ψ_{ck} denotes the k-th eigenvector of the c-th class auto-correlation matrix Γ_c:

$$\Gamma_c = \frac{1}{n_c} \sum_{x \in \omega_c} x x^\top. \tag{5}$$

ψ_{ck} is not an eigenvector of the covariance matrix, but obviously has information about the distribution of samples of the class c (because this is an eigenvector of auto-correlation). Our key idea is to use ψ_{ck} to compute dispersions between classes: intuitively, a classifier that separates ψ_{ck}s from ψs of other classes is a good classifier because it segregates "shapes" of class distributions. Our criterion is described as maximization of

$$\Sigma_{B2} = \sum_c \sum_{d \neq c} \sum_{k=1}^{d_u} \sum_{l=1}^{d_u} (\psi_{ck} - \psi_{dl})(\psi_{ck} - \psi_{dl})^\top. \tag{6}$$

d_u denotes a number of eigenvectors of auto correlation matrix used for computation, which must be predefined by users. Based on this criterion, we define a new between scatter matrix as

$$\Sigma_{B_{new}} = \Sigma_B + \Sigma_{B2}. \tag{7}$$

If the rank of $\Sigma_{B_{new}}$, $r = \text{rank}(\Sigma_{B_{new}})$ is greater than $C - 1$, then we can expect improvement of classification accuracy. We call this extension of FLDA as Detailed FLDA (DFDA), which maximizes $\Sigma_{B_{new}}$. A concept sketch of DFDA illustrate in Fig.1[1].

Figure 1(a) illustrates when classes share the same covariance as assumed implicitly in FLDA. In such a case, the FLDA provides optimal projections. However, FLDA is not optimal in the case (b) because classes have different covariances (distributions, or "shapes" of shaded regions). On the other hand, DFDA projection also tries to separate eigenvectors of class auto-correlation matrices ψ and there is no assumption of shared covariances among classes. Thus we can expect improvement of classification accuracy when covariances among classes are different.

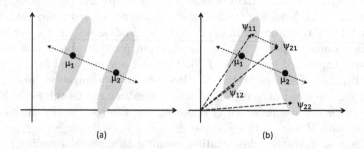

(a)   (b)

Fig. 1. Conceptual sketch of the proposed method. (a)conventional FLDA (b)proposed DFDA. The vectors ψ implicitly reveal the distributions (shapes) of the classes. The proposed DFDA tries to take the class distribution away from the others by incorporating ψ into the original FLDA.

---

[1] Though this figure is not truly correct, we expect the figure helps readers to understand the concept of the proposed method.

4   Experiments

In this section, we present the experimental evaluations. As the first experiment, we employ MNIST handwritten digit database [2]. The MNIST dataset is known as a standard benchmark data for statistical pattern recognition and machine learning researches. We would like to understand and present behaviors of proposed DFDA by this experiment. Second experiment employs a few datasets taken from UCI machine learning repository. A goal of the second experiment is to confirm effectiveness of the proposed DFDA over various datasets[3].

4.1   Experiments with MNIST Dataset

MNIST database consists of 10 digits (=classes, $C = 10$) handwritten character images. Images are 28×28 real valued matrices. We used vectorized matrices as feature vectors, thus feature space dimensionality is $D = 784$.

One of the characteristics of MNIST dataset is "FLDA hard." [15] So far many researchers have evaluated various classifiers and feature extractors with MNIST dataset. However the accuracy score is relatively low if we employ FLDA: for example classification accuracy of 1-NN classifier in original pixel value space is 97.3% that in FLDA space is 90.5%. We guess this difficulty is caused by dimension limitation of FLDA.

Since we would like to understand behaviors of FLDA and DFDA as feature extractors, we employ a 1-NN classifier in reduced spaces induced by FLDA or DFDA. We used PRTools 4.3[4] and a Matlab implementation of proposed method. We used 60000 samples as training data and 10000 samples as test data. We tested several values of d_u, the number of used eigenvectors ψ of auto-correlation matrices.

The results are shown in Table 1, Fig. 2, and Fig. 3.

Table 1. Classification accuracy of FLDA and DFDA

Method	d_u	r	Classification Accuracy (%)
FLDA	0	9	90.5
DFDA	50	51	94.3
DFDA	100	77	95.2
DFDA	200	120	95.3
DFDA	400	229	93.5

From the Table 1, it is obvious that the rank of augmented between-class scatter matrix $\Sigma_{B_{new}}$, r, grows as the number of used ψ, d_u, increases. This indicates that the information from auto-correlation matrices actually augments the information for class separations. Figure 2 illustrates the evolution of classification

---

[2] http://yann.lecun.com/exdb/mnist

[3] http://archive.ics.uci.edu/ml

[4] http://www.prtools.org

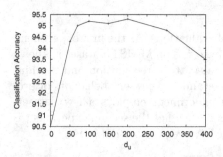 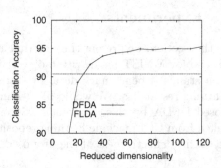

Fig. 2. Classification accuracy of proposed method. The horizontal axis denotes d_u, the number of used eigenvectors ψ of auto-correlation matrices. The vertical axis denotes the classification accuracy.

Fig. 3. Comparison of classification accuracy between FLDA and DFDA on MNIST dataset ($d_u = 200$). The horizontal axis denotes the dimensionality of the extracted features.

accuracies against d_u, the number of used eigenvectors ψ of auto-correlation matrices in DFDA. The classification accuracy score hits the highest at $d_u = 200$. This result indicates that there is an balancing point in adding ψs, possibly because of the ranks of auto-correlations matrices: these auto-correlations matrices may not be full-rank, too.

Finally, Fig.3 shows a comparison of classification accuracy of the original FLDA and the proposed DFDA ($d_u = 200$). The horizontal axis denotes the dimensionality of the extracted features. In other words, the number of eigenvectors (projections) obtained by DFDA. Note that the dimensionality of extracted features by FLDA is $9 = C - 1$. If we employ very few number of eigenvectors (less than 20), the DFDA performs poor, even worse than FLDA. However, as the number of eigenvectors increases, the performance of DFDA outperforms the FLDA, and saturates around 60 dimensions.

4.2   Experiments with Dataset from UCI Machine Learning Repository

To confirm effectiveness of DFDA, we evaluated the proposed DFDA with several datasets from UCI machine learning repository. The selection of datasets is based on the following conditions:

- The number of classes, C, is smaller than dimensionality of the feature vectors, D.
- The number of samples in each class, n_c, is larger than the dimensionality of the feature vectors, D.

Table 2 summarizes the computed classification accuracies. As evident from the table, DFDA surprisingly performs better than the FLDA in all the datasets.

Table 2. Evaluation of proposed method on UCI MLR data

Data	D	C	# of training samples ($n_c \times C$)	r	FLDA	DFDA
Breast cancer	9	2	200	8	78.0%	79.9%
magic	10	2	200	8	55.8%	69.6%
wine	13	3	60	13	40.0%	72.9%
spambase	8	2	200	8	55.8%	62.7%
image segmentation	19	7	210	19	48.9%	86.0%
ionosphere	34	2	100	8	56.6%	75.7%
statlog(Landsat)	36	6	1800	36	26.1%	75.3%
statlog(Shuttle)	9	7	43500	9	91.4%	99.7%
statlog (vehicle)	18	4	400	18	37.2%	69.7%
madelon	500	2	2000	500	54.2%	60.7%
optdigits	64	10	3823	64	45.4%	97.9%
Cardiotocography	21	3	1000	21	73.5%	78.3%

5 Conclusion

This paper proposed an extension of Fisher's Linear Discriminant Analysis (FLDA) by injecting inherent differences of distributions among classes. The proposed method exploited the auto-correlation matrix of each class samples inspired by CLAFIC. The proposed Detailed Fisher Discriminant Analysis (DFDA) integrates the subspace spanned by eigenvectors obtained from the auto-correlation matrix into the between-class scatter matrix of FLDA. Experimental evaluations with MNIST dataset and several dataset in UCI machine learning repository demonstrated the effectiveness of proposed method. Our proposed method is composed of only simple matrix operations, and therefore it can be naturally applied to multi-class categorization. The method might provide some new direction of FLDA.

The major weakness of the proposed method is its theoretical foundations. Since our idea is intuitively sound, we need more theoretical justification for this extension. It is also interesting to compare the proposed method with other extensions of FLDA such as [13,9,10,8]. Finally, some extensions of canonical correlation analysis can be achieved in a similar way, which might be fruitful for many applications.

References

1. Fisher, R.A.: The use of multiple measurements in taxonomic problems. Annals of Eigenics 7, 179–188 (1936)
2. Duda, R.O., Hart, P.E.: Pattern Classification and Scene Analysis. John Willey and Sons (1973)
3. Belhumeur, P.N., et al.: Eigenfaces vs Fisherfaces: Recognition using class specific linear projection. IEEE Transaction of Pattern analysis and Machine PAMI 19, 711–720 (1997)

4. Hastie, T., Buja, A., Tibshirani, R.: Panalized Discriminant Analysis. The Annals of Statistics 23(1), 73–102 (1995)
5. Fukunaga, K.: Introduction to Statistical Pattern Recognition, 2nd edn. Academic Press (1990)
6. Baudat, G., Anouar, F.: Generalized Discriminant Analysis Using a Kernel Approach. Neural Computation 12(10), 2385–2404 (2006)
7. Sierra, A.: High-order Fishers discriminant analysis. Pattern Recognition 35(6), 1291–1302 (2002)
8. Hastie, T., Tibshirani, R.: Discriminant Analysis by Gaussian Mixture. J. Royal Society of Statistical. Soc. B. 58, 155–176 (1996)
9. Zhu, M., Martinez, A.M.: Subclass discriminant analysis. IEEE Transactions on Pattern Analysis and Machine Intelligence 28(8), 1274–1286 (2006)
10. Gkalelis, N., Mezaris, V., Kompatsiaris, I.: Mixture subclass discriminant analysis. IEEE Signal Processing Letters 18(5), 319–322 (2011)
11. Sakano, H.: A Brief History of the Subspace Methods. In: Koch, R., Huang, F. (eds.) ACCV Workshops 2010, Part II. LNCS, vol. 6469, pp. 434–435. Springer, Heidelberg (2011)
12. Decell, H.P., Mayekar, S.M.: Feature Combinations and the Divergence Criterion. Computers and Math. with Applications 3, 71–76 (1977)
13. Loog, M., Duin, R.P.W.: Linear dimensionality reduction via a heteroscedastic extension of LDA: the Chernoff criterion. IEEE Transactions on Pattern Analysis and Machine Intelligence 26(6), 732–739 (2004)
14. Watanabe, S., Lambert, P.F., Kulikowski, C.A., Buxton, J.L., Walker, R.: Evaluation and selection of variables in pattern recognition. Comp. & Info. Sciences 2, 91–122 (1967)
15. Lim, G., Park, C.H.: Semi-supervised Dimension Reduction Using Graph-Based Discriminant Analysis. In: 2009 Ninth IEEE International Conference on Computer and Information Technology, pp. 9–13 (2009)

Poisoning Adaptive Biometric Systems

Battista Biggio, Giorgio Fumera, Fabio Roli, and Luca Didaci

Department of Electrical and Electronic Engineering, University of Cagliari
Piazza d'Armi 09123, Cagliari, Italy
{battista.biggio,fumera,roli,luca.didaci}@diee.unica.it
http://prag.diee.unica.it/

Abstract. Adaptive biometric recognition systems have been proposed to deal with natural changes of the clients' biometric traits due to multiple factors, like *aging*. However, their adaptability to changes may be exploited by an attacker to compromise the stored templates, either to impersonate a specific client, or to deny access to him. In this paper we show how a carefully designed attack may gradually poison the template gallery of some users, and successfully mislead a simple PCA-based face verification system that performs self-update.

Keywords: Biometric recognition, Adaptive biometric systems, Template self-update, Principal component analysis, Poisoning attack.

1 Introduction

Adaptive biometric recognition systems have been proposed to deal with changes of the clients' biometric traits over time, like aging. Biometric data acquired over time during system operation can be exploited to account for the natural temporal variations of biometric traits. One of the proposed approaches, inspired by semi-supervised learning techniques, is template self-update. It consists of periodically updating the template gallery of a user, using samples assigned with high confidence to the corresponding identity during operation. Adaptation may allow a biometric system to maintain a good performance over time. However, an *attacker* may exploit it to compromise the stored templates, either to impersonate a specific client or to deny access to him, violating system security.

In this paper we present a preliminary investigation on how to exploit the above discussed vulnerability in the context of adaptive biometric systems, using as a case study a simple PCA-based face verification system that performs self-update. We show that an attacker can submit a carefully designed set of fake faces to the camera while claiming the identity of another user (i.e., the *victim*), to gradually compromise the stored templates of the victim. The fake faces can be obtained by printing a face image on paper. This is a well-known procedure in the literature of *spoofing* of biometric traits (see, e.g., [2]). The goal of the attacker is to eventually be able to impersonate the victim without presenting any fake face to the sensor, i.e., to include at least one of her templates into the victim's gallery.

G.L. Gimel' farb et al. (Eds.): SSPR & SPR 2012, LNCS 7626, pp. 417–425, 2012.
© Springer-Verlag Berlin Heidelberg 2012

We derive the optimal attack, i.e., the one that minimizes the number of fake faces to be submitted to the sensor, under two distinct template update policies. To this end, we exploit the results reported in [4] about poisoning attacks against a different, but related application (online anomaly detection). Our results show that an attacker may effectively compromise the system with relatively small effort, i.e., by submitting a few, carefully designed fake faces. We also highlight a trade-off between the ability of a system to adapt to changes, and its security.

In Sect. 2 we summarize background concepts on adaptive biometric systems. Poisoning attacks and our application example are described in Sect. 3 and 4, respectively. Conclusions are drawn in Sect. 5.

2   Adaptive Biometric Recognition Systems

One of the issues that affect the performance of biometric systems in real operational scenarios is that biometric data can exhibit a large intra-class variability due to multiple factors, like illumination changes, pose variations and aging. This can make the templates stored for each client during enrollment not representative of the biometric traits submitted during verification (or identification) [10]. In particular, it is very difficult to deal with *temporal* changes of biometric patterns, like the ones due to aging. To this end, the exploitation of biometric data acquired over time during system operation has recently been proposed [3,8,7]. The reason is that such data stream naturally contains temporal variations of the considered biometric trait, which may allow one to implement *adaptive* systems that improve with use. In the following, we focus on the template self-update technique, that will be considered in the rest of this work.

Template Self-update. Template self-update is a semi-supervised learning technique that can be easily implemented in many biometric recognition systems, to enable adaptation to temporal changes. It consists of updating the stored templates of each enrolled client over time, exploiting unlabelled biometric data acquired during system operation [7]. In this work we consider a simple biometric verification system that stores one template for each client, computed by averaging the set of n enrolled images of the same client. It will thus be referred to as *centroid*. Denoting the feature vectors of the enrolled images of a given client c as $\{\mathbf{x}_{c,1}, \ldots, \mathbf{x}_{c,n}\}$, their centroid is $\mathbf{x}_c = \frac{1}{n}\sum_{k=1}^{n}\mathbf{x}_{c,k}$. During verification, a user submits a sample $\mathbf{x}$ and claims an identity c. A matching score $s(\mathbf{x}, \mathbf{x}_c)$ is then computed, e.g.:

$$s(\mathbf{x}, \mathbf{x}_c) = 1/(1 + \|\mathbf{x} - \mathbf{x}_c\|) ,\qquad(1)$$

where $\|\cdot\|$ is the Euclidean distance. The user is accepted as genuine, if $s(\mathbf{x}, \mathbf{x}_c) \geq t_c$, otherwise it is rejected as an impostor, where t_c is a predefined, client-dependent acceptance threshold. Template self-update can be implemented by updating $\mathbf{x}_c$ using $\mathbf{x}$, if $s(\mathbf{x}, \mathbf{x}_c) \geq \theta_c$, where θ_c is an update threshold, usually more conservative, i.e., $\theta_c > t_c$. The centroid $\mathbf{x}_c$ can be updated to $\mathbf{x}'_c$ according to different policies, more or less adaptive. We will consider two policies discussed in [4], which can be expressed as:

$$\mathbf{x}'_c = \mathbf{x}_c + (\mathbf{x} - \mathbf{x}_c)/n . \tag{2}$$

The *infinite window* policy updates $\mathbf{x}_c$ without discarding any of the past n samples [4,6]. Thus, n increases by 1 *before* each update, and the impact of new samples reduces as n grows. A more adaptive policy is *finite window (average-out)*, that discards the current centroid at each iteration, and keeps n fixed to its initial value.

3 Poisoning Attacks

Biometric recognition is an example of the use of *machine learning in adversarial environments*, in which a human "adversary" can be interested in subverting a recognition system, e.g., to impersonate a given client [5,1]. In particular, if the adversary has some degree of control on training data (e.g., in scenarios like template self-update), she may "contaminate" it by adding carefully designed attack samples. This attack, known as *poisoning*, has been investigated in [4,6] for online anomaly detection tasks. Since their results apply also to biometric template self-update, we summarize them below.

In template self-update, a poisoning attack exploits adaptation to gradually compromise the template $\mathbf{x}_c$ of the targeted client, until it is replaced by a sample $\mathbf{x}_a$ chosen by the attacker. To this end, the adversary may be required to iteratively submit to the system a carefully designed sequence of attack samples. We consider the case when the attacker aims to gain access with the identity of user c without using any fake trait. In this case, $\mathbf{x}_a$ must be a representative sample of the attacker's biometric trait. The type and number of attack samples depend on the template update policy, and on the capability and knowledge of the attacker. The analysis of [4,6] was made under the worst-case assumption that the attacker perfectly knows the targeted system, which is typical in security problems. In our case, this amounts to knowing the feature vector representation of samples, the initial template gallery of the targeted client and the template updating policy, the matching score function $s(\cdot, \cdot)$, and the thresholds t_c and θ_c of the victim. The optimal attack can be derived, in terms of the minimum number of attack samples required to replace $\mathbf{x}_c$ with $\mathbf{x}_a$, as well as a lower bound on such number, depending on the update policy.

The optimal poisoning attack against the policies mentioned in Sect. 2 is depicted in Fig. 1. At iteration i, it amounts to place the attack sample $\mathbf{x}_a^{(i)}$ on the line joining $\mathbf{x}_a$ and the initial centroid $\mathbf{x}_c$, in the so-called *attack direction* $\mathbf{a} = \frac{\mathbf{x}_a - \mathbf{x}_c}{\|\mathbf{x}_a - \mathbf{x}_c\|}$, at the maximum distance from the current centroid $\mathbf{x}_c^{(i)}$ that satisfies the update condition $s(\mathbf{x}_a^{(i)}, \mathbf{x}_c^{(i)}) \geq \theta_c$. Given the matching score of Eq. 1, this distance is $d_c(\theta_c) = \|\mathbf{x}_a^{(i)} - \mathbf{x}_c^{(i)}\| = 1/\theta_c - 1$. This leads to: $\mathbf{x}_a^{(i)} = \mathbf{x}_c^{(i)} + d_c \cdot \mathbf{a}$.

The minimum number of attack samples needed to replace $\mathbf{x}_c$ with $\mathbf{x}_a$, for the *infinite* and *finite window* policies, is respectively lower bounded by:

$$n \left[\exp\left(\|\mathbf{x}_a - \mathbf{x}_c\|/\theta_c\right) - 1\right], \quad n \left(\|\mathbf{x}_a - \mathbf{x}_c\|/\theta_c\right) . \tag{3}$$

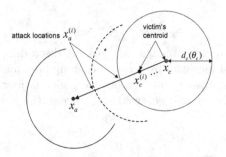

Fig. 1. Illustration of a poisoning attack, similar to [4]

It is worth noting that: (i) in both cases the number of attack samples scales linearly with the initial number of averaged samples n; (ii) in the *infinite window* case, the number of attack samples increases exponentially as $\|\mathbf{x}_a - \mathbf{x}_c\|$ grows; (iii) in the *finite window* case, such number scales linearly with $\|\mathbf{x}_a - \mathbf{x}_c\|$. Therefore, although more adaptive to changes, the latter policy may be misled by a poisoning attack with a significantly lower number of attack samples. This quantifies the intuitive trade-off between the ability of the system to adapt to changes and its *security* to poisoning attacks. To better characterize this trade-off in the context of biometric systems, one should also evaluate the probability for an attacker (and the victim) to be accepted as the targeted victim, as the attack proceeds. However, this can be only done empirically, as shown in the next section.

4   Application Example

In this section we describe the case study, related to PCA-based face verification, that we used to investigate the vulnerability of template self-update techniques.

PCA-Based Face Verification. The standard PCA-based face recognition method works as follows [9,11]. During *enrollment*, a set of face images is acquired for each user and pre-processed (e.g., the background is removed by applying a specific mask to each image, and face images are normalized to have the same size and eye position). Each image is then stored as a column vector of d pixels to constitute the training set $Z = \{\mathbf{z}_1, \ldots, \mathbf{z}_n\} \in R^{d \times n}$, and the PCA is applied as follows. (i) The average face image and the covariance matrix are respectively computed as $\mathbf{z}_\mu = \frac{1}{n} \sum_{k=1}^{n} \mathbf{z}_k$, and $C = (Z - \mathbf{z}_\mu)(Z - \mathbf{z}_\mu)^T$. (ii) The eigenvalues and eigenvectors of C can be more efficiently computed from the matrix $K = (Z - \mathbf{z}_\mu)^T(Z - \mathbf{z}_\mu)$ instead of C, as explained in [11,9]. Usually only a subset of them (those associated to the highest eigenvalues) is retained for computational efficiency. (iii) Samples in Z can be now projected onto the eigenspace as $\mathbf{x}_i = V^t(\mathbf{z}_i - \mathbf{z}_\mu)$, $i = 1, \ldots, n$, where V is the matrix of the eigenvectors (one per column). (v) For each user c, a face template $\mathbf{x}_c$ is stored. Such template is often computed as the mean (or centroid) of the projected faces of that user.

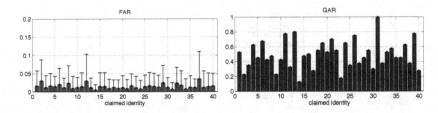

Fig. 2. FAR and GAR for each client under no attack. The FAR is averaged over all possible attackers. Standard deviation is also shown as error bars.

During *verification*, the input image $\mathbf{z}$ is pre-processed and projected onto the eigenspace as $\mathbf{x} = V^t(\mathbf{z} - \mathbf{z}_\mu)$. Then, it is compared to the centroid of the claimed identity through the matching score $s(\mathbf{x}, \mathbf{x}_c)$, either to accept or reject the user (see Sect. 2).[1] Template self-update was implemented as explained in Sect. 2, using the two update policies of Sect. 3. In our experiments we set the initial template gallery size to $n = 10$. We also assume that the PCA projection is not updated during verification, since it is too computationally expensive [7].

Data Set. We collected a data set consisting of 40 different clients with 60 images each, for a total of 2,400 face images. The face images of each client were collected into two sessions, using a commercial webcam, with a time interval of about two weeks between them, under different lighting conditions and facial expressions. This induced a high intra-class variability of the face images, which makes face recognition particularly challenging. The data set is available under request to the authors, and it was also used in [2].

Experimental Setup. We split the face images as follows. We randomly selected 10 images for each client as training data, to compute the PCA eigenvectors and the clients' centroids. A further set of 10 images for each client was used as validation data, to tune the acceptance threshold t_c and update threshold θ_c for each client. We set θ_c by computing the 0% FAR operational point for the corresponding client, and t_c to a less conservative value, namely, at the 1% FAR operational point. The remaining 40 images per client were retained as testing set. We observed that randomly choosing different data splits do not substantially affect our results. For the sake of ease of interpretation, we thus chose not to average them on different data splits.

Performance under No Attack. We first computed the performance of the considered face verification system on the testing set, when no attack is considered, in terms of the Genuine Acceptance Rate (GAR) and False Acceptance Rate (FAR), namely, the fraction of clients correctly verified and of impostors wrongly accepted as genuine clients. For each client (i.e., claimed identity), the

---

[1] Note that more than one template per user may be also used, to better capture the high intra-class variability typical of biometric images. This would however slightly complicate the verification process, and the consequent poisoning attack; thus, we only consider here the simplest case in which only one centroid per user is stored.

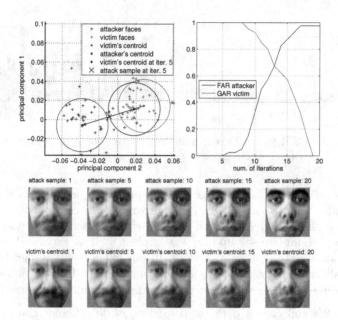

Fig. 3. Poisoning attack under finite window (average-out) update policy. *Top-left plot:* illustration of the attack in the space spanned by the first two principal components. *Top-right plot:* Variation of the attacker's FAR and victim's GAR during the attack progress. *Bottom plot:* attack samples and victim's centroids during the attack progress, at iterations 1, 5, 10, 15, 20.

corresponding GAR and FAR are shown in Fig. 2. The FAR of each client was estimated as the average FAR of the other 39 users, considered as attackers. As expected from the choice of t_c, the average FAR is around 1% for most of the clients. A rather low GAR is attained for several users (lower than 50%), due to the high-intra class variability. Further, we observed that for most clients no update actually occurred, due to the very conservative choice of θ_c.

Poisoning under the Finite Window (average-out) Update Policy. We implemented the poisoning attack as described in Sect. 3. We simulated the simplest scenario in which the template gallery of the targeted client (victim) is updated by a sequence of attack samples only, in a given period of time.[2]

In Fig. 3, we report the results attained when considering a specific attacker (user 13) and victim (user 31). The attack progress is depicted in the top-left plot, where it can be noted how the victim's centroid drifts toward the attacker's centroid. In particular, we report the initial victim's and attacker's centroids, and the drifted victim's centroid after 5 iterations (i.e., after submitting 5 attack samples). The top-right plot shows how the victim's GAR decreases, and,

---

[2] The scenario when the gallery is updated with interleaved attack and genuine samples (i.e., samples coming from genuine verification attempts by the targeted client) can be however investigated in a similar manner, as done in [4].

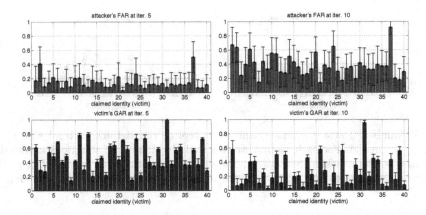

Fig. 4. Poisoning attack under the finite window (average-out) update policy

simultaneously, the FAR relative to the attacker identity increases, as a function of the number of attack iterations. Although the number of iterations required to replace the victim's centroid with the attacker's centroid is 20, the attacker's FAR raises quite quickly, being equal to 40% after only 10 iterations. The bottom plot shows some attack samples, and the corresponding change in the victim's centroid. Note how the initial victim's face (victim's centroid at iteration 1) is eventually replaced by the attacker's face (victim's centroid at iteration 20).

Fig. 4, and Fig. 5 (left plot) summarize the results obtained considering all possible pairs of attacker and victim ($39 \times 40 = 1560$). The latter depicts the average number of iterations and the standard deviation, over the 39 possible attackers, required to replace the victim's centroid with the attacker's one. Note that most of the attacks are successfully completed after 10 to 20 iterations.

Since in many cases it is not realistic for an attacker to perform more than 10 attempts without being caught, we focused on the GAR of each victim, and the FAR relative to the corresponding attacker, after 5 and 10 iterations, which are reported in Fig. 4. As in the previous case, for each victim we average the GAR and FAR with respect to all possible 39 attackers. In other words, the FAR represents the probability that a randomly chosen attacker cracks a specific victim account. Similarly, the GAR represents the probability of a specific victim being correctly accepted as a genuine user, under a poisoning attack carried out by a randomly chosen attacker. It can be seen that the FAR is relatively high even at the early stages of the attack: it ranges from 10% to 20% after only 5 iterations, and approaches 50% for most of the targeted victims after 10 iterations. The GAR remains instead almost the same after 5 iterations, but significantly decreases after 10 iterations for most of the victims. This means that an attacker may significantly increase the chance of being accepted after few iterations (e.g., from 1% FAR to 10% FAR with just 5 iterations) without causing a substantial denial of access to the victim. Lastly, note that after 10 iterations, the FAR is much higher than the GAR, although most of the poisoning attacks are not complete at this stage.

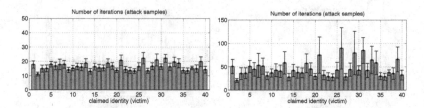

Fig. 5. Number of attack samples required to replace the victim's centroid with the attacker's centroid. For each victim, we reported the number of iterations averaged over all possible attackers. Standard deviation is also shown as error bars. *Left plot:* finite window (average-out) update policy. *Right plot:* infinite window update policy.

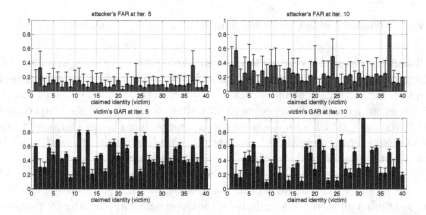

Fig. 6. Poisoning attack under the infinite window update policy

Poisoning under the Infinite Window Update Policy. Poisoning is much harder under this policy, since it requires an exponential number of attack samples with respect to the relative displacement (see Eq. 3). We report the evaluation involving all pairs of attacker and victim, as above, in Fig. 6, and Fig. 5 (right plot). As expected, the latter plot shows that the number of iterations required to complete a poisoning attack is much higher in this case, and its effectiveness (in terms of FAR at the same number of iterations) is lower. However, Fig. 6 shows that the increase in FAR is still significant, even after few iterations.

Finally, we repeated the experiments reducing the initial number of templates per client to $n = 5$. As predicted by Eq. 3, the number of attack iterations scaled linearly with n for both policies, without substantially changing the attack effectiveness in terms of FAR and GAR. In particular, almost the same values were attained after half of the iterations.

5   Conclusions and Future Work

In this work, we demonstrated that adaptive biometric recognition systems can be vulnerable to poisoning attacks, namely, carefully designed attacks that exploit system adaptation. Such attacks can significantly violate system security from the early stages, i.e., with few attack iterations. To our knowledge, this is the first time that such vulnerability is highlighted in the context of biometric adaptive systems. Further, we observed that more adaptive update policies (e.g., finite window), which may be more beneficial in the standard scenario without attacks, can be more vulnerable to poisoning than less adaptive policies (e.g., infinite window). This highlights that a trade-off between security and ease of adaptation is required in adaptive biometric systems, and that it should be investigated more in detail in future work; e.g., considering different adaptive systems and more update policies.

Acknowledgment. This work has been partly supported by the project CRP-18293 funded by Regione Autonoma della Sardegna, L.R. 7/2007, Bando 2009.

References

1. Barreno, M., Nelson, B., Sears, R., Joseph, A.D., Tygar, J.D.: Can machine learning be secure? In: Proc. Symp. on Information, Computer and Comm. Sec. (ASIACCS), pp. 16–25. ACM (2006)
2. Biggio, B., Akhtar, Z., Fumera, G., Marcialis, G.L., Roli, F.: Security evaluation of biometric authentication systems under real spoofing attacks. IET Biometrics 1(1), 11–24 (2012)
3. Jiang, X., Ser, W.: Online fingerprint template improvement. IEEE Trans. Pattern Analysis and Machine Intell. 24(8), 1121–1126 (2002)
4. Kloft, M., Laskov, P.: Online anomaly detection under adversarial impact. In: Proc. 13th Int'l Conf. on AI and Statistics (AISTATS), pp. 405–412 (2010)
5. Laskov, P., Lippmann, R.: Machine learning in adversarial environments. Machine Learning 81, 115–119 (2010)
6. Nelson, B., Joseph, A.D.: Bounding an attack's complexity for a simple learning model. In: Proc. 1st Workshop on Tackling Computer Systems Problems with ML Techniques, SysML (2006)
7. Roli, F., Marcialis, G.L.: Semi-supervised PCA-Based Face Recognition Using Self-training. In: Yeung, D.-Y., Kwok, J.T., Fred, A., Roli, F., de Ridder, D. (eds.) SSPR 2006 and SPR 2006. LNCS, vol. 4109, pp. 560–568. Springer, Heidelberg (2006)
8. Ryu, C., Kim, H., Jain, A.K.: Template adaptation based fingerprint verification. In: Proc. 18th Int'l Conf. Pattern Rec., vol. 04, pp. 582–585. IEEE CS (2006)
9. Turk, M., Pentland, A.: Eigenfaces for recognition. J. Cogn. Neuroscience 3(1), 71–86 (1991)
10. Uludag, U., Ross, A., Jain, A.K.: Biometric template selection and update: a case study in fingerprints. Pattern Recognition 37(7), 1533–1542 (2004)
11. Yambor, W.S.: Analysis of PCA-based and Fisher discriminant-based image recognition algorithms. Technical Report, Colorado State University (2000)

Modified Divergences for Gaussian Densities

Karim T. Abou–Moustafa[1] and Frank P. Ferrie[2]

[1] Robotics Institute, Carnegie Mellon University
5000 Forbes Avenue, Pittsburgh, PA 15213, U.S.A
karimt@andrew.cmu.edu
[2] Dept. of Electrical & Computer Engineering, McGill University
3480 University Street, Montréal, QC, H3A 0E9, Canada
ferrie@cim.mcgill.ca

Abstract. Multivariate Gaussian densities are pervasive in pattern recognition and machine learning. A central operation that appears in most of these areas is to measure the difference between two multivariate Gaussians. Unfortunately, traditional measures based on the Kullback–Leibler (KL) divergence and the Bhattacharyya distance do not satisfy all metric axioms necessary for many algorithms. In this paper we propose a modification for the KL divergence and the Bhattacharyya distance, for multivariate Gaussian densities, that transforms the two measures into distance metrics. Next, we show how these metric axioms impact the unfolding process of manifold learning algorithms. Finally, we illustrate the efficacy of the proposed metrics on two different manifold learning algorithms when used for motion clustering in video data. Our results show that, in this particular application, the new proposed metrics lead to significant boosts in performance (at least 7%) when compared to other divergence measures.

1   Introduction

There are various applications in machine learning and pattern recognition in which the data of interest $\mathcal{D}$ are represented as a family or a collection of sets $\mathcal{D} = \{\mathcal{S}_i\}_{i=1}^n$, where $\mathcal{S}_i = \{\mathbf{x}_j^i\}_{j=1}^{n_i}$, and $\mathbf{x}_j^i \in \mathbb{R}^p$. For some of these applications, it is reasonable to model each $\mathcal{S}_i$ as a Gaussian distribution $\mathcal{G}_i(\boldsymbol{\mu}_i, \boldsymbol{\Sigma}_i)$ with mean vector $\boldsymbol{\mu}_i$ and a covariance matrix $\boldsymbol{\Sigma}_i$.[1] In these settings, a natural measure for the (dis)similarity between two Gaussians, $\mathcal{G}_1$ and $\mathcal{G}_2$ say, is the divergence measure of probability distributions [3,6]. For instance, some of the well known divergence measures with closed form expressions for Gaussian densities are the symmetric Kullback–Leibler (KL), or Jeffreys, divergence $d_J(\mathcal{G}_1, \mathcal{G}_2)$ [12], the Bhattacharyya distance $d_B(\mathcal{G}_1, \mathcal{G}_2)$ and the Hellinger distance $d_H(\mathcal{G}_1, \mathcal{G}_2)$ [9].

---

[1] Notations: Bold small letters $\mathbf{x}, \mathbf{y}$ are vectors. Bold capital letters $\mathbf{A}, \mathbf{B}$ are matrices. Calligraphic and double bold capital letters $\mathcal{X}, \mathcal{Y}, \mathbb{X}, \mathbb{Y}$ denote sets and/or spaces. Positive (semi-)definite matrices, PD (and PSD) are denoted by $\mathbf{A} \succ 0$ and $\mathbf{A} \succeq 0$ respectively. $\mathrm{tr}(\cdot)$ is the matrix trace. $|\cdot|$ is the matrix determinant. $\mathbf{I}$ is the identity matrix.

G.L. Gimel' farb et al. (Eds.): SSPR & SPR 2012, LNCS 7626, pp. 426–436, 2012.

When considering a learning problem such as classification, clustering, or low dimensional embedding for the family of sets $\mathcal{D}$, via its representation as the set of Gaussians $\{\mathcal{G}_i\}_{i=1}^n$, a natural question that arises is that of *which divergence measure will yield a better performance?* At first glance, one can consider an answer along two main dimensions: 1) the learning algorithm that shall be used for the sought task, and 2) the data set under consideration. In this research, however, we show that the metric properties of these divergence measures form a third crucial dimension that has a direct impact on the algorithm's performance. In particular, we show that when modifying the closed form expressions for $d_J(\mathcal{G}_1, \mathcal{G}_2)$ and $d_B(\mathcal{G}_1, \mathcal{G}_2)$ such that both measures satisfy all metric axioms[2], the resulting new measures yield significant improvements in the discriminability of the embedding spaces obtained from two different manifold learning algorithms, classical Multidimensional Scaling (cMDS) [21] and Laplacian Eigenmaps (LEM) [4]. These improvements in discriminability, in turn, result in consistent and significant boosts in clustering accuracy. For the application considered in this paper, motion clustering in video data, an improvement in discriminability of at least 7% is observed.

Starting with the closed form expressions for $d_J(\mathcal{G}_1, \mathcal{G}_2)$ and $d_B(\mathcal{G}_1, \mathcal{G}_2)$, in Section (2) we take, a closer look on how each term in these expressions violate the desired metric axioms. Then, we propose modifications for these expressions that result in new distances that satisfy all metric axioms. In Section (3), we show how the metric properties of divergence can impact the unfolding process of manifold learning algorithms such as cMDS and LEM. In Section (4), we compare the performance of cMDS and LEM using the proposed divergence measures in the context of clustering human motion in video data. Finally, conclusions are drawn in Section (5).

2   Characteristics of $d_J(\mathcal{G}_1, \mathcal{G}_2)$ and $d_B(\mathcal{G}_1, \mathcal{G}_2)$

Our discussion begins with the characteristics of $d_J(\mathcal{G}_1, \mathcal{G}_2)$ and $d_B(\mathcal{G}_1, \mathcal{G}_2)$ in terms of structure and metric properties. Let $\mathbb{G}_p$ be the family of p–dimensional Gaussian densities, where the density $\mathcal{G}(\mu, \Sigma) \in \mathbb{G}_p$ is defined as:

$$\mathcal{G}(\mathbf{x}; \mu, \Sigma) = (2\pi)^{-\frac{p}{2}} |\Sigma|^{-\frac{1}{2}} \exp\{-\tfrac{1}{2}(\mathbf{x} - \mu)^\top \Sigma^{-1}(\mathbf{x} - \mu)\},$$

$\mathbf{x}, \mu \in \mathbb{R}^p$, $\Sigma \in \mathbb{S}_{++}^{p \times p}$, and $\mathbb{S}_{++}^{p \times p}$ is the manifold of symmetric positive definite (PD) matrices. For $\mathcal{G}_1, \mathcal{G}_2 \in \mathbb{G}_p$, Jeffreys divergence has the closed form expression:

$$d_J(\mathcal{G}_1, \mathcal{G}_2) = \tfrac{1}{2}\mathbf{u}^\top \Psi \mathbf{u} + \tfrac{1}{2}\mathrm{tr}\{\Sigma_1^{-1}\Sigma_2 + \Sigma_2^{-1}\Sigma_1 - 2\mathbf{I}\}, \qquad (1)$$

---

[2] A metric space [11, p. 3] is an ordered pair $(\mathcal{X}, d)$, where $\mathcal{X}$ is a non-empty abstract set (of any objects/elements whose nature is left unspecified), and d is a distance function, or a metric, defined as: $d : \mathcal{X} \times \mathcal{X} \mapsto \mathbb{R}$, and $\forall\, a, b, c \in \mathcal{X}$, the following axioms hold : (i) $d(a, b) \geq 0$, (ii) $d(a, a) = 0$, (iii) $d(a, b) = 0$ iff $a = b$, (iv) Symmetry : $d(a, b) = d(b, a)$, and (v) The triangle inequality : $d(a, c) \leq d(a, b) + d(b, c)$. Semi-metrics satisfy axioms (i), (ii), and (iv) only. Note that the axiomatic definition of metrics and semi-metrics, in particular axioms (i) and (ii), produce the positive semi-definiteness of d. Hence metrics and semi-metrics are PSD.

where $\boldsymbol{\Psi} = (\boldsymbol{\Sigma}_1^{-1} + \boldsymbol{\Sigma}_2^{-1})$, and $\mathbf{u} = (\boldsymbol{\mu}_1 - \boldsymbol{\mu}_2)$. The Bhattacharyya coefficient ρ, which is a measure of similarity between probability distributions, is defined as:

$$\rho(\mathcal{G}_1, \mathcal{G}_2) = |\boldsymbol{\Gamma}|^{-\frac{1}{2}} |\boldsymbol{\Sigma}_1|^{\frac{1}{4}} |\boldsymbol{\Sigma}_2|^{\frac{1}{4}} \exp\{-\tfrac{1}{8} \mathbf{u}^\top \boldsymbol{\Gamma}^{-1} \mathbf{u}\}, \tag{2}$$

where $\boldsymbol{\Gamma} = (\tfrac{1}{2}\boldsymbol{\Sigma}_1 + \tfrac{1}{2}\boldsymbol{\Sigma}_2)$. From $\rho(\mathcal{G}_1, \mathcal{G}_2)$, the Hellinger distance d_H is defined as $\sqrt{2[1 - \rho(\mathcal{G}_1, \mathcal{G}_2)]}$, while the Bhattacharyya distance d_B is $-\log \rho(\mathcal{G}_1, \mathcal{G}_2)$, which also yields an interesting closed form expression:

$$d_B(\mathcal{G}_1, \mathcal{G}_2) = \tfrac{1}{8} \mathbf{u}^\top \boldsymbol{\Gamma}^{-1} \mathbf{u} + \tfrac{1}{2} \ln \left\{ |\boldsymbol{\Sigma}_1|^{-\frac{1}{2}} |\boldsymbol{\Sigma}_2|^{-\frac{1}{2}} |\boldsymbol{\Gamma}| \right\}. \tag{3}$$

It is well known that the KL divergence is not a metric since it does not satisfy the triangle inequality [12], and hence d_J in (1) is not a metric. Similarly, d_B in (3) is not a metric for the same reason, however, d_H is indeed a metric [9].

The two closed form expressions in Equations (1) and (3) have the same structure which is a summation of two components in terms of their first and second order moments. The first term in Equations (1) and (3) measures the difference between the means $\boldsymbol{\mu}_1$ and $\boldsymbol{\mu}_2$ weighted by the covariance matrices $\boldsymbol{\Sigma}_1$ and $\boldsymbol{\Sigma}_2$. The second term, however, measures the difference or discrepancy between the covariance matrices $\boldsymbol{\Sigma}_1$ and $\boldsymbol{\Sigma}_2$ only, and is independent from the means $\boldsymbol{\mu}_1$ and $\boldsymbol{\mu}_2$.

The first term in Equations (1) and (3), up to a scale factor and a square root, is equivalent to the generalized quadratic distance (GQD) between $\mathbf{x}, \mathbf{y} \in \mathbb{R}^p$: $d(\mathbf{x}, \mathbf{y}; \mathbf{A}) = \sqrt{(\mathbf{x} - \mathbf{y})^\top \mathbf{A} (\mathbf{x} - \mathbf{y})}$, where $\mathbf{A} \in \mathbb{S}_{++}^{p \times p}$. If $\boldsymbol{\Sigma}_1 = \boldsymbol{\Sigma}_2 = \boldsymbol{\Sigma}$, then Equations (1) and (3) reduce to:

$$d_J(\mathcal{G}_1, \mathcal{G}_2) = \tfrac{1}{2} \mathbf{u}^\top \boldsymbol{\Psi} \mathbf{u}, \text{ and} \tag{4}$$

$$d_B(\mathcal{G}_1, \mathcal{G}_2) = \tfrac{1}{8} \mathbf{u}^\top \boldsymbol{\Gamma}^{-1} \mathbf{u}. \tag{5}$$

Note that the squared GQD $d^2(\mathbf{x}, \mathbf{y}; \mathbf{A})$ is a semi-metric, and if $\mathbf{A}$ is PSD, then $d(\mathbf{x}, \mathbf{y}; \mathbf{A})$ is a pseudo-metric. Both, semi-metrics and pseudo metrics, do not satisfy the triangle inequality, and hence Equations (4) and (5) are semi-metrics. Further, if $\boldsymbol{\Sigma}_1 = \boldsymbol{\Sigma}_2 = \mathbf{I}$, then Equations (4) and (5), up to a scale factor, reduce to the squared Euclidean distance. Note that the squared Euclidean distance is also a semi-metric. The second term in Equations (1) and (3) is the distance or discrepancy measure between $\boldsymbol{\Sigma}_1$ and $\boldsymbol{\Sigma}_2$, and is independent of $\boldsymbol{\mu}_1$ and $\boldsymbol{\mu}_2$. If $\boldsymbol{\mu}_1 = \boldsymbol{\mu}_2 = \boldsymbol{\mu}$ then:

$$d_J(\mathcal{G}_1, \mathcal{G}_2) = \tfrac{1}{2} \mathrm{tr}\{\boldsymbol{\Sigma}_1^{-1} \boldsymbol{\Sigma}_2 + \boldsymbol{\Sigma}_2^{-1} \boldsymbol{\Sigma}_1 - 2\mathbf{I}\}, \text{ and} \tag{6}$$

$$d_B(\mathcal{G}_1, \mathcal{G}_2) = \tfrac{1}{2} \ln \left\{ |\boldsymbol{\Gamma}| |\boldsymbol{\Sigma}_1|^{-\frac{1}{2}} |\boldsymbol{\Sigma}_2|^{-\frac{1}{2}} \right\}. \tag{7}$$

Since Equations (1) and (3) by definition, do not satisfy the triangle inequality, and hence are semi-metrics, then Equations (6) and (7) are also semi-metrics between $\boldsymbol{\Sigma}_1$ and $\boldsymbol{\Sigma}_2$.

We note that it is easy to satisfy all the metric properties for Equations (4) and (5) by taking their square root, and ensuring that $\boldsymbol{\Psi}$ and $\boldsymbol{\Gamma}^{-1}$ are PD.

In practice, the positive definiteness of $\boldsymbol{\Psi}$ and $\boldsymbol{\Gamma}^{-1}$ can be achieved by ensuring that $\boldsymbol{\Sigma}_1$ and $\boldsymbol{\Sigma}_2$ are PD. For high dimensional data, shrinkage estimators for covariance matrices [5] are usually used to estimate regularized versions of $\boldsymbol{\Sigma}_1$ and $\boldsymbol{\Sigma}_2$. These estimates are statistically efficient, PD, and well conditioned[3].

The problem, however, remains with Equations (6) and (7). Covariance matrices $\boldsymbol{\Sigma}_1$ and $\boldsymbol{\Sigma}_2$ are elements of $\mathbf{S}_{++}^{p \times p}$, which is a metric space with a defined metric for its elements. The semi-metrics in Equations (6) and (7), although naturally derived from divergence measures [3,6], do not define proper metrics for $\mathbf{S}_{++}^{p \times p}$, and hence violate its geometric properties.

The set of symmetric PD matrices is a set of geometric objects that define the Riemannian manifold $\mathbb{S}_{++}^{p \times p}$. A Riemannian manifold is a differentiable manifold equipped with an inner product that induces a natural distance metric, or a Riemannian metric between all its elements. Förstner and Moonen [7] and independently X. Pennec [17] derived this metric for $\mathbb{S}_{++}^{p \times p}$, however its history goes back to C. R. Rao in 1945 [18]. For $\boldsymbol{\Sigma}_1, \boldsymbol{\Sigma}_2 \in \mathbb{S}_{++}^{p \times p}$, the Riemannian metric is defined as:

$$d_{\mathcal{R}}(\boldsymbol{\Sigma}_1, \boldsymbol{\Sigma}_2) = \left(\sum_{j=1}^{p} \log^2 \lambda_j \right)^{\frac{1}{2}}, \tag{8}$$

where $\operatorname{diag}(\lambda_1, \ldots, \lambda_p) = \boldsymbol{\Lambda}$ is the generalized eigenvalue matrix for the generalized eigenvalue problem (GEP): $\boldsymbol{\Sigma}_1 \mathbf{V} = \boldsymbol{\Lambda} \boldsymbol{\Sigma}_2 \mathbf{V}$, and $\mathbf{V}$ is the column matrix of its generalized eigenvectors. Note that $d_{\mathcal{R}}$ satisfies all metric axioms and is invariant to inversion and to affine transformations of the coordinate system [7].

2.1   Modifying $d_J(\mathcal{G}_1, \mathcal{G}_2)$ and $d_B(\mathcal{G}_1, \mathcal{G}_2)$

Modifying the divergence measures $d_J(\mathcal{G}_i, \mathcal{G}_j)$ and $d_B(\mathcal{G}_i, \mathcal{G}_j)$ in Equations (1) and (3) respectively, will rely on (i) their special structure which decomposes the difference between two Gaussian densities into the difference between their first and second order moments, and (ii) the fact that the second term in Equations (1) and (3) is independent from the means $\boldsymbol{\mu}_1$ and $\boldsymbol{\mu}_2$. This split of the Gaussian parameters encourages us to exchange the second term in $d_J(\mathcal{G}_1, \mathcal{G}_2)$ and $d_B(\mathcal{G}_1, \mathcal{G}_2)$, i.e. the semi-metrics for covariance matrices in Equations (6) and (7), with the Riemannian metric $d_{\mathcal{R}}$ in Equation (8). More specifically, we propose the following metrics as measures for the difference between two Gaussians:

$$d_{J\mathcal{R}}(\mathcal{G}_1, \mathcal{G}_2) = (\mathbf{u}^{\top} \boldsymbol{\Psi} \mathbf{u})^{\frac{1}{2}} + d_{\mathcal{R}}(\boldsymbol{\Sigma}_1, \boldsymbol{\Sigma}_2), \quad \text{and} \tag{9}$$

$$d_{B\mathcal{R}}(\mathcal{G}_1, \mathcal{G}_2) = (\mathbf{u}^{\top} \boldsymbol{\Gamma}^{-1} \mathbf{u})^{\frac{1}{2}} + d_{\mathcal{R}}(\boldsymbol{\Sigma}_1, \boldsymbol{\Sigma}_2), \tag{10}$$

where $\boldsymbol{\Psi} \succ 0$, and $\boldsymbol{\Gamma}^{-1} \succ 0$. Note that each term of the proposed measures satisfy all metric axioms. Further, Equations (9) and (10) keep the same structure and characteristics of Equations (1) and (3); in particular the second term is independent from $\boldsymbol{\mu}_1$ and $\boldsymbol{\mu}_2$. If $\boldsymbol{\mu}_1 = \boldsymbol{\mu}_2 = \boldsymbol{\mu}$ then Equations (9) and (10)

---

[3] See for instance [5] and its affiliated references for a nice overview on these methods, and some recent developments in this direction.

reduce to the Riemannian metric $d_\mathcal{R}$ in Equation (8). If $\Sigma_1 = \Sigma_2 = \Sigma$, then Equations (9) and (10) will yield the exact GQD with symmetric PD matrices Ψ and Γ^{-1} respectively, and if $\Sigma = I$, then the two metrics will yield the Euclidean distance. In the case when $\mu_1 \neq \mu_2$ and $\Sigma_1 \neq \Sigma_2$, an α–weighted version of (9) and (10) can be expressed as:

$$d_{J\mathcal{R}}(\mathcal{G}_1, \mathcal{G}_2; \alpha) = \alpha(u^\top \Psi u)^{\frac{1}{2}} + (1-\alpha)d_\mathcal{R}(\Sigma_1, \Sigma_2), \quad \text{and}$$
$$d_{B\mathcal{R}}(\mathcal{G}_1, \mathcal{G}_2; \alpha) = \alpha(u^\top \Gamma^{-1} u)^{\frac{1}{2}} + (1-\alpha)d_\mathcal{R}(\Sigma_1, \Sigma_2),$$

where $\alpha \in (0,1)$ weights the contribution (or importance) of each term in $d_{J\mathcal{R}}$ and $d_{B\mathcal{R}}$. Note that when the α–weighted version of the measures are plugged in a learning algorithm, α can be optimized by methods of cross validation, or jointly optimized with the intensity/shrinkage parameters used to regularize the covariance matrices Σ_1 and Σ_2.

3   Manifold Learning with Divergence Measures

Given a set vectors $\mathcal{X} = \{x_i\}_{i=1}^n$, $x_i \in \mathbb{R}^p$, manifold learning algorithms [20,4] construct a neighbourhood graph in which the input points x_i act as its vertices. This graph is an estimate for the topology of an underlying low dimensional manifold on which the data are assumed to lie on. The learning algorithm then, tries to unfold this manifold – while preserving some local information – to partition the graph (as in clustering), or to redefine metric information (as in dimensionality reduction). The algorithm's output is the set $\mathcal{Y} = \{y_i\}_{i=1}^n$ that lives in a subspace of dimensionality $p_0 \ll p$, where $y_i \in \mathbb{R}^{p_0}$ is the embedding of the input x_i.

A different setting occurs when each vertex v_i on the graph represents a set $\mathcal{S}_i$, where $\mathcal{S}_i = \{x_j^i\}_{j=1}^{n_j}$ is a set of vectors. For instance, $\mathcal{S}_i$ can be the feature vectors describing a multimedia file [16], an image [10], or a short video clip [1]. In these settings, each $\mathcal{S}_i$ is modelled as a Gaussian distribution $\mathcal{G}_i$, and the pairwise dissimilarity between all the Gaussians $\{\mathcal{G}_i\}_{i=1}^n$ is measured using divergence measures. This, however, turns the problem into obtaining a low dimensional embedding for the family of Gaussians $\{\mathcal{G}_i\}_{i=1}^n$. Again, the algorithm's output is the set $\mathcal{Y} = \{y_i\}_{i=1}^n$, with $y_i \in \mathbb{R}^{p_0}$ being the low dimensional embedding (representation) of the Gaussian $\mathcal{G}_i$.

Before proceeding to obtain such an embedding, it is important to understand how the metric properties of divergence measures can affect the graph embedding process of these algorithms. To illustrate these properties, we pick two different types of algorithms: cMDS [21] and LEM [4].

It turns out that the metric properties of divergence measures are intimately related to the positive semi-definiteness of the affinity matrix $A \in \mathbb{R}^{n \times n}$ extracted from the graph's adjacency matrix. Let $D \in \mathbb{R}^{n \times n}$ be the matrix of pairwise divergences where $D_{ij} = div(\mathcal{G}_i, \mathcal{G}_j)$, $\forall i, j$, and div is a symmetric divergence measure.

For cMDS, the affinity matrix A is defined as $A_{ij} = -\frac{1}{2}D_{ij}^2$, $\forall i, j$. The matrix A is guaranteed to be PSD *if and only if* $div(\mathcal{G}_i, \mathcal{G}_j)$ is a metric; in particular satisfies the triangle inequality. This result is due to Theorem (3) in [21] and Theorem (4) in [8]. Therefore, div in the case of cMDS can be d_H, d_{JS}, d_{JR}, or d_{BR} since they are all metrics.

For LEM, and for input vectors $\mathbf{x}_i, \mathbf{x}_j$, the affinity matrix $\mathbf{A}$ is defined as $\mathbf{A}_{ij} = K(\mathbf{x}_i, \mathbf{x}_j)$, $\forall i, j$, where K is a symmetric PSD kernel that measures the similarity between $\mathbf{x}_i$ and $\mathbf{x}_j$. From Mercer kernels [15], it is known that $\mathbf{A}$ is PSD *if and only if* K is symmetric and PSD. Note that for any two probability distributions P_1 and P_2, and by definition of divergence [3], $div(P_1, P_2) \geq 0$, and equality only holds when $P_1 = P_2$. Hence $div(P_1, P_2)$ is PSD by definition and it can also be symmetric as d_J, d_B, d_H, d_{JS}, d_{JR}, and d_{BR}.

A possible kernel for $\mathcal{G}_i$ and $\mathcal{G}_j$ using a symmetric div is: $K(\mathcal{G}_i, \mathcal{G}_j) = \exp\{-\frac{1}{\sigma} div(\mathcal{G}_i, \mathcal{G}_j)\} = \exp\{-\frac{1}{\sigma}\mathbf{D}_{ij}\}$, where $\sigma > 0$ is a parameter that scales the affinity between two densities. Since div is PSD and symmetric, then $K(\mathcal{G}_i, \mathcal{G}_j)$ is PSD and symmetric as well. This simple fact is due to Theorems (2) and (4) in [19], and a discussion on these particular kernels can be found in [2]. Further, if div is a metric, then the isometric embedding $\exp\{-div\}$ will result in a metric space (see footnote in pp. 525 of [19]), and the resulting embedding of LEM will be isometric as well. Therefore, for LEM, a symmetric PSD affinity matrix can be defined as $\mathbf{A}_{ij} = K(\mathcal{G}_i, \mathcal{G}_j)$, $\forall i, j$, and using any symmetric div to define the kernel K. Note that LEM is more flexible than cMDS since it only requires a symmetric divergence, while cMDS needs all metric axioms to be satisfied.

4 Experiments

To test the validity and efficacy of the proposed measures d_{JR} and d_{BR}, and to compare their performance to d_J, d_B, and d_H, we conduct a set of experiments in the context of clustering human motion from video sequences. Our main objective from these experiments is to show that the proposed measures d_{JR} and d_{BR} can consistently outperform other divergence measures in a nontrivial and rather challenging task such as human motion clustering in video data.

For the purpose of our experiments, we use the KTH data set for human action recognition[4]. The data set consists of video clips for 6 types of human actions (boxing, hand clapping, hand waving, jogging, running, and walking) performed by 25 subjects in 4 different scenarios (outdoors, outdoors with scale variation, outdoor with different clothes, and indoors), resulting in a total number of video clips $n = 6 \times 25 \times 4 = 600$. All sequences were taken over homogeneous backgrounds with a static camera with a frame rate of 25 fps. The spatial resolution of the videos is 160×120, and each clip has a length of 20 seconds on average.

4.1 Representing Motion as Sets of Vectors

In these experiments, a long video sequence $V = \{\mathbf{F}_t\}_{t=1}^{\tau}$ with intensity frames $\mathbf{F}_t$ is divided into very short video clips $VClip$ of equal length k where it is assumed that an apparent smallest human action can occur; i.e. $V = \{VClip_i\}_{i=1}^{n}$.

---

[4] http://www.nada.kth.se/cvap/actions/

Depending on the video sampling rate, $k = \{20, 25, 30, 35\}$ frames/clip. This is the first column in Tables in (1) and (2).

To extract the motion information, a dense optical flow is computed for each video clip using the Lucas-Kanade algorithm [13][5], resulting in a large set of spatio-temporal gradients vectors describing the motion of pixels in each frame. To capture the motion information encoded in the gradient direction, first we apply an adaptive threshold based on the norm of the gradient vectors to eliminate all vectors resulting from slight illumination changes and camera jitter. Second, each video frame is divided into $h \times w$ blocks – typically 3×3 and 4×4 – and the motion in each block is encoded by an m–bins histogram of gradient orientations. In all our experiments, m is set to 4 and 8 bins. The histograms of all blocks for one frame are concatenated to form one vector of dimensionality $p = m \times h \times w$. Therefore, a video clip $VClip_i$ with k frames is finally represented as a set $S_i = \{\mathbf{x}_1^i, \ldots, \mathbf{x}_k^i\}$, where $\mathbf{x}_j^i$ is a p-dimensional vector of the concatenated histograms of frame j. Last, for each subject, the video clips for the 6 actions from one scenario were concatenated to form one long video sequence. This resulted in $25 \times 4 = 100$ long video sequences that were used in our experiments. To validate the accuracy of clustering, each video frame was labeled with the type of action it contains.

4.2   Experimental Setting

Once the motion information in video V is represented as a family of sets $\{S_i\}_{i=1}^n$, motion clustering tries to group together video clips (or sets) with similar motion vectors. To this end, we use a recently proposed framework for learning over sets of vectors [1] to obtain such a clustering for the S_i's. In this framework, each $S_i = \{\mathbf{x}_j^i\}_{j=1}^{n_i}$ is modelled as a Gaussian distribution $\mathcal{G}_i$ with mean vector $\hat{\boldsymbol{\mu}}_i = \frac{1}{n_i} \sum_{j=1}^{n_i} \mathbf{x}_j^i$, and a covariance matrix $\hat{\boldsymbol{\Sigma}}_i = \frac{1}{n_i-1} \sum_{j=1}^{n_i} (\mathbf{x}_j^i - \hat{\boldsymbol{\mu}}_i)(\mathbf{x}_j^i - \hat{\boldsymbol{\mu}}_i)^\top + \gamma\mathbf{I}$, where γ is a regularization parameter. This forms the family of Gaussians $\{\mathcal{G}_i\}_{i=1}^n$ which represents the motion in V. Note that regularization is necessary for high dimensional data especially when $n_i \leq p$ (rank deficient covariance matrix) to avoid over fitting, leverage noise effect in the data, and outlier reliance[6].

Using cMDS and LEM together with the divergence measures discussed here, d_J, d_B, d_H, d_{JR} and d_{BR}, we obtain a low dimensional embedding for the family of Gaussians as the set $\{\mathbf{y}_i\}_{i=1}^n$, where $\mathbf{y}_i \in \mathbb{R}^{p_o}$, and $p_0 \ll p$. Finally, the k-Means clustering is run on the data set $\{\mathbf{y}_i\}_{i=1}^n$. To summarize, a video sequence goes through the following transformations:
$$V \longmapsto \{VClip_i\}_{i=1}^n \longmapsto \{S_i\}_{i=1}^n \longmapsto \{\mathcal{G}_i\}_{i=1}^n \longmapsto \{\mathbf{y}_i\}_{i=1}^n.$$

The dimensionality p_0 of the embedding space is a hyperparameter for cMDS and LEM. For cMDS this is allowed to change from 2 up to 100 dimensions, while for LEM it is usually set equal to the number of clusters which is 6 in this case [14]. This is due to our *a priori* knowledge that there are 6 types of motion

---

[5] Implemented in Piotr's Image and Video Toolbox for Matlab
http://vision.ucsd.edu/~pdollar/toolbox/doc/

[6] In all our experiments $\gamma = 1$.

Table 1. Average clustering accuracy (with standard deviations) over 100 video sequences in 4 different embedding spaces obtained using cMDS+d_J, cMDS+d_B, cMDS+d_H, and cMDS+$d_{J\mathcal{R}}$

frames/clip	$p = m \times h \times w = 8 \times 3 \times 3$			
	cMDS+d_J	cMDS+d_B	cMDS+d_H	cMDS+$d_{J\mathcal{R}}$
20	70.9 (11.9)	71.0 (12.0)	75.5 (12.1)	**80.3 (10.9)**
25	62.8 (10.9)	62.8 (11.0)	68.2 (12.3)	**75.5 (13.1)**
30	66.7 (11.7)	66.7 (11.8)	71.5 (12.7)	**77.4 (12.7)**
35	62.8 (10.9)	62.8 (11.1)	68.2 (12.3)	**75.3 (13.1)**
frames/clip	$p = m \times h \times w = 8 \times 4 \times 4$			
	cMDS+d_J	cMDS+d_B	cMDS+d_H	cMDS+$d_{J\mathcal{R}}$
20	68.3 (12.1)	68.9 (11.6)	74.2 (12.0)	**79.5 (11.7)**
25	66.5 (12.0)	66.5 (12.4)	72.5 (12.2)	**78.6 (12.1)**
30	61.9 (10.9)	63.0 (10.6)	68.9 (11.4)	**75.5 (12.3)**
35	71.3 (12.1)	71.8 (12.3)	76.5 (11.8)	**80.7 (10.1)**

in each video. Another hyperparameter to optimize for LEM is the kernel width σ which was allowed to take 4 different values from all the pairwise divergences; the median, 0.25, 0.75, and 0.9 of the quantile.

For the k-Means algorithm, the number of clusters k was set to 6, and to avoid local minima, the algorithm was run with 30 different initializations and the run with the minimum sum of squared distances was selected as the final result for clustering. The clustering accuracy here is measured using the Hungarian score used in [22] which finds the maximum matching between the true labeling of each video clip and the labeling produced by the clustering algorithm. Note that this is the accuracy for clustering one and only one long video sequence. The values recorded in Columns 2, 3, 4, and 5 in Tables (1) and (2) are the average accuracies (with standard deviations) over the 100 video sequences created for these experiments (§4.2). During these experiments, it was noted that the performance for $d_{J\mathcal{R}}$ and $d_{B\mathcal{R}}$ are very similar under both algorithms, and hence, due to space limitations, we show the results of cMDS+$d_{J\mathcal{R}}$ in Table (1) and the results for LEM+$d_{B\mathcal{R}}$ in Table (2).

4.3   Analysis of the Results

Our hypothesis, before running the experiments, is that clustering accuracy in the embedding space obtained through the modified divergences $d_{J\mathcal{R}}$ and $d_{B\mathcal{R}}$ will be higher than the clustering accuracy in the embedding spaces obtained by other divergence measures. Note that the k-Means accuracy here is a quantitative indicator on the quality of the embedding and its capability to define clusters, or regions of high density (manifolds), which correspond to clusters of different motion types. Therefore, each embedding space is optimized to maximize the clustering accuracy, and then the highest accuracy obtained is compared against all other highest accuracies of other embedding spaces.

Table 2. Average clustering accuracy (with standard deviations) over 100 video sequences in 4 different embedding spaces obtained using LEM+d_J, LEM+d_B, LEM+d_H, and LEM+$d_{B\mathcal{R}}$

frames/clip	$p = m \times h \times w = 8 \times 3 \times 3$			
	LEM+d_J	LEM+d_B	LEM+d_H	LEM+$d_{B\mathcal{R}}$
20	55.7 (11.2)	56.0 (10.9)	60.1 (11.5)	**65.1 (13.2)**
25	58.2 (12.0)	58.1 (11.9)	63.6 (13.1)	**69.6 (13.6)**
30	60.0 (12.7)	59.9 (12.6)	64.8 (12.9)	**70.3 (13.4)**
35	63.0 (13.3)	62.9 (13.3)	67.4 (13.1)	**71.8 (13.6)**
frames/clip	$p = m \times h \times w = 8 \times 4 \times 4$			
	LEM+d_J	LEM+d_B	LEM+d_H	LEM+$d_{B\mathcal{R}}$
20	54.0 (12.5)	54.6 (12.7)	60.8 (12.2)	**66.3 (12.7)**
25	57.7 (14.0)	57.7 (13.9)	64.7 (13.2)	**69.5 (13.2)**
30	59.5 (13.4)	59.5 (13.2)	66.3 (12.5)	**70.5 (12.6)**
35	59.5 (13.4)	59.5 (13.2)	66.3 (12.5)	**70.5 (12.6)**

Tables (1) and (2) show that, under the embeddings of cMDS and LEM with $d_{J\mathcal{R}}$ and $d_{B\mathcal{R}}$, the clustering accuracy is consistently superior to the accuracy of both algorithms with other divergence measures. This implies that the embedding spaces obtained via the new proposed measures can better characterize the cluster structure in the data, and hence the high clustering accuracies in Tables (1) and (2). Another observation to note from Tables (1) and (2) is that the clustering accuracies under the embedding of cMDS and LEM with d_H (which is a metric) are higher than the accuracies obtained with the same algorithms but using d_J and d_B. Again, this implies that the obtained embedding space via d_H can better characterize the cluster structure in the data. However, when comparing d_H on one hand, versus $d_{J\mathcal{R}}$ and $d_{B\mathcal{R}}$ on the other, we note that the embeddings obtained via d_H yield consistently lower performance than $d_{J\mathcal{R}}$ and $d_{B\mathcal{R}}$ do. In our understanding, this is due to its measure for the difference between covariance matrices[7], $(2 - 2|\mathbf{\Gamma}|^{-\frac{1}{2}}|\mathbf{\Sigma}_1|^{\frac{1}{4}}|\mathbf{\Sigma}_2|^{\frac{1}{4}})^{\frac{1}{2}}$, which is not a metric on $\mathbb{S}_{++}^{p \times p}$ and hence it violates its geometry.

The low performance for d_J and d_B with both algorithms when compared to the other divergence measures is again due to their lack of metric properties (in particular the triangle inequality), which in turn impacts the characteristics preserved (or relinquished) by the embedding procedure. Note that the difference in performance is more clear for the cMDS case in Table (1). None of d_J and d_B them is a true metric, and hence, they can result in embeddings that do not preserve the relative dissimilarities among all objects assigned to the graph's vertices. This can easily collapse a group of objects to be very close to each other in the embedding space thereby misleading the k–Means clustering algorithm.

In summary, it can be seen that, on the same data sets, and despite the differences between cMDS and LEM as dimensionality reduction algorithms, both algorithms showed consistent and identical behaviour in terms of relative

---

[7] By setting $\boldsymbol{\mu}_1 = \boldsymbol{\mu}_2 = \boldsymbol{\mu}$ in $d_H(\mathcal{G}_1, \mathcal{G}_2)$.

responses to the different divergence measures discussed here which validates our hypothesis with regards to the proposed metrics d_{JR} and d_{BR}.

5   Concluding Remarks

Our research presented here is motivated by the following question: Do metric properties of divergence measures have an impact on the output hypothesis of a learning algorithm, and hence on its performance? In this paper, we tried to answer this question through the following: First, we analyzed some well known divergence measures for the particular case of multivariate Gaussian densities since they are pervasive in machine learning and pattern recognition. Second, based on our analysis, we proposed a simple modification to two well known divergence measures for Gaussian densities. The modification led to two new distance metrics between Gaussian densities in which their constituting elements respect the geometry of their corresponding spaces. Next, we showed how the metric properties can impact the graph embedding process of manifold learning algorithms, and demonstrated empirically how the proposed new metrics yield better embedding spaces in a totally unsupervised manner.

Our study suggests that metric properties of divergence measures constitute an important aspect of the model selection question for divergence based learning algorithms. Further, the proposed metrics developed here are not restricted to manifold learning algorithms, and they can be used in various contexts, such as metric learning, discriminant analysis, and feature selection to mention a few.

References

1. Abou–Moustafa, K., Ferrie, F.: A framework for hypothesis learning over sets of vectors. In: Proc. of 9th SIGKDD Workshop on Mining and Learning with Graphs, pp. 335–344. ACM (2011)
2. Abou-Moustafa, K., Shah, M., De La Torre, F., Ferrie, F.: Relaxed Exponential Kernels for Unsupervised Learning. In: Mester, R., Felsberg, M. (eds.) DAGM 2011. LNCS, vol. 6835, pp. 184–195. Springer, Heidelberg (2011)
3. Ali, S.M., Silvey, S.D.: A general class of coefficients of divergence of one distribution from another. J. of the Royal Statistical Society, Series B 28(1), 131–142 (1966)
4. Belkin, M., Niyogi, P.: Laplacian eigenmaps and spectral techniques for data representation. Neural Computation 15, 1373–1396 (2003)
5. Cao, G., Bachega, L., Bouman, C.: The sparse matrix transform for covariance estimation and analysis of high dimensional signals. IEEE. Trans. on Image Processing 20(3), 625–640 (2011)
6. Csiszár, I.: Information–type measures of difference of probability distributions and indirect observations. Studia Scientiarium Mathematicarum Hungarica 2, 299–318 (1967)
7. Förstner, W., Moonen, B.: A metric for covariance matrices. Tech. rep., Dept. of Geodesy and Geo–Informatics, Stuttgart University (1999)
8. Gower, J., Legendre, P.: Metric and Euclidean properties of dissimilarity coefficients. J. of Classification 3, 5–48 (1986)

9. Kailath, T.: The divergence and Bhattacharyya distance measures in signal selection. IEEE Trans. on Communication Technology 15(1), 52–60 (1967)
10. Kondor, R., Jebara, T.: A kernel between sets of vectors. In: ACM Proc. of ICML (2003)
11. Kreyszig, E. (ed.): Introductory functional Analysis with Applications. Wiley Classics Library (1989)
12. Kullback, S.: Information Theory and Statistics – Dover Edition. Dover, New York (1997)
13. Lucas, B., Kanade, T.: An iterative image registration technique with an application to stereo vision. In: Proc. of IJCAI, pp. 674–679 (1981)
14. von Luxburg, U.: A tutorial on spectral clustering. Statistics and Computing 17(4), 395–416 (2007)
15. Mercer, J.: Functions of positive and negative type, and their connection with the theory of integral equations. Philosophical Trans. of the Royal Society of London, Series A 209, 415–446 (1909)
16. Moreno, P., Ho, P., Vasconcelos, N.: A Kullback–Leibler divergence based kernel for svm classification in multimedia applications. In: NIPS, vol. 16 (2003)
17. Pennec, X., Fillard, P., Ayache, N.: A Riemannian Framework for Tensor Computing. Tech. Rep. RR-5255, INRIA (July 2004)
18. Rao, C.R.: Information and the accuracy attainable in the estimation of statistical parameters. Bull. Calcutta Math. Soc. (58), 326–337 (1945)
19. Schoenberg, I.: Metric spaces and positive definite functions. Trans. of the American Mathematical Society 44(3), 522–536 (1938)
20. Tenenbaum, J., de Silva, V., Langford, J.: A global geometric framework for non-linear dimensionality reduction. Science 290(5500), 2319–2323 (2000)
21. Young, G., Householder, A.: Discussion of a set of points in terms of their mutual distances. Psychometrika 3(1), 19–22 (1938)
22. Zha, H., Ding, C., Gu, M., He, X., Simon, H.: Spectral relaxation for k–means clustering. In: NIPS, vol. 13. MIT Press (2001)

Graph Database Retrieval Based on Metric-Trees[*]

Francesc Serratosa, Xavier Cortés, and Albert Solé-Ribalta

Universitat Rovira i Virgili, Departament d'Enginyeria Informàtica i Matemàtiques, Spain
{francesc.serratosa,xavier.cortes,albert.sole}@urv.cat

Abstract. M-trees are well-know structures used to speed-up queries in databases. In this paper, we evaluate the applicability of m-trees to graph databases. In classical schemes based on metric-trees, the routing information kept in a metric-tree node is a selected element from the sub-cluster that represents. Nevertheless, defining a graph that represents a set of graphs is not a trivial task. We evaluate different graphs-class prototype as routing nodes in the metric tree. The considered prototypes are: Median Graphs, Closure Graphs, First-Order Random Graphs, Function-Described Graphs and Second-Order Random Graphs.

Keywords: Metric-tree, Graph Indexing, Median Graph, First-Order Random Graph, Function-Described Graph, Second-Order Random Graph.

1 Introduction

Indexing structures are fundamental tools in database technology; they are used to obtain efficient access to large collections of elements. Traditional image database systems manage global properties of images, such as histograms [1]. Many techniques for indexing one-dimensional data sets have been defined. Since a total order function over a particular attribute domain always exists, this ordering can be used to partition the data and moreover it can be exploited to efficiently support queries. Several multi-dimensional indexes have appeared, such as, colour, texture, shape, with the aim of increasing the efficiency in executing queries on sets of objects characterized by multi-dimensional features.

Effective access to image databases requires queries addressing the expected appearance of searched images [2]. To this end, it is needed to represent the image as a set of entities and relations between them. The effectiveness of retrieval may be improved by registering images as structural elements rather than global features [3, 4]. In the most practiced approach to content-based image retrieval, the visual appearance of each spatial entity is represented independently by a vector of features. Mutual relationships between entities can be taken into account in this retrieval process. Thus, local entities and mutual relationships may be considered to have the same relevance and to be defined as parts of a global structure that captures mutual

---

[*] This research is supported by Consolider Ingenio 2010: project CSD2007-00018 & by the CICYT project DPI 2010-17112.

G.L. Gimel'farb et al. (Eds.): SSPR & SPR 2012, LNCS 7626, pp. 437–447, 2012.

dependencies [5]. In this case, the model of content takes the structure of an Attributed Graph.

While the distance between two sets of independent features can be computed in polynomial time [6,7], the exact distance between two graphs is computed in exponential time with respect to the number of nodes of the graphs. Although some sub-optimal solutions have been presented to compare a pair of graphs, in which, the computational complexity is reduced to polynomial cost, few contributions of practical interest have been proposed supporting the application of graphs to content-based retrieval from image databases [8, 9].

Out of the specific context of content-based image retrieval, the problem of comparing an input graph against a large number of model graphs has been addressed in several approaches. In some applications, the classes of objects are represented explicitly by a set of graphs, which means that a huge amount of model graphs must be matched with the input graph and so the conventional error-tolerant graph matching algorithms must be applied to each model-input pair sequentially. As a consequence, the total computational cost is linearly dependent on the number of model graphs and exponential (or polynomial if suboptimal methods are used) with the size of the graphs. For applications dealing with large databases, this may be prohibitive. To alleviate these problems, some attempts have been designed with the aim of reducing the computational time of matching the unknown input patterns to the whole set of models from the database. Those approaches assume that the graphs that represent a cluster or class are not completely dissimilar in the database and, in this way, only one structural model is defined from the graphs that represent the cluster. These structures are called *Graph-Class Prototypes*. In the classification process, only one comparison is needed for each cluster.

In this paper, we evaluate an indexing scheme, modelled by an m-tree, in which the cluster knowledge embedded in each node of the m-tree is represented by one of the six Graph-Class Prototypes presented in the literature. The different representations of Graph-Class Prototypes are: 1) Set Median Graph [8]; 2) Generalise Median Graph [10, 11, 12, 13] synthesised through a hierarchical method [14], synthesised through a genetic algorithm [15] or synthesised through an extension of the Graduated Assignment algorithm [16]; 3) First-Order Random Graphs [17]; 4) Function-Described Graphs [18, 19]; 5) Second-Order Random Graphs [20]; 6) Closure Graphs [21]. Moreover, we evaluate two types of graph queries; the ones that the user imposes the number of graphs to be queried and the ones that the user imposes the maximum distance between the query graph and the returned graphs. It is not the aim of this paper to explain the structural representation of each graph prototype but to evaluate its representational power in metric trees. Some of the methods presented in this paper have been presented in [14, 23] but only applied to Median Graphs. The aim of this paper is to evaluate the representational power of the Graph-Class Prototypes presented in the literature.

In this paper, we have performed more experiments with more databases and we have put together both types of queries and we have used more Graph-Class Prototypes with the aim of obtaining a more general results and conclusions.

The rest of the paper is organised as follows. In section 2, we comment the few approaches that have been presented for indexing Attributed Graphs. In chapter 3, we introduce metric-trees. In section 4, we explain the methods used to synthesise the graph prototypes. In section 5, we experimentally evaluate the graph prototypes as routing elements of m-trees. We finish the paper drawing some conclusions.

2 Indexing Databases of Graphs

Some indexing techniques have been developed for graph queries. We divide these techniques into two categories. In the first ones, the index is based on several tables and filters [25, 26]. In the second ones, the index structure is based on m-trees [8,21,27].

In the first group of techniques, the ones that are not based on trees, we emphasize the method developed by Shasha et. al. [26] called GraphGrep. GraphGrep is based on a table in which each row stands for a path inside the graph (up to a threshold length) and each column stands for a graph. Each entry in the table compounds to the number of occurrences of a particuar path in the graph. Queries are processed in two phases. The filtering phase generates a set of candidate graphs for which the count of each path is at least that of the query. Since indexing schemes based on paths do not ensure graph isomorphism, in a verification phase, each candidate is strictly compared to the query graph and only isomorphic graphs are returned. More recently, Yan et. al. [25] proposed G_{Index} that uses frequent patterns as indexing features. These frequent patterns reduce the index space as well as improve the filtering rate. The main drawback of these models is that the construction of the indices requires an exhaustive enumeration of the paths or fragments that increases the memory and time requirements of the model. Moreover, since paths or fragments carry little information about a graph, the lost of information at the filtering step seems to be unavoidable.

Considering the second group, the first time that metric trees were applied to graph databases was done by Berretti et. al. [8]. Attributed graphs were clustered hierarchically according to their mutual distances and indexed by m-trees [22]. Queries are processed in a top-down manner by routing the query along the index tree. Each node of the index tree represents a cluster and it has one of the graphs of the cluster as a representative. The graph matching problem, in the tree construction and at query time, was solved by an extension of the A* algorithm that uses a look-ahead strategy plus a stopping threshold. A drawback of this method is that the computational cost is exponential respect the number of nodes in the graphs. Lee et. al. [27] used this technique to model graphical representations of foreground and background scenes in videos. The resulting graphs were clustered using the edit-distance metric, and similarity queries were answered using a multi-level index structure.

More recently, He and Singh [21] proposed what they called a Closure-tree. It uses a similar structure than the one presented by Berretti [8] but, the representative of the cluster was not one of the graphs but a graph prototype (called closure graph) that could be seen as the union of the Attributed Graphs that compose the cluster. The structurally similar nodes that have different attributes in the graphs are represented in the Closure graph with only one node but with more than one attribute. Closure trees have two main drawbacks. First, they can only represent discrete attributes at nodes of the attributed graphs. Second, they tend to generalize too much the set of graphs they represent, allowing graphs that have not been used to synthesize the closure graph.

Finally, Median Graphs have been used as a new prototype to represent Attributed Graphs in [14, 15]. More specifically, in [14], they defined queries in which a maximum distance between the query and the graphs was considered. And in [23], they performed k-nearest neighbour queries.

3   Database Indexing Based on Metric-Trees

A metric-tree (m-tree) [22] is a scheme to partition a database in a hierarchical set of clusters, collecting similar objects. Each cluster has a routing object and a radius providing an upper bound for the maximum distance between the reference object and any other object in the cluster. Triangle inequality can be used during the access to the database to prune clusters that are bound out of an assigned range from the query.

Formally, a metric-tree is a tree of nodes. Each node contains a fixed maximum number of m entries, $< node > := \{< entry >\}^m$. In turn, each entry is constituted by a routing element M; a reference to the father r^H of a sub-index containing the element in the so-called covering region of M; and a radius d^M providing an upper bound for the distance between M and any element in its covering region, $< entry > := \{M, r^M, d^M\}$. During retrieval of an element Q, triangular inequality is used to support efficient processing of queries. To this end, the distance between Q and any element in the covering region of a routing element M can be max-bounded using the radius d^M plus the distance between Q and M.

Two different types of queries can be performed to databases organised by m-trees: *k-Nearest-Neighbour* queries [23] and *Similarity* queries [14]. The aim of the k-Nearest Neighbour Queries is to retrieve the k elements in a database that have minimum distance between them and the query element. On the contrary, the aim of the Similarity queries is to retrieve all the elements in the database which its distance to the queried element is lower than a threshold d_{max}.

The m-tree can be constructed using different schemes for the insertion of a new element and the selection of the routing element [22]. In this paper, we use a general construction methodology from which we are able to construct an m-tree independently of the type of the routing element. We use a non-balanced tree constructed through a hierarchical clustering algorithm and complete linkage clustering [24]. In this way, given a set of graphs, the distance matrix over the whole set is computed and then a dendogram is constructed. Using the dendogram and some horizontal cuts, a set of partitions that clusters the graphs in the database is obtained. With these partitions the m-tree is generated. Finally, the information on the routing elements in the m-tree is inserted, M and d^M.

In our case, M is a Graph-Class Prototype and d^M is the maximum distance between the Graph Prototype and any of the graphs in the covering region. Figure 1 shows an example of a dendogram. Elements G^i are placed on the leaves of the

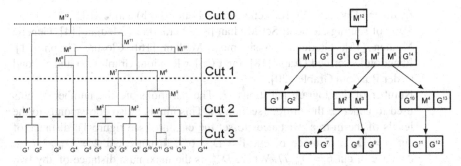

Fig. 1. Example of a dendogram   **Fig. 2.** The obtained m-tree

dendogram and the routing elements M^j are placed on the junctions between the cuts and the horizontal lines of the dendograms. Dendogram of figure 1 defines 4 different partitions. Figure 2 shows the obtained m-tree. Note that in some tree nodes, there are Class-Graph Prototypes (M^j) together with original graphs (G^i).

4 Synthesis of Graph Prototypes Related to Metric-Trees

Two types of methods exist to generate Graph Prototypes from a given set of graphs [18, 20]. We assume the structure of the metric-tree has been computed (section 3) and we have to compute the Graph Prototype and the radius of the cluster d^M. The first method is based on a hierarchical synthesis. The second one is based on a Global Synthesis based on a Common Labelling [28, 29, 30].

In the Hierarchical method, each Graph Prototype is computed only using two Graph Prototypes or Attributed Graphs at a time. Therefore, a Common Labelling is not needed and Graph Prototypes are computed as pairwise consecutive computations of other Graph Prototypes obtained in lower levels of the tree.

In the Global Synthesis, each Graph Prototype is computed using the whole set of Attributed Graphs in the cluster that the m-tree node represents, independently of whether the m-tree node has other nodes as descendants in the tree. The first step of this method computes a Common Labelling from the Attributed Graphs of the sub-cluster and the second step obtains the Graph Prototype. In this paper, we have used two different Common Labelling algorithms, which have the main feature that are independent of the prototype graph to be synthesised. The first one is based on the Graduated Assignment [16] and the second one is based on a genetic algorithm [15].

Note that the Set Median is a special prototype since it does not need to be synthesised. The Set Median is the graphs of the cluster that has the minimum distance between it and the other graphs.

5 Practical Evaluation

Test Parameters: In each test, only one m-tree is constructed with 50 graphs of the reference set. The parameters used to construct each m-tree are:

- Evaluated Datasets: COIL, Letter (low), Letter (high) and GREC.
- Type of routing element: Set Median [8], Hierarchical Median [14], Genetic Median [15], Graduated Assignment Median [16], Closure Graph [21], Function Described Graph [18], First Order Random Graph [17] and Second Order Random Graphs [20].
- Number of dendogram partitions: 7. The partitions are the number of cuts used to generate the m-tree (section 3). This number also corresponds to the levels of the m-tree. Distances to set the cuts are (see figure 1); distance of cut $0 = D_{max}$, distance of cut $1 = D_{max} \cdot 6/7$, distance cut $2 = D_{max} \cdot 5/7$, ... distance of cut $6 = D_{max}/7$. Where D_{max} is the maximum distance of any two graphs of the m-tree.

Parameters for each query are:
- Graph query. The graph has been extracted from the test set.
- Number of queries: 50. Results values are the mean of these 50 queries.
- Metric-tree.
  - o If k-NN query: Number of elements to be retrieved: $\mathbf{k} = 3$.
  - o If similarity query: Range of the query: $\mathbf{d_{max}} = 0.6 \cdot D_{max}$.

Evaluation Indices: Three indices have been used: Access ratio, Precision and Recall. Access ratio evaluates the capacity of the m-tree to properly route the queries [14, 23]. It is obtained as the normalised number of accessed nodes and leaves of the m-tree given a query. Precision is the fraction of retrieved documents that are relevant to the search and Recall is the fraction of the documents that are relevant to the query that are successfully retrieved. Ground truth of precision and recall are computed by exhaustive search of the elements in the dataset. Note that Precision and Recall values depend on the construction of the m-tree due to we use sub-optimal algorithms to synthesise the prototypes and to compute d^M. In addition, at query time, since distances are also sub-optimally computed, the algorithm may violate triangle inequality restrictions and so return not accurate results.

Datasets: COIL, Letter (low), Letter (high) and GREC (presented in [31]). The 72 images of each element of COIL dataset have been clustered in 4 classes instead of one class. Each class is composed by 18 consecutive images.

Results: Tables 1 and 2 show the access ratio in nearest neighbour and similarity queries. Lower is the access ratio faster is the query. Besides, if the access ratio is greater than 1, the number of comparisons done using the m-tree is higher than if there was no m-tree and the graphs of the whole database where all compared. This situation does not appear in the K-nn queries, which means that it is worth to structure the database in an m-tree. On the contrary, some values of table 2 (similarity query) are greater than 1. To reduce this problem, d_{max} whole have to be reduced but we preferred to use the same value for all the experiments for the compactness of the result values. Besides, some cells of table 2 have value 0.02. This is because, given a query, only the root node of the m-tree is explored. Considering that the m-tree has been built using 50 graphs, the value comes from 0.02 = 1/50. Again, this problem could be solved by adapting d_{max} value to each Graph-Class Prototype.

In general, Median Graphs are the prototypes with better access ratio. So, they obtain faster queries (except for the commented extreme values, 0.02).

Table 1. Access ratio on nearest neighbour queries

Access Ratio (k=3)	Synthesis	COIL	Letter L	Letter H	GREC
Set Median	----	0.54	0.34	0.35	0.45
Generalise Median	Hierarchical	0.38	0.31	0.33	0.37
Generalise Median	Genetic	0.40	0.38	0.36	0.57
Generalise Median		0.44	0.34	0.37	0.40
Closure Graph		0.65	0.33	0.57	0.80
FORG	Graduated	0.42	0.35	0.42	0.47
SORG	Assignment	0.35	0.33	0.36	0.38
FDG		0.45	0.36	0.45	0.56

Table 2. Access Ratio on similarity queries

Access Ratio (similarity)	Synthesis	COIL	Letter L	Letter H	GREC
Set Median	----	0.02	1.03	1.46	0.06
Generalise Median	Hierarchical	0.47	0.99	1.34	1.07
Generalise Median	Genetic	0.92	1.65	1.66	0.78
Generalise Median		0.89	0.98	1.30	0.94
Closure Graph		0.02	0.02	0.02	0.02
FORG	Graduated	0.02	0.02	0.02	0.02
SORG	Assignment	0.02	2.27	2.78	0.02
FDG		0.02	0.47	0.04	0.02

Tables 3 and 4 show the mean precision. SORGs and Closures are the prototypes that obtain the best results although there are other prototypes with similar values. In general, prototypes computed using the Graduated Assignment obtains better results than the Hierarchical and Genetic synthesis. Considering values on tables 1 and 3 (k-NN), we can conclude that the probabilistic prototypes are slower but obtain greater precision. And considering values on tables 2 and 4 (similarity), we realise that the fact that some queries only explore the root node penalises the obtained precision.

Table 3. Precision on nearest neighbour queries

Precision (k=3)	Synthesis	COIL	Letter L	Letter H	GREC
Set Median	----	0.46	0.76	0.42	0.48
Generalise Median	Hierarchical	0.37	0.62	0.34	0.22
Generalise Median	Genetic	0.11	0.16	0.22	0.34
Generalise Median		0.32	0.94	0.46	0.32
Closure Graph		0.62	0.98	0.38	0.59
FORG	Graduated	0.58	0.86	0.42	0.57
SORG	Assignment	0.65	0.81	0.50	0.37
FDG		0.48	0.84	0.44	0.53

Table 4. Precision on similarity queries

Precision (similarity)	Synthesis	COIL	Letter L	Letter H	GREC
Set Median	----	0.75	0.99	0.99	0.78
Generalise Median	Hierarchical	0.81	0.99	0.99	0.98
Generalise Median	Genetic	0.89	1	0.99	0.92
Generalise Median		0.99	0.99	0.99	0.97
Closure Graph		0.75	0.62	0.74	0.77
FORG	Graduated	0.75	0.62	0.74	0.77
SORG	Assignment	0.75	1	1	0.77
FDG		0.75	0.82	0.78	0.77

Finally, tables 5 and 6 show the recall results. In general, probabilistic prototypes obtain greater recall than non-probabilistic ones, except in some cases. Note that in cases that the access ratio is 0.02, the recall is always 1. This is because, if all graphs of the database are accepted, then the recall has to be 1 by definition.

Table 5. Recall on nearest neighbour queries

Recall (k=3)	Synthesis	COIL	Letter L	Letter H	GREC
Set Median	----	0.26	0.58	0.32	0.50
Generalise Median	Hierarchical	0.24	0.48	0.26	0.22
Generalise Median	Genetic	0.06	0.12	0.18	0.34
Generalise Median		0.18	0.72	0.36	0.34
Closure Graph		0.38	0.76	0.30	0.60
FORG	Graduated	0.38	0.66	0.32	0.58
SORG	Assignment	0.40	0.62	0.38	0.38
FDG		0.30	0.64	0.32	0.54

Table 6. Recall on similarity queries

Recall (similarity)	Synthesis	COIL	Letter L	Letter H	GREC
Set Median	----	1	0.99	1	0.99
Generalise Median	Hierarchical	1	1	0.98	0.90
Generalise Median	Genetic	0.92	1	1	0.96
Generalise Median		1	0.98	0.98	0.90
Closure Graph		1	1	1	1
FORG	Graduated	1	1	1	1
SORG	Assignment	1	1	1	1
FDG		1	0.60	0.90	1

Table 7 summarises the results presented in the last 6 tables. Each value is the average of the eight corresponding values. Statistically best values are bolded. FORGs obtain the fastest queries (lower access ratio). Generalise Median (with Graduated Assignment) and SORGs obtain the greatest Precision. Closure Graphs, FORGs and SORGs obtain the greatest Recall. Finally, SORGs obtain the best F-measure.

Table 7. Average results of Access Ratio, Precision, Recall and F-measure

	Synthesis	Access	Precision	Recall	F-measure
Set Median	----	0.53	0.70	0.70	0.70
Generalise Median	Hierarchical	0.65	0.66	0.63	0.64
Generalise Median	Genetic	0.84	0.57	0.57	0.57
Generalise Median		0.70	**0.74**	0.68	0.71
Closure Graph		0.30	0.68	**0.75**	0.71
FORG	Graduated	**0.21**	0.66	**0.74**	0.70
SORG	Assignment	0.81	**0.73**	0.72	**0.72**
FDG		0.29	0.64	0.32	0.67

6 Conclusions

We have evaluated a graph indexing technique based on metric-trees and several Graph-Class Prototypes. Specifically, we have studied the behaviour of Graph-Class Prototypes as routing elements of m-trees. Several papers have been published that compare the accuracy of the evaluated prototypes. In this paper, we evaluated the goodness of those prototypes on speeding-up queries on graph databases. The evaluation has been performed using four different datasets with different characteristics. We see from the practical validation that probabilistic prototypes seem to achieve better results on k-nn queries. On the contrary, the Generalise Median together with the Set Median seem to give better results on similarity queries at the cost of giving a larger access ratio. Up to now, Set Median Graphs and Closure Graphs where the only prototypes used as routing elements of metric trees. The general conclusion of this work is that other existing Graph-Class Prototypes can also be successfully used as routing elements of metric trees in graph databases.

References

1. Konstantinidis, K., Gasteratos, A., Andreadis, I.: Image retrieval based on fuzzy colour histogram processing. In: Optics Communications, vol. 248, pp. 375–386 (2005)
2. Datta, R., Joshi, D., Li, J., Wang, J.Z.: Image Retrieval: Ideas, Influences, and Trends of the New Age. ACM Computing Surveys 40(2), Article 5 (2008)
3. Jouili, S., Tabone, S.: Hypergraph-based image retrieval for graph-based representation. Pattern Recognition 45(11), 4054–4068 (2012)
4. Lebrun, J., Gosselin, P., Philipp, S.: Inexact graph matching based on kernels for object retrieval in image databases. Image and Vision Computing 29(11), 716–729 (2011)

5. Le Saux, B., Bunke, H.: Feature Selection for Graph-Based Image Classifiers. In: Marques, J.S., Pérez de la Blanca, N., Pina, P. (eds.) IbPRIA 2005. LNCS, vol. 3523, pp. 147–154. Springer, Heidelberg (2005)

6. Gold, S., Rangarajan, A.: A Graduated Assignment Algorithm for Graph Matching. Transactions on Pattern Analysis and Machine Intelligence 18(4), 377–388 (1996)

7. Neuhaus, M., Riesen, K., Bunke, H.: Fast Suboptimal Algorithms for the Computation of Graph Edit Distance. In: Yeung, D.-Y., Kwok, J.T., Fred, A., Roli, F., de Ridder, D. (eds.) SSPR 2006 and SPR 2006. LNCS, vol. 4109, pp. 163–172. Springer, Heidelberg (2006)

8. Berretti, S., Del Bimbo, A., Vicario, E.: Efficient Matching and Indexing of Graph Models in Content-Based Retrieval. IEEE Transactions on Pattern Analysis and Machine Intelligence 23(10), 1089–1105 (2001)

9. Zhao, J.L., Cheng, H.K.: Graph Indexing for Spatial Data Traversal in Road Map Databases. Computers & Operations Research 28, 223–241 (2001)

10. Jiang, X., Münger, A., Bunke, H.: On median graphs: Properties, algorithms and applications. IEEE Trans. on PAMI 23(10), 1144–1151 (2001)

11. Ferrer, M., Valveny, E., Serratosa, F., Riesen, K., Bunke, H.: Generalized Median Graph Computation by Means of Graph Embedding in Vector Spaces. Pattern Recognition 43(4), 1642–1655 (2010)

12. Ferrer, M., Valveny, E., Serratosa, F.: Median graphs: A genetic approach based on new theoretical properties. Pattern Recognition 42(9), 2003–2012 (2009)

13. Ferrer, M., Valveny, E., Serratosa, F.: Median graph: A new exact algorithm using a distance based on the maximum common subgraph. Pattern Recognition Letters 30(5), 579–588 (2009)

14. Serratosa, F., Solé-Ribalta, A., Vidiella, E.: Graph Indexing and Retrieval Based on Median Graphs. In: Martínez-Trinidad, J.F., Carrasco-Ochoa, J.A., Kittler, J. (eds.) MCPR 2010. LNCS, vol. 6256, pp. 311–321. Springer, Heidelberg (2010)

15. Bunke, H., Munger, A., Jiang, X.: Combinatorial search versus genetic algorithms: A case study based on the generalized median graph problem. Pattern Recognition Letter 20, 1271–1277 (1999)

16. Solé-Ribalta, A., Serratosa, F.: Graduated Assignment Algorithm for Finding the Common Labelling of a Set of Graphs. In: Hancock, E.R., Wilson, R.C., Windeatt, T., Ulusoy, I., Escolano, F. (eds.) SSPR&SPR 2010. LNCS, vol. 6218, pp. 180–190. Springer, Heidelberg (2010)

17. Wong, A.K.C., You, M.: Entropy and distance of random graphs with application to structural pattern recognition. IEEE Transactions on Pattern Analysis and Machine Intelligence 7, 599–609 (1985)

18. Serratosa, F., Alquézar, R., Sanfeliu, A.: Function-described graphs for modeling objects represented by attributed graphs. Pattern Recognition 36(3), 781–798 (2003)

19. Serratosa, F., Alquézar, R., Sanfeliu, A.: Synthesis of Function-Described Graphs and clustering of Attributed Graphs. International Journal of Pattern Recognition and Artificial Intelligence 16(6), 621–655 (2002)

20. Sanfeliu, A., Serratosa, F., Alquézar, R.: Second-Order Random Graphs for modeling sets of Attributed Graphs and their application to object learning and recognition. Intern. Journal of Pattern Recognition and Artificial Intelligence 18(3), 375–396 (2004)

21. He, H., Singh, A.K.: Closure-Tree: An Index Structure for Graph Queries. In: Proc. International Conference on Data Engineering, p. 38 (2006)

22. Ciaccia, P., Patella, M., Zezula, P.: M-tree: An Efficient Access Method for Similarity Search in Metric Spaces. In: Proc. 23rd VLDB Conference, pp. 426–435 (1997)

23. Serratosa, F., Solé-Ribalta, A., Cortés, X.: K-nn Queries in Graph Databases Using M-Trees. In: Real, P., Diaz-Pernil, D., Molina-Abril, H., Berciano, A., Kropatsch, W. (eds.) CAIP 2011, Part I. LNCS, vol. 6854, pp. 202–210. Springer, Heidelberg (2011)

24. Fernández, A., Gómez, S.: Solving Non-uniqueness in Agglomerative Hierarchical Clustering Using Multidendrograms. Journal of Classification (25), 43–65 (2008)

25. Yan, X., Yu, P.S., Han, J.: Graph indexing: a frequent structure-based approach. In: ACM SIGMOD International Conference on Management of Data, pp. 335–346 (2004)

26. Shasha, D., Wang, J.T.L., Giugno, R.: Algorithmics and applications of tree and graph searching. In: ACM SIGMOD-SIGACT-SIGART, pp. 39–52 (2002)

27. Lee, S.Y., Hsu, F.: Spatial Reasoning and Similarity Retrieval of Images using 2D C-Strings Knowledge Representation. Pattern Recognition 25(3), 305–318 (1992)

28. Lozano, M.A., Escolano, F., Bonev, B., Suau, P., Aguilar, W., Saez, J.A., Cazorla, M.A.: Region and constellations based categorization of images with unsupervised graph learning. Image and Vision Computing 27, 960–978 (2009)

29. Solé-Ribalta, A., Serratosa, F.: Graduated Models and Algorithms for computing the Common Labelling of a set of Attributed Graphs. In: CVIU, vol. 115 (7), pp. 929–945 (2011)

30. Solé-Ribalta, A., Serratosa, F.: A Probabilistic Framework to Obtain a Common Labelling between Attributed Graphs. In: Vitrià, J., Sanches, J.M., Hernández, M. (eds.) IbPRIA 2011. LNCS, vol. 6669, pp. 516–523. Springer, Heidelberg (2011)

31. Riesen, K., Bunke, H.: IAM Graph Database Repository for Graph Based Pattern Recognition and Machine Learning. In: da Vitoria Lobo, N., Kasparis, T., Roli, F., Kwok, J.T., Georgiopoulos, M., Anagnostopoulos, G.C., Loog, M. (eds.) S+SSPR 2008. LNCS, vol. 5342, pp. 287–297. Springer, Heidelberg (2008)

Validation of Network Classifiers

James Li, Abdullah Sonmez, Zehra Cataltepe, and Eric Bax

Cornell University, Istanbul Technical University, and Yahoo! Inc.
jyl73@cornell.edu, abdullah.sonmez@gmail.com,
cataltepe@itu.edu.tr, baxhome@yahoo.com

Abstract. This paper develops PAC (probably approximately correct) error bounds for network classifiers in the transductive setting, where the network node inputs and links are all known, the training nodes class labels are known, and the goal is to classify a working set of nodes that have unknown class labels. The bounds are valid for any model of network generation. They require working nodes to be selected independently, but not uniformly at random. For example, they allow different regions of the network to have different densities of unlabeled nodes.

Keywords: network classifier, collective classification, validation, error bound, worst likely assignment.

1 Introduction

Networks play fundamental roles in our lives. A network of interactions among genes determines how our bodies grow. Neural networks enable us to think and learn. We participate in social networks. The ecosystem we live in is a complex web of interactions between all living things and our shared environment. Electrical grids supply power; transportation systems bring us goods, and the internet supports information sharing around the world.

Data analysis based on networks has a deep history in social science and telecommunications. Networks are emerging as a basis for data analysis in many other fields, including biology, economics, and engineering [10]. Network data analysis builds on well-established foundations in graph theory [4], including study of small-world [17,16] and other random graphs [3], and statistics [7], including analysis of Markov chains.

Classification of network data [12,9,18], sometimes called collective classification [14], is an emerging sub-field of machine learning. Collective classification techniques have been developed for these transductive network classification problems. These techniques include Loopy Belief Propagation or the Iterative Classification Algorithm [13], which assign an initial set of labels to working nodes, then iteratively apply a model to re-label working nodes based on their network neighbors' input data and labels. Other collective classification techniques, such as Weighted-Vote Relational Classifiers [11], can also be applied in the transductive setting. For a study comparing some collective classification techniques, refer to [14].

In machine learning, data usually consists of examples, each containing input data and output data. In classification problems, the output is a label that

G.L. Gimel' farb et al. (Eds.): SSPR & SPR 2012, LNCS 7626, pp. 448–457, 2012.
© Springer-Verlag Berlin Heidelberg 2012

takes one of a finite set of values. To illustrate, consider a medical application. Each example corresponds to a person. The input data include the person's age, gender, and levels of some proteins measured in their blood. The output label indicates whether the person has a certain type of infectious disease.

In classification, there is a set of in-sample training examples with known labels and a set of out-of-sample test examples with unknown labels. The training examples are used to develop a classifier, which is used to label the test examples. (The classifier is a function that maps from example inputs to labels.) The primary goal of classification is accuracy – developing a classifier that has a low error rate on the test examples. A secondary goal is to validate classifier accuracy – producing probabilistic error bounds for the classifier. Since accuracy is the primary goal, we prefer to use all available examples for training, introducing the challenge that we lack independent data for validation.

This paper addresses classification problems with network data, where the inputs include links. In our example, a link between two people could indicate they share a water source. In traditional machine learning, examples are assumed to be drawn i.i.d. from an underlying joint input-output distribution. With network data, this is often not the case. For example, medical researchers may have invited an initial "seed" set of people to join the study and be represented by network nodes, and then participants may have invited people they know to join the study, growing the network by non-uniform sampling.

In the transductive setting for machine learning [15,6], the test example inputs are available for training; only the test labels are unknown. In the transductive setting, the test examples are called working examples. This paper focuses on collective classifier validation in the transductive setting, where all node input data and links are known, the training node labels are known, and only the working node labels are unknown. The classifier may take advantage of training node inputs and labels, all links, and working node inputs.

This paper develops worst likely assignment error bounds [2] for network classifiers. These error bounds validate a classifier by computing how much an assignment of labels to the working nodes may disagree with the classifier's labels and still be likely to be the actual unknown labels, according to some statistical criteria. The statistics are based on the fact that training and working nodes are interchangeable a priori, so it is unlikely that the unknown labels for the working examples have very different statistics than the known labels for the training examples.

This paper is organized as follows. Section 2 develops an error bound algorithm and some variations. Section 3 presents experimental results showing bound effectiveness. Then Section 4 concludes with ideas for future work.

2   Algorithms

This section presents algorithms to compute error bounds for network classifiers based on worst likely assignments. Subsection 2.1 develops a basic algorithm that is effective for problems with small working sets but requires too much computation for larger problems. Subsection 2.2 develops an algorithm for larger problems, using the basic algorithm as a component. Subsection 2.3 shows how to modify the algorithm to control the relationship between component and whole-network bound failure probabilities. Subsection 2.4 shows how to adapt the algorithms for network labeling processes where different nodes have different probabilities of being selected for the working set.

2.1   Basic Algorithm

Let n be the number of nodes in the network. Let $(X_1, Y_1), \ldots, (X_n, Y_n)$ be the input-label pairs for the nodes. The inputs may be from any domain. The labels are drawn from a finite set.

Assume all network nodes and links are known, and all node inputs X_i are known. Assume some node labels Y_i are known and others are unknown. Refer to nodes with known labels as training nodes and nodes with unknown labels as working nodes. (Working nodes are sometimes called test nodes.) Define T to be the set of training nodes and W to be the set of working nodes. Let $t = |T|$ and $w = |W|$.

For now, suppose that the working set was selected by applying i.i.d. Bernoulli trials to the nodes, with success indicating the node is a working node and failure indicating the node is a training node. Then each training-working split with t training and w working nodes is equally likely. (The constraint that all nodes are equally likely to be working nodes will be removed later, in Subsection 2.4.)

The algorithm uses a concept of *continuous rank* to assess likelihood. Define the continuous rank $r(s_0, \{s_0, \ldots, s_{m-1}\})$ of a value s_0 among values $s_0, \ldots, s_{m-1}$ to be the sum of

- The number of values in $s_1, \ldots, s_{m-1}$ less than s_0,
- a value drawn uniformly at random from the integers zero to the number of values in $s_1, \ldots, s_{m-1}$ equal to s_0, and
- a value drawn uniformly at random from the real numbers in $[0, 1]$.

Note that if $s_0, \ldots, s_{m-1}$ are drawn i.i.d., then $r(s_0, \{s_0, \ldots, s_{m-1}\})$ is uniformly distributed over the real numbers in $[0, m]$.

The basic algorithm is:

1. Select bound failure probability $\delta > 0$. Select sample size $m > 0$. Choose $m - 1$ comparison training-working splits $(T_1, W_1), \ldots, (T_{m-1}, W_{m-1})$ uniformly at random from the $\binom{t+w}{t}$ splits that have t training and w working nodes. Either choose them without replacement and prohibit the actual training-working split (T, W), or choose them with replacement. Let $(T_0, W_0) = (T, W)$.
2. Use a set M to store error counts for likely assignments. Initially, $M = \emptyset$.
3. Let A be the set of assignments a' that agree with the labels on nodes in T and assign any labels to the nodes in W. $\forall a' \in A$:
   (a) Assign a' to the node labels.

(b) Train $g_0, \ldots, g_{m-1}$, using $T_0, \ldots, T_{m-1}$ as the training nodes and $W_0, \ldots, W_{m-1}$ as the working nodes. (The training procedure for g_i may use all node inputs, all links, and the labels on T_i, but not the labels on W_i.)

(c) For $i \in \{0, \ldots, m-1\}$, let score s_i be the number of errors g_i makes when classifying the nodes in W_i under assignment a':

$$s_i = \sum_{x \in W_i} I(g_i(x) \neq a'(x)),$$

where x is a node, $I()$ is the indicator function – one if the argument is true and zero otherwise, and $a'(x)$ is the label a' assigns to x.

(d) If

$$r(s_0, \{s_0, \ldots, s_{m-1}\}) \leq (1 - \delta)m,$$

then insert s_0 into set M, because a' is a likely assignment: $a' \in L$.

4. Return $\max M$ as the error bound, because it is the maximum error count for a likely assignment.

Theorem 1. *With probability at least $1 - \delta$, the basic algorithm returns a valid error bound:*

$$\mathbb{P}\{\sum_{x \in W} I(g(x) \neq a(x)) \leq \max M\} \geq 1 - \delta,$$

where $\max M$ is the value returned by the basic algorithm, g is the algorithm's g_0 when $a' = a$, and $a(x)$ is the actual (unknown) label for x.

Proof. Let $s_i(a')$ be the value of s_i for assignment a' in the algorithm. Then the theorem states:

$$\mathbb{P}\{s_0(a) \leq \max M\} \geq 1 - \delta.$$

At the end of the algorithm,

$$M = \{s_0(a') | r(s_0(a'), \{s_0(a'), \ldots, s_{m-1}(a')\}) \leq (1 - \delta)m\}.$$

Since each training-working split with t training and w working nodes is equally likely to be each of $(T_0, W_0), \ldots, (T_{m-1}, W_{m-1})$, the values $s_0(a), \ldots, s_{m-1}(a)$ are i.i.d. So

$$r(s_0(a), \{s_0(a), \ldots, s_{m-1}(a)\}) \sim U(0, m).$$

Hence

$$\mathbb{P}\{r(s_0(a), \{s_0(a), \ldots, s_{m-1}(a)\}) \leq (1 - \delta)m\} = 1 - \delta.$$

So

$$\mathbb{P}\{s_0(a) \in M\} \geq 1 - \delta,$$

which implies

$$\mathbb{P}\{s_0(a) \leq \max M\} \geq 1 - \delta.$$

The basic algorithm requires $O(w^c m)$ classifier trainings, where c is the number of different class labels, and $O(w^c mw)$ node classifications. So the basic algorithm requires too much computation to be feasible for large working sets.

2.2   Partition the Working Set to Speed Computation

Since the basic algorithm has running time exponential in the size of the working set, one way to reduce computation is to partition the working set, apply the basic algorithm to each partition, and combine error bounds over partitions to produce an error bound over the whole working set. When applying the basic algorithm to a subset of the working nodes, the remaining working nodes are also unlabeled. Modify the basic algorithm to accommodate these *reserved nodes* as follows:

- Withhold the reserved nodes from all training-working splits, so all comparison splits have the same reserved nodes.
- Leave the reserved nodes unlabeled in assignments a'.
- Allow the training procedure to use the reserved nodes, but not their labels, which are unknown.

The partitioning algorithm is:

1. Select bound failure probability $\delta > 0$.
2. Partition the working set W into subsets $W^1, \ldots, W^p$.
3. For each partition W^i, apply the basic algorithm, using $\delta_p = \frac{\delta}{p}$ as the bound failure probability and $W - W^i$ as the reserved nodes. Let b_i be the bound returned.
4. Return $b_1 + \ldots + b_p$ as an error bound.

Theorem 2. *With probability at least $1 - \delta$, the partitioning algorithm returns a valid error bound:*

$$\mathbb{P}\{\sum_{i=1}^{p} \sum_{x \in W^i} I(g^i(x) \neq a(x)) \leq b_1 + \ldots + b_p\} \geq 1 - \delta,$$

where g^i is the classifier trained as g_0 in the modified basic algorithm when $a' = a$, with working set W^i and reserved nodes $W - W^i$.

Note that for most (non-random) training procedures, $g^1, \ldots, g^p$ are the same classifier, because each g_0 classifier training in the modified basic algorithm ignores all labels in W.

Proof. By Theorem 1,

$$\forall i \in \{1, \ldots, p\} : \mathbb{P}\{\sum_{x \in W^i} I(g^i(x) \neq a(x)) \leq b_i\} \geq 1 - \frac{\delta}{p}.$$

So

$$\forall i \in \{1, \ldots, p\} : \mathbb{P}\{\sum_{x \in W^i} I(g^i(x) \neq a(x)) > b_i\} \leq \frac{\delta}{p}.$$

Using sum bounds on the probability of a union:

$$\mathbb{P}\{\exists i \in \{1, \ldots, p\} : \sum_{x \in W^i} I(g^i(x) \neq a(x)) > b_i\} \leq p\frac{\delta}{p} = \delta.$$

So
$$\mathbb{P}\{\forall i \in \{1,\dots,p\} : \sum_{x \in W^i} I(g^i(x) \neq a(x)) \leq b_i\} \geq 1 - \delta,$$

and this implies
$$\mathbb{P}\{\sum_{i=1}^{p} \sum_{x \in W^i} I(g^i(x) \neq a(x)) \leq b_1 + \dots + b_p\} \geq 1 - \delta.$$

The partitioning algorithm requires $O((|W^1|^c + \dots + |W^p|^c)m)$ classifier trainings and $O((|W^1|^{c+1} + \dots + |W^p|^{c+1})m)$ node classifications. So as the number of partitions increases, the computation required decreases. The tradeoff is weaker bounds, because δ is divided by the number of partitions to produce δ_p, the bound failure probability in the basic algorithm for each partition.

2.3 Use Nearly Uniform Validation to Strengthen Bounds

One way to prevent slicing δ too thinly over partitions is to employ *nearly uniform* error bounds [1]. Instead of insisting that all partition bounds b_i hold to ensure that the sum bound $b_1 + \dots + b_p$ holds, nearly uniform bounds allow some partition bound failures. For example, suppose one partition bound failure is allowed. Then the error bound $b_1 + \dots + b_p$ becomes

$$b_1 + \dots + b_p + \max_i(|W^i| - b_i),$$

since when a partition bound fails, all nodes in the partition may be misclassified. Since this bound accounts for a single failure, it is invalid only if there are two or more partition bound failures. The distribution that maximizes the probability of two or more failures has each single bound failure accompanied by exactly one more. So the worst-case probability of two or more failures is at most $\frac{p}{2}\delta_p$, where δ_p bounds each single-bound failure probability. (Recall that the worst-case probability of one or more failures is $p\delta_p$.) So

$$\mathbb{P}\{\sum_{i=1}^{p} \sum_{x \in W^i} I(g^i(x) \neq a(x)) \leq b_1 + \dots + b_p + \max_i(|W^i| - b_i)\} \geq 1 - \frac{p}{2}\delta_p,$$

making it possible to set partition bound failure probabilities $\delta_p = 2\frac{\delta}{p}$ and still achieve a valid overall bound with probability at least $1 - \delta$.

Similarly, allowing k partition bound failures produces an error bound

$$b_1 + \dots + b_p + max_{i_1 \neq \dots \neq i_k}(\sum_{j=1}^{k} |W^{i_j}| - b_{i_j}).$$

Since this bound is valid with probability at least

$$1 - \frac{p}{k+1}\delta_p,$$

setting $\delta_p = (k+1)\frac{\delta}{p}$ for each partition achieves a valid bound with probability at least $1 - \delta$.

2.4   Cohorts and Non-identical Selection for the Working Set

The error bounds and algorithms in the previous subsections require that each node is equally likely to be unlabeled. Consider the case when there are subsets of nodes that are equally likely to be unlabeled, but the probability varies between subsets. Refer to each subset as a *cohort*. Within each cohort, nodes may be selected for the working set by i.i.d. Bernoulli trials or by uniform selection over subsets of a fixed size.

To illustrate, recall the example problem from Section 1. Nodes represent people, and each label indicates whether a person has an infectious disease. Suppose the study began with 100 volunteers, who all received testing for the disease. Then suppose those volunteers recruited another 250 volunteers, of whom 100 were selected at random for testing. In this case, the initial 100 volunteers are one cohort, and the next 250 are another.

To apply the algorithms to cohorts, limit the selection of comparison training-working splits $(T_1, W_1), \ldots, (T_{m-1}, W_{m-1})$ to splits that have the same number of nodes from each cohort as T in each T_i and as W in each W_i. Select the comparison splits uniformly among such splits. Then each such split is equally likely a priori to be (T, W). So the error bounds are valid.

Now consider the general case where nodes are selected for the working set based on independent Bernoulli trials with different probabilities. In the algorithms, select working sets for the comparison training-working splits by independent Bernoulli trials over $T \cup W$, using each node's a priori probability to determine whether it is in the working set. Since the working sets in different comparison splits may have different sizes, consider ranking by error rate (error count divided by size of working set) rather than error count to determine which assignments are likely. Whether ranking by error rate or error count, when $a' = a$, the scores for (T, W) and the comparison splits are i.i.d. over selection of (T, W) and the comparison splits. So the error bounds are valid.

3   Experiments

This section describes experiments using the bounds developed in this paper. The experiments are based on a network generated at random using the procedure outlined in Section 5.1 of [5] (with their $m_s = 32$). The network has 1000 nodes and 7882 edges. Each node has one of two labels.

For each of 100 trials, 400 nodes are selected to be the training set and another 16 nodes are selected to be the working set. The selections are uniformly at random and without replacement. The classifier assigns each working node the label of the majority of its neighboring training nodes, with ties broken at random. The test error rate is the fraction of working set examples misclassified.

For each trial, error bounds are computed for $\delta \in \{0.1, 0.2, 0.3\}$, for numbers of comparison working-training sets $m \in \{100, 200\}$, and with and without partitioning. With partitioning, the working set is divided in half, making 8 examples in each partition, and the bounds are computed using the algorithm in Subsection 2.2. Without partitioning, the algorithm in Subsection 2.1 is used.

Over the 100 trials, the test error rate averages 0.216, with standard deviation 0.098. (We state all results to three decimals.) Table 1 shows the averages and standard deviations of differences between bounds and test error rates for the various bound computation methods and values of δ. The table shows sample standard deviations; standard deviations of the estimates of the means are one tenth of these values, indicating that the differences among means are statistically significant.

The unpartitioned bounds are stronger than their partitioned counterparts – the decreased computation bears a price in bound strength. This is partly because δ is halved for each partition, but also because having smaller working sets in the partitions produces weaker bounds. To see this, observe that the partitioned bounds with $\delta = 0.2$ are weaker than the unpartitioned bounds with $\delta = 0.1$. Increasing the sample size m increases bound strength for all but one bound, but the effect is not as large as for partitioning.

Table 1. Bound - Test Error Rate

		unpartitioned		partitioned
m	δ	difference	δ	difference
	0.1	0.170 ± 0.094	0.1	0.284 ± 0.097
100	0.2	0.113 ± 0.094	0.2	0.212 ± 0.099
	0.3	0.075 ± 0.099	0.3	0.171 ± 0.098
	0.1	0.166 ± 0.095	0.1	0.286 ± 0.102
200	0.2	0.108 ± 0.096	0.2	0.206 ± 0.106
	0.3	0.069 ± 0.101	0.3	0.168 ± 0.096

4   Discussion

This paper presents a method to validate network classifiers, computing PAC error bounds for a classifier over working examples in a transductive setting. The method is based on the idea that training and working sets of nodes are interchangeable a priori, so the likelihood of substantial differences between the two sets is small. The method does not depend on which underlying process generated the network. The only requirement is that whether each node is labeled is determined independently from the labeling determination of other nodes. The bounds do not require that labeling has the same probability for each node – using cohorts produces error bounds when nodes in some sets are more likely than others to be in the working set a priori.

One direction for future research is to remove the constraint that nodes are selected independently for the working set. In practice, this constraint may not hold. For example, in a social network, if one person provides the data to label her node, that may encourage her friends to do the same. For some types of correlation, it may be possible to alter the selection of comparison training-working splits, as we do to accommodate cohorts. If a sampling technique, such as snowball sampling [8], is known to mimic how the working set is selected, then perhaps the technique can be used to generate comparably likely training-working splits.

Another challenge for future research is to develop faster algorithms. There is an efficient algorithm to compute error bounds for 1-nearest neighbor classifiers, based on dynamic programming [2]. The method depends on avoiding long cycles of dependency, where the label for one example influences the classification of another, which influences the classification of another, and so on, returning to influence the first example. It may be possible to develop a similar error bound algorithm for network classifiers by partitioning the examples in the working set into subsets in which the examples do not influence one another's classifications. Then the error bounds over subsets could be combined to produce an error bound over the whole working set, as in this paper.

The error bounds developed in this paper apply to the transductive setting, where the working nodes and their links are known. In the future, it would be interesting to extend these bounds to cases where the working nodes are unknown. If the working nodes have a known distribution, that information could be used to sample working nodes to develop samples of bounds. Probabilistic error bounds can be based on the statistics over these samples. Even without a known distribution, the training nodes may be used to estimate a distribution for working nodes, if they are known to be drawn from the same distribution. In some cases, with growing networks, only the most recent training nodes may be drawn from a similar distribution to the next nodes to be drawn, which are the nodes of interest for error bounds.

References

1. Bax, E.: Nearly uniform validation improves compression-based error bounds. Journal of Machine Learning Research 9, 1741–1755 (2008)
2. Bax, E., Callejas, A.: An error bound based on a worst likely assignment. Journal of Machine Learning Research 9, 581–613 (2008)
3. Bollobas, B.: Random Graphs, 2nd edn. Cambridge University Press (2001)
4. Bondy, J.A., Murty, U.: Graph Theory. Springer (2008)
5. Cataltepe, Z., Sonmez, A., Baglioglu, K., Erzan, A.: Collective classification using heterogeneous classifiers. In: 7th International Conference on Machine Learning and Data Mining, MLDM 2011 (2011)
6. Cristianini, N., Shawe-Taylor, J.: An Introduction to Support Vector Machines and Other Kernel-Based Learning Methods. Cambridge University Press (2000)
7. Feller, W.: An Introduction to Probability Theory and Its Applications. John Wiley & Sons, New York (1968)
8. Frank, O.: Survey sampling in graphs. Journal of Statistical Planning and Inference 1, 235–264 (1977)
9. Getoor, L., Friedman, N., Koller, D., Taskar, B.: Learning probabilistic models of link structure. Journal of Machine Learning Research 3, 679–707 (2002)
10. Kolaczyk, E.D.: Statistical Analysis of Network Data. Springer (2010)
11. Macskassy, S., Provost, F.: A simple relational classifier. In: Proceedings of the Multi-Relational Data Mining Workshop (MRDM) at the Ninth ACM SIGKDD International Conference on Knowledge Discovery and Data Mining, Washington, DC, USA, pp. 64–76 (2003)
12. Macskassy, S.A., Provost, F.: Classification in networked data: A toolkit and a univariate case study. Journal of Machine Learning Research 8, 935–983 (2007)

13. Sen, P., Getoor, L.: Empirical comparison of approximate inference algorithms for networked data. In: ICML Workshop on Open Problems in Statistical Relational Learning, SRL 2006 (2006)
14. Sen, P., Namata, G., Bilgic, M., Getoor, L., Gallagher, B., Eliassi-Rad, T.: Collective classification in network data. AI Magazine 29(3) (2008)
15. Vapnik, V.: Statistical Learning Theory. John Wiley & Sons (1998)
16. Watts, D.: Six Degrees: The Science of a Connected Age. Norton & Company (2003)
17. Watts, D., Strogatz, S.: Collective dynamics of 'small-world' networks. Nature 393(6684), 440–442 (1998)
18. Zheleva, E., Getoor, L.: To join or not to join: The illusion of privacy in social networks with mixed public and private user profiles. In: 18th International World Wide Web Conference, pp. 531–531 (April 2009)

Alignment and Morphing
for the Boundary Curves of Anatomical Organs

Keiko Morita[1], Atsushi Imiya[2], Tomoya Sakai[3],
Hidetaka Hontan[4], and Yoshitaka Masutani[5]

[1] School of Advanced Integration Sciences, Chiba University, Japan,
Yayoi-cho 1-33, Inage-ku, Chiba, 263-8522, Japan
[2] Institute of Media and Information Technology, Chiba University, Japan,
Yayoi-cho 1-33, Inage-ku, Chiba, 263-8522, Japan
[3] Department of Computer and Information Science, Nagasaki University,
Bunkyo-cho 1-44, Nagasaki, 852-8521, Japan
[4] Department of Computer Science, Nagoya Institute of Technology, Gokiso,
Showa-ku, Nagoya 466-8555, Aichi, Japan
[5] Department of Radiology, The University of Tokyo Hospital,
Division of Radiology and Biomedical Engineering, Graduate School of Medicine
The University of Tokyo,
Hongo 7-3-1, Bunkyo-ku 113-8655, Tokyo, Japan

Abstract. In this paper, we develop a tracking method for the deformable boundary curves of biological organs using variational registration method. We first define the relative distortion of a pair of curves using curvatures of curves. This minimum distortion aligns corresponding points of a pair of curves. Then, we derive the mean of curves as the curve which minimises the total distortion of a collection of shapes. We compute the intermediate boundary curve of a pair of curves as the mean of these curves.

1 Introduction

Morphing is a fundamental technique in computer graphics to interpolate and generate shapes and objects. In medical application, morphing is used for the description of deformation process of biological organs. This process predicates deformable motion of biological organs in human torso such as beating heart, deformation of lungs during blessing. Follow up analysis of tumors in censer diagnosis tracks and predicates deformation of censer. In this paper, we develop a tracking method for the deformable boundary curves of biological organs using variational registration method. This registration process between images clarifies the difference between images which is used for medical diagnosis. This registration process is mainly achieved by the matching process, which is an established fundamental idea in pattern recognition. In both structure pattern recognition [4, 5] and variational registration [1, 3], the mean shape of a collection of given shapes is interested.

G.L. Gimel' farb et al. (Eds.): SSPR & SPR 2012, LNCS 7626, pp. 458–466, 2012.
© Springer-Verlag Berlin Heidelberg 2012

Shape retrieval categorises and classifies shapes, and finds shapes from portions of shapes. In shape retrieval, the matching of shapes based on the defomorphorism of shapes [8, 9, 16, 19] and descriptor of shape boundary contours [10] are used. In the matching process for discrete shapes, the string edit-distance [4, 6] computed by dynamic programming is a fundamental tool. Moreover, in the matching process of images, the variational registration strategy [1, 2] is a typical tool. Since, in registration of grey-valued images, the deformation is assumed to be relatively small, the point correspondences between the target and reference images are estimated as a local deformation of images [1–3]. For the matching of planar curves, we are required to estimate both alignments and local deformation of curves. In this paper, we separate this problem into alignment estimation [16, 17] in the normalised set of curves and deformation of curves.

2   Alignment of Curves

For a pair of planar curves $c(s) = (c_1(s), c_2(s))^\top$ and $\bar{c}(s) = (\bar{c}_1(s), \bar{c}_2(s))^\top$, whose lengths are C and $\bar{C}$, respectively, assuming $c(s+C) = c(s)$ and $\bar{c}(s+\bar{C}) = \bar{c}(s)$, the alignment of curves is obtained as

$$\text{Align}(c, \bar{c}) = \min_{t,\psi} \int_0^C |\bar{c}(\psi(s) - t) - c(s)|^2 ds, \qquad (1)$$

where $\psi(\tau)$ is a monotone function from the interval $[0, C]$ to the interval $[0, \bar{C}]$ [13, 14, 16]. The function $\tau = \psi(s)$ and the displacement t define the correspondences of points on a pair of curves $c(s)$ and $\bar{c}(\tau)$.

The dynamic time warping (DTW) is a fundamental procedure to achieve curve alignment employing dynamic programming [13]. The time warping sometimes maps a point on a curve to a relatively long interval of another curve. The derivative dynamic time warping technique (DDTW) [14], which computes alignment of derivative curves, solves this pathological mapping. Therefore, we can also use

$$\text{Align}(c, \bar{c}) = \min_{t,\psi} \int_0^C |\dot{\bar{c}}(\psi(s) - t) - \dot{c}(s)|^2 ds, \qquad (2)$$

for the derivative of curves [14] $\dot{c}$ and $\dot{\bar{c}}$.

For a planar curve S_i, the normal curve s_i is the curve whose length are normalised to unity is a normalised curve. For the normal curve $x(s)$ the unit normal vector is $n(s) = (-\sin\theta(s), \cos\theta(s))^\top$, if $\dot{x}(s)/|\dot{x}(s)| = (\cos\theta(s), \sin\theta(s))^\top$. We call $\theta(s)$ the p-expression of the curve. The p-expression is invariant for Euclidean motion, that is, for curve $x(s)$, $y(s) = x(s) + a$ derives the same p-expression.

We define the log measure between two normal curves as

$$H(\theta_1, \theta_2) = \int_0^1 \left| \ln \frac{\exp(i\theta_1(s - t_1))}{\exp(i\theta_2(s - t_2))} \right|^2 ds, \qquad (3)$$

using p-expression of each curve of a pair. Using p-expression of each curve of a pair, the p-distance of a pair of simple polygonal curve S_i and S_j is defined by

$$D(S_i, S_j)^2 = \min H(\theta_i, \theta_j) = \min_t \int_0^1 |\theta_i(s-t) - \theta_j(s)|^2 ds. \qquad (4)$$

where θ_i and θ_j are the p-expressions of the normalised curve s_i and s_j of S_i and S_j, respectively. Then, setting

$$t_{ij} = \arg\left(\min_t \left(\int_0^1 |\theta_i(s-t_{ij}) - \theta_j(s)|^2 ds\right)\right), \qquad (5)$$

we define the alignment of s_i and s_j as $\theta_i(s-t_{ij})$ and $\theta_j(s)$.

The p-expression $\theta(k)$ of a normalised polygonal curve, whose vertices are $\{x_k\}_{k=1}^n$, is computed as

$$\frac{x_k - x_{k-1}}{|x_k - x_{k-1}|} = (\cos\theta(k), \sin\theta(k))^\top. \qquad (6)$$

Furthermore, setting $\theta_i^k = \theta_i(k\Delta)$, the distance between a pair of normalised curves is approximately computed by

$$d_{ij} \approx \min_p \sum_{k=1}^m |\theta_i^k - \theta_j^{k-p}|^2 \qquad (7)$$

for an appropriately large m such that $\Delta m = 1$.

Next, we define the mean $\phi_{ij}(s)$ of a pair of p-expressions $\theta_i(s)$ and $\theta_j(s)$ as the minimiser of the functional

$$\begin{aligned}
J(\phi_i, \phi_j, \phi_{ij}, t_i, t_j) = \int_0^1 &\Big\{ |(\theta_i(s-t_i) - \phi_i(s)) - \phi_{ij}(s)|^2 \\
&+ |(\theta_j(s-t_2) - \phi_j(s)) - \phi_{ij}(s)|^2 \\
&+ \lambda|\dot\theta_{ij}(s)|^2 + \mu|\dot\phi_i(s)|^2 + \mu|\dot\phi_j(s)|^2 \Big\} ds.
\end{aligned} \qquad (8)$$

Equation (8) is converted to the problem,

$$\begin{aligned}
J_2(\phi_{ij}) = \int_0^1 &\Big\{ |(\theta_i(s-t_{ij}) - \phi_i(s) - \phi_{ij}(s))|^2 \\
&+ |(\theta_j(s) - \phi_j(s) - \phi_{ij}(s)|^2 \\
&+ \lambda|\dot\phi_{ij}(s)|^2 + \mu|\dot\phi_i(s)|^2 + \mu|\dot\phi_j(s)|^2 \Big\} ds
\end{aligned} \qquad (9)$$

since t_{ij} aligns a pair of p-expressions, for a generalisation of eq. (9), the initial points of a collection of curves are required to be aligned.

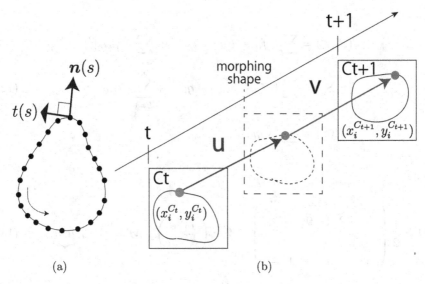

Fig. 1. Geometric property of the image boundary of curve. (a) Configuration of the normal and tangent vectors on a planar simple curve. (b) Morphing of temporal curves.

3   The Mean of Polygonal Curves

Assuming that the correspondence of the vertices of curves S_i and S_j in a collection of curves $\{S_\alpha\}_\alpha^n$ are established by minimising eq. (3), we define the distance between a pair of polygonal curves $S_i = \{f_{ik}\}_{k=1}^n$ and $S_j = \{f_{jk}\}_{k=1}^n$ where $f_{\cdot k} = (x_{\cdot k}, y_{\cdot k})^\top$ is the vertex of the curve $S_\cdot$ with the condition $f_{\cdot m+k} = f_{\cdot k}$, as

$$
\begin{aligned}
d(S_i, S_j) &= \min_{\boldsymbol{u}_k^{ij}} \left\{ \sum_{k=1}^m \{(f_{ik} - \boldsymbol{u}_k^{ij}) - f_{jk}\}^2 + \mu \sum_{k=1}^m (\overline{\nabla} \boldsymbol{u}_k^{ij})^2 \right\} \\
&= \min_{\boldsymbol{u}_k^{ij}} \left\{ \sum_{k=1}^m \{f_{ik} - (f_{jk} - \boldsymbol{u}_k^{ji})\}^2 + \mu \sum_{k=1}^m (\overline{\nabla} \boldsymbol{u}_k^{ji})^2 \right\} \\
&= d(S_j, S_i),
\end{aligned}
\tag{10}
$$

where $\boldsymbol{u}_k^{ij} = -\boldsymbol{u}_k^{ji}$ is the displacement between f_{ik} and f_{jk} and $\overline{\nabla}$ stands for the discrete differential operation along a polygonal curve [1].

Definition 1. *Setting $\boldsymbol{u}_{ki}$ to be deformation of the vertex f_{ki} of the shape S_i, the vertices $\boldsymbol{g}$ of the mean curve S of S_i and S_j is the minimiser of the discrete variational problem*

$$
J(S_i(f_i), S(g), S_j(f_j)) = d(S_i, S) + d(S, S_j) + \lambda P(S)
\tag{11}
$$

---

[1] $\overline{\nabla} g_k = g_{k+\frac{1}{2}} - g_{k-\frac{1}{2}}$ and $\overline{\nabla}^2 g_k = \frac{1}{2}(g_{k+1} - 2g_k + g_{k-1})$.

where

$$d(S_i, S) = \sum_{k=1}^{n} \{(f_{ik} - u_{ik}) - g_k\}^2 + \sum_{k=1}^{n} |\overline{\nabla}|u_{ik}|^2,$$

$$d(S, S_j) = \sum_{k=1}^{n} \{(f_{jk} - u_{jk}) - g_k\}^2 + \sum_{k=1}^{n} |\overline{\nabla}|u_{jk}|^2, \tag{12}$$

$$P(S) = \sum_{k=1}^{n} |\overline{\nabla} g_k|^2.$$

Setting

$$D = \frac{1}{2} \begin{pmatrix} -2 & 1 & 0 & \cdots & 0 & 1 \\ 1 & -2 & 1 & 0 & \cdots & 0 \\ \vdots & & & & & \\ 1 & 0 & \cdots & 0 & 1 & -2 \end{pmatrix}, \quad A = \begin{pmatrix} I_2 & e \\ e^\top & m \end{pmatrix}, \quad M = Diag(\underbrace{\mu, \mu, \lambda}). \tag{13}$$

where $e = (1,1)^\top$, we have the Euler-Lagrange equation of eq. (13)

$$(I_{m+1} \otimes D)s = (M^{-1}A \otimes I_2)s - (M^{-1} \otimes I_n)c, \tag{14}$$

where I_k is the $k \times k$ identity matrix and

$$s = (u_i^\top, u_j^\top, g^\top)^\top, \quad c = (f_i^\top, f_j^\top, f_i^\top + f_j^\top)^\top. \tag{15}$$

Rewriting eq. (14) as

$$Bs = Ks - c, \tag{16}$$

the semi-implicit discretisation of the diffusion equation

$$\frac{\partial s}{\partial t} = Bs - Ks + c \tag{17}$$

derives the iteration form

$$(I + \tau K)s^{(k+1)} = (I + \tau B)s^{(k)} + \tau c. \tag{18}$$

This iteration form implies that

$$g = (\underbrace{0_n, \cdots, 0_n}_{m}, I_n)s^{(\infty)}, \quad s^{(\infty)} = \lim_{k \to \infty} s^{(k)}. \tag{19}$$

Setting u_{ti} and $u_{(t+1)i}$ to be deformations from the vertices f_{ti} of the mean curve $S_{t+\frac{1}{2}}$ to vertices f_{ti} and $f_{(t+1)i}$ of S_t and S_{t+1}, respectively, we define an interframe curve. Fig. 2 shows morphing and tracking of a temporal curve sequence using variational mean curve.

Definition 2. *The vertices* $\boldsymbol{f}_{t+\frac{1}{2}}$ *of the intermediate shape* $S_{t+\frac{1}{2}}$ *is the minimiser of the discrete variational problem*

$$J(S_t(\boldsymbol{f}_t), S_{t+\frac{1}{2}}(\boldsymbol{f}_{t+\frac{1}{2}}), S_{t+1}(\boldsymbol{f}_{t+1})) = D(S_t, S_{t+\frac{1}{2}}) + D(S_{t+\frac{1}{2}}, S_{t+1}) + \lambda P(S_{t+\frac{1}{2}}) \tag{20}$$

where

$$D(S_t, S_{t+\frac{1}{2}}) = \sum_{k=1}^{n} \{(f_{ik} - \boldsymbol{u}_{ik}) - \boldsymbol{g}_k\}^2 + \sum_{k=1}^{n} |\overline{\nabla}|\boldsymbol{u}_{ik}|^2,$$

$$D(S_{t+\frac{1}{2}}, S_{t+1}) = \sum_{k=1}^{n} \{(f_{jk} - \boldsymbol{u}_{jk}) - \boldsymbol{g}_k\}^2 + \sum_{k=1}^{n} |\overline{\nabla}|\boldsymbol{u}_{jk}|^2, \tag{21}$$

$$P(S_{t+\frac{1}{2}}) = \sum_{k=1}^{n} |\overline{\nabla}\boldsymbol{g}_k|^2.$$

4   Numerical Examples

Fig. 2 shows Images of the boundary curves of tumors observed in the years, 1996, 19980, 2000, 2002, and 2004.

Fig. 3 comparative results between the means and the linear averages of the years Y. Y^*, and Y° express the images of the year Y computed by our method and linear average of corresponding vertices of the curve. In Fig. 4, we have evaluated the distances $D(Y, Y^*)$ and $D(Y, Y^\circ)$. These results show our method derives smooth intermediate curves, by computing the alignment of corresponding vertices and by minimising the relative distortion of curves.

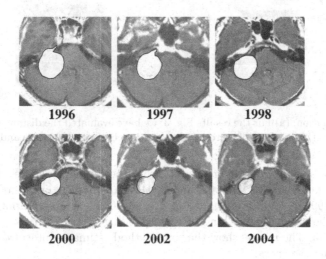

Fig. 2. Images of the boundary curves of tumors. The boundary curves observed in the years, 1996, 19980, 2000, 2002, and 2004.

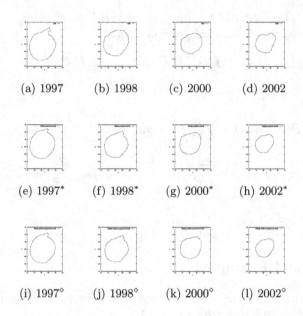

(a) 1997   (b) 1998   (c) 2000   (d) 2002

(e) 1997*   (f) 1998*   (g) 2000*   (h) 2002*

(i) 1997°   (j) 1998°   (k) 2000°   (l) 2002°

Fig. 3. Performance evaluation. from top to bottom the original images, the means, and the linear averages of the years Y. Y^*, and Y° express the images of the year Y computed by our method and linear average of corresponding vertices of the curve.

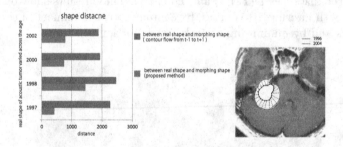

Fig. 4. Evaluation. (a) For the results Fig. 3, we have evaluated the distances $D(Y, Y^*)$ and $D(Y, Y^\circ)$. (b) Tracking of corresponding vertices between 1996 data and 2004 data.

Fig.5 shows shape morphing results. From the boundary curves observe by the years, 1996, 1998, 2000, 2002, and 2004, the boundary curves of the years $1996 + \frac{1}{2}$, $1997 + \frac{1}{2}$, $1998 + 1$, $2000 + 1$, and $2002 + 1$ are computed. Fig. 4 (b) shows the result for tracking of corresponding vertices between 1996 data and 2004 data. The results show that our method estimates intermediate tumor shapes during the therapy.

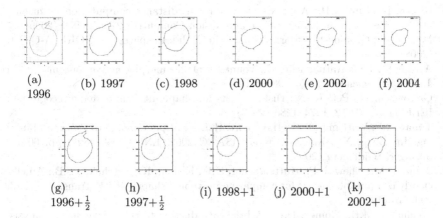

(a)
1996   (b) 1997   (c) 1998   (d) 2000   (e) 2002   (f) 2004

(g)
$1996+\frac{1}{2}$ (h)
$1997+\frac{1}{2}$ (i) 1998+1   (j) 2000+1   (k)
2002+1

Fig. 5. Shape morphing results. from the boundary curves observe by the years, 1996, 19980, 2000, 2002, and 2004, the boundary curves of the years $1996+\frac{1}{2}$, $1997+\frac{1}{2}$, $1898+1$, $2000+1$, and $2002+1$ are computed.

5   Conclusions

We first define the relative distortion of a pair of curves using curvatures of curves. This minimum distortion aligns corresponding points of a pair of curves. Then, we derive the mean of curves as the curve which minimises the total distortion of a collection of shapes. We compute the intermediate boundary curve of a pair of curves as the mean of these curves.

Our method automatically detects and tracks corresponding vertices of the temporal-deformation curves.

This research was supported by "Computational anatomy for computer-aided diagnosis and therapy: Frontiers of medical image sciences" funded by the Grant-in-Aid for Scientific Research on Innovative Areas, MEXT, Japan, the Grants-in-Aid for Scientific Research funded by Japan Society of the Promotion of Sciences and the Grant-in-Aid for Young Scientists (A), NEXT, Japan.

References

1. Hill, D.L.G., et al.: Medical image registration. Phys. Med. Biol. 46, R1–R45 (2001)
2. Fischer, B., Modersitzki, J.: Ill-posed medicine- an introduction to image registration. Inverse Problem 24, 1–17 (2008)
3. Rumpf, M., Wirth, B.: A nonlinear elastic shape averaging approach. SIAM Journal on Imaging Sciences 2, 800–833 (2009)
4. Sebastian, T.B., Klein, P.N., Kimia, B.B.: On aligning curves. IEEE Trans. PAMI 25, 116–125 (2003)
5. Baeza-Yates, R., Valiente, G.: An image similarity measure based on graph matching. In: Proc. 7th Int. Symp. String Processing and Information Retrieval, pp. 8–38 (2000)

6. Riesen, K., Bunke, H.: Approximate graph edit distance computation by mean s of bipartite graph matching. Image and Vision Computing 27, 950–959 (2009)
7. Mémoli, F.: Gromov-Hausdorff distances in Euclidean spaces. In: NORDIA-CVPR (2008)
8. Arrate, F., Tilak Ratnanather, J., Younes, L.: Diffeomorphic active contours. SIAM J. Imaging Sciences 3, 176–198 (2010)
9. Grigorescu, C., Petkov, N.: Distance sets for shape filters and shape recognition. IEEE Trans. IP 12, 1274–1286 (2003)
10. Tănase, M., Veltkamp, R.C., Haverkort, H.J.: Multiple Polyline to Polygon Matching. In: Deng, X., Du, D.-Z. (eds.) ISAAC 2005. LNCS, vol. 3827, pp. 60–70. Springer, Heidelberg (2005)
11. Arkin, E.M., Chew, L.P., Huttenlocher, D.P., Kedem, K., Mitchell, J.S.B.: An efficiently computable metric for comparing polygonal shapes. IEEE Trans. PAMI 13, 209–216 (1991)
12. Stegmann, M.B., Gomez, D.D.: A brief introduction to statistical shape analysis. Informatics and Mathematical Modelling, Technical University of Denmark (2002), http://www2.imm.dtu.dk/pubdb/p.php?403
13. Müller, M.: In: Information Retrieval for Music and Motion, ch. 4. Springer (2007)
14. Keogh, E.J., Pazzani, M.J.: Derivative dynamic time warping. In: First SIAM International Conference on Data Mining, SDM 2001 (2001), http://www.cs.ucr.edu/~eamonn/
15. Srivastava, A., Joshi, S., Mio, W., Liu, X.: Statistical shape analysis: Clustering, learning, and testing. IEEE Trans. PAMI 27, 590–602 (2005)
16. Sebastian, T.B., Klein, P.N., Kimia, B.B.: On aligning curves. PAMI 25, 116–125 (2003)
17. Marques, J.S., Abrantes, A.J.: Shape alignment? optimal initial point and pose estimation. Pattern Recognition Letters 18 (1997)
18. Kass, M., Witkin, A., Terzopoulos, D.: Snakes: Active contour models. IJCV 1, 321–331 (1988)
19. Sharon, E., Mumford, D.: 2D-shape analysis using conformal mapping. IJCV 70, 55–75 (2006)
20. Mumford, D., Shah, J.: Boundary detection by minimizing functionals. In: Proc. CVPR 1985, pp. 22–26 (1985)
21. Mumford, D., Shah, J.: Optimal approximations by piecewise smooth functions and associated variational problems. Comm. on Pure and Applied Math. bXLII, 577–684 (1989)

Unsupervised Clustering of Human Pose
Using Spectral Embedding

Muhammad Haseeb and Edwin R. Hancock*

Department of Computer Science, The University of York, UK

Abstract. In this paper we use the spectra of a Hermitian matrix and the coefficient of the symmetric polynomials to cluster different human poses taken by an inexpensive $3D$ camera, the Microsoft 'Kinect' for XBox 360. We construct a Hermitian matrix from the joints and the angles subtended by each pair of limbs using the three-dimensional 'skeleton' data delivered by Kinect. To compute the angles between a pair of limbs we construct the line graph from the given skeleton. We construct pattern vectors from the eigenvectors of the Hermitian matrix. The pattern vectors are embedded into a pattern-space using Principal Component Analysis (PCA). We compere the results obtained with the Laplacian spectra pattern vectors. The empirical results show that using the angular information can be efficiently used to clusters different human poses.

1 Introduction

Graph partitioning/clustering and classification is one of the most extensively studied topics in computer vision and machine learning community. Clustering is closely related to unsupervised learning in pattern recognition systems. Graphs are structures formed by a set of vertices called nodes and a set of edges that are connections between pairs of nodes. Graph clustering is grouping similar graphs based on structural similarity within clusters. Bunke et al. [1] proposed a structural method referred to as the Weighted Minimum Common Supergraph (WMCS), for representing a cluster of patterns. There has been significant amount of work aimed at using spectral graph theory [2] to cluster graphs. This work shows the common feature of using graph representations of the data for the graph partitioning. Luo et al. [3] have used the discriminatory qualities of a number of features constructed from the graph spectra. Using the leading eigenvalues and eigenvectors of the adjacency matrix they found that the leading eigenvalues have the best capabilities for structural comparison. There are a number of examples of applying pairwise clustering methods to graph edit distances [4]. Recently, the properties of the eigenvectors and eigenvalues of the Laplacian matrix of graph have been exploited in many areas of computer vision. For instance, Shi and Malik [5] used the eigenvector corresponding to second smallest (none zero) eigenvalue (also called Fielder vector) of the Laplacian matrix to iteratively bi-partition the graph for image segmentation. The information encoded in the eigenvectors of the Laplacian has been used for shape registration [6] and clustering. Veltkamp et al. [7] developed a shape retrieval method using a complex Fielder vector of a Hermitian property matrix. Recent spectral approaches

---

* Edwin R. Hancock is supported by a Royal Society Wolfson Research Merit Award.

use the eigenvectors corresponding to the k smallest eigenvalues of the Laplacian matrix to embed the graph onto a k dimensional Euclidian space [8], [9].

In this paper we propose a clustering method using the angular information and the distance between each pair of joints, from the skeleton extracted from the Microsoft Kinect $3D$ sensor [10]. We construct a Hermitian matrix using the distance as real part and the angles between each pair of limb as imaginary part. We use the spectra of the Hermitian matrix to cluster similar human poses. We construct a feature vector from the eigenvalues and eigenvectors of the Hermitian matrix of the graph. Once the feature-vectors for all the poses are to hand, we subject these vectors to Principal Component Analysis (PCA).

The remainder of the paper is organized as follows. In Section 2 the Hermitian matrix is defined. The symmetric polynomials are briefly reviewed in Section 3. Section 4 details the construction of the feature vectors. Experimental results are provided in Section 5 and finally Section 6 concludes the paper.

2   Complex Laplacian (Hermitian) Matrix

A Hermitian matrix H (or self-adjoint matrix) is a square matrix with complex elements that remains unchanged under the joint operation of transposition and complex conjugation of the elements. That is, the element in the ith row and j^{th} column is equal to the complex conjugate of the element in the jth row and i^{th} column, for all indices i and j, i.e. $a_{i,j} = \overline{a}_{j,i}$. Complex conjugation is denoted by the dagger operator † i.e. $H^\dagger = H$. Hermitian matrices can be viewed as the complex number extension of the symmetric matrix for real numbers. The on-diagonal elements of a Hermitian matrix are necessarily real quantities. Each off-diagonal element is a complex number which has two components, and can therefore represent a 2-component measurement.

To create a positive semi-definite Hermitian matrix of a graph, there should be some constraints applied on the measurement representations. Let $\{x_1, x_2, ..., x_n\}$ be a set of measurements for the node-set $\mathcal{V}$ and $\{y_{1,2}, y_{1,2}, ..., y_{n,n}\}$ be the set of measurements associated with the edges of the graph, in addition to the graph weights. Each edge then has a pair of observations $(\mathcal{W}_{a,b}, y_{a,b})$ associated with it. There are a number of ways in which the complex number $H_{a,b}$ could represent this information, for example with the real part as $\mathcal{W}$ and the imaginary part as y. However, here we follow Wilson, Hancock and Luo [11] and construct the complex property matrix so as to reflect the Laplacian. As a result the off-diagonal elements of H are chosen to be $H_{a,b} = -\mathcal{W}_{a,b}e^{\iota y_{a,b}}$. The edge weights are encoded by the magnitude of the complex number $H_{a,b}$ and the additional measurement by its phase. By using this encoding, the magnitude of the number is the same as the original Laplacian matrix. This encoding is suitable when measurements are angles, satisfying the conditions $-\pi \leq y_{a,b} < \pi$ and $y_{a,b} = -y_{a,b}$ to produce a Hermitian matrix. To ensure a positive definite matrix, H_{aa} should be greater than $-\Sigma_{b \neq a}|H_{ab}|$. This condition is satisfied if $H_{aa} = x_a + \Sigma_{b \neq a}\mathcal{W}_{a,b}$ and $x_a \geq 0$. When defined in this way the property matrix is a complex analogue of the weighted Laplacian matrix for the graph.

For a Hermitian matrix there is an orthogonal complete basis set of eigenvectors and eigenvalues i.e. $H\phi = \lambda\phi$. The eigenvalues λ_i of Hermitian matrix are real while the eigenvectors ϕ_i are complex. There is a potential ambiguity in the eigenvectors, in that any multiple of an eigenvector is a solution of the the eigenvector equation $H\phi = \lambda\phi$. i.e. $H\alpha\phi = \lambda\alpha\phi$. Therefore, we need two constraints for them. Firstly, make each eigenvector of unit length vector i.e. $|\phi_i| = 1$, and secondly impose the condition $\arg \sum_i \phi_{ij} = 0$.

3   Symmetric Polynomials

A symmetric polynomial is a polynomial $P(x_1, x_2, \ldots, x_n)$ in n variables, such that if any of the variables are interchanged, the same polynomial is obtained. A symmetric polynomial is invariant under permutation of the variable indices. There is a special set of symmetric polynomials referred to as the *elementary symmetric polynomial (S)* that form a basis set for symmetric polynomial. Any symmetric polynomial can be expressed as a polynomial function of the elementary symmetric polynomials. For a set of variables $x_1, x_2, \ldots, x_n$ the elementary symmetric polynomials can be defined as:

$$S_1(v_1, v_2, \ldots, v_n) = \sum_{i=1}^{n} v_i$$

$$S_2(v_1, v_2, \ldots, v_n) = \sum_{i=1}^{n} \sum_{j=i+1}^{n} v_i v_j$$

$$\vdots$$

$$S_n(v_1, v_2, \ldots, v_n) = \prod_{i=1}^{n} v_i$$

The power symmetric polynomial functions (P) defined as

$$P_1(v_1, v_2, \ldots, v_n) = \sum_{i=1}^{n} v_i$$

$$P_2(v_1, v_2, \ldots, v_n) = \sum_{i=1}^{n} v_i^2$$

$$\vdots$$

$$P_n(v_1, v_2, \ldots, v_n) = \sum_{i=1}^{n} v_i^n$$

The elementary symmetric polynomials can be efficiently computed using the power symmetric polynomials using the Newton-Girard formula

$$S_r = \frac{(-1)^{r+1}}{r} \sum_{k=1}^{r} (-1)^{k+r} P_r S_{r-k} \tag{1}$$

here the shortcut S_r is used for $S_r(v_1, v_2, \ldots, v_n)$ and P_r is used for $P_r(v_1, v_2, \ldots, v_n)$.

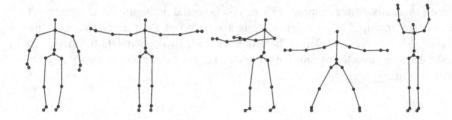

Fig. 1. Poses for Experiments

4   Feature Vectors

The skeleton of human body with twenty, 3-dimensional points representing the joints connected by the lines representing the limbs, is acquired using the Microsoft Kinect SDK. Kinect provides the skeletal data with the rate of 30 frames per second. Figure 1 shows examples of the skeletons captured with the Kinect sensor. Each point in the skeleton is represented by a three dimensional vector $w_i = (x_i, y_i, z_i)^T$.

We used the limb joint angles and the limb length assigned by the Microsoft Kinect SKD. We convert the skeleton to its equivalent line graph. The line graph of undirected graph G is another graph that represents the adjacency between edges of G. The nodes in the line graph represents the edges of the original graph G. We construct a Hermitian matrix from the difference between the lengths of each pair of edges and the angles subtended by those edges. Given two adjacent edges e_i and e_j, with the nodes w_{k-1}, w_k and w_{k+1}, where w_k is the common node. The angle between the edges e_i and e_j is given by

$$\theta_{ij} = \frac{\cos((w_k - w_{k-1})^T(w_k - w_{k+1}))}{||(w_k - w_{k-1})|| \times ||(w_k - w_{k+1})||} \tag{2}$$

The Hermitian matrix H has element with row index i and column index j is

$$H(i,j) = -\mathcal{W}_{i,j}e^{\iota\theta_{i,j}} \tag{3}$$

where $\mathcal{W}_{i,j}$ is the difference of the lengths of the edges e_i and e_j and $\theta_{i,j}$ is the angle between the edges e_i and e_j. To obey the antisymmetric condition $\theta_{i,j} = -\theta_{j,i}$ we multiply $\theta_{i,j}$ with -1 if length of edge $e_i > e_j$.

With the complex matrix H to hand, we compute its eigenvalues and eigenvectors. The eigenvector of a Hermitian matrix are complex and the eigenvalues are real. We order the eigenvectors according to the decreasing magnitude of the eigenvalues i.e. $|\lambda_1| > |\lambda_2| > \ldots > |\lambda_n|$. From the eigenvectors the symmetric polynomial coefficients are computed by first computing the power symmetric polynomial. From the power symmetric polynomials elementary symmetric polynomials are computed using the Newton-Girard formula [11] (equation 1) as described in Section 3. We take only the first ten coefficients as the rest of the coefficients approach to zero because of the product terms appearing in the higher order polynomials. Since the components of the eigenvector are complex numbers, therefore each symmetric polynomial coefficient is also complex. The real and imaginary components of symmetric polynomials are inter-leaved and stacked to form a long feature vector F_i for the graph representing the pose frame.

5   Experimental Results

In this section, we provide some experimental investigations of the clustering of differ-ent human poses. We focus on its use in two different settings. In the first setting we choose five different poses for the experiment which are shown in Figure 1. We take 200 different instances of each pose. We construct the feature vectors using a complex Laplaican property matrix detailed in Section 2. We embed the graph feature vectors into a three dimensional pattern-space by performing the PCA for visualization. Figure 3(a) shows the result of the clustering using the first three eigenvectors. We compare our clustering result with the clustering result of the Laplacian spectral pattern vectors [12]. Figure 3(c) shows the result of the clustering using the Laplacian spectral pattern vectors.

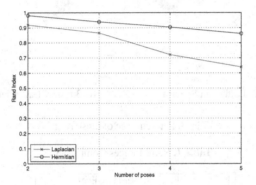

Fig. 2. Rand Indices Comparison

Under the second setting we choose first three poses shown in Figure 1 and take 100 different instances of each pose. We construct the feature vectors according to the steps mentioned in Section 4. We then embed the feature vectors into a three dimensional pattern-space by performing the PCA. Figure 3(b) shows the result of the clustering using the fist three eigenvectors. We compare the result with the result of the Lapla-cian spectral pattern vectors. Figure 3(d) shows the result of the clustering using the Laplacian spectral pattern vectors.

The Laplacian spectral pattern vectors are formed by taking the second smallest through to the nineteenth smallest eigenvalues of graph Laplacian as components.

Table 1 shows the Rand indices obtained when clustering is attempted using differ-ent number of poses. The first row shows the Rand indices obtained using the Laplacian spectral pattern vectors (referred to as Laplacian), while the second row shows the Rand indices obtained using the pattern vectors detailed in Section 4 (referred to as Hermi-tian). The same statistics have been shown in the Figure 2 visually which shows that the clustering results using the angular information is better than the that of the Laplacian spectral pattern vectors.

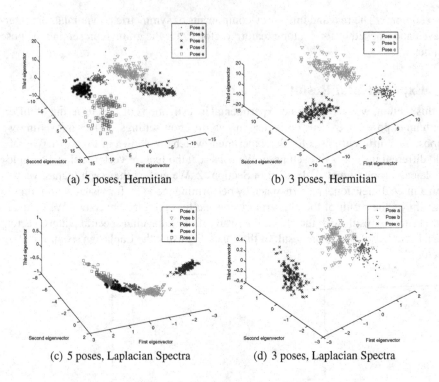

(a) 5 poses, Hermitian   (b) 3 poses, Hermitian

(c) 5 poses, Laplacian Spectra   (d) 3 poses, Laplacian Spectra

Fig. 3. Performance of clustering

Table 1. Rand Indices Comparison

	Rand Indices			
# of poses	2	3	4	5
Laplacian	0.9189	0.8659	0.7212	0.6393
Hermitian	0.9816	0.9395	0.9042	0.8617

6 Conclusion and Future Work

In this paper we construct feature vectors for the Microsoft Kinect skeletal data from the spectra of a Hermitian property matrix employing the angle between the limbs and the lengths of the limbs. The empirical results show that the angular information clusters different poses efficiently. In future, we would like to extend the Hermitian property matrix to four components complex number representation known as quaternion.

Acknowledgement. Edwin R. Hancock was supported by a Royal Society Wolfson Research Merit Award.

References

1. Bunke, H., Foggia, P., Guidobaldi, C., Vento, M.: Graph Clustering Using the Weighted Minimum Common Supergraph. In: Hancock, E.R., Vento, M. (eds.) GbRPR 2003. LNCS, vol. 2726, Springer, Heidelberg (2003)
2. Chung, F.R.K.: Spectral Graph Theory. American Mathematical Society (1997)
3. Luo, B., Wilson, R.C., Hancock, E.R.: Spectral Feature Vectors for Graph Clustering. In: Hancock, E.R., Vento, M. (eds.) GbRPR 2003. LNCS, vol. 2726, pp. 190–201. Springer, Heidelberg (2003)
4. Pavan, M., Pelillo, M.: Dominant sets and hierarchical clustering. In: Proceedings 9th IEEE Conference on Computer Vision and Pattern Recognition I, pp. 362–369 (2003)
5. Shi, J., Malik, J.: Normalized cuts and image segmentation. IEEE Transactions on Pattern Analysis and Machine Intelligence 22, 888–905 (2000)
6. Mateus, D., Cuzzolin, F., Horaud, R.P., Boyer, E.: Articulated shape matching using laplacian eigenfunctions and unsupervised point registration. In: Proceedings of the IEEE Conference on Computer Vision and Pattern Recognition, CVPR (2008)
7. van Leuken, R.H., Symonova, O., Veltkamp, R.C., De Amicis, R.: Complex Fiedler Vectors for Shape Retrieval. In: da Vitoria Lobo, N., Kasparis, T., Roli, F., Kwok, J.T., Georgiopoulos, M., Anagnostopoulos, G.C., Loog, M. (eds.) S+SSPR 2008. LNCS, vol. 5342, pp. 167–176. Springer, Heidelberg (2008)
8. Ng, A.Y., Jordan, M.I., Weiss, Y.: On spectral clustering: Analysis and an algorithm. In: Advances in Neural Information Processing Systems (2002)
9. Yu, S.X., Shi, J.: Multiclass spectral clustering. In: International Conference on Computer Vision (2003)
10. Microsoft: MS Kinect for XBOX 360, http://www.xbox.com/kinect
11. Wilson, R.C., Hancock, E.R., Luo, B.: Pattern vectors from algebraic graph theory. IEEE Transactions on Pattern Analysis and Machine Intelligence 27, 1112–1124 (2005)
12. Luo, B., Wilson, R.C., Hancock, E.R.: Spectral embedding of graphs. Pattern Recognition 36, 2213–2223 (2002)

Human Action Recognition in Video
by Fusion of Structural and Spatio-temporal Features

Ehsan Zare Borzeshi[1], Oscar Perez Concha[2], and Massimo Piccardi[1]

[1] School of Computing and Communications, Faculty of Engineering and IT,
University of Technology, Sydney (UTS), Sydney, Australia
{ehsan.zareborzeshi, massimo.piccardi}@uts.edu.au
[2] Centre for Health Informatics, Australian Institute of Health Innovation,
University of New South Wales, Sydney (UNSW), Australia
o.perezconcha@unsw.edu.au

Abstract. The problem of human action recognition has received increasing attention in recent years for its importance in many applications. Local representations and in particular STIP descriptors have gained increasing popularity for action recognition. Yet, the main limitation of those approaches is that they do not capture the spatial relationships in the subject performing the action. This paper proposes a novel method based on the fusion of global spatial relationships provided by graph embedding and the local spatio-temporal information of STIP descriptors. Experiments on an action recognition dataset reported in the paper show that recognition accuracy can be significantly improved by combining the structural information with the spatio-temporal features.

Keywords: Graph, Graph embedding, Human action recognition, STIP, Markov models.

1 Introduction and Related Work

Human action recognition has been the focus of much recent research for its increasing importance in applications such as video surveillance, human-computer interaction, multimedia and others. Recognising human actions is a challenging task, especially when the background is not fixed or known and the lighting conditions are changeable. Local representations and in particular appearance descriptors centred around spatio-temporal interest points (STIPs) [1] have gained increasing popularity for action recognition since they describe salient points in space and time and have demonstrated strong recognition performance. Nevertheless, spatio-temporal features may fail when the activities become complex since they are unable to capture the global spatial relationships in the subject performing the action [2]. Conversely, graphs are a powerful tool to represent structured objects and as such have been used for action recognition in a recent work from Ta *et al* [3]. Nevertheless, in [3] graphs are directly compared to assess the similarity of two action instances, a procedure that is prone to significant noise. An efficient alternative to the direct comparison of action graphs is offered by graph embedding [4]: in each frame, the graph representing the actor's shape can be converted to a finite set of distances from prototype graphs, and the distance vector then used as

G.L. Gimel' farb et al. (Eds.): SSPR & SPR 2012, LNCS 7626, pp. 474–482, 2012.

a feature vector with conventional statistical classifiers. Other approaches leveraging on a graphical representation of the actor are based on models akin to Pictorial Structures [5]. Such models were originally proposed for limb motion tracking and require higher resolution imagery to ensure accurate fitting. In all cases, purely structural approaches do not take advantage of the useful information offered by spatio-temporal appearance descriptors.

In this paper, we introduce a novel framework for the fusion of the structural information provided by graph embedding and the spatio-temporal information given by STIP descriptors, thus benefitting from both powerful representations and overcoming their respective limitations. Experiments are performed over the popular dataset KTH [6].

The remainder of this paper is organised as follows. Firstly, in section 2 we define the feature set used in our framework. The proposed approach is then described in section 3. In section 4, we present an experimental evaluation of the proposed method on the KTH action dataset. Finally, conclusions and discussion of future work (section 5).

2 Features

The following section provides a description of the *structural* and *spatio-temporal* features provided by graph embedding and typical descriptors such as those extracted from STIPs, respectively.

2.1 Structural Features

Graphs can represent many patterns very effectively by adjusting the graphs' complexity to that of the patterns. However, their main limitation is that they are computationally cumbersome for pattern analysis. One method of circumventing this problem is that of transforming the graphs into a vector space by means of graph embedding. This section briefly provides an overview of prototype-based graph embedding and then describes its use for incorporating structural information into feature vectors.

Overview of Prototype-Based Graph Embedding
In this work, we avail of the definition of *attributed graph*, noted as $g = (V, E, \alpha, \beta)$ with:

- $V = \{1, 2, ..., M\}$, a set of vertices (nodes),
- $E \subseteq (V \times V)$, a set of edges,
- $\alpha : V \to L_V$, a vertex labeling function, and
- $\beta : E \to L_E$, an edge labeling function.

Vertex and edge labels are restricted to fixed-size tuples, ($L_V = \mathbb{R}^p$, $L_E = \mathbb{R}^q$, p, $q \in \mathbb{N} \cup \{0\}$). When attributed graphs are used to represent objects, the problem of pattern recognition changes to that of graph matching. One of the most widely used methods for error-tolerant graph matching is the graph edit distance (GED), defined as the cost of a transformation "morphing" a given graph into another [7]. GED measures the (dis)similarity of arbitrarily structured and arbitrarily labeled graphs and is flexible thanks to its ability to cope with any kind of structural errors [7]. The edit transformation is usualy broken up into atomic edit operations which can be of six basic types: insertion, deletion and substitution, for either nodes or edges, and noted as $(e^{i,n}, e^{d,n}, e^{s,n}, e^{i,e}, e^{d,e}, e^{s,e})$. It can be proven that every arbitrary graph can be transformed into another, equally arbitrary graph by applying a finite sequence of edit operations (also called an *edit path*). The distance between the two graphs is defined as the minimum cost amongst all edit paths transforming the first graph into the other. Let $g_i = (V_i, E_i, \alpha_i, \beta_i)$ and $g_j = (V_j, E_j, \alpha_j, \beta_j)$ be a pair of graphs in a set. The graph edit distance of such graphs is formally defined as:

$$d(g_i, g_j) = \min_{(e_1,...,e_k) \in E(g_i, g_j)} \sum_{l=1}^{k} C(e_l) \tag{1}$$

where $E(g_i, g_j)$ denotes the set of edit paths between the two graphs, C denotes the edit cost function and e_l denotes the individual edit operation. Based on (1), the problem of evaluating the structural similarity of two graphs is changed into that of finding a minimum-cost edit path between them. Among the various methods, the *probabilistic graph edit distance* (P-GED) proposed in [8] is capable of automatically inferring the cost function from a training set of manually-paired graphs. P-GED measures the similarity of two graphs by a learned probability, $p(g_i, g_j)$, and defines the dissimilarity measure as: $d(g_i, g_j) = -\log p(g_i, g_j)$. A further advantage of P-GED is its claimed ability to learn from large sets of graphs with huge distortion between samples of the same class, which makes it suitable for application to vision problems [8].

Graph Embedding. In the literature, "graph embedding" refers interchangeably to the embedding of a graph as a whole into a point in vector space, or the embedding of its set of nodes into a set of corresponding points in vector space. In this work, we assume the former meaning, although similar embedding techniques can be applied in the two cases and for other types of non-vectorial objects such as strings or trees [9]. The embedding assumes that a set of objects is given alongside distance values between any two objects in the set. The goal is that of converting the set of objects into a set of points in a vector space of given dimensionality while ensuring certain properties or constraints. Well-known embedding techniques include Laplacian eigenmaps, commute times, symmetric polynomials, amongst others [10], [11], [12]. After the embedding of the initial set of objects, it is also possible to embed new, out-of-sample objects, albeit not always straightforward. An alternative embedding approach is to make use of a given set of "prototype" objects (or prototypes, for short) which can equally embed in-sample and out-of-sample data, in a way not unlike that of eigenvectors in principal component analysis. Let $G = \{g_1, g_2, ..., g_m\}$ be a set of graphs, $P = \{p_1, p_2, ..., p_n\}$ be a set of prototype graphs with $n < m$, and d be a dissimilarity measure. For embedding any graph $g_j \in G$ by way of P, the dissimilarity measure $d_{ji} = d(g_j, p_i)$ of graph g_j to prototype $p_i \in P$ is computed $\forall i$. Then, an n-dimensional vector $(d_{j1}, ..., d_{jn})$ is assembled from all the n dissimilarities. With this procedure, any graph can be individually transformed into a vector of real numbers [13]. Prototype-based embedding is certainly the simplest and fastest embedding approach and for these reasons is adopted hereafter.

Prototype Selection. Selecting informative prototypes from the underlying graph domain plays a vital role in graph embedding [13]. In order to obtain a meaningful as well as class-discriminative vector representation in the embedding space, a set of selected prototypes $P = \{p_1, p_2, ..., p_n\}$ should be adequately distributed over the whole graph domain, at the same time avoiding redundancies in terms of selection of similar graphs [13] ,[14]. Among various prototype selection algorithms [13], [15], [16], the *discriminative prototype selection* method [15] was chosen in this study. This approach select prototypes from a graph set by adequately balancing within-class and between-class scattering.

Structural Features Extraction

The approach used for extracting structural features consists of the following main steps:

1. Use of a modified tracker [17] to extract a bounding box of each actor in each frame, and detection of the scale-invariant feature transform (SIFT) keypoints [18] within such a bounding box by using the software of Vedaldi and Fulkerson [19]. Based on the chosen threshold, this number for the selected dataset (KTH [6]; details provided in section 4) typically varies between 5 and 8. After detection, the location of each SIFT keypoint, (x, y), is expressed relatively to the actor's centroid and employed as a node label for an attributed graph describing the human's shape. In a preliminary study, we found that graphs with only labeled nodes granted comparable accuracy to graphs with both labeled nodes and labeled edges, yet resulted in faster processing. We therefore decided to employ graphs consisting only of labeled nodes (labeled edgeless graphs).

2. Next, in order to identify a prototype set which could lead to meaningful feature vectors in the embedded space, a number of different reference postures was chosen to describe all human shapes in the action dataset. For the dataset at hand (KTH [6]), we arbitrarily chose a set of 16 different reference postures across all human actions (running, walking, boxing, jogging, hand-waving, hand-clapping). Such selected postures should prove adequate for recognising human actions also in any other dataset where the actors are approximately in full view such as UCF Sports [20] and MuHAVi [21]. For training purposes, we manually selected a number of different frames varying in scenario (e.g. outdoor, outdoor with different clothes, indoor), action (e.g. hand waving, hand clapping, jogging) and actor (e.g. person01, person25, person12).

3. Finally, the graph is embedded into the feature vectors by means of P-GED with the prototype set of choice.

2.2 Spatio-Temporal Features

In this paper, in order to establish a fair comparison and focus the scope on the benefits of structural information, we have chosen to adopt the same features - STIP descriptors - of a deservedly much-cited paper from Laptev [1]. STIP descriptors have gained increasing popularity for action recognition since they describe salient points in space and time and do not require a preliminary step of foreground extraction which is generally regarded as inaccurate. A STIP descriptor consists of the concatenation of a histogram of quantised gradient (HOG) and a histogram of quantised optical flow (HOF) computed over a small spatio-temporal volume of pixels [1]. In this paper, we have used a combination of HOG and HOF for an overall dimensionality of 145. The main difference with [1] is that we do not convert descriptors into codewords; rather, we use each descriptor individually as an observation for our model (details in section 3).

3   Graphical Model

In this paper, we have used the *hidden Markov model with multiple, independent observations (HMM-MIO)* [22], a modified hidden Markov model (HMM) [23] capable of dealing with sequences of observations that include outlier, high-dimensional, and sparse measurements typical of action recognition.

Robustness to outliers is obtained by modelling the observation densities with Student's t distributions [24]. Dimensionality reduction is implemented by using the probabilistic principal component framework [25], and multimodality is taken into account by using a mixture distribution. Finally, modifications to the Baum-Welch algorithm allow for a variable number of observations per frame (single, multiple or none) by the assumption of independence and identical distribution of observations given the state of the HMM. This is a simplifying assumption given that in reality dependencies between these observations may exist. By noting as $O_t \equiv O_t^{1:N_t}$ the set of observations at time t, N_t their number, and Q_t the corresponding hidden state, we define:

$$P(O_t^{1:N_t}|Q_t) \equiv P(O_t^1, ..., O_t^{N_t}|Q_t) = \prod_{n=1}^{N_t} P(O_t^n|Q_t), if N_t > 1 \qquad (2)$$

and

$$P(O_t^{1:N_t}|Q_t) \equiv 1, if N_t = 0 \qquad (3)$$

Posing $P(O_t^{1:N_t}|Q_t) = 1$ in the case of no observations is equivalent to a missing observation and has neutral effect in the chain evaluation of the HMM-MIO.

In this study, the probability for all the observations in a frame, t, is calculated by the fusion of two likelihoods which model two types of measures:

- Spatio-Temporal Texture or Appearance Observations ($O_{a,t}$) provided by the STIP descriptors: the different numbers of STIP points per frame introduced a scale problem in the resulting probability that is solved in HMM-MIO by means of the following normalization:

$$P_a(O_{at}^{1:N_t}|Q_t) = \sqrt[N_t]{\prod_{n=1}^{N_t} P(O_{a,t}^n|Q_t)} \qquad (4)$$

- Structural Observations ($O_{s,t}$) provided by graph embedding: In our experiments, the embedding of a graph with 16 different selected prototypes leads to a 16-dimensional feature vector describing the shape of a single actor in each frame. This feature vector is modelled statistically by likelihood $P(O_{s,t}|Q_t)$.

The combination of the two likelihoods (equation 5) is performed as a weighted sum of weights W_a and W_s, such that $W_a + W_s = 1$.

$$P(O_t|Q_t) = W_a \cdot P_a(O_{a,t}^{1:N_t}|Q_t) + W_s \cdot P(O_{s,t}|Q_t) \qquad (5)$$

The graphical model for the modified HMM-MIO can be seen in Figure 1. The generative model is then obtained as the joint probability $P(O_{1:T}, Q_{1:T}|\lambda)$ of a sequence of observations, $O_{1:T} \equiv \{O_1, ..., O_t, ..., O_T\}$, and a sequence of corresponding hidden states, $Q_{1:T} \equiv \{Q_1, ..., Q_t, ..., Q_T\}$.

$$P(O_{1:T}, Q_{1:T}|\lambda) \equiv p(O_1, Q_1, ..., O_t, Q_t, ..., O_T, Q_T|\lambda) \qquad (6)$$

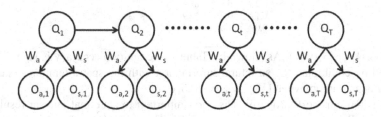

Fig. 1. Modified HMM-MIO (hidden Markov model with multiple, independent observations); O_t are the observations at time t (appearance observations provided by the STIP descriptors, O_a, and the structural observation provided by graph embedding, O_s); Q_t is the corresponding hidden state; W_a and W_s are the two weights for computing the total observation probability $P(O_t|Q_t) = W_a \cdot P_a(O_{a,t}^{1:N_t}|Q_t) + W_s \cdot P(O_{s,t}|Q_t); W_a + W_s = 1$

4 Experiments

This section provides the experimental evaluation of the proposed approach and shows the advantages of combining the structural information provided by graph embedding with the spatio-temporal information provided by STIPs. As dataset, we have chosen the KTH human action dataset containing 2,391 video sequences (from 25 different actors) and acquisition conditions, inclusive of four different scenarios and mild camera movements. The action classes include walking, jogging, running, boxing, hand waving and hand clapping [6]. Although KTH is becoming saturated in recent years with results reporting high accuracies, it still offers the widest platform for comparison with previous work [26]. For accuracy evaluation, we have used the evaluation procedure proposed by Schuldt *et al.* in [6]. In this procedure, all sequences are divided into three sets with respect to the actors: training (8 actors), validation (8 actors) and test (9 actors). Each classifier is then tuned using the first two sets (training and validation sets), and the accuracy on the test set is measured "blindly" by using the parameters selected on the validation set, without any further tuning. In order to assess the individual contribution of the features and show the advantages of the proposed fusion, we have conducted experiments with different weights (Table 1). A value of $(W_a, W_s) = (1, 0)$ means that only appearance features are used, whereas structural features are solely utilised

Table 1. Accuracy (%) of our approach over the KTH dataset with variable weights over the appearance and structural components.

W_a	W_s	Test accuracy (%)
1.0	0.0	85.7 [22]
0.9	0.1	85.9
0.8	0.2	86.8
0.7	0.3	87.9
0.6	0.4	88.9
0.5	**0.5**	**89.8**
0.4	0.6	87.9
0.3	0.7	85.8
0.2	0.8	82.2
0.1	0.9	77.9
0.0	1.0	48.7 [4]

when $(W_a, W_s) = (0, 1)$. As shown by Table 1, recognition accuracy is significantly improved by combining the structural information with the spatio-temporal features, reaching its maximum when $(W_a, W_s) = (0.5, 0.5)$.

To position our work properly, it is very important to state that current results on KTH are well in excess of 90% accuracy [27]. The goal of our paper is not that of proposing a more accurate action recognition method; rather, assessing the fusion of structural information with spatio-temporal features in a significant classification exercise. As for what action recognition is concerned, we have gathered empirical evidence that the graphs built by using SIFT keypoints as their nodes are rather unstable and noisy, and we are working on the use of graph-cut techniques to substantially improve nodes' extraction [28]. However, we believe that the work conducted to date already provides evidence that the fusion of structural information obtained by graph embedding with spatio-temporal information provided by STIPS is capable of encoding the human action to a significant extent.

5   Conclusions and Future Work

In this paper, we have presented a novel approach for human action recognition based on the fusion of structural and spatio-temporal information. To this aim, the structural information provided by graph embedding and the local spatio-temporal information provided by STIP descriptors are jointly modelled by a modified hidden Markov model with multiple, independent observations (HMM-MIO) [22]. Although our approach does not yet outperform the state-of-the-art accuracy, it shows that structural and spatio-temporal features can be fused constructively to obtain higher accuracy than from either separately. In the near future, we plan to further investigate other keypoint sets to improve the stability of the graph-based representation along the frame sequence and extend our study to other challenging action datasets.

Acknowledgments. The authors wish to thank the Australian Research Council and its industry partners that have partially supported this work under the Linkage Project funding scheme - grant LP 0990135 "Airport of Future".

References

1. Laptev, I.: On space-time interest points. International Journal of Computer Vision 64(2), 107–123 (2005)
2. Niebles, J., Chen, C.W., Fei-Fei, L.: Modeling Temporal Structure of Decomposable Motion Segments for Activity Classification. In: Daniilidis, K., Maragos, P., Paragios, N. (eds.) ECCV 2010, Part II. LNCS, vol. 6312, pp. 392–405. Springer, Heidelberg (2010)
3. Ta, A.-P., Wolf, C., Lavoue, G., Baskurt, A.: Recognizing and localizing individual activities through graph matching, pp. 196–203. IEEE Computer Society, Los Alamitos (2010)
4. Borzeshi, E.Z., Xu, R.Y.D., Piccardi, M.: Automatic Human Action Recognition in Videos by Graph Embedding. In: Maino, G., Foresti, G.L. (eds.) ICIAP 2011, Part II. LNCS, vol. 6979, pp. 19–28. Springer, Heidelberg (2011)
5. Fischler, M., Elschlager, R.: The representation and matching of pictorial structures. IEEE Transactions on Computers 22(1), 67–92 (1973)
6. Schuldt, C., Laptev, I., Caputo, B.: Recognizing human actions: a local SVM approach. In: Proceedings of the 17th International Conference on Pattern Recognition, ICPR 2004, vol. 3 (2004)
7. Gao, X., Xiao, B., Tao, D., Li, X.: A survey of graph edit distance. Pattern Analysis & Applications 13(1), 113–129 (2010)
8. Neuhaus, M., Bunke, H.: Automatic learning of cost functions for graph edit distance. Information Sciences 177(1), 239–247 (2007)
9. Rieck, K., Laskov, P.: Linear-Time Computation of Similarity Measures for Sequential Data. Journal of Machine Learning Research 9, 23–48 (2007)
10. Belkin, M., Niyogi, P.: Laplacian Eigenmaps for Dimensionality Reduction and Data Representation. Neural Computation 15(6), 1373–1396 (2003)
11. Qiu, H., Hancock, E.R.: Clustering and embedding using commute times. IEEE Transactions on Pattern Analysis and Machine Intelligence 29(11), 1873–1890 (2007)
12. Wilson, R.C., Hancock, E.R., Luo, B.: Pattern vectors from algebraic graph theory. IEEE Transactions on Pattern Analysis and Machine Intelligence, 1112–1124 (2005)
13. Riesen, K., Neuhaus, M., Bunke, H.: Graph Embedding in Vector Spaces by Means of Prototype Selection. In: Escolano, F., Vento, M. (eds.) GbRPR. LNCS, vol. 4538, pp. 383–393. Springer, Heidelberg (2007)
14. Hjaltason, G.R., Samet, H.: Properties of embedding methods for similarity searching in metric spaces. IEEE Transactions on Pattern Analysis and Machine Intelligence 25(5), 530–549 (2003)
15. Borzeshi, E.Z., Piccardi, M., Xu, R.Y.D.: A discriminative prototype selection approach for graph embedding in human action recognition. In: 2011 IEEE International Conference on Computer Vision Workshops (ICCV Workshops), pp. 1295–1301. IEEE (2011)
16. Riesen, K., Bunke, H.: Graph classification by means of Lipschitz embedding. IEEE Transactions on Systems, Man, and Cybernetics, Part B: Cybernetics 39(6), 1472–1483 (2009)
17. Chen, T.P., Haussecker, H., Bovyrin, A., Belenov, R., Rodyushkin, K., Kuranov, A., Eruhimov, V.: Computer vision workload analysis: case study of video surveillance systems. Intel Technology Journal 9(2), 109–118 (2005)
18. Lowe, D.G.: Distinctive image features from scale-invariant keypoints. International Journal of Computer Vision 60(2), 91–110 (2004)

19. Vedaldi, A., Fulkerson, B.: Vlfeat: An open and portable library of computer vision algorithms. In: Proceedings of the International Conference on Multimedia, pp. 1469–1472. ACM (2010)
20. Rodriguez, M.D., Ahmed, J., Shah, M.: Action mach a spatio-temporal maximum average correlation height filter for action recognition. In: IEEE Conference on Computer Vision and Pattern Recognition, CVPR 2008, pp. 1–8. IEEE (2008)
21. Singh, S., Velastin, S.A., Ragheb, H.: Muhavi: A multicamera human action video dataset for the evaluation of action recognition methods. In: 2010 Seventh IEEE International Conference on Advanced Video and Signal Based Surveillance (AVSS), pp. 48–55. IEEE (2010)
22. Concha, O.P., Xu, D., Yi, R., Moghaddam, Z., Piccardi, M.: Hmm-mio: an enhanced hidden markov model for action recognition. In: 2011 IEEE Computer Society Conference on Computer Vision and Pattern Recognition Workshops (CVPRW), pp. 62–69. IEEE (2011)
23. Rabiner, L., Juang, B.: An introduction to hidden markov models. IEEE ASSP Magazine 3(1), 4–16 (1986)
24. Liu, C., Rubin, D.B.: Ml estimation of the t distribution using em and its extensions, ecm and ecme. Statistica Sinica 5(1), 19–39 (1995)
25. Archambeau, C., Delannay, N., Verleysen, M.: Mixtures of robust probabilistic principal component analyzers. Neurocomputing 71(7), 1274–1282 (2008)
26. Gao, Z., Chen, M., Hauptmann, A., Cai, A.: Comparing Evaluation Protocols on the KTH Dataset. In: Salah, A.A., Gevers, T., Sebe, N., Vinciarelli, A. (eds.) HBU 2010. LNCS, vol. 6219, pp. 88–100. Springer, Heidelberg (2010)
27. Guo, K., Ishwar, P., Konrad, J.: Action recognition using sparse representation on covariance manifolds of optical flow. In: 2010 Seventh IEEE International Conference on Advanced Video and Signal Based Surveillance (AVSS), pp. 188–195. IEEE (2010)
28. Rother, C., Kolmogorov, V., Blake, A.: Grabcut: Interactive foreground extraction using iterated graph cuts. ACM Transactions on Graphics (TOG) 23, 309–314 (2004)

An Incremental Structured Part Model
for Image Classification

Huigang Zhang[1], Xiao Bai[1], Jian Cheng[2], Jun Zhou[3], and Huijie Zhao[1]

[1] School of Computer Science and Engineering, Beihang University, Beijing, China
baixiao.buaa@googlemail.com
[2] Institute of Automation Chinese Academy of Sciences, Beijing 100190, China
[3] School of Information and Communication Technology, Griffith University, Nathan,
QLD 4111, Australia

Abstract. The state-of-the-art image classification methods usually require many training samples to achieve good performance. To tackle this problem, we present a novel incremental method in this paper, which learns a part model to classify objects using only a small number of training samples. Our model captures the inherent connections of the semantic parts of objects and builds structural relationship between them. In the incremental learning stage, we use high entropy images that have been accepted by users to update the learned model. The proposed approach is evaluated on two datasets, which demonstrates its advantages over several alternative classification methods in the literature.

Keywords: Image classification, semantic parts, structural relationship, incremental learning.

1 Introduction

Image classification is one of the most important tasks in computer vision and pattern recognition. A number of methods based on the bag-of-words (BOW) model [1] have been proposed to fulfill this task and have shown to be effective for object and scene classification [2,3]. The BOW method represents an image as a histogram of its local features. It is robust against spatial translations of features, and has demonstrated decent performance in the whole-image classification. However, the BOW method does not sufficiently characterize the spatial relationship between features. Therefore, it is incapable of capturing structural shapes or locating objects in an image.

Structure based methods extract invariant structures to characterize objects in an image [4]. One popular solution is to use graph structure because graph can be used to represent high level vision information. This property has made the graph based methods capable of bridging the low-level local invariant feature with the high-level vision information in images [5,6]. More recently, part based models have been proposed [7,8], which operate on image structure rather than solely extracting discrete features.

Learning frameworks have been introduced to further improve the adaptability of statistical and structural image classification methods. Of particular interest

G.L. Gimel' farb et al. (Eds.): SSPR & SPR 2012, LNCS 7626, pp. 483–491, 2012.
© Springer-Verlag Berlin Heidelberg 2012

is the spatial pyramid matching (SPM) method [9]. It partitions an image into increasingly finer spatial subregions and computes a histogram of local features from each subregion. The same rationale has been employed by several methods, such as sparse coding for linear spatial pyramid matching (ScSPM) [10] and locality-constrained linear coding (LLC) [11]. Similar work also includes the coarse-to-fine learning framework presented by Li et al [12]. In this work, a novel automatic dataset collecting and model learning approach, OPTIMOL, has been developed to refine online picture selection in an incremental way.

Enlightened by these work, we propose an approach to improve the image classification performance via learning semantic parts of objects and exploring their structural relationship. It includes a feature learning method [13] to enrich the part description, and an incremental framework to iteratively update the learned model. Figure 1 illustrates the framework of this classification method. To validate the effectiveness of this method, we have compared it against several state-of-the-art methods in the literature.

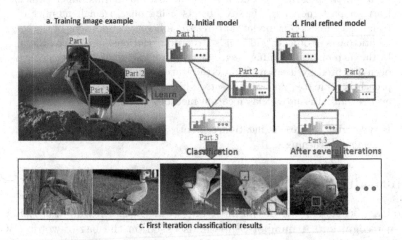

Fig. 1. Framework of the incremental structured part model for image classification. (a) Extracted relevant semantic parts. (b) Training an SVM classifier for each semantic part and building the structured part model. (c) Initial classification results. (d) Iterative model updating using selected images. After several iterations, the model is then updated to a refined model.

The main contribution of this paper is three-fold. Firstly, we propose a part description method that provides abundant mid-level features for image classification. Secondly, the structured part model combines both appearance and structure information of objects in images, which leads to improved classification performance. Thirdly, the incremental learning algorithm can adapt to novel image features and structures introduced from unseen testing objects. This has greatly reduced the number of training images required.

2   Incremental Structured Part Model

The proposed approach is a combination of both statistical and structural pattern recognition methods. It is based on the observation that different parts of objects in the same class normally share similar spatial relationship. For example, all birds have beaks, legs, and tails, and they follow similar spatial layout. Therefore, we only need to recognize these three parts and model their spatial relationship in order to distinguish birds from other objects (Figure 1(a)).

2.1   Semantic Part Learning

We commence by semantic part learning which allows the treatment of each part as mid-level semantic attribute. We first define the part classes that are important to object classification, then image patches for these parts are manually selected from the training set. From each of these patches, SIFT, texture, color, and edge direction features are extracted. The SIFT features [14] are extracted in a grid-based manner, while the texture descriptors [15] are computed at each pixel using a set of filter banks. To extract the color feature, we use the LAB values [16] of densely sampled pixels. Edges are generated via standard Canny edge detector [17]. Using the bag-of-words model, these four types of features are quantized into vectors with 1000, 256, 128 and 8 dimensions, respectively, and are concatenated into a vector of length 1392.

Using the part feature vectors, a multi-class support vector machines (SVMs) can be learned. Let M be the number of part classes, x_n denotes the n-th training sample and y_n denotes its part class label. The multi-class SVM generates an M-dimensional weight vector $\{w_m^*\}_{m=1}^M$, with one weight for each class. Let W denote a matrix whose columns are w_m. To estimate W, we minimize the following loss function:

$$W^* = \arg\min \sum_n \sum_{t=1}^M d(w_t^T x_n, y_{nt}) + \gamma \sum_m \|w_m\|_2^2 \qquad (1)$$

where $\gamma \geq 0$ is a tradeoff parameter that regularizes the model complexity, and is set to 0.8 by threefold cross-validation. $d(\cdot, \cdot)$ is the loss function.

After solving this optimization problem, we get a semantic part classifier. When an unlabeled image is given, this classifier can be applied to detect relevant parts in the image. In the next section, we explore the structural relationship between these parts.

2.2   Structured Part Model Matching

In this step, we effectively arrange the semantic parts in a deformable configuration to represent an object. The structure model here is inspired by the pictorial structure method [7].

Given an image, let $p_i(l_i)$ be a function measuring the degree of part similarity when part v_i is placed at location l_i. Let $p_{ij}(l_i, l_j)$ be a function measuring the degree of deformation when part v_i is placed at location l_i and part v_j is placed at location l_j. We define the problem of matching a structured part model to an image as a statistical function to be maximized

$$L^* = \arg\max_L (\sum_{i=1}^{n} p_i(l_i) + \lambda \sum_{(v_i,v_j)\in E} p_{ij}(l_i,l_j)) \tag{2}$$

This function maximizes the sum of the matching probabilities $p_i(l_i)$ of each individual part and the deformation similarities $p_{ij}(l_i,l_j)$ for connected pairs of parts. Therefore, it can be decomposed into two equations as follows:

$$L_1^* = \arg\max_L \sum_{i=1}^{n} p_i(l_i) \tag{3}$$

$$L_2^* = \arg\max_L \sum_{(v_i,v_j)\in E} p_{ij}(l_i,l_j) \tag{4}$$

where Eq. 3 is a standard part model and Eq. 4 is a structure model. λ is a parameter that adjust the contribution from the part model and the structure model. It leads to the extension of [7] to a more flexible setting and is self-adaptive through the incremental process to be described later.

We use a sliding window method to detect parts in an unseen image and to compute $p_i(l_i)$. This is achieved by searching the testing images at three different scales, i.e., 0.7, 1, 1.3 times the reference part scale (50×50 pixels), respectively. Using the learned multi-class SVM classifier, we can compute the probability of these candidate patches by fitting a sigmoid function to the original SVM decision values [18]. To compute $p_{ij}(l_i,l_j)$, we use the same method as [7] to calculate the degree of deformation, and fit it to $(0,1]$ via an exponential function.

The proposed structured part model is robust to missing parts in an image. In Eq. 2, even if one or two p_i is incorrect, high probability still can be achieved on parts from object in the same class due to the contribution from the structure model.

2.3   Coarse-To-Fine Updating

Given a very small number of training images of an object class, our algorithm learns the optimal structured part model L^* that best describes this class using the steps introduced above. Now we introduce a coarse-to-fine process to iteratively update L^*, which further improves the robustness of the proposed method.

We randomly separate testing images into several batches and feed them sequentially into the system. Each batch is treated as an iteration. Our incremental process is performed when a new batch comes in. It continuously classifies the images while learning a more robust model. On each image batch, we compute the probability that the current optimal structured part model matches the images using Eq. 2. The model update is dependent on the image matching results.

Images with low matching probability are discarded, while the rest are divided into two sets based on the entropy value generated from the following equation

$$H(I) = -\sum_i p_i \ln p_i - \lambda \sum_E p_{ij} \ln p_{ij} \tag{5}$$

According to Shannon's entropy theory, Eq. 5 relates to the amount of uncertainty of an image I. High entropy indicates high uncertainty of an image, which, in turn, suggests possible new structures. Thus, we choose those images with high probability and high entropy for model updating. Images with high likelihood and low entropy are classified to be positive images. The model updating follows the method introduced in the previous two subsections. It allows refinement of the part classifiers and the corresponding structure model.

At the same time, the weight parameter λ is updated iteratively to make the learned model more robust. In each iteration, the image probabilities are calculated using L_1^* and L_2^*. This can be achieved by setting λ to 0 and 100 (a large enough number) respectively. Let $\varphi_i = \{x|x$ be an image belongs to the positive part using $L_i^*\}$; $\varphi = \{x|x$ is an image belongs to the positive part using $L^*\}$; con_i represents the contribution of model L_i^* to L^*. Then

$$con_i = \frac{\#\{\varphi_i \cap \varphi\}}{\#\{\varphi\}}, i = 1, 2 \tag{6}$$

$$\lambda = \frac{con_2}{con_1} = \frac{\#\{\varphi_2 \cap \varphi\}}{\#\{\varphi_1 \cap \varphi\}} \tag{7}$$

Eq. 7 determines the weights of the part model and the structure model. By calculating λ in each batch, more refined model can be achieved. The proposed coarse-to-fine framework is an iterative process that continuously classifies an image data set with high accuracy while learning a more robust object model. We summarize the steps of our algorithm in Algorithm 1.

Algorithm 1. Incremental Structured Part Model for Classification

---

Input: Set of N positive images (N is a small number), set of novel unlabeled images, part number n, and weight $\lambda=1$.

Output: Set of classified positive images, and the final Structured Part Model

Initialize Manually select n parts in each training image

Repeat

  Learn Calculate the features of each part in the latest input images and train SVM models. (Sec. 2.1)

  Learn the Structured Part Model. (Sec. 2.2)

  Classify Classify images using the current Structured Part Model. (Sec. 2.2)

  Incremental Use the images with high probability and high entropy for model updating. (Sec. 2.3)

until User satisfied or images exhausted

---

3   Experimental Results

We evaluate the performance of the proposed incremental structured part model on two widely used datasets, Caltech-256 [19] and Pascal VOC 2007 [20], and show that only a small number of training images is required for the proposed model (Section 3.1). We also compare our method with other classification methods such as the model by Gritfin et al. [19], ScSPM [10], and LLC [11].

The Caltech-256 dataset contains 30,607 images in 256 categories, with each class containing at least 80 images. The Pascal VOC 2007 dataset consists of 9,963 images from 20 classes. Objects in this dataset reside in cluttered scenes with a high degree of variation in viewing angle, illumination and object appearance. Before the experiments, each image is resized to less than 300×300 pixels with the aspect ratio unchanged. We used all classes in these two datasets for the experiments.

3.1   Incremental Structured Part Model Evaluation

In the first experiment, we randomly chose 5, 10, 15, 20, 25 and 30 training images per class respectively to validate the effectiveness of the proposed method. We consider three baselines to compare our system with: 1) a standard part model L_1^* as in Eq. 3, 2) a structure model L_2^* of Eq. 4, and 3) our structured part model without a coarse-to-fine process. The results are shown in Figure 2. It can be seen that our incremental structured part model outperforms the the baselines by nearly 10 percent. The proposed model is very stable on both datasets when different training sizes are used. At the 5% level, our method achieves classification accuracies that are nearly 10 and 20 percent higher than the alternatives, respectively.

The reason that our model can achieve good performance under small number of training images is due to the effect of the coarse-to-fine process. By choosing those images with high entropy, large amount of novel information can be acquired for model updating. The effect of the incremental process is three-fold.

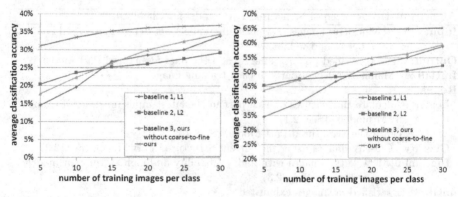

Fig. 2. The average classification results of all the categories in the Caltech-256 dataset (*left*) and Pascal VOC 2007 dataset (*right*), when different training sizes is used.

Firstly, it can refine the multi-SVM part model. As illustrated in Figure 1, part 2 of the first model is actually a coarse model, as marked by bars in different colors. After several incremental iterations, this model is well refined, which is represented by bars in the same color. Secondly, this process can refine the structural model both in shape and in edge relationships. Take the last model in Figure 1 for example, the dotted line between part 2 and 3 shows that this relationship should be week compared with others, because it's changes in accordance with different birds' postures. Thirdly, the iteration refines the parameter λ in Eq. 2, which leads to a refined global model.

Figure 3 shows some example images with high classification accuracy in the Caltech-256 dataset. We have also tracked those image data with missing parts. The results show that most of them can be classified correctly, which proves the robustness of the proposed method.

Fig. 3. Example images from categories with high classification accuracy in the Caltech-256 dataset. The percentages in the brackets represent the corresponding classification accuracy.

3.2   Comparison with other Classification Methods

In this experiment, we first compared the proposed method with several state-of-the-art classification methods on the Pascal dataset. The classification performance is evaluated using the Average Precision (AP) measure. It computes the area under the Precision/Recall curve, in which higher score means better performance. Table 1 shows the classification accuracy on all 20 classes compared against several other classification methods [11,21,20]. Our method has achieved the highest accuracy in most classes, especially those with similar shapes such as *bicycle and motorbike, cat and dog, cow and sheep*. The results show that our semantic part model is capable of extracting features and their structural relationships in order to distinguish similar objects. We also tested the method on Caltech-256 dataset, in which we used 5, 15, and 30 training images per class. Detailed results are shown in Table 2. It suggests that our method leads the performance with a small number of training images.

Table 1. Image classification results on Pascal VOC 2007 dataset

Category	aero	bicyc	bird	boat	bottle	bus	car	cat	chair	cow
PASCAL 07 Best [20]	**77.5**	63.6	56.1	71.9	33.1	60.6	78	58.8	53.5	42.6
LLC [11]	74.8	65.2	50.7	70.9	28.7	68.8	**78.5**	61.7	54.3	48.6
Su [21]	76.2	66.4	**59.2**	70.3	35.4	63.6	79.4	62.4	59.5	47.9
ours	77.1	**73.0**	54.8	**75.2**	**37.2**	**70.3**	72.4	**65.7**	**60.6**	**50.8**
Category	table	dog	horse	mbike	person	plant	sheep	sofa	train	tv
PASCAL 07 Best [20]	54.9	45.8	77.5	64.0	85.9	36.3	44.7	50.9	79.2	53.2
LLC [11]	51.8	44.1	76.6	66.9	83.5	30.8	44.6	53.4	78.2	53.5
Su [21]	**58.8**	44.9	**78.3**	67.4	**87.9**	32.9	46.9	**53.8**	78.6	**58.9**
ours	57.5	**49.3**	75.7	**72.9**	77.2	**42.1**	**47.9**	51.5	**80.6**	58.6

Table 2. Image classification results on Caltech-256 dataset

Algorithms	5 training	15 training	30 training
Gritfin et al. [19]	18.40	28.30	34.10
ScSPM [10]	-	27.73	34.02
LLC [11]	-	34.36	**41.19**
ours	**31.15**	**35.22**	36.87

4   Conclusion

In this paper we have proposed a novel incremental structured part model for image classification. This method first builds image classification models by incorporating both advantages from semantic parts and their structural relation description. Then an incremental framework is employed to refine the model iteratively, which makes the proposed method more robust. This method requires only a small number of training images to achieve good classification performance. Future work will explore the use of hierarchical segmentations to find the semantic parts at the training stage. We will also investigate other features to train the part classifier.

References

1. Fei-Fei, L., Fergus, R., Torralba, A.: Recognizing and learning object categories. In: ICCV Short Course (2005)
2. Yang, L., Jin, R., Sukthankar, R., Jurie, F.: Unifying discriminative visual codebook generation with classifier training for object category recognition. In: IEEE Conference on Computer Vision and Pattern Recognition, CVPR 2008, pp. 1–8 (2008)
3. Fergus, R., Fei-Fei, L., Perona, P., Zisserman, A.: Learning object categories from google's image search. In: IEEE International Conference on Computer Vision, ICCV 2005, vol. 2, pp. 1816–1823 (2005)
4. Bunke, H., Sanfeliu, A.: Syntactic and structural pattern recognition: theory and applications, vol. 7. World Scientific Pub. Co. Inc. (1990)

5. Xiao, B., Hancock, E., Wilson, R.: Graph characteristics from the heat kernel trace. Pattern Recognition 42(11), 2589–2606 (2009)

6. Wilson, R., Hancock, E., Luo, B.: Pattern vectors from algebraic graph theory. IEEE Transactions on Pattern Analysis and Machine Intelligence 27(7), 1112–1124 (2005)

7. Felzenszwalb, P., Huttenlocher, D.: Pictorial structures for object recognition. International Journal of Computer Vision 61(1), 55–79 (2005)

8. Felzenszwalb, P., Girshick, R., McAllester, D., Ramanan, D.: Object detection with discriminatively trained part-based models. IEEE Transactions on Pattern Analysis and Machine Intelligence 32(9), 1627–1645 (2010)

9. Grauman, K., Darrell, T.: The pyramid match kernel: Discriminative classification with sets of image features. In: IEEE International Conference on Computer Vision, ICCV 2005, pp. 1458–1465 (2005)

10. Yang, J., Yu, K., Gong, Y., Huang, T.: Linear spatial pyramid matching using sparse coding for image classification. In: IEEE Conference on Computer Vision and Pattern Recognition, CVPR 2009, pp. 1794–1801 (2009)

11. Wang, J., Yang, J., Yu, K., Lv, F., Huang, T., Gong, Y.: Locality-constrained linear coding for image classification. In: IEEE Conference on Computer Vision and Pattern Recognition, CVPR 2010, pp. 3360–3367 (2010)

12. Li, L., Fei-Fei, L.: Optimol: Automatic online picture collection via incremental model learning. International Journal of Computer Vision 88(2), 147–168 (2010)

13. Farhadi, A., Endres, I., Hoiem, D., Forsyth, D.: Describing objects by their attributes. In: IEEE Conference on Computer Vision and Pattern Recognition, CVPR 2009, pp. 1778–1785 (2009)

14. Lowe, D.: Distinctive image features from scale-invariant keypoints. International Journal of Computer Vision 60(2), 91–110 (2004)

15. Varma, M., Zisserman, A.: A statistical approach to texture classification from single images. International Journal of Computer Vision 62(1), 61–81 (2005)

16. Wyszecki, G., Stiles, W.: Color Science: Concepts and Methods, Quantitative Data and Formulae. Wiley, New York (1982)

17. Canny, J.: A computational approach to edge detection. IEEE Transactions on Pattern Analysis and Machine Intelligence (6), 679–698 (1986)

18. Platt, J., et al.: Probabilistic outputs for support vector machines and comparisons to regularized likelihood methods. Advances in Large Margin Classifiers 10(3), 61–74 (1999)

19. Griffin, G., Holub, A., Perona, P.: Caltech-256 object category dataset (2007)

20. Everingham, M., Van Gool, L., Williams, C., Winn, J., Zisserman, A.: The pascal visual object classes challenge 2007 (voc 2007) results (2007)

21. Su, Y., Allan, M., Jurie, F.: Improving object classification using semantic attributes. In: Proceedings of the British Machine Vision Conference, BMVC 2010, pp. 26–21 (2010)

Top-Down Tracking
and Estimating 3D Pose of a Die

Fuensanta Torres* and Walter G. Kropatsch

PRIP, Vienna University of Technology, Austria
http://www.prip.tuwien.ac.at

Abstract. Real-time 3D pose estimation from monocular image sequences is a challenging research topic. Although current methods are able to recover 3D pose, they are severely challenged by the computational cost. To address this problem, we propose a tracking and 3D pose estimation method supported by three main pillars: a pyramidal structure, an aspect graph and the checkpoints. Once initialized the systems performs a top-down tracking. At a high level it detects the position of the object and segments its time-space trajectory. This stage increases the stability and the robustness for the tracking process. Our main objective is the 3D pose estimation, the pose is estimated only in relevant events of the segmented trajectory, which reduces the computational effort required. In order to obtain the 3D pose estimation in the complete trajectory, an interpolation method, based on the aspect graph describing the structure of the object's surface, can be used to roughly estimate the poses between two relevant events. This early version of the method has been developed to work with a specific type of polyhedron with strong edges, texture and differentiated faces, a die.

Keywords: tracking, 3D pose estimation, pyramid, checkpoints, aspect graph.

1 Introduction

The proliferation of high speed videos, high-end computers and the need for automated video analysis have generated an increasing interest in visual tracking and pose estimation algorithms. The higher resolution of the images and the higher frame rate increase the data rate by a higher factor than the increase in computing power. This paper addresses the challenging problem of real-time tracking and 3D pose estimation exploring the efficient use of knowing the past for predicting and for verifying the future. Selecting the right features for tracking plays a critical role [3]. Nowadays, the illumination changes, the partial occlusion and the matching errors are simple to achieve with localized features [7]. However, computation of descriptors that are invariant across large view changes is usually

---

* Thanks to Doctoral College on Computational Perception (Vienna University of Technology, Austria) for funding.

G.L. Gimel' farb et al. (Eds.): SSPR & SPR 2012, LNCS 7626, pp. 492–500, 2012.

expensive [15]. To overcome this weakness, the state of the art feature descriptors, detect and match points in successive images, in a non-recursive way, [10], [1], [8]. SIFT [8] is known to be a strong, but computationally expensive feature descriptor and on the contrary Ferns [10] classification is fast, but requires large amounts of memory. Therefore, our work investigates the applicability of a new markerless tracking method based on the checkpoints. Checkpoints are a small group of 3D points on the known object surface, which allow reliable tracking and preserve the structure. These are robust to illumination changes, computationally cheap and do not require large amount of memory. Moreover, they are 3D points. Therefore, once initialized their positions in the next frame can be predicted assuming smooth movement, without the need to back-project the 2D locations to obtain the 3D pose.

Top-down tracking encodes the current frame into a hierarchical structure, a pyramid, which reduces the search cost and allow large view changes. The use of a hierarchical approach for tracking have been widely used in the literature [16], [5], [9]. The main drawback of theses approaches have been the high computational cost to build a pyramid per frame. To overcome this weakness, we can use the computational power and increasing programmability of the graphics processing unit (GPU) present in modern graphics hardware that provides great scope for acceleration of computer vision algorithms which can be parallelized [13].

In order to reduce the computational effort our method distinguishes relevant (frame with only one visible face of the die) and normal events (with two or three visible faces) of the time-space trajectory of the object. The changes between two relevant events are handle by the aspect graph [12], [11].

The rest of the paper is organized as follows: Sec.2 describes the different structures and processes of this approach. Sec. 3 presents the top-down tracking and 3D pose estimation method. The experimental results revealing the efficacy of the method are shown in Section 4. Finally, the paper concludes along with discussions and future work in Section 5.

2   Definitions

We begin by providing some necessary definitions.

2.1   Checkpoints

Checkpoints are a small group of points characterizing local and salient features embedded in the object's surface and allowing to detect and correct displacements in the image frame. They require to distinguish between the background and the foreground of the object. This early version of the method is based on a strong foreground-background contrast, the background and the foreground points are differentiated by their gray values.

Let S= $(I_t, I_{t+1}..., I_{t+k})$ be an image sequence. The initial estimation of a group of checkpoints location in time t $(x1_t = (x1_t^1, x1_t^2, x1_t^3, x1_t^4$ and $x1_t^5))$ can be found by giving some correspondences between 3D points in the object model and their projections in I_t [7]. Checkpoints are projected into the current image $x1'_{t+1}$ frame and the corresponding pixel values checked whether they belong to the object or the background. Based on the result the correction is estimated that brings the object back to a location where the checkpoints are appropriately placed in the image $x1_{t+1}$. The correction (C= (s or/and T or/and R)) is the uniform scale (s), the translation (T) and the rotation (R) to get $x1_{t+1}$ from $x1'_{t+1}$ in the current frame I_{t+1}. For this purpose, considering a circular target, five checkpoints ($x1'_{t+1}$), which preserve the order, placed as shown Fig. 1 a), $x1^1$, $x1^2$, $x1^4$, $x1^5$ in the background and $x1^3$ in center of the object, in the foreground, are enough to detect the translation and the scale error. However, to estimate the rotation error, at least two groups of checkpoints are needed ($x1_t$ and $x2_t$) (Fig. 1 b)).

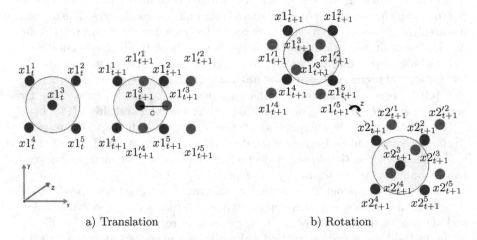

a) Translation   b) Rotation

Fig. 1. Predicting and correcting translations and rotations of checkpoints

2.2   Prediction-Estimation-Correction

This section defines the Prediction-Estimation-Correction method (PEC) of the checkpoints' positions. Let I_t be the current frame and x_t be the checkpoints' locations in time t. Using a motion model[1], the checkpoints are predicted forward for one frame, x'_{t+1}. First, it checks if x'^1_{t+1}, x'^2_{t+1}, x'^4_{t+1}, x'^5_{t+1} are placed in the background(0) and x'^3_{t+1} in the foreground(1) of the image, otherwise the prediction is incorrect. When an error has been detected, it estimates the location where the checkpoints are appropriately placed in the image x'_{t+1}. The estimation method is based on a table with the possible cases of prediction errors and their respective estimation (Tab. 1). The table has been built considering all the possible movements of the prediction with respect to the real projection and their optimal improvement. Moreover, its efficacy has been demonstrated in the experimental results. In the table there are five columns (x'^1, x'^2, x'^3,

---

[1] here is used the 3D affine motion model.

x'^4, x'^5), which represent one group of checkpoints illustrated in Fig. 1 a) and the last column (x") is the translation or the scale needed to get the estimated appropiate position from x'.

The zeros in the table mean that the value of x'^i is close to the background, the one appears when it is more similar to the foreground and the * means that this checkpoint does not have any effect in the estimation. For instance, the case of Fig. 1 a) corresponds to the box in Tab. 1, x'^1, x'^3, x'^4 are equal to 1 while x'^2 and x'^5 are equal to 0. Therefore, the estimation(x") is a translation of the prediction to the left. The direction and the sense of the arrows describe the translations for correction. The correction step finds the relationship between the estimated position of all groups ($x''1$, $x''2$ ...$x''i$) of the current frame and their prediction ($x'1$, $x'2$ ...$x'i$).

This least-squares problem in the 3D space is solved by using Horn [4], which returns the uniform scale factor (s), the rotation matrix (R_{3x3}) and the translation vector (T_{3x1}) needed to get the correction (x) from the prediction (x')(eq. 1)

Table 1.

| \multicolumn{6}{c}{values at prediction} |
x'^1	x'^2	x'^3	x'^4	x'^5	x''
0	0	*	0	1	↘
0	0	*	1	0	↙
0	0	*	1	1	↓
0	1	*	0	0	↗
0	1	*	0	1	→
0	1	*	1	1	↘
1	0	*	0	0	↖
1	0	*	1	0	←
1	0	*	1	1	↙
1	1	*	0	0	↑
1	1	*	0	1	↗
1	1	*	1	0	↖
0	0	1	0	0	s
1	1	1	1	1	$1/s$

$$x = (s \cdot R_{3x3} + T_{3x1}) \cdot x'; \qquad (1)$$

2.3 Recall of the Maximum Pyramid

The structure of a regular pyramid can be described as an array hierarchy in which each level l^t is at least defined by a set of nodes N_l. A node of a regular pyramid can be determined by its position (i, j, l) in the hierarchy, being l the level of the pyramid and (i, j) its (x, y) coordinates within the level. On the base level of the pyramid, the nodes are the pixels of the input image. Each pyramid level is recursively obtained by processing the level below. The children-parent relationships are fixed and for each node in level l+1, there is a reduction window of children at level l. We have selected a 2x2/4 pyramid [2] other types are also under investigation. To detect bright spots in images we use the Maximum pyramid, which uses the maximum as reduction function. The top of the pyramid receives the maximum gray value of the base image. There is a closed chain of links between the maximum in the base and the top. This can effectively be used to find its location top-down. Small non-maxima holes disappear quickly [6].

2.4 Aspect Graph

An aspect describes the appearance of an object from a specific view point. Views of one aspect may differ by continuous deformations but they all have the same topology. The appearance from one aspect to another aspect changes, i.e.

a new surface patch becomes visible, another one disappears. The aspect graph is a graph with a node for every aspect and edges connecting adjacent aspects. Therefore, it allows us to know the relationship between each aspect (Fig. 2).

3   Top-Down Tracking and Pose Estimation

The novel approach for target localization and 3D pose estimation is described in this section. The first step of tracking is to obtain a hierarchical representation of the current frame. In order to decrease the computational cost, we assume that the object does not move very much from one frame to the next one as well as their backgrounds are quite similar. Having the pyramid for the previous frame, it subtracts two consecutive frames to update in this pyramid only the information corresponding to their differences. Once the pyramid is available, the system performs the top-down tracking method (Fig. 3). This method segments the time-space trajectory of the object. It recovers the object position in all the frames, which increases the stability and the robustness for the tracking process. Although, the pose estimation is obtained only in the relevant events. We can use an interpolation method, based on the aspect graph, between two pose estimations to roughly estimate the pose in the intermediate frames. The top-down process is the following:

1. Target localization: works at the top level of the pyramid l_T^t. In this level the target region has approximately homogeneous color. It selects the nodes with this color, where the object is placed $N_{l_T}^t$.
2. Trajectory Segmentation: Each node below the target object in the top level $N_{l_T}^t$ is linked to its children. This top-down process continues until the method estimates if the current frame is a relevant or a normal event. In the case of a normal event, the object position is estimated. Otherwise, it determines its 3D pose estimation.

3.1   Object Position

The position of an object is determined in normal events and at the first level where the number of nodes of the target is bigger than a given threshold. It

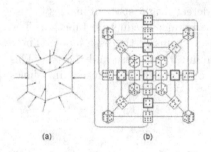

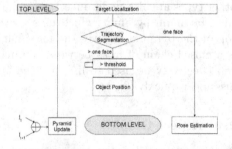

Fig. 2. a) Different viewing angles. b) Aspect graph of a die.

Fig. 3. Illustration of the Top-down tracking and pose estimation algorithm

is chosen in such a way that the completed target region has approximately homogeneous color, which is a compromise between homogeneous color in $ROI_{l_i}^t$ and precision in the PEC method. If the threshold raises, the precision increases but the homogeneity in the color decreases. At the highest level where $ROI_{l_i}^t <$ threshold, a group of checkpoints $(x1_{t,l})$ and the PEC method are used to to estimate the position of the object in the current frame.

3.2   Pose Estimation

This method works with relevant events. Two groups of checkpoints $(x1_{t,0}$ and $x2_{t,0})$ and the PEC method are used at the base level to estimate the 3D pose[14].

4   Experiments

In this section we demonstrate the effectiveness of our approach using a video sequence S= $(I_t, I_{t+1}..., I_{t+k})$ of a die. Figs. 5 a) and 5 b) have the same nine frames of the video sequence $(I_1, I_2..., I_9)$. They show the prediction of the checkpoints's positions (green points) and the result of the PEC method (red points).

The object position method (Sec. 3.1) allows abrupt displacements and large view changes (Fig. 5 a)). Although, it does not detect rotation changes and its prediction is not very accurate, as can be seen in Tab. 2. This shows the biggest error in pixels between the estimated position of the center point of the die and its real position in the base level. We calculated the biggest error in the prediction and also in the correction in the frames I_1, I_{10}, I_{20}, I_{30}, I_{40}, I_{50}.

Otherwise, the isolated pose estimation method (Sec. 3.2) is not robust to large view changes and translations (Fig. 5 b)). But this refinement step increases the accuracy of the method. As shown in Tab. 2 the errors are smaller than 5 after of the PEC method, except in the case of I_{50}, where the die is lost.

We have observed that the size of the target in ROI at the different levels of the pyramid strongly depends of the number of visible faces of the die in the current frame. Fig. 4 shows a graph with the size of the die in ROI for the different frames of a video sequence at a given level. As can be seen in the minimum values of the graph there are frames with only one visible face and in the maximum values there are views with three visible faces. Our current method

Table 2. Errors in pixels at the base level with two methods

object position method (Sec. 3.1)			pose estimation method (Sec. 3.2)		
Frame	Prediction error	Correction error	Frame	Prediction error	Correction error
I_1	8.6	4.4	I_1	6.9	1.5
I_{10}	12.6	12.6	I_{10}	9	1
I_{20}	13	10	I_{20}	13.29	0
I_{30}	10	13.1	I_{30}	5.5	2.7
I_{40}	14.5	19	I_{40}	11	5
I_{50}	10.5	8	I_{50}	28.9	42.3(lost)

Fig. 4. Size of the target in ROI for each frame of a video sequence

to segment the time-space trajectory fails in some cases, which strongly depend on the position of the die in the frame (related to the shift variance problem of non overlapping pyramids). We are working to overcome this weakness.

Finally, the strengths of our method have been proven with different experiments:

– Robustness to illumination changes: We changed the illumination in the training sequence Fig. 6 a). As can be seen in the bottom row the checkpoints handle a very abrupt lighting changes.
– Insensitivity to large view changes: Thanks to the object position method (Sec. 3.1), the algorithm can handle large view changes and it also updates the motion model. Fig. 6 b) shows in the top row the frames I_t, I_{t+12}, I_{t+13} and I_{t+14} of a video sequence. As can be seen in the bottom row, it localizes the die in the frame I_{t+12} and updates the motion model. Therefore, the prediction in the frame I_{t+14} is quite accurate.
– Computationally Cheap: Once initialized, the pyramid of the current frame I_{t+1} is the same pyramid as the previous frame I_t, where only the differ-

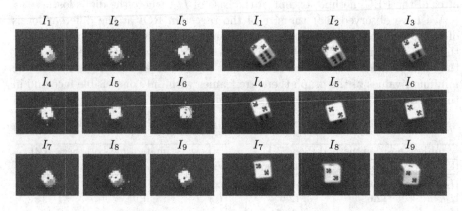

a) Object position   b) Pose estimation

Fig 5. Prediction and correction of checkpoints

I_t $\quad$ I_{t+1} $\quad$ I_{t+2} $\quad$ I_t $\quad$ I_{t+12} $\quad$ I_{t+13} $\quad$ I_{t+14}

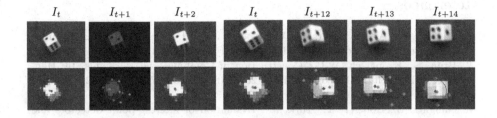

a) Robustness to illumination changes.   b) Insensitivity to large view changes.

Fig 6. Experiments to prove the strengths of our method

Fig 7. Two consecutive frames and their differences at the base level (l^0) and at l^1 respectively

ences between I_{t+1} and I_t have been updated. Fig. 7 shows two consecutive frames and their differences on the top row, while the row below shows the differences at the higher level(l^1). In this particular example, the dimensions of I_t and I_{t+1} are equal to 640x480= 307200 pixels, there are 5020 nodes different at the base level (l^0), 17 at l^1, 10 at l^2 , 2 at l^3 and 0 in the rest of levels.

5   Conclusions and Future Work

This paper has proposed a novel approach to track and to estimate the 3D pose of a (partially) known object. To demonstrate the new concept we have chosen a die because of it's simple structure: six well distinguished faces. We have developed a marker-less 3D tracking, which extracts the checkpoints with a top-down method and matches them across images, in a recursive way. This is robust to changes to illumination, computationally cheap and do not require large amount of memory. In order to reduce the search cost and allow large view changes, the method is based on the Maximum Pyramid. Moreover, the time-space trajectory of the object was divided into relevant and normal events, that reduces the computational effort and allows us to focus only on those relevant frames of the video stream. The 3D pose was estimated only in the relevant events. Although, the target was localized in all the events to increase the stability and the robustness for the tracking process. Finally, in order to obtain the 3D pose estimation in the complete trajectory, the future work will be an interpolation method, based on the aspect graph describing the structure of the object's surface, can be used to roughly estimate the poses between two relevant events [11].

References

1. Bay, H., Ess, A., Tuytelaars, T., Van Gool, L.J.: Surf: Speeded up robust features. Computer Vision and Image Understanding (CVIU) 110(3) (2008)
2. Brun, L., Kropatsch, W.G.: Construction of Combinatorial Pyramids. In: Hancock, E.R., Vento, M. (eds.) GbRPR 2003. LNCS, vol. 2726, pp. 1–12. Springer, Heidelberg (2003)
3. Gauglitz, S., Höllerer, T., Turk, M.: Evaluation of interest point detectors and feature descriptors for visual tracking. International Journal of Computer Vision, 1–26 (2011)
4. Horn, B.K.P.: Closed-form solution of absolute orientation using unit. J. Optical Society of America 4(4), 629–642 (1987)
5. Klein, G., Murray, D.: Parallel tracking and mapping for small ar workspaces. In: Proc. of ISMAR (2007)
6. Kropatsch, W.G., Bischof, H., Englert, R.: Hierarchies. In: Digital Image Analysis: Selected Techniques and Applications, ch. III Robust and Adaptive Image Understanding (2001)
7. Lepetit, V., Fua, P.: Monocular model-based 3d tracking of rigid objects: A survey. Foundations and Trends in Computer Graphics and Vision 1(1) (2005)
8. Lowe, D.: Distinctive image features from scale-invariant keypoints. Computer Vision and Image Understanding 20 (2004)
9. Marfil, R., Molina-Tanco, L., Rodríguez, S.F.: Real-time object tracking using bounded irregular pyramids. Pattern Recognition Letters (2007)
10. Ozuysal, M., Calonder, M., Lepetit, V., Fua, P.: Fast keypoint recognition using random ferns. IEEE Transactions on Pattern Analysis and Machine Intelligence 32 (2010)
11. Ramachandran, G., Kropatsch, W.: Using aspect graphs for view synthesis. In: Proceedings of Computer Vision Winter Workshop, CVWW (2012)
12. Ravela, S., Draper, B., Lim, J., Weiss, R.: Adaptive tracking and model registration across distinct aspects. In: International Conference on Intelligent Robots and Systems (1995)
13. Sinha, S.N., Frahm, J.M., Pollefeys, M., Genc, Y.: Gpu-based video feature tracking and matching. In: EDGE, Workshop on Edge Computing Using New Commodity Architectures (2006)
14. Torres, F., Kropatsch, W.G., Artner, N.M.: Predict pose and position of rigid objects in video sequences. In: Proceedings of International Conference on Systems, Signals and Image Processing, IWSSIP (2012)
15. Wagner, D., Reitmayr, G., Mulloni, A., Drummond, T., Schmalstieg, D.: Pose tracking from natural features on mobile phones. In: International Symposium on Mixed and Augmented Reality, Cambridge, UK (2008)
16. Wagner, D., Schmalstieg, D., Bischof, H.: Multiple target detection and tracking with guaranteed framerates on mobile phones. In: Proceedings of Int. Symposium on Mixed and Augmented Reality (2009)

Large Scale Experiments
on Fingerprint Liveness Detection

Gian Luca Marcialis[1], Luca Ghiani[1], Katja Vetter[2], Dirk Morgeneier[2], and Fabio Roli[1]

[1] Department of Electrical and Electronic Engineering, University of Cagliari, Italy
{marcialis,luca.ghiani,roli}@diee.unica.it
[2] Crossmatch Technologies Inch.
{katja.vetter,dirk.morgeneier}@crossmatch.com

Abstract. Fingerprint liveness detection consists in extracting measurements, from a fingerprint image, allowing to distinguish between an "alive" fingerprint image, that is, an image coming from the fingertip of the claimed identity, and an artificial replica. Several algorithms have been proposed so far, but the robustness of their performance has not yet been compared when varying several environmental conditions. In this paper, we present a set of experiments investigating the performance of several feature sets designed for fingerprint liveness detection. In particular we assessed the decrease of performance when varying the pressure and the environmental illumination as well as the size of the region of interest (ROI) used for extracting such features. Experimental results on a large data set show the different dependence of some features sets on the investigated conditions.

1   Introduction

Identification of a person based on the so-called biometrics, namely physical (fingerprints, face, iris) or behavioural (gait, signature) attributes is an alternative paradigm to those relying on what he/she possesses (e.g. a card that can be lost or stolen) or remembers (e.g. a password that can be forgotten) [1]. Nowadays, more than ever, it is very important to be able to tell if an individual is authorized to perform actions like entering a facility, access privileged information or even cross a border. Therefore, biometric systems are considered to be more reliable for the recognition of a person than traditional methods.

A biometric system is a pattern recognition system that acquires biometric data from an individual, extracts a features set from the data, compares these features against those stored in a database and executes an action based on the comparison result.

Fingerprints are the most used, oldest and well-known biometric measurements [2]. Fingerprints exhibit important properties as uniqueness and permanence. They are composed of epidermic ridges and valleys flow, which smoothly varies around two or more singular points named *core* and *delta*.

Although fingerprints were often claimed difficult to be steal and reproduced, it has been recently shown that artificial replication is possible [3]. Furthermore, the

G.L. Gimel' farb et al. (Eds.): SSPR & SPR 2012, LNCS 7626, pp. 501–509, 2012.
© Springer-Verlag Berlin Heidelberg 2012

related image obtained by electronic sensors can be difficult to distinguish from "alive" ones, even by visual inspection. Therefore, the development of "liveness" detection techniques is important to try to distinguish if a fingerprint image is coming from an alive person or from a replica. Liveness detection seeks additional data to verify if a biometric measure is authentic. Fingerprint liveness detection, with either hardware-based or software-based systems, is used to check if a presented fingerprint originates from a live person or an artificial finger [4]. It is based on the principle that additional information can be obtained from the data acquired by a standard verification system. This additional data can be used to verify if an image is authentic.

To detect liveness, hardware-based systems use additional sensors to gain measurements outside of the fingerprint image itself while the software-based ones use image processing algorithms to gather information directly from the collected fingerprint. These systems classify images as either live or fake [4-9].

Software-based approaches are cheaper than hardware-based, since these require additional and invasive hardware to measure the liveness directly from the fingertip of people. Instead, software-based must detect liveness from features extracted from the fingerprint images captured by the sensor. In other words, the liveness detection problem is treated as a pattern recognition problem, where a set of features must be selected in order to train an appropriate classifier.

Although several feature sets have been proposed to this aim, it is difficult to assess the state-of-the-art appropriately. Moreover, the variables to be taken into account are so much, that it is often impossible to perform an exhaustive and fair comparison among methods: for example, the sensor type, materials used for fabricating the fingerprint replicas, the environmental conditions as temperature, illumination, the ability of the attacker in pressing the replica on the sensor surface, the fingerprint region used to extract liveness features (ROI), and so on.

Therefore, in this paper we assess a fair comparison of several state-of-the-art approaches to fingerprint liveness detection on a large data set made up of live and fake fingerprint images acquired by the Crossmatch sensor LSCAN Guardian USB. In particular, after analyzing the baseline performance of such algorithms, we focus on three environmental conditions: illumination, pressure and selected ROI. In all cases, we measure the effect on the system performance and point out some countermeasures in order to improve the system robustness.

This paper is organized as follows. Section 2 briefly describes the investigated algorithms. Section 3 describes data set, protocol and experiments performed al results and performance obtained. Section 4 concludes the paper.

2   Investigated Algorithms and Open Issues

In this paper, we reported experimental results on several state-of-the-art fingerprint liveness detection algorithms. We briefly describe them in the following. Further details can be found in the related references.

Local Binary Patterns (LBP) [5]: local binary patterns were first employed for two-dimensional textures analysis and excellent results were obtained due to their invariance with respect to grey level, orientation and rotation. It extracts certain uniform patterns corresponding to micro-features in the image. The histogram of these uniform patterns occurrence is capable of characterize the image as it combines structural (it identify structures like lines and borders) and statistical (micro-structures distribution) approaches. According to [5], a 54-sized feature vector has been obtained.

Power Spectrum [6]: Coli *et al.* analyzed fingerprints images in terms of high frequency information loss. In the artificial fingerprint creation, the ridge-valley periodicity is not altered by the reproduction process but some micro-characteristics are less defined. Consequently, high frequency details can be removed or strongly reduced. It is possible to analyze these details by computing the image Fourier transform modulus also called "power spectrum". We selected twenty sub-bands on the power spectrum, so obtaining a 20-sized feature vector.

Wavelet Energy Signature and Gray-Level Co-occurrence Matrix (GLCM) [7]: The gray-level co-occurrence matrix (GLCM) takes account of how often a pixel with gray-level (grayscale intensity) value i is adjacent to a pixel with the value j. Actually, the element (i, j, d, θ) represents the probability that a couple of pixels x, y at distance d and orientation θ have gray levels i and j respectively.We considered a distance d = 1, so the GLMC matrix is related of local characteristic of the image, and four orthogonal directions for θ, as done in [7]. Therefore we computed four matrices $C_\theta(i, j)$, and, for each of them, a group of ten features, so obtaining a 40-sized feature vector.

Wavelet 2D [8]: wavelet decomposition of an image lead to the creation of four sub-bands: the approximation sub-band containing global low frequency information, and three detail sub-bands containing high frequency information. The image is decomposed in four levels, three sub-bands for each one, and three different wavelet filters (Haar, Daubechies (db4) and Biorthogonal (bior2.2)), so obtaining a 70-sized feature vector.

Curvelet [9] decomposition is very efficient for representing edges and other singularities along fingerprint ridges due to his high directional sensitivity and his high anisotropy. We consider two different sets of features, also called "signatures" in [9]:

- Curvelet energy signature: the energies of the 18 sub-bands are measured by computing means and variances of curvelet coefficients.
- Curvelet co-occurrence signature: for each of the 18 sub-bands, the GLCM (Gray Level Co-occurrence Matrix) is calculated together with 10 corresponding features.

In order to test the accuracy of these algorithms, Refs. [5-9] use appropriate data sets, but different, one each others, with respect to the size, the materials used for replicating fingerprints. Classifiers used are different too. Therefore, methods cannot be compared by simply considering results reported in those papers.

Moreover, no experimental investigation has been done on the environmental conditions affecting the liveness detection performance, especially if these algorithms must be intergrated in real fingerprint verification systems [8]. From this point of view, we can consider characteristics "intrinsic" to the feature set chosen, like the location of the region of interest selected for feature extraction, and "intrinsic" to the sensor adopted that, like the pressure of the attacker on the fingerprint sensor surface, and the environmental illumination. These points may impact on the quality of the liveness feature set extracted, thus being crucial to analyze the "robustness" of such feature sets against attacks based on fake fingerprints. In fact, if the system can be less robust where environmental illumination changes, a person can take advantage of this, by choosing the best moment to attack the system or modifying the environmental light. The same holds for the pressure. Finally, if the features are sensitive to the ROI position, a wrong ROI extraction could lead to misclassification errors. However, it is unknown at which extent they are important, and, eventually, which countermeasures can be adopted to reduce their impact. This is the scope of the present paper, where these characteristics are analyzed by experiments, and some preliminary observations are drawn from the obtained results.

3   Experimental Results

3.1   Data Sets and Experimental Protocol

We used a four data sets for our investigations:

1) **D.1.** This data set is made up of 1816 live fingerprints and 1624 fake fingerprints created with commonly used materials, uniformly distributed along replicas: silicone, gelatin, wood glue and latex. Molds are made up of plastiline-like material which allows to replicate the 2d contour of the fingertip. Fingerprint images have been acquired by the Crossmatch LSCAN Guardian USB electronic sensor. The data set has been subdivided in two parts, namely, training set and test set, according to the protocol adopted in the recent Second edition of Fingerprint Liveness Detection Competition (LivDet2011) [10]. A multi layer perceptron (MLP) has been trained on the first part of data, so obtaining the *baseline fingerprint liveness detector*. The MLP output is interpreted, as usual, as liveness detection score in the range [0,1]. Features sets are extracted from ROIs located on the core of fingerprint images (the core is centre of the fingerprint image according to [1]), as shown in Fig. 1(a). Such ROIs are quadrangular regions. Two sizes has been used: 80x80 pixels and 160x160 pixels.

2) **D.2.** This data set has been built to test the variability of the feature set performance when ROIs are not correctly located. Four different location errors are studied, as reported in Fig. 1(b). We tested the performance on the baseline system with 80x80 pels ROIs, but also the performance which can be obtained by adding to the training set also patterns extracted from wrongly located ROIs.

3) **D.3.** This data set has been built for evaluating the impact of the pressure of the fake fingerprint on the sensor surface when baseline system is used with 80x80 pels for ROIs. In order to generate novel images, we put a increasing weight from 500 g to 4000 g on the fake fingerprint, thus simulating the different pressure. Obtained data set is thus made of 500 fake fingerprint frames (with increasing weight over frames) per three different types of silicon-like materials, gelatin and latex for replicating fingerprints. On overall, 2,100 test images have been used (not overlapped with the training/test set used for the baseline system). Effect is studied by evaluating the variation and correlation degree of the liveness detection score with the related weight on the fake fingerprint.

4) **D.4.** This data set has been built for evaluating the impact of the environmental illumination on the features sets related to fake fingerprints. It has been organized as follows: 103 fake fingerprint images simulate device initialization in dark room without any enviromental illumination (condition 1); 103 fake fingerprint images simulate device initialization with directed light (condition 2). Influence of environmental illumination has been tested by evaluating average and standard deviation of the liveness score for conditions 1-2.

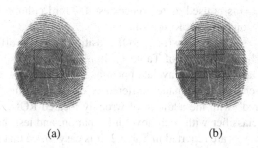

(a)   (b)

Fig. 1. ROI positions. (a) Baseline ROI. (b) Wrong locations: up, down, left, right.

3.2 Baseline Results

Figs. 2(a-b) show the ROC curves of the baseline system according to D.1 data set.

It can be seen that LBP feature set leads to the best performance, whilst the feature set Power spectrum one leads to the worst one. Experiments show that the error slightly depends on the size of the ROI. Doubling the size of the ROI, that is, from 80x80 to 160x160 pixels, the error decreases of about 3% on average. It is worth noting that, using a ROI of 160x160 pixels, a fraction of background could be present during feature extraction, but this does not seem relevant on the basis of reported results.

Moreover, the rank of investigated feature sets, from the best one to the worst one, is: LBP, Wavelet2D, Curvelet Energy, Curvelet Glcm, GLCM, PS. It is independent on the ROI size. Therefore, all feature sets are sensitive to the ROI size, in a very similar manner.

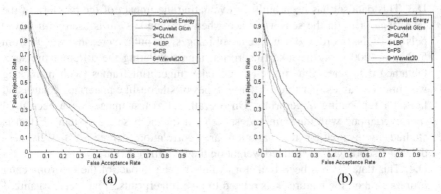

Fig. 2. ROC curves showing the expected performance on the D.1 test set. (a) ROI size: 80x80 pixel. (b) ROI size: 160x160 pixel.

3.3 Performance on Wrongly Located ROIs

Table 1 summarizes results as follows. For each feature set (first column), EER is reported when training the classifier on centered images (second column), that is, on the baseline system. This is used as reference result. Third column reports the same classifier when tested on different ROI positions.

It is possible to see that a wrong ROI position weakly affects the system performance (second-third column of Table 1). In all cases, a loss of performance is about 1%. In particular, LBP and Wavelet appear as the preferred feature sets.

In order to recover this performance difference, one could think that training set should be "empowered" with the addition of wrongly located ROIs. Therefore, we re-trained the baseline classifier with such novel information, and test the performance of the same test sets. Results are reported in Table 2. It is easy to see that recovering above performance variation by adding to the training set bad centered images is not possible.

These results allow to observe that ROI location does not appear a very crucial point for all feature sets, so estimation errors of the ROI do not impact on the final system performance, independently on features sets adopted among the ones investigated here.

Table 1. EER for all feature sets and different ROI positions

Feature set	Baseline test	Wrongly located ROIs test
LBP	10.90	12.45
GLCM	27.32	28.03
Wavelet2D	20.94	22.07
Curvelet Energy	23.34	24.56
Curvelet GLCM	27.32	27.99
Power Spectrum	28.84	28.59

Table 2. EER for all features sets on the classifier "empowered" with patterns extracted from wrongly located ROIs

Feature set	Empowered classifier Wrongly located ROIs test	Empowered classifier Baseline test
LBP	12.66	11.84
GLCM	26.07	25.68
Wavelet2D	21.36	20.21
Curvelet Energy	24.15	23.74
Curvelet GLCM	27.31	26.66
Power Spectrum	29.24	29.76

3.4   Effect of Pressure Variation

We summarise basic results in Table 3, where we report, for each material and each feature set, the sign of the correlation between the system output (1 – liveness score) and the applied pressure. If this correlation is more than 0.5 we write '+' on the related cell in Table 3; if it is less than -0.5, '-' is written; if correlation is averagely "low", that is, between -0.5 and 0.5, we indicate this fact by '*'.

The most promising feature set is still LBP, because is positive in almost every case. This means that, if a fingerprint is spoof, the more the pressure, the more this evidence. The same holds for the live class. This can be also desired in optical sensors as the one adopted for these tests, because the more the pressure, the more the sharpness of related image. Worth noting, all feature sets are positively correlated with pressure when live fingerprints are submitted. This is good, but the decrease of the posterior probability for the fake class may lead to an increase of false acceptance rate (see for example Wave and PS columns).

Table 3. Positive ('+'), negative ('-') or no correlation ('*') between system output (posterior probability of the correct class) and the applied pressure

	LBP	GLCM	Wavelet2D	Curv. Energy	Curv. GLCM	P.S.
Silicone 1	+	+	-	*	*	-
Silicone 2	*	-	-	-	*	-
Silicone 3	+	+	-	+	-	*
Gelatine	+	-	-	+	*	-
Latex	+	-	-	+	*	-

We reported in Table 4 if the system outputs, for each feature set, fall into the related average range of spoof and fingers on D.1 Test. Symbols in Table 4 can be interpreted as follows:

"+": object is always in range from 500g-4000g;
"*": object is not in range for each weight, but there is a visible tendency (correlation exists);

"-": object is not in range for each weight, but there is not a visible tendency (no correlation exists).

Reported results largely confirm observations from Table 3. LBP, Curvelet GLCM and Curvelet Energy related outputs fall in the standard range in the most of cases, thus they cannot considered as preferred feature sets.

Table 4. Positive ('+'), negative ('-') or no correlation ('*') between system output (posterior probability of the correct class) and the applied pressure when considering only standard output range (D1.1. Test) and related outputs

	LBP	GLCM	Wavelet2D	Curv. Energy	Curv. GLCM	P.S.
Silicone 1	+	+	*	-	+	+
Silicone 2	+	*	*	+	+	*
Silicone 3	*	*	*	+	+	-
Gelatine	+	*	*	+	-	+
Latex	*	-	*	+	+	+

3.5   Effect of Environmental Illumination

Results are shown in Table 5. Conditions 1-2 are the ones explained in Section 3.1. In all cases, similar output values, that is, liveness score, are obtained, thus illumination does not appear as a relevant environmental conditions for the system output.

On the basis of available data set and reported experiment, LBP and Wavelet feature sets appear as the preferred ones.

Table 5. Posterior probabilities of spoof samples by varying the illumination conditions, and related standard deviation (in brackets)

	Liveness score average and standard deviation		
	Baseline	Condition (1)	Condition (2)
LBP	0.960(0.019)	0.962(0.066)	0.969(0.028)
GLCM	0.706(0.008)	0.660(0.066)	0.666(0.081)
Wavelet2D	0.941(0.020)	0.934(0.022)	0.932(0.019)
Curv. Energy	0.788(0.024)	0.747(0.043)	0.762(0.043)
Curv. GLCM	0.684(0.064)	0.630(0.097)	0.659(0.084)
PS	0.765(0.021)	0.805(0.028)	0.813(0.019)

4   Conclusions

In this preliminary set of large scale experiments on fingerprint liveness detection, we focused on the baseline system performance of several state-of-the-art feature sets, with respect to some of variability elements. In particular, we have studied an intrinsic characteristic of the feature sets, namely, the choice of the ROI (size and location), and two external characteristics of the fingerprint sensor, that is, the pressure of the fake fingerprint on the sensor surface, and the environmental illumination, which may impact on the captured image, and, thus on the feature set extracted.

On the basis of reported experiments, we noticed that, in almost all cases, the larger is ROI size, the better is the system performance, but the location of the ROI is not relevant. It has also been obtained that the most of features sets are sensitive to the pressure, thus improving or worsening the liveness detection result depending on how much an individual tend to press the fake fingerprint (see in particular the LBP and Curvelet Energy cases). Finally, environmental illumination is not crucial, since the system output is substantially stable independently on rough changes of the light intensity.

These results point out that performance is not yet acceptable for their integration in standard fingerprint verification algorithms, but also that they have some invariant characteristics to some settings and environmental conditions which make them worthy of further theoretical and experimental investigations.

Acknowledgments. This work has been supported by the sponsored research agreement between University of Cagliari and Crossmatch Technologies Inch.

References

1. Jain, A.K., Flynn, P., Ross, A.: Handbook of Biometrics. Springer (2007) ISBN 9780387710402
2. Maltoni, D., Maio, D., Jain, A.K., Prabhakar, S.: Handbook of Fingerprint Recognition. Springer, New York (2003) ISBN 0387954317
3. Matsumoto, T., Matsumoto, H., Yamada, K., Hoshino, H.: Impact of artificial 'gummy' fingers on fingerprint systems. In: Proceedings of SPIE, vol. 4677 (2002)
4. Coli, P., Marcialis, G.L., Roli, F.: Vitality Detection from Fingerprint Images: A Critical Survey. In: IEEE/IAPR 2nd International Conference on Biometrics ICB (2007), doi:10.1007/978-3-540-74549-5_76
5. Nikam, S.B., Aggarwal, S.: Local binary pattern and wavelet-based spoof fingerprint detection. International Journal of Biometrics 1(2), 141–159 (2008)
6. Coli, P., Marcialis, G.L., Roli, F.: Power spectrum-based fingerprint vitality detection. In: IEEE Int. Work. on Automatic Identification Advanced Technologies AutoID 2007, pp. 169–173 (2007)
7. Nikam, S.B., Aggarwal, S.: Wavelet energy signature and GLCM features-based fingerprint anti-spoofing. In: IEEE Int. Conf. On Wavelet Analysis and Pattern Recognition (2008), doi:10.1109/ICWAPR.2008.4635872
8. Abyanka, A., Schuckers, S.: Integrating a wavelet based perspiration liveness check with fingerprint recognition. Pattern Recognition 42, 452–464 (2009)
9. Nikam, S.B., Agarwal, S.: Fingerprint Liveness Detection Using Curvelet Energy and Co-occurrence Signatures. In: IEEE Fifth International Conference on Computer Graphics, Imaging and Visualization (2008) doi:10.1109/CGIV.2008.9
10. Yambay, D., et al.: LivDet 2011 - Fingerprint Liveness Detection Competition 2011. In: 5th IAPR/IEEE Int. Conf. on Biometrics (ICB 2012), pp. 208–215 (2012), doi:10.1109/ICB.2012.6199810

Implicit and Explicit Graph Embedding: Comparison of Both Approaches on Chemoinformatics Applications

Benoit Gaüzère[1], Makoto Hasegawa[2], Luc Brun[1], and Salvatore Tabbone[2]

[1] Université de Caen - Basse Normandie-GREYC CNRS UMR 6072, Caen, France
[2] Université de Lorraine-LORIA CNRS UMR 7503, Nancy, France

Abstract. Defining similarities or distances between graphs is one of the bases of the structural pattern recognition field. An important trend within this field consists in going beyond the simple formulation of similarity measures by studying properties of graph's spaces induced by such distance or similarity measures . Such a problematic is closely related to the graph embedding problem. In this article, we investigate two types of similarity measures. The first one is based on the notion of graph edit distance which aims to catch a global dissimilarity between graphs. The second family is based on comparisons of bags of patterns extracted from graphs to be compared. Both approaches are detailed and their performances are evaluated on different chemoinformatics problems.

1 Introduction

Graphs allow to encode not only the elementary features of a set of objects but also the relationships between these objects. Graphs constitute thus an efficient tool to model complex objects or phenomenons. Classification, regression or clustering operations applied on graphs constitute an important sub field of the structural pattern recognition framework, all these operations being based either implicitly or explicitly on a distance or a similarity measure.

Definition of graph distances or graph similarity measures constitute an active field within the structural pattern recognition framework. Main distance definitions are based on one hand on the size of the minimum common super graph or the maximum common sub graph and on the other hand on the minimal number of vertex/edge insertion/removal/relabeling required to transform one graph into an other. This last measure called the edit distance is related to the notion of maximum common sub graph [2] and provides a nicely interpretable measure of distance between two graphs. Moreover, assuming basic properties on the elementary edit costs, one can show that this distance satisfies the 4 properties of a distance (positivity, separation, symmetry and triangular inequality). However, the number of calculus required by edit distance computation grows exponentially with the number of nodes of both input graphs and several heuristics have been proposed to obtain efficient but sub optimal edit distances [8,14].

G.L. Gimel' farb et al. (Eds.): SSPR & SPR 2012, LNCS 7626, pp. 510–518, 2012.

Restricting structural pattern recognition to pairwise comparisons of graphs leads to restrict the field to efficient but often basic classification or clustering algorithms such as the k-nearest neighbor or the k-median algorithms. Computing efficiently more global feature on a set of graphs requires additional properties of the topology of the graph's space implicitly defined by a distance measure between graphs. Such a problem may be solved by defining a natural embedding of graphs. Such an embedding leads to associate explicitly or implicitly a vector to each graph and to define a metric between these vectors which corresponds to the metric defined by the graph distance. However, the fact that a distance satisfies the 4 usual distance's axioms does not insure that an embedding within an Hilbert space may be associated to graphs [3]. More precisely, given a set of n graphs, and a matrix D encoding all pairwise distances between the graphs of the set, the type of space induced by D is provided by the spectrum of the matrix $S^c = -\frac{1}{2}(I - \frac{1}{n}ee^t)D(I - \frac{1}{n}ee^t)$ where e is the vector of ones (Section 2). The metric space encoding similarities between graphs is a Krein space if this spectrum contains negative eigen values and an Hilbert space otherwise. Krein spaces have unusual properties such as possibly negative distances between graphs. In order to avoid to use such spaces, several authors [8] regularize the matrix S^c in order to remove its negative eigen values hereby slightly modifying the original metric defined by D. An alternative approach consists in associating a vector to each graph using for example spectral analysis [12]. The approach is in this case slightly different since the metric defined between vectors does not correspond to a metric initially defined in the graph's space. A last approach consists in defining a symmetric similarity measure between graphs. The matrix encoding all pairwise similarities between the graphs of a set is called the Gram matrix of this set. If for some sets of graphs the Gram matrix is non definite positive the embedding space associated to this similarity measure is a Krein space. Otherwise, the embedding space corresponds to an Hilbert space and the similarity measure is called a kernel. In this last case the similarity function corresponds to a scalar product between the vectors associated to both input graphs. One may note the symmetry between embeddings based on distances and similarity measures. Both problems are indeed related, since within an Hibert space or a Krein space a distance measure may be defined from scalar products and conversely.

This paper provides a comparison of both distance and similarity approaches. We first present two important methods within the distance based embedding framework in Section 2. Then we provide an overview of graph kernels methods in Section 3. Both approaches are finally compared in Section 4 on several chemoinformatics data sets.

2   Graph Embedding

Embedding graph in vector space aims to define points in a vector space such that their mutual distances is as close as possible to the initial graph dissimilarity matrix wrt a cost function (eg. graph edit distance). More precisely, let $G=\{g_1, ..., g_n\}$ be a set of graphs and $d\colon G \times G \to \mathbb{R}$ a graph distance function between pairs of its elements and let $D = D_{ij} = d(g_i, g_j) \in \mathbb{R}^{n \times n}$ be the dissimilarity matrix. The aim in the graph embedding is to provide n p-dimensional vectors x_i such that the distance between x_i and x_j is as close as possible to the similarity D_{ij} between g_i and g_j. Thus, embedding graph into a vector space make the graph available to numerous machine learning techniques which require vectorial representation.

Numerous approaches [4,8,12,18] have been proposed in the literature. In this paper we recall the approach proposed in [8] and which is based on the constant shift embedding [15]. Originally, the constant shift embedding was introduced in order to embed pairwise data into Euclidean vector spaces. In [8], the authors adapt this method to the domain of graphs. The key issue is to convert general dissimilarity data into metric data.

Constant Shift Embedding. We briefly describe the method of Roth et al. [15] to embed D (restricted by the constraint that self-dissimilarities are equal to zero) into a Euclidian space, without influencing the distribution of the initial data. The aim of this approach is to determine a matrix $\tilde{D}$ close to D such that it exists a set of vectors $(x_i)_{i \in \{1,\dots,n\}}$ with $\tilde{D}_{ij} = \|x_i - x_j\|^2$. The solution of this problem is of course not unique since any translation of vectors x_i would provide a same distance matrix. In order to overcome this problem we perform a centralization of matrix D by considering $S^c = -\frac{1}{2}D^c$, where $D^c = QDQ$ is the definition of the centralization and $Q = I_n - \frac{1}{n}e_n e_n^\mathsf{T}$ is the projection matrix on the orthogonal complement of $e_n = (1, \dots, 1)$. Such a matrix S^c satifies:

$$D_{ij}^c = S_{ii}^c + S_{jj}^c - 2S_{ij}^c \tag{1}$$

If S^c is semidefinite positive, its singular value decomposition is equal to $S^c = V \Lambda V^t$ where columns of V encode the eigen vectors of S^c and Λ is a diagonal matrix encoding its positive eigen values. Setting $X = V(\Lambda)^{\frac{1}{2}}$, we obtain $S^c = XX^t$. Hence, each element S_{ij}^c of S^c is equal to a scalar product $< x_i, x_j >$ between the lines i and j of X. Equation 1 may thus be intepreted as a classical result on Euclidean norms stating that the squared distance between two vectors is equal to the sum of the squared norms of these vectors minus twice their scalar product. The scaled eigen vectors $(x_i)_{i \in \{1,\dots,n\}}$ provide thus a natural embedding of matrix D when matrix S^c is definite positive.

Following the constant shift embedding S^c can be transformed into a positive semidefinite matrix (see Lemma 2 in [15]):

$$\tilde{S} = S^c - \lambda_n(S^c)I_n$$

where $\lambda_n(S^c)$ is the minimal eigenvalue of the matrix S^c. The diagonal shift of the matrix S^c transforms the dissimilarity matrix D in a matrix representing squared Euclidean distances. The resulting embedding of D is defined by (minimal shift theorem):

$$\tilde{D}_{ij} = \tilde{S}_{ii} + \tilde{S}_{jj} - 2\tilde{S}_{ij} \iff \tilde{D} = D - 2\lambda_n(S^c)(e_n e_n^\mathsf{T} - I_n)$$

Setting Dimension. In PCA it is known that small eigenvalues contain noise. Therefore, the dimensionality p can reduced by choosing $t \leq p$. Consequently, a $n \times t$ map matrix $X_t = V_t(\Lambda_t)^{1/2}$ will be computed where V_t is the column-matrix of the selected eigenvectors and Λ_t the diagonal matrix of the corresponding eigenvectors.

Graph Similarity Measure. Let us recall how the similarity (or dissimilarity) in the domain of graphs can be computed. Similarity between two graphs is almost always referred as a graph matching problem. Graph matching is the process of finding a correspondence between nodes and edges of two graphs that satisfies some constraints ensuring that similar substructures in one graph are mapped to similar substructures in the other. Many approaches have been proposed to solve the graph matching problem. Among these, the graph edit distance has been widely used as the most appropriate similarity measure for representing the distance between graphs. In this paper we use two approaches [7, 14] based both on an approximation of the graph edit distance as an instance of an assignment problem where the edit distance between two graphs is based on a bipartite graph matching. In both approaches, the authors formulate the assignment problem by cost matrix where the optimal match is solved by the Hungarian algorithm.

In [14], each entry of the cost matrix encodes the cost of a node substitution, deletion or insertion. Substitution costs are defined using the Hungarian algorithm on the set of incident edges of both vertices. The substitution cost of two incident edges takes into account the label of the edges and the label of the incident vertices.

In [7], the cost matrix is encoded differently using a distance (HEOM distance) between node signatures. A signature describes the node (degree, attributes), the incident edges attributes but also the degrees of the adjacent nodes. The main differences with the previous approach is that no prior computation (learning phase) of the edit cost function are needed and more global information are taken into account on the graph in the signature.

3   Graph Kernels Methods

Graph embedding methods aim to associate coordinates to graphs. Such an embedding allows us to define similarity or distance measures from graph's coordinates. An alternative strategy consists in computing directly a similarity measure between graphs. Graph kernels can be understood as symmetric graph similarity measures. Using a semi definite positive kernel, the value $K(G, G')$, where G and G' encode two input graphs corresponds to a scalar product between two vectors $\phi(G)$ and $\phi(G')$ in some Hilbert space, called feature space. Distance between two graphs G and G' can be retrieved from kernel function by the relation (Equation 1) $d^2(G, G') = K(G, G) + K(G', G') - 2K(G, G')$. Thanks to this possibly implicit embedding of graphs into an Hilbert space, graph kernels can be combined with machine learning methods based on scalar products between input data, such as the well-known SVM. This use of kernels into statistical machine learning method, called kernel trick, provides a natural connection between structural pattern recognition and graph theory on one hand and statistical pattern recognition on the other hand.

A large family of graph kernels are based on the extraction of a bag of patterns from each graph. Methods corresponding to this family consists in three key steps. First, bags of pattern are built from graphs by enumerating a given set of patterns $\mathcal{P}$ within graphs. This enumeration, possibly implicit, defines an embedding of graphs into a feature space where each dimension is associated to a pattern. Second, global similarity between graphs is defined by the similarity of their bags of patterns. Finally, this similarity between bags is based on a sub kernel between pattern $k_p : \mathcal{P} \times \mathcal{P} \rightarrow \mathbb{R}$. This sub kernel k_p encodes the similarity of two patterns extracted from graphs.

A common approach defines the set of patterns as all possible walks included within a graph. A first method, defined by Gärtner and al., proposes a formulation of a kernel based on graph product and powers of adjacency matrix [5] which computes the number of common walks of the two graphs to be compared. A second method proposed by Kashima and al. [9] defines a random walk kernel by considering the probability $p(w|G)$ of encountering a random walk w within a graph G. Using such probabilities, the kernel is defined as:

$$k_{rw}(G, G') = \sum_{w \in \mathcal{W}(G)} \sum_{w' \in \mathcal{W}(G')} p(w|G)p(w'|G')k(w, w') \qquad (2)$$

with $\mathcal{W}(G)$ denoting the set of walks extracted from G. Vishwanathan [16] has proposed an unified and efficient computation of both methods by means of Sylvester equations. However, comparison of graphs based on random walks suffers from tottering. Tottering corresponds to possible infinite oscillations between two nodes which leads to artificially long walks not representative of the structure of the graphs.

The major drawback of methods based on linear patterns is that linear structures can not represent most of the structural information encoded within complex and non linear structures such as molecular graphs. In order to tackle this limitation, Ramon and Gärtner [11] and Mahé and Vert [10] have proposed a kernel based on the comparison of non linear patterns. This set of non linear patterns is defined as the set of tree patterns, denoted $\mathcal{T}_P$, i.e. trees where a same node can appears more than once. This kernel maps each tree pattern having a different labeling to a specific dimension in an infinite feature space representing all possible tree patterns. This embedding may be encoded by projection $\phi_{\mathcal{T}_P}(G)$ and graph kernel is defined as an inner product between these projections: $K_{\mathcal{T}_P}(G, G') = \langle \phi_{\mathcal{T}_P}(G), \phi_{\mathcal{T}_P}(G') \rangle$. Computation of this kernel is based on a recursive comparison of neighborhood matching sets up to a given depth [10].

Mahé and Vert have proposed in [10] an extension of tree pattern kernel which weights each tree pattern according to its structural complexity. This measure of structural complexity may be encoded by the branching cardinality or the ratio between number of nodes and depth of tree patterns. However, since the

number of occurrences of each tree pattern is not explicitly computed during
kernel computation, only an a priori weighting of tree patterns can be applied to
each tree pattern. In addition, as observed on walks, tree patterns suffers from
tottering. However, Mahé and Vert [10] have proposed an extension to prevent
tottering based on a transformation of input graphs.

Another method based on non linear patterns computes an explicit distribution of each pattern within a graph. This method, called treelet kernel [6],
explicitly enumerates treelets included within a graph, the set of treelets being
defined as the 14 trees having a size lower than or equals to 6 nodes. Thanks
to the limited number of different patterns encoding treelets, an efficient algorithm allows to enumerate the number of occurrences of each pattern within a
graph. Given this first enumeration, a first kernel on unlabeled graphs can be
defined. When applying this method to set of labeled graphs, labeling information included within treelets is encoded by a canonical key. This canonical key is
defined such that given two treelets with a same structure, their canonical key is
similar if and only if the two treelets are isomorphic. Each treelet being uniquely
identified by the index of its pattern and its canonical key, any graph G can
be associated to a vector $f(G)$ which explicitly encodes the number of occurrences of each treelet t by $f_t(G)$. Using this vector representation, treelet kernel
between graphs is defined as a sum of sub kernels between common treelets of
both graphs:

$$K_T(G, G') = \sum_{t \in T(G) \cap T(G')} k(f_t(G), f_t(G')) \tag{3}$$

where $k(.,.)$ defines any positive definite kernel between real numbers such as
linear, Gaussian or polynomial kernel. In the same way as tree pattern kernel,
each pattern can be weighted in order to improve kernel accuracy as follows:

$$K_T(G, G') = \sum_{t \in T(G) \cap T(G')} w(t) k(f_t(G), f_t(G')) \tag{4}$$

However, conversely to tree pattern kernel, the explicit enumeration of each sub
structure provided by treelet kernel method allows to weight each pattern according to a property to predict and not only according to an a priori function.
This weighting may be computed using variable selection algorithms [6] or multiple kernel learning [1].

4   Experiments

Our first experiment is based on two regression problems[1] which consist in predicting molecule boiling points. The first dataset is composed of 150 alkanes, an
alkane corresponding to an acyclic molecule solely composed of carbons and hy-
drogens. A common encoding is to implicitly encode hydrogen atoms using the

---

[1] These databases are available on the IAPR TC15 Web page:
   http://www.greyc.ensicaen.fr/iapr-tc15/links.html#chemistry

Table 1. Boiling point prediction

Method	RMSE (°C)		Computation Time (s)
	Alkane	Acyclic	
(1) Gaussian edit distance	10.01	10.27	1.35
(2) Random Walks Kernel	16.28	18.72	19.10
(3) Treelet Kernel	**1.92**	**8.10**	**0.07**
(4) Tree Pattern Kernel	3.48	11.02	4.98
(5) Graph Embedding	6.15	12.3	7.21

valency of carbon atoms. Such an encoding allows to represent alkanes as unlabeled graphs. The second dataset is composed of 183 acyclic molecules, each molecule being composed of heteroatoms and thus encoded as acyclic labeled graphs. We evaluate the boiling point of each molecule using several test sets composed of 10% of the database, the remaining 90% being used as training set.

First, we can note that linear patterns (Table 1, Line 2) do not encode enough structural information to correctly predict boiling points of molecules. Conversely, methods based on bags of non linear patterns obtain better results (Table 1, Lines 3 and 4). Differences between Treelet Kernel and Tree Pattern Kernel may be explained by the use of a Gaussian kernel for Treelet kernel, which is not possible within the tree pattern computational scheme. In addition, limitation on the size of patterns induced by explicit enumeration of treelets does not have a large influence on these problems since molecules have a low number of atoms. Second, Table 1 shows results obtain by graph embedding method (Line 5) and a Gaussian kernel applied on the approximate edit distance as defined by [14] (Line 1). Graph embedding results have been computed using different subsets of eigenvalues obtained by applying a threshold on variance encoded within the matrix.We can note that the improvement on edit distance approximation leads to better results approximation defined in [14] when applied to unlabeled graphs. Finally, the last column of Table 1 shows the time required to compute the Gram matrix on acyclic dataset. Note that while most of the methods are computed within the same order of magnitude (seconds), Treelet Kernel can be computed in 0.07 seconds thanks to the efficient enumeration of a limited set of patterns.

The second experiment consists of two classification problems. The first one is taken from the Predictive Toxicity Challenge [17] which aims to predict carcinogenicity of 416 chemical compounds applied to female (F) and male (M) rats (R) and mice (M). This experiment consists of ten different datasets for each class of animal, each of them being composed of one train set of about 310 molecules and one test set of about 35 molecules. The second dataset is provided by [13]. This database defined from the AIDS Antiviral Screen Database of Active Compounds is composed of 2000 chemical compounds. These chemical compounds have been screened as active or inactive against HIV and they are split into three different sets. A train set composed of 250 compounds used to train SVM, a validation set composed of 250 compounds used to find parameters giving the best prediction accuracy and a test set composed of remaining 1500 compounds.

Table 2. Classification accuracy on the two classification experiments

Method	PTC				AIDS
	MM	FM	MR	FR	
(1) Gaussian Edit Distance	**223**	212	194	234	**99.7%**
(2) Random Walks Kernel	216	221	201	232	98.5%
(3) Treelet Kernel (TK)	208	205	209	212	99.1%
(4) TK with variable weighting	217	224	**223**	**250**	**99.7%**
(5) Graph Embedding	218	**227**	206	239	**99.7%**

Table 2 shows the amount of correctly classified molecules over the ten test sets for each class of animal for the first dataset and the accuracy obtained by differents methods on AIDS dataset. Note however that results obtained by tree pattern kernel are not displayed since the source code provided by the authors is restricted to molecules with a degree bounded by 4. First, we can note that method based on graph embedding (Table 2, Line 5) leads to globally better results than Gaussian kernel applied on an approximation of the graph edit distance (Table 2, Line 1). In the same way, graph embedding methods outperforms Random Walks Kernel (Table 2, Line 2) and Treelet Kernel (Table 2, Line 3). However, combination of a variable weighting scheme with Treelet Kernel (Table 2, Line 4) improves the prediction accuracy of Treelet Kernel and obtains the best results on 3 over 5 datasets a slightly lower prediction accuracy than graph embedding methods on the two others. However, weighting each treelet according to a property to predict requires about 30 minutes for each train set of PTC dataset whereas computational time of graph embedding is performed in about 74 seconds for each PTC dataset. The accuracy provided by variable weighting can thus be obtained at the cost of an high computational time.

5 Conclusion

As shown in our experiments graph kernels and graph embedding methods provide close results in most of experiments. This last point is expected since as stressed in this paper both approaches are closely related. The main difference of both approaches should rather be determined from their potential usage. On one hand, Graph embedding methods provide an explicit embedding in a finite dimensional space for each input data sets. Hence, this approach is not restricted to kernel methods but can use explicitly the coordinates associated to graphs. On the other hand, this approach requires the whole data set to compute an embedding. Graph kernels based on bag of patterns, only require to compute the similarity between an input graph and the one of the training set. These methods may thus be used on unbounded data sets. The choice between both approaches should thus be determined from the ability for a given application to obtain the whole data set and from the ability of algorithms applied on graphs to be kernelized.

Acknowledgments. The authors thanks Salim Jouili for providing the graph embedding code.

References

1. Villemin, D., Gaüzére, B., Brun, L., Mokhtari-Brun, M.: Graph kernels based on relevant patterns and cycle information for chemoinformatics. In: Proceedings of ICPR 2012 (to be published, 2012)
2. Bunke, H.: Error correcting graph matching: On the influence of the underlying cost function. IEEE Transactions on Pattern Analysis and Machine Intelligence 21(9), 917–922 (1999)
3. Dattorro, J.: Convex Optimization and Euclidean Distance Geometry. Meboo Publishing, USA (2005)
4. de Mauro, C., Diligenti, M., Gori, M., Maggini, M.: Similarity learning for graph-based image representations. Pattern Recognition Letters 24(8), 1115–1122 (2003)
5. Gärtner, T., Flach, P.A., Wrobel, S.: On graph kernels: Hardness results and efficient alternatives. In: Proceedings of the 16th Annual Conference on Computational Learning Theory and the 7th Kernel Workshop (2003)
6. Gaüzére, B., Brun, L., Villemin, D.: Two new graphs kernels in chemoinformatics. Pattern Recognition Letters (in Press, 2012)
7. Jouili, S., Tabbone, S.: Graph Matching Based on Node Signatures. In: Torsello, A., Escolano, F., Brun, L. (eds.) GbRPR 2009. LNCS, vol. 5534, pp. 154–163. Springer, Heidelberg (2009)
8. Jouili, S., Tabbone, S.: Graph Embedding Using Constant Shift Embedding. In: Ünay, D., Çataltepe, Z., Aksoy, S. (eds.) ICPR 2010. LNCS, vol. 6388, pp. 83–92. Springer, Heidelberg (2010)
9. Kashima, H., Tsuda, K., Inokuchi, A.: Marginalized Kernels Between Labeled Graphs. Machine Learning (2003)
10. Mahé, P., Vert, J.-P.: Graph kernels based on tree patterns for molecules. Machine Learning 75(1), 3–35 (2009)
11. Ramon, J., Gärtner, T.: Expressivity versus efficiency of graph kernels. In: First International Workshop on Mining Graphs, Trees and Sequences, Citeseer, pp. 65–74 (2003)
12. Ren, P., Wilson, R., Hancock, E.: Graph Characteristics from the Ihara Zeta Function. In: da Vitoria Lobo, N., Kasparis, T., Roli, F., Kwok, J.T., Georgiopoulos, M., Anagnostopoulos, G.C., Loog, M. (eds.) S+SSPR 2008. LNCS, vol. 5342, pp. 257–266. Springer, Heidelberg (2008)
13. Riesen, K., Bunke, H.: IAM Graph Database Repository for Graph Based Pattern Recognition and Machine Learning. In: da Vitoria Lobo, N., Kasparis, T., Roli, F., Kwok, J.T., Georgiopoulos, M., Anagnostopoulos, G.C., Loog, M. (eds.) S+SSPR 2008. LNCS, vol. 5342, pp. 287–297. Springer, Heidelberg (2008)
14. Riesen, K., Bunke, H.: Approximate graph edit distance computation by means of bipartite graph matching. Image Vision Comput. 27(7), 950–959 (2009)
15. Roth, V., Laub, J., Kawanabe, M., Buhmann, J.M.: Optimal cluster preserving embedding of nonmetric proximity data. IEEE Trans. Pattern Anal. Mach. Intell. 25(12), 1540–1551 (2003)
16. Vishwanathan, S.V.N., Schraudolph, N., Kondor, I.R., Borgwardt, K.: Graph kernels. Journal of Machine Learning Research 11 (April 2010)
17. Toivonen, H., Srinivasan, A., King, R., Kramer, S., Helma, C.: Statistical evaluation of the predictive toxicology challenge 2000-2001. Bioinformatics 19(10), 1183–1193 (2003)
18. Torsello, A., Hancock, E.R.: Graph embedding using tree edit-union. Pattern Recognition 40(5), 1393–1405 (2007)

Modeling Spoken Dialog Systems under the Interactive Pattern Recognition Framework

M. Inés Torres, Jose Miguel Benedí, Raquel Justo, and Fabrizio Ghigi

Dpto Electricidad y Electrónica, Universidad del País Vasco UPV/EHU, Spain
{manes.torres,raquel.justo,fabrizio.ghigi}@ehu.es
Instituto Tecnológico de Informática, Universidad Politécnica de Valencia, Spain
jbenedi@iti.upv.es

Abstract. The new Interactive Pattern Recognition (IPR) framework has been recently proposed. This proposal lets a human interact with a Pattern Recognition system allowing the system to learn from the interaction as well as adapt it to the human behavior. The aim of this paper is to apply the principles of IPR to the design of Spoken Dialog Systems (SDS). We propose a new formulation to present SDS as an IPR problem. To this end some extensions to the IPR approach are proposed. Additionally a user model based on the IPR paradigm is also defined. We applied the proposed formulation to compose a preliminary graphical model that has been experimentally developed to deal with a Spanish dialog task. An initial maximum likelihood strategy for the dialog manager actions along with a stochastic simulation of user behavior have allowed to get new dialogs. The preliminary evaluation of these results allowed us to consider this formulation as a promising framework to deal with SDS.

1 Introduction

Interacting with machines has proved to help many human activities. But machines can also take advantage of the human feedback to improve their performances. In this context the new Interactive Pattern Recognition (IPR) framework has been recently proposed [1]. This proposal lets a human to interact with a Pattern Recognition (PR) system allowing the system to learn from the interaction as well as adapt it to the human behavior. IPR has been applied to some classical PR problems such as interactive transcription of handwritten and spoken documents, computer assisted translation, interactive text generation and parsing, among others [1].

Speech-based human-computer interaction seems to be a straightforward application of the IPR framework. However the management of a SDS is a very complex task that involves many other problems to be solved like the Automatic Speech Recognition (ASR), semantic representation and understanding, answer generation, etc. The Dialog Manager (DM) is the main component of a SDS. It is devoted to manage the state of the dialog as well as the dialog strategy. According to the information provided by the user the DM must decide the action to be taken. Due to its complexity the design of DM has been traditionally related to rules based methodologies, sometimes combined with some statistical knowledge [2] [3]. However during the last few years some proposal based on classical pattern recognition

G.L. Gimel' farb et al. (Eds.): SSPR & SPR 2012, LNCS 7626, pp. 519–528, 2012.

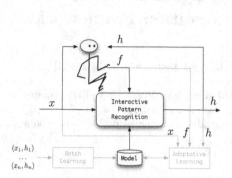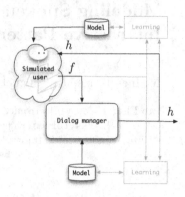

Fig. 1. a) Diagram of an Interactive Pattern Recognition system that provides an hypothesis h given a stimulus x and a user feedback f. b) Diagram of an SDS where a DM provides an hypothesis h given the previous hypothesis and the user feedback f. In the next interaction step a *simulated* user provides the feedback f given its previous feedback and the system hypothesis h.

methodologies can be found in the literature [4] [5] [6] [3] [7]. Some of them are based on Markov decision process and reinforcement learning. However, only small problems can be addressed in this framework up to now, since global optimization is still a hard computational problem. This problem is addressed by factorization of the states space [5] and partition of the dialog state distributions [8].

The aim of this paper is to apply the principles of IPR to the management of SDS. We propose a new formulation to present SDS as an IPR problem. To this end some extensions to the IPR approach presented in [1] are also proposed (Section 2). We deal with both speech and text-based dialog systems, decoding as well as with the relationship between SDS and decision theory. Additionally a user model based on the IPR paradigm is also defined (Section 3). We have applied the proposed formulation to compose a preliminary graphical model that deals with both manager and user behavior (Section 4). The preliminary evaluation of these models over a Spanish Dialog Task (Section 5) allowed us to consider this formulation as a promising framework to deal with SDS.

2   Spoken Dialog Systems in the IPR Framework

Let x be an input stimulus, observation or signal and h an hypothesis or output, which a classical PR system has to derive from x. Let $\mathcal{M}$ be a *model* or set of models used by the system to derive its hypotheses. In general, $\mathcal{M}$ is obtained through a *batch* learning procedure from a given set of training pairs (x_i, h_i) from the task being considered. Under the IPR framework [1] the user of the system provides some (perhaps null) feedback signals, f, which may iteratively help the system to refine or to improve its hypothesis until it is finally accepted, as diagram in Figure 1a) shows. The interaction allows to consider the human feedback f. Thus, an adaptive *on-line* procedure can also be now considered.

Under the decision theory point of view, after I iterations the system has received $F = f^1, f^2, \ldots, f^I$ user feedbacks and has produced $H = h^1, h^2, \ldots, h^I$ hypotheses. The loss function $l(x, h, h^\star, H, F)$ defines the cost incurred by the system due to an erroneous hypothesis, being $h^\star$ the *correct* one. A best hypothesis is now given by:

$$\hat{h} = \arg\min_{h \in \mathcal{H}} R_l(h|x, H, F) = \arg\min_{h \in \mathcal{H}} \sum_{h^\star \in \mathcal{H}} l(x, h, h^\star, H, F) Pr(h^\star|x, H, F) \quad (1)$$

where $R_l(h|x, H, F)$ is the risk, or cost of proposing a hypothesis h. A basic simplification is to ignore the user feedback except for the last interaction and/or hypothesis; that is define the loss function as $l(x, h, h^\star, h', f)$. Then the classical PR *minimum-error criterion* corresponds to a $0/1$ *loss function* defined to be 0 if $h = h^\star$ and 1 otherwise. In such a case the Baye's decision rule is simplified to maximize the posterior $Pr(h|h', f)$, and a best hypothesis $\hat{h}$ is obtained as follows:

$$\hat{h} = \arg\max_{h \in \mathcal{H}} P(h|x, h', f) \quad (2)$$

Equation 2 corresponds to a zero-order approach, where $\hat{h}$ is derived using only the feedback obtained in the previous iteration step and h' is the history. In a first-order approach h' can be represented by the optimal hypothesis $\hat{h}$ obtained by the system in its previous interaction step for the given x.

Let now apply the IPR paradigm to provide a new formal framework for SDS. We first assume that the system interacts with the user providing a first hypothesis through a greeting turn that acts as unique stimulus x. So, we can ignore it from now on. Then, the probability to be maximized in Equation 2 is now $P(h|h', f)$. This maximization procedure defines the way the Dialog Manager of a SDS choose the best hypothesis, i.e. the best action at each interaction step, given the previous hypothesis h' and the user feedback f. However, alternative criteria can also be considered to make this decision. In fact, the $0/1$ *loss function* may be substituted by a *loss function* proportional to the number of user turns in a dialog or to an estimation of the number of turns required to successfully ending a dialog, at each interaction step. Thus, the estimation of the best hypothesis $\hat{h}$ given at each interaction step may not be based on the classical *minimum-error criterion* criterium. Moreover, in SDS this decision is usually taken according with a DM strategy that maximizes the probability of achieving the unknown user goal at the end of the interaction procedure while minimizing the cost of getting them [5] [9].

In a SDS, the interpretation of the user feedback can not be considered a deterministic process. Let now $\mathcal{D}$ be the space of decoded feedback signals and $d \in \mathcal{D}$ the decoding of f. Considering d as a hidden variable we can rewrite Equation 2 as follows:

$$\hat{h} = \arg\max_{h \in \mathcal{H}} P(h|h', f) = \arg\max_{h \in \mathcal{H}} \sum_{d} P(h, d|h', f) \quad (3)$$

Approximating the sum with the value of the mode, applying basic probabilities rules, ignoring terms which do not depend on the optimization variables h, d and then assuming independence of $P(h|d, h', f)$ on f given h', d and of $P(f|d, h')$ on h' given d, Equation 3 can be rewritten as follows [1]:

$$\hat{h} \approx \arg \max_{h \in \mathcal{H}} \max_{d} P(h|d, h')P(f|d)P(d|h') \tag{4}$$

where f is the user turn, d is the decoding of the user turn, h is the hypothesis or output produced by the system and h' is the *history of the dialog*.

The optimizing problem to be solved is to find $\hat{h}$ according to Equation 4. A suboptimal approach is a two step decoding. Find first an optimal user feedback:

$$\hat{d} = \arg \max_{d} P(f|d)P(d|h') \tag{5}$$

Then, use $\hat{d}$ to decode $\hat{h}$ as follows:

$$\hat{h} \approx \arg \max_{h \in \mathcal{H}} P(h|\hat{d}, h') \tag{6}$$

A Particular Case: A Text-Based Dialog System Let now consider a deterministic feedback that can be specified as a function $d : \mathcal{F} \to \mathcal{D}$ mapping each user turn signal into its corresponding unique decoding $d = d(f)$. For instance, if f is a sequence of written words then $d(f)$ is a deterministic decoding of f in terms of semantic units, i.e. an unambiguous semantic tagging procedure. In such a particular case Equation 4 becomes:

$$\hat{h} \approx \arg \max_{h \in \mathcal{H}} \max_{d} P(h|d, h')P(d|h') \tag{7}$$

Equation 7 stands for a text-based dialog system whereas Equation 4 stands for a SDS. In both equations, $P(d|h')$ represents the semantic model of the task that is constrained by the *history* h'. Finally $P(h|d, h')$ includes both the task and dialog manager models since this distribution provides the hypotheses, i.e. outputs of the system, given the *history* h' and the user intervention d.

3   A Simulated User

Equation 3 summarizes a system that provides an hypothesis h given its previous hypothesis h' and a user feedback f, according with the distribution $P(h|h', f)$. In fact, this system is the Dialog Manager of the SDS that needs to take decisions at each interaction step. The probability distribution $P(h|h', f)$ can be approached by some system model $\mathcal{M}_S$ whose parameters need to be estimated from data trough a learning process. Thus, corpora consisting of sets of (h, h', f) can be used to train $\mathcal{M}_S$. However, *loss functions* that take into account the success in achieve the user goals and the system cost minimization, which is measured in terms of number of turns, are not very well supported. The final goal of a Dialog Manager is to achieve the user goals and expectations, which are absolutely unknown for the system [5]. Thus, online learning, i.e., learning from the interaction, is the only way in this case for the system to be trained by users. Therefore, a large amount of dialogs as well as real users with different goals, expectations and behavior are required.

This is the reason that statistical dialog managers are usually trained by simulated users [3]. Accurate training of DM includes a first *batch* training step using large dialog corpora and a second training step with simulated users. On line learning algorithms would also allow the system to be adapted to the task and to the real user behavior, when running.

No user model is considered up to now in the IPR framework [1]. A simulated user must provide to the system the feedback f at each interaction step. Let now the user feedback f depend on its previous feedback f' according to some unknown distribution $P(f|f', h)$, which represents the user response to the history of system hypotheses and user' feedbacks. This distribution stands for user behavior and represents, to some extend, the user model defined in classical statistical frameworks proposed for spoken dialog systems [5].

Let us define a model of *user behavior* $\mathcal{M}_U$ that is applied by the user to produce the feedback f. Such a model can also be defined under the IPR framework considering now the user point of view. Thus, after I iterations the user has received $H = h^1, h^2, \ldots, h^I$ hypotheses from the system and has generated $F = f^1, f^2, \ldots, f^I$ feedbacks to the system. The *loss function* can now be defined as $l(f, f^\star, F, H)$ such that the estimation of the user *best* feedback is given by:

$$\hat{f} = \arg\min_{f \in \mathcal{F}} R_l(f|F, H) \tag{8}$$

where $R_l(f|F, H)$ is now the user interactive conditional risk. Ignoring the history of system hypotheses except for the last user feedback and considering again a $0/1$ *loss function*, a *best* user feedback $\hat{f}$ is the one that maximizes the posterior $P_{\mathcal{M}_U}(f|f', h)$.

$$\hat{f} = \arg\max_{f \in \mathcal{F}} P(f|f', h) \approx \arg\max_{f \in \mathcal{F}} P_{\mathcal{M}_U}(f|f', h) \tag{9}$$

where $\hat{f}$ is estimated using only the hypothesis produced by the system and the optimal feedback produced by the user in the previous interaction step according with its *user model*. Figure 1b) shows a Simulated User (SU) interacting with a Dialog Manager according with a model of the *user behavior*.

Equation 9 represents the way the user decides the feedback f. As the case of the system model, alternative criteria could be also considered to simulate the user behavior. In fact, many simulated user models can be found in the SDS bibliography [7][10].

Feedback f' produced by user in the previous interaction is not corrupted by any noisy channel, such as an ASR system, before arriving to the user again. Thus, a deterministic decoding $d : \mathcal{F} \to \mathcal{D}$ maps each user turn signal into its corresponding unique decoding $d = d(f)$. If f is a sequence of acoustic observations then $d(f)$ is a deterministic decoding in terms of semantic units. We are now representing $d(f)$ just by d and $d(f')$ by d'. Then Equation 9 can now be rewritten as

$$\hat{d} = \arg\max_{d \in \mathcal{D}} P(d|d', h) \approx \arg\max_{d \in \mathcal{D}} P_{\mathcal{M}_U}(d|d', h) \tag{10}$$

where $\mathcal{D}$ represents the set $d(\mathcal{F})$.

4 Modelling the DM and the User Behavior

In this section we are providing a preliminary approach to model both the dialog manager hypothesis probability distributions $P(h|d, h')$ and the user feedback probability distribution $P(d|h, d')$. We are defining a graphical model consisting of sets of states representing (h, d) pairs. Some of these states correspond to the DM and are labelled by (d, h'), being d the output of the Speech understanding system given the user feedback f and h' the system hypothesis at the previous interaction. Then, states corresponding to the user are labelled by pairs (h, d') where h is the system hypothesis and d' is the deterministic decoding of the previous user feedback f'. The DM generates a system hypothesis h at each machine turn and the simulated user provides a feedback f at each user turn. Figure 2 shows a diagram of a machine and a user turn. Additionally, each

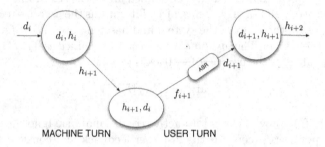

Fig. 2. Machine state at interaction turn i is labelled by the pair (d_i, h_i) where d_i has been updated. The system then generates a hypothesis h_{i+1} that updates the user model state labelled by (h_{i+1}, d_i). In the user turn a feedback f_{i+1}, decoded as d_{i+1} by the speech understanding module, is provided. d_{i+1} updates the machine state for interaction turn $i + 1$.

state needs to be labelled by the values of all the relevant internal variables, thus leading to an *attributed* model. Then, an additional alphabet appears to represent variables and internal attributes.

The parameters of the model can be estimated in a three step learning procedure as follows:

1. Get a dialog corpus consisting of pairs of user and machine turns. Then get an initial maximum likelihood estimation of the parameters of both models.
2. Define a Dialog Manager strategy and several simulated user model behaviors. Define also error recovery strategies. Run the system until desired dialog goals are successfully achieved for different simulated user behaviors.
3. Run the SDS with real users while adapting the Dialog Manager using real interaction feedbacks.

5   Experimental Application

This work mainly focusses on the formulation of a SDS as an IPR problem. Thus, the aim of this section is to put into practice the approach described in Sec. 4, for a Spanish dialog task.

DIHANA Corpus. [11]. It is a set of spoken dialogs in Spanish, providing information related to the Spanish railway system. The corpus is composed by 900 dialogs acquired by the Wizard of Oz technique. It consists of 225 speakers (153 males and 72 females) asking for information about long-distance train timetables, fares, destinations and services. In order to obtain more realistic dialogs, the speakers had to reach a certain goal in each dialog, while they were entirely free to express themselves as desired. The corpus consists of 9.133 system turns and 6.280 user turns with a vocabulary of 823 words.

This corpus has been annotated in terms of Dialog Acts (DA), according to an udapted version of the Interchange Format defined in the C-STAR project. Each minimum segment of a turn is labelled with a single label composed by three hierarchical levels [12]. The first one is the most generic and represent the action of the segment: *affirmation, opening, closing, confirmation, wait, undefined, negation, not_understood, new_question, question, answer.* The second level stores the information directly related to the first level, and the third level contains other data present in the segment. In DIHANA corpus second and third level labels are combinations of the 13 variables that define the task: *ticket_class, destination, day, arrival_hour, departure_hour, nil, origin, price, service, duration, train_type, relative_order_number* and *number_of_trains.* Note that these variable labels are just descriptors of the data and do not store real values, i.e. there are not attributes in the sense defined in previous section. An example of a labelled user turn in DIHANA corpus is:

| (U:Question:Price:Day) *I would like to know the fare on next Monday* |

where "U" indicates a user turn, "Question" is the first level that represents the action performed in the segment, "Price" the second level as the user is asking about the fare and "Day" is the third level because it is the additional information provided by the user.

Building the Model. A graph like the one shown in Fig 2 has been obtained for DIHANA corpus. This preliminary model only considers a list of 13 attributes consisting of the values of the variables defining the task. The information provided by some of these attributes is required to successfully complete a dialog. In the user turn *I would like to know the fare on next Monday*, the user is providing an attribute value to the variable *day* and expects the system to prompt the attributed value of the variable *price*. To this end the dialog manager may need additional values such as *destination* and *hour*. Thus, it needs to take into account which attributes have been already provided by the user and which of them are not still filled, throughout the dialog. A more sophisticated model

Table 1. Sizes of the Dialog Manager and User Model (number of nodes and number of edges) which have been trained with two subsets (equal size) of the corpus.

	DIHANA	
	# nodes	# no. edges
system	7,466	5,049
user	7,387	4,289

would include a larger list of attributes for each variable including, for instance, confident measures provided by the ASR and other modules of the SDS system.

Each node associated to a system turn would be determined by a couple of three-level DA labels associated to the previous system action and to the previous user feedback, (d_i, h_i) in Fig. 2, as well as the list of variables that has been already provided. In the same way, each user turn node is determined by two three-level DA labels associated to the previous user feedback and to the system action, (h_{i+1}, d_i) in Fig. 2, as well as the list of attributes.

We then wanted to get a separated maximum likelihood estimation of both the system model probability distribution $P_{\mathcal{M}_S}(h|h', d)$ defined in Section 2 and the *user model* probability distribution $P_{\mathcal{M}_U}(d|d', h)$ defined in Section 3. To this end, the DIHANA corpus has been split into two equal subsets to train the DM and the user model respectively. The sizes of both models are summarized in Table 1. This table shows that the graph has an affordable size for vocabularies associated to restricted domain tasks, which are the most frequent dialog applications.

We then put both models to interact with the aim to obtain new dialogs. For these preliminary experiments the 0/1 *loss function* was selected as a DM strategy, whereas a random behavior was set as simulated user behavior. Additionally, a smoothing strategy was required for unseen events. Let us note that given a node in the user turn, the simulated user provides an output that may lead to an unknown node of the graph. In the same way, given a node in the system turn, the system model may provide an action that has not an associated edge and a target node in the graph. In this work a simple smoothing strategy was adopted; it consists on looking for the most similar state in the graph that is closest to the end when an unseen node has to be reached. An example of a dialog obtained with these preliminary approaches is given below:

(**M:Open:Nil:Nil**) *Welcome to the railway information system. How can I help you?*
(**U:Question:Nil:Orig:U:Question:Nil:Dest:U:Question:Nil:Day:CITY,ORIG**)
I'd like to know trains from $ORIG to $CITY to travel on a working day next week.
(**M:Confirm:DepartHour:Dest:Day:Orig:M:Wait:Nil:Nil:M:Answer:**
DepartHour:DepartHour:RelOrderNumber:TrainsNumber:M:NewQuestion:
Nil:Nil:CITY,DEST,ORIG,DATE,HOUR,DEPARTHOUR,
ORDNUMBER) *I'm searching for trains from $ORIG to $DEST on $DATE*
Wait a moment. There are several trains. $ORDNUMBER leave at $DEPARTHOUR
and the $ORDNUMBER leave at $HOUR. Do you need something more?
(**U:Closing:Nil:Nil:CITY,DEST,ORIG,DATE,HOUR,DEPARTHOUR,**
ORDERNUMBER) *No, thank you.*
(**M:Closing:Nil:Nil:CITY,DEST,ORIG,DATE,HOUR,DEPARTHOUR,**
ORDERNUMBER) *Thank you for using this service. Have a nice journey.*

The previous example shows that coherent dialogs that successfully finish can be generated with the formulation presented in previous sections. These preliminary results give rise to an in depth study of the proposed strategies that could lead to a promising approach in the field of DMs.

6   Concluding Remarks and Future Work

We have proposed a new formulation to model SDS in IPR problem framework. To this end some extensions to the IPR approach have been proposed. We dealt with both speech and text-based dialog systems, decoding as well as the relationship between SDS and decision theory. Additionally a user model based on the IPR paradigm is also defined. We have applied the proposed formulation to compose a preliminary graphical model dealing with both manager strategy and user behavior. The preliminary evaluation of these models over a Spanish Dialog task allowed us to consider this formulation as a promising framework to deal with SDS. Future work includes choosing alternative smoothing schemas to deal with more complex error recovering strategies. In the same way adequate loss functions will define more sophisticated Dialog Manager strategies and simulated user model behaviors.

Acknowledgments. Work supported by Spanish Ministry of Science under Consolider Ingenio program MIPRCV CSD2007-00018 and grant TIN2011-28169-C05-04 and by Basque Government under grant GIC10/158 IT375-10

References

1. Toselli, A.H., Vidal, E., Casacuberta, F. (eds.): Multimodal Interactive Pattern Recognition and Applications. Springer (2011)
2. Lemon, O., Pietquin, O.: Machine learning for spoken dialogue systems. In: Proceedings of the 10th European Conference on Speech Communication and Technology. Interspeech, Antwerp, Belgium, August 27-31, pp. 2685–2688 (2007)
3. Griol, D., Hurtado, L.F., Segarra, E., Sanchis, E.: A statistical approach to spoken dialog systems design and evaluation. Speech Communication 50, 666–682 (2008)
4. Meng, H., Wai, C., Pieraccini, R.: The use of belief networks for mixed-initiative dialog modeling. IEEE Trans. Speech and Audio Processing 11(6), 757–773 (2003)
5. Williams, J.D., Young, S.: Partially observable markov decision processes for spoken dialog systems. Computer Speech and Language 21, 393–422 (2007)
6. Sarigaya, R., Gao, Y., Picheney, M.: A comparison of rule-based and statistical methods for semantic language modeling and confidence measurement. In: Proceedings of the Human Language Technology Conference. North American Chapter of the Association for Computational Linguistics Annual Meeting. HLT-NAACL, Boston, pp. 65–68 (2007)
7. Cuayáhuitl, H., Renals, S., Lemon, O., Shimodaira, H.: Evaluation of a hierarchical reinforcement learning spoken dialogue system. Computer, Speech and Language 25, 395–429 (2010)

8. Williams, J.D.: Incremental partiten recombination for efficient tracking of multiple dialog states. In: Proceedings of International Conference on Acoustics, Speech and Signal Processing (ICASSP), Dallas, USA (2010)
9. Hajdinjak, M., Mihleič, F.: The pradise evaluation framework: Issues and findings. Computational Linguistics 32(2), 263–272 (2006)
10. Lee, S., Eskenazi, M.: An unsupervised approach to user simulation: toward self-improving dialog systems. In: Proceedings of the SIGDIAL Conference, Seoul, Korea, pp. 50–59 (July 2012)
11. Benedí, J., Lleida, E., Varona, A., Castro, M., Galiano, I., Justo, R., López, I., Miguel, A.: Design and acquisition of a telephone spontaneous speech dialogue corpus in Spanish: DIHANA. In: Proceedings of LREC 2006, Genoa (May 2006)
12. Alcácer, N., Benedí, J.M., Blat, F., Granell, R., Martínez, C.D., Torres, F.: Acquisition and labelling of a spontaneous speech dialogue corpus. In: Proceeding of 10th International Conference on Speech and Computer (SPECOM), Patras, Greece, pp. 583–586 (2005)

Hierarchical Graph Representation for Symbol Spotting in Graphical Document Images

Klaus Broelemann[1], Anjan Dutta[2], Xiaoyi Jiang[1], and Josep Lladós[2]

[1] Institute for Computer Science, University of Münster, Germany
[2] Computer Vision Center, Universitat Autonoma de Barcelona, Spain

Abstract. Symbol spotting can be defined as locating given query symbol in a large collection of graphical documents. In this paper we present a hierarchical graph representation for symbols. This representation allows graph matching methods to deal with low-level vectorization errors and, thus, to perform a robust symbol spotting. To show the potential of this approach, we conduct an experiment with the SESYD dataset.

Keywords: hierarchical graph representation, graph matching, maximal clique finding, symbol spotting, graphics recognition.

1 Introduction

Symbol spotting has experienced a growing interest among the graphics recognition community. It can be defined as locating a given query graphical symbol into a set of graphical document images. Example applications of symbol spotting are finding a mechanical part in a database of engineering drawings or retrieving invoices of a provider from a large database of documents by querying a particular logo. The problem of symbol spotting in documents for real world situation is difficult as the documents often suffer from different noises. Graphs are very effective tool to represent any graphical elements, especially line drawings. Hence, in line drawings represented by graphs, the problem of symbol spotting can be formulated as a subgraph matching problem, where graph theory offers robust approaches to solve it. This explains our motivation to work with graphs.

The list of approaches proposed for spotting symbols in graphical documents is long [10]. The current paper only mentions the recent works dealing with the graph representations: Nayef and Breuel [7] proposed a branch and bound algorithm for spotting symbols in documents, where they used geometric primitives of images as features. Luqman *et al.* [6] proposed a graph embedding based subgraph spotting method applied to symbol spotting. Here the candidate regions containing symbols are filtered out beforehand using some criteria of loop. Recently Dutta *et al.* [5] proposed graph factorization based symbol spotting methods for architectural floorplans. Of course, the above set of algorithms deal with some kind of error tolerance when matching the subgraphs but they seldom can handle disconnection between nodes i.e. when two nodes are disconnected but supposed to belong to the same graph. In case of such disconnection usually

G.L. Gimel' farb et al. (Eds.): SSPR & SPR 2012, LNCS 7626, pp. 529–538, 2012.
© Springer-Verlag Berlin Heidelberg 2012

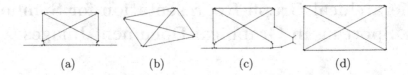

(a)   (b)   (c)   (d)

Fig. 1. Examples for low-level segmentation errors

the methods just loose the connectivity which reduce some topological feature of the graph. So handling these kinds of distortions is the inspiration of proposing a hierarchical representation of graph where we deal with different kind of errors propagated from the lower level to the graph level.

The construction of graph representations of documents is followed by some inter-dependent pre-processing steps viz. binarization, skeletonization, polygonal approximation. These low-level pre-processing steps result in the vectorized documents which often contain some structural errors. In this work our graph representation considers the critical points as the nodes and the lines joining them as the edges. So often the graph representation contains spurious nodes, edges, disconnection between nodes etc (see Figure 1). Our present work deals with these kinds of distortion in the graph level. To do that we propose hierarchical representation of graphs. The hierarchical representation of graphs allows to incorporate the various segmentation errors hierarchically. The main motivation of our work comes from [1], where the authors introduced a hierarchical representation of the segmented image regions to support an approximated maximal clique finding algorithm in matching objects in natural images.

The rest of the paper is organized into four sections. In Section 2 we present the hierarchical representation of graphs to represent a database in terms of the descriptors of graph paths. Section 3 describes the hierarchical graph matching methods we used. Section 4 contains the detailed experimental results. After that, in Section 5, we conclude the paper and discuss future work.

2   Hierarchical Graph Representation

An essential part for graph-based symbol spotting methods is the representation of symbols. This representation often contains low-level vectorization errors that will affect later graph matching methods. In this section we present a hierarchical representation that overcomes these problems by covering different possible vectorizations.

First we will give a brief overview of the initial vectorization and some errors that can occur due to it. Afterwards we will describe our hierarchical representation and how this representation overcomes the vectorization errors.

2.1   Vectorization

Graph representation of documents follows some pre-processing steps, vectorization is one of them. Here vectorization can be defined as approximating the binary images to a polygonal representation. In our method we have done it with the Rosin-West algorithm [8] which is implemented in the Qgar package[1]. This particular algorithm works without any parameter except one to prune the isolated components. The algorithm produces a set of critical points and the information whether they are connected. Our graph representation considers the critical points as the nodes and the lines joining them as the edges.

Vectorization Errors. The resulting graph can contain vectorization errors. Reasons for that can be inaccurate drawings, artefacts in the binarization or errors in the vectorization algorithm. There are different kinds of vectorization errors that can occur. Among these, we concentrated on the following ones:

Gaps. In the drawing there can be small gaps between lines that ought to be connected. Reasons for that are inaccurate drawings and mistakes in the binarization. The result are either two unconnected nodes at the border of the gap or a node on one and an edge on the other side of the gap. Beside being caused by errors, gaps can also be drawn intentionally to separate nearby symbols.

Split nodes. On the other hand, one original node can be split into two or more nodes. This can happen, if lines in the drawing do not intersect exactly at one point. Another reason are artefacts from the skeletonization step. Nearby nodes that seem to be a split node can be the result of fine details instead of vectorization errors.

Dispensable nodes. The vectorization can create nodes of order two that divide a straight edge into two or more parts. One reason for these nodes are small inaccuracies in the drawing that cause a local change in direction. For a later symbol spotting, these nodes are often undesired and should be removed. Nevertheless, in some cases such structures reflect details of the symbol.

Though all these errors can be corrected in a post-processing step, a simple post-processing causes other problems: often it is not clear for the system whether a situation is an error or intentional. To deal with this uncertainty, we introduce a hierarchical representation that will be described in the next part.

2.2   Hierarchical Graph Construction

This section describes the construction of hierarchical graph that is able to cover different possible vectorizations. This enables a later graph matching algorithm to deal with the uncertainties whether a part of the graph is intentional or caused by a vectorization error.

The basic idea of our approach is to extend a given graph G so that it contains the different possibilities. These possibilities are connected hierarchically. This allows us to embed the constraint not to match two hierarchically connected nodes into the graph matching and, thus, only accept one alternative. In Section 3 we will give further details for the graph matching and this constraint.

---

[1] http://www.qgar.org/

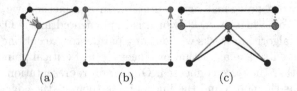

Fig. 2. Three cases for simplification. Displayed are the original nodes and edges (black) and the simplified nodes and their edges (gray): (a) Merge nodes (b) Remove dispensable node (c) Merge node and edge.

In order to create different possible vectorizations, we take the initial vectorization represented in G and simplify it step by step. For this purpose, we identify three cases that allow a simplification. These three cases will be motivated in the following. Afterwards, a formal definition of these cases is given.

Nearby nodes. Both gaps in drawing as well as split nodes result in nodes near to each other and can be solved by merging these nodes. Since nearby nodes can also be the result of correct vectorization, e.g. due to two nearby symbols, we store both versions and hierarchically connect the merged node with the basic nodes. The merged node inherits all connection of its basic nodes. Figure 2 (a) shows an example for such a merging step.

Dispensable nodes. In case of dispensable nodes, the vectorization error can be solved be removing the node. Again, a hierarchical structure can store both versions. As described before, we only consider dispensable nodes that have two neighbors. The simplified versions of these neighbors are directly connected. This is shown in Figure 2 (b). Applying this rule multiple times allows us to remove chains of dispensable nodes.

Nodes near to edges. The third simplification is the merging of nodes with nearby edges. In this way the second kind of gaps can be corrected. To merge a node with an edge, the edge has to be divided into two edges by a copy of the node. This can be seen for an example in Figure 2 (c).

Recursive Definition. Based on the previous motivation we will give a recursive definition of our hierarchical graphs that reflects the construction algorithm based on the vectorization outcome.

The result of the vectorization is an undirected graph $G = (V_G, E_G, \sigma_G)$ where V_G it the set of nodes, $E_G \subseteq V_G \times V_G$ is the set of edges and $\sigma_G : V_G \to \mathbb{R}^2$ is a labeling function that maps the nodes to their coordinates in the plane.

A hierarchical graph has two kinds of edges: undirected neighborhood edges and directed hierarchical edges. Hierarchical edges represent simplification operations, i.e. they link nodes from the original graph arising from the vectorization to successor nodes representing simplified vectorizations. Formally, we define a hierarchical graph H as a tuple $H = (V, E_N, E_H, \sigma)$ with the neighborhood edges $E_N \subseteq V \times V$ and the hierarchical edges $E_H \subseteq V \times V$.

To detect the three previously described cases, we define:

1. function $\delta_1 : V \times V \to \{0,1\}$ to test for pairs of nearby nodes.
2. function $\delta_2 : V \to \{0,1\}$ to test for dispensable nodes.
3. function $\delta_3 : V \times E \to \{0,1\}$ to test for nodes near to edges.

Furthermore, given two nodes $u, v \in V$, let $u \rightsquigarrow v$ denote that v is a hierarchical successor of u and $L(u)$ denote the set of all predecessors of u that belong to G: $L(u) = \{v \in V_G | v \rightsquigarrow u\}$. Based on these functions and formulations we can define the hierarchical simplification $H = \mathcal{H}(G) = (V, E_N, E_H, \sigma)$ of G by the following rules:

Initial. As initialization for the recursion, G is a subgraph of H, i.e. $V_G \subseteq V$ and for $u, v \in V_G : (u, v) \in E_G \Leftrightarrow (u, v) \in E_N$

Merging. For $u, v \in V$ with $\delta_1(u, v) = 1$ there is a merged node $w \in V$ with

 − w is a hierarchically successor of u and v:
   $\forall s \in V : s \rightsquigarrow w \Leftrightarrow s \rightsquigarrow u \vee s \rightsquigarrow v \vee s \in \{u, v\}$
 − w has all neighbors of u and v except u and v:
   $\forall s \in V : (s, w) \in E_N \Leftrightarrow ((s, u) \in E_N \vee (s, v) \in E_N) \wedge s \notin \{u, v\}$
 − w lies in the center of its leaf nodes: $\sigma(w) = \frac{1}{|L(w)|} \sum_{s \in L(w)} \sigma(s)$

Removing. For a dispensable node $u \in V$ with $\delta_2(u) = 1$ there exist two nodes $v, w \in V_G$ with $(u, v), (u, w) \in E_N$. Since v and w can have hierarchical successors, these have to be included in the definition: for all $v_i : (v_i, u) \in E_N \wedge v \in L(v_i)$ there exists a $\bar{v}_i$. In the same way a set of $\bar{w}_j$ is defined.

 − $\bar{v}_i$ hierarchical successor of v_i: $(v_i, \bar{v}_i), (w_j, \bar{w}_j) \in E_H$
 − to cover all possibilities, there is neighborhood connection between all of $\bar{v}_i$ and all $\bar{w}_j$. Furthermore, the $\bar{v}_i$ has the same connections as v_i with exception of the removed node u:
   $(s, \bar{v}_i) \in E_N \Leftrightarrow ((s, v_i) \in E_N \wedge s \neq u) \vee \exists j s = w_j$. (analogous for w_j)
 − The coordinates do not change: $\sigma(v_i) = \sigma(\bar{v}_i)$, $\sigma(w_j) = \sigma(\bar{w}_j)$

Node/Edge merging. For $u \in V, e = (v, w) \in E$ with $\delta_3(u, e) = 1$ there exist simplifications $\bar{u}, \bar{v}, \bar{w}$ with

 − $\bar{u}, \bar{v}, \bar{w}$ are hierarchical successors of u, v, w:
   $\forall s \in V : s \rightsquigarrow \bar{u} \Leftrightarrow s \rightsquigarrow u \vee s = u$ (analog for v, w)
 − $\bar{u}$ intersects the edge between $\bar{v}$ and $\bar{w}$:
   $\forall s \in V : (s, \bar{u}) \in E_N \Leftrightarrow ((s, u) \in E_N \vee s \in \{\bar{v}, \bar{w}\}$
 − The coordinates do not change: $\sigma(u) = \sigma(\bar{u})$, $\sigma(v) = \sigma(\bar{v})$ and $\sigma(w) = \sigma(\bar{w})$

Based on these recursive rules, we construct the smallest hierarchical graph that satisfies these rules, i.e. no additional nodes are added. For our hierarchical graph we defined the testing functions $\delta_1, \delta_2, \delta_3$ by using thresholds: for δ_1 define an upper bound for the distance between two nodes, for δ_3 we do the same for the distance between edge and node. We define δ_2 by a threshold for the relative distance of the dispensable node from the direct line between it's neighbors. In contrast to other definitions like the angle at the dispensable node, this can easily be extended to chains of dispensable nodes.

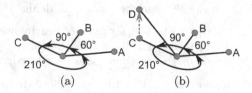

Fig. 3. Example for node labels for graphs based on angles between edges: (a) for plane graphs and (b) for hierarchical graphs. Both will be labeled with $(90, 210, 60)$.

Pre-processing. Depending on the chosen thresholds there can be a huge number of possibilities and, thus, a large hierarchical graph. To reduce the size of the hierarchy, we perform a pre-processing step. The idea is that in some cases the confidence in the simplification is strong enough not to store both versions, e.g. it is not very likely that a one-pixel gap is intentional. For that purpose we perform merging and removing steps on the graph with stricter thresholds. With these thresholds we do not create hierarchically connected possibilities, but change the original graph structure.

3   Graph Matching

In this section we will describe how to make use of the hierarchical graph representation described in the previous section for subgraph matching in order to spot symbols for vectorial drawings. Graph matching has a long history in pattern recognition and there exist several algorithms for this problem [3]. Our approach is based on solving the maximal weighted clique problem in association graphs [2]. In this section we will first give a brief overview over the graph matching algorithm. This method relies on similarities between nodes. Hence, we will present a geometric node similarity for hierarchical graphs afterwards.

Given two hierarchical graphs $H^i = (V^i, E^i_N, E^i_H, \sigma^i)$, $i = 1, 2$, we construct the association graph A. Each node of A consists of a pair of nodes of H^1 and H^2, representing the matching between these nodes. Two nodes (u_1, u_2), $(v_1, v_2) \in H_1 \times H_2$ are connected in A, if the matchings are consistent with each other. For hierarchical graphs, we define the constraints for edges in A: u_i and v_i are different, not hierarchically connected and if u_1 and v_1 are neighbor, this also holds for u_2 and v_2. By blocking the matching of hierarchically connected nodes, we force the matching algorithm to select a version of the vectorization. The first and the third constraint ensure that both subgraphs have the same structure.

We use replicator dynamics [2] to find the maximal weighted clique of the association graph and, hence, the best matching subgraphs of H^1 and H^2. Based on the results of this, we perform the following steps to spot symbols. Let us consider H^1 be the query graph or the model graph and H^2 be the input graph where we want to spot the instances of H^1. First of all, we perform n iterations and in each iteration we perform the replicator dynamics to find the correspondences of the H^1 to H^2. Since the replicator dynamics only provide a one-to-one

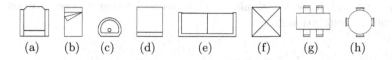

(a) (b) (c) (d) (e) (f) (g) (h)

Fig. 4. Model symbols in the SESYD dataset used for our experiment

matching, in each iteration we obtain the correspondences from the nodes of H^1 to the nodes of H^2. So for m nodes in H^1 we get m nodes in H^2. But it is not constrained that these m nodes in H^2 will belong to the same instance of H^1. So to obtain the different instances of the H^1 we consider each of the m nodes in the H^2 and all the neighborhood nodes of a node which can be reached within a k graph path distance. The graph path distance between two nodes is calculated as the minimum total number of nodes between the two nodes. Let us denote this set of nodes as V_s^1 and consider all the hierarchical and normal edges connecting the nodes in V_s^1 as in H^1, this forms a subgraph which we can denote as $H_s^1 = (V_s^1, E_{s_N}^1, E_{s_H}^1, \sigma_s^1)$. We again apply the replicator dynamics to get the best matching subgraph and compute the bounding box around the nodes of best correspondences. The bounding box gives the best matching region of interest expected to contain instance of a query symbol.

The complexity of replicator dynamics is $\mathcal{O}(|A|^2)$ (see [1]). Since we perform n iterations, we get a complexity of $\mathcal{O}(n \cdot |A|^2)$

Node Attributes. The graph matching algorithm operates on the association graph with similarity labels for the nodes. To use this algorithm, we have to define the similarity between two nodes of the hierarchical graph. Since the matching reflects geometric structures, we use geometric attributes for the similarity.

In a non-hierarchical plane graph, a single node can be labeled by the sequence of adjacent angles which sum up to 360°. Figure 3 (a) gives an example for such a labeling. This naive approach will cause some problems for hierarchical graphs since nodes can have several hierarchically connected neighbors. Thus, the number of possible vectorizations has a strong influence on the node description. Because the number of possibilities is also affected by the level of distortion of the original image, such an approach is not robust to distortion.

To reduce the influence of the hierarchical structure and the distortion on the node labeling, we use only edges to nodes that have no predecessor connected with the central node. An example for that can be seen in Figure 3 (b): though the central node is connected to four nodes, only three edges are used to compute the node label, because D has the predecessor C and, thus, is not used.

To compute the similarity between two node labels, we define an editing distance on these labels. The editing operations are rotating one edge, i.e. lowering one angle and rising another one, removing one edge, i.e. merging two angles, and rotating the whole description. The last operation is cost-free and makes the similarity rotation-invariant. The cost for rotating an edge is set to the angle of rotation. The cost for removing an edge is set to a fixed value. Using this editing distance, we can define the similarity between nodes that is used to weight the nodes of the association graph.

4   Experimental Results

We have evaluated the performance of our method on the SESYD (floorplans)[2] database which is a synthetically generated graphical document benchmark [4]. Actually, this dataset contains 10 different subdatasets, each of which consists of 100 different synthetically generated floorplans and 16 model symbols (see Figure 4). For this work we have considered one such subdataset and eight randomly chosen query symbols. All the floorplans in a subdataset are created on a same floorplan template by putting different model symbols in different places in random orientation and scale. The query symbol is always ideal and does not contain any distortion. The average number of nodes in the query graph and the input graph are 12 and 1500 respectively. Since we are focused on the document retrieval aspect of the problem, we use the standard performance measures of precision, recall, and F-measure for evaluating the performance of our system. For a more detailed discussion on performance evaluation of spotting systems we refer to [9]. Even though the set of images are synthetically generated, the vectorization algorithm generates some distortion like disconnection between nodes, insertion of spurious nodes etc. The aim of this experiment is to see how the algorithm performs for these distortions in a moderately sized database.

The results obtained by our system are presented in Table 1 in a symbol wise manner, which shows that the method is not equally successful for all the symbols, in particular for the simple symbols with trivial nodes, for example sofa1 (Figure 4(d)). This is because the nodes of the graph representing those symbols contain similar attributes with the nodes from the floorplans that do not belong to the symbol. In general, the precision of the algorithm is quite good which ensures the confidence of the system for retrieving the system. The recall values vary depending on the symbol but in most of the cases it is quite satisfactory. This ensures that most of the instances of the query symbols can be retrieved by the system. To get an idea about the results obtained the system, in Figure 5 we present the symbol spotting results of querying armchair (Figure 4(a)) and table1 (Figure 4(f)). The average processing time for spotting a symbol with number of nodes 12 into a floorplan with number of nodes 1500 is 0.9 min on an Intel i5 processor with GB memory.

Table 1. Results with SESYD dataset

Symbol	Precision	Recall	F-measure	Symbol	Precision	Recall	F-measure
armchair	92.71	83.86	88.06	sofa1	32.65	77.45	45.94
bed	23.67	87.17	37.23	sofa2	47.98	81.87	60.50
table1	98.56	97.23	97.89	table2	32.76	79.98	46.48
sink1	82.85	78.98	80.87	table3	23.51	78.23	36.15

---

[2] http://mathieu.delalandre.free.fr/projects/sesyd/index.html

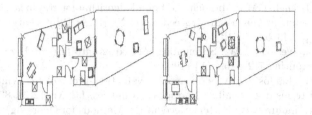

Fig. 5. Qualitative results of retrieving armchair (Figure 4(a)) and table1 (Figure 4(f)).

5   Conclusion and Future Work

In this paper we have presented a new hierarchical graph representation that enables us to store different possibilities for the vectorization of a drawing in one graph. With this representation, symbol spotting by graph matching can deal with typical vectorization errors. We could show the efficiency in an experiment.

Though our method performs well for most symbols, we still have some problems with too simple symbols. In the future we want to improve the efficiency for simple symbols and apply the approach to free-hand sketches, which have a higher level of distortion.

Acknowledgment. K. Broelemann is supported by the International Research Training Group 1498 "Semantic Integration of Geospatial Information" funded by DFG (German Research Foundation). This was done during a research stay of A. Dutta at the University of Münster, Germany. His research stay was supported by the BE (2011_BE_100469) and FI (2011_FI_B01022) scholarship provided by the Catalan research agency (AGAUR).

References

1. Ahuja, N., Todorovic, S.: From Region Based Image Representation to Object Discovery and Recognition. In: Hancock, E.R., Wilson, R.C., Windeatt, T., Ulusoy, I., Escolano, F. (eds.) SSPR&SPR 2010. LNCS, vol. 6218, pp. 1–19. Springer, Heidelberg (2010)
2. Bomze, I.R., Pelillo, M., Stix, V.: Approximating the maximum weight clique using replicator dynamics. IEEE TNN 11(6), 1228–1241 (2000)
3. Conte, D., Foggia, P., Sansone, C., Vento, M.: Thirty Years Of Graph Matching In Pattern Recognition. IJPRAI 18(3), 265–298 (2004)
4. Delalandre, M., Pridmore, T., Valveny, E., Locteau, H., Trupin, E.: Building Synthetic Graphical Documents for Performance Evaluation, pp. 288–298. Springer, Heidelberg (2008)
5. Dutta, A., Lladós, J., Pal, U.: Symbol spotting in line drawings through graph paths hashing. In: Proceedings of 11th ICDAR, pp. 982–986 (2011)
6. Luqman, M.M., Ramel, J., Llados, J., Brouard, T.: Subgraph spotting through explicit graph embedding: An application to content spotting in graphic document images. In: Proceedings of 11th ICDAR, pp. 870–874 (2011)

7. Nayef, N., Breuel, T.M.: A branch and bound algorithm for graphical symbol recognition in document images. In: Proceedings of Ninth IAPR International Workshop on DAS, pp. 543–546 (2010)
8. Rosin, P.L., West, G.A.W.: Segmentation of edges into lines and arcs. Image and Vision Computing 7(2), 109–114 (1989)
9. Rusiñol, M., Lladós, J.: A performance evaluation protocol for symbol spotting systems in terms of recognition and location indices. IJDAR 12(2), 83–96 (2009)
10. Tombre, K., Lamiroy, B.: Pattern Recognition Methods for Querying and Browsing Technical Documentation. In: Ruiz-Shulcloper, J., Kropatsch, W.G. (eds.) CIARP 2008. LNCS, vol. 5197, pp. 504–518. Springer, Heidelberg (2008)

Compact Form of the Pseudo–inverse Matrix in the Approximation of a Star Graph Using the Conductance Electrical Model (CEM) *

Manuel Igelmo[1] and Alberto Sanfeliu[1,2]

[1] Universitat Politècnica de Catalunya (UPC)
[2] Institut de Robòtica i Informàtica Industrial (UPC–CSIC)
migelmo@xtec.cat, sanfeliu@iri.upc.edu
http://www.upc.edu, http://www.iri.upc.edu

Abstract. The Conductance Electrical Model (CEM) transforms a graph into a circuit and can be use to do "inexact graph isomorphism" as it was shown in [13]. In second stage of this process, we transform the circuit r_{eq} in a star circuit, using the Moore–Penrose pseudo–inverse of a matrix for which there is a general formula that requires transpose, multiply and invert matrices with a time complexity of $O(N^4)$, where N is the number of nodes of the graph. However, due to the special structure of the star transformation, we are able to exploit this special structure to compute the pseudo–inverse without using the general Moore–Penrose formula. We have developed a closed formula that can compute the elements of the pseudo–inverse without using that formula, that means without multiplying matrices neither doing the matrix inversion and that moreover can be computed in $O(N^3)$. This method also eliminates the problems due to computer rounding and due to bad–conditioned problems in mathematical terms.

1 Introduction

Graphs have been successfully applied in various fields such as chemistry and biochemistry; transportation, telephony and computers networks, speech recognition and computer vision [1]. Examples of graphs in computer vision can be seen in [7] and they usually have a large number of nodes and/or edges.

The methods for graph and sub-graph matching are based on enumerative techniques [2,3], edit operations [4,5,6], spectral methods [8], expectation-maximization [9], random walks [10], genetics algorithms [11] and probabilistic approximations [12]. The time complexity in the enumerative and edit operation methods is NP–complete while in the other inexact methods it is polynomially bounded. Only in the enumerative solutions there exist an exact solution, in the other cases only graph matching approximations can be obtained.

---

* This research was conducted at the Institut de Robòtica i Informàtica Industrial (CSIC-UPC). It was partially supported by the CICYT project RobTaskCoop (DPI2010-17112)and the MIPRCV Ingenio Consolider 2010 (CSD2007-018)

G.L. Gimel' farb et al. (Eds.): SSPR & SPR 2012, LNCS 7626, pp. 539–547, 2012.
© Springer-Verlag Berlin Heidelberg 2012

In [13], we proposed a model to replace a graph by a circuit and we use the methods of Circuit Theory to solve the graph isomorphism. In order to compare two graphs, the method transform a graph G into a star circuit G^* using the following stages (see article [13]):

First Stage: Computation of the equivalent circuit resistances r_{eq}^G
1. Obtain the adjacency matrix, A^G
2. Compute the Laplacian matrix, Y^G obtained from A^G
3. Eliminate one row and one column; the one that belong to the node that will be consider the reference node in the electrical circuit (any node can be consider the reference node) and obtain the new matrix X^G
4. Apply the Ohm Law and compute the r_{eq}^G

Second Stage: Computation of the star circuit G^*
5. Obtain the branch resistances of the star circuit r, by computing $r = B^+ r_{eq} = (B^t B)^{-1} B^t r_{eq}$

In the first stage, each undirected weighted graph, G, of N nodes is transformed in the circuit CEM model, a passive resistive circuit where the weights of the edges are the conductance in siemens, and then in the adjacency matrix A^G. In the second stage, the CEM model is transformed in a star circuit, G^*, with $N+1$ nodes and N branches by minimizing the mean square error of the $N(N-1)/2$ equivalent resistances. This operation requires the calculation of the Moore–Penrose pseudo–inverse (hereinafter simply pseudo–inverse) of the B matrix by the general formula involving the product and inversion of matrices. In this work we have developed a new close form that computes the pseudo–inverse of B without the need of matrix multiplication and inversion.

2   Star Approximation Using CEM

By using the first described stage, we obtain the $N(N-1)/2$ equivalent resistances $(r_{eq_{ij}})$ which values can be represented by a column vector

$$r_{eq} = (r_{eq1,2}, r_{eq1,3}, \dots, r_{eq1,N}, r_{eq2,3}, \dots, r_{eq2,N}, \dots \dots, r_{eqN-3,N-1}, r_{eqN-2,N-1}, r_{eqN-1,N})^t$$

In the work [13] we proposed to approximate the original circuit by a star circuit (with N branches and $N+1$ nodes including one node in the center of the star) with one resistance (r_i) for each branch. These values can be written as a column vector

$$r = (r_1, r_2, \dots, r_N)^t$$

Also there are $N(N-1)/2$ equivalent resistances in the star circuit. Note the central node is not involved in the calculation of the equivalent resistances. The equivalent resistances can be written as a column vector

$$r'_{eq} = (r'_{eq1,2}, r'_{eq1,3}, \dots, r'_{eq1,N}, r'_{eq2,3}, \dots, r'_{eq2,N}, \dots \dots, r'_{eqN-3,N-1}, r'_{eqN-2,N-1}, r'_{eqN-1,N})^t$$

It easy to see that $r'_{eq_{ij}} = r_i + r_j$, since the equivalent resistance between two nodes in the star circuit, is the association of two serial resistances. Then we have $r'_{eq} = Br$ where B is the matrix show in (1)

$$
B = \begin{pmatrix}
1\,1\,0\,0\cdots0\,0\,0 \\
1\,0\,1\,0\cdots0\,0\,0 \\
1\,0\,0\,1\cdots0\,0\,0 \\
1\,0\,0\,0\cdots0\,0\,0 \\
\vdots\;\vdots\;\vdots\;\vdots\;\ddots\;\vdots\;\vdots\;\vdots \\
1\,0\,0\,0\cdots0\,1\,0 \\
1\,0\,0\,0\cdots0\,0\,1 \\
\hline
0\,1\,1\,0\cdots0\,0\,0 \\
0\,1\,0\,1\cdots0\,0\,0 \\
0\,1\,0\,0\cdots0\,0\,0 \\
\vdots\;\vdots\;\vdots\;\vdots\;\ddots\;\vdots\;\vdots\;\vdots \\
0\,1\,0\,0\cdots0\,1\,0 \\
0\,1\,0\,0\cdots0\,0\,1 \\
\hline
0\,0\,1\,1\cdots0\,0\,0 \\
0\,0\,1\,0\cdots0\,0\,0 \\
\vdots\;\vdots\;\vdots\;\vdots\;\ddots\;\vdots\;\vdots\;\vdots \\
0\,0\,1\,0\cdots0\,1\,0 \\
0\,0\,1\,0\cdots0\,0\,1 \\
\hline
0\,0\,0\,1\cdots0\,0\,0 \\
\vdots\;\vdots\;\vdots\;\vdots\;\ddots\;\vdots\;\vdots\;\vdots \\
0\,0\,0\,1\cdots0\,1\,0 \\
0\,0\,0\,1\cdots0\,0\,1 \\
\vdots\;\vdots\;\vdots\;\vdots\;\ddots\;\vdots\;\vdots\;\vdots \\
\vdots\;\vdots\;\vdots\;\vdots\;\ddots\;\vdots\;\vdots\;\vdots \\
0\,0\,0\,0\cdots0\,1\,1
\end{pmatrix}
\begin{matrix}
\left.\vphantom{\begin{matrix}a\\a\\a\\a\\a\\a\\a\end{matrix}}\right\}N-1\text{ rows} \\
\left.\vphantom{\begin{matrix}a\\a\\a\\a\\a\\a\end{matrix}}\right\}N-2\text{ rows} \\
\left.\vphantom{\begin{matrix}a\\a\\a\\a\\a\end{matrix}}\right\}N-3\text{ rows} \\
\left.\vphantom{\begin{matrix}a\\a\\a\\a\end{matrix}}\right\}N-4\text{ rows} \\
\left.\vphantom{a}\right\}1\text{ row}
\end{matrix}
\tag{1}
$$

The approximation discussed above must be understood as the search for the values of r such that r'_{eq} is approximately equal to r_{eq}, in the sense of minimizing the mean square error between r_{eq} and r'_{eq} is given by

$$r = (B^t B)^{-1} B^t r_{eq}$$

where

$$B^+ = (B^t B)^{-1} B^t \tag{2}$$

is known as the pseudo-inverse of B, note that B^+ has N rows and $N(N-1)/2$ columns. The above equation we can finally be written

$$r = B^+ r_{eq}$$

3 Compact Form of the Pseudo–inverse

To obtain the pseudo–inverse (B^+) of any matrix B by the (2) expression it is necessary to make a matrix inversion, two products of matrices and matrix transpose. But for the particular case that the matrix B has the form given in (1), it is not necessary to use the (2) expression. This substantially simplifies calculations as discussed in the following theorem.

Theorem 1. *Let B be the matrix with the structure shown in (1) with $N \neq 1$ and $N \neq 2$, then its pseudo–inverse is*

$$B^+ = \frac{1}{(N-1)(N-2)} \left[(N-1)B^t - \mathbb{1}_{N,N(N-1)/2} \right] \qquad (3)$$

where N is the number of columns of the matrix B and $\mathbb{1}_{N,N(N-1)/2}$ is a matrix full of ones with N rows and $N(N-1)/2$ columns.

Proof. We call M the result of $B^t B$ then it is easy to see that

$$M = \begin{pmatrix} 1\,1\,1 & \cdots & 1\,1\,0\,0 & \cdots & 0\,0\,0 & \cdots & 0\,0\,0 & \cdots & 0\,0 & \cdots & 0 \\ 1\,0\,0 & \cdots & 0\,0\,1\,1 & \cdots & 1\,1\,0 & \cdots & 0\,0\,0 & \cdots & 0\,0 & \cdots & 0 \\ 0\,1\,0 & \cdots & 0\,0\,1\,0 & \cdots & 0\,0\,1 & \cdots & 1\,1\,0 & \cdots & 0\,0 & \cdots & 0 \\ 0\,0\,1 & \cdots & 0\,0\,0\,1 & \cdots & 0\,0\,1 & \cdots & 0\,0\,1 & \cdots & 1\,1 & \cdots & 0 \\ \vdots & \ddots & \vdots & \ddots & \vdots & \ddots & \vdots & \ddots & \vdots & \ddots & \vdots \\ 0\,0\,0 & \cdots & 1\,0\,0\,0 & \cdots & 1\,0\,0 & \cdots & 1\,0\,0 & \cdots & 1\,0 & \cdots & 1 \\ 0\,0\,0 & \cdots & 0\,1\,0\,0 & \cdots & 0\,1\,0 & \cdots & 0\,1\,0 & \cdots & 0\,1 & \cdots & 1 \end{pmatrix}$$

$$= \begin{pmatrix} N-1 & 1 & 1 & \cdots & 1 \\ 1 & N-1 & 1 & \cdots & 1 \\ 1 & 1 & N-1 & \cdots & 1 \\ \vdots & \vdots & \vdots & \ddots & \vdots \\ 1 & 1 & 1 & \cdots & N-1 \end{pmatrix} \qquad (4)$$

where M is a square matrix of order N. To calculate M^{-1} we will use that

$$M^{-1} = \frac{M^*}{|M|}$$

where M^* is the adjugate matrix and $|M|$ is the determinant that must necessarily be non-zero so that the inverse matrix does exist. Applying the formula (8) of the Theorem 2 obtained in the Annex and substituting n by N and k by $N-1$ then

$$|M| = 2(N-1)(N-2)^{N-1} \tag{5}$$

Performing the same substitutions in (9) of the Theorem 3 for the adjugate matrix we obtain the following

$$M^* = (N-2)^{N-2} \begin{pmatrix} 2N-3 & -1 & -1 & \cdots & -1 \\ -1 & 2N-3 & -1 & \cdots & -1 \\ -1 & -1 & 2N-3 & \cdots & -1 \\ \vdots & \vdots & \vdots & \ddots & \vdots \\ -1 & -1 & -1 & \cdots & 2N-3 \end{pmatrix} \tag{6}$$

Dividing the expressions (5) and (6) we obtain

$$M^{-1} = \frac{1}{2(N-1)(N-2)} \begin{pmatrix} 2N-3 & -1 & -1 & \cdots & -1 \\ -1 & 2N-3 & -1 & \cdots & -1 \\ -1 & -1 & 2N-3 & \cdots & -1 \\ \vdots & \vdots & \vdots & \ddots & \vdots \\ -1 & -1 & -1 & \cdots & 2N-3 \end{pmatrix}$$

Note that M^{-1} can be written as

$$M^{-1} = \frac{1}{2(N-1)(N-2)} \left[2(N-1)\mathbb{I}_N - \mathbb{1}_{N,N} \right]$$

where $\mathbb{I}_N$ is the identity matrix of order N. We finally have

$$B^+ = M^{-1}B^t = \frac{1}{2(N-1)(N-2)} \left[2(N-1)\mathbb{I}_N - \mathbb{1}_{N,N} \right] B^t =$$

$$= \frac{1}{2(N-1)(N-2)} \left[2(N-1)\mathbb{I}_N B^t - \mathbb{1}_{N,N}B^t \right] =$$

$$= \frac{1}{2(N-1)(N-2)} \left[2(N-1)B^t - 2\mathbb{1}_{N,N(N-1)/2} \right] =$$

$$= \frac{1}{(N-1)(N-2)} \left[(N-1)B^t - \mathbb{1}_{N,N(N-1)/2} \right]$$

The last step is due to the fact that all the columns of B^t add the constant 2.   □

4   Conclusions and Advantages of the Compact Form of the Pseudo-inverse

We have shown in this article that there is a way of computing the pseudo–inverse of the second stage of the inexact isomorphism computation, without requiring matrix transpose, inversion and the multiplication of matrices, because we can built the pseudo-inverse in a direct way.

The advantages of the calculation of pseudo-inverse by the compact formula (3) versus the general formula (2) are:

1) The computational complexity is reduced from $O(N^4)$ to $O(N^3)$.
2) This improvement avoids the problem of numerical resolution in matrix pseudo-inversion on a computer (numerical stability is ensured) and also avoids potential bad–conditioned problems in mathematical terms.
3) The elements of matrix B^+ may be obtained on the fly. It is not necessary to work in memory with the entire matrix, therefore this improvement is important in systems with low memory.

5   Annex

Let be Q a matrix of order n

$$Q = \begin{pmatrix} k & 1 & 1 & \cdots & 1 \\ 1 & k & 1 & \cdots & 1 \\ 1 & 1 & k & \cdots & 1 \\ \vdots & \vdots & \vdots & \ddots & \vdots \\ 1 & 1 & 1 & \cdots & k \end{pmatrix} \tag{7}$$

The following two theorems are fulfilled:

Theorem 2. *The determinant of the matrix Q is*

$$|Q| = (k+n-1)(k-1)^{n-1} \tag{8}$$

Proof. Let us going to obtain the upper triangular matrix. For each row it has to be added all the columns to first column

$$|Q| = \begin{vmatrix} k+n-1 & 1 & 1 & \cdots & 1 \\ k+n-1 & k & 1 & \cdots & 1 \\ k+n-1 & 1 & k & \cdots & 1 \\ \vdots & & \vdots & \ddots & \vdots \\ k+n-1 & 1 & 1 & \cdots & k \end{vmatrix} \quad \text{then} \quad |Q| = (k+n-1) \begin{vmatrix} 1 & 1 & 1 & \cdots & 1 \\ 1 & k & 1 & \cdots & 1 \\ 1 & 1 & k & \cdots & 1 \\ \vdots & \vdots & \vdots & \ddots & \vdots \\ 1 & 1 & 1 & \cdots & k \end{vmatrix}$$

Each row is replaced, except the first row that is obtained by subtracting the first row

$$|Q| = (k+n-1) \begin{vmatrix} 1 & 1 & 1 & \cdots & 1 \\ 0 & k-1 & 0 & \cdots & 0 \\ 0 & 0 & k-1 & \cdots & 0 \\ \vdots & \vdots & \vdots & \ddots & \vdots \\ 0 & 0 & 0 & \cdots & k-1 \end{vmatrix}$$

Since the determinant of a triangular matrix is the product of the diagonal elements, then

$$|Q| = (k+n-1)(k-1)^{n-1} \qquad \square$$

Corollary 1. *The determinant of Q is not zero if and only if $k \neq 1$ and $k \neq 1 - n$.*

Theorem 3. *The adjoint matrix (Q^*) of the matrix Q is*

$$
Q^* = (k-1)^{n-2}
\begin{pmatrix}
k+n-2 & -1 & -1 & \cdots & -1 \\
-1 & k+n-2 & -1 & \cdots & -1 \\
-1 & -1 & k+n-2 & \cdots & -1 \\
\vdots & \vdots & \vdots & \ddots & \vdots \\
-1 & -1 & -1 & \cdots & k+n-2
\end{pmatrix}
\tag{9}
$$

Proof. To proof this theorem we have to divide the problem in two parts: *(i) the calculation of the diagonal adjoints (Q_{ii}) and (ii) the calculation of the off-diagonal adjoints (Q_{ij}).*

(i) Calculation of the Diagonal Adjoints (Q_{ii}).
The adjoint of a diagonal element (all the adjoints of the diagonal elements are identical) will be a determinant of order $n - 1$. Applying the formula (8) of Theorem 2 we obtain the following expression

$$
Q_{ii} =
\begin{vmatrix}
k & 1 & 1 & \cdots & 1 \\
1 & k & 1 & \cdots & 1 \\
1 & 1 & k & \cdots & 1 \\
\vdots & \vdots & \vdots & \ddots & \vdots \\
1 & 1 & 1 & \cdots & k
\end{vmatrix}
= (k+n-2) \cdot (k-1)^{n-2}
$$

(ii) Calculation of the Off-Diagonal Adjoints (Q_{ij})
As it was shown in (10) to calculate the adjoint Q_{ij} ($i \neq j$) it must be removed the row i and column j (solid line) of Q.

$$
Q =
\left(
\begin{array}{ccccccc|c|ccc}
k & \cdots & 1 & 1 & 1 & \cdots & 1 & 1 & 1 & \cdots & 1 \\
\vdots & \ddots & \vdots & \vdots & \vdots & \ddots & \vdots & \vdots & \vdots & \ddots & \vdots \\
1 & \cdots & k & 1 & 1 & \cdots & 1 & 1 & 1 & \cdots & 1 \\ \hline
1 & \cdots & 1 & k & 1 & \cdots & 1 & 1 & 1 & \cdots & 1 \\
1 & \cdots & 1 & 1 & k & \cdots & 1 & 1 & 1 & \cdots & 1 \\
\vdots & \ddots & \vdots & \vdots & \vdots & \ddots & \vdots & \vdots & \vdots & \ddots & \vdots \\
1 & \cdots & 1 & 1 & 1 & \cdots & k & 1 & 1 & \cdots & 1 \\
1 & \cdots & 1 & 1 & 1 & \cdots & 1 & k & 1 & \cdots & 1 \\
1 & \cdots & 1 & 1 & 1 & \cdots & 1 & 1 & k & \cdots & 1 \\
\vdots & \ddots & \vdots & \vdots & \vdots & \ddots & \vdots & \vdots & \vdots & \ddots & \vdots \\
1 & \cdots & 1 & 1 & 1 & \cdots & 1 & 1 & 1 & \cdots & k
\end{array}
\right)
\tag{10}
$$

Thereafter, it easy to see in (11) that only appears one row and one column with all elements with value one (solid line) in the adjoint of Q_{ij} (with $i \neq j$)

$$C_{ij} = (-1)^{i+j} \begin{vmatrix} k & \cdots & 1 & 1 & 1 & \cdots & 1 & 1 & \cdots & 1 \\ \vdots & \ddots & \vdots & \vdots & \vdots & \ddots & \vdots & \vdots & \ddots & \vdots \\ 1 & \cdots & k & 1 & 1 & \cdots & 1 & 1 & \cdots & 1 \\ 1 & \cdots & 1 & 1 & k & \cdots & 1 & 1 & \cdots & 1 \\ \vdots & \ddots & \vdots & \vdots & \vdots & \ddots & \vdots & \vdots & \ddots & \vdots \\ 1 & \cdots & 1 & 1 & 1 & \cdots & k & 1 & \cdots & 1 \\ 1 & \cdots & 1 & 1 & 1 & \cdots & 1 & 1 & \cdots & 1 \\ 1 & \cdots & 1 & 1 & 1 & \cdots & 1 & k & \cdots & 1 \\ \vdots & \ddots & \vdots & \vdots & \ddots & \vdots & \vdots & \vdots & \ddots & \vdots \\ 1 & \cdots & 1 & 1 & \cdots & 1 & 1 & 1 & \cdots & k \end{vmatrix} \tag{11}$$

This row will be permuted to the first row. We will proceed in a similar way for the column. For each permutation the determinant changes its sign.

Suppose that $i < j$ then the column filled with ones will appear at position i meanwhile the row with all ones will appear at position $j - 1$. Therefore the number of permutations (and consequent changes of sign) of the row and column with all ones is $j - 2$ and $i - 1$ respectively, and the determinant is affected by $(-1)^{i+j-3}$. Analogous results are obtained assuming $j < i$.

Summarizing, the coefficient that multiplies the determinant is $(-1)^{i+j}$ $(-1)^{i+j-3}$. It will always have the value -1, because the exponent is always odd, as it can be seen in

$$(-1)^{i+j}(-1)^{i+j-3} = (-1)^{2i+2j-3} = (-1)^{2(i+j)-3} = -1$$

Then the adjoint is as follows ($i \neq j$)

$$Q_{ij} = - \begin{vmatrix} 1 & 1 & 1 & 1 & \cdots & 1 \\ 1 & k & 1 & 1 & \cdots & 1 \\ 1 & 1 & k & 1 & \cdots & 1 \\ 1 & 1 & 1 & k & \cdots & 1 \\ \vdots & \vdots & \vdots & \vdots & \ddots & \vdots \\ 1 & 1 & 1 & 1 & \cdots & k \end{vmatrix}$$

The calculation of this determinant is similar to that of Theorem 2, for $i \neq j$.

$$C_{ij} = -(k-1)^{n-2}$$

Finally we will have

$$Q^* = (k-1)^{n-2} \begin{pmatrix} k+n-2 & -1 & -1 & \cdots & -1 \\ -1 & k+n-2 & -1 & \cdots & -1 \\ -1 & -1 & k+n-2 & \cdots & -1 \\ \vdots & \vdots & \vdots & \ddots & \vdots \\ -1 & -1 & -1 & \cdots & k+n-2 \end{pmatrix}$$

$\square$

References

1. Bunke, H., Sanfeliu, A.: Syntactic and Structural Pattern Recognition – Theory and Applications. Series in Computer Science, vol. 7. World Scientific Publishing Co. Pte. Ltd, Singapore (1990)
2. Ullman, J.R.: An algorithm for subgraph isomorphism. Journal of the Association for Computing Machinery 23(1), 31–42 (1976)
3. Messmer, B.T., Bunke, H.: A new algorithm for error–tolerant subgraph isomorphism detection. IEEE Transactions on Pattern Analysys and Machine Intelligence 20(5), 493–504 (1998)
4. Sanfeliu, A., Fu, K.S.: A distance measure between attributed relational graphs for pattern recognition. IEEE Trans. Syst. Man Cybern. SMC-13(3), 353–362 (1983)
5. Neuhaus, M., Bunke, H.: Automatic learning of cost function for graph edit distance. Information Sciences 177, 239–247 (2007)
6. Neuhaus, M., Bunke, H.: Edit distance–based kernel functions for structural pattern classification. Pattern Recognition 39(10), 1852–1863 (2006)
7. Sanfeliu, A., Alquézar, R., Andrade, J., Climent, J., Serratosa, F., Vergés, J.: Graph-based representations and techniques for image processing and image analysis. Pattern Recognition 35, 639–650 (2002)
8. Umeyama, S.: An Eigendecomposition approach to weighted graph matching problems. IEEE Trans. PAMI 10, 695–703 (1998)
9. Robles–Kelly, A., Hancock, E.R.: An expectation–maximisation framework for segmentation and grouping. Image and Vision Computing 20, 725–738 (2002)
10. Gori, M., Maggini, M., Sarti, L.: Exact and Approximate Graph Matching Using Random Walks. Pattern Anal. and Mach. Intelligence 27(7), 1100–1111 (2005)
11. Cross, A.D.J., Wilson, R.C., Hancock, E.R.: Inexact Graph Matching Using Genetic Search. Pattern Recognition 30, 953–970 (1997)
12. Wilson, R.C., Hancock, E.R.: Structural matching by discrete relaxation. Pattern Analysis and Machine Intelligence 19(6), 634–648 (1997)
13. Igelmo, M., Sanfeliu, A., Ferrer, M.: A Conductance Electrical Model for Representing and Matching Weighted Undirected Graphs. In: Proceedings of the International Conference on Pattern Recognition (ICPR 2010), pp. 958–961 (2010)

A Heuristic Based on the Intrinsic Dimensionality for Reducing the Number of Cyclic DTW Comparisons in Shape Classification and Retrieval Using AESA

Vicente Palazón-González and Andrés Marzal*

Dept. Llenguatges i Sistemes Informàtics and Institute of New Imaging Technologies,
Universitat Jaume I de Castelló, Spain
{palazon,amarzal}@lsi.uji.es

Abstract. Cyclic Dynamic Time Warping (CDTW) is a good dissimilarity of shape descriptors of high dimensionality based on contours, but it is computationally expensive. For this reason, to perform recognition tasks, a method to reduce the number of comparisons and avoid an exhaustive search is convenient. The Approximate and Eliminate Search Algorithm (AESA) is a relevant indexing method because of its drastic reduction of comparisons, however, this algorithm requires a metric distance and that is not the case of CDTW. In this paper, we introduce a heuristic based on the intrinsic dimensionality that allows to use CDTW and AESA together in classification and retrieval tasks over these shape descriptors. Experimental results show that, for descriptors of high dimensionality, our proposal is optimal in practice and significantly outperforms an exhaustive search, which is the only alternative for them and CDTW in these tasks.

Keywords: Cyclic strings, cyclic sequences, cyclic dynamic time warping, shape classification, shape retrieval, intrinsic dimensionality, metric spaces, AESA.

1 Introduction

Shape classification and retrieval are very important problems with applications in several areas such as industry, medicine, biometrics and even entertainment.

Among the methods to solve this problem the ones related to Dynamic Time Warping (DTW) [1] and descriptors of the contour with sequences of components of several dimensions have had a significant presence [2–8]. In general, these shape descriptors aim to have information from all of the contour with respect to each point, that is the reason for their large size (see Figure 1 for an example of the shape descriptor used in [6]). These methods offer very competitive results because of their full description and the properties that DTW has as a dissimilarity (DTW is able to align parts instead of points and it is robust with elastic deformations). Nevertheless, this combination has a high computational cost. Besides, the problem of the starting point invariance appears, i.e., where we have to start the comparison in the sequence. Although there are many

* Work partially supported by the Spanish Government (TIN2010-18958 and *Consolider Ingenio 2010* CSD2007-00018), the Generalitat Valenciana (*Prometeo*/2010/028), and *Fundació Caixa Castelló-Bancaixa* (P11B2009-48).

G.L. Gimel'farb et al. (Eds.): SSPR & SPR 2012, LNCS 7626, pp. 548–556, 2012.

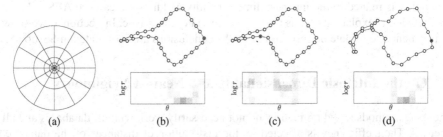

Fig. 1. Shape context computation. Given a set of landmark points from the contour, for each point is defined a histogram of the relative coordinates of the remaining points. (a) Diagram of log-polar histogram bins used in computing the shape contexts. Five bins for $\log r$ and 12 bins for θ, 60 dimensions. (b) Landmark points and the corresponding histogram of the point marked by a black dot. (c) The same shape but using a different point. (d) Another shape similar to (b).

heuristic methods to obtain this invariance, they are not suitable in most of the domains. Therefore, the literature accepts that to obtain a good starting point we must make the comparison between every possible starting point of the sequence [2, 3, 9, 4]. Hence the necessity to use cyclic sequences and then CDTW (Cyclic DTW) arises.

In [10], an algorithm is proposed to calculate the CDTW in time $O(n^2 \log n)$ (being n the size of the sequences). Although this algorithm considerably reduces the cost, with the shape descriptors mentioned before, the local distance or dissimilarity [10] between the components of the sequence has too much weight on the final cost, due to its dimensionality. Thus, in recognition tasks to use solutions that avoid the computation of CDTW over all the prototypes of the database is necessary, i.e., to avoid an exhaustive search.

In [9], the authors, using a method similar to their previous work with DTW [11], try to speed up the CDTW as well. In this work, they do not use the algorithm of [10], but they make clusters of sequences based on their similarity, treating every possible starting point as a different sequence and using indexing methods with lower bounds of these clusters. This solution seems to be suitable just for shape descriptors with only one dimension (such as the curvature) and not for much more dimensions [2–8]. For instance, in [5], 60 dimensions are required for each point (Figure 1). Another problem is that it cannot use more sophisticated local distances (in CDTW between elements of the shape descriptors) such as χ^2 [6], due to their lower bound.

AESA [12, 13] is characterised by a drastic reduction of the computation of distances. It is then specially interesting when the distance has a high cost and that is precisely our case. However, CDTW is not a metric because it does not satisfy the triangular inequality, which is an indispensable property for using AESA. In [14–17], the authors used AESA to speed up a speech recognition task based on DTW with good results in spite of not satisfying this property. In the current work, we improve their heuristic adding an important factor: the intrinsic dimensionality [18]. As far as we know the heuristic presented here is the only alternative to an exhaustive search in the context of shape classification and retrieval with cyclic sequences of high dimensionality.

The paper is organized as follows: The next section describes how the intrinsic dimensionality is affected in the search of nearest neighbours. In Section 3, the triangular inequality is related to the intrinsic dimensionality and how we can use AESA due to this relation is explained, that is to say, we present our heuristic. In Section 4, we show experiments to validate our proposal. Finally, conclusions are formulated in Section 5.

2   On the Intrinsic Dimensionality and Nearest Neighbours

Indexing methods based on metrics do not necessarily work with all databases and all metrics. Their efficiency is affected by the distribution of distances of the database. From this distribution we can obtain the intrinsic dimensionality. According to [18], given a database D and a metric m, the intrinsic dimensionality, ϱ, is: $\varrho(D, m) = \frac{\mu^2}{2\sigma^2}$, where μ and σ^2 are the mean and the variance of the distribution of distances.

In [18], it is shown, in an analytical and experimental way, that all the algorithms based on metrics degrade in a systematic way as the dimensionality increases, i.e., the computational cost is getting close to the one of an exhaustive search.

We can observe that the intrinsic dimensionality increases because of the two next reasons: the variance decreases and/or the mean of the distribution of distances increases. In Figure 2, we can see two distributions of distances showing a low and high intrinsic dimensionality. Two extreme cases, where both variance and mean vary. If the variance decreases, it means that the most distances have similar value, then we are going to have less information for pruning (in the case of AESA, bounds are going to be worse). On the other hand, if the mean increases, to obtain the nearest neighbours we will have to explore more prototypes (in AESA, we will take more time to find a prototype for a good pruning).

But, in our problem, what determines the distribution of distances?, i.e., what provokes that ϱ increases?. We can consider two causes. One is the sequence or the shape descriptor, especially affecting the number of points and the number of dimensions for each point. For instance, the BAS descriptor [2] uses 4 dimensions for each point and the shape contexts [5] 60 dimensions for each point. The second cause is the distance for comparing the sequences. Even though, if we set as the distance the CDTW, the local distance gains importance, which for the BAS descriptor is the euclidean distance and for the shape contexts is χ^2 [5].

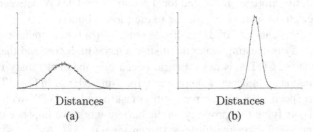

Distances   Distances
(a)   (b)

Fig. 2. Synthetic example of two distributions of distances. (a) With a low intrinsic dimensionality. (b) With a high intrinsic dimensionality.

3 Improving the Heuristic with the Intrinsic Dimensionality

The only problematic property for DTW to be a metric, is the triangular inequality:

$$d(x, z) \leq d(x, y) + d(y, z),$$

since it is possible to find counterexamples where DTW does not satisfy it [14, 19] (thus, CDTW is not a metric either). The correction of algorithms such as AESA (Figure 3) depends on having a metric distance and then it has to satisfy this property.

In [14–17], a study was performed with a task of speech recognition with isolated words using DTW. They aimed to see how not to satisfy the triangular inequality by DTW affected in samples of the real world. These samples were speech frames that were represented by sequences of components of eight dimensions. In [15], in 15 millions of triplets there were no cases where the triangular inequality was violated. In [19], the authors made experiments with synthetic time series (sequences of one dimension) of three types: *white-noise, random-walk* and *cylinder-bell-funnel*. The most problematic was *random-walk* where 20% of triplets violated the triangular inequality.

To see how many triplets x, y, z violate the triangular inequality we can use the next formula:

$$H = d(x, y) + d(y, z) - d(x, z). \tag{1}$$

All the triplets that have an H less than zero do not satisfy the triangular inequality. In [14], distributions (or histograms) of the frequencies of triplets for each H are shown. These distributions seem to have a gaussian form and when $H = 0$ the frequency is very low.

In (1), we can observe that the distribution of H has a relation to the distribution of distances (Section 2). That is to say, H is a composition of three random variables with the same distribution (the distribution of distances). The greater the mean, μ, of the distribution of distances, the greater the value of H of most of triplets, therefore, there will be more positive values because we will be adding two distances of the same distribution and subtracting another one of the same distribution too. In the case of the variance, σ^2, a similar thing will happen but with a lower variance, since the distances will be similar, and then, there will be more values of H that are greater or equal to zero. Therefore, we can say that, when the intrinsic dimensionality, ϱ, is greater, we will very probably find a lower number of triplets, x, y, z, that violate the triangular inequality. In practice, and in the case of CDTW, this statement shows that it will be easier to find triplets that violate triangular inequality in sets of sequences whose components have one dimension, like the curvature descriptor, than in sequences with 60 dimensions, like the shape contexts descriptor [5]. Thus, we can apply AESA with greater chances of success the greater the dimensionality of our cyclic sequences.

In our experiments with real world data (Section 4) we obtained few cases that violate the triangular inequality. However, in the curvature descriptor it arrives to almost a 3%. For the other types of descriptor the amount is very low as we expected and, given the characteristics of AESA (Figure 3), the recognition rates are not going to be significantly affected in practice. The fact that the intrinsic dimensionality increases is good for the triangular inequality but not for AESA, since it degrades the search [18], as we

Input: P: prototypes, x: sample to classify, $D \in \mathbb{R}^{|P| \times |P|}$: distances between prototypes
Output: $nn \in P$: nearest neighbour
begin

> **for** $p \in P$ **do**
> > $\llcorner\ G[p] = 0$
>
> $nn =$ unknown; $d_{nn} = \infty$; $s =$ any element from P
> **while** $|P| > 0$ **do**
> > $d_s = d(x, s)$; $P = P - s$
> > **if** $d_s < d_{nn}$ **then**
> > > $\llcorner\ nn = s$; $d_{nn} = d_s$
> >
> > $next =$ unknown; $gmin = \infty$
> > **for** $p \in P$ **do**
> > > $G[p] = \max(G[p], |D[s, p] - d_s|)$ // lower bound based on the
> > >   // triangular inequality
> > >
> > > **if** $G[p] > d_{nn}$ **then**
> > > > $\llcorner\ P = P - p$
> > >
> > > **else**
> > > > **if** $G[p] < gmin$ **then**
> > > > > $\llcorner\ gmin = G[p]$; $next = p$
> >
> > $\llcorner\ s = next$

end

Fig. 3. AESA. In our case the distance d is the CDTW.

mentioned before. Even so, as we will see in the next section the results are satisfactory both in time and in classification and retrieval rates.

4 Experiments

In order to assess the behaviour of our proposal, we performed experiments on an Intel i7 2.66GHz machine running under linux 3.2.0. The real world databases used were the MPEG-7 Core Experiment CE-Shape-1 (part B) [20] and the Silhouette database [21]. The shape descriptors were: curvature (as an example descriptor of one dimension for each point), BAS [2] (four dimensiones) and the shape contexts (SC) [5, 6] (60 dimensions). The results achieved with these descriptors, and in particular the ones with the shape contexts, can be applied to other ones of similar characteristics from the bibliography [3, 4, 6–8]. We also used a synthetic corpus of sequences of several dimensions (1, 5, 10, 20 and 60). We generated 1000 sequences (for each number of dimensions) with a random walk (for each dimension of the sequence) defined by $x_i = x_{i-1} + N(0, 1)$ and $x_1 = 0$ as in [19].

In the following, we will observe how the intrinsic dimensionality affects CDTW to satisfy the triangular inequality property. Subsequently, speeding up results are shown with AESA with respect to an exhaustive search. Finally, we will see how using AESA affects classification and retrieval rates.

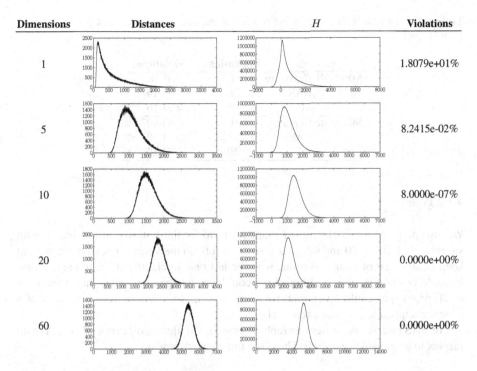

Fig. 4. Dimensions of the sequences, histograms of the distribution of distances, the distribution of H and the percentage of triplets that violate the triangular inequality, for the experiments with the random-walk synthetic corpus.

4.1 Intrinsic Dimensionality and Triangular Inequality

In Figure 4, we can see by means of histograms the relation between the distribution of distances and the distribution of H (Section 3). We performed an experiment similar to the one in [19], with random-walk sequences, but varying the number of dimensions (in [19] this experiment was done for just one dimension). We generated 1000 sequences for each number of dimensions, then we checked 1000000000 triplets. The fact of having the distribution of distances near to 0 (as it happens with sequences of one dimension) makes more probable to find triplets that violate the triangular inequality in the distribution of H. On the other hand, if the distribution of distances is far from the value 0 (as it happens with sequences of 20 dimensions), the percentage considerably decreases.

With respect to real world data, Table 1 shows the dimensionality and the corresponding percentage of triplets that violate the triangular inequality for each shape descriptor. As we can be observe, the greater the dimensionality the lower the percentage of triplets. As it is commented in Section 3, a great value in the intrinsic dimensionality makes the violation of the triangular inequality less probable.

Table 1. Comparison of the dimensionality with the percentage of triplets that violate the triangular inequality

		Dimensions	Violations
MPEG7B	Curvature	1	2.95 %
	BAS	4	$7.25 \cdot 10^{-3}$ %
	SC	60	$7.07 \cdot 10^{-5}$ %
Silhouette	Curvature	1	$3.93 \cdot 10^{-1}$ %
	BAS	4	$2.61 \cdot 10^{-5}$ %
	SC	60	$6.54 \cdot 10^{-7}$ %

4.2   Time

We also performed experiments of shape retrieval for the k most similar shapes, with values of k: 1, 5, 10, 20 and 40. To use AESA to obtain the k nearest neighbours we can keep a sorted list of them and prune with the last one. In classification, in many cases, it would be enough $k = 1$, although we could also use greater values. In retrieval, 10 or 20 prototypes could be enough for a first answer (or even unique) for a user of a concrete application of shape retrieval.

For BAS and SC we present a graph (Figure 5) with the average time of AESA, with respect to an exhaustive search. There is a huge improvement.

4.3   Classification and Retrieval Rates Using AESA and CDTW

Finally, we need to mention the classification and retrieval rates for the k nearest neighbours (Table 2). The only results that change are the ones of the curvature, but the difference is not so great.

Table 2. Recognition rates for an exhaustive search and AESA

		Curvature		BAS		SC	
	k	Exhaustive	AESA	Exhaustive	AESA	Exhaustive	AESA
MPEG7B	1	90.50	**90.00**	97.64	97.64	98.78	98.78
	5	83.21	**83.23**	94.91	94.91	96.67	96.67
	10	70.94	**70.74**	88.44	88.44	91.63	91.63
	20	55.39	**55.29**	74.61	74.61	79.20	79.20
	40	63.83	**63.43**	82.85	82.85	86.73	86.73
Silhouette	1	91.87	**91.77**	96.91	96.91	98.59	98.59
	5	88.25	**88.20**	94.66	94.66	97.77	97.77
	10	80.27	80.27	90.75	90.75	95.81	95.81
	20	69.47	69.47	83.13	83.13	90.60	90.60
	40	61.20	61.20	73.85	73.85	83.03	83.03

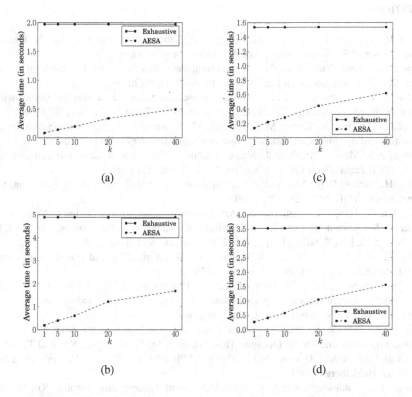

Fig. 5. Average time of an exhaustive search and AESA for (a) BAS, and (b) the shape contexts with the MPEG7B database. For (c) BAS, and (d) the shape contexts with the Silhouette database.

5   Discussion

From the experiments presented in the last section, it is clear that when the dimensionality of the cyclic sequences is sufficiently high we can obtain a very low percentage of triplets that violate the triangular inequality. In real tasks of shape classification and retrieval with AESA and CDTW, we have studied three shape descriptors with different number of dimensions. In particular the curvature, despite having a significant percentage of violations of the triangle inequality, surprisingly obtains quite acceptable rates with AESA (without being the same ones). With BAS and SC descriptors the rates are the same with respect to an exhaustive search. But if the results of the curvature are acceptable, with descriptors of higher dimensionality we can be more confident that AESA will have a good behaviour. We want to remark as well that our proposal significantly speeds up the classification and retrieval of these shape descriptors [2–8] and that our heuristic is the only alternative to an exhaustive search for them.

Of course, this proposal can be applied to other contexts based on DTW, not just the one of shape recognition and obviously it is possible to use other indexing methods based on metric spaces. In posterior work we aim to explore these contexts.

References

1. Sankoff, D., Kruskal, J. (eds.): Time Warps, String Edits, and Macromolecules: the Theory and Practice of Sequence Comparison. Addison-Wesley, Reading (1983)
2. Arica, N., Yarman-Vural, F.T.: BAS: a perceptual shape descriptor based on the beam angle statistics. Pattern Recognition Letters 24(9-10), 1627–1639 (2003)
3. Adamek, T., O'Connor, N.E.: A multiscale representation method for nonrigid shapes with a single closed contour. IEEE Trans. Circuits Syst. Video Techn. 14(5), 742–753 (2004)
4. Alajlan, N., Rube, I.E., Kamel, M.S., Freeman, G.: Shape retrieval using triangle-area representation and dynamic space warping. Pattern Recognition 40(7), 1911–1920 (2007)
5. Belongie, S., Malik, J., Puzicha, J.: Shape matching and object recognition using shape contexts. IEEE Trans. Pattern Anal. Mach. Intell. 24(4), 509–522 (2002)
6. Ling, H., Jacobs, D.W.: Shape classification using the inner-distance. IEEE Trans. Pattern Anal. Mach. Intell. 29(2), 286–299 (2007)
7. Gopalan, R., Turaga, P., Chellappa, R.: Articulation-Invariant Representation of Non-planar Shapes - Supplementary Material. In: Daniilidis, K., Maragos, P., Paragios, N. (eds.) ECCV 2010, Part III. LNCS, vol. 6313, pp. 286–299. Springer, Heidelberg (2010)
8. Wang, J., Bai, X., You, X., Liu, W., Latecki, L.: Shape matching and classification using height functions. Pattern Recognition Letters (2011)
9. Keogh, E., Wei, L., Xi, X., Vlachos, M., Lee, S., Protopapas, P.: Supporting exact indexing of arbitrarily rotated shapes and periodic time series under Euclidean and warping distance measures. The VLDB Journal 18(3), 611–630 (2009)
10. Marzal, A., Palazón, V., Peris, G.: Shape Retrieval Using Normalized Fourier Descriptors Based Signatures and Cyclic Dynamic Time Warping. In: Yeung, D.-Y., Kwok, J.T., Fred, A., Roli, F., de Ridder, D. (eds.) SSPR 2006 and SPR 2006. LNCS, vol. 4109, pp. 208–216. Springer, Heidelberg (2006)
11. Keogh, E.J., Ratanamahatana, C.A.: Exact indexing of dynamic time warping. Knowl. Inf. Syst. 7(3), 358–386 (2005)
12. Vidal, E.: An algorithm for finding nearest neighbours in (approximately) constant average time. Pattern Recognition Letters 4(3), 145–157 (1986)
13. Vidal, E.: New formulation and improvements of the nearest-neighbour approximating and eliminating search algorithm (AESA). Pattern Recognition Letters 15(1), 1–7 (1994)
14. Vidal, E., Casacuberta, F., Rulot, H.M.: Is the DTW distance really a metric? An algorithm reducing the number of DTW comparisons in isolated word recognition. Speech Communication 4(4), 333–344 (1985)
15. Casacuberta, F., Vidal, E., Rulot, H.: On the metric properties of dynamic time warping. IEEE Trans. Acoustics, Speech and Signal Processing ASSP-35(11), 1631 (1987)
16. Vidal, E., Casacuberta, F., Benedi, J., Lloret, M.: On the verification of triangle inequality by dynamic time-warping dissimilarity measures. Speech Commun. 7(1), 67–79 (1988)
17. Vidal, E., Rulot, H.M., Casacuberta, F., Benedi, J.M.: On the use of a metric-space search algorithm (AESA) for fast DTW-based recognition of isolated words. IEEE Trans. Acoustics, Speech and Signal Processing ASSP-36(5), 651 (1988)
18. Chávez, E., Navarro, G., Baeza-Yates, R., Marroquín, J.: Searching in metric spaces. ACM Computing Surveys (CSUR) 33(3), 273–321 (2001)
19. Lemire, D.: Faster retrieval with a two-pass dynamic-time-warping lower bound. Pattern Recognition 42(9), 2169–2180 (2009)
20. Bober, M.: MPEG-7 visual shape descriptors. IEEE Trans. Circuits Syst. Video Techn. 11(6), 716–719 (2001)
21. Sharvit, D., Chan, J., Tek, H., Kimia, B.B.: Symmetry-based indexing of image databases. In: Workshop on Content-Based Access of Image and Video Libraries, pp. 56–62 (1998)

Support Vector Machines Training Data Selection Using a Genetic Algorithm

Michal Kawulok and Jakub Nalepa[*]

Institute of Informatics, Silesian University of Technology
Akademicka 16, 44-100 Gliwice, Poland
{michal.kawulok,jakub.nalepa}@polsl.pl

Abstract. This paper presents a new method for selecting valuable training data for support vector machines (SVM) from large, noisy sets using a genetic algorithm (GA). SVM training data selection is a known, however not extensively investigated problem. The existing methods rely mainly on analyzing the geometric properties of the data or adapt a randomized selection, and to the best of our knowledge, GA-based approaches have not been applied for this purpose yet. Our work was inspired by the problems encountered when using SVM for skin segmentation. Due to a very large set size, the existing methods are too time-consuming, and random selection is not effective because of the set noisiness. In the work reported here we demonstrate how a GA can be used to optimize the training set, and we present extensive experimental results which confirm that the new method is highly effective for real-world data.

1 Introduction

Support vector machines (SVM) [1] is a widely adopted classifier which has been found highly effective for a variety of pattern recognition problems. Based on a labeled training set, it determines a hyperplane that linearly separates two classes in a higher-dimensional kernel space. The hyperplane is defined by a small subset of the vectors from the entire training set, termed *support vectors* (SV). Afterwards, the hyperplane is used to classify the data of the same dimensionality as the training set data.

SVM training is a constrained quadratic programming problem of $O(n^3)$ time and $O(n^2)$ memory complexity, where n is the number of samples in the training set. This is one of the most important shortcomings of SVM, as it makes it virtually inapplicable in case of huge amounts of training samples. Therefore, some attempts have been made to refine the training sets and use only those samples, from which the support vectors are selected. Existing techniques are focused either on random selection or analysis of the data geometry.

Our contribution lies in using a genetic algorithm (GA) for selecting the relevant data from the entire available set of training samples. From the work

---

[*] This work has been supported by the Polish Ministry of Science and Higher Education under research grant no. IP2011 023071 from the Science Budget 2012-2013.

G.L. Gimel' farb et al. (Eds.): SSPR & SPR 2012, LNCS 7626, pp. 557-565, 2012.

reported here we conclude that in certain cases it is better to use only a small portion of the available data for training SVM. Moreover, we demonstrate that the data must be selected carefully as it has a crucial impact on the obtained classification score, and the selection process can be effectively managed using a GA. Our work was motivated by the problems related to skin detection. SVM have been already used for this purpose [2], however training set selection was not investigated there. It is worth noting that due to huge amount of available training data, proper data selection is very important in this case, which was confirmed by obtained experimental results.

The paper is organized as follows. Existing training set reduction techniques are outlined in Section 2. The details of proposed method are presented in Section 3, while the validation results are shown and discussed in Section 4. Conclusions and directions for our future work are given in Section 5.

2 Related Literature

Initial approaches towards dealing with large training sets were aimed at decomposing the optimization problem into a number of sub-problems that can be easily solved, reducing the overall training time [3]. However, for very large training sets this is insufficient, and the number of training samples must be significantly decreased. The simplest method for reducing large training sets is to select a smaller subset randomly [4]. Such an approach was the basis for reduced support vector machines (RSVM) [5]. Not only does random sampling help reduce the training time, but the classification is accelerated as well. This is because the classification time is linearly dependent on the number of SV, and generally for smaller training sets there are less SV determined.

Random sampling may be extended by analyzing the geometry of the training data in the input space. In particular, k-means clustering has been found effective here [6]. Another approach is to find crisp clusters with safety regions [7]. This method rejects the vectors inside single-class clusters, preserving those positioned at clusters' boundaries. Recently, the clustering-based approach has also been applied for one-class SVM [8]. The entire training set must be processed using these methods, which increases the computation time.

In order to achieve better performance, the clustering can be performed only in proximity of the decision boundary [9]. As the boundary is unknown before the SVM is trained, it is estimated using heterogeneity analysis based on entropy measure. Another approach to estimate the boundary is to classify the training data based on their mutual Mahalanobis distances and use only the misclassified vectors for training [10]. Mahalanobis distance-based data clustering was also studied in [11]. The points that are closest to the decision boundary are selected from every cluster. This process is well-demonstrated using artificial 2D data. Another method that operates in the kernel space rather than in the input space, applied to two-teachers-one-student problem was recently presented in [12].

There is also a group of methods which use alternative techniques to the clustering to analyze the data geometry. In [13] the convex hulls are determined which embed the training data. Later, the vectors are selected using Hausdorff distance between the convex hulls of opposite classes. It was presented there that appropriate reduction of the training set makes it possible to achieve almost as good results as using the entire set. In [14] the points from the training set are interpreted as a graph and subject to β-skeleton algorithm. This makes it possible to reduce both training and testing time while being almost as effective as using the entire training set. Other geometry-based approaches include minimum enclosing ball [15] and smallest enclosing ball with a ring region [16].

Huge training sets can also be reduced using active learning techniques [17, 18].They operate based on a large unlabeled set, and labels for the individual samples are acquired dynamically. According to [17], these algorithms determine the points near the decision boundary, similarly to the clustering methods.

The aforementioned methods report similar conclusions. Classification accuracy for reduced training sets is comparable to that obtained using the entire training set. In some of the referenced works it is indicated that the results are slightly better than using random sampling.

3   Genetic Training Set Optimization

It must be noted that the methods which analyze the data geometry or perform clustering need to process the entire training set, and therefore their execution time depends on the total number of samples. Contrary to these methods, random sampling is applicable regardless of the number of available samples, but it is not reliable for noisy sets or when the data may be mislabeled. In such cases, it is difficult to select "good" vectors based on random drawing. In the work reported here we have successfully solved this problem using a GA to select appropriate subset of training samples. Our approach is based on the iterative random sampling, during which different draws (i.e. *individuals*) are verified, and optimal training set is selected using a GA process. This approach combines the advantages of RSVM and geometry-based methods.

A GA, firstly introduced by Holland [19], is a heuristic search approach inspired by the biological mechanism of evolution and natural selection. Encoded solutions belonging to the solution space S are called *chromosomes*. The initial population is a subset of N chromosomes, and it is successively improved during the subsequent *generations*. The chromosomes p_A and p_B are selected and recombined using the crossover operator to generate one or more offspring solutions. Selected individuals are mutated with a certain probability to avoid premature convergence of the search. The quality of each chromosome is assessed by the *fitness function* corresponding to the objective function of the problem. These with a high fitness survive and form the next generation.

3.1   Genetic Operators

For the problem reported here, a chromosome defines the content of a single subset from the entire training set T, which consists of labeled samples belonging to

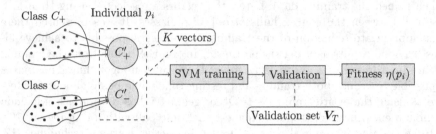

Fig. 1. Creation and validation of an individual

two classes C_+ and C_-. The chromosome's length $(2K)$ is equal to the number of samples that are used for training after the reduction. The first generation of N individuals is created based on random sampling, which is illustrated in Fig. 1. From each class, K vectors are selected randomly to create a new individual p_i. This initial selection is independent from the cardinality of T, which means that the genetic operations are independent from the training set size. Afterwards, SVM is trained using p_i and its fitness $\eta(p_i)$ is determined based on the classification score obtained for the validation set V_T.

A set of individuals from every i-th generation are used for reproduction to create the $(i+1)$-th generation. This process is similar to generating a new individual. First, two individuals p_A and p_B create an initial training set consisting of $4K$ samples, from which $2K$ samples are selected randomly as individual p_{A+B}. Then, the new individual is subject to mutation with the probability P_m. It is performed by random changes to the training subsets of the individual. Some samples are randomly substituted with others from the entire training set T. At every step it is reassured that the chromosome contains unique samples, and the same sample cannot be selected twice to the same chromosome.

3.2 Operator Strategies

The performance of a GA depends on the genetic operators including parents selection, crossover and mutation. The selection strategies address the problem of choosing two individuals from the population for recombination. The offspring solutions inherit the features of both parents p_A and p_B, thus the well-adapted individuals should be drawn from the population with a larger probability. However, recombining only the best individuals may cause saturating the population with the chromosomes of similar configurations, which in turn leads to the *diversity crisis* [20]. Four selection strategies are discussed here, namely: *high-low fit*, *AB-selection*, *truncation* and *enhanced truncation*.

1. **High-low fit**— this selection method was proposed in [21]. The population is sorted according to the fitness. The parent p_A is selected from the $c_h \cdot N$ fittest individuals, where c_h is the high-low coefficient. The parent p_B is drawn from the less-fitted part of the population. The offspring solutions are appended

to the population forming a new population of size $2N$. The N individuals with the highest fitness survive to maintain the constant population size.

2. **AB-selection**– this selection strategy was successfully used in the memetic algorithms to solve the vehicle routing problem with time windows [22, 23]. Each individual is selected for reproduction twice: first as p_A, then as p_B. If the offspring solution p_i generated for a pair of parents has higher fitness than the parent p_A then it replaces the parent p_A.

3. **Truncation.** At first, the population is sorted according to the fitness. Both parents p_A and p_B are selected from the $c_t \cdot N$ fittest individuals, where c_t is the truncation coefficient. The new population is composed of the offspring solutions generated for N pairs of parents.

4. **Enhanced truncation.** At first, the population is sorted according to the fitness. The $c_r \cdot N$ pairs of parents p_A and p_B are selected from the $c_e \cdot N$ fittest individuals, where c_r is the reproduction coefficient and c_e is the enhanced truncation coefficient. To maintain the constant population size N, the $N - c_r \cdot N$ individuals are generated randomly. The randomization simulates additional mutation for the search diversification.

The individuals of the child population are mutated with a certain probability as described in Section 3.1. In case of the AB-selection the best individuals will survive the recombination. However, they may be mutated and their fitness can decrease. Similarly, it is not guaranteed that the best chromosomes will survive for the other selection and replacement strategies. In order to keep the well-adapted individuals, the $c_c \cdot N$ best chromosomes replace a set of randomly chosen chromosomes with lower fitness, where c_c is the restoring coefficient.

The best fitness $\eta(p_b^i)$ and the average fitness $\bar{\eta}(p^i)$ in subsequent generations determine the necessity of regenerating the population. More formally, if $\eta(p_b^i) - \eta(p_b^{i-1}) < \epsilon$ for s_b consecutive steps and $\bar{\eta}(p^i) - \bar{\eta}(p^{i-1}) < \epsilon$ for s_a consecutive steps, where ϵ is the minimal improvement threshold, then the population is regenerated. The regeneration is based on copying $c_g \cdot N$ best individuals and drawing $N - c_g \cdot N$ individuals randomly, where c_g is the regeneration coefficient. The GA finishes after r regenerations.

4   Experimental Validation

The proposed method (termed GASVM) has been validated using two data sets, namely: 1) real-world data derived from ECU skin image database [24], and 2) artificial set of 2D points. ECU database consists of 4000 images coupled with binary ground-truth skin masks. The training set T was formed out of 6938255 pixels from 100 images. Every pixel was represented by a three-dimensional vector, indicating its color in YC_bC_r. Two validation sets were created, namely: V_T for evaluating the individual's fitness during the GA optimization and V, which was not fed back to the GA process (all the results are presented for V). The validation sets were created by sampling pixels from the remaining images. As a result, 560732 pixels were selected to every validation set. The sets are available at http://sun.aei.polsl.pl/~mkawulok/spr.

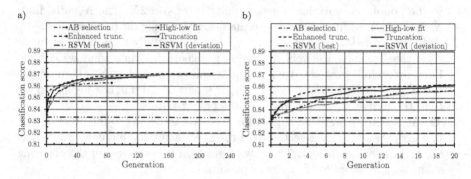

Fig. 2. Optimization process using different GA strategies compared with random sampling: a) whole process, b) first 20 generations

The GA was implemented in C++ and the experiments were performed using Intel Core i7 2.3 GHz with 16 GB RAM. We used LIBSVM [25], which is a popular SVM implementation, with RBF kernel: $K\left(\boldsymbol{u},\boldsymbol{v}\right) = \exp\left(-\|\boldsymbol{u}-\boldsymbol{v}\|^2/\sigma^2\right)$, where σ is the kernel width. SVM parameters (i.e. σ and C) were selected based on a grid search approach [25] using ranges $0.1 \leq \sigma \leq 10$ and $0.1 \leq C \leq 1000$ with a dynamic step. This simple approach was sufficient in the analyzed case and more sophisticated methods [26] were not exploited here. For skin detection we used $\sigma = 1$ and $C = 10$, and for 2D points $\sigma = 0.26$ and $C = 100$. The GA parameters were tuned experimentally in a similar manner. The following values were used: $N = 50$, $P_m = 0.3$, $c_h = 0.5$, $c_t = 0.5$, $c_r = 0.9$, $c_e = 0.2$, $c_c = 0.1$, $c_g = 0.1$, $\epsilon = 10^{-5}$, $s_a = s_b = r = 3$. In order to verify performance of RSVM [5], 20 independent tests were performed for every configuration, and within each test $N = 50$ subsets were drawn and validated to make it comparable to a single GA generation. Hence, a total number of 1000 random draws were executed to validate each setting. The best result out of each test, averaged over all the tests, is referred to as RSVM (best), while a global average result – RSVM (average). Minimal and maximal scores for all the draws are presented as RSVM (deviation) in Fig. 2 and as error bars for RSVM (best) in Fig. 3.

For each GA strategy discussed in Section 3.2, five optimization processes were run. Average maximal fitness obtained in subsequent generations for $K = 50$ samples in each class is presented in Fig. 2 for the skin data. GA strategies are compared here with RSVM. It can be seen from the graphs that after just a few generations GASVM outperforms RSVM. Enhanced truncation offers the fastest improvement, however it is the high-low fit strategy which delivers the best final score, and it has been chosen for further validation. The premature convergence of the search occurs in case of AB-selection strategy and after a relatively small number of generations the best individual cannot be further improved.

For high-low fit strategy we ran extensive tests to validate performance for various number of samples (K) in each class of the training set. In Fig. 3 our method is compared with RSVM. Error bars present maximal and minimal value.

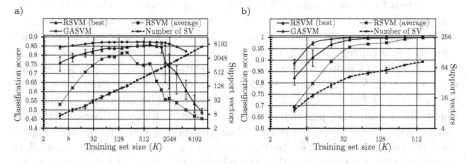

Fig. 3. GA and random sampling results depending on the training set size for skin segmentation set (a) and for artificial 2D data (b)

a) RSVM, $\eta = 0.882$ b) RSVM, $\eta = 0.883$ c) GASVM, $\eta = 0.985$ d) GASVM, $\eta = 1$

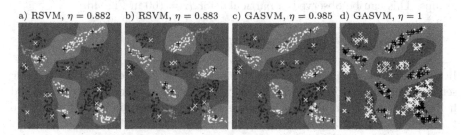

Fig. 4. Examples of training set selection using RSVM (a, b) and GASVM (c) for $K = 10$ vectors in each class, and GASVM for $K = 160$ vectors in each class (d)

For RSVM (average) the error bars were skipped as RSVM (best) indicates the maximal scores, and the minimal scores are irrelevant here. In addition, the dependence between the training set size and the number of SV is presented. For small value of K, GASVM selects definitely better training sets than those generated using random sampling, and this influences the final classification score. It is less dependent on K than RSVM, and the scores achieved in different runs are very similar. It is worth to note that the number of SV is linearly dependent on K, which induces linear dependence between K and the classification time. Theoretically, it is possible that using random sampling the same set is drawn as in case of GASVM, but for huge training sets this is little probable and has not been observed during our experiments– the best score achieved using RSVM was always worse than the worst obtained after the GA optimization.

For the skin data (Fig. 3a), the best RSVM score drops drastically after exceeding a certain threshold (ca. $K = 1500$), and the score variance increases. GASVM is more stable, but the decrease is observed as well. This can be explained by the fact that for larger sets it is hard to eliminate noisy data, which seriously affects the effectiveness. However, it is still easier to eliminate them using GASVM. We have not run GASVM for K greater than 5000 due to the required computation time. For $K = 5000$ the GA process required 4800 min to reach the stop condition, but for smaller sets the times were definitely shorter (e.g. 80 min for $K = 30$ and 210 min for $K = 200$). Due to the SVM training complexity it would be virtually impossible to use the entire training set.

Contrary to the skin data, the artificial set of 2D points can be classified without any error using the whole set for training, which is possible due to small data set size. For smaller K, the classification error appears, however it is smaller using GASVM. For $K = 160$ GASVM eliminated the classification error, which has not been achieved using RSVM for $K < 320$. The data are visualized in Fig. 4. Black and white points indicate the vectors from the entire set, and those marked with white and black crosses show the data selected to the training set (here the colors are altered for better visualization). Also, the decision boundary is presented. It can be noticed that the selected points do not follow any specific geometric pattern as proposed in [11]. In some cases they are located near the decision boundary, but in others they are positioned in the centers of the point groups. This can be observed in particular for $K = 160$ in Fig. 4d.

5   Conclusions and Future Work

In this paper we proposed to use a genetic algorithm for selecting SVM training sets. Presented experimental results show that while in some cases our method helps reduce the training set size, which means shorter training and validation times, it also makes it possible to achieve higher classification scores for noisy or mislabeled data. Although the GA process may require many generations to converge, it is independent from the total number of available samples, which cannot be offered by existing geometry-based approaches. Furthermore, after just a few generations it manages to select better training sets than those found using random sampling, so the optimization process can be terminated earlier, if it is critical to reduce the training time.

Our ongoing research includes comparing GASVM with the geometry-based methods using benchmark data sets. This should allow us to design a memetic approach, which would combine a GA with the data structure analysis to further improve the classification results. Also, our aim is to design a parallel GA to accelerate the computations. Finally, we want to use the method for selecting the training data from unlabeled data sets.

References

1. Cortes, C., Vapnik, V.: Support-Vector Networks. Machine Learning 20(3), 273–297 (1995)
2. Khan, R., Hanbury, A., Stöttinger, J., Bais, A.: Color based skin classification. Pattern Recogn. Lett. 33(2), 157–163 (2012)
3. Joachims, T.: Making large-scale SVM learning practical. In: Schölkopf, B., Burges, C.J.C., Smola, A.J. (eds.) Advances in kernel methods, pp. 169–184. MIT Press, USA (1999)
4. Balc'azar, J., Dai, Y., Watanabe, O.: A Random Sampling Technique for Training Support Vector Machines. In: Abe, N., Khardon, R., Zeugmann, T. (eds.) ALT 2001. LNCS (LNAI), vol. 2225, pp. 119–134. Springer, Heidelberg (2001)

5. Lee, Y.J., Huang, S.Y.: Reduced support vector machines: A statistical theory. IEEE Trans. on Neural Networks 18(1), 1–13 (2007)
6. Chien, L.J., Chang, C.C., Lee, Y.J.: Variant methods of reduced set selection for reduced support vector machines. J. Inf. Sci. Eng. 26(1), 183–196 (2010)
7. Koggalage, R., Halgamuge, S.: Reducing the number of training samples for fast support vector machine classification. Neural Information Process. Lett. and Reviews 2(3), 57–65 (2004)
8. Li, Y.: Selecting training points for one-class support vector machines. Pattern Recogn. Lett. 32(11), 1517–1522 (2011)
9. Shin, H., Cho, S.: Neighborhood property–based pattern selection for support vector machines. Neural Comput. 19(3), 816–855 (2007)
10. Abe, S., Inoue, T.: Fast Training of Support Vector Machines by Extracting Boundary Data. In: Dorffner, G., Bischof, H., Hornik, K. (eds.) ICANN 2001. LNCS, vol. 2130, pp. 308–313. Springer, Heidelberg (2001)
11. Wang, D., Shi, L.: Selecting valuable training samples for SVMs via data structure analysis. Neurocomputing 71, 2772–2781 (2008)
12. Chang, C.C., Pao, H.K., Lee, Y.J.: An RSVM based two-teachers-one-student semi-supervised learning algorithm. Neural Networks 25, 57–69 (2012)
13. Wang, J., Neskovic, P., Cooper, L.N.: Training Data Selection for Support Vector Machines. In: Wang, L., Chen, K., S. Ong, Y. (eds.) ICNC 2005. LNCS, vol. 3610, pp. 554–564. Springer, Heidelberg (2005)
14. Zhang, W., King, I.: Locating support vectors via β-skeleton technique. In: Int. Conf. on Neural Information Process, pp. 1423–1427 (2002)
15. Tsang, I.W., Kwok, J.T., Cheung, P.M.: Core vector machines: Fast SVM training on very large data sets. J. of Machine Learning Research 6, 363–392 (2005)
16. Zeng, Z.Q., Xu, H.R., Xie, Y.Q., Gao, J.: A geometric approach to train SVM on very large data sets. Intell. System and Knowledge Eng. 1, 991–996 (2008)
17. Schohn, G., Cohn, D.: Less is more: Active learning with support vector machines. In: 17th Int. Conf. on Machine Learning, pp. 839–846. Morgan Kaufmann Publishers Inc., USA (2000)
18. Musicant, D.R., Feinberg, A.: Active set support vector regression. IEEE Trans. on Neural Networks 15(2), 268–275 (2004)
19. Holland, J.H.: Adaptation in Natural and Artificial Systems. The University of Michigan Press (1975)
20. Corne, D., Dorigo, M., Glover, F., Dasgupta, D., Moscato, P., Poli, R., Price, K.V. (eds.): New ideas in optimization. McGraw-Hill Ltd., UK (1999)
21. Elamin, E.E.A.: A proposed genetic algorithm selection method. In: 1st National Symposium, NITS (2006)
22. Nagata, Y., Bräysy, O., Dullaert, W.: A penalty-based edge assembly memetic algorithm for the vehicle routing problem with time windows. Computers & OR 37(4), 724–737 (2010)
23. Nalepa, J., Czech, Z.J.: A parallel heuristic algorithm to solve the vehicle routing problem with time windows. Studia Informatica 33(1), 91–106 (2012)
24. Phung, S.L., Chai, D., Bouzerdoum, A.: Adaptive skin segmentation in color images. In: Proc. IEEE Int. Conf. on Acoustics, Speech and Signal, pp. 353–356 (2003)
25. Chang, C.C., Lin, C.J.: LIBSVM: A library for support vector machines. ACM Trans. on Intell. Systems and Technology 2, 27:1–27:27 (2011)
26. Staelin, C.: Parameter selection for support vector machines. Technical Report HPL-2002-354. HP Laboratories, Israel (2002)

A Unified View of Two-Dimensional Principal Component Analyses

Kohei Inoue*, Kenji Hara, and Kiichi Urahama

Department of Communication Design Science, Kyushu University,
4-9-1, Shiobaru, Minami-ku, Fukuoka, 815-8540 Japan
{k-inoue,hara,urahama}@design.kyushu-u.ac.jp

Abstract. Recently, two-dimensional principal component analysis (2D-PCA) and its variants have been proposed by several researchers. In this paper, we summarize their 2DPCA variants, show some equivalence among them, and present a unified view in which the non-iterative 2DPCA variants are interpreted as the non-iterative approximate algorithms for the iterative 2DPCA variants, i.e., the non-iterative 2DPCA variants are derived as the first iterations of the iterative algorithm started from different initial settings. Then we classify the non-iterative 2DPCA variants on the basis of their algorithmic patterns and propose a new non-iterative 2DPCA algorithm based on the classification. The effectiveness of the proposed algorithm is experimentally demonstrated on three publicly accessible face image databases.

Keywords: Dimensionality reduction, Principal component analysis, Two-dimensional principal component analysis.

1 Introduction

Principal component analysis (PCA) and linear discriminant analysis (LDA) are well-known techniques for dimensionality reduction. Since they are based on vectors, matrices such as 2D face images must be transformed into 1D image vectors in advance. However, the resultant vectors usually lead to a high-dimensional vector space, where it is difficult to solve the (generalized) eigenvalue problems for PCA and LDA.

Recently, Yang and Yang [1] and Yang et al. [2] have proposed two-dimensional PCA (2DPCA) which can handle matrices directly without vectorizing them. However, 2DPCA is approximately equivalent to the conventional PCA operated only on the row vectors of matrices [3–5], and needs many more coefficients for image representation than PCA. To overcome this problem, several variants of 2DPCA have been proposed recently. Since they have been proposed almost independently and simultaneously, the relationship among them is not clear. Therefore, the systematization of them is desired for a deeper understanding of 2DPCA variants.

---

* This work was partially supported by the Japan Society for the Promotion of Science under the Grant-in-Aid for Scientific Research (23700212).

G.L. Gimel' farb et al. (Eds.): SSPR & SPR 2012, LNCS 7626, pp. 566–574, 2012.

In this paper, we summarize the variants of 2DPCA [2] and show some equivalence among them. Moreover, we present a unified view of 2DPCA variants, where the non-iterative 2DPCA variants are interpreted as the non-iterative approximate algorithms for the iterative ones, i.e., we show that the non-iterative ones are derived as the first iterations of the iterative algorithm started from different initial settings. Then we classify the non-iterative ones on the basis of their algorithmic patterns and present a new non-iterative 2DPCA algorithm based on the classification result.

The rest of this paper is organized as follows: Section 2 briefly surveys the related work. Section 3 summarizes the original 2DPCA and its variants and shows some equivalence among them. Section 4 presents a unified view of 2DPCA variants and classifies the non-iterative 2DPCA variants on the basis of their algorithmic patterns. From the classification result, a new non-iterative algorithm is derived. Section 5 shows experimental results which demonstrate the effectiveness of the derived algorithm compared with the conventional non-iterative 2DPCA variants. Section 6 concludes this paper.

2 Related Work

Yang and Yang [1] and Yang et al. [2] proposed two-dimensional principal component analysis (2DPCA) which is based on 2D image matrices rather than 1D vectors so the image matrix does not need to be transformed into a vector prior to feature extraction. Ye et al. [6] proposed generalized PCA (GPCA) which is formulated as an optimization problem and derived an iterative procedure for GPCA. Kong et al. [7] proposed a framework of generalized 2DPCA to extend the original 2DPCA in two perspectives: a bilateral-projection-based 2DPCA (B2DPCA) and a kernel-based 2DPCA. Zhang and Zhou [8] proposed two-directional 2DPCA, i.e., $(2D)^2$PCA which combines 2DPCA and alternative 2DPCA. Zhang et al. [9] proposed a method for representing 2D image matrices using eigenimages, which are 2D matrices with the same size as the original images and can be directly computed from the original image matrices. Benito and Peña [10] proposed a method for dimensionality reduction based on the projection of images as matrices. Xu et al. [11] proposed complete 2DPCA (C2DPCA) in which two image covariance matrices are constructed and their eigenvectors are derived for image feature extraction. Xu et al. [12] proposed a two-stage strategy, parallel image matrix compression (PIMC), to compress the image matrix redundancy among both row vectors and column ones. Zuo et al. [13] proposed bi-directional PCA (BDPCA) and an assembled matrix distance metric to calculate the distance between two feature matrices. Wen and Shi [14] proposed image PCA (IPCA) in which a family of projective feature vectors, which is called the projective feature image, is obtained by 2DPCA and then the transpose of the projective feature image is processed by 2DPCA again. Lu et al. [15] proposed doubleside 2DPCA (D2DPCA) and the constructive method for incrementally adding observation to the existing eigen-space model, called incremental D2DPCA. Xi and Ramadge [16] proposed separable PCA (SPCA) and unified 2DPCA [2], BDPCA [13] and generalized low rank approximations of matrices (GLRAM) [17]. Yang et al. [18] proposed Bi-2DPCA which performs 2DPCA [2] twice: the first one is in horizontal direction and the second is in vertical direction.

The above 2DPCA variants proposed almost independently and simultaneously. Therefore, the theoretical relationship among them is not clear. In the following, we will discuss the relationship theoretically.

3   2DPCA and Its Variants

In this section, we summarize the original 2DPCA [2] and its variants which are roughly divided into two categories: iterative and non-iterative algorithms, and present some equivalence among them.

3.1   2DPCA

Suppose that there are M training image samples, the kth training image is denoted by an $m \times n$ matrix $A_k \in \Re^{m \times n}$ where $\Re$ denotes the set of real numbers, and the average image of all training samples is denoted by $\bar{A} = \frac{1}{M} \sum_{k=1}^{M} A_k$. Then the image covariance (scatter) matrix [2] is defined by

$$G = \frac{1}{M} \sum_{k=1}^{M} \left(A_k - \bar{A} \right)^T \left(A_k - \bar{A} \right) \in \Re^{n \times n}, \tag{1}$$

where ^{T} denotes the transpose of a matrix, and the generalized total scatter criterion [2] is expressed by

$$J(X) = \text{tr} \left(X^T G X \right), \tag{2}$$

where tr denotes the matrix trace and $X \in \Re^{n \times \tilde{n}}$ for $\tilde{n} < n$ is subject to $X^T X = I_{\tilde{n}}$ where $I_{\tilde{n}}$ is the $\tilde{n} \times \tilde{n}$ identity matrix. The optimal X that maximize $J(X)$ is obtained by $X = [x_1, \ldots, x_{\tilde{n}}]$ where x_j $(j = 1, \ldots, \tilde{n})$ denotes the eigenvector of G corresponding to the jth largest eigenvalue. Finally, each A_k is transformed into

$$B_k = A_k X \in \Re^{m \times \tilde{n}}. \tag{3}$$

Although the above 2DPCA can reduce the number of columns from n to $\tilde{n}$ in (3), the number of rows, m, is unchanged. Therefore, 2DPCA needs many more coefficients for image representation than PCA [2, 9]. To overcome this problem, several variants of 2DPCA have been proposed recently. They can be classified into two categories: iterative and non-iterative algorithms.

3.2   Non-iterative 2DPCA Variants

In this subsection, we summarize the non-iterative 2DPCA variants, which are further divided into two sub-categories: parallel and serial methods.

Parallel Method. Instead of the image covariance (scatter) matrix in (1), another one can be defined as follows:

$$\tilde{G} = \frac{1}{M} \sum_{k=1}^{M} \left(A_k - \bar{A} \right) \left(A_k - \bar{A} \right)^T \in \Re^{m \times m}, \tag{4}$$

from which another criterion is obtained by

$$\tilde{J}(Y) = \mathrm{tr} \left(Y^T \tilde{G} Y \right), \tag{5}$$

where $Y \in \Re^{m \times \tilde{m}}$ for $\tilde{m} < m$ is subject to $Y^T Y = I_{\tilde{m}}$. The optimal Y that maximize $\tilde{J}(Y)$ is obtained by $Y = [y_1, \ldots, y_{\tilde{m}}]$ where y_i $(i = 1, \ldots, \tilde{m})$ denotes the eigenvector of $\tilde{G}$ corresponding to the ith largest eigenvalue. Zhang and Zhou [8] called this method the alternative 2DPCA. Finally, each A_k is transformed into

$$\tilde{B}_k = Y^T A_k X \in \Re^{\tilde{m} \times \tilde{n}}. \tag{6}$$

We call this type of 2DPCA variant the parallel method because X and Y are calculated in a parallel manner. Essentially, $(2D)^2$PCA [8], eigenimages [9], C2DPCA [11], BDPCA [13], and D2DPCA [15] are equivalent to this parallel method.

Serial Method. As opposed to the above parallel method, we can consider the serial method as follows. First, 2DPCA [2] described in Subsection 3.1 is conducted to obtain $\{B_k\}_{k=1}^{M}$. Next, the image covariance (scatter) matrix is constructed for the set of the transposed matrices $\{B_k^T\}_{k=1}^{M}$ as follows:

$$\hat{G} = \frac{1}{M} \sum_{k=1}^{M} \left(B_k^T - \bar{B}^T \right)^T \left(B_k^T - \bar{B}^T \right) \in \Re^{m \times m}, \tag{7}$$

where $\bar{B} = \frac{1}{M} \sum_{k=1}^{M} B_k$. Then the total scatter criterion

$$\hat{J}(\hat{Y}) = \mathrm{tr} \left(\hat{Y}^T \hat{G} \hat{Y} \right) \tag{8}$$

for $\hat{Y} \in \Re^{m \times \tilde{m}}$ which is subject to $\hat{Y}^T \hat{Y} = I_{\tilde{m}}$ is maximized by $\hat{Y} = [\hat{y}_1, \ldots, \hat{y}_{\tilde{m}}]$ where $\hat{y}_i$ $(i = 1, \ldots, \tilde{m})$ denotes the eigenvector of $\hat{G}$ corresponding to the ith largest eigenvalue. Each A_k is transformed into

$$\hat{B}_k = \hat{Y}^T B_k = \hat{Y}^T A_k X \in \Re^{\tilde{m} \times \tilde{n}}. \tag{9}$$

Essentially, PIMC [12], IPCA [14], and Bi-2DPCA [18] are equivalent to this serial method.

3.3   Iterative 2DPCA Variants

In the above non-iterative 2DPCA variants, two matrices X and Y (or $\hat{Y}$) are derived from different criteria. On the other hand, the iterative 2DPCA variants are formulated as an optimization of a single criterion as follows [19]:

$$\max_{X,Y} \frac{1}{M} \sum_{k=1}^{M} \left\| Y^T \tilde{A}_k X \right\|_F^2 \tag{10}$$

$$\text{subj.to } X^T X = I_{\tilde{n}}, \ Y^T Y = I_{\tilde{m}}, \tag{11}$$

where $\tilde{A}_k = A_k - \bar{A}$. Let $F(X,Y)$ be the objective function in (10). Then the Lagrange function for (10)-(11) is given by

$$\mathcal{L} = F(X,Y) - \text{tr}\left[\Lambda_X \left(X^T X - I_{\tilde{n}} \right) \right] - \text{tr}\left[\Lambda_Y \left(Y^T Y - I_{\tilde{m}} \right) \right], \tag{12}$$

where Λ_X and Λ_Y are symmetric matrices of which the elements are the Lagrange multipliers. From $\partial\mathcal{L}/\partial X = 0$ and $\partial\mathcal{L}/\partial Y = 0$, we have

$$G_Y X = X \Lambda_X, \tag{13}$$

$$G_X Y = Y \Lambda_Y, \tag{14}$$

respectively, where $G_Y = \frac{1}{M} \sum_{k=1}^{M} \tilde{A}_k^T Y Y^T \tilde{A}_k$ and $G_X = \frac{1}{M} \sum_{k=1}^{M} \tilde{A}_k X X^T \tilde{A}_k^T$. Hence, for a fixed Y, the optimal X is obtained by $X = [x_1, \ldots, x_{\tilde{n}}]$ where x_j ($j = 1, \ldots, \tilde{n}$) denotes the eigenvector of G_Y corresponding to the jth largest eigenvalue, and similarly, for the obtained X, the optimal Y is obtained by $Y = [y_1, \ldots, y_{\tilde{m}}]$ where y_i ($i = 1, \ldots, \tilde{m}$) denotes the eigenvector of G_X corresponding to the ith largest eigenvalue. This procedure is repeated until the convergence. Essentially, GPCA [6], B2DPCA [7], Benito's method [10], and SPCA [16] are equivalent to this method.

4   A Unified View of 2DPCA Variants

In this section, we present a unified view of the 2DPCA variants described in the above section. That is, we show that the non-iterative 2DPCA variants including the original 2DPCA [2] can be interpreted as the non-iterative approximate algorithms for the iterative algorithm in Subsection 3.3.

First, the original 2DPCA [2] is derived from the iterative algorithm in Subsection 3.3 as follows: if we initialize $Y = I_m$, then $F(X,Y)$ becomes

$$F(X,Y) = \text{tr}\left[X^T \left(\frac{1}{M} \sum_{k=1}^{M} \tilde{A}_k^T Y Y^T \tilde{A}_k \right) X \right] = \text{tr}\left(X^T G X \right) = J(X). \tag{15}$$

Therefore, X obtained by the first iteration coincides with that of 2DPCA [2].

On the other hand, if we initialize $X = I_n$, then we have $F(X,Y) = \tilde{J}(Y)$. Therefore, Y obtained by the first iteration in this setting coincides with that of the parallel method in Subsection 3.2. That is, the parallel method uses X and Y obtained by the first iterations of the iterative algorithm started from different initial settings: $Y = I_m$ and $X = I_n$, respectively.

The serial method in Subsection 3.2 can be derived by initializing $Y = I_m$. Since X obtained by the serial method coincides with that of 2DPCA [2], it can be obtained by the first iteration. Next, the obtained X is used for computing Y, i.e., we have

$$F(X,Y) = \text{tr} \left[Y^T \left(\frac{1}{M} \sum_{k=1}^{M} \tilde{A}_k X X^T \tilde{A}_k^T \right) Y \right] = \text{tr} \left(Y^T \hat{G} Y \right). \qquad (16)$$

Therefore, Y obtained by the first iteration coincides with that of the serial method; $Y = \hat{Y}$.

Thus, the conventional non-iterative 2DPCA variants can be derived as the first iterations of the iterative algorithm started from different initial settings. Furthermore, this viewpoint suggests the existence of the other non-iterative variant, i.e., we can consider another (alternative) serial method which is initialized as $X = I_n$. Then we obtain a pair of Y and $\hat{X}$ which is the solution to $\max_X F(X,Y)$. Finally, we can combine the two serial methods to obtain the selective method as follows:

$$(X^*, Y^*) = \arg \max_{(X,Y) \in \{(X,\hat{Y}),\, (\hat{X},Y),\, (\hat{X},\hat{Y})\}} F(X,Y). \qquad (17)$$

Each A_k is transformed into

$$B_k^* = (Y^*)^T A_k X^* \in \Re^{\tilde{m} \times \tilde{n}}. \qquad (18)$$

This method will achieve better performance than the conventional non-iterative 2DPCA variants because it is guaranteed that the objective function value obtained by the selective method is greater than or equal to that of the serial methods. The superiority of the proposed method to the other methods will be experimentally demonstrated in the next section. Table 1 shows the classification of the non-iterative 2DPCA variants. The proposed selective method in (17) fills up the blank in Table 1.

Table 1. Classification of non-iterative 2DPCA variants

	Renew X or Y	Renew X and Y
Initialize X or Y	(alternative) 2DPCA	(alternative) serial
Initialize X and Y	parallel	**selective** (proposed)

5   Experimental Results

In this section, we show experimental results on the ORL face image database [20], the Caltech Faces [21] and the UMIST face database [22]. The ORL database [20]

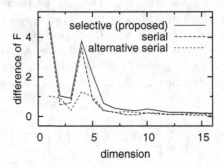

Fig. 1. Difference of F for the ORL face image database [20]

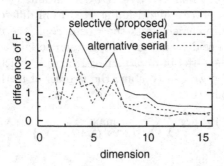

Fig. 2. Difference of F for the Caltech Faces [21]

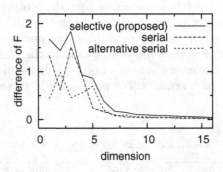

Fig. 3. Difference of F for the UMIST face database [22]

contains face images of 40 persons. For each person, there are 10 different face images. In our experiment, we used the first 5 images per person, i.e., $M = 5 \times 40 = 200$. The height and width of each image are $m = 112$ and $n = 92$, respectively. Fig. 1 shows the differences of the objective function values: $F(\xi, \eta) - F(X, Y)$, where $\xi = X$, $\eta = \hat{Y}$ for broken line (the serial method), $\xi = \hat{X}$, $\eta = Y$ for dotted line (the alternative serial method) and $\xi = \hat{X}$, $\eta = \hat{Y}$

for solid line (the proposed selective method), and $F(X, Y)$ denotes the objective function value for the parallel method, i.e., X and Y are obtained by maximizing (2) and (5), respectively. The horizontal axis denotes $\tilde{m}$ ($= \tilde{n}$). Since all lines lie in the positive region, it is clear that the serial, alternative serial and selective methods achieve higher objective function values than the parallel method. Furthermore, among the three methods, the proposed selective method achieves the highest objective function value.

Figs. 2 and 3 show the results for the Caltech Faces [21] and the UMIST face database, respectively. In the Caltech Faces [21], we used 445 cropped face images. The height and width of each image are $m = 165$ and $n = 122$, respectively. In the UMIST face database, we used 380 face images. The height and width of each image are $m = 112$ and $n = 92$, respectively. Figs. 2 and 3 also demonstrate the superiority of the proposed selective method to the other methods.

6 Conclusion

In this paper, we summarized the 2DPCA variants which have been proposed by several researchers recently, and presented a unified view of the 2DPCA variants. We discussed some equivalence of the 2DPCA variants and classified them on the basis of their algorithmic patterns. Then we proposed a new non-iterative 2DPCA algorithm based on the classification result. The proposed method achieved higher objective function value than the other non-iterative 2DPCA variants. Future work will include the summarization of the variants of two-dimensional linear discriminant analysis.

References

1. Yang, J., Yang, J.-Y.: From Image Vector to Matrix: A Straightforward Image Projection Technique - IMPCA vs. PCA. Pattern Recognition 35(9), 1997–1999 (2002)
2. Yang, J., Zhang, D., Frangi, A.F., Yang, J.-Y.: Two-Dimensional PCA: A New Approach to Appearance-Based Face Representation and Recognition. IEEE Trans. PAMI 26(1), 131–137 (2004)
3. Wang, L., Wang, X., Zhang, X., Feng, J.: The equivalence of two-dimensional PCA to line-based PCA. Pattern Recogn. Lett. 26(1), 57–60 (2005)
4. Zhang, D., Chen, S., Liu, J.: Representing Image Matrices: Eigenimages Versus Eigenvectors. In: Wang, J., Liao, X.-F., Yi, Z. (eds.) ISNN 2005. LNCS, vol. 3497, pp. 659–664. Springer, Heidelberg (2005)
5. Gao, Q.: Is two-dimensional PCA equivalent to a special case of modular PCA? Pattern Recogn. Lett. 28(10), 1250–1251 (2007)
6. Ye, J., Janardan, R., Li, Q.: GPCA: An Efficient Dimension Reduction Scheme for Image Compression and Retrieval. In: KDD, pp. 354–363. ACM, New York (2004)
7. Kong, H., Li, X., Wang, X., Teoh, E.K., Wang, J.-G., Venkateswarlu, R.: Generalized 2D Principal Component Analysis. In: Proc. IJCNN, pp. 108–113 (2005)
8. Zhang, D., Zhou, Z.-H. (2D)²PCA: Two-Directional Two-Dimensional PCA for Efficient Face Representation and Recognition. Neurocomputing 69(1-3), 224–231 (2005)

9. Zhang, D., Chen, S., Liu, J.: Representing Image Matrices: Eigenimages Versus Eigenvectors. In: Wang, J., Liao, X.-F., Yi, Z. (eds.) ISNN 2005. LNCS, vol. 3497, pp. 659–664. Springer, Heidelberg (2005)
10. Benito, M., Peña, D.: A Fast Approach for Dimensionality Reduction with Image Data. Pattern Recognition 38(12), 2400–2408 (2005)
11. Xu, A., Jin, X., Jiang, Y., Guo, P.: Complete Two-Dimensional PCA for Face Recognition. In: Proc. ICPR, vol. (3), pp. 481–484 (2006)
12. Xu, D., Yan, S., Zhang, L., Li, M., Ma, W., Liu, Z., Zhang, H.: Parallel Image Matrix Compression for Face Recognition. In: MMM 2005, pp. 232–238. IEEE Computer Society, Washington (2005)
13. Zuo, W., Zhang, D., Wang, K.: Bidirectional PCA with Assembled Matrix Distance Metric for Image Recognition. IEEE trans. SMC-B36(4), 863–872 (2006)
14. Wen, Y., Shi, P.: Image PCA: A New Approach for Face Recognition. In: ICASSP 2007, vol. (1), pp. 1241–1244. IEEE (2007)
15. Lu, C., Liu, W., Liu, X., An, S.: Double Sides 2DPCA for Face Recognition. In: Huang, D.-S., Wunsch II, D.C., Levine, D.S., Jo, K.-H. (eds.) ICIC 2008. LNCS, vol. 5226, pp. 446–459. Springer, Heidelberg (2008)
16. Yongxin Taylor Xi, Y.T., Ramadge, P.J.: Separable PCA for Image Classification. In: ICASSP, pp. 1805–1808. IEEE (2009)
17. Ye, J.: Generalized Low Rank Approximations of Matrices. Mach. Learn. 61(1-3), 167–191 (2005)
18. Yang, J., Xu, Y., Yang, J.-Y.: Bi-2DPCA: A Fast Face Coding Method for Recognition. In: Pattern Recognition Recent Advances, pp. 313–340. InTech (2010)
19. Inoue, K., Urahama, K.: Equivalence of Non-Iterative Algorithms for Simultaneous Low Rank Approximations of Matrices. In: Proc. CVPR, pp. 154–159. IEEE (2006)
20. Samaria, F.S., Harter, A.C.: Parameterisation of a Stochastic Model for Human Face Identification. In: Proc. 2nd IEEE Workshop on Applications of Computer Vision, pp. 138–142 (1994)
21. Caltech Faces, http://www.vision.caltech.edu/html-files/archive.html
22. Graham, D.B., Allinson, N.M.: Characterizing Virtual Eigen Signatures for General Purpose Face Recognition. Face Recognition: From Theory to Applications 163, 446–456 (1998)

Automatic Dimensionality Estimation for Manifold Learning through Optimal Feature Selection

Fadi Dornaika[1,2], Ammar Assoum[3], and Bogdan Raducanu[4]

[1] University of the Basque Country UPV/EHU, San Sebastian, Spain
[2] IKERBASQUE, Basque Foundation for Science, Bilbao, Spain
[3] LaMA Laboratory, Lebanese University, Tripoli, Lebanon
[4] Computer Vision Center, 08193 Bellaterra, Spain

Abstract. A very important aspect in manifold learning is represented by automatic estimation of the intrinsic dimensionality. Unfortunately, this problem has received few attention in the literature of manifold learning. In this paper, we argue that feature selection paradigm can be used to the problem of automatic dimensionality estimation. Besides this, it also leads to improved recognition rates. Our approach for optimal feature selection is based on a Genetic Algorithm. As a case study for manifold learning, we have considered Laplacian Eigenmaps (LE) and Locally Linear Embedding (LLE). The effectiveness of the proposed framework was tested on the face recognition problem. Extensive experiments carried out on ORL, UMIST, Yale, and Extended Yale face data sets confirmed our hypothesis.

1 Introduction

In recent years, a new family of non-linear dimensionality reduction techniques for manifold learning has emerged. The most known ones are: Kernel Principal Component Analysis (KPCA) [1], Locally Linear Embedding (LLE) [2], Isomap [3], Supervised Isomap [4], Laplacian Eigenmaps (LE)[5,6]. This family of non-linear embedding techniques appeared as an alternative to their linear counterparts which suffer of severe limitation when dealing with real-world data: i) they assume the data lie in an Euclidean space, and ii) they may fail when the number of samples is too small. On the other hand, the non-linear dimensionality techniques are able to discover the intrinsic data structure by exploiting the local topology. In general, they attempt to optimally preserve the local geometry around each data sample while using the rest of the samples to preserve the global structure of the data. Most of existing works on non-linear manifold learning techniques are focused either on the graph design [7] or on the objective function that should be optimized. However, to the best of our knowledge, there is no work attempting to automatically estimate the dimensionality of the non-linear embedding. For classification tasks, the common way was to plot the performance over a validation (test) data set as a curve from which the optimal dimension can be estimated. This assumes that all dimensions below the found one will be considered as relevant and all dimensions beyond it should be irrelevant. This assumption seems to be very simplistic and does not take into account the effect of subsets of dimensions.

G.L. Gimel' farb et al. (Eds.): SSPR & SPR 2012, LNCS 7626, pp. 575–583, 2012.
© Springer-Verlag Berlin Heidelberg 2012

For this reason, we address this problem in the current paper. The main contribution of our work is represented by a generic framework associated with manifold learning which allows the extraction and selection of optimal features (dimensions) in the embedded subspace (from the perspective of pattern classification). Our approach for feature selection is guided by a Genetic Algorithm (GA). The advantage of the proposed framework is twofold. First, by selecting the most relevant features (dimensions), the classification performance is enhanced (as proven by the experimental results). Second, by retaining only the most relevant dimensions, pattern classification task becomes much more efficient[1] (from the point of view of computational complexity).

The remainder of the paper is organized as follows. Section 2 reviews two non-linear manifold learning techniques. Section 3 briefly describes the feature selection paradigm. Section 4 provides some experimental results obtained with four public face data sets. Finally, section 5 contains our conclusions.

2 Non-linear Embedding Techniques

2.1 Laplacian Eigenmaps

Laplacian Eigenmaps is a recent non-linear dimensionality reduction technique that aims to preserve the local structure of data [5]. Using the notion of the Laplacian of the graph, this non-supervised algorithm computes a low-dimensional representation of the data set by optimally preserving local neighborhood information in a certain sense. We assume that we have a set of N samples $\{\mathbf{x}_i\}_{i=1}^{N} \subset \mathbb{R}^D$. Let's define a neighborhood graph on these samples, such as a K-nearest-neighbor or ϵ-ball graph, or a full mesh, and weigh each edge $\mathbf{x}_i \sim \mathbf{x}_j$ by a symmetric affinity function $W_{ij} = K(\mathbf{x}_i; \mathbf{x}_j)$, typically Gaussian $W_{ij} = \exp(-\frac{\|\mathbf{x}_i - \mathbf{x}_j\|^2}{\beta})$

where β is usually set to the average of squared distances between all pairs. LE seeks latent points $\{\mathbf{y}_i\}_{i=1}^{N} \subset \mathbb{R}^L$ that minimize $\frac{1}{2}\sum_{i,j}\|\mathbf{x}_i - \mathbf{x}_j\|^2 W_{ij}$, which discourages placing far apart latent points that correspond to similar observed points. If $\mathbf{W} \equiv W_{ij}$ denotes the symmetric affinity matrix and $\mathbf{D}$ is the diagonal weight matrix, whose entries are column (or row, since $\mathbf{W}$ is symmetric) sums of $\mathbf{W}$, then the Laplacian matrix is given $\mathbf{L} = \mathbf{D} - \mathbf{W}$. The objective function can also be written as:

$$\frac{1}{2}\sum_{i,j}\|\mathbf{y}_i - \mathbf{y}_j\|^2 W_{ij} = tr(\mathbf{Z}^T \mathbf{L}\mathbf{Z}) \tag{1}$$

where $\mathbf{Z}^T = \mathbf{Y} = [\mathbf{y}_1, \ldots, \mathbf{y}_N]$ is the $N \times L$ embedding matrix and $tr(.)$ denotes the trace of a matrix. The ith row of the matrix $\mathbf{Z}$ provides the vector $\mathbf{y}_i$—the embedding coordinates of the sample $\mathbf{x}_i$.

The embedding matrix $\mathbf{Z}$ is the solution of the optimization problem:

$$\min_{\mathbf{Z}} tr(\mathbf{Z}^T \mathbf{L}\mathbf{Z}) \ \ s.t. \ \ \mathbf{Z}^T \mathbf{D}\mathbf{Z} = \mathbf{I}, \ \ \mathbf{Z}^T \mathbf{L}\mathbf{e} = \mathbf{0} \tag{2}$$

---

[1] This is a clear advantage for large data sets for which the dimensionality of the non-linear embedded space is equal to the size of the data set.

where $\mathbf{I}$ is the identity matrix and $\mathbf{e} = (1, \ldots, 1)^T$. The first constraint eliminates the trivial solution $\mathbf{Z} = \mathbf{0}$ (by setting an arbitrary scale) and the second constraint eliminates the trivial solution $\mathbf{e}$ (all samples are mapped to the same point). Standard methods show that the embedding matrix is provided by the matrix of eigenvectors corresponding to the smallest eigenvalues of the generalized eigenvector problem,

$$\mathbf{L}\,\mathbf{z} = \lambda \mathbf{D}\,\mathbf{z} \tag{3}$$

Let the column vectors $\mathbf{z}_0, \ldots, \mathbf{z}_{N-1}$ be the solutions of (3), ordered according to their eigenvalues, $\lambda_0 = 0 \leq \lambda_1 \leq \ldots \leq \lambda_{N-1}$. The eigenvector corresponding to eigenvalue 0 is left out and only the next eigenvectors for embedding are used. The embedding of the original samples is given by the row vectors of the matrix $\mathbf{Z}$, that is, $\mathbf{Y} = [\mathbf{y}_1, \mathbf{y}_2, \ldots, \mathbf{y}_N] = \mathbf{Z}^T$.

2.2 Locally Linear Embedding

One important geometric intuition behind the LLE algorithm is that each data point and its neighbors lie on or are close to a locally linear patch of the manifold. LLE tries to characterize the geometry of the local patches by finding the linear coefficients that reconstruct each data point from its neighbors. In the first step, each sample is approximated by a weighted linear combination of its K nearest neighbors, making use of the assumption that neighboring samples will lie on a locally linear patch of the nonlinear manifold. To find the reconstruction weight matrix $\mathbf{W}$, where the entry W_{ij} contains the weight of neighbor j in the reconstruction of sample $\mathbf{x}_i$. The reconstruction error is minimized subject to the constraint that the rows of the weight matrix sum to one: $\sum_{j=1}^{N} w_{ij} = 1$.

Let $\mathbf{Y}$ be the non-linear embedding of the original data $\mathbf{Y} = [\mathbf{y}_1, \mathbf{y}_2, \ldots, \mathbf{y}_N]$. Then $\mathbf{Y}$ can be computed by minimizing the following embedding cost function:

$$\sum_{i=1}^{N} \|\mathbf{y}_i - \sum_{j=1}^{N} w_{ij}\mathbf{y}_j\|^2 = tr(\mathbf{Y}\,\mathbf{M}\,\mathbf{Y}^T) \tag{4}$$

where $\mathbf{M}$ is given by $\mathbf{M} = (\mathbf{I} - \mathbf{W})(\mathbf{I} - \mathbf{W})^T$. The eigenvectors of the matrix $\mathbf{M}$ corresponding to the smallest eigenvalues then form the final embedding $\mathbf{Y}$.

3 Feature Selection

3.1 Overview

In many fields, including pattern recognition and machine learning, the input data are represented by a very large number of features, but only few of them are relevant for classification task. Many algorithms become computationally intractable when the dimensionality of the data is too high. On the other hand, once an optimal set of selected features has been chosen, even the basic classifiers (e.g., K-nearest neighbor) can achieve desirable performance. Therefore, the process of feature selection, i.e. the task of choosing a small subset of features which is statistically relevant, can be critical to minimize the classification error. At the same time, feature selection also reduces training and inference time and leads to a better data visualization as well as to a reduction of measurement and storage requirements. Roughly speaking, feature selection algorithms have two key problems [8,9]: (i) search strategy and (ii) evaluation criterion. The first key problem refers to the strategy of the search in the space of all possible solutions. Roughly speaking, the search strategies can be optimal or heuristic. Regarding the second key problem, feature selection algorithms can be categorized into filter model and wrapper model. In the wrapper model, the feature selection method tries to directly optimize the performance of a specific predictor (classification or clustering algorithm). The main drawback of this method is its computational deficiency. In the filter model, the feature selection is done as a preprocessing, without trying to optimize the performance of any specific predictor directly [10,11,12]. A comprehensive discussion of feature selection methodologies can be found in [13].

3.2   Optimal Feature Subset Using a Genetic Algorithm

We adopt here a wrapper technique for feature selection. The adopted evaluation strategy will attempt to maximize the recognition accuracy over a given validation set.

The adopted search strategy will be carried out using a Genetic Algorithm (GA). Genetic Algorithms (GAs) are biologically motivated adaptive systems based on natural selection and genetic recombination [14] whose main application is for optimization problems. In the standard GA, candidate solutions are encoded as fixed length vectors–strings. We use a bit string representation whose length is determined by the number of eigenvectors obtained as a result of the embedding process. Thus, each eigenvector is associated with one bit in the string. If the ith bit is 1, then the ith eigenvector is selected. Otherwise, that component is discarded. Each string thus represents a different subset of eigenvectors. The initial population of solutions is chosen randomly. These candidate solutions are allowed to evolve over a certain number of generations. At each generation, the fitness of each string is set to the recognition rate over a fixed validation set.

4   Performance Study

To verify the effectiveness of our proposed framework, we applied it to the face recognition problem. Four public face data sets are considered.

4.1   Data Sets

1. The ORL face data set[2]. There are 10 images for each of the 40 human subjects, which were taken at different times, varying the lighting, facial expressions (open/closed eyes, smiling/not smiling) and facial details (glasses/no glasses). The images were taken with a tolerance for some tilting and rotation of the face up to 20°. Some samples are shown in figure 1.

---

[2] http://www.cl.cam.ac.uk/research/dtg/attarchive/
   facedatabase.html

2. The UMIST face data set[3]. The UMIST data set contains 575 gray images of 20 different people. The images depict variations in head pose. Some samples are show in figure 2.
3. The Yale face data set[4]. It contains 11 grayscale images for each of the 15 individuals. The images demonstrate variations in lighting condition (left-light, center-light, right-light), facial expression (normal, happy, sad, sleepy, surprised, and wink), and with/without glasses. Figure 3 shows some instances from this dataset.
4. The Extended Yale Face Database B[5]. It contains 16128 images of 28 human subjects under 9 poses and 64 illumination conditions. In our study, a subset of 1800 images has been used. Figure 4 shows some face samples in the extended Yale Face Database B.

Fig. 1. Some samples in ORL data set

4.2 Experimental Results

The experiments consisted of two stages. In the first stage, the selection paradigm was run over a fixed validation set. For every face data set, the validation set was randomly set to 40% of the whole data set. In the second stage, we evaluated the generalization capacity of the obtained features (generalization tests). For each face data set and for every method, we conducted three groups of experiments for which the percentage of training samples was set to 30%, 50% and 70% of the whole data set. The remaining data was used for testing. The partition of the data set was done randomly. In all our experiments the classification in the embedded spaces (selected or unselected) was carried out by the Nearest Neighbor Classifier. Figure 5 shows the results of the first stage,

---

[3] http://www.shef.ac.uk/eee/research/vie/research/face.html
[4] http://see.xidian.edu.cn/vipsl/database_Face.html
[5] http://vision.ucsd.edu/~leekc/ExtYaleDatabase/ExtYaleB.html

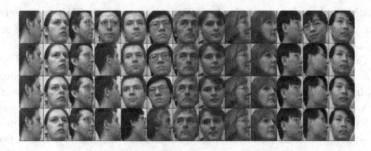

Fig. 2. Some samples in UMIST data set

Fig. 3. Some samples in YALE data set

Fig. 4. Some samples in Extended Yale data set

where the plots depict the validation results before (blue line) and after feature selection (red line), for each data set.

In table 1, we summarize the face recognition performance using the Laplacian Eigenmaps embedding on the four data sets (output of second stage). For every data set and for every training percentage, three schemes were used: the original features/ dimensions (Orig.), the selected features using the GA (GA), and the sorted features using the Fisher Score[6] (FS). This table illustrates the average best recognition rate (%) over 5 random splits. The number in parenthesis is the mean recognition over the available dimensions. We can observe that: (i) the best recognition rate remains almost the same for all schemes, however, the GA scheme got these results with a fraction of the original features which varies between 30% and 40% of the total dimensions, and

---

[6] This scheme re-ranks the features according to their Fisher Score.

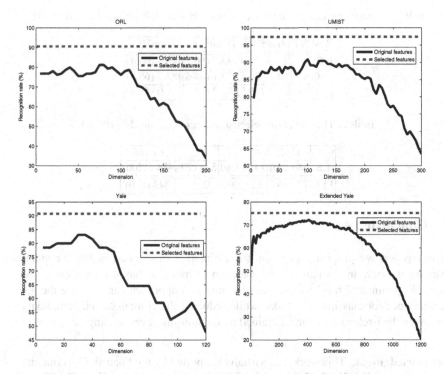

Fig. 5. Feature selection for LE embedding using a GA for four face data sets

Table 1. Comparison of recognition rates between the three schemes: maximum and average (in parenthesis)

30%	ORL	UMIST	Yale	Ex.Yale
Orig.	67.8 (52.5)	90.7 (73.8)	**75.9** (51.7)	67.8 (56.0)
GA	68.1 (**66.9**)	**91.3** (**83.8**)	74.1 (**61.7**)	68.2 (**65.2**)
FS	**71.8** (48.9)	89.7 (60.3)	73.3 (44.2)	**69.1** (54.7)
50%				
Orig.	73.5 (65.4)	94.1 (83.1)	**79.2** (57.7)	73.5 (63.2)
GA	74.0 (**79.8**)	**94.5** (**90.0**)	79.0 (**66.2**)	**74.0** (**70.8**)
FS	**84.0** (61.8)	94.0 (70.2)	77.5(49.5)	73.8 (61.4)
70%				
Orig.	87.1 (72.9)	**95.8** (88.9)	**78.8** (64.0)	75.5 (64.2)
GA	87.3 (**83.8**)	95.6 (**92.2**)	78.4 (**71.9**)	**76.0** (**72.2**)
FS	**88.1** (69.1)	94.9 (77.7)	76.7 (54.7)	75.5(60.4)

(ii) the GA method provided the most stable recognition rate as a function of the features used. Table 2 shows the face recognition performance using the LLE embedding on UMIST data set. Table 3 provides a comparison between the original dimensionality of the embedded space and the dimensionality discovered by our feature selection approach (in bold).

Table 2. Average best recognition rate (%) over 5 random splits using the LLE method

UMIST	Train30%	Train50%	Train70%
Orig.	**64.3** (47.4)	**73.7** (60.1)	**78.5** (66.4)
GA	61.6 (**54.5**)	73.1 (**64.6**)	77.7 (**68.8**)
FS	60.3 (45.5)	72.3 (58.2)	76.0 (63.9)

Table 3. The size of the selected features as obtained by the GA

	ORL	UMIST	Yale	Ex.Yale
LE	**68** (200)	**141** (300)	**54** (120)	**486** (1200)
LLE	**77** (200)	**82** (200)	**32** (80)	**243** (600)

5 Conclusion

In this paper, we proposed an automatic estimation of dimensionality for manifold learning through an optimal feature selection framework, based on a Genetic Algorithm. Experimental results show that the proposed approach can enhance the global performance for classification tasks, while reducing the computational complexity by removing the irrelevant features obtained by the non-linear embeddings.

Acknowledgment. This work was partially supported by the Spanish Government under the project TIN2010-18856 and the Lebanese National Council for Scientific Research (LCNRS) under the project 03-10-11.

References

1. Schölkopf, B., Smola, A., Müller, K.R.: Nonlinear component analysis as a kernel eigenvalue problem. Neural Computation 10, 1299–1319 (1998)
2. Saul, L.K., Roweis, S.T., Singer, Y.: Think globally, fit locally: Unsupervised learning of low dimensional manifolds. Journal of Machine Learning Research 4, 119–155 (2003)
3. Tenenbaum, J.B., de Silva, V., Langford, J.C.: A global geometric framework for nonlinear dimensionality reduction. Science 290(5500), 2319–2323 (2000)
4. Geng, X., Zhan, D., Zhou, Z.: Supervised nonlinear dimensionality reduction for visualization and classification. IEEE Transactions on Systems, Man, and Cybernetics-Part B: Cybernetics 35, 1098–1107 (2005)
5. Belkin, M., Niyogi, P.: Laplacian eigenmaps for dimensionality reduction and data representation. Neural Computation 15(6), 1373–1396 (2003)
6. Jia, P., Yin, J., Huang, X., Hu, D.: Incremental Laplacian Eigenmaps by preserving adjacent information between data points. Pattern Recognition Letters 30(16), 1457–1463 (2009)
7. Zhan, L., Qiao, L., Chen, S.: Graph-optimized locality preserving projections. Pattern Recognition 43, 1993–2002 (2010)
8. Dy, J.G., Brodley, C.E.: Feature selection for unsupervised learning. Journal of Machine Learning Research 5, 845–889 (2004)
9. Zhao, Z., Liu, H.: Spectral feature selection for supervised and unsupervised learning. In: Int. Conference on Machine Learning (2007)

10. Mitra, P., Murthy, C., Pal, S.: Unsupervised feature selection using feature similarity. IEEE Trans. Pattern Analysis and Machine Intelligence 24, 301–312 (2002)
11. He, X., Cai, D., Niyogi, P.: Laplacian score for feature selection. In: Advances in Neural Information Processing Systems 18 (2005)
12. Cai, D., Zhang, C., He, X.: Unsupervised feature selection for multi-cluster data. In: 16th ACM SIGKDD Conference on Knowledge Discovery and Data Mining, KDD 2010 (2010)
13. Liu, H., Yu, L.: Toward integrating feature selection algorithms for classification and clustering. IEEE Trans. Knowledge Data Engineering 17, 494–502 (2005)
14. Srinivas, M., Patnaik, L.: Genetic algorithms: a survey. IEEE Computer 27(6), 17–26 (1994)

Novel Gabor-PHOG Features
for Object and Scene Image Classification

Atreyee Sinha*, Sugata Banerji, and Chengjun Liu

Department of Computer Science,
New Jersey Institute of Technology,
Newark, NJ 07102, USA
{as739,sb256,chengjun.liu}@njit.edu
http://cs.njit.edu/liu

Abstract. A new Gabor-PHOG (GPHOG) descriptor is first introduced in this paper for image feature extraction by concatenating the Pyramid of Histograms of Oriented Gradients (PHOG) of all the local Gabor filtered images. Next, a comparative assessment of the classification performance of the GPHOG descriptor is made in six different color spaces, namely the RGB, HSV, YCbCr, oRGB, DCS and YIQ color spaces, to propose the novel YIQ-GPHOG and the YCbCr-GPHOG feature vectors that perform well on different object and scene image categories. Third, a novel Fused Color GPHOG (FC-GPHOG) feature is presented by integrating the PCA features of the six color GPHOG descriptors for object and scene image classification, with applications to image search and retrieval. Finally, the Enhanced Fisher Model (EFM) is applied for discriminatory feature extraction and the nearest neighbor classification rule is used for image classification. The effectiveness of the proposed feature vectors for image classification is evaluated using two grand challenge datasets, namely the Caltech 256 dataset and the MIT Scene dataset.

Keywords: Gabor-PHOG (GPHOG), YIQ-GPHOG, YCbCr-GPHOG, FC-GPHOG, PCA, EFM, color spaces, image search.

1   Introduction

Color images provide powerful discriminating information than grayscale images [1], and color based image search can be very effective for face, object, scene, and texture image classification [2], [3], [4]. Some desirable properties of the descriptors defined in different color spaces include relative stability over changes in photographic conditions such as varying illumination. Global color features such as the color histogram and local invariant features provide varying degrees of success against image variations such as rotation, viewpoint and lighting changes, clutter and occlusions [5]. Shape and local features also provide important cues for content based image classification and retrieval. Local object shape and the spatial layout of the shape within an image can be described

---

* Corresponding author.

G.L. Gimel' farb et al. (Eds.): SSPR & SPR 2012, LNCS 7626, pp. 584–592, 2012.
© Springer-Verlag Berlin Heidelberg 2012

by the Pyramid of Histograms of Oriented Gradients (PHOG) descriptor [6]. Several researchers have described the biological relevance and computational properties of Gabor wavelets for image analysis [7], [8]. Lately, Donato et al. [9] showed experimentally that the Gabor wavelet representation is optimal for classifying facial actions.

We subject the image to a series of Gabor wavelet transformations, whose kernels are similar to the 2D receptive field profiles of the mammalian cortical simple cells [7]. In this paper, we design several novel feature vectors based on Gabor filters. Specifically, we first introduce a novel Gabor-PHOG (GPHOG) descriptor by concatenating the Pyramid of Histograms of Oriented Gradients (PHOG) of the components of the images produced by the result of applying a combination of Gabor filters in different orientations. We then measure the classification performance of our GPHOG descriptor on six different color spaces and propose the novel YIQ-GPHOG and the YCbCr-GPHOG features. We further extend this concept by integrating the Principal Component Analysis (PCA) features of the six color GPHOG vectors to produce the novel Fused Color GPHOG (FC-GPHOG) descriptor. Feature extraction applies the Enhanced Fisher Model (EFM) [10], and image classification is based on the nearest neighbor classification rule. Finally, the effectiveness of the proposed descriptors for image classification is evaluated using two datasets: the Caltech 256 grand challenge dataset and the MIT Scene dataset.

2   Novel Gabor-PHOG Features for Object and Scene Image Classification

This section briefly reviews the color spaces in which our new descriptors are defined, and then discusses the proposed novel descriptors and classification methodology for image classification.

2.1   Color Spaces

A color image contains three component images. The commonly used color space is the RGB color space, from which other color spaces are derived by means of either linear or nonlinear transformations. The HSV color space is motivated by human vision system as humans describe color by means of hue, saturation, and brightness. Hue and saturation define chrominance, while intensity or value specifies luminance [1]. The YIQ color space is adopted by the NTSC (National Television System Committee) video standard in reference to RGB NTSC. The I and Q components are derived from the U and V counterparts of the YUV color space via a clockwise rotation (33°) [3]. The YCbCr color space is developed for digital video standard and television transmissions. In YCbCr, the RGB components are separated into luminance, chrominance blue, and chrominance red. The oRGB color space [11] has three channels L, C_1 and C_2. The primaries of this model are based on the three fundamental psychological opponent axes: white-black, red-green, and yellow-blue. The color information is contained in

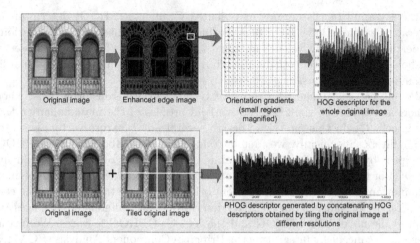

Fig. 1. Generation of the PHOG descriptor for a color image

C_1 and C_2. The value of C_1 lies within $[-1, 1]$ and the value of C_2 lies within $[-0.8660, 0.8660]$. The L channel contains the luminance information and its values ranges between $[0, 1]$. The Discriminating Color Space (DCS) [12], is derived from the RGB color space by means of discriminant analysis [13]. In the RGB color space, a color image with a spatial resolution of $m \times n$ contains three color component images R, G, and B with the same resolution. Each pixel (x,y) of the color image thus contains three elements corresponding to the red, green, and blue values from the R, G, and B component images. The DCS defines discriminating component images via a linear transformation $W_D \in \mathbb{R}^{3 \times 3}$ from the RGB color space. The transformation matrix $W_D \in \mathbb{R}^{3 \times 3}$ may be derived through a procedure of discriminant analysis [13] and has been discussed in [12].

2.2 The Color Gabor-PHOG (GPHOG) and FC-GPHOG Image Descriptors

A Gabor filter is obtained by modulating a sinusoid with a Gaussian distribution. In a 2D scenario such as images, a Gabor filter is defined as:

$$g_{\nu,\theta,\phi,\sigma,\gamma}(x, y) = \exp(-\frac{x'^2 + \gamma^2 y'^2}{2\sigma^2}) \exp(i(2\pi\nu x' + \phi)) \qquad (1)$$

where $x' = x \cos\theta + y \sin\theta$, $y' = -x \sin\theta + y \cos\theta$, and ν, θ, ϕ, σ, γ denote the spatial frequency of the sinusoidal factor, orientation of the normal to the parallel stripes of a Gabor function, phase offset, standard deviation of the Gaussian kernel and the spatial aspect ratio specifying the ellipticity of the support of the Gabor function respectively. For a grayscale image $f(x, y)$, the Gabor filtered image is produced by convolving the input image with the real and imaginary components of a Gabor filter [14].

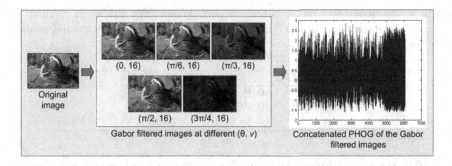

Fig. 2. The generation of the proposed Gabor-PHOG descriptor

The Pyramid of Histograms of Oriented Gradients (PHOG) [6] descriptor, inspired from the Histograms of Oriented Gradients (HOG) [15] and the image pyramid representation of Lazebnik et al. [16], represents local image shape and its spatial layout, together with a spatial pyramid kernel. The local shape is captured by the distribution over edge orientations within a region, and the spatial layout by tiling the image into regions at multiple resolutions. The distance between two PHOG image descriptors then reflects the extent to which the images contain similar shapes and correspond in their spatial layout [6]. Figure 1 illustrates the generation of the PHOG feature vector.

We used the Gabor wavelet representation for subsequent extraction of our feature vectors as it captures the local structure corresponding to spatial frequency (scale), spatial localization, and orientation selectivity. We subject each of the three color components of the image to different combinations of Gabor filters. For our experiments, we choose the parameter values as $\phi = 0$, $\sigma = 2$, $\gamma = 0.5$, and $\theta = [0, \pi/6, \pi/3, \pi/2, 3\pi/4]$. We derive the novel Gabor-PHOG (GPHOG) feature vector by concatenating the PHOG of the components of the Gabor filtered images and normalize it to zero mean and unit standard deviation. It should be noted that we computed the PHOG with two levels, and used

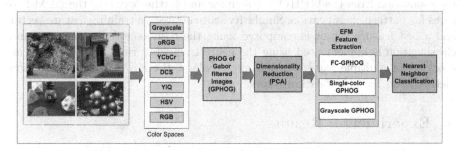

Fig. 3. An overview of multiple features fusion methodology, the EFM feature extraction method, and the classification stages

Table 1. Comparison of the classification performance (%) with other methods on Caltech 256 dataset. Note that [17] used 250 of the 256 classes with 30 training samples per class.

#train	#test	GPHOG		[4]		[17]
		YCbCr	**30.4**	oRGB-SIFT	23.9	
12800	6400	YIQ	**30.5**	CSF	30.1	
		FC	33.2	CGSF	**35.6**	SPM-MSVM 34.1

$\nu = 16$ as the spatial frequency of the Gabor filters for generating our GPHOG descriptor. Figure 2 illustrates the creation of the proposed GPHOG descriptor.We assess the performance of the GPHOG descriptor on six different color spaces, namely RGB, HSV, oRGB, YCbCr, YIQ and DCS as well as on grayscale and propose two new color feature vctors - the YIQ-GPHOG and the YCbCr-GPHOG descriptors. For fusion, we first use PCA for the optimal representation of our color GPHOG vectors with respect to minimum mean square error. We then combine the PCA features of the six normalized color GPHOG descriptors to form the novel Fused Color GPHOG (FC-GPHOG) descriptor which outperforms the classification results of the individual color GPHOG features.

2.3   The EFM-NN Classifier

Learning and classification are performed using the Enhanced Fisher Linear Discriminant Model (EFM) [10] and the nearest neighbor classification rule. The EFM method first applies Principal Component Analysis (PCA) to reduce the dimensionality of the input pattern vector. The Fisher Linear Discriminant (FLD) is a popular classification method that achieves high separability among the different pattern classes. However, the FLD method, if implemented in an inappropriate PCA space, may lead to overfitting. The EFM method hence applies an eigenvalue spectrum analysis criterion to choose the number of principal components to avoid overfitting and improves the generalization performance of the FLD. The EFM method thus derives an appropriate low dimensional representation from the GPHOG descriptor and further extracts the EFM features for pattern classification. Similarity score between a training feature vector and a test feature vector is computed using the cosine similarity measure and classification is implemented using the nearest neighbor rule. Figure 3 gives an overview of multiple feature fusion methodology, the EFM feature extraction method, and the classification stages.

3   Experimental Results

3.1   Caltech 256 Dataset

The Caltech 256 dataset [17] holds 30,607 images divided into 256 object categories and a clutter class. The images have high intra-class variability and high

Fig. 4. Some sample images from the Caltech 256 dataset

object location variability. Each category contains at least 80 images and at most 827 images. The mean number of images per category is 119. The images represent a diverse set of lighting conditions, poses, backgrounds, and sizes. Images are in color, in JPEG format with only a small percentage in grayscale. The average size of each image is 351x351 pixels. Figure 4 shows some sample images from this dataset.

For each class, we choose 50 images for training and 25 images for testing. The data splits are the ones provided on the Caltech website [17]. In this dataset, YIQ-GPHOG performs the best among single-color descriptors giving 30.5% success followed by YCbCr-GPHOG with 30.4% classification rate. Figure 5 shows the success rates of the GPHOG descriptors for this dataset. The FC-GPHOG descriptor here achieves a success rate of 33.2%. Table 1 compares our results with other methods.

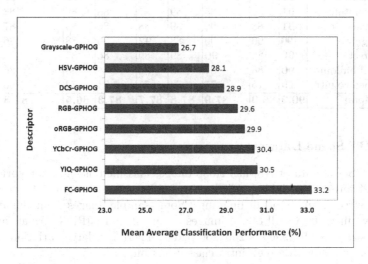

Fig. 5. The mean average classification performance of the proposed color GPHOG and FC-GPHOG descriptors on the Caltech 256 dataset

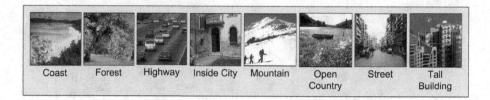

| Coast | Forest | Highway | Inside City | Mountain | Open Country | Street | Tall Building |

Fig. 6. Some sample images from the MIT Scene dataset

Table 2. Comparison of the classification performance (%) with other methods on the MIT Scene dataset

#train	#test	GPHOG		[2]		[18]
		YCbCr	**87.9**	CLF	86.4	-
2000	688	YIQ	**88.0**	CGLF	86.6	
		FC	**90.2**	CGLF+PHOG	89.5	
		YIQ	**84.6**	CLF	79.3	
800	1888	YCbCr	**84.7**	CGLF	80.0	
		FC	**86.6**	CGLF+PHOG	84.3	83.7

Table 3. Category wise descriptor performance (%) on the MIT Scene dataset. Note that the categories are sorted on the FC-GPHOG results

Category	FC	YIQ	YCbCr	RGB	DCS	oRGB	HSV	Grayscale
forest	97	96	96	96	96	96	**98**	97
coast	**94**	91	89	91	89	90	88	87
mountain	**91**	88	85	89	87	88	88	83
inside city	**91**	88	90	89	88	86	86	87
highway	90	90	90	88	**92**	90	88	86
street	90	88	**90**	89	89	86	85	84
tall building	90	88	87	88	86	88	88	86
open country	**79**	75	75	73	75	74	72	68
Mean	**90.2**	**88.0**	**87.9**	**87.8**	**87.7**	**87.3**	**86.5**	**84.8**

3.2   MIT Scene Dataset

The MIT Scene dataset [18] has 2,688 images classified as eight categories: 360 coast, 328 forest, 260 highway, 308 inside of cities, 374 mountain, 410 open country, 292 streets, and 356 tall buildings. Some sample images from this dataset are shown in figure 6. All of the images are in color, in JPEG format, and the average size of each image is 256x256 pixels. There is a large variation in light and angles along with a high intra-class variation.

From each class, we use 250 images for training and the rest of the images for testing the performance, and we do this for five random splits. Here too, YIQ-GPHOG is the best single-color descriptor at 88.0% followed closely by

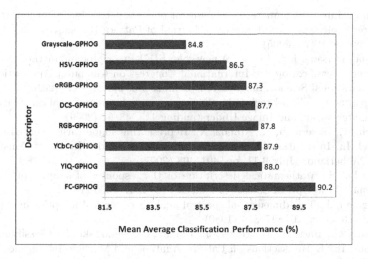

Fig. 7. The mean average classification performance of the proposed GPHOG descriptor in individual color spaces as well as after fusing them on the MIT Scene dataset

YCbCr-GPHOG at 87.9%. The combined descriptor FC-GPHOG gives a mean average performance of 90.2%. See Figure 7 for details. Table 2 compares our result with that of other methods. Table 3 shows the class wise classification rates for this dataset on applying the proposed GPHOG descriptors.

4   Conclusion

The contributions of this paper are in the generation of several novel descriptors for object and scene image classification based on Gabor wavelet transformation. We have introduced a new Gabor-PHOG descriptor and further proposed the robust YIQ-GPHOG and YCbCr-GPHOG features. The six color GPHOG features beat the recognition performance of the Grayscale-GPHOG descriptor which show information contained in color images can be significantly more useful than that in grayscale images for classification. Experimental results using two datasets, the Caltech 256 object categories dataset and the MIT Scene dataset, show that the proposed novel FC-GPHOG image descriptor exceeds or achieves comparable performance to some of the best performance reported in the literature for object and scene image classification.

References

1. Gonzalez, R., Woods, R.: Digital Image Processing. Prentice-Hall (2001)
2. Banerji, S., Verma, A., Liu, C.: Novel color LBP descriptors for scene and image texture classification. In: 15th International Conference on Image Processing, Computer Vision, and Pattern Recognition, Las Vegas, Nevada, July 18-21 (2011)

3. Shih, P., Liu, C.: Comparative assessment of content-based face image retrieval in different color spaces. International Journal of Pattern Recognition and Artificial Intelligence 19(7) (2005)
4. Verma, A., Banerji, S., Liu, C.: A new color SIFT descriptor and methods for image category classification. In: International Congress on Computer Applications and Computational Science, Singapore, December 4-6, pp. 819–822 (2010)
5. Burghouts, G., Geusebroek, J.M.: Performance evaluation of local color invariants. Computer Vision and Image Understanding 113, 48–62 (2009)
6. Bosch, A., Zisserman, A., Munoz, X.: Representing shape with a spatial pyramid kernel. In: International Conference on Image and Video Retrieval, Amsterdam, The Netherlands, July 9-11, pp. 401–408 (2007)
7. Marcelja, S.: Mathematical description of the responses of simple cortical cells. Journal of the Optical Society of America 70, 1297–1300 (1980)
8. Daugman, J.: Two-dimensional spectral analysis of cortical receptive field profiles. Vision Research 20, 847–856 (1980)
9. Donato, G., Bartlett, M., Hager, J., Ekman, P., Sejnowski, T.: Classifying facial actions. IEEE Transactions on Pattern Analysis and Machine Intelligence 21(10), 974–989 (1999)
10. Liu, C., Wechsler, H.: Robust coding schemes for indexing and retrieval from large face databases. IEEE Transactions on Image Processing 9(1), 132–137 (2000)
11. Bratkova, M., Boulos, S., Shirley, P.: oRGB: A practical opponent color space for computer graphics. IEEE Computer Graphics and Applications 29(1), 42–55 (2009)
12. Liu, C.: Learning the uncorrelated, independent, and discriminating color spaces for face recognition. IEEE Transactions on Information Forensics and Security 3(2), 213–222 (2008)
13. Fukunaga, K.: Introduction to Statistical Pattern Recognition, 2nd edn. Academic Press (1990)
14. Lee, H., Chung, Y., Kim, J., Park, D.: Face Image Retrieval Using Sparse Representation Classifier with Gabor-LBP Histogram. In: Chung, Y., Yung, M. (eds.) WISA 2010. LNCS, vol. 6513, pp. 273–280. Springer, Heidelberg (2011)
15. Dalal, N., Triggs, B.: Histograms of oriented gradients for human detection. In: Proceedings of the 2005 IEEE Computer Society Conference on Computer Vision and Pattern Recognition (CVPR 2005), Washington, DC, USA, vol. 1, pp. 886–893 (2005)
16. Lazebnik, S., Schmid, C., Ponce, J.: Beyond bags of features: Spatial pyramid matching for recognizing natural scene categories. In: Proceedings of the 2006 IEEE Computer Society Conference on Computer Vision and Pattern Recognition, Washington, DC, USA, vol. 2 (2006)
17. Griffin, G., Holub, A., Perona, P.: Caltech-256 object category dataset. Technical Report 7694, California Institute of Technology (2007)
18. Oliva, A., Torralba, A.: Modeling the shape of the scene: A holistic representation of the spatial envelope. International Journal of Computer Vision 42(3), 145–175 (2001)

Binary Gabor Statistical Features
for Palmprint Template Protection

Meiru Mu*, Qiuqi Ruan, Xiaoying Shao,
Luuk Spreeuwers, and Raymond Veldhuis

Institute of Information Science, Beijing Jiaotong University, Beijing, China
Systems and Signals Group, University of Twente, Enschede, The Netherlands
{m.mu,x.shao,L.J.Spreeuwers,R.N.J.Veldhuis}@utwente.nl

Abstract. The biometric template protection system requires a high-quality biometric channel and a well-designed error correction code (ECC). Due to the intra-class variations of biometric data, an efficient fixed-length binary feature extractor is required to provide a high-quality biometric channel so that the system is robust and accurate, and to allow a secret key to be combined for security. In this paper we present a binary palmprint feature extraction method to achieve a robust biometric channel for template protection system. The real-valued texture statistical features are firstly extracted based on Gabor magnitude and phase responses. Then a bits quantization and selection algorithm is introduced. Experimental results on the HongKong PloyU Palmprint database verify the efficiency of our method which achieves low verification error rate by a robust palmprint binary representation of low bit error rate.

Keywords: Palmprint verification, Binary feature extraction, Feature template protection, Gabor filtering.

1 Introduction

It has been widely known that the typical biometrics system encounters some security and privacy problems such as identity fraud, limited-renewability, cross-matching, and leaking sensitive personal information [1]. Biometric template protection system, as a countermeasure to these security and privacy threats, has become an important issue, which requires that biometric data is firstly quantized into a fixed-length binary string as template. For typical (unprotected) palmprint verification systems, there have been many feature representation approaches reported achieving high verification accuracy. Among them there are some coding based methods which generate binary features [2,3]. However, most of them require a template registration during the matching stage, which might be not allowed for template protection system. Further, most of the current reported binary representations give the bit error rates as high as 40% from the genuine matching, which are too noisy to be corrected by the current error correcting coding schemes [4,5].

---

* Corresponding author.

G.L. Gimel' farb et al. (Eds.): SSPR & SPR 2012, LNCS 7626, pp. 593–601, 2012.
© Springer-Verlag Berlin Heidelberg 2012

In this paper, we present a method of binary feature extraction from palm-print image, aiming to achieve robust biometric channel, *i.e.* low bit error rate (BER) for matching channel, and low verification error rate, which we indicate by false rejection rate (FRR), false acceptance rate (FAR) and equal error rate (EER). The robust binary representation gives a solution of combing the typical biometric verification and the template protection. Besides, a fixed-length binary representation also has additional advantages such as small template storage and high matching speed. Since the over-complete information of Gabor filtering responses contributes to the discriminating ability, our method chooses to filter the palmprint image by a group of two-dimensional Gabor functions firstly [2]. Figure 1 shows the flow chart of our proposed binary feature extraction method. As can be seen from it, instead of employing the filtered Gabor magnitude (GM) and Gabor phase (GP) responses directly which are generally high-dimensional, we extract the statistic features from them respectively for real-valued representation (denoted by V_{GM} and V_{GP}), then based on which the one-bit quantization and reliable bits selection are subsequently processed. Finally a binary sequence (denoted by $B^{1 \cdots L}$) length of L is achieved as palmprint template for storage and matching. From GM information the global statistical features are extracted which we denote by LogGM [6], while from GP information the local statistical features are extracted after the local XOR pattern (LXP) operating which we denote by LxpGP [7]. After quantizing the global and local statistical features respectively, we fuse the obtained binary bits together and then from them some reliable bits are selected to construct the final binary Gabor statistical features. For verification the Hamming distance is employed as the dissimilarity measurement.

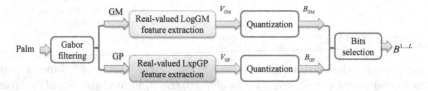

Fig. 1. Flow chart of the proposed method

The main contributions of this paper are: a highly efficient scheme is proposed to extract the binary features from the Gabor filtering responses by considering the fusion of the local and global texture statistical information, which provides a high-quality biometric channel achieving low bit error rate for the genuine template matching and the corresponding low verification error rate so as to give a solution of combining the typical palmprint verification system and template protection scheme.

In this paper, we will firstly present the scheme of real-valued Gabor texture statistical representation in Section 2, which includes the statistical analysis of Gabor magnitude and phase responses respectively. Next, the quantization and reliable bits selection algorithm will be introduced in Section 3. The experimental results will be given in Section 4. Finally, it is the conclusion.

2   Gabor Statistical Feature Extraction

In order to represent a palmprint by a binary string for template protection, we need to firstly extract some real-valued features from it. Since the Gabor filtered representation can provide the optimal localization of image details, we choose a group of Gabor functions to perform a joint spatial-frequency multi-channel transform on the palmprint image, which can be expressed as following [2]:

$$g_{m,n}(x,y) = \frac{1}{2\pi\sigma^2} \exp\left\{\frac{-(x^2+y^2)}{2\sigma^2}\right\}$$
$$\times \exp\{2\pi i(u_m x \cos\theta_n + u_m y \sin\theta_n)\}. \tag{1}$$

u_m is the frequency of sinusoidal wave along directional θ_n from x-axis, and σ specifies the Gaussian envelope along x and y axes, which determines the bandwidth of the Gabor filter. Each Gabor function $g_{m,n}(x,y)$ with the parameters (u_m, θ_n, σ) is commonly transformed into a discrete Gabor filter and its direct current is turned to zero, which can be denoted by $\tilde{g}_{m,n}(x,y)$. Given an image $I(x,y)$, its Gabor-filtered images can be defined as follow: $J_{m,n}(x,y) = \tilde{g}_{m,n}(x,y) * I(x,y)$, where $J_{m,n}(x,y)$ is a complex number. The Gabor magnitude and phase angle responses can be respectively denoted by $GM_{m,n}(x,y) = \| J_{m,n}(x,y) \|$, and $\phi_{m,n}(x,y) = \arctan(J_{m,n}(x,y))$, where $\| \bullet \|$ denotes the modulus operator of a complex number. To alleviate the sensitivity of Gabor phase to the varying positions, we transform $\phi_{m,n}(x,y)$ into four different ranges as the Gabor phase (GP) information $GP_{m,n}(x,y)$ by the following expression:

$$GP_{m,n}(x,y) = p, \text{ if } 90*(p-1) \le \phi_{m,n}(x,y) < 90*p; \quad p \in \{1,2,3,4\} \tag{2}$$

$GM_{m,n}(x,y)$ and $GP_{m,n}(x,y)$ will be respectively further processed to construct the real-valued statistical features.

By investigating the histogram distribution of each GM, it has been found that the lognormal densities fit the GMs very well, and the sub-blocks of each GM are also close to lognormal distribution [6]. Figure 2 gives an example of GM histogram fitting. After the lognormal transformation of each GM, some Gaussian distributions are obtained, which can be expressed as $LogGM_{m,n}(x,y) = \log(GM_{m,n}(x,y))$. Since a Gaussian sequence can be represented specifically by its mean ν and standard deviation ρ, the palmprint feature representation can be constructed by these Gaussian parameters. Assuming the Gabor filter bank has E scales and F orientations, and each GM is partitioned into A sub-blocks, we will get $(G \times R \times A)$ pairs of ν and ρ values. By concatenating them together, the final real-valued feature vector can be formulated as $[\nu_1 \cdots \nu_{(E \times F \times A)}, \rho_1 \cdots \rho_{(E \times F \times A)}]$. Following the experimental results in Ref. 6, the best verification performance is achieved when $E = 5, F = 8, A = 21$. To obtain more discriminating feature components from which we expect to extract

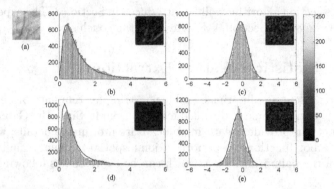

Fig. 2. Examples of histogram fitting. (a) The original palmprint image. (b) and (d) hist Gabor magnitudes from two different Gabor parameters. (c) shows the logarithmic transform of (b). (e) is the logarithmic transform of (d).

Fig. 3. Flow chart of real-valued LogGM feature extraction

bits, the linear discriminant analysis (LDA) projection is applied to get the final real-valued representation V_{GM} of GM, the process of which we denote by real-valued LogGM feature extraction as Fig. 3 illustrates.

To extract some statistical features from GP, we firstly encode $GP_{m,n}(x, y)$ by local XOR pattern (LXP) operator [7], which is expressed as $LxpGP_{m,n}(x, y) = LXP(GP_{m,n}(x, y))$. As shown in Fig. 4, for each pixel P_0, its eight neighborhoods with one pixel interval $P_i \in \{1, 2, 3, 4\}$, $i = 0, 1, \ldots 8$ need to be encoded into $B_i \in \{0, 1\}$, $i = 0, 1, \ldots 8$ by computing $B_i = \texttt{if}\{P_i \neq P_0\}$. Then P_0 is mapped into a decimal number $S_0 \in \{0, 1, \ldots 255\}$ by calculating $S_0 = \sum_{i=1}^{8} B_i \times 2^i$. Finally all the mapped S_0 forms $LxpGP_{m,n}(x, y)$. Assuming the Gabor filter bank has E scales and F orientations, and each GM is partitioned into D sub-blocks, the histograms of each sub-block at all the scales and orientations are concatenated. Then D histograms of $E \times F \times 256$ bins will be obtained. LDA is applied for each histogram respectively. In this paper, we set $E = 5$, $F = 8$, $D = 5$ by taking into account the computational complexity and the performance. Finally the features from the D LDA modulus are concatenated into a vector V_{GP} as the real-valued representation of GP information, the flow chart of which is shown in Fig. 5.

3   Bits Quantization and Selection

Feature quantization and bits selection procedures strongly affect the verification performance of the template protection system. Based on the real-valued features described in Section 2, we introduce our proposed bit quantization and selection

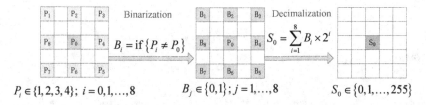

$P_i \in \{1,2,3,4\}; \ i = 0,1,\ldots,8$ $B_j \in \{0,1\}; j = 1,\ldots,8$ $S_0 \in \{0,1,\ldots,255\}$

Fig. 4. An illustration of local XOR pattern (LXP) operating

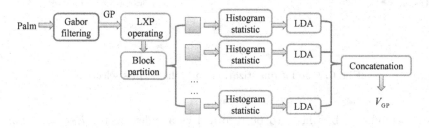

Fig. 5. Flow chart of real-valued LxpGP feature extraction

method in this section, which is illustrated in Fig. 6. Since V_{GM} and V_{GP} are processed in the same way, here we take V_{GM} as an illustration of the main steps as follows:

(1) Compute the averaged real-valued feature from the enrolled N samples of one palm;

Assuming V_t palms are captured for training and N samples from one palm are enrolled, we will get a feature set $\{V_{GM,Enroll}^i, \ i = 1 \ldots N\}$ to represent each palm after the LDA projection. The averaged feature vector for each palm needs to be computed as expressed following:

$$\overline{V}_{GM,Enroll} = \frac{1}{N} \sum_{i=1}^{N} V_{GM,Enroll}^{i,j}, \ j = 1 \ldots (V_t - 1) \qquad (3)$$

(2) Quantize the real-valued features by a threshold;

Here the threshold is set to 0. Each component of the real-valued features $\overline{V}_{GM,Enroll}$ is quantized into one bit $B_{GM,Enroll}^j$, $j = 1 \ldots V_t - 1$ which can be expressed as following:

$$(B_{GM,Enroll}^j \mid j = 1 \ldots V_t - 1) = \begin{cases} 1, \overline{V}_{GM,Enroll}^j \geq 0; \\ 0, \text{otherwise.} \end{cases} \qquad (4)$$

By the same process, $V_{GM,Test}$ from each test sample is quantized into a binary string $B_{GM,Test}$.

(3) Select some reliable bits as the final binary feature template.

During the enrollment phase, we assume that the larger $|\overline{V}_{GM,Enroll}^j|$, the more reliable of its corresponding bit $B_{GM,Enroll}^j$. In the meanwhile, the positions of

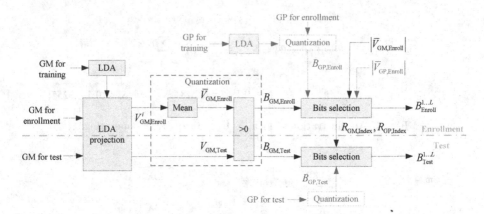

Fig. 6. One-bit quantization and reliable bits selection

the most reliable bits need to be recorded as a index vector $R_{GM,Enroll}$. For a test sample, the bits are selected by following the position index of $R_{GM,Enroll}$. Assuming L bits are selected then our proposed binary Gabor statistical features are $B_{Enroll}^{1...L}$ for enrollment palmprint and $B_{Test}^{1...L}$ for test palmprint respectively.

4 Experimental Results

The HongKong Polytechnic University (PolyU) palmprint database is used to test our proposed method [8]. They were captured by a CCD camera from 386 different palms and collected in two sessions with two different illumination conditions. There are 3889 images in session one and 3863 palms in session two respectively. Because there is one palm which has only one sample captured in session 2, we use the other 385 palms for our experiments. The resolution of original captured images is 384×284 pixels at 75 dpi. By preprocessing each captured image (as shown in Fig. 7), the central region size of 128×128 is used for feature extraction.

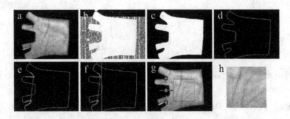

Fig. 7. (a) A typical palmprint image from HongKong PolyU Palmprint Database; (b)-(f) show our used registration and region-crop method

Experiment I: Since the verification performance is heavily depended on the number of the selected bits, we test our algorithm by varying the bits number from 20 to 400. For this experiment, the images in session one are used, which are randomly split as the training set (185 palms) and the evaluation set (the remaining 200 palms for enrollment and test), which can be referred as training-evaluation-set split. For training, all the samples are used. For enrollment, five samples are randomly selected and the remaining ones for test. The training-evaluation-set split is performed six times. For matching genuine pairs, we test all the samples in the test set. For matching imposter pairs, the first sample of each palm in the test set is chosen for all the possible imposter combinations. Thus, we have totally 6082 genuine scores and 238,800 imposter scores.

Figure 8 (Left) plots the comparison of the verification equal error rate (EER) among the methods of LogGM (only GM information is used), LxpGP (only GP information is used), Fusion (the proposed method by fusing GM and GP information). As can be seen from it, the EER decreases as the bits number increases. However, since more bits lead to be more noisy string, we determine 120 as the optimal bits number for the proposed fusion method. Figure 8 (Right) plots the comparison of ROC curves. As can be seen, the verification performance can be greatly improved by fusing the bits generated from GM and GP information respectively. Figure 9 shows the percentage distributions of bit error rate (BER) for genuine and imposter matching respectively by the proposed method. As we can see that the proposed method is able to achieve good verification performance by adjusting the BER threshold of error correcting coding (ECC) module to around 30% assuming the system works under the template protection framework.

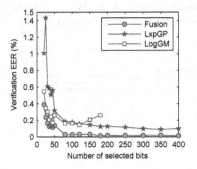 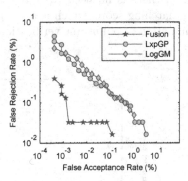

Fig. 8. Verification performance. (Left) Verification EER (%) comparison when the number of selected bits varies from 20 to 400; (Right) ROC curves when the number of selected bits is set to 120.

Experiment II: Here the proposed method is compared with the binary co-occurrence vector (BOCV) method on the verification error rate and the corresponding BER threshold which is depended on the used ECC module [3,5]. For

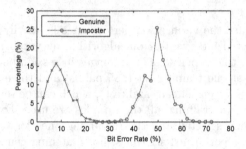

Fig. 9. Percentage distribution of Bit Error Rate (%) for genuine and imposter matching respectively by the proposed method

this experiment, the samples for training (185 palms) are from session 1 while the ones for evaluation (200 palms) from session 2. Other protocols are the same as those in experimental I. For the BOCV method, no training procedure is needed. Here we set the evaluation protocol the same as that for ours. In total, we have 6013 genuine scores and 236,412 imposter scores. BOCV requires to shift the whole image by several pixel horizontally and vertically and then matches multiple times to get the final matching score. Here each bits plane is down-sampled into a binary matrix size of 32×32. The range of shift is considered as [-2, 2].

Figure 10 plots the comparison of verification error rate which is indicated by false acceptance rate (FAR) and false rejection rate (FRR) when adjust the BER threshold of ECC module. As can be seen from it, the proposed method greatly outperforms the coding based method BOCV when the applied ECC module has an error correcting capability of lower than 25% . Besides, for the proposed method there is no need to shift the feature templates multiple times for the final matching score, which is not only time-consuming, but also challenges the combination of palmprint verification and template protection system.

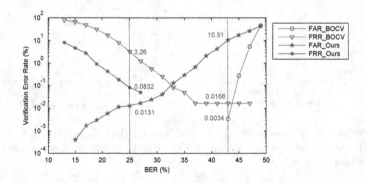

Fig. 10. Comparisons of verification error rate (%) and their corresponding BER thresholds between the binary co-occurrence vector (BOCV) method and the proposed method (Ours)

5   Conclusion and Discussion

In this paper, we present a binary feature extraction method for palmprint verification under consideration of feature template protection which requires a robust biometric channel. Experimental results demonstrate that fusion of the global and local statistical features extracted from Gabor magnitude and phase responses respectively outperforms that of global or local features represented separately on the verification accuracy. Compared with the popular coding based methods, the proposed approach achieves comparable verification error rate while much lower bit error rate of the genuine matching which gives a solution of combining palmprint verification and feature template protection scheme so that the system could be robust and accurate.

However, the extracted discriminative binary string is still not long enough to be secure. To alleviate the intra-class variations, how to extract the robust binary palmprint features of more bits so that a long key can be combined for security will be the point of our future work.

Acknowledgments. This work is supported partly by the National Grand Fundamental Research 973 Program of China (Grant No. 2004CB318005), the Fundamental Research Funds for the Central Universities (Grant No. KKJB110 34536), and China Scholarship Council.

References

1. Jain, A., Nandakumar, K., Nagar, A.: Biometric Template Security. EURASIP Journal on Advances in Signal Processing 113 (2008)
2. Kong, W., Zhang, D., Li, W.: Palmprint Feature Extraction Using 2-D Gabor Filters. Pattern Recognition 36, 2339–2347 (2003)
3. Guo, Z., Zhang, D., Zhang, L., Zuo, W.: Palmprint Verification Using Binary Orientation Co-Occurrence Vector. Pattern Recognition Letters 30, 1219–1227 (2009)
4. Richard, R.E.: Theory and Practice of Error Control Codes. Addison-Wesley Publishing Company, Inc. (1983)
5. Shao, X., Xu, H., Veldhuis, R.N.J., Slump, C.H.: A Concatenated Coding Scheme for Biometric Template Protection. In: 2012 IEEE International Conference on Acoustics, Speech and Signal Processing (ICASSP). IEEE Press, Japan (2012)
6. Mu, M., Ruan, Q.: Mean and standard deviation as features for palmprint recognition based on gabor filters. Int. J. Patt. Recog. Art. Intel. 25(4), 491–512 (2011)
7. Xie, S., Shan, S., Chen, X., Chen, J.: Fusing Local Patterns of Gabor Magnitude and Phase for Face Recognition. IEEE Transactions on Image Processing 19(5), 1349–1361 (2010)
8. Biometrics Research Centre (BRC) in HongKong,
   http://www.comp.polyu.edu.hk/~biometrics/

Class-Dependent Dissimilarity Measures
for Multiple Instance Learning

Veronika Cheplygina, David M.J. Tax, and Marco Loog

Pattern Recognition Laboratory, Delft University of Technology
{v.cheplygina,d.m.j.tax,m.loog}@tudelft.nl

Abstract. Multiple Instance Learning (MIL) is concerned with learning from sets (bags) of feature vectors (instances), where the individual instance labels are ambiguous. In MIL it is often assumed that positive bags contain at least one instance from a so-called concept in instance space, whereas negative bags only contain negative instances. The classes in a MIL problem are therefore not treated in the same manner. One of the ways to classify bags in MIL problems is through the use of bag dissimilarity measures. In current dissimilarity approaches, such dissimilarity measures act on the bag as a whole and do not distinguish between positive and negative bags. In this paper we explore whether this is a reasonable approach and when and why a dissimilarity measure that is dependent on the bag label, might be more appropriate.

1 Introduction

Multiple-instance learning (MIL) [6] extends traditional supervised learning methods in order to learn from objects that are described by a set (*bag*) of feature vectors (*instances*), rather than a single feature vector only. MIL problems are often considered to be two-class problems, i.e., a bag of instances can belong either to the positive or the negative class. The bag labels are available, but the labels of the individual instances are not defined. Often assumptions are made about the instance labels and their relationship with the bag labels.

Traditional MIL problems assume that positive bags contain one or more positive instances from a so-called *concept*, whereas negative bags contain only negative instances [6,9]. E.g. when classifying images represented by a bag of image segments as "tiger" or "no tiger", a segment containing black stripes could be seen as a positive instance for the "tiger" concept, whereas segments containing grass, sky e.t.c. would be considered negative, or background, instances.

Many traditional, "instance-based" MIL approaches try to model the concept by identifying the "most positive" instances in bags, and classify new bags as positive if they appear to have instances within this concept [6,9]. Other, "bag-based" MIL approaches compare bags directly, using distances[18], kernels[7] or dissimilarities [17,14]. It is possible to define a dissimilarity measure between bags, represent each bag by its dissimilarities to other bags, and use these dissimilarity values as features for supervised classifiers. A number of such dissimilarities are investigated in [14,17], where it is shown that some bag dissimilarities can be effective even when a concept is not clearly defined.

G.L. Gimel' farb et al. (Eds.): SSPR & SPR 2012, LNCS 7626, pp. 602–610, 2012.

The instance-based methods explicitly use the assumption that positive bags are different from negative bags, whereas the bag-based methods typically do not differentiate between classes. This may not be completely natural for a MIL problem, because we have some information about how positive bags are different from negative bags. In supervised learning problems where classes are expected to behave differently, class-dependent distances[10,5,19] or features[2,8] have been suggested. In this work we examine whether a similar approach might be reasonable for MIL problems.

2   Related Work

Using a class-dependent distance measure, rather than a fixed distance measure, is not a new idea. Quadratic Discriminant Analysis already allows different classes to have different covariance matrices. More attention to class-dependent distances is given in [10], where the goal is to learn weights for each feature/class combination, and to use these weights in a Mahalanobis-type metric. A similar approach is taken in [5] to improve performance in speech recognition. In [19], the authors propose learning different metrics for different classes and show that this improves classification results. In all cases, the goal is obtain high nearest neighbor performance on the learnt distances.

Other authors have examined the importance of class-dependent features rather than distances. In [2] several examples are provided where such class-dependent features are important: classification of handwritten characters, textures and documents. For instance, in a bag of words approach to document classification, it might be better to represent documents based on words that frequently occur in a particular class, as opposed to words that frequently occur in all documents. The same motivation is given in [8], where a weight is associated with each (word, class) pair. Although here, the term "dissimilarity" is used rather than distance, the learned dissimilarities are still used in a nearest neighbor setting.

For MIL, the only example of using a class-dependent dissimilarity we are aware of is from [20]. Here, bag dissimilarities are used for feature selection. The authors propose to use different dissimilarity measures for two positive bags, two negative bags, or a positive and a negative bags, to best capture the properties of the classes, such as the presence of a concept. Because the purpose is feature selection, only the dissimilarities between bags in the training set are computed. The same class-dependent dissimilarity cannot be used for the purpose of classification, because the labels of test bags are not available.

3   Review of MIL and Bag Dissimilarities

In Multiple Instance Learning, an object is represented by a bag $B_i = \{x_{ik}|k = 1,...,n_i\} \subset \mathbb{R}^d$ of n_i feature vectors or instances. The training set $T = \{(B_i, y_i)|i = 1,...N\}$ consists of positive ($y_i = +1$) and negative ($y_i = -1$) bags. The traditional assumption for MIL is that there are instance labels y_{ik}

which relate to the bag labels as follows: a bag is positive if and only if it contains at least one positive instance[6]. In this case we can speak of concept (positive) instances, which are assumed to be close together in a region of the feature space called the concept $C \subset \mathbb{R}^d$, and which directly affect the bag label by their presence.

Alternatively, we can represent an object (and therefore, also a bag in a MIL problem) by its dissimilarities to prototype objects in a representation set R[11]. Often, R is taken to be the training set T, and each bag is represented as $\mathbf{d}(B_i, T) = [d(B_i, B_1), ...d(B_i, B_N)]$: a vector of dissimilarities. Therefore, each bag is represented by a single feature vector and the MIL problem can be viewed as a standard supervised learning problem.

There are various ways of defining the bag dissimilarity measure $d(B_i, B_j)$. Here we focus on defining $d(B_i, B_j)$ through the pairwise instance dissimilarities $D = [d(\mathbf{x}_{ik}, \mathbf{x}_{jl})]_{N_i \times N_j}$. We use the squared Euclidean distance for the instance dissimilarity, but other choices are also possible. In all the dissimilarities considered here, the first step is to find, for each instance in B_i, the distance to its closest instance in B_j. Using these minimum instance distances, we can define the following dissimilarities:

- Overall minimum or *minmin*: $d_{minmin}(B_i, B_j) = \min_k \min_l d(\mathbf{x}_{ik}, \mathbf{x}_{jl})$
- Average minimum or *meanmin*: $d_{meanmin}(B_i, B_j) = \frac{1}{n_i} \sum_{k=1}^{n_i} \min_l d(\mathbf{x}_{ik}, \mathbf{x}_{jl})$
- Maximum minimum or *maxmin*: $d_{maxmin}(B_i, B_j) = \max_k \min_l d(\mathbf{x}_{ik}, \mathbf{x}_{jl})$

Note that these dissimilarities are very similar to (variants) of the Hausdorff distance. However, in literature, the name "modified Hausdorff distance" has been used for a number of different distances (see [21] for some examples), so we prefer to use these more straightforward names instead. Furthermore, the Hausdorff distance is generally not symmetric, i.e. $d(B_i, B_j) \neq d(B_j, B_i)$, and often a symmetric version is obtained by taking the average or the maximum of the two values. In this paper we refrain from doing so for reasons that will become apparent in the next section.

The three dissimilarities above have their advantages and disadvantages for particular types of datasets. For instance, *minmin* performs well with a very tight concept, whereas *meanmin* is more appropriate for cases where instances from positive and negative bags arise from different distributions. A more detailed explanation is available in [4][1].

4   Class-Dependent Dissimilarity

We argue that, in a MIL problem, it may be advantageous to exploit the bag label information when defining a dissimilarity between two bags. Let's assume we are dealing with a MIL problem with a well-defined concept, such as in

---

[1] In press, available online from
http://prlab.tudelft.nl/sites/default/files/icpr2012.pdf

Figure 1(a). In this problem, if we consider all instances in a bag, any two bags may be similar or dissimilar overall. However, in MIL problems with a concept, we could speculate that the positive bags are similar at the concept level. Figure 1(b) illustrates the a dissimilarity matrix corresponding with this intuition. Each square here is a dissimilarity value between two bags, where the color of the square represents the dissimilarity value (black = 0, i.e. similar bags, white=1, i.e. dissimilar bags). Notice the difference between treating these values as distances, or as dissimilarities. In terms of distances, this representation is quite poor, because each bag has several neighbors in the opposite class. However, in terms of dissimilarities, the situation is quite different: the positive bags are clearly represented in a different way than the negative bags, so the classes are well separated.

By using the same dissimilarity to compare positive and negative bags, we risk overlooking an important difference between positive and negative bags, producing a dissimilarity matrix where all values are nearly equal. It seems that using the class information could help us capture the correct aspects of dissimilarity between bags. Ideally, we would want to have the class information of both bags when determining their dissimilarity (e.g. using the overall minimum distance for two positive bags, as in [20], but for classification purposes, it is obvious that only the labels of the prototypes are available.

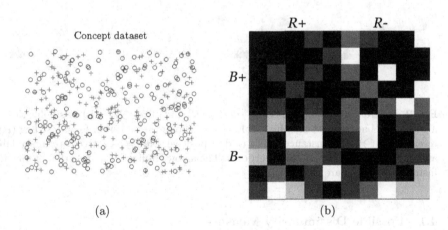

Fig. 1. (a) Artificial concept dataset, + and ◯ represent instances from positive and negative bags respectively. (b) Dissimilarity matrix reflecting the intuition we have about the positive and negative bags. The intensity value reflects the dissimilarity of two bags (black = 0, white = 1).

For positive prototypes, we want to find out something about the presence of a concept in the test bag (denoted by B), i.e. the concept instance of the prototype bag (denoted by $R+$ or $R-$ depending on the prototype label) needs to be involved. As illustrated in Figure 2, the asymmetry of the bag dissimilarities becomes important here. If we measure the dissimilarity of the test bag to the prototype bag, denoted by $d(B \to R+)$, it may happen that none of the instances in B are matched to the positive instance in $R+$. If we measure $d(R+ \to B)$ instead, which measures the dissimilarity from the prototype to the test bag, the positive instance of $R+$ has to be matched to an instance in B. In other words, the distance from a positive prototype $d(R+ \to B)$ should be more informative. For negative prototypes, we want to highlight the absence of the concept in the prototype. In this case, we are interested in the dissimilarity to the prototype $d(B \to R-)$ because this ensures that incorrect matches (of concept instances in the test bag to background instances in the prototype) will be present.

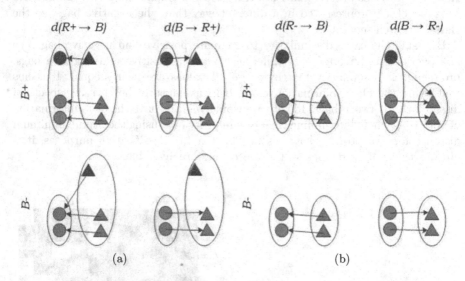

Fig. 2. Difference between the "from" $d(R \to B)$ and "to" $d(B \to R)$ dissimilarities for a prototype bag (represented by $\triangle$). Each row shows the situation for a test bag (represented by $\bigcirc$). The instance labels (red = positive, blue = negative) are unavailable and only shown for explanation purposes. The arrows indicate the direction of how the instance distances are measured.

4.1   Possible Dissimilarity Measure

Just the direction of measuring the dissimilarity does not yet provide us with a way to produce a single dissimilarity, but with a vector of minimum instance distances between a bag and a prototype. If the bags were very large, we could see these vectors as distributions of distances. Assuming that these distributions would be somehow different for positive and negative test bags, we could define a dissimilarity value between two bags by comparing the distributions directly.

However, in real applications, some bags may be very small (e.g. in the Musk datasets, bags with just one instance are present), so such comparisons would not always be feasible. Instead we try to define cheap approximations for the overall bag distance, given only finite samples from the instance distance distributions.

Given our previous experiences with *meanmin*, we propose to approximate both directions with the following dissimilarity:

$$d_{cd}(\cdot, R) = \begin{cases} d_{meanmin}(R \to \cdot) & \text{if } R \text{ is positive} \\ d_{meanmin}(\cdot \to R) & \text{if } R \text{ is negative} \end{cases} \quad (1)$$

Table 1. AUC performance and standard error (x100), 5x10-fold cross-validation for 1-NN classifier in the dissimilarity space. The numbers in bold indicate which dissimilarity is best (or not significantly worse than best) per dataset.

dataset	d_{to}	d_{from}	d_{cd}	d_{avg}
Musk1	**92.6 (1.1)**	**92.6 (1.1)**	**92.9 (1.2)**	**93.4 (1.1)**
Musk2	**89.7 (1.8)**	87.2 (1.7)	**89.7 (1.6)**	88.5 (1.6)
Fox	57.2 (1.4)	65.3 (1.7)	**68.7 (1.7)**	66.2 (1.7)
Tiger	**78.2 (1.6)**	**79.7 (1.3)**	75.1 (1.3)	75.6 (1.6)
Elephant	83.0 (1.3)	87.1 (1.2)	**90.8 (1.0)**	88.9 (1.0)
Protein	62.1 (2.5)	61.0 (3.1)	63.0 (3.1)	64.2 (2.7)
Mutagen easy	**89.6 (1.0)**	**89.0 (1.1)**	88.4 (1.0)	88.8 (1.0)
Mutagen hard	**79.5 (3.1)**	73.3 (3.6)	**79.7 (3.2)**	77.3 (3.6)
African	**89.9 (0.7)**	88.5 (0.7)	**90.8 (0.7)**	**89.3 (0.7)**
Beach	82.2 (0.8)	**82.1 (1.0)**	**83.3 (0.8)**	**82.4 (0.8)**
Historical	**87.0 (0.7)**	85.9 (0.8)	**87.1 (0.7)**	**87.6 (0.7)**
Buses	96.6 (0.3)	**97.2 (0.3)**	**97.1 (0.3)**	96.9 (0.3)
Dinosaurs	99.2 (0.2)	**99.7 (0.0)**	99.0 (0.2)	**99.3 (0.1)**
Elephants	**92.9 (0.5)**	92.6 (0.8)	92.8 (0.6)	**92.9 (0.5)**
Flowers	98.0 (0.2)	97.7 (0.3)	**98.1 (0.2)**	97.6 (0.2)
Horses	**98.9 (0.1)**	96.1 (0.4)	**98.9 (0.1)**	97.8 (0.2)
Mountains	82.7 (0.9)	82.6 (0.8)	82.5 (0.8)	**85.7 (0.7)**
Food	95.9 (0.3)	**97.0 (0.2)**	96.8 (0.3)	**97.3 (0.2)**
Dogs	84.5 (0.9)	85.4 (0.8)	**86.8 (0.8)**	**86.6 (0.7)**
Lizards	**93.0 (0.5)**	91.7 (0.7)	92.0 (0.6)	**92.2 (0.6)**
Fashion	90.0 (0.5)	**90.2 (0.6)**	**90.9 (0.5)**	**90.2 (0.5)**
Sunset	**94.3 (0.6)**	93.3 (0.5)	94.1 (0.5)	**94.7 (0.4)**
Cars	88.8 (0.7)	87.9 (0.7)	**89.8 (0.6)**	88.2 (0.6)
Waterfalls	**93.8 (0.4)**	91.1 (0.6)	93.5 (0.4)	**93.4 (0.4)**
Antique	**92.8 (0.7)**	**93.7 (0.5)**	**93.2 (0.6)**	**93.4 (0.6)**
Battleships	92.7 (0.5)	92.8 (0.4)	93.9 (0.4)	**94.5 (0.4)**
Skiing	87.0 (0.8)	**91.4 (0.6)**	87.3 (0.7)	89.9 (0.6)
Desserts	**72.0 (1.4)**	68.6 (1.1)	71.0 (1.4)	**72.7 (1.0)**
AjaxOrange	81.6 (1.5)	**85.2 (1.5)**	**86.7 (1.3)**	**86.7 (1.1)**
Apple	**69.3 (1.3)**	62.4 (1.7)	**68.6 (1.3)**	66.3 (1.5)
Banana	65.4 (1.4)	61.5 (1.7)	**66.6 (1.7)**	65.4 (1.8)
BlueScrunge	72.3 (1.5)	**81.2 (1.2)**	76.2 (1.4)	**81.6 (1.2)**
CandleWithHolder	80.6 (1.4)	79.7 (1.3)	85.4 (1.2)	**87.1 (1.0)**
CardboardBox	72.3 (1.5)	83.4 (1.0)	76.7 (1.4)	**86.4 (1.1)**
CheckeredScarf	95.1 (0.4)	94.2 (0.3)	95.7 (0.4)	**96.7 (0.3)**
CokeCan	85.0 (1.2)	81.8 (1.2)	**87.9 (1.1)**	88.5 (1.1)
DataMiningBook	84.7 (1.1)	78.6 (1.3)	**86.9 (1.1)**	85.7 (1.2)
DirtyRunShoes	90.9 (0.9)	90.6 (0.9)	**92.2 (0.8)**	**91.6 (0.9)**
DirtyWorkGloves	75.5 (1.6)	75.9 (1.5)	78.5 (1.5)	**82.8 (1.4)**
FabricSoftener	89.0 (1.1)	88.9 (1.1)	**95.7 (0.7)**	89.7 (1.0)
FeltFlowerRug	83.7 (1.4)	86.3 (0.9)	88.4 (1.1)	**90.2 (0.8)**
GlazedWoodPot	58.7 (1.2)	63.5 (1.7)	59.9 (1.3)	**68.1 (1.6)**
GoldMedal	75.6 (1.5)	75.1 (1.3)	80.8 (1.2)	**83.5 (1.1)**
GreenTeaBox	81.8 (1.3)	79.6 (1.4)	**85.9 (1.1)**	83.4 (1.2)
JuliesPot	68.1 (1.5)	60.3 (1.7)	**70.9 (1.5)**	64.7 (1.4)
LargeSpoon	79.0 (1.4)	71.1 (1.7)	**82.7 (1.2)**	**81.7 (1.5)**
RapBook	70.3 (1.5)	71.4 (1.5)	71.5 (1.4)	**74.1 (1.4)**
SmileyFaceDoll	**79.0 (1.4)**	59.6 (2.0)	78.2 (1.4)	75.7 (1.4)
SpriteCan	77.7 (1.2)	71.0 (1.5)	**79.5 (1.2)**	80.2 (1.2)
StripedNotebook	76.6 (1.2)	**78.1 (1.6)**	**78.8 (1.2)**	**77.8 (1.2)**
TranslucentBowl	67.4 (1.4)	62.1 (1.4)	**68.8 (1.5)**	63.8 (1.5)
WD40Can	86.3 (1.3)	85.6 (1.1)	**90.3 (1.0)**	89.0 (1.1)
WoodRollingPin	78.8 (1.4)	79.2 (1.3)	**82.6 (1.4)**	82.2 (1.2)

5   Experiments

We test our approach on several benchmark MIL datasets:

- Musk 1, Musk 2 [6], molecule activity prediction.
- Trx Protein [16], protein function prediction.
- Mutagenesis easy, Mutagenesis hard [15], drug activity prediction.
- Fox, Tiger, Elephant [1], image classification.
- 20 Corel datasets [3], image classification.
- 25 SIVAL datasets [13], image classification.

We compare the performance of the 1-nearest neighbor classifier in the dissimilarity space using the class-dependent dissimilarity d_{cd} and its "ingredients", which we denote by d_{to} and d_{from} for brevity. In addition, we provide results of the symmetric mean $d_{avg} = \frac{1}{2}(d_{from} + d_{to})$ because before considering asymmetric dissimilarities, we have achieved good results with this symmetrized measure.

The results are given in Table 1. The class-dependent d_{cd} is performing better than d_{from} and d_{to}, which is in line with our intuition about it being able to capture more class differences. Overall, the performance of d_{avg} is comparable to that of d_{cd}, which might mean that averaging d_{from} and d_{to} captures some of the same information as in d_{cd}. For several datasets, one of these dissimilarities does significantly better than the other, although it is not entirely clear what these datasets have in common. However, it seems that in many cases where d_{to} outperforms d_{from}, d_{cd} also outperforms d_{avg} (e.g. SmileyFaceDoll).

The difference in the results of d_{from} and d_{to} is another interesting observations. For some datasets, these dissimilarities have comparable results, while for others, especially SIVAL datasets, one outperforms the other greatly. Although d_{to} is often better than d_{from}, for instance for the Apple dataset, in other datasets, such as CardboardBox, the situation is reversed.

6   Discussion and Conclusion

We have emphasized that in Multiple Instance Learning problems, it might be appropriate to treat the classes differently due to an important difference between positive and negative bags: the presence of concept instances. Most MIL approaches which compare bags directly disregard this difference. Therefore, we proposed to use a class-dependent dissimilarity based on the average minimum instance distance, which adapts itself based on the labels of the prototype bags. Experimental results showed that this class-dependent dissimilarity is indeed more informative than the independent versions, and that it is comparable to averaging of these two dissimilarities.

In several datasets, we have noticed large differences between measuring dissimilarities from bags to prototypes (d_{to}), or from prototypes (d_{from}) to the test bags. We believe these differences may be related to the class imbalance in the Corel and SIVAL datasets, where only 4 to 5% of the bags are positive. Therefore, the representation d_{to} is actually very similar to the class-dependent representation d_{cd}. This also explains why the successes of d_{to} and d_{from} are related. In fact, the correlation coefficient between the difference of performances of d_{to} and d_{from}, and the difference of performances of d_{cd} and d_{avg}, is equal to 0.55. This suggests that d_{from} contains some information which negatively affects the performance of d_{avg}, but which can be avoided when using the dissimilarities in a class-dependent manner.

To better understand the obtained results, we also examined the performances of the individual "to" and "from" dissimilarities using only the positive, or only the negative bags as prototypes. The results were surprising, because the performances were comparable to the dissimilarities where prototypes of both classes are available. In about half of the datasets, the dissimilarity from positive prototypes outperformed all the dissimilarities in Table 1. This provides opportunities for investigating how prototype selection [12] or assigning weights to the prototype classes can further improve performance. Furthermore, this result might be of interest in MIL problems with class imbalance such as in medical image diagnosis, and is worth investigating further.

References

1. Andrews, S., Hofmann, T., Tsochantaridis, I.: Multiple instance learning with generalized support vector machines. In: Proc. of the National Conference on Artificial Intelligence, pp. 943–944. AAAI Press, MIT Press, Menlo Park, Cambridge (2002)
2. Bailey, A.: Class-dependent features and multicategory classification. Ph.D. thesis, Citeseer (2001)
3. Chen, Y., Bi, J., Wang, J.: Miles: Multiple-instance learning via embedded instance selection. IEEE Transactions on Pattern Analysis and Machine Intelligence 28(12), 1931–1947 (2006)
4. Cheplygina, V., Tax, D., Loog, M.: Does one rotten apple spoil the whole barrel? In: International Conference on Pattern Recognition (in press)
5. De Wachter, M., Demuynck, K., Wambacq, P., Van Compernolle, D.: A locally weighted distance measure for example based speech recognition. In: International Conference on Acoustics, Speech, and Signal Processing, vol. 1, pp. I–181. IEEE (2004)
6. Dietterich, T., Lathrop, R., Lozano-Pérez, T.: Solving the multiple instance problem with axis-parallel rectangles. Artificial Intelligence 89(1-2), 31–71 (1997)
7. Gärtner, T., Flach, P., Kowalczyk, A., Smola, A.: Multi-instance kernels. In: Proc. of the 19th Int. Conf. on Machine Learning, pp. 179–186 (2002)
8. Kummamuru, K., Krishnapuram, R., Agrawal, R.: On learning asymmetric dissimilarity measures. In: International Conference on Data Mining, p. 4. IEEE (2005)
9. Maron, O., Lozano-Pérez, T.: A framework for multiple-instance learning. In: Advances in Neural Information Processing Systems, pp. 570–576. Morgan Kaufmann Publishers (1998)
10. Paredes, R., Vidal, E.: A class-dependent weighted dissimilarity measure for nearest neighbor classification problems. Pattern Recognition Letters 21(12), 1027–1036 (2000)
11. Pękalska, E., Duin, R.P.W.: The dissimilarity representation for pattern recognition: foundations and applications, vol. 64. World Scientific Pub. Co. Inc. (2005)

12. Pękalska, E., Duin, R., Paclík, P.: Prototype selection for dissimilarity-based classifiers. Pattern Recognition 39(2), 189–208 (2006)
13. Rahmani, R., Goldman, S., Zhang, H., Krettek, J., Fritts, J.: Localized content based image retrieval. In: Proc. of the 7th ACM SIGMM International Workshop on Multimedia Information Retrieval, pp. 227–236. ACM (2005)
14. Sørensen, L., Loog, M., Tax, D., Lee, W., de Bruijne, M., Duin, R.: Dissimilarity-Based Multiple Instance Learning. In: Hancock, E.R., Wilson, R.C., Windeatt, T., Ulusoy, I., Escolano, F. (eds.) SSPR&SPR 2010. LNCS, vol. 6218, pp. 129–138. Springer, Heidelberg (2010)
15. Srinivasan, A., Muggleton, S., King, R.: Comparing the use of background knowledge by inductive logic programming systems. In: Proceedings of the 5th International Workshop on Inductive Logic Programming, pp. 199–230 (1995)
16. Tao, Q., Scott, S., Vinodchandran, N., Osugi, T.: Svm-based generalized multiple-instance learning via approximate box counting. In: Proc. of the 21st Int. Conf. on Machine learning, p. 101. ACM (2004)
17. Tax, D., Loog, M., Duin, R., Cheplygina, V., Lee, W.: Bag dissimilarities for multiple instance learning. Similarity-Based Pattern Recognition, 222–234 (2011)
18. Wang, J.: Solving the multiple-instance problem: A lazy learning approach. In: Proc. of the 17th Int. Conf. on Machine Learning (2000)
19. Weinberger, K., Saul, L.: Distance metric learning for large margin nearest neighbor classification. The Journal of Machine Learning Research 10, 207–244 (2009)
20. Zafra, A., Pechenizkiy, M., Ventura, S.: Reducing dimensionality in multiple instance learning with a filter method. Hybrid Artificial Intelligence Systems, 35–44 (2010)
21. Zhao, C., Shi, W., Deng, Y.: A new hausdorff distance for image matching. Pattern Recognition Letters 26(5), 581–586 (2005)

Bidirectional Language Model
for Handwriting Recognition

Volkmar Frinken[1], Alicia Fornés[1], Josep Lladós[1], and Jean-Marc Ogier[2]

[1] Computer Vision Center, Dept. of Computer Science
Edifici O, UAB, 08193 Bellaterra, Spain
{vfrinken,afornes,josep}@cvc.uab.cat
[2] L3i Laboratory, Université de La Rochelle
Av. M. Crépeau, 17042 La Rochelle Cédex 1, France
jean-marc.ogier@univ-lr.fr

Abstract. In order to improve the results of automatically recognized handwritten text, information about the language is commonly included in the recognition process. A common approach is to represent a text line as a sequence. It is processed in one direction and the language information via n-grams is directly included in the decoding. This approach, however, only uses context on one side to estimate a word's probability. Therefore, we propose a bidirectional recognition in this paper, using distinct forward and a backward language models. By combining decoding hypotheses from both directions, we achieve a significant increase in recognition accuracy for the off-line writer independent handwriting recognition task. Both language models are of the same type and can be estimated on the same corpus. Hence, the increase in recognition accuracy comes without any additional need for training data or language modeling complexity.

Keywords: handwriting recognition, language models, neural networks.

1 Introduction

The recognition of handwritten text is a very active research field among researchers on pattern recognition [12]. Promising approaches for handwriting recognition are segmentation-free and learning-based, such as hidden Markov models (HMM) [2,13], neural networks (NN) [6], or combinations thereof [3].

Still, the recognition of unconstrained text can not be considered a solved problem. The main reason is the difficulty in dealing with the high variability encountered in different handwriting styles. Often, a semantic understanding of the text is necessary to be able to read a text. In case of automatic recognition systems, contextual understanding is usually emulated by estimating word probabilities, such as n-grams [5,7]. Yet, despite their simplicity and inability to capture any long-term relationships between words, n-gram approaches perform remarkably well and are still state-of-the-art.

Current handwriting recognition systems represent the text line as a sequence and perform the recognition usually in the direction of writing, i.e., left to right

G.L. Gimel' farb et al. (Eds.): SSPR & SPR 2012, LNCS 7626, pp. 611–619, 2012.
© Springer-Verlag Berlin Heidelberg 2012

for Roman scripts. This allows to directly include n-gram language model information in the decoding. In this form of language probability estimation, however, only a limited context is used to estimate the occurrence probability of a word. As a result, recognizers face the problem of error propagation. Correct word that are required in a larger context might be dropped due to pruning. Instead a wrong hypotheses propagates wrong language model information to the following words and may disturb their recognition, hence creating a form of decoding direction dependent error.

As a consequence, one-directional decoding seems to be an unnecessary restriction, especially when the input data are off-line text images. A word's probability can be estimated more robustly by considering both n-grams, the one considering the words on the left, and the one considering the words on the right side. Thus, taking also the reversed decoding direction into account could reduce the recognition error-propagation.

In this paper we propose the use of bi-directional n-grams for improving the recognition performance of unconstrained handwritten text. In order to do this, N-best lists are created for both directions separately, using a distinct forward and a backward language model. Then, these lists are combined to produce the final recognition output. Note that the system used in this paper is based on Neural Networks [6], but it could easily be extended to HMM-based approaches as well.

The rest of the paper is structured as follows. In Section 2 the proposed bidirectional language model approach is introduced and explained in detail. The experimental evaluation is presented in Section 3 and conclusions are drawn in Section 4.

2   Bidirectional Language Models

The ambiguity of different handwritten text and the huge variances in different writing styles require an integration of contextual information for an automatic transcription. The standard way of doing this is to integrate a statistical language model in the decoding process. However, language modeling using bi-grams do not capture the language sufficiently well. One option is to increase the complexity of the language model by using higher order n-grams, however, the number of distinct n-grams increases exponentially with n. Hence, even in a large training corpus, many word combinations do not occur at all or they occur with a frequency not high enough for a robust occurrence probability estimation.

Another challenge to handwriting recognition is the error propagation of a mis-recognized word. As a sequential decoding problem, common recognition methods process the text line in one direction, left-to-right or right-to-left. Hence, any mis-recognition propagates in the direction of recognition due to the language model which takes the current recognition result to estimate the next word's probability.

To address both issues, the challenge to estimate sophisticated language models on sparse data as well as the problem of error propagation, we propose in this paper to decode the text from both directions and combine the results. Forward and backward decoding require to different language models which can still be estimated on the same corpus and the combination can successfully increase the recognition accuracy.

The proposed approach is a step towards holistic language models to better capture syntactic and semantic information. While such models have been proposed for speech recognition in a sophisticated way [14], our approach does not increase the language modeling and hence the computational complexity.

2.1 Contribution

From a mathematical point of view, continuous handwriting recognition systems map a text image to a sequence of words $w_1^S = w_1 w_2 \ldots w_S$. This is done by using both, an observation model ϑ that assigns a probability value to a character sequence according to the observed image and a language model LM that assigns a probability value to a given character sequence according to the language at hand. The character sequence that maximizes the combined score is then selected as the final output.

In this paper we focus on the language model probability score which can be factorized as

$$p(w_1^S) = p(w_1) \cdot p(w_2|w_1) \cdots p(w_S|w_1^{S-1}) \tag{1}$$

$$= p(w_1) \prod_{i=2}^{S} p(w_i|w_1^{i-1}) \tag{2}$$

$$= p(w_S) \cdot p(w_{S-1}|w_S) \cdots p(w_1|w_2^S) \tag{3}$$

$$= p(w_S) \prod_{i=1}^{S} p(w_i|w_{i+1}^S) . \tag{4}$$

Note that we define $w_i = \varepsilon$ for $i \leq 0$ and $i > S$ to make the Equations more readable. Following from the rules of probability, it does not matter whether the LM probability is factorized such that the probability for a word w_i is conditioned on its left context w_1^{i-1} (see Eqn. 2) or its right context w_{i+1}^S. However, keeping track of the entire context is unfeasible for real word applications, hence state-of-art recognition systems use n-gram models which only take a limited number of words into account. Usually this is done in the direction of text processing, i.e., for languages that are written and recognized from left-to-right, the left hand side context of a word is considered to estimate its probability. Here, we will indicate this with $LM_\rightarrow$ and call it *forward LM*

$$p_\rightarrow(w_1^S|LM_\rightarrow) = p(w_1|LM_\rightarrow) \prod_{i=2}^{S} p(w_i|w_{i-n+1}^{i-1}, LM_\rightarrow) . \tag{5}$$

Obviously, every text image can also be recognized in the reversed direction, requiring different n-grams, indicated here with $LM_\leftarrow$ (*backward LM*)

$$p_\leftarrow(w_1^S|LM_\leftarrow) = p(w_S|LM_\leftarrow) \prod_{i=1}^{S-1} p(w_i|w_{i+1}^{i+n-1}, LM_\leftarrow) . \tag{6}$$

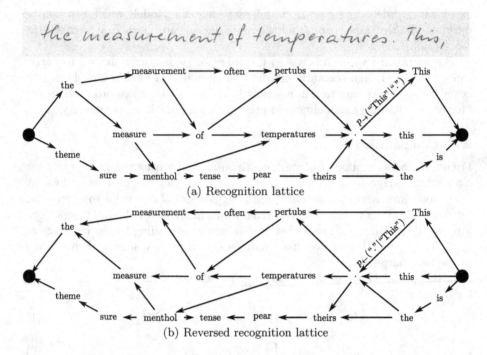

Fig. 1. In (a) the recognition lattice for the left-to-right decoding direction is given for a sample text line. In (b) the reversed lattice with the modified language model information is shown. Note that all node labels but only one edge label is shown for the sake of readability. In the left-to-right decoding, the bi-gram information that the word "This" occurs after the symbol "." is used. In the right-to-left decoding, the corresponding edge contains the probability of the symbol "." occurring before the word "This".

Although the Equations (2) and (4) are a factorization of the same probability, their n-gram simplification in Equations (5) and (6) is expected to produce different results. Yet, both can be estimated on the same corpus. Thus, we propose in this paper to exploit this fact. We show that a significant improvement of the recognition rate can be achieved by combining the recognition output of the two systems using the forward LM and the backward LM of the same n-gram order.

2.2 Approach

We propose to generate two different N-best lists of recognition hypotheses, one generated by a left-to-right and one generated by a right-to-left decoding. Afterwards, these N-best lists can be combined to generate a new output.

A straightforward way to build both lists is to use a recognizer for handwritten text that produces a recognition lattice, such as HMMs or BLSTM Neural Networks in conjunction with a Token Passing algorithm. A recognition lattice (see Fig. 1), is a directed graph with node and edge labels and constitutes a comprehensive way of storing various decoding paths. From this, N-best lists can easily be generated by searching the most likely paths across the lattice using A^*-search. The exact specifications, what information is stored in the nodes and labels may vary, but usually a node represents a word and an edge indicates a transition between two words. In our approach, nodes are labeled with the position where the word ends. Edges are labeled with two probability scores, the bi-gram transition probability between the word at the starting node and the word at the ending node and the observation probability of the word at the ending node. From this, we generate the N-best lists of the forward direction.

Next, we reverse the directions of the edges and adjust the bi-gram probabilities. That is, an edge $e = (u, v) \in V \times V$ from node u to v labeled with $p_{\rightarrow}(v|u, LM)$ and $p_{obs}(v)$ is changed into an edge $e = (v, u)$ with labeling $p_{\leftarrow}(u|v, LM)$ and $p_{obs}(v)$. A path in the new lattice now represents a decoding using the reversed bi-gram language model and an N-best list is also generated. Note that the word ordering of the hypotheses in this list is in reversed order and needs to be changed back.

To make use of higher order N-grams, the bi-gram word transition probabilities on the edges are ignored. Instead, an A^*-search on the lattices is done using an external language model file to generate the forward and backward N-best lists.

Finally, the two N-best lists can be combined using a generalized *recognizer output voting error reduction* (ROVER) scheme [4,16]. In this system, the N-best output word strings are first aligned and then combined in a weighted voting scheme. The weights of the word hypotheses in the combination are based on their posterior probabilities which are estimated from the N-best lists of the recognizers. The combination was done using the SRILM toolkit [15]. The toolkit allows the use of different weight parameters, which were optimized on the validation set.

3   Experimental Evaluation

3.1   Setup

For the experiments, we have used the IAM off-line database [11], which contains forms of unconstrained handwritten English text. The database is composed of 1,539 pages (13,353 text lines, 115,320 words) written by 657 writers. In our experiments we have followed the benchmark defined by the authors, which consists in 6,161 lines in the training set, and 920 lines in the validation set, and 2,781 lines in the test set.

First of all, each text line has been binarized and normalized in order to cope with different handwriting styles. The normalization consists in correcting the skew and slant, and normalizing the size and width of the text. The result of the text line normalization process can be seen in Fig. 2. Once the text lines are normalized, a sliding window moving from left to right over the text image. At each column of width one pixel, the following nine features are extracted: the 0th, 1st and 2nd moment of the black pixels' distribution within the window, the position of the top-most and bottom-most black pixel, the inclination of the

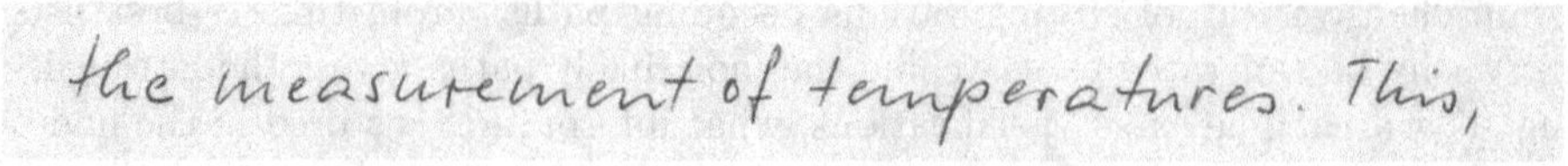

the measurement of temperatures. This,

(a) The original text line image.

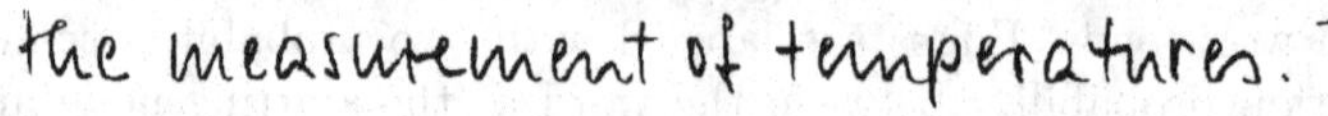

the measurement of temperatures. This,

(b) The normalized text line.

Fig. 2. The text line preprocessing

top and bottom contour of the word at the actual window position, the number of vertical black/white transitions, and the average gray scale value between the top-most and bottom-most black pixel. For a more detailed description of the normalization and feature extraction, we refer to [10].

As a recognizer, a bidirectional LSTM neural network (BLSTM NN) is used, i.e., the sequence of feature vectors is fed into the network from both directions, left-to-right and right-to-left. The output layer consists of one node for each possible character. By normalizing the output activations, the result is a matrix of posterior probabilities for each letter and each position. Given that matrix and a bi-gram language model, a token passing algorithm can be used to generate the recognition lattices. For details about BLSTM networks and the CTC token passing algorithm, we refer to [6].

Both the forward and the backward N-grams with $N = 2, 3, 4$ are estimated on the union of the Brown and Wellington corpus [1,8,9] as well as the part of the LOB corpus not used in the validation or testing. The total amount of text is 3.34M words in 162.6K sentences.

We chose the dictionary to be the 20,000 most frequent English words. Since we consider the open vocabulary recognition task, some words in the training, validation, and test set do not occur in the dictionary and can not be recognized. This imposes an upper bound to the word recognition rate of 93.74%.

3.2   Results

In Fig. 3, the impact of the bidirectional language model on the handwriting recognition task can be seen. The solid line indicate the standard left-to-right language model and it can be seen that the recognition accuracy increases from 75.08% using a 2-gram LM, up to 75.47% (3-grams) and 75.50% (4-grams). The results using bi-grams are comparable to the ones found in [6,3].

Using a right-to-left language model, the recognition rates are consistently higher by reaching 75.25% (2-grams), 75.80% (3-grams), and 75.82% (4-grams). The lack of significant increase when switching from a 3-gram to the 4-gram LM can be explained by the size of the language corpus. Obviously the limit of the generalization capability is reached.

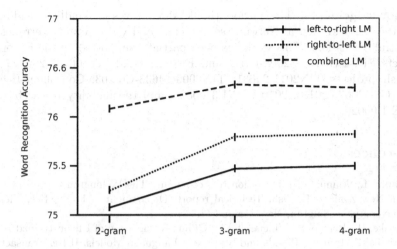

Fig. 3. Word level recognition accuracies of the different systems

The proposed, combined language model, however, achieves a significant increase by combining the left-to-right and right-to-left models. With bi-gram model, a recognition accuracy of 76.08% is reached, outperforming even the 4-gram recognition with an unidirectional model. The performance using the 3-gram combined model is 76.33% and slightly better than using the 4-gram (76.29%) combined model. However, this difference is not statistically significant, while all increases from the uni-directional models to the proposed bi-directional model are statistically significant at $\alpha = 0.05$ for every N.

4   Conclusion

The recognition of unconstrained handwritten text is still considered an open problem mainly due to the high variability in the handwriting styles. Since state-of-the-art handwriting recognition systems decode a text line sequentially, the contextual information used for solving ambiguities is only taken from one side of a word. To increase the robustness of estimating a word's language model probability one the one hand and to reduce the effect of error-propagation of mis-recognized words, we propose bidirectional language models. In considering contextual information from both sides of a word, our approach may be seen as a step towards full sentence language models that capture the meaning of a text holistically.

The experimental results obtained with bidirectional n-grams have shown a significant improvement over current state-of-the-art approaches. The improvement has been achieved without increasing the amount of training data, language corpus, or the complexity of the language model.

Thus, we can conclude that bidirectional language models are promising approaches. Therefore, further work could be focused on investigating holistic whole sentence analysis with bidirectional grammars and context-free grammars.

Acknowledgements. The authors thank Alex Graves for kindly providing us with the BLSTM Neural Network source code and Oriol Ramos Terrades for insightful discussions. This work has been partially supported by the European projects FP7-PEOPLE-2008-IAPP and ERC-2010-AdG-20100407-269796, the Spanish projects TIN2011-24631, TIN2009-14633-C03-03, Consolider-Ingenio 2010 (CSD2007-00018), 2010 CONE3 00029, and the mobility research grant 10 BE-1 00020.

References

1. Bauer, L.: Manual of Information to Accompany The Wellington Corpus of Written New Zealand English. Technical report, Department of Linguistics, Victoria University, Wellington, New Zealand (1993)
2. Bunke, H., Bengio, S., Vinciarelli, A.: Offline Recognition of Unconstrained Handwritten Texts using HMMs and Statistical Language Models. IEEE Transactions on Pattern Analysis and Machine Intelligence 26(6), 709–720 (2004)
3. Espana-Boquera, S., Castro-Bleda, M.J., Gorbe-Moya, J., Zamora-Martínez, F.: Improving Offline Handwritten Text Recognition with Hybrid HMM/ANN Models. IEEE Transactions on Pattern Analysis and Machine Intelligence 33(4), 767–779 (2011)
4. Fiscus, J.: A Post-processing System to Yield Reduced Word Error Rates: Recognizer Output Voting Error Reduction (ROVER). In: Workshop on Automatic Speech Recognition and Understanding, pp. 347–354. IEEE (December 1997)
5. Goodman, J.T.: A Bit of Progress in Language Modeling - Extended Version. Technical Report MSR-TR-2001-72, Microsoft Research, One Microsoft Way Redmond, WA 98052, 8 (2001)
6. Graves, A., Liwicki, M., Fernández, S., Bertolami, R., Bunke, H., Schmidhuber, J.: A novel Connectionist System for Unconstrained Handwriting Recognition. IEEE Transactions on Pattern Analysis and Machine Intelligence 31(5), 855–868 (2009)
7. Jelinek, F.: Stochastic Analysis of Structured Language Modeling. In: Mathematical Foundations of Speech and Language Processing, vol. 138, pp. 37–71. Springer,
8. Johansson, S., Atwell, E., Garside, R., Leech, G.: The tagged lob corpus: Users' manual. Technical report, The Norwegian Computing Centre for the Humanities (1986)
9. Kucera, H., Francis, W.N.: Manual of Information to accompany A Standard Corpus of Present-Day Edited American English, for use with Digital Computers. Brown University, Department of Linguistics, Providence, Rhode Island, 1964. Revised 1971. Revised and amplified (1979)
10. Marti, U.-V., Bunke, H.: Using a Statistical Language Model to Improve the Performance of an HMM-Based Cursive Handwriting Recognition System. Int. Journal of Pattern Recognition and Artificial Intelligence 15, 65–90 (2001)
11. Marti, U.V., Bunke, H.: The iam-database: An English Sentence Database for Offline Handwriting Recognition. Int'l Journal on Document Analysis and Recognition 5(1), 39–46 (2002)
12. Plamondon, R., Srihari, S.N.: Online and Off-Line Handwriting Recognition: A Comprehensive Survey. IEEE Transactions on Pattern Analysis and Machine Intelligence 22(1), 63–84 (2000)

13. Plötz, T., Fink, G.A.: Markov Models for Offline Handwriting Recognition: A Survey. Int'l Journal on Document Analysis and Recognition 12(4), 269–298 (2009)
14. Rosenfeld, R., Chen, S.F., Zh, X.: Whole-Sentence Exponential Language Models: A Vehicle for Linguistic-Statistical Integration. Computers, Speech and Language 15, 55–73 (2001)
15. Stolcke, A.: SRILM: An Extensible Language Modeling Toolkit, pp. 901–904 (2002)
16. Stolke, A., König, Y., Weintraub, M.: Explicit Word Error Minimization in N-Best List Rescoring. In: EUROSPEECH, pp. 163–166 (1997)

Hypergraph Spectra for Unsupervised Feature Selection

Zhihong Zhang and Edwin R. Hancock*

Department of Computer Science,
University of York, UK

Abstract. Most existing feature selection methods focus on ranking individual features based on a utility criterion, and select the optimal feature set in a greedy manner. However, the feature combinations found in this way do not give optimal classification performance, since they neglect the correlations among features. In an attempt to overcome this problem, we develop a novel unsupervised feature selection technique by using hypergraph spectral embedding, where the projection matrix is constrained to be a selection matrix designed to select the optimal feature subset. Specifically, by incorporating multidimensional interaction information (MII) for higher order similarities measure, we establish a novel hypergraph framework which is used for characterizing the multiple relationships within a set of samples. Thus, the structural information latent in the data can be more effectively modeled. Secondly, we derive a hypergraph embedding view of feature selection which casting the feature discriminant analysis into a regression framework that considers the correlations among features. As a result, we can evaluate joint feature combinations, rather than being confined to consider them individually, and are thus able to handle feature redundancy. Experimental results demonstrate the effectiveness of our feature selection method on a number of standard datasets.

Keywords: Hypergraph representation, Hypergraph subspace learning.

1 Introduction

In order to render the analysis of high-dimensional data tractable, it is crucial to identify a smaller subset of features that are informative for classification and clustering. Dimensionality reduction aims to reduce the number of variables under consideration, and the process can be divided into feature extraction and feature selection. Feature extraction usually projects the features onto a low-dimensional and distinct feature space, e.g., kernel PCA [1], Locality preserving Projection (LPP) [2] and Laplacian eigenmap [3]. Unlike feature extraction, feature selection identifies the optimal feature subset in the original feature space. By maintaining the original features, feature selection improves the interpretability of the data, which is preferred in many real world applications, such as face recognition and text mining. Feature selection algorithms can be roughly classified into two groups, namely a) supervised feature selection and b) unsupervised feature selection.

---

* Edwin Hancock is supported by a Royal Society Wolfson Research Merit Award.

G.L. Gimel' farb et al. (Eds.): SSPR & SPR 2012, LNCS 7626, pp. 620–628, 2012.

While the labeled data required by supervised feature selection can be scarce, there is usually no shortage of unlabeled data. Hence, there are obvious attractions in developing unsupervised feature selection algorithms which can utilize this data. The typical examples in unsupervised learning are graph-based spectral learning algorithms. Examples include the Laplacian score [6], SPEC [5], Multi-Cluster Feature Selection (MCFS) [8] and Unsupervised Discriminative Feature Selection (UDFS) [10]. Given d features, and a similarity matrix S for the samples, the idea of spectral feature selection algorithms is to identify features that align well with the leading eigenvectors of S. The leading eigenvectors of S contain information of concerning the structure of the sample distribution and group similar samples into compact clusters. Consequently, features that align closely to them will better preserve sample similarity [5]. For example, the Laplacian score [6] uses a nearest neighbor graph to model the local geometric structure of the data, using the pairwise similarities between features are calculated using the heat kernel. In this framework, the features are evaluated individually and are selected one by one. The SPEC [5] algorithm is an extension of the Laplacian score that render it more robust to noise. The method selects the features most consistent with the graph structure. Note that SPEC also evaluates features independently.

However, there are two limitations to the above graph-based spectral feature selection methods. Firstly, they evaluate features individually, and hence cannot handle redundant features. Redundant features increase the dimensionality unnecessarily, and worsen learning performance when faced with a shortage of data. It is also shown empirically that removing redundant features can result in significant performance improvement. The second weakness is that in many situations the graph representation for relational patterns can lead to substantial loss of information. This is because in real-world problems objects and their features tend to exhibit multiple relationships rather than simple pairwise ones. For example, consider the problem of classifying faces which are under different lighting conditions [7]. Therefore, the higher order relations cannot be suitably characterized by pairwise similarity measures.

A natural way for remedying the misleading representation described above is to represent the dataset as a hypergraph instead of a graph. Hypergraph representations allow vertices to be multiply connected by hyperedges and can hence capture multiple or higher order relationships between features. Due to their effectiveness in representing multiple relationships, for the task of feature selection addressed in this paper, we introduce a hypergraph embedding view of feature selection by subspace learning. The method jointly evaluates the utility sets of features rather than individual feature. There are three novel ingredients. The first is that by incorporating hypergraph representation into feature selection, we can be more effective capture the higher order relations among samples. Secondly, inspired from the recent works on mutual information [16], we determine the weight of a hyperedge using an information measure referred to as multidimensional interaction information (MII) which precisely preserves the higher order relations captured by the hypergraph. The advantage of MII is that it is sensitive to the relations between sample combinations, and as a result can be used to seek third or even higher order dependencies among the relevant samples. Thus, the structural information latent in the data can be more effectively modeled. Finally, we describe a new feature selection strategy through hypergraph embedding, which casts the feature discriminant analysis into a regression framework that considers the correlations among features. As a result, we can evaluate joint feature combinations, rather than being confined to consider them individually, thus it is able to handle feature redundancy.

2   Hypergraph Construction

In this section, we establish a novel hypergraph framework which is used for characterizing the multiple relationships within a set of samples. To this end, we commence by introducing a new method for measuring higher order similarities among samples based on information theory. According to Shannon's study, the uncertainty of a random variable X can be measured by the entropy $H(X)$. For two random variables X and Y, the conditional entropy $H(Y|X)$ measures the remaining uncertainty about Y when X is known. The mutual information $I(X;Y)$ of X and Y quantifies the information gain about Y provided by X. The relationship between $H(Y)$, $H(Y|X)$ and $I(X;Y)$ is $I(X;Y) = H(Y) - H(Y|X)$. As defined by Shannon, the initial uncertainty for X is $H(X) = -\sum_{x \in Y} P(x) \log P(x)$, where $P(x)$ is the prior probability density function over $x \in X$. The remaining uncertainty for Y if X is known is defined by the conditional entropy $H(Y|X) = -\int_x p(x)\{\sum_{y \in Y} p(y|x) \log p(y|x)\}dx$, where $p(y|x)$ denotes the posterior probability for $y \in Y$ given $x \in X$. After observing x, the amount of additional information gain is given by the mutual information

$$I(X;Y) = \sum_{y \in Y} \int_x p(y,x) log \frac{p(y,x)}{p(y)p(x)} dx . \tag{1}$$

The mutual information (1) quantifies the information which is shared by X and Y. When the $I(X;Y)$ is large, it implies that x and y are closely related. Otherwise, when $I(X;Y)$ is equal to 0, it means that two variables are totally unrelated. Analogically, the conditional mutual information of X and Y given Z, denoted as $I(X;Y|Z) = H(X|Z) - H(X|Y,Z)$, represents the quantity of information shared by X and Y when Z is known. The conditioning on a third random variable may either increase or decrease the original mutual information. In this context, the Interaction Information $I(X;Y;Z)$ is defined as the difference between the conditional mutual information and the simple mutual information, i.e.

$$I(X;Y;Z) = I(X;Y|Z) - I(X;Y) . \tag{2}$$

The interaction information $I(X;Y;Z)$ measures the influence of the variable Z on the amount of information shared between variables X and Y. Its value can be positive, negative, or zero. Zero valued Interaction Information $I(X;Y;Z)$ implies that the relation between X and Y entirely depends on Z. A positive value of $I(X;Y;Z)$ implies that X and Y are independent of each other themselves, but are correlated with each other when combined with Z. A negative value of $I(X;Y;Z)$ indicates that Z can account for or explain the correlation between X and Y. The generalization of Interaction Information to K variables is defined recursively as follow

$$I(\{X_1, \cdots, X_K\}) = I(\{X_2, \cdots, X_K\}|X_1) - I(\{X_2, \cdots, X_K\}) . \tag{3}$$

Based on the higher order similarity measure, we establish a hypergraph framework for characterizing a set of high dimensional samples. A hypergraph is defined as a triplet $H = (V, E, w)$. Here V denotes the vertex set, E denotes the hyperedge set in which each hyperedge $e \in E$ represents a subset of V, and w is a weight function which assigns a real value $w(e)$ to each hyperedge $e \in E$. We only consider K-uniform hypergraphs (i.e. those for which the hyperedges have identical cardinality K) in our work. Given a set of high dimensional samples $\mathbf{X} = [x_1, \cdots x_N]^T$ where $x_i \in \mathbb{R}^d$, we establish a K-uniform hypergraph, with each hypergraph vertex representing an individual sample and each hyperedge representing the Kth order relations among a K-tuple of participating samples. A K-uniform hypergraph can be represented in terms of Kth order matrix, i.e. a tensor $\mathcal{W}$ of order K, whose element $W_{i_1,\cdots,i_K}$ is the hyperedge weight associated with the K-tuple of participating vertices $\{v_{i_1}, \cdots, v_{i_K}\}$. In our work, the hyperedge weight associating with $\{x_{i_1}, x_{i_2}, \cdots, x_{i_K}\}$ is computed as follows

$$W_{i_1,\cdots,i_K} = K \frac{I(x_{i_1}, x_{i_2}, \cdots, x_{i_K})}{H(x_{i_1}) + H(x_{i_2}) + \cdots H(x_{i_K})} . \tag{4}$$

It is clear that $W_{i_1,\cdots,i_K}$ is a normalized version of Kth order Interaction Information. The greater the value of $W_{i_1,\cdots,i_K}$ is, the more relevant the K samples are. On the other hand, if $W_{i_1,\cdots,i_K} = 0$, the K samples are totally unrelated.

3   Hypergraph Representation

Unlike matrix eigen-decomposition, there has not yet been a widely accepted method for spanning a rationale eigen-space for a tensor [13]. Therefore, it is hard to directly embed a hypergraph into a feature space spanned by its tensor representation through eigen-decomposition. In our work, we consider the transformation of a K-uniform hypergraph into a graph. Accordingly, the associated hypergraph tensor $\mathcal{W}$ is transformed to a graph adjacency matrix A, and the higher order information exhibited in the original hypergraph can be encoded in an embedding space spanned by the related matrix representation. In this scenario, one straightforward way for the transformation is marginalization which computes the arithmetic average over all the hyperedge weights $W_{i_1,\cdots,i_{K-2},i,j}$ associated with the edge weight $A_{i,j}$

$$\tilde{A}_{i,j} = \sum_{i_1=1}^{|V|} \cdots \sum_{i_{K-2}=1}^{|V|} W_{i_1,\cdots,i_{K-2},i,j} \tag{5}$$

The edge weight $\tilde{A}_{i,j}$ for edge ij is generated by a uniformly weighted sum of hyperedge weights $W_{i_1,\cdots,i_{K-2},i,j}$. However, the form appearing in (5) behaves as a low pass filter, and thus results in information loss through marginalization.

To make the process of marginalization more comprehensive, we use marginalization to constrain the sum of edge weights and then estimate their values through solving an over-constrained system of linear equations. Our idea is motivated by the so called *clique average* introduced in the higher order clustering literature [11]. We characterize the relationships between A and $\mathcal{W}$ as follows

$$W_{i_1,\cdots,i_K} = \sum_{\{i,j\}\subseteq\{i_1,\cdots,i_K\}} A_{i,j} \qquad (6)$$

There are $\binom{|V|}{2}$ variables and $\binom{|V|}{K}$ equations in the system of equations described in (5). When $K > 2$, the linear system (5) is over-determined and cannot be solved analytically. We thus approximate the solution to (5) by minimizing the least squares error

$$\hat{A} = \underset{A}{\mathrm{argmax}} \sum_{i_1,\cdots,i_K} \left(\sum_{\{i,j\}\subseteq\{i_1,\cdots,i_K\}} A_{i,j} - W_{i_1,\cdots,i_K} \right)^2 \qquad (7)$$

In practical computation, we normalize the compatibility tensor $\mathcal{W}$ by using the extended Sinkhorn normalization scheme [14], and constrain the element of A to be in the interval $[0, 1]$ to avoid unexpected infinities. Effective iterative numerical methods are used to compute the approximated solutions [15].

The adjacency matrix A computed through (7) is one effective representation for a K-uniform hypergraph, because it naturally avoids the operation of arithmetic average and thus to a certain degree overcomes the low pass information loss arising in (5). Furthermore, the Laplacian matrix L for a hypergraph can be defined as $L = D - A$, where D is the diagonal matrix with its ith diagonal element being $A_{ii} = \sum_j A_{ij}$. In this context, a hypergraph can be easily embedded into a feature space spanned by its Laplacian matrix, which will be explained in detail in the next Section.

4   Feature Selection through Hypergraph Embedding

In this section, we formulate the procedure of feature extraction on a basis of hypergraph spectral embedding. One goal of spectral embedding is to represent the high dimensional data $\mathbf{X} \in \mathbb{R}^{N \times d}$ by a low dimensional representation $\mathbf{Y} \in \mathbb{R}^{N \times C}$ $(C \ll d)$ in the low dimensional feature space such that the structural characteristics of the high dimensional data are well preserved or are more "obvious". Here we use the representations $\mathbf{X} = [x_1, \cdots x_N]^T$ and $\mathbf{Y} = [y_1, \cdots, y_k, \cdots, y_C]$, where y_k is a N-dimensional vector and its N elements represent the N samples $x_1, \cdots x_N$ separately in the kth dimension of the low dimensional feature space.

Based on the hypergraph transformation described in Section 3 and the scheme of Laplacian eigen-decomposition [3], the hypergraph spectral embedding can be easily conducted as follows

$$D^{-1}LY = \lambda Y . \qquad (8)$$

The hypergraph embedding procedure can be viewed as feature extraction, and can be expressed as $\mathbf{Y} = \mathbf{X}\Phi$ where $\Phi \in \mathbb{R}^{d \times C}$ is a column-full-rank projection matrix. However, unlike feature extraction, feature selection attempts to select the optimal feature subset in the original feature space. Therefore, for the task of feature selection, the projection matrix $\Phi = [\Phi_1, \ldots \Phi_C]$ can be constrained to be a selection matrix which contains the combination coefficients for different features in approximating $\mathbf{Y} = [y_1, \ldots, y_C]$. That is, given the kth column of $\mathbf{Y}$, i.e y_k, we aim to find a subset of features, such that their linear span is close to y_k. This idea can be formulated as the minimization problem

$$\widehat{\Phi} = \underset{\Phi}{\operatorname{argmin}} \sum_{k=1}^{C} \|y_k - X\Phi_k\|^2 . \tag{9}$$

where $\Phi = [\Phi_1, \cdots, \Phi_k, \cdots, \Phi_C]$ and Φ_k is a d dimensional vector that contains the combination coefficients required to compute for different features in approximating y_k. However, feature selection requires to locate a optimal subset of features that are close to y_k. This is a combinatorial problem which is NP-hard. Thus we approximate the problem in (9) subject to the constraint

$$|\Phi_k| \leq \gamma \tag{10}$$

where $|\Phi_k|$ is the ℓ_1-norm and $|\Phi_k| = \sum_{j=1}^{d} |\Phi_{j,k}|$. When applied in regression, the ℓ_1-norm constraint is equivalent to applying a Laplace prior on Φ_k. This tends to force some entries in Φ_k to be zero, resulting in a sparse solution. Therefore, the representation $\mathbf{Y}$ is generated by using only a small set of selected features in $\mathbf{X}$.

In order to efficiently solve the optimization problem in Equations (9) and (10), we use the Least Angle Regression (LARs) algorithm [9]. Instead of setting the parameter γ, LARs allow us to control the sparseness of Φ_k. This is done by specifying the cardinality of the number of nonzero subset of Φ_k, which is particularly convenient for feature selection.

We consider selecting m features from the d feature candidates. For a dataset containing C clusters, we can compute C selection vectors $\{\Phi_k\}_{k=1}^{C} \in R^d$. The cardinality of each Φ_k is m and each entry in Φ_k corresponds to a feature. Here, we use the following computationally effective method for selecting exactly m features based on the C selection vectors. For every feature j, we define the HG score for the feature as

$$HGscore(j) = \max_{k} |\Phi_{j,k}| . \tag{11}$$

where $\Phi_{j,k}$ is the jth element of vector Φ_k. We then sort the features in descending order according to their HG scores, and then select the top m features.

5   Experiments and Comparisons

We test the performance of our proposed algorithm on one publicly available face database (ORL) and one handwritten digit databases (MNIST). Table. 1 summarizes the coverage and properties of the two benchmark datasets.

Data Transformation: We compare the data transformation performance of our proposed method using hypergraph embedding (HG embedding) with alternative methods, including kernel PCA [1], the Laplacian eigenmap [3] and LPP [2]. In order to visualize the results, we have used five randomly selected subjects from each dataset, and these are shown in Fig. 1 and Fig. 2. In each figure, we have shown the projections onto the leading two most significant eigenmodes from different spectral embedding methods,

Table 1. Summary of benchmark datasets

Dataset	Examples	Features	Classes
ORL	400	1024	40
MNIST	4000	784	10

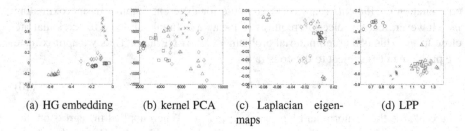

(a) HG embedding   (b) kernel PCA   (c) Laplacian eigen-maps   (d) LPP

Fig. 1. Distribution of samples of five subjects in ORL dataset

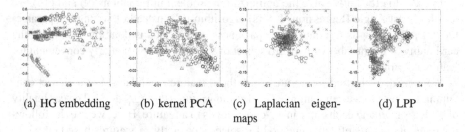

(a) HG embedding   (b) kernel PCA   (c) Laplacian eigen-maps   (d) LPP

Fig. 2. Distribution of samples of five subjects in MNIST dataset

ordered according to their eigenvalues. This provides a low-dimensional representation for the images. From the above figures, it is clear that our hypergraph spectral embedding method demonstrates much clearer cluster structure than alternative spectral clustering methods. This implies that the hypergraph representation is more appropriate and more complete in describing feature relations and structures existing in these datasets.

Classification Accuracy: In order to explore the discriminative capabilities of the information captured by our method, we use the selected features for further classification. We compare the classification results from our proposed method (UFSHE) with five alternative feature selection algorithms. For unsupervised learning, three alternative feature selection algorithms are selected as baselines. These methods are the Laplacian score [6], SPEC [5] and UDFS [10]. We also compare our results with two state-of-the-art supervised feature selection methods, namely a) the Fisher score [4] and b) the MRMR algorithm [12]. We use 5-fold cross-validation for the SVM classifier on the feature subsets obtained by the feature selection algorithms to verify their classification performance. Here we use the linear SVM with LIBSVM.

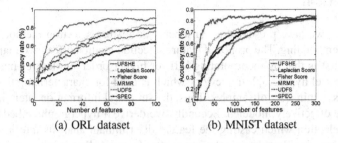

(a) ORL dataset   (b) MNIST dataset

Fig. 3. Accuracy rate vs. the number of selected features on two benchmark image datasets

Table 2. The best result of all methods and their corresponding size of selected feature subset on two benchmark image datasets

Dataset	MRMR	Fisher Score	Laplacian Score	SPEC	UDFS	UFSHE
ORL	83.5%(95)	80%(99)	65.25%(99)	64.5%(95)	76.5%(99)	**91%(75)**
MNIST	82.5%(284)	81.25%(293)	82.05%(291)	82.1%(292)	81.3%(293)	**84.33%(90)**

The classification accuracies obtained with different feature subsets are shown in Fig. 3. From the figure, it is clear that our proposed method UFSHE is, by and large, superior to the alternative feature selection methods. Specifically, it selects both a smaller and better performing (in terms of classification accuracy) set of discriminative features on both datasets. Moreover, UFSHE rapidly converges, with typically around 30 features. Each of the alternative unsupervised methods, usually require more than 100 features to achieve a comparable result. The reason for this improvement is that the hypergraph representation is effective in capturing the higher order relations among samples and thus the structural information latent in the data can be effectively preserved. Additionally, our hypergraph based feature selection method casts the feature discriminant analysis into a regression framework which suitably characterizes the correlations among features. As a result, the optimal feature combinations can be located so as to remove redundant features.

The best result for each method together with the corresponding size of the selected feature subset are shown in Table. 2. In this table, the classification accuracy is shown first and the optimal number of features selected is reported in brackets. Overall, UFSHE achieves the highest degree of dimensionality reduction, i.e. it selects a smaller feature subset compared with those obtained by the alternative methods. For example, in the MNIST dataset, the best result obtained by the alternative feature selection methods is 82.5% with the MRMR algorithm and 284 features. However, our proposed method (UFSHE) gives a better accuracy of 84.33% when only 90 features are used. The results further verify that our feature selection method can guarantee the optimal size of the feature subset, as it not only achieves a higher degree of dimensionality reduction but it also gives better discriminability.

6 Conclusion

In this paper, we have presented an unsupervised feature selection method based on hypergraph embedding. The proposed feature selection method offers two major advantages. The first is that by incorporating MII for higher order similarities measure, we establish a novel hypergraph framework which is used for characterizing the multiple relationships within a set of samples. Thus, the structural information latent in the data can be more effectively modeled. Secondly, we derive a hypergraph embedding view of feature selection which casting the feature discriminant analysis into a regression framework that considers the correlations among features. As a result, we can evaluate joint feature combinations, rather than being confined to consider them individually. These properties enable our method to be able to handle feature redundancy effectively.

References

1. Scholkopf, B., Smola, A., Muller, K.R.: Nonlinear component analysis as a kernel eigenvalue problem. Neural Computation 10(5), 1299–1319 (1998)
2. He, X., Niyogi, P.: Locality preserving projections (LPP). In: Proc. NIPS (2004)
3. Belkin, M., Niyogi, P.: Laplacian eigenmaps and spectral techniques for embedding and clustering. In: Proc. NIPS, pp. 585–592 (2002)
4. Duda, R.O., Hart, P.E., Stork, D.G.: Pattern Classification. John Wiley & Sons, New York (2001)
5. Zhao, Z., Liu, H.: Spectral feature selection for supervised and unsupervised learning. In: Proc. ICML, pp. 1151–1157 (2007)
6. He, X., Cai, D., Niyogi, P.: Laplacian score for feature selection. In: Proc. NIPS (2005)
7. Belhumeur, P.N., Kriegman, D.J.: What is the set of images of an object under all possible illumination conditions? IJCV 28(3), 245–260 (1998)
8. Cai, D., Zhang, C., He, X.: Unsupervised feature selection for multi-cluster data. In: Proc. ACM SIGKDD, pp. 333–342 (2010)
9. Efron, B., Hastie, T., Johnstone, I., Tibshirani, R.: Least angle regression. The Annals of Statistics 32(2), 407–499 (2004)
10. Yang, Y., Shen, H.T., Ma, Z., Huang, Z., Zhou, X.: L21-norm regularized discriminative feature selection for unsupervised learning. In: Proc. IJCAI, pp. 1589–1594 (2011)
11. Agarwal, S., Lim, J., Zelnik-Manor, L., Perona, P., Kriegman, D., Belongie, S.: Beyond pairwise clustering. In: Proc. CVPR, pp. 838–845 (2005)
12. Peng, H., Long, F., Ding, C.: Feature selection based on mutual information: Criteria of max-dependency, max-relevance, and min-redundancy. IEEE Trans. on PAMI 27(8), 1226–1238 (2005)
13. Kolda, T.G., Bader, B.W.: Tensor decompositions and applications. SIAM Review 51(3), 455–500 (2009)
14. Shashua, A., Zass, R., Hazan, T.: Multi-way Clustering Using Super-Symmetric Non-negative Tensor Factorization. In: Leonardis, A., Bischof, H., Pinz, A. (eds.) ECCV 2006. LNCS, vol. 3954, pp. 595–608. Springer, Heidelberg (2006)
15. Björck, A.: Numberical methods for least squares problems. In: Proc. SIAM (1996)
16. Zhang, Z., Hancock, E.R.: Hypergraph based Information-theoretic Feature Selection. Pattern Recognition Letters (2012)

Feature Selection Using Counting Grids: Application to Microarray Data

Pietro Lovato[1], Manuele Bicego[1], Marco Cristani[1],
Nebojsa Jojic[2], and Alessandro Perina[2]

[1] Computer Science Department, University of Verona (ITALY)
[2] Microsoft Research (US)

Abstract. In this paper a novel feature selection scheme is proposed, which exploits the potentialities of a recent probabilistic generative model, the Counting Grid. This model is able to cluster together similar observations, highlighting the compactness of a class and its underlying structure. The proposed feature selection scheme is applied to the expression microarray scenario, a peculiar context with very few patterns and a huge number of features. Experiments on benchmark datasets show that the proposed approach is effective and stable, assessing state-of-the-art classification accuracies.

Keywords: feature selection, gene selection, generative models.

1 Introduction

Feature selection techniques definitely represent an important class of prepro-cessing tools in many Pattern Recognition applications: such methods, by elim-inating uninformative features, can reduce the dimension of the problem space, thus alleviating the curse of dimensionality issue [1]. Further, there are appli-cation fields – like biology, where everyday lab procedures generate enormous amount of data to be processed – where it is inconceivable to devise an analysis procedure which does not comprise a feature selection step. A clear example can be found in the analysis of expression microarray data, where the expression level of thousands of genes is simultaneously measured. A typical classification task implies few dozens of samples, each one characterized by the expression level of thousands of genes (i.e. few points in a huge dimensional space). In this con-text, feature selection techniques are even more important, since they can help the medical/biological researchers in identifying a stable and informative set of biomarkers for cancer diagnosis, prognosis, and therapeutic targeting [2, 3].

A large amount of approaches have been introduced in the past in the feature selection field. Broadly, they can be divided in three major classes, depending on how they interact with the classification technique. Filter approaches do not interact with the classifier system, and perform selection just by looking at the intrinsic properties of data. Usual examples are ranking of the features according to criteria which spans from simple variance up to complicates statistics [2, 4].

G.L. Gimel' farb et al. (Eds.): SSPR & SPR 2012, LNCS 7626, pp. 629–637, 2012.

Wrapper methods interact with a specific model trained on the subset of features, using metrics such as the classifier performance / error estimate to assess the quality of the selected features. Finally, in embedded techniques, the search for an optimal subset of features is built into the classifier construction. In the popular SVM-RFE algorithm [5], the weight given to each feature by the SVM classifier is used as a score to rank features, from the most important to the less important. In the specific field of expression microarray – where the feature selection is called gene selection – a common problem of most methods proposed in the past is the stability of the extracted features/genes: actually, datasets which differ by a few samples can lead to complete different sets of genes selected by the feature selection algorithm, still guaranteeing good classification performances [6]. This issue has been often disregarded and has been addressed only recently [7, 8].

This paper presents a novel feature selection scheme, which is based on the Counting Grid (CG) model [9] – a probabilistic model which clusters together similar observations, highlighting the compactness of a class and its underlying structure. The proposed approach is specifically thought for the microarray scenario, which is characterized by the presence of few points in a very high dimensional space. In fact, in [9] it has been show that CGs provide a rich and powerful description of a microarray dataset: samples and gene expressions can be placed on an N-dimensional grid; samples coming from the same class are placed close together in this grid, allowing easy and interpretable visualization of the transition from one class to the other, which turns out to be smooth in most of the cases. In this paper we make one step ahead along this direction, proposing a method which starts from the embedding of the data into the grid and permits to gain insights into which genes characterize a particular class. In fact, starting from the dense embedding of the data provided by the CG, *i)* we embed the class label on the grid, *ii)* we highlight the directions of maximum variation between classes by means of directional derivatives, and finally *iii)* we rank the genes based on how much they vary along these directions. Eventually the ranking is used to extract a stable set of genes for classification or biomarker identification. A further important note concerns the assumption made by most of the gene selection techniques about the independence between genes (actually the typical approach is to rank individually the genes): actually this assumption oversimplifies the complex relationship between genes – which are well known to interact with each other through gene regulative networks. Therefore, models like Counting Grid which can measure and consider the relation and the influence between genes should be preferred.

The experimental evaluation, performed on well-known datasets and compared with state-of-the-art methodologies, shows the suitability of the proposed approach in terms of classification accuracy. Furthermore, to assess the stability of the selected genes, we show that slight alterations in the composition of the training set do not change the selected features, giving confidence that the genes may be somehow involved in the pathology of interest.

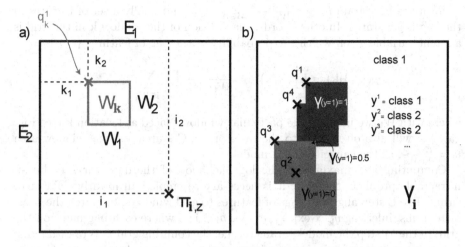

Fig. 1. a) An example of a counting grid geometry. b) Label Embedding γ_i.

2 Background: Counting Grid Model

In Pattern Recognition, data samples are often represented as bags of features without particular order; each t-th observation is characterized by a vector – often called count vector $\{c_z^t\}$ – containing the number of occurrences of each feature z [10, 11]. For example, a text document may be described by the number of occurrences of the different words it contains (or an image with the number of occurrences of different visual features it contains). This choice is often motivated by the difficulty or computational efficiency of modeling the known structure of the data. Concerning microarray, it has been shown in [12–14] that the bag-of-features representation is well-suited also for microarray data, providing interpretable and descriptive signatures. Each sample can be seen as an independent observation; the gene expression value is then interpreted as the "count" of that gene in the sample: the higher the expression level, the "more present" the gene is in such experiment.

The counting grid model, recently introduced in [9], is a generative model for such representations. Formally, the basic counting grid $\pi_{i,z}$ is a set of normalized counts of features indexed by z on the 2-dimensional[1] discrete grid indexed by $\mathbf{i} = (i, j)$ where $i \in [1 \ldots E_1]$, $j \in [1 \ldots E_2]$ and $\mathbf{E} = [E_1, E_2]$ describes the extent of the counting grid. Since π is a grid of distributions, $\sum_z \pi_{i,z} = 1$ everywhere on the grid (see Fig.1a for an illustration).

A given bag of features, represented by counts $\{c_z\}$ is assumed to follow a count distribution found in a patch of the counting grid. In particular, using a window of dimensions $\mathbf{W} = [W_1, W_2]$, each bag can be generated by first selecting a position $\mathbf{k}$ on the grid and then by placing the window in the grid such that $\mathbf{k}$ is its upper left corner. Then, all counts in this patch are averaged

---

[1] N-dimensional in general, here we focus on 2 dimensions.

to form the histogram $h_{\mathbf{k},z} = \frac{1}{W_1 \cdot W_2} \sum_{i \in W_\mathbf{k}} \pi_{i,z}$, and finally a set of features in the bag is generated. In other words, the position of the window $\mathbf{k}$ in the grid is a latent variable given which the probability of the bag of features $\{c_z\}$ is

$$p(\{c_z\}|\mathbf{k}) = \prod_z (h_{\mathbf{k},z})^{c_z} = \frac{1}{W_1 \cdot W_2} \prod_z (\sum_{i \in W_\mathbf{k}} \pi_{i,z})^{c_z}$$

where with $W_\mathbf{k}$ we indicate the particular window placed at location $\mathbf{k}$ (see Fig. 1a). We will also often refer to the ratio of the CG area and the window area $\kappa = \frac{E_1 \cdot E_2}{W_1 \cdot W_2}$, as the capacity of the model.

Computing and maximizing the log likelihood of the data turns to be an intractable problem; therefore it is necessary to employ an iterative EM algorithm. The E step aligns all bags of features to grid windows, to match the bags' histograms, inferring $q_\mathbf{k}^t \propto \exp \sum_z c_z^t \cdot \log h_{i,z}$, i.e., where each bag maps on the grid. In the M-step the model parameter, i.e. the counting grid π, is re-estimated. To avoid severe local minima it is important to consider the Counting Grid as a torus, and perform all windowing operation accordingly. For details on the learning algorithm and on its efficiency see [9].

3   The Proposed Approach

Once a Counting Grid is learned, each sample can be mapped on it through $q_\mathbf{k}^t$, which represents a map telling which part of the CG has more likely generated the pattern t. As a first step of our procedure, we can map all samples belonging to the same class to the CG, trying to obtain a class-related averaged map. This step in [9] has been called class labels embedding, where the goal was to embed the samples' class labels $y^t = l, l = [1, \ldots, L]$ to obtain a posterior probability of each class $p(l|\mathbf{i}) = \gamma_l(\mathbf{i})$ in each position $\mathbf{i}$: this indicates which positions of the CG better "explain" that class. This is achieved using the posterior probabilities $q_\mathbf{k}^t$ already inferred like illustrated in Fig.1b and described by Eq.1

$$\gamma_l(\mathbf{i}) = \frac{\sum_t \sum_{\mathbf{k}|i \in W_\mathbf{k}} q_\mathbf{k}^t \cdot [y^t = l]}{\sum_t \sum_{\mathbf{k}|i \in W_\mathbf{k}} q_\mathbf{k}^t} \tag{1}$$

where $[\cdot]$ is the indicator function, which returns 1 if sample t belongs to class l and 0 otherwise. Roughly speaking, the main idea is to "average" all the mappings $q_\mathbf{k}^t$ of the training samples belonging to a given class. If the CG is able to capture the underlying behaviour of a specific class, then all the mappings will be more or less coherent, and only a part of this averaged map will be different than zero, possibly in a spatially coherent small region – the region which more likely "explains" the training patterns of that class. In order to clarify this concept, in Fig. 2a we show the label embedding for the prostate cancer dataset [15], which comprises two classes. In the figure the tumoral class is embedded. Please observe that the active (non zero) locations are all grouped in spatially coherent zones of the averaged map. Therefore, even if the labels are not used

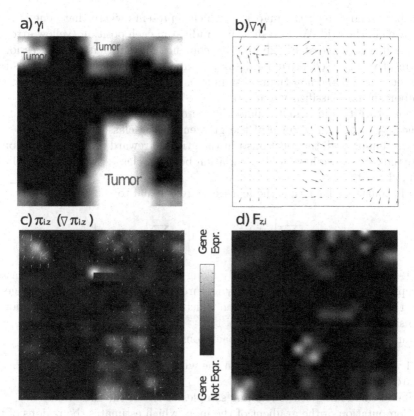

Fig. 2. a) Label embedding γ_i. b) Gradient of the embedding. c) Counting grid for a particular gene (π_z) and its gradient. d) $F_{z,i}$.

during the learning of the CG, tumoral and non-tumoral samples are naturally separated (since we are in a two class problem, the embedding of non tumoral clas is simply obtained by reversing this image); this suggests that indeed CGs are suitable to describe the latent structure which generates the data.

As a second step, we compute the gradient of the embedding, $\nabla\gamma_i$, which returns information about where and how the classes separates (see Fig.2b). In this case the idea is to find which are the regions in the CG where the first class "translates" to the second class or vice versa. Please note that, in the two class case, we only need to compute the gradient on one map, since the map of the second class is just the complementary of the first. Even if the generalization to the multiclass case is somehow straightforward (for example 1 versus all embeddings, or others), for simplicity here we present the two class case.

As a final step, to get the feature score F_z, upon which we will base our feature selection strategy, we rank the genes depending on how much their expression vary along the borders between the classes. The idea is straightforward: to discriminate between the two classes the most useful features are the ones which vary most where we have the class transition. For example in Fig.2c we show for a particular gene $\hat{z}$ the map $\pi_{\hat{z},\mathbf{i}}$, which represents where that gene is more expressed in the grid. We also show its gradient in each position (yellow arrows). After a quick glance at Fig.2b one can convince himself that the expression of $\hat{z}$ is mostly expressed in tumoral samples and often varies where a transition between tumoral and non-tumoral samples is present; that suggets that the gene is important for classification and related to the disease.

To capture this idea mathematically we compute the directional derivatives of the $\pi_{z,\mathbf{i}}$ in the direction v of the gradient of the class embedding $v = \nabla \gamma_{\mathbf{i}}$ and we sum over all the locations $\mathbf{i}$ in the grid. To reward more the variation in expression where we have a high variation between classes, we also multiply by the module of v.

In formulae we have that the feature score is equal to:

$$F_z = \sum_{\mathbf{i}} \left| |v| \cdot \frac{v}{|v|} \cdot \nabla \pi_{z,\mathbf{i}} \right| = \sum_{\mathbf{i}} \left| v \cdot \nabla \pi_{z,\mathbf{i}} \right| \tag{2}$$

In Fig.2d we show that $F_{\hat{z},\mathbf{i}} \neq 0$ only along the borders between the 2 classes. F_z represents the rank score of every feature, which permits to order the genes from the most prominent (i.e. the one which varies the most in the direction of "transition" of the classes) to the least.

Summarizing, the proposed approach consists in the following steps:

1. Training of the Counting Grid on the whole dataset (generative step, labels are not used)
2. Label embedding of the training samples of one of the two classes
3. Computation of the gradient of the map, which estimates the regions of the maps where there is the transition from one class to the other
4. Computation in such zones of the gradient of the genes
5. As a final score, each gene is ranked by its averaged variation in the direction where the two classes vary most.

4   Experimental Evaluation

We tested the proposed approach on two well-known microarray benchmark datasets for two-class problems; a brief description can be found on table 1.

Table 1. Summary of the datasets used

Name	N. Features (genes)	N. Samples	Reference
Colon	2000	62 (40-22)	[16]
Prostate	6033	102 (50-22)	[15]

Since, as a base level, we are mostly interested in the quality of unsupervised learning of the distributions over the microarray samples, the whole dataset

Table 2. Classification results (AUC) for the dataset used

Colon dataset

Sel. Method	Gene Signature Size				
	10	50	100	150	200
SVM-RFE [8]	76.4	77.5	79.2	79.4	80.1
Ens.SVM-RFE [8]	80.3	79.4	78.6	78.6	79.4
SW SVM-RFE [8]	79.5	81.2	78.4	76.2	76.2
ReliefF [8]	78.8	80.1	78.5	77.5	76.1
Ens. ReliefF [8]	78.9	80.2	79.1	77.3	76.1
SW ReliefF [8]	78.3	79.6	78.1	76.4	75.4
[7]	**85.0**	86.0	87.0	87.5	86.5
Our method	81.38	**89.53**	**89.64**	**89.25**	**88.97**

Prostate dataset

Sel. Method	Gene Signature Size				
	10	50	100	150	200
SVM-RFE [8]	89.8	91.3	92.1	92.1	92.2
Ens.SVM-RFE [8]	92.9	92.0	92.0	92.6	92.7
SW SVM-RFE [8]	93.4	91.3	90.0	90.7	91.2
ReliefF [8]	93.3	93.0	91.4	91.4	91.7
Ens. ReliefF [8]	93.4	92.4	91.4	91.0	91.9
SW ReliefF [8]	93.3	92.7	91.4	91.3	91.4
[7]	**95.5**	**96.0**	**95.0**	94.0	94.0
Our method	78.21	88.30	92.45	**94.99**	**95.73**

has been used to train a CG (of course labels are ignored in this phase), in a transductive way [14, 17]. Then, in order to have a fair comparison with the state-of-the-art, we adopted the testing protocol of [8]: the data set was randomly split 2:3/1:3 (training/testing). Labels have been embedded in the Counting Grid, the score F_z has been calculated for each gene z and the top-ranked genes have been extracted, ranging in the values [10 50 100 200]. In order to have a fair evaluation, the gene ranking has been calculated using only the training samples, and applied to the testing samples. The classification is performed using a linear SVM with the parameter C = 1, using the area under the ROC curve (AUC) as an estimate for the classification performance. The test has been repeated 100 times, and the mean of the computed AUCs is shown in table 2, along with comparative state-of-the-art results (see the references between brackets). As for the Counting Grid size, we varied its dimensions by selecting κ between 5 and 40, reporting in the table the mean of the obtained AUCs.

From table 2 it is evident that the proposed approach produces results comparable, and in many cases superior, with state-of-the-art techniques. Furthermore, we assessed the stability of the selected features using the Kuncheva index [18]. The idea is to compare the subsets of genes extracted while varying the training/testing splitting. Given two sets of features $\mathbf{f}_1$ and $\mathbf{f}_2$, the stability index is defined as follows:

Table 3. Stability of the proposed approach

Colon dataset

Feat. Sel.	Gene Signature Size				
	10	50	100	150	200
Best [8]	0.78	0.75	0.70	0.69	0.67
[7]	0.65	0.59	0.58	0.61	0.62
Our method	**0.94**	**0.92**	**0.92**	**0.91**	**0.91**

Prostate dataset

Feat. Sel.	Gene Signature Size				
	10	50	100	150	200
Best [8]	0.68	0.65	0.68	0.68	0.69
[7]	0.72	0.72	0.73	0.72	0.71
Our method	**0.90**	**0.94**	**0.96**	**0.96**	**0.96**

$$KI(\mathbf{f}_1, \mathbf{f}_2) = \frac{r - (s^2/N)}{s - (s^2/N)} \tag{3}$$

where s denotes the signature size, $r = |\mathbf{f}_1 \cap \mathbf{f}_2|$ and N is the total number of genes in the dataset. The Kuncheva index takes values in [-1, 1], and the higher its value, the larger the number of commonly selected genes in both signatures. The index is shown in Table 3, for our approach and other methods. Since the proposed approach is aimed at explaining the data through a generative model, and labels are used later on, the stability index is very high: for both datasets and all different signature sizes, it is always above 0.9, while the best result found in the references we used for comparison is 0.78.

5   Conclusions

In this paper we presented a filter algorithm to perform feature selection, which is based on the recently proposed Counting Grid generative model. The representation given by this model in terms of patterns placed on a 2-dimensional grid has been tailored to derive a new feature selection algorithm. We applied the proposed approach to expression microarray data validating through a series of experiments on benchmark microarray datasets found in the literature. Obtained results were satisfactory.

References

1. Duda, R., Hart, P., Stork, D.: Pattern Classification, 2nd edn. John Wiley & Sons (2001)
2. Guyon, I., Elisseeff, A.: An introduction to variable and feature selection. Journal of Machine Learning Research 3, 1157–1182 (2003)

3. Saeys, Y., Inza, I., Larraaga, P.: A review of feature selection techniques in bioinformatics. Bioinformatics 23(19), 2507–2517 (2007)
4. Thomas, J., Olson, J., Tapscott, S., Zhao, L.: An efficient and robust statistical modeling approach to discover differentially expressed genes using genomic expression profiles. Genome Research 11, 1227–1236 (2001)
5. Guyon, I., Weston, J., Barnhill, S., Vapnik, V.: Gene selection for cancer classification using support vector machines. Machine Learning 46, 389–422 (2002)
6. Li, T., Zhang, C., Ogihara, M.: A comprehensive study on feature selection and multiclass classification methods for tissue classifcation based on gene expression. Bioinformatics 20, 2429–2437 (2004)
7. Abeel, T., Helleputte, T., de Peer, Y.V., Dupont, P., Saeys, Y.: Robust biomarker identification for cancer diagnosis with ensemble feature selection methods. Bioinformatics 26, 392–398 (2010)
8. Yu, L., Han, Y., Berens, M.: Stable gene selection from microarray data via sample weighting. IEEE Transaction on Computational Biology and Bionformatics 9, 262–272 (2012)
9. Jojic, N., Perina, A.: Multidimensional counting grids: Inferring word order from disordered bags of words. In: Uncertainty in Artificial Intelligence (2011)
10. Salton, G., McGill, M.: Introduction to Modern Information Retrieval. McGraw-Hill, New York (1983)
11. Blei, D.M., Ng, A.Y., Jordan, M.I.: Latent dirichlet allocation. J. Mach. Learn. Res. 3, 993–1022 (2003)
12. Rogers, S., Girolami, M., Campbell, C., Breitling, R.: The latent process decomposition of cdna microarray datasets. IEEE/ACM Transactions on Computational Biology and Bioinformatics (2005)
13. Bicego, M., Lovato, P., Oliboni, B., Perina, A.: Expression microarray classification using topic models. In: SAC, pp. 1516–1520 (2010)
14. Perina, A., Lovato, P., Cristani, M., Bicego, M.: A Comparison on Score Spaces for Expression Microarray Data Classification. In: Loog, M., Wessels, L., Reinders, M.J.T., de Ridder, D. (eds.) PRIB 2011. LNCS, vol. 7036, pp. 202–213. Springer, Heidelberg (2011)
15. Singh, D., Febbo, P., Ross, K., Jackson, D., Manola, J., Ladd, C., Tamayo, P., Renshaw, A., D'Amico, A., et al.: Gene expression correlates of clinical prostate cancer behavior. Cancer Cell 98, 203–209 (2002)
16. Alon, U., Barkai, N., Notterman, D., Gish, K., Ybarra, S., Mack, D., Levine, A.: Broad patterns of gene expression revealed by clustering analysis of tumor and normal colon tissues probed by oligonucleotide arrays. Proc. Natl. Acad. Sci. 96, 6745–6750 (1999)
17. Vapnik, V.: Statistical Learning Theory. Wiley, New York (1998)
18. Kuncheva, L.: A stability index for feature selection. In: IASTED International Multi-Conference Artificial Intelligence and Applications, pp. 390–395 (2007)

Infinite Sparse Factor Analysis for Blind Source Separation in Reverberant Environments

Kohei Nagira, Takuma Otsuka, and Hiroshi G. Okuno

Graduate School of Informatics, Kyoto University, Kyoto, Japan
{knagira,ohtsuka,okuno}@kuis.kyoto-u.ac.jp

Abstract. Sound source separation in a real-world indoor environment is an ill-formed problem because sound source mixing is affected by the number of sounds, sound source activities, and reverberation. In addition, blind source separation (BSS) suffers from a permutation ambiguity in a frequency domain processing. Conventional methods have two problems: (1) impractical assumptions that the number of sound sources is given, and (2) permutation resolution as a post processing. This paper presents a non-parametric Bayesian BBS called permutation-free infinite sparse factor analysis (PF-ISFA) that solves the two problems simultaneously. Experimental results show that PF-ISFA outperforms conventional complex ISFA in all measures of BSS_EVAL criteria. In particular, PF-ISFA improves Signal-to-Interference Ratio by 14.45 dB and 5.46 dB under $RT_{60} = 30$ ms and $RT_{60} = 460$ ms conditions, respectively.

Keywords: Blind source separation, Reverberant mixtures, Infinite sparse factor analysis, Non-parametric Bayes.

1   Introduction

Machine listening functions, e.g. a robot audition system [1] or a distant speech recognition [2], cannot dispense with a sound source separation technique because we often observe a mixture of sound sources. For instance, HARK [1],a robot audition software, provides functions of source localization, separation, and recognition of separated speech signals. Since HARK may be deployed to various kinds of acoustic environments, parameter tuning is critical to avoid performance degradation.

In order to maximize the availability of a source separation function, the following requirements should be fulfilled for the application to practical environments:

1. source separation under an unknown mixing process dependent on the locations of sources and microphones,
2. separation under the condition of unknown number of sources,
3. robustness against the reverberation.

Many source separation methods need prior information such as the number of sources or the mixing process. Since prior information is usually difficult to obtain in advance, source separation methods should work without prior information, or at least with minimal prior information. Such a separation method is called **blind source separation** (BSS).

G.L. Gimel' farb et al. (Eds.): SSPR & SPR 2012, LNCS 7626, pp. 638–647, 2012.

For sound source separation in a practical environment, the system should separate mixtures of reverberant speeches. This is because the mixed signals captured by microphones are affected by room reverberation.

Frequency domain processing is effective to separate reverberant mixed signals, and a lot of frequency domain BSS systems are proposed. One of the problems of frequency domain processing is a permutation problem [3]. Conventional frequency domain BSS systems separates signals for all frequency bins independently, and consequently, a permutation ambiguity arises in output orders for all frequency bins. The source separation system should resolve this permutation ambiguity to reconstruct separated signals.

Independent component analysis (ICA) [4] is a well known BSS method. Frequency domain ICA [5] fulfills the first and the third requirements. However, ICA assumes the number of sources because ICA cannot detect source activities. This means that ICA does not satisfy the second requirement. In addition, Frequency domain ICA suffers from the permutation problem. Independent vector analysis (IVA) [6] and permutation free ICA [7] are BSS methods avoiding permutation problem. These methods are based on ICA and also assume the number of sources. Thus they do not satisfy the second requirement. In our previous work, frequency domain infinite sparse factor analysis (FD-ISFA) is proposed [8]. This method achieves all the three requirements, but the separation quality of FD-ISFA may be deteriorated by the subsequent permutation resolution process.

This paper presents permutation free ISFA (PF-ISFA), a BSS method which meets all the requirements and offers permutation resolution. PF-ISFA is based on nonparametric Bayesian framework, which allows BSS under the uncertainty of source numbers. The key idea of our method is that all frequency bins of signals are processed at a time by introducing a unified source activity variable for the joint optimization of the separation and permutation resolution.

2 BSS in Frequency Domain

2.1 Problem Statement of BSS

The problem of BSS is stated as below:

Input: Sound mixtures of K sources captured by D microphones.
Output: Estimated K source signals
Assumption: K is not more than D.
The locations of microphones and those of sources are fixed.

The system extracts K source signals from the mixture signal captured with D microphones without prior information of mixing process such as the location of sources, the location of microphones, and impulse responses between microphone and sound sources.

2.2 Frequency Domain Processing and Permutation Problem

In real environment with reverberation, the mixing process of speech signal is convolutive. The observed signals consist of a mixture of sources and they are contaminated by their reverberations. To model these time-delayed signals, the convoluted mixture is often employed.

$$\overline{\mathbf{x}}(t) = \sum_{j=0}^{J} \overline{\mathbf{A}}(j)\overline{\mathbf{s}}(t-j) \tag{1}$$

where $\overline{\mathbf{x}}(t)$, $\overline{\mathbf{s}}(t)$, and $\overline{\mathbf{A}}(j)$ are observed signals, source signals, and transfer function coefficients in the time domain, respectively. BSS problem aims to retrieve that constituent sound sources $\overline{\mathbf{s}}(t)$ only given the observation $\overline{\mathbf{x}}(t)$ where the mixing process including the reverberation $\overline{\mathbf{A}}(j)$ is unknown.

When solving a BSS problem involving convoluted mixtures of signals, short time Fourier transform (STFT) is often applied in order to convert a convoluted mixture in a time domain into an instantaneous mixture in a frequency domain. In the case that signals are separated for each frequency bin independently in frequency domain, the permutation ambiguity of output order of separated signals has to be solved. This is called "permutation problem" [3]. The permutation problem is one of the well-known problems of frequency-domain BSS.

Some methods are proposed to solve this problem. One method is based on the direction of arrival estimation and the inter-frequency correlation of signal envelopes [3], and another uses power ratio of signals as dominance measure [9]. However, existing permutation resolution as a post-processing of a frequency-wise separation process, the resulting sound sources may severely be affected by the preceding separation quality. For example, if the frequency-wise separation is deteriorated by the reverberation, the permutation resolution by the signal envelopes fails, which results in the failure of BSS as well.

3   Permutation-Free ISFA

3.1   Outline of Our System

The flow of PF-ISFA is depicted in Fig. 1. After STFT, the complex spectra are whitened in each frequency bin, and PF-ISFA is applied to these whitened signals. The output order of PF-ISFA is already aligned, but the amplitude of the output signals may not equals to that of original sources. This is called scaling ambiguity, and this is another well-known problem of frequency domain BSS. The projection back method [10] is an effective solution for this problem. After projection back processing, the separated signals are reconstructed by inverse STFT.

3.2   Generative Model and Likelihood of PF-ISFA

Let K, D, F, and T be the number of sources, the number of microphones, the number of frequency bins, and the length of the source signals, respectively. The ISFA model is based on the instantaneous mixture model:

$$\mathbf{X}_f = \mathbf{A}_f(\mathbf{Z}_f \odot \mathbf{S}_f) + \mathbf{E}_f \ (f = 1, \cdots, F), \tag{2}$$

where $\mathbf{Z}_f = [\mathbf{z}_{f1}, \cdots \mathbf{z}_{fT}]$, $\mathbf{X}_f = [\mathbf{x}_{f1}, \cdots \mathbf{x}_{fT}]$, $\mathbf{S}_f = [\mathbf{s}_{f1}, \cdots \mathbf{s}_{fT}]$, $\mathbf{E}_f = [\varepsilon_{f1}, \cdots \varepsilon_{fT}]$, $\mathbf{x}_{ft} = [x_{1ft}, x_{2ft}, \cdots, x_{Dft}]^T$ is a mixed signal vector at time t, $\mathbf{s}_{ft} = [s_{1ft}, s_{2ft}, \cdots, s_{Kft}]^T$

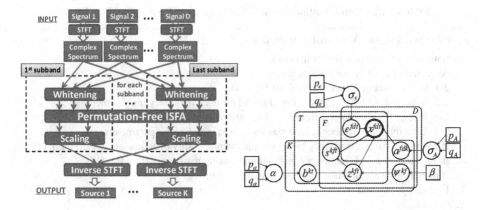

Fig. 1. Schematic overview for our method   **Fig. 2.** Graphical model of PF-ISFA

is the source signal vector, and $\varepsilon_{ft} = [\varepsilon_{1ft}, \varepsilon_{2ft}, \cdots, \varepsilon_{Dft}]^{\mathrm{T}}$ is the Gaussian noise vector. Here, $\mathbf{A}_f$ is the $D \times K$ mixing matrix, $\mathbf{z}_{ft} = [z_{1ft}, z_{2ft}, \cdots, z_{Kft}]^{\mathrm{T}}$ is the activity of each source at time t in f-th frequency bin, and source activity z_{kft} is a binary variable: $z_{kft} = 1$ if source k is active at time t in f-th frequency bin, otherwise $z_{kft} = 0$. Operator $\odot$ indicates the element-wise product. PF-ISFA deals with F-tuple frequency bins at the same time. $\mathbf{Z}$, $\mathbf{X}$, $\mathbf{S}$. $\mathbf{E}$, and $\mathbf{A}$ are defined as $\mathbf{Z} = [\mathbf{Z}_1, \cdots \mathbf{Z}_F]$, $\mathbf{X} = [\mathbf{X}_1, \cdots \mathbf{X}_F]$, $\mathbf{S} = [\mathbf{S}_1, \cdots \mathbf{S}_F]$, $\mathbf{E} = [\mathbf{E}_1, \cdots \mathbf{E}_F]$, and $\mathbf{A} = [\mathbf{A}_1, \cdots \mathbf{A}_F]$, respectively.

To unify activities of all frequency bins, the following model is introduced.

$$z_{kft} = b_{kt}\phi, \quad \phi \sim \mathrm{Bernoulli}(\psi_{kf}), \tag{3}$$

where Bernoulli(x) is the Bernoulli distribution with parameter x. b_{kt} is the unified source activity of source k at time t, and ψ_{kf} is a activation probability of source k in the f-th frequency bin. $\mathbf{B}$ represents $K \times T$ matrix of b_{kt} and Ψ means $K \times F$ matrix of ψ_{kf}.

PF-ISFA estimates the source signals $\mathbf{S}$, their time-frequency activities $\mathbf{Z}$, mixing matrix $\mathbf{A}$, unified activities $\mathbf{B}$, activation probability Ψ, and other parameters by using only the observed signal $\mathbf{X}$.

The prior distributions of the variables are assumed as follows:

$$\varepsilon_{ft} \sim \mathcal{N}_C(0, \sigma_\varepsilon^2 \mathbf{I}), \quad \sigma_\varepsilon^2 \sim \mathcal{IG}(p_\varepsilon, q_\varepsilon), \tag{4}$$

$$s_{kft} \sim \mathcal{N}_C(0, 1), \tag{5}$$

$$\mathbf{a}_{kf} \sim \mathcal{N}_C(0, \sigma_\mathbf{A}^2 \mathbf{I}), \quad \sigma_\mathbf{A}^2 \sim \mathcal{IG}(p_A, q_A), \tag{6}$$

$$\mathbf{B} \sim \mathrm{IBP}(\alpha), \quad \alpha \sim \mathcal{G}(p_\alpha, q_\alpha), \text{ and} \tag{7}$$

$$\Psi \sim \mathrm{Beta}(\beta/K, \beta(K-1)/K). \tag{8}$$

Table 1. Algorithm for estimating model parameters of Permutation-Free ISFA

Input: Observed signals $\mathbf{X}$, Output: Source signals $\mathbf{S}$.

1. Initialize parameters using their priors.
2. At each time t, carry out the following:

   2-1 In each source k, sample b_{kt} from Eq. (14).

   2-2 If $b_{kt} = 1$, sample z_{kft} from Eq. (11) and for each frequency bin f; otherwise $z_{kft} = 0$.

   2-3 If $z_{kft} = 1$, sample s_{kft} from Eq. (10); otherwise $s_{kft} = 0$.

   2-4 Determine the number of new classes κ_t, and initialize the parameters.
3. In each source k and frequency bin f, sample the probability of activation ψ_{kf} from Eq. (16).
4. In each source k and frequency bin f, sample mixing matrix $\mathbf{a}_{kf}$ from Eq. (17).
5. If there is a source that is always inactive, remove it.
6. Update σ_ε^2, σ_A^2, and α.
7. Go to 2.

Here, $\mathbf{a}_{fk}$ is the kth row of $\mathbf{A}_f$, and p_ε, q_ε, p_A, q_A, p_α, q_α, and β are the hyperparameters. $\mathcal{N}_C$, $\mathcal{G}$, $\mathcal{IG}$ are the univariate complex normal, gamma, inverse gamma distributions, respectively. The prior for the variance of each parameter is the inverse Gamma since the inverse Gamma distribution is conjugate to the normal distribution. IBP(α) is the Indian buffet process (IBP) [11] with parameter α. IBP is a stochastic process that provides the probability distribution over sparse binary matrices with infinite number of columns. Therefore, IBP can deal with a potentially infinite number of signals.

The likelihood function of PF-ISFA is written as follows.

$$P(\mathbf{X}|\mathbf{A},\mathbf{S},\mathbf{Z}) = \prod_{f=1}^{F}\prod_{t=1}^{T} P(\mathbf{x}_{ft}|\mathbf{A}_f,\mathbf{s}_{ft},\mathbf{z}_{ft}) = \prod_{f=1}^{F}\prod_{t=1}^{T} \mathcal{N}_C(\mathbf{x}_{ft};\mathbf{A}_f(\mathbf{z}_{ft}\odot\mathbf{s}_{ft}),\sigma_\varepsilon^2\mathbf{I})$$

$$= \prod_{f=1}^{F}\frac{1}{(\pi\sigma_\varepsilon^2)^{TD}}\exp\left(-\frac{\mathrm{tr}(\mathbf{E}_f^H\mathbf{E}_f)}{\sigma_\varepsilon^2}\right). \tag{9}$$

where $\mathbf{E}_f = \mathbf{X}_f - \mathbf{A}_f(\mathbf{Z}_f\odot\mathbf{S}_f)$. Here, each data point is assumed to be independent and identically distributed.

3.3 Source Separation through the Inference of Latent Variables

The model parameters of PF-ISFA are estimated using an iterative algorithm. The algorithm is given in Table 1, and a graphical model of PF-ISFA is shown in Fig. 2. This method is based on the Metropolis-Hastings algorithm. Posterior distributions of latent variables are derived from Bayes' theorem by multiplying priors by the likelihood function.

Sound Sources. When z_{kft} is active, s_{kft} is sampled by the following posterior.

$$P(s_{kft}|\mathbf{A}_f,\mathbf{s}_{-kft},\mathbf{x}_{ft}\mathbf{z}_{ft}) \propto P(\mathbf{x}_{ft}|\mathbf{A}_f,\mathbf{s}_{ft},\mathbf{z}_{ft},\sigma_\varepsilon^2)P(s_{kft}) = \mathcal{N}_C\left(s_{kft};\mu_{s,f},\sigma_{s,f}^2\right), \tag{10}$$

where

$$\sigma_{s,f}^2 = \sigma_\varepsilon^2/\left(\sigma_\varepsilon^2 + \mathbf{a}_{kf}^H\mathbf{a}_{kf}\right), \quad \mu_{s,f} = \mathbf{a}_{kf}^H\varepsilon_{-kft}/\left(\sigma_\varepsilon^2 + \mathbf{a}_{kf}^H\mathbf{a}_{kf}\right).$$

$\mathbf{s}_{-kft}$ means $\mathbf{s}_{ft}$ except for s_{kft}, and ε_{-kft} means $\varepsilon|_{z_{kft}=0}$.

Source Activity of Each Time-Frequency Frame. If $b_{kt} = 1$, z_{kft} is sampled from its posterior distribution. The posterior of z_{kft} is calculated as follows.

$$P(z_{kft}|b_{kt}, \psi_{kf}, z_{-kft}, \mathbf{x}_{ft}, \mathbf{s}_{ft}, \mathbf{A}_f) \propto P(z_{kft}|b_{kt}, \psi_{kf})P(x_{ft}|\mathbf{A}_f, \mathbf{s}_{ft}, \mathbf{z}_{ft}, \sigma_\varepsilon^2)$$
$$= \text{Bernoulli}(p_1/(p_0 + p_1)), \tag{11}$$

where

$$\log(p_1) = \log(\psi_{kf}) + (2\,\text{Re}(s_{kft}^* \mathbf{a}_{kf}^H \varepsilon_{-kft}) + |s_{kft}|^2 \mathbf{a}_{kf}^H \mathbf{a}_{kf})/\sigma_\varepsilon^2$$
$$\log(p_0) = \log(1 - \psi_{kf})$$

Unified Activity for Each Time Frame. The ratio of the probability that b_{kt} becomes active to the probability that b_{kt} becomes inactive is calculated by Eq. (12). This ratio r consists of the ratio of prior r_p and the ratio of likelihood of each frequency bin $r_{l,f}$.

$$r = \frac{P(b_{kt} = 1|\mathbf{A}, \mathbf{S}_{-kt}, \mathbf{X}_t, \mathbf{S}_{-kt})}{P(b_{kt} = 0|\mathbf{A}, \mathbf{S}_{-kt}, \mathbf{X}_t, \mathbf{Z}_{-kt})} = r_p \prod_{f=1}^F r_{l,f}. \tag{12}$$

where

$$r_p = \frac{P(b_{kt} = 1|\mathbf{b}_{kt})}{P(b_{kt} = 0|\mathbf{b}_{kt})} = \frac{m_{k,-t}}{T - m_{k,-t}}, \text{and}$$

$$r_{l,f} = \frac{P(\mathbf{x}_{ft}|\mathbf{A}_f, \mathbf{s}_{-kft}, \mathbf{z}_{-kft}, \mathbf{b}_{-kt}, b_{kt} = 1, \psi_{kf})}{P(\mathbf{x}_{ft}|\mathbf{A}_f, \mathbf{s}_{-kft}, \mathbf{z}_{-kft}, \mathbf{b}_{-kt}, b_{kt} = 0, \psi_{kf})} = \psi_{kf}\sigma_{s,f}^2 \exp\left(\frac{|\mu_{s,f}|^2}{\sigma_{s,f}^2}\right) + (1 - \psi_{kf}). \tag{13}$$

where $m_{k,-t} = \sum_{t' \neq t} b_{kt'}$. Here, $\mathbf{X}_t$ is $\mathbf{x}_{1t}, \cdots, \mathbf{x}_{Ft}$, and $\mathbf{S}_{-kt}$ and $\mathbf{Z}_{-kt}$, are $\mathbf{S}$ and $\mathbf{Z}$ except for $s_{k1t}, \cdots, s_{kFt}$ and $z_{k1t}, \cdots, z_{kFt}$, respectively. The ratio of prior r_p is derived from the priors of source activity based on IBP [11].

The posterior probability of $z_{kt} = 1$ is calculated using ratio r.

$$P(b_{kt} = 1|\mathbf{A}, \mathbf{S}_{-kt}, \mathbf{X}_t, \mathbf{Z}_{-kt}, \mathbf{b}_{-kt}) = r/(1+r) \tag{14}$$

To decide whether or not b_{kt} is active, we sample u from Uniform(0,1) and compare it to $r/(1+r)$. If $u \leq r/(1+r)$, b_{kt} becomes active; otherwise it is not.

Number of New Sources. Some source signals that were not active at the beginning are active at time t for the first time. Let κ_t be the number of these sources.

First, the prior distribution of κ_t is $P(\kappa_t|\alpha) = \text{Poisson}\left(\frac{\alpha}{T}\right)$. After sampling κ_t, we initialize new sources and their activities. Next, we decide whether this update is acceptable or not. The acceptance probability of the transition is $\min(1, r_{\xi \to \xi^*})$. According to Meeds [12] and Knowles [13], $r_{\xi \to \xi^*}$ becomes the ratio of the likelihood of the current state to that of the next state. Let $\mathbf{A}_f^*$ be the $D \times \kappa_t$ matrix of the additional part of $\mathbf{A}_f$. The ratio can be calculated as follows.

$$r_{\xi \to \xi^*} = \prod_{f=1}^F (\det \Lambda_{\xi,f})^{-1} \exp\left(\mu_{\xi,f}^H \Lambda_{\xi,f} \mu_{\xi,f}\right), \tag{15}$$

where

$$\Lambda_{\xi,f} = \mathbf{I} + \mathbf{A}_f^{*H} \mathbf{A}_f^* / \sigma_\varepsilon^2, \quad \Lambda_{\xi,f} \mu_{\xi,f} = \mathbf{A}_f^{*H} \varepsilon_{ft} / \sigma_\varepsilon^2.$$

Probability of Activation for Each Frequency Bin. ψ_{kf} is sampled by the following posterior.

$$
\begin{aligned}
P(\psi_{kf}|\mathbf{z}_{kf},\Psi_{-kf},\mathbf{B}_{-kt}) &\propto P(\psi_{kf}|\beta)\prod_{t=1}^{T}P(z_{kft}|\psi_{kf},b_{kt}) \\
&= \mathrm{Beta}\left(n_{kf}+\beta/K, m_k-n_{kf}+\beta(K-1)/K\right),
\end{aligned} \tag{16}
$$

where $n_{kf}=\sum_{t=1}^{T}z_{kft}$ is the number of active time-frequency frames of source k in f-th frequency bin, and $m_k=\sum_{t=1}^{T}b_{kt}$ is the number of active time frames of source k.

Mixing Matrix. The mixing matrix is estimated in each column. The posterior distribution is

$$
\begin{aligned}
P(\mathbf{a}_{kf}|\mathbf{A}_{f,-k},\mathbf{S}_f,\mathbf{X}_f,\mathbf{Z}_f) &\propto P(\mathbf{X}_f|\mathbf{A}_f,\mathbf{S}_f,\mathbf{Z}_f,\sigma_\varepsilon^2)P(\mathbf{a}_{kf}|\sigma_A^2) \\
&= \mathcal{N}_C(\mathbf{a}_{kf};\mu_{\mathbf{A}},\Lambda_{\mathbf{A}}^{-1}),
\end{aligned} \tag{17}
$$

where

$$
\Lambda_{\mathbf{A}} = \left(\frac{\mathbf{s}_{kf}^{H}\mathbf{s}_{kf}}{\sigma_\varepsilon^2}+\frac{1}{\sigma_A^2}\right)\mathbf{I}_D,\quad \mu_{\mathbf{A}} = \frac{\sigma_A^2}{\mathbf{s}_{kf}^{H}\mathbf{s}_{kf}\sigma_A^2+\sigma_\varepsilon^2}\mathbf{E}_f|_{\mathbf{a}_{kf}=0}\mathbf{s}_{kf}.
$$

Variance of Noise and Mixing Matrix. The variance of noise corresponds to the noise level of the estimated signals, and the variance of the mixing matrix affects the scale of the estimated signals. Their posteriors are as follows.

$$
P(\sigma_\varepsilon^2|\mathbf{E}) \propto P(\mathbf{E}|\sigma_\varepsilon^2)P(\sigma_\varepsilon^2|p_\varepsilon,q_\varepsilon) = \mathscr{IG}\left(\sigma_\varepsilon^2;p_\varepsilon+FTD,\frac{q_\varepsilon}{(1+q_\varepsilon\sum_{f=1}^{F}\mathrm{tr}(\mathbf{E}_f^{H}\mathbf{E}_f))}\right). \tag{18}
$$

$$
P(\sigma_A^2|\mathbf{A}) \propto P(\mathbf{A}|\sigma_A^2)P(\sigma_A^2|p_A,q_A) = \mathscr{IG}\left(\sigma_A^2;p_A+FDK,\frac{q_A}{1+q_A\sum_{f=1}^{F}\mathrm{tr}(\mathbf{A}_f^{H}\mathbf{A}_f)}\right). \tag{19}
$$

Parameter of IBP. Since the IBP parameter α can be updated in the same way as FD-ISFA [8], the detailed explanation is omitted here.

4   Experimental Results

We test our method in a separation experiment using speech signals in order to evaluate the separation performance of our method. In this experiment, our method is compared with the baseline method, complex ISFA [8]. We use two kinds of mixed signals for this experiment: convoluted mixture with impulse responses measured in anechoic chamber, and convoluted mixture with impulse responses measured in meeting room ($RT_{60} = 460$ ms). Figure 4 shows the locations of the microphones and sources, and Table 2

Table 2. Experimental conditions

Number of sources K	2
Number of microphones D	2
Test set	ASJ-JNAS
Sampling rate	16000 Hz
Window length	64 ms
Shift length	32 ms
Iterations	300

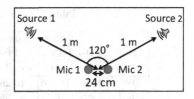

Fig. 3. Locations of microphones and sources in experiment

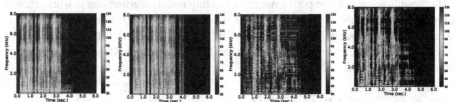

Fig. 4. Spectrogram of original source

Fig. 5. Spectrogram of PF-ISFA result

Fig. 6. Spectrogram of FD-ISFA result

Fig. 7. Spectrogram of permutation-aligned FD-ISFA result

lists the conditions for this experiment. We used 200 utterances from JNAS phoneme balanced sentences on each condition.

First, an example of experimental results of separation experiment using mixed signals in the anechoic chamber is shown. Figures 4–7 show the spectrograms of a source signal, a signal separated with PF-ISFA, a signal separated with conventional FD-ISFA, and a permutation-aligned signal separated with FD-ISFA. In the results of FD-ISFA, many horizontal lines are seen in Figure 6, but in Figure 7, the number of these lines decrease. These lines are the spectrogram of the other separated signal. This means that the output orders of FD-ISFA result are not aligned for all frequency bins. In contrast, there is no horizontal line in the spectrogram of PF-ISFA (Figure 5). This shows that the output order is aligned, in other words the permutation problem is solved by PF-ISFA.

We also evaluate our method in terms of the Signal to Distortion Ratio (SDR), the Image to Spatial distortion Ratio (ISR), the Source to Interference Ratio (SIR), and the Source to Artifacts Ratio (SAR) [14]. Table 3 summarizes the results. "Non-Perm" is calculated by output signals themselves, in other words, their permutations are not aligned. "Perm" means that output signals are aligned their permutations using the correlation between outputs and original sources. In other word, permutation is aligned by using original source signals as reference. Our proposed method outperforms FD-ISFA by all criteria in Non-Perm case. Especially, proposed method improves SIR by 14.45 dB in anechoic chamber reverberations and 5.46 dB in meeting room reverberations.

One of the reasons of poor performance of FD-ISFA is caused by the permutation problem, because the difference between the performance of permutation-aligned results of FD-ISFA and that of FD-ISFA results without aligning permutations is large. In contrast, that of PF-ISFA results is smaller. This means that the permutations of outputs are automatically aligned when PF-ISFA is applied.

Table 3. Average separation performance [dB]

	Anechoic chamber				Meeting room ($RT_{60} = 460$ ms)			
	FD-ISFA		PF-ISFA		FD-ISFA		PF-ISFA	
	Non-Perm	Perm	Non-Perm	Perm	Non-Perm	Perm	Non-Perm	Perm
SDR	0.38	11.96	**10.26**	**12.59**	0.35	5.85	**3.56**	**5.31**
ISR	4.98	18.23	**15.96**	**18.75**	4.73	10.41	**8.08**	**9.88**
SIR	1.38	18.58	**15.83**	**19.20**	1.12	9.86	**6.58**	**9.22**
SAR	5.22	14.39	**13.91**	**15.16**	5.72	10.36	**9.30**	10.36

This results show that the performance in meeting room reverberation is worse than that in anechoic chamber reverberation. This is because the reverberation time of meeting room ($RT_{60} = 460$ ms) is longer than STFT window length (64 ms). If the reverberation time is longer than STFT window length, reverberation affects multiple time frames, and this degrades the performance.

5   Conclusion and Future Work

This paper presented PF-ISFA based on a non-parametric Bayesian framework for reverberant environments. PF-ISFA achieves BSS without the assumptions about observations such as the number of sources, reverberation, and the mixing process. This method is processed in frequency domain to separate reverberant speeches without prior information, and it can avoid permutation problem. Experimental results show that PF-ISFA outperforms conventional FD-ISFA.

Future work includes the following. We focus on the source activity accuracy, and achieve voice activity detection using the source activity estimated by PF-ISFA for an effective speech recognition system. In addition, the time complexity of PF-ISFA should be reduced for an accelerated separation system. If we attain a real-time processing, PF-ISFA can be applied to many applications including robot audition.

Acknowledgement. This study was partially supported by the Grant-in-Aid for Scientific Research (S), and Honda Research Institute Japan Inc., Ltd.

References

1. Nakadai, K., et al.: Design and Implementation of Robot Audition System "HARK" Open Source Software for Listening to Three Simultaneous Speakers. Advanced Robotics 24(5-6), 739–761 (2010)
2. Wölfel, M., et al.: Distant Speech Recognition. Wiley (2009)
3. Sawada, H., et al.: A robust and precise method for solving the permutation problem of frequency-domain blind source separation. IEEE Trans. on Speech and Audio Processing 12(5), 530–538 (2004)
4. Hyvärinen, A., et al.: Independent component analysis. Wiley Interscience (2001)
5. Sawada, H., et al.: Polar coordinate based nonlinear function for frequency-domain blind source separation. In: Proc. of IEEE Intl. Conf. on Acoustics, Speech, and Signal Processing (ICASSP 2002), pp. 1001–1004 (2002)

6. Lee, I., et al.: Fast fixed-point independent vector analysis algorithms for convolutive blind source separation. Signal Processing 87(8), 1859–1871 (2007)
7. Hiroe, A.: Solution of Permutation Problem in Frequency Domain ICA, Using Multivariate Probability Density Functions. In: Rosca, J.P., Erdogmus, D., Príncipe, J.C., Haykin, S. (eds.) ICA 2006. LNCS, vol. 3889, pp. 601–608. Springer, Heidelberg (2006)
8. Nagira, K., Takahashi, T., Ogata, T., Okuno, H.G.: Complex Extension of Infinite Sparse Factor Analysis for Blind Speech Separation. In: Theis, F., Cichocki, A., Yeredor, A., Zibulevsky, M. (eds.) LVA/ICA 2012. LNCS, vol. 7191, pp. 388–396. Springer, Heidelberg (2012)
9. Sawada, H., et al.: Measuring dependence of bin-wise separated signals for permutation alignment in frequency-domain BSS. In: IEEE Intl. Symposium on Circuits and Systems, ISCAS 2007, pp. 3247–3250. IEEE (2007)
10. Murata, N., et al.: An approach to blind source separation based on temporal structure of speech signals. Neurocomputing 41(1-4), 1–24 (2001)
11. Griffiths, T., et al.: Infinite latent feature models and the Indian buffet process. Advances in Neural Information Processing Systems 18, 475–482 (2006)
12. Meeds, E., et al.: Modeling dyadic data with binary latent factors. Advances in Neural Information Processing Systems 19, 977–984 (2007)
13. Knowles, D., Ghahramani, Z.: Infinite Sparse Factor Analysis and Infinite Independent Components Analysis. In: Davies, M.E., James, C.J., Abdallah, S.A., Plumbley, M.D. (eds.) ICA 2007. LNCS, vol. 4666, pp. 381–388. Springer, Heidelberg (2007)
14. Vincent, E., Sawada, H., Bofill, P., Makino, S., Rosca, J.P.: First Stereo Audio Source Separation Evaluation Campaign: Data, Algorithms and Results. In: Davies, M.E., James, C.J., Abdallah, S.A., Plumbley, M.D. (eds.) ICA 2007. LNCS, vol. 4666, pp. 552–559. Springer, Heidelberg (2007)

Sparse Discriminant Analysis Based on the Bayesian Posterior Probability Obtained by L1 Regression

Akinori Hidaka and Takio Kurita

[1] Tokyo Denki University
[2] Hiroshima University

Abstract. Recently the kernel discriminant analysis (KDA) has been successfully applied in many applications. However, kernel functions are usually defined a priori and it is not known what the optimum kernel function for nonlinear discriminant analysis is. Otsu derived the optimum nonlinear discriminant analysis (ONDA) by assuming the underlying probabilities similar with the Bayesian decision theory. Kurita derived discriminant kernels function (DKF) as the optimum kernel functions in terms of the discriminant criterion by investigating the optimum discriminant mapping constructed by the ONDA. The derived kernel function is defined by using the Bayesian posterior probabilities. We can define a family of DKFs by changing the estimation method of the Bayesian posterior probabilities. In this paper, we propose a novel discriminant kernel function based on L1-regularized regression, called L1 DKF. L1 DKF is given by using the Bayesian *posterior* probabilities estimated by L1 regression. Since L1 regression yields a sparse representation for given samples, we can naturally introduce the sparseness into the discriminant kernel function. To introduce the sparseness into LDA, we use L1 DKF as the kernel function of LDA. In experiments, we show sparseness and classification performance of L1 DKF.

1   Introduction

Recently the kernel discriminant analysis (KDA), a non-linear extension of linear discriminant analyasis (LDA), has been successfully applied in many applications [1, 8]. KDA constructs a nonlinear discriminant mapping by using kernel functions. Usually the kernel function is defined a priori, and it is not known what the best kernel function for nonlinear discriminant analysis (NDA) is. Also the class information is usually not introduced in kernel functions.

On the other hand, Otsu derived the optimum nonlinear discriminant analysis (ONDA) by assuming the underlying probabilities [9–11] similar with the Bayesian decision theory [2]. He showed that the optimum nonlinear discriminant mapping was obtained by using variational calculus and was closely related to Bayesian decision theory (The *posterior* probabilities). The optimum nonlinear discriminant mapping can be defined as a linear combination of the Bayesian *posterior* probabilities and the coefficients of the linear combination are obtained

G.L. Gimel' farb et al. (Eds.): SSPR & SPR 2012, LNCS 7626, pp. 648–656, 2012.

by solving the eigenvalue problem of the matrices defined by using the Bayesian *posterior* probabilities.

Kurita showed that the best kernel function is derived from the optimum discriminant mapping constructed by ONDA by investigating the dual problem of the eigenvalue problem of ONDA [7]. The derived kernel function, called the discriminant kernel function (DKF), is also given by using the *posteriori* probabilities. This means the class information is naturally introduced in the kernel function. As like ONDA, the DKF is also optimum in terms of the discriminant criterion. Kurita also showed that a family of DKFs can be defined by changing the estimation method of the Bayesian *posterior* probabilities [7].

Recently, many researchers have actively studied about *sparseness* of features or classifiers [3][13]. It is known that the sparse representation often brings several good properties for classification problems; lower computational load, higher classification accuracy or a feature representation which is easy to interpret.

One of the approach to give sparse representation to existing methods is to introduce the L1-regularized penalty into optimization problems. Based on this approach, Sparse principal component analysis (PCA) by Zou et al. [13] and sparse LDA by Clemmensen et al. [3] were proposed.

In this paper, we propose a novel discriminant kernel function based on L1-regularized regression, called L1 DKF. L1 DKF is obtained by using the Bayesian *posterior* probabilities estimated by L1 regression. Since L1 regression yields a sparse representation for given samples, we can naturally introduce the sparseness into DKF.

We use L1 DKF as the kernel function of LDA to introduce the sparseness into LDA indirectly. Our approach is different from Clemmensen's approach which brings the sparseness into LDA directly [3]. In experiments, we show sparseness and classification performance.

In Sec. 2, we briefly summarize LDA and its nonlinear extensions, KDA and ONDA. In Sec. 3, we describe about discriminant kernels. In Sec. 4, we propose L1 regression based discriminant kernel function. The experiments are shown in Sec. 5. The conclusions are described in Sec. 6.

2   Optimal Nonlinear Discriminant Analysis

2.1   Linear Discriminant Analysis

Linear Discriminant Analysis (LDA) [4] is defined as a method to find the linear combination of features which best separates two classes of objects. LDA is regarded as one of the well known methods to extract the best discriminating features for multi-class classification.

Let an $m-$D feature vector be $\boldsymbol{x} = (x_1, \ldots, x_m)^T$. Consider K classes denoted by $\{C_1, \ldots, C_K\}$. Assume that we have N feature vectors $\{\boldsymbol{x}_i | i = 1, \ldots, N\}$ as training samples and they are labeled as one of the K classes. Then LDA constructs a dimension reducing linear mapping from the input feature vector $\boldsymbol{x}$ to a new feature vector $\boldsymbol{y} = A^T \boldsymbol{x}$ where $A = [a_{ij}]$ is the coefficient matrix.

The objective of LDA is to maximize the discriminant criterion,

$$J = \mathrm{tr}(\hat{\Sigma}_T^{-1} \hat{\Sigma}_B) \tag{1}$$

where $\hat{\Sigma}_T$ and $\hat{\Sigma}_B$ are respectively the total covariance matrix and the between-class covariance matrix of the new feature vectors y.

The optimal coefficient matrix A is then obtained by solving the following generalized eigenvalue problem

$$\Sigma_B A = \Sigma_T A \Lambda \quad (A^T \Sigma_T A = I) \tag{2}$$

where $\Lambda = \mathrm{diag}(\lambda_1, \ldots, \lambda_L)$ is a diagonal matrix of eigen values and I shows the unit matrix. The matrices Σ_T and Σ_B are respectively the total covariance matrix and the between-class covariance matrix of the input feature vectors x.

2.2 Kernel Discriminant Analysis

The kernel discriminant analysis (KDA) is one of the nonlinear extensions of LDA. Consider a nonlinear mapping Φ from a input feature vector x to the new feature vector $\Phi(x)$. For the case of $1-D$ feature extraction, the discriminant mapping can be given as $y = a^T \Phi(x)$. Since the coefficient vector a can be expressed as a linear combinations of the training samples as $a = \sum_{i=1}^{N} \alpha_i \Phi(x_i)$, the discriminant mapping can be rewritten as

$$y = \sum_{i=1}^{N} \alpha_i \Phi(x_i)^T \Phi(x) = \sum_{i=1}^{N} \alpha_i K(x_i, x) = \alpha^T k(x), \tag{3}$$

where $K(x_i, x) = \Phi(x_i)^T \Phi(x)$ and $k(x) = (K(x_1, x), \ldots, K(x_N, x))$ are the kernel function defined by the nonlinear mapping $\Phi(x)$ and the vector of the kernel functions, respectively.

Then the discriminant criterion is given as

$$J = \frac{\sigma_B^2}{\sigma_T^2} = \frac{\alpha^T \Sigma_B^{(K)} \alpha}{\alpha^T \Sigma_T^{(K)} \alpha}, \tag{4}$$

where σ_T^2 and σ_B^2 are respectively the total variance and the between-class variance of the discriminant feature y, and $\Sigma_T^{(K)}$ and $\Sigma_B^{(K)}$ are respectively the total covariance matrix and the between-class covariance matrix of the kernel feature vector $k(x)$ (details are denoted in [7]).

The optimum coefficient vector α can be obtained by solving the generalized eigenvalue problem $\Sigma_B^{(K)} \alpha = \Sigma_W^{(K)} \alpha \lambda$.

For the multi-dimension case, the kernel discriminant mapping is given by $y = A^T k(x)$, where the coefficinet matrix A is defined by $A^T = (\alpha_1, \ldots, \alpha_N)$. The optimum coefficient matrix A is obtained by solving the eigenvalue problem

$$\Sigma_B^{(K)} A = \Sigma_W^{(K)} A \Lambda. \tag{5}$$

Usually the kernel function is defined a priori in KDA. However it is not noticed what the best kernel function for nonlinear discriminant analysis is. Also the class information is usually not introduced in these kernel functions.

2.3   Optimal Nonlinear Discriminant Analysis

Otsu derived the optimal nonlinear discriminant analysis (ONDA) by assuming the underlying probabilities [9–11]. This assumption is similar with the Bayesian decision theory. Similar with LDA, ONDA constructs the dimension reducing optimum nonlinear mapping which maximizes the discriminant criterion J. Namely ONDA finds the optimum nonlinear mapping in terms of the discriminant criterion J.

By using Variational Calculus, Otsu showed that the optimal nonlinear discriminant mapping is obtained as

$$y = \sum_{k=1}^{K} P(C_k|x)u_k \tag{6}$$

where $P(C_k|x)$ is the Bayesian *posterior* probability of the class C_k given the input x. The vectors $u_k(k = 1, \ldots, K)$ are class representative vectors which are determined by the following generalized eigenvalue problem

$$\Gamma U = PU\Lambda \tag{7}$$

where $\Gamma = [\gamma_{ij}]$ is a $K \times K$ matrix whose elements are defined by

$$\gamma_{ij} = \int (P(C_i|x) - P(C_i))(P(C_j|x) - P(C_j))p(x)dx \tag{8}$$

and the other matrices are defined as $U = [u_1, \ldots, u_K]^T$, $\Lambda = \text{diag}(\lambda_1, \ldots, \lambda_L)$, $P = \text{diag}(P(C_1), \ldots, P(C_K))$. It is important to notice that the optimal nonlinear mapping is closely related to Bayesian decision theory, namely the *posterior* probabilities $P(C_k|x)$.

By using the eigen vectors obtained by solving the generalized eigenvalue problem (7), we can construct the optimum nonlinear discriminant mapping from a given input feature x to the new discriminant feature y as shown in the equation (6) if we can know or estimate all the *posterior* probabilities. This means that we have to estimate the *posterior* probabilities for real applications. It also implies a family of nonlinear discriminant mapping can be defined by changing the estimation method of the *posterior* probabilities.

3   Discriminant Kernel Functions

3.1   Dual Problem of ONDA

In the KDA, usually the kernel function is defined a priori. The polynomial functions or the Radial Basis functions are often used as the kernel functions but such kernel functions are general and are not related to the discrimination. Thus the class information is usually not introduced in these kernel functions. Also it is not known what the optimum kernel function for nonlinear discriminant analysis is.

Kurita showed the optimum kernel function, called discriminant kernel function (DKF), can be derived by investigating the dual problem of the eigenvalue problem of ONDA [7]. The DKF is also optimum in terms of the discriminant criterion.

The eigenvalue problem of ONDA given by the equation (7) is the generalized eigenvalue problem. By multiplying $P^{-1/2}$ from the left, this eigen equations can be rewritten as the usual eigenvalue problem as

$$P^{-1/2}\Gamma P^{-1/2}P^{1/2}U = P^{1/2}U\Lambda. \tag{9}$$

By denoting $\tilde{U} = P^{1/2}U$, we have the following usual eigenvalue problem as

$$(P^{-1/2}\Gamma P^{-1/2})\tilde{U} = \tilde{U}\Lambda. \tag{10}$$

Then the optimum nonlinear discriminant mapping of ODNA is rewritten as

$$\boldsymbol{y} = U^T \tilde{\boldsymbol{B}}(\boldsymbol{x}) = \tilde{U}^T P^{-1/2}\tilde{\boldsymbol{B}}(\boldsymbol{x}) = \tilde{U}^T \phi(\boldsymbol{x}) \tag{11}$$

where $\phi(\boldsymbol{x}) = P^{-1/2}\tilde{\boldsymbol{B}}(\boldsymbol{x})$ and $\tilde{\boldsymbol{B}}(\boldsymbol{x}) = (P(C_1|\boldsymbol{x}) - P(C_1), \ldots, P(C_K|\boldsymbol{x}) - P(C_K))^T$.

For the case of N training samples, the eigenvalue problem to determine the class representative vectors (10) is given by

$$(\Phi^T \Phi)\tilde{U} = \tilde{U}\Lambda, \tag{12}$$

where $\Phi = (\phi(\boldsymbol{x}_1), \ldots, \phi(\boldsymbol{x}_N))^T$.

The dual eigenvalue problem of (12) is then given by

$$(\Phi\Phi^T)V = V\Lambda. \tag{13}$$

From the relation on the singular value decomposition of the matrix Φ, these two eigenvalue problems (12) and (13) have the same eigenvalues and there is the following relation between the eigenvectors $\tilde{U}$ and V as $\tilde{U} = \Phi^T V\Lambda^{-1/2}$.

By inserting this relation into the nonlinear discriminant mapping (11), we have

$$\boldsymbol{y} = \Lambda^{-1/2}V^T\Phi\phi(\boldsymbol{x}) = \sum_{i=1}^{N}\Lambda^{-1/2}\boldsymbol{v}_i\phi(\boldsymbol{x}_i)^T\phi(\boldsymbol{x}) = \sum_{i=1}^{N}\alpha_i K(\boldsymbol{x}_i, \boldsymbol{x}) - \alpha_0 \tag{14}$$

where

$$
\begin{aligned}
K(\boldsymbol{x}_i, \boldsymbol{x}) &= \phi(\boldsymbol{x}_i)^T\phi(\boldsymbol{x}) + 1 \\
&= \sum_{k=1}^{K}\frac{P(C_k|\boldsymbol{x}_i) - P(C_k)(P(C_k|\boldsymbol{x}) - P(C_k))}{P(C_k)} + 1 \\
&= \sum_{k=1}^{K}\frac{P(C_k|\boldsymbol{x}_i)P(C_k|\boldsymbol{x})}{P(C_k)}.
\end{aligned}
\tag{15}
$$

This shows that the kernel function of the optimum nonlinear discriminant mapping is given by

$$K(\boldsymbol{x}, \boldsymbol{y}) = \sum_{k=1}^{K} \frac{P(C_k|\boldsymbol{x})P(C_k|\boldsymbol{y})}{P(C_k)}. \tag{16}$$

This is called the discriminant kernel function (DKF).

The derived DKF is defined by using the Bayesian *posterior* probabilities $P(C_k|\boldsymbol{x})$. This means that the class information is explicitly introduced in this kernel function. Also there is no kernel parameters. This means that we do not need to estimate the kernel parameters.

4   Sparse LDA Based on L1 DKF

There are many ways to estimate the Bayesian *posterior* probabilities. Depending on the estimation method, we can define the corresponding DKF [6][7].

In this paper, we propose L1-regularized discriminant kernel function which is defined by using the Bayesian posterior probability obtained from L1-regularized regression. We use L1 DKF as the kernel function for LDA.

4.1   L1-Regularized Regression

Given training samples $\{\mathbf{x}_n, t_n\}_{n=1}^{N}$ where $\mathbf{x}_n$ is n-th observation and t_n is the corresponding target value of $\mathbf{x}_n$, the objective of regression is to estimate the value t for a new data $\mathbf{x}$.

The simplest estimation model is given as the linear model:

$$y(\mathbf{x}, \mathbf{w}) = \mathbf{w}^{\mathrm{T}}\mathbf{x} \tag{17}$$

where $\mathbf{x} = (1, x_1, \cdots, x_D)$, $\mathbf{w} = (w_0, w_1, \cdots, w_D)$ and $y(\mathbf{x}, \mathbf{w})$ is the predicted value of t. An appropriate cofficient vector $\mathbf{w}$ is obtained by minimizing a certain error function $E_D(\mathbf{w})$. A sum-of-squares error function is commonly used:

$$E_D(\mathbf{w}) = \frac{1}{2} \sum_{n=1}^{N} (t_n - \mathbf{w}^T \mathbf{x}_n)^2. \tag{18}$$

To control over-fitting, we can add a regularization term $E_W(\mathbf{w})$ with a regularization parameter λ into the error function. Given general regularizer $E_W(\mathbf{w}) = \sum_{j=1}^{M} |w_j|^q$, we obtain a regularized error function,

$$\frac{1}{2} \sum_{n=1}^{N} (t_n - \mathbf{w}^T \mathbf{x}_n)^2 + \frac{1}{\lambda} \sum_{j=1}^{M} |w_j|^q. \tag{19}$$

The case of $q = 1$ is known as the L1-regularized regression [12]. L1 regression has the property that if $\lambda(> 0)$ is sufficiently large, some of the coefficients w_j

Table 1. Classification accuracy for the test set (agerage of 10 trials)

dataset	bre	dna	ger	hea	iri	seg	sem	spl	veh	win
# of classes	2	3	2	2	3	7	10	2	4	3
# of samples	683	2586	1000	270	4	2310	1593	3175	846	178
# of features	10	180	24	13	150	19	256	60	18	13
LDA	96.3%	93.0%	72.1%	83.9%	86.1%	88.9%	81.9%	83.9%	76.3%	98.5%
sparse LDA	96.4%	93.5%	72.8%	85.2%	89.6%	89.9%	85.7%	84.4%	77.3%	99.3%

are driven to zero. It leads to a sparse model in which the corresponding basis functions play no role.

for K class classification problems, L1 regression can be used as the Baysian *posterior* probability estimator by the *one-vs-all* manner. Let $\mathbf{x}_n$ denote n-th d-dimensional feature vector ($n = 1, \cdots, N$). For all $k = 1, \cdots, K$, the regression about *k-th class vs other classes* is performed between independent varible $\mathbf{x}_n$ and following dependent variable t_n^k,

$$t_n^k = \begin{cases} 1 & \text{if sample } \mathbf{x}_n \text{ belongs to class } k, \\ 0 & \text{otherwise.} \end{cases} \tag{20}$$

After K times regression, we obtain summarized projection vector of $\mathbf{x}_n$,

$$\mathbf{p}(\mathbf{x}_n) = (p_1, \cdots, p_K) = (\mathbf{w}_1^T \mathbf{x}_n, \cdots, \mathbf{w}_K^T \mathbf{x}_n) \tag{21}$$

where $\mathbf{w}_k$ is k-th projection cofficients. Ideally, if $\mathbf{x}_n$ belongs to class k, the k-th component of $\mathbf{p}$ should be 1, and the others should be 0. By normalizing $\mathbf{p}(\mathbf{x}_n)$ to satisfy the condition $\forall k(p_k \geq 0)$ and $p_1 + \cdots + p_K = 1$, we can consider p_k as the estimation of the Bayesian *posterior* probabilities $P(C_k|\boldsymbol{x})$.

4.2   L1 DKF and Sparse LDA

In this paper, we use L1 regression for the estimator of the Baysian *posterior* probability in the K class problems. For the input vector $\mathbf{x}$, the regression outputs probabilistic vector $(p_1, \cdots, p_K)$ in Eq. (21) as the estimation of the Bayesian *posterior* probabilities $(P(C_1|\mathbf{x}), \cdots, P(C_K|\mathbf{x}))$.

Then the corresponding discriminant kernel function, L1 DKF, is given as

$$K(\mathbf{x}, \mathbf{y}) = \sum_{k=1}^{K} \frac{p_k(\mathbf{x})p_k(\mathbf{y})}{p(C_k)}. \tag{22}$$

We use L1 DKF as the kernel function of LDA to introduce the sparseness into LDA indirectly.

5   Experiments

We confirmed the performance of L1 DKF for the kernel of LDA, by using several data sets in UCI machine learning repository [5]: Breast-cancer (bre),

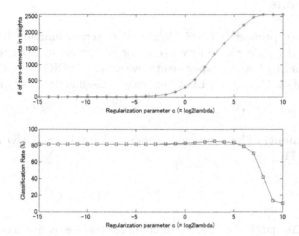

Fig. 1. Results for semeion data. The top figure shows the sparseness of the regression coefficients. The bottom figure shows the classification rate for the test set. The horizontal line shows LDA's performance. The curve shows sparse LDA's performance. Both graphs show the average of 10 trials' results.

dna, german (ger), heart (hea), iris (iri), segment (seg), semeion (sem), splice (spl), vehicle (veh) and wine (win) data. Each data set was divided into a training set (2/3 of all samples) and a test set (remaining samples), at random. For classification experiments, we made 10 different divisions of the training and test sets. For all experiments, we used class prior $P(C_k) = N_k/N$ where N_k is the number of samples in C_k. We use a nearest mean classifier for usual LDA and sparse LDA.

We train L1 DKF by using different regularization parameter $\lambda = 2^{-15}, 2^{-14}$, $\cdots$, 2^{10}. Fig. 1 shows the training result for the semeion data. Note that the figure shows the average of results of 10 trials.

The top figure shows the sparseness of the L1 DKF. The semeion data has 10 classes and 256 features, therefore the summarized projection matrix $[\mathbf{w}_1, \cdots, \mathbf{w}_K]$ has totally 2560 elements. The vertical axis shows the number of zero elements in 2560 elements. The number of zero elements is increasing in proportion to the regularization parameter λ.

The bottom figure shows the classification accuracy for the test set. As the baseline performance, LDA has 81.9% accuracy. The accuracy of sparse LDA is better than LDA in some part. The highest averaged accuracy of sparse LDA is 85.7 % ($\lambda = 2^3$). In this case, about 1,400 features in 2,560 original features did not be used in the classification task. It is considered that the features which are not suitable for the classification task were removed by L1 regression.

Tab.1 shows the classification performances of LDA and sparse LDA for each data set. In all cases, the highest performance of sparse LDA was better than the performance of LDA.

6   Conclusions

In this paper, we propose a novel discriminant kernel function based on L1 regression (called L1 DKF), and we use it for the kernel of LDA to introduce the sparseness into LDA. In experiments, we show L1 DKF is appropriate as the kernel for LDA. Our sparse LDA has better classification performance than usual LDA.

Acknowledgement. This work was supported by JSPS KAKENHI Grant Number 23500211.

References

1. Baudat, G., Anouar, F.: Generalized discriminant analysis using a kernel approach. Neural Computation 12(10), 2385–2404 (2000)
2. Chow, C.K.: An optimum character recognition system using decision functions. IRE Trans. EC-6, 247–254 (1957)
3. Clemmensen, L., Hastie, T., Witten, D., Ersboll, B.: Sparse discriminant analysis (2011)
4. Fisher, R.A.: The Use of Multiple Measurements in Taxonomic Problems. Annals of Eugenics 7, 179–188 (1936)
5. Frank, A., Asuncion, A.: UCI Machine Learning Repository. University of California, School of Information and Computer Science, http://archive.ics.uci.edu/ml
6. Hidaka, A., Kurita, T.: Discriminant Kernels based Support Vector Machine. In: The First Asian Conference on Pattern Recognition (ACPR 2011), Beijing, China, November 28-30, pp. 159–163 (2011)
7. Kurita, T.: "Discriminant Kernels derived from the Optimum Nonlinear Discriminant Analysis. In: Proc. of 2011 International Joint Conference on Neural Networks, San Jose, California, USA, July 31-August 5 (2011)
8. Mika, S., Ratsch, G., Weston, J., Scholkopf, B., Smola, A., Muller, K.: Fisher discriminant analysis with kernels. In: Proc. IEEE Neural Networks for Signal Processing Workshop, pp. 41–48 (1999)
9. Otsu, N.: Nonlinear discriminant analysis as a natural extension of the linear case. Behavior Metrika 2, 45–59 (1975)
10. Otsu, N.: Mathemetical Studies on Feature Extraction In Pattern Recognition. Researches on the Electrotechnical Laboratory 818 (1981) (in Japanease)
11. Otsu, N.: Optimal linear and nonlinear solutions for least-square discriminant feature extraction. In: Proceedings of the 6th International Conference on Pattern Recognition, pp. 557–560 (1982)
12. Tibshirani, R.: Regression shrinkage and selection via the lasso. J. Royal. Statist. Soc B. 58(1), 267–288 (1996)
13. Zou, H., Hastie, T., Tibshirani, R.: Sparse principal component analysis. Journal of Computational and Graphical Statistics 15(2), 262–286 (2006)

Conditional Variance of Differences: A Robust Similarity Measure for Matching and Registration

Atsuto Maki and Riccardo Gherardi

Cambridge Research Laboratory, Toshiba Research Europe
Cambridge, United Kingdom

Abstract. This paper presents a new similarity measure, the *sum of conditional variance of differences* (SCVD), designed to be insensitive to highly non-linear intensity transformations such as the ones occurring in multi-modal image registration and tracking. It improves on another recently introduced statistical measure, the sum of conditional variances (SCV), which has been reported to outperform comparable information theoretic similarity measures such as mutual information (MI) and cross-cumulative residual entropy (CCRE). We also propose two additional extensions that further increase the robustness of SCV(D) by relaxing the quantisation process and making it symmetric. We demonstrate the benefits of SCVD and improvements on image matching and registration through experiments.

1 Introduction

A robust similarity measure between different regions of images plays a fundamental role in several image analysis applications such as stereo matching, motion estimation, registration and tracking. Similarity measures commonly used in these tasks, SSD or NCC for example, can at most cope with linear variations of intensity, such as global changes in gain and bias. Matching and registration techniques in general need to be robust to a wider range of transformations that can arise from non-linear illumination changes caused by anisotropic radiance distribution functions, occlusions or different acquisition processes [1] (e.g. visible light and infrared, those employed in medical imaging). These more challenging contexts, which represent the main focus of this article, have been extensively explored in the literature.

Most of the existing methods for computing similarity measures across multimodal images are based on information theoretic approaches and make use of the probability of the intensity co-occurrence. The seminal works on mutual information (MI) [2,3] introduced the use of joint intensity distributions, recognising the statistical dependence between intensities of corresponding pixels. Other statistical dependencies have also been explored: cross-cumulative residual entropy (CCRE) [4] for example measures the entropy defined using cumulative distributions. The increased resilience to non-linear intensity transformations however comes at the cost of a higher computational complexity than conventional sum-comparing metrics, whose complexity is linear with respect to the number of elements.

G.L. Gimel' farb et al. (Eds.): SSPR & SPR 2012, LNCS 7626, pp. 657–665, 2012.

A recently proposed method, called the *sum of conditional variances* (SCV) [5], also uses the joint distribution of image intensities, but generates it directly as a histogram. Intuitively, SCV exploits a statistical property assuming that a group of pixels clustered by neighbouring intensities in the first image should be similarly clustered in the second, even if their mapped ranges are very different. SCV was originally developed in the context of medical image registration [6] and therefore aimed at being robust against non-linear intensity variations such as those occurring when capturing images through different acquisition modalities. It has been shown to have a larger convergence basin than MI's Parzen window approach in medical alignment tasks [7]. These results have been confirmed in the context of visual tracking [8], showing SCV to have better performance than several competing approaches in terms of convergence radius, computational complexity and stability (quantified by the number of iteration necessary for convergence). SCV is closely related to the correlation ratio [9], but has a lower computational complexity and is therefore more amenable to efficient optimization strategies.

In this paper, we introduce a new similarity measure called the *sum of conditional variance of differences* (SCVD). In the original SCV formulation, the reference image is used solely in its quantized form for generating a partition to be applied to the second image (i.e. the set of *conditions*). This process discards a significant amount of information. Assuming the intensity map to be weakly order preserving, whether directly or inversely, we show that the information loss can be mitigated employing the variance of *intensity differences*, leading to a more discriminative measure without increase in the computational complexity. We also generalise the computation of conditions, improving both our matching measure SCVD and the original SCV implementation.

The contribution of this paper is thus two-fold:

1. we introduce a novel similarity measure, the *sum of conditional variance of differences* (SCVD) and show its superior performance in comparison to other metrics designed against non-linear intensity variations,
2. we generalise the definitions of conditions, leading to improvements for both our formulation and the original SCV approach.

The rest of the paper is organized as follows: the next section will contain a brief description of the SCV algorithm, followed by the description of our proposal (SCVD) and its extensions respectively in section 3 and 4. We will evaluate the performance of the novel matching measure in section 5, focusing on matching and registration tasks. Finally, section 6 will report our conclusions.

2   The Sum of Conditional Variances

Given a pair of images X and Y, the sum of conditional variances (SCV) matching measure [5] prescribes to partition the pixels of Y into n_b disjoint bins $Y(j)$ with $j = 1, ..., n_b$, corresponding to bracketed intensity regions $X(j)$ of X (called

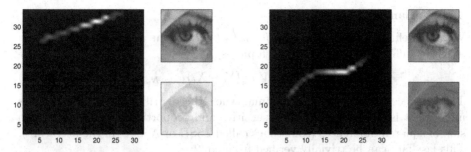

Fig. 1. Joint intensity histograms. A joint histogram H_{XY} can be interpreted as non-injective relation that maps the ranges of two images. **On the left:** the resulting joint histogram after linearly reducing the contrast of the reference image. **On the right:** the joint histogram for a non-linear intensity map. Hotter (brighter) colors correspond to more frequently occurring values.

the *reference* image). The value of the matching measure is then obtained summing the variances of the intensities within each bin $Y(j)$.

$$\mathcal{S}_{SCV}(X,Y) = \sum_{j=1}^{n_b} E[(Y_i - E(Y_i))^2 \mid X_i \in X(j)] \tag{1}$$

where X_i and Y_i with $i = 1, ..., N_p$ indicate the pixel intensities of X and Y respectively, N_p being the total number of pixels. The *conditions* that appear in the sum are obtained uniformly partitioning the intensity range of X.

The behaviour of SCV can be characterised by the joint histogram H_{XY} of X and Y. As shown in figure 1, the joint histogram can be interpreted as non-injective relation that maps the range of the first image to the second one. The set of pixels that contributed to the non zero entry of each column (row) corresponds to one of the regions selected by the j-th condition.

The number of discretisation levels n_b is problem specific; for images quantised at byte precision, a typical choice is usually $n_b = 32$ or 64 [8]. Larger intervals can help in achieving a wider convergence radius and offer more resilience to noise (the matching measure will not change as long as the pixels do not cross the current bin boundaries). On the other hand, narrow ranges will boost the matching accuracy and reduce the information that is lost during the quantisation step.

3   Sum of Conditional Variance of Differences

According to the SCV algorithm, the reference image is used solely to determine the subregions in which the variances of equation 1 should be computed. In this section, we present a new similarity measure based on the *conditional variance of differences*, which uses all the information present in both images leading to a more discriminative matching measure. We also propose two generalisations of the conditionals computation, which further increase the robustness of our approach.

3.1   Variance of Differences

We first define the *variance of differences* ($\mathcal{VD}$) as the second moment of the intensity differences between two templates:

$$\mathcal{VD}(X,Y) = Var[\{Y_i - X_i\}_{i=1...N_p}] \tag{2}$$

The variance of differences is minimal when the distribution of differences is uniform. It is bias invariant, scale sensitive and proportional to the zero-mean sum of squared differences (sometimes called ZSSD or ASSSD in the literature). This last fact can be trivially verified from eq. 2:

$$\mathcal{VD}(X,Y) = E[(Y - X - E(Y - X))^2] \tag{3}$$

$$\propto \sum_i [(Y_i - E(Y_i)) - (X_i - E(X_i))]^2, \tag{4}$$

where the mean of an image is understood to indicate its element-wise mean.

3.2   Sum of Conditional Variance of Differences

Given two images X and Y, we define the *sum of the conditional variance of differences* (SCVD) as the sum of the variances over a partition of their difference. As before, the subsets are selected bracketing the range of the reference image to produce a set of bins $X(j)$. In order for the difference to be meaningful, the two signals should be in direct relation; since the matching measure need be insensitive to changes in scale and bias, we maximise direct relation by adjusting the sign of one of them in accordance with eq. 6. In symbols:

$$\mathcal{S}_{SCVD}(X,Y) = \sum_{j=1}^{n_b} \mathcal{VD}(X_i, \Phi Y_i \mid X_i \in X(j)), \tag{5}$$

$$\Phi = \Gamma \left(\sum_{j=2}^{n_b} \Gamma\left(E(Y_i \mid X_i \in X(j)) - E(Y_i \mid X_i \in X(j-1))\right) \right), \tag{6}$$

where Γ indicates the step function mapping $\mathbb{R}$ to $\{-1, 1\}$. Φ encodes a cumulative result of comparisons between a pair of $E(Y_i)$ in the adjacent histogram bins, so that the sign is properly adjusted. Hence, the requirement for the mapping from X and Y is to be weakly order preserving (the function should be monotonic but is not required to be injective). This restriction, not present in the original SCV formulation, makes it possible to make better use of the available information and largely valid, e.g. between signals captured for the same target with different modes.

4   Generalising the Conditions

Uniformly partitioning the intensity range of X into equally sized bins $X(j)$ can lead subpar performances when the intensity distribution is uneven: poorly sampled intensity ranges are noisy and their variance unreliable. Overly sampled regions of the spectrum conversely lead to compressing many pixels into a single bin,

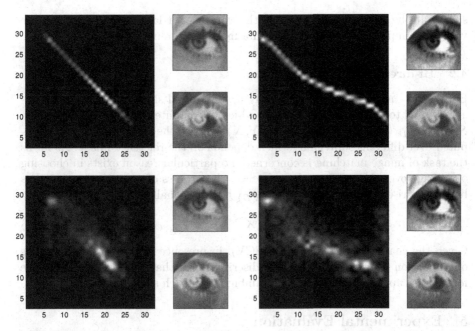

Fig. 2. Effects of quantisation and displacement. On the top row: H_{XY} for a pair of aligned images. Bottom row: H_{XY} for the same pair plus displacement. Left column: H_{XY} using uniform quantisation of intensity range. Right column: H_{XY} by using histogram equalised intensities for the reference image.

discarding a large amount of useful information in the process. The procedure is also inherently asymmetric, producing in general different results when swapping the images involved. In this section we discuss two non-mutually exclusive modifications of our proposal in order to deal with these issues. Each one of them provides an independent performance boost to the baseline approach described.

4.1 Uniform Quantizations

In fig. 2 (top-left) is shown the joint histogram between an image and its gray scale inverse. As it can be seen, the bins corresponding to the low and high end of the intensity spectrum are not receiving any vote, thus compressing the image information into a smaller number of regions.

To achieve a uniform bin utilisation, we perform histogram equalisation on the reference image X. Figure 2 (right) shows an H_{XY} generated by replacing the input reference image X with its histogram equalized version, achieving full utilisation of the entire dynamic range.

On the bottom row of fig. 2 are shown the original and histogram equalised version after applying a 5 pixel displacement to one of the images. As a result, the entries are more scattered and less sharp. As in the previous case, the non equalised version does not make full use of the available bins; the equalized one, shown at the bottom right spreads the vote over a larger area, affecting the variance computation and resulting in a more discriminative measure.

4.2 Bi-directional Quantisations

Both SCV and SCVD are structurally asymmetrical since only one of the images is used to define the partitions in which to compute the variance. Generally, $\mathcal{S}_{\{SCV,SCVD\}}(X,Y) \neq \mathcal{S}_{\{SCV,SCVD\}}(Y,X)$ because the two quantities are computed over different subregions which depends on the reference image. As far as the task of image matching is concerned, no particular reason exists in choosing one image over the other as the reference; the process of quantization can thus be symmetrised computing $\mathcal{S}_{\{SCV,SCVD\}}$ bi-directionally:

$$\mathcal{S}^B_{\{SCV,SCVD\}} = \left(\mathcal{S}_{\{SCV,SCVD\}}(X,Y) + \mathcal{S}_{\{SCV,SCVD\}}(Y,X)\right) / 2 . \tag{7}$$

Given the characteristics of SCVD (SCV), in presence of uneven quantizations one direction is usually much more discriminative than the other. The above formula is capable of successfully disambiguating such situations.

5 Experimental Evaluation

Experiment I. In order to compare our proposal, its variations and the original SCV approach, in this first experiment we study the discriminativeness of each one of them for increasing, isotropic displacements. We selected an image location, a direction and a displacement all at random, computing the measure between the selected reference window and the template after applying the translation. Notice that the template is negated in order to simulate multi-modal inputs. The size of the region was fixed to 50×50 pixels while the maximum distance was set to be half of its edge length, i.e. 25 pixels. The results are shown for a single image (the *peppers* image included in Matlab) but the plot of figure 3 is similar for any non-periodic, non-uniform picture.

Figure 3 was produced averaging 20,000 iterations of this procedure, to remove the effects of noise (each single trial is roughly monotonic). As it can be seen, all $\mathcal{S}_{SCVD}$ versions are better at discriminating the minimum. Histogram equalized and symmetric variants obtain steeper gradients for both SCV and SCVD. When utilising both improvements, SCVD shows a nearly constant slope, a crucial property in order to use optimization algorithms based on implicit derivatives.

Experiment II. We now compare the performance of different similarity measures on a synthetic registration task using a gradient descent search; given a random location and displacement as before, we optimize the cost function following the direction of the steepest gradient. The procedure terminates when reaching a local minima or the maximum number of allowed iterations (set to 50 in our experiments). Figure 4 was obtained averaging 4000 different trials; as it can be seen, each SCVD version beats the equivalent SCV measure using the same set of variants, which provide a non negligible performance boost.

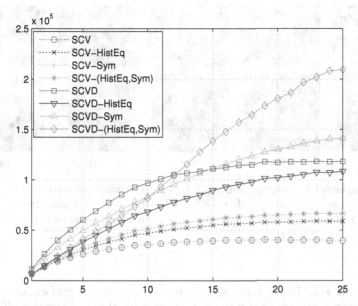

Fig. 3. Matching measure vs. displacement. We compare our proposal, its variations and the original SCV approach over random displacements within an image. SCVD plus both extensions results in the most discriminative measure, with a nearly constant slope across the entire search domain.

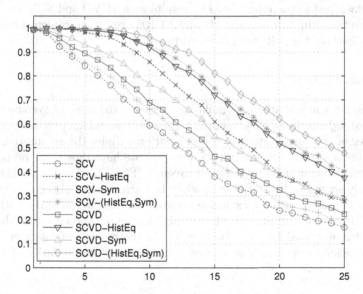

Fig. 4. Convergence vs. displacement. The plots show the convergence rate as a function of the distance between the reference and displaced window. We compare our proposal, its variations and the original SCV approach over random displacements within an image.

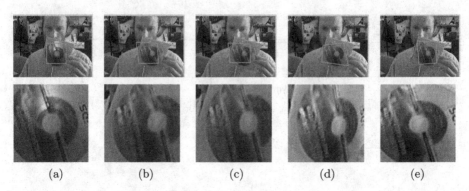

(a)   (b)   (c)   (d)   (e)

Fig. 5. Registration experiment. (a) Input frame with reference region marked green. (b-e) Registrations by MI, CCRE, SCV and SCVD. On the second row are shown the registered regions backwarped to the template (sequence part of the ESM project, http://esm.gforge.inria.fr).

Experiment III. In our final experiment we compare the performance of several similarity measures on a tracking task over a real image sequence. Figure 5 (a) shows one of the frames of the sequence, and its reference template. The subsequent frame has both photometric and geometric deformations; in figure 5 (b-e) we display the registration results respectively for MI, CCRE, SCV and SCVD, showing both the best matching quadrilateral on the frame and the regions backwarped to the reference. The results with SCVD and SCV are by our implementation while those with MI and CCRE are by an implementation by [8] on the basis of the software presented in [10].

6   Conclusions

We presented a new statistical similarity measure, the *sum of the conditional variance of difference* (SCVD), tailored for robustly matching two image regions in presence of non-linear intensity transformations. Under the assumption of the transfer function being weakly order preserving, we have shown our proposal to outperform the *sum of the conditional variance* (SCV), a recent algorithm that was already shown to be competitive with the current state of the art. We also developed two non mutually exclusive improvements that can make both SCV and SCVD more discriminative at a negligible computational cost. Although we have demonstrated the benefit of SCVD in the context of image matching and registration, its principle is applicable to measure the similarity of two 3D volumes.

References

1. Irani, M., Anandan, P.: Robust multi-sensor image alignment. In: ICCV, pp. 959–966 (1998)

2. Viola, P.A., Wells III, W.M.: Alignment by maximization of mutual information. International Journal of Computer Vision 24, 137–154 (1997)
3. Maes, F., Collignon, A., Vandermeulen, D., Marchal, G., Suetens, P.: Multimodality image registration by maximization of mutual information. IEEE Trans. Med. Imaging 16, 187–198 (1997)
4. Wang, F., Vemuri, B.C.: Non-rigid multi-modal image registration using cross-cumulative residual entropy. International Journal of Computer Vision 74, 201–215 (2007)
5. Pickering, M.R., Muhit, A.A., Scarvell, J.M., Smith, P.N.: A new multi-modal similarity measure for fast gradient-based 2d-3d image registration. In: Int. Conf. of IEEE Engineering in Medicine and Biology Society, pp. 5821–5824 (2009)
6. Zitová, B., Flusser, J.: Image registration methods: a survey. Image Vision Comput. 21, 977–1000 (2003)
7. Pickering, M.R.: A new similarity measure for multi-modal image registration. In: International Conference of Image Processing, pp. 2273–2276 (2011)
8. Richa, R., Sznitman, R., Taylor, R.H., Hager, G.D.: Visual tracking using the sum of conditional variance. In: IROS, pp. 2953–2958 (2011)
9. Roche, A., Malandain, G., Pennec, X., Ayache, N.: The Correlation Ratio as a New Similarity Measure for Multimodal Image Registration. In: Wells, W.M., Colchester, A.C.F., Delp, S.L. (eds.) MICCAI 1998. LNCS, vol. 1496, pp. 1115–1124. Springer, Heidelberg (1998)
10. Baker, S., Matthews, I.: Lucas-kanade 20 years on: A unifying framework. International Journal of Computer Vision 56, 221–255 (2004)

A Class Centric Feature and Classifier Ensemble Selection Approach for Music Genre Classification

Hasitha Bimsara Ariyaratne, Dengsheng Zhang, and Guojun Lu

Monash University, Gippsland School of IT, Churchill, Australia
{hasitha.ariyaratne,dengsheng.zhang,guojun.lu}@monash.edu

Abstract. Music genre classification has attracted a lot of research interest due to the rapid growth of digital music. Despite the availability of a vast number of audio features and classification techniques, genre classification still remains a challenging task. In this work we propose a class centric feature and classifier ensemble selection method which deviates from the conventional practice of employing a single, or an ensemble of classifiers trained with a selected set of audio features. We adopt a binary decomposition technique to divide the multiclass problem into a set of binary problems which are then treated in a class specific manner. This differs from the traditional techniques which operate on the naive assumption that a specific set of features and/or classifiers can perform equally well in identifying all the classes. Experimental results obtained on a popular genre dataset and a newly created dataset suggest significant improvements over traditional techniques.

Keywords: music retrieval, feature selection, classifier ensemble, music genre classification.

1 Introduction

Recent advancements in digital media encoding, storing and delivering technologies have led to a significant increase in the number of digital audio files. As a result, managing music has become a challenging task. Genre is the most widely used descriptor in organizing and searching large music collections [1]. It has been shown that existing audio features and classifiers have reached a "glass ceiling" [2], because most new features and classifiers only show a marginal improvement. Further investigation of existing literature reveals that majority of the methods use a single classifier trained with a collection of different audio features. Some of the recent works have investigated the effectiveness of employing attribute selection methods for feature combination [3], while others[4,5] have demonstrated the effectiveness of using classifier ensembles to improve classification accuracy. However, these techniques operate on the assumption that a specific set of features and/or a specific set of classifiers can perform well for all music classes. But studies [5] have shown class-specific feature selection can produce better performance. In this paper we take the existing research one

G.L. Gimel' farb et al. (Eds.): SSPR & SPR 2012, LNCS 7626, pp. 666–674, 2012.

step further by proposing a class centric feature and classifier ensemble selection method.

Music genre classification has two steps: audio feature extraction and classification. Audio features are usually designed only to capture specific qualities of sound; therefore, they are limited in terms of generalization. For example, features like Zero Crossing Rate (ZCR), and Linear Predictive coefficients (LPC) can be used to discriminate instrumental music (e.g. Classical) from vocal music (e.g. Country) due to their ability in capturing certain characteristics of human voice. However, they perform poorly at discriminating between pure instrumental genres (e.g. classical or jazz). The most common way to solve this problem is to combine different features. However, fusing features should be done carefully, since not all features contribute to the classification task equally.

Moreover, large feature vectors can also lead to high computational complexities and complications (generally known as the "curse of dimensionality"). Therefore feature selection based on individual classes can provide an effective way to reduce the number of features while preserving the discrimination power. Most traditional techniques rely on multiclass classifiers working on all the classes; therefore, class specific feature selection is not performed. An alternative to this approach is the use of hierarchical classifiers where a multiclass problem can be split into a set of smaller problems which can be optimized individually. Zhang et.al [6] proposed the use of a manually generated classification hierarchy and a set of manually selected features for each node, while Silla et.al [7] proposed a node specific classifier and feature selection technique based on a predefined genre taxonomy, however neither have provided empirical results on the effectiveness of hierarchical classification against traditional techniques. Ariyaratne et.al [8] has proposed a hierarchical classification technique based on an auto generated tree which utilizes node specific feature and classifier selection, their empirical results have shown the hierarchical approach to perform better, however despite high accuracies obtained at higher levels, the leaf level accuracies were not significantly improved due to errors at higher levels propagating downward. Therefore in this paper we focus on a nonhierarchical approach to class centric classification based on a binary decomposition technique.

There are two main techniques for breaking down a multiclass problem into a set of binary classification problems: One-Vs-One (OVO) and One-Vs-All (OVA). OVA is the most simplest of the two, this technique creates n binary classifiers for each n classes where one class is considered positive while all other classes are considered negative. In comparison OVO technique builds n(n-1)/2 classifiers for each possible combination of class pairs. These two approaches adopt different techniques to aggregate the results of multiple classification problems to decide the final outcome [9].

The rest of the paper is organized as follows: Section 2 presents the proposed method which utilizes a binary decomposition technique for dissembling the multiclass problem in order to perform class centric feature and classifier selection. Section 3 presents experimental results and the paper is concluded in section 4.

2   The Proposed Method

In this section we present an overview of our proposed class centric feature and classifier ensemble selection method. As indicated earlier, in order to perform class centric feature and classifier selection, we need to address each class separately. Both OVA and OVO binary decomposition techniques let us focus on each class individually. Galar et.al [9] has compared the performance of several different classifiers for solving multiclass problems using OVA and OVO decomposition methods on number of publicly available nonmusical datasets. They have concluded that OVO methods generally perform better. Similar evidence can be seen in Silla et.als work [10] where a new feature selection technique was introduced for music genre classification. They have evaluated the performance on a set of classifiers under OVA and OVO schemes. Both previous work have used a relatively low number of features (less than 40), therefore we conducted our own experiments to validate these claims with large number of features(183 features). We adopted a voting aggregation method for OVO and maximum confidence level for OVA. Experimental results are shown in Table 3 under section 3.1. As indicated in previous work and as confirmed by our own experiments, OVO noticeably outperforms OVA, hence we adopted the one-vs-one decomposition technique for our class centric approach. Fig. 1 shows an overview of our proposed method. We start by disassembling the multiclass genre classification problem into a set of binary problems. Training data sets for each problem are constructed of instances belonging to the corresponding class pairs. These datasets are then preprocessed by performing feature selection to reduce dimensionality and improve predictive accuracy (by removing irrelevant features that can introduce noise into the data).

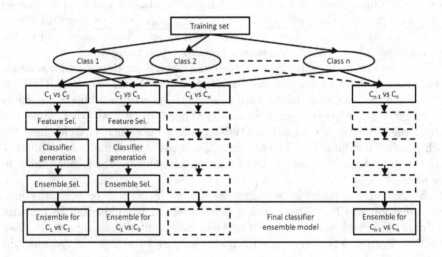

Fig. 1. Overview of the proposed method

Afterwards, we generate a set of classifiers for each problem and then choose the best classifier ensemble for each task through a classifier ensemble selection method outlined in following section.

2.1   Classifier Ensemble Approach

There are many classifier ensemble techniques in the literature such as boosting, bagging, random forests and stacking [11]. The main idea behind the use of classifier ensembles is to combine multiple classifiers to improve classification accuracy. However, as with feature selection, we hypothesize that selecting the best set of classifiers out of a multitude of classifiers for each classification problem is more important than randomly combining a set of classifiers in a multiclass setting. Therefore we propose the use of a classifier ensemble selection technique to choose the best performing set of classifiers (classifier ensemble) for each classification task. Caruana et.al [12] have proposed a classifier ensemble selection technique to select a set of classifiers from a multitude of model libraries. We have adopted this approach to choose the best classifier ensemble. The basic steps of the algorithm are as follows:

- Separate training instances into individual classes
- Construct n(n-1)/2 datasets for OVO comparisons with each pair of classes
- Carry out feature selection for each dataset
- Generate a collection of classifiers with bagging and selection with replacement
- Perform feature selection for each dataset based on the classifier optimum feature selection method.
- Train the classifiers on the classifier optimal set of features
- Select the best classifier ensemble for each binary class by:
  - Add to the ensemble, the classifier model which maximizes the ensembles' overall performance (measured using a certain error metric: i.e accuracy,RMS etc.) in a hill climbing fashion. The overall performance is computed by averaging the performance of each classifier.
  - Repeat the above step until all the models have been examined.
  - Return the ensemble that has maximum performance.

3   Experimental Results

In this section we present the details of the experiments carried out to measure the impact of class-centric feature and classifier selection. As mentioned earlier, previous work in MIR research has produced a wide variety of features, different authors use different sets of audio features, however there are no "correct" set of features for any particular MIR task. In our work, we chose a set of 13 widely used features which are robust and computationally efficient low level features as listed in Table 1. They comprise of six MPEG7 features and seven other most commonly used audio features. We derive 4 statistical properties: i.e. mean,

Table 1. Audio features used for the experiments

Feature Name	No.of features
1. Total Energy	4
2. Fundamental Frequency	4
3. Loudness Sensation	32
4. Integral Loudness	4
5. Audio Spectral Centroid	4
6. Spectral Rolloff	4
7. Audio Spectrum Spread	4
8. Audio Spectrum Flatness	16
9. Audio auto correlation	13
10. Log Attack Time	1
11. Temporal Centroid	1
12. Zero Crossing Rate	4
13. Mel Frequency Cepstral Coefficients	96
Total	183

Table 2. Details of the HBA and GTZAN datasets

	HBADS	GTZAN
No. of genres	15	10
Genres	Blues, Classical, Country, Disco, Hiphop, Indian, Jazz, Metal, Opera, Pop, Reggae, Rock, Salsa, Techno, Ambient	Blues, Classical, Country, Disco, Hiphop, Jazz, Metal, Pop, Reggae, Rock
No. of files	7500 (500 per genre)	1000 (100 per genre)
Format	22,000KHz, 64kbps MP3	22,000KHz, 16bit PCM
Size	2.6GB	1.2GB

variance, covariance, and their numerical partial derivatives (i.e. the differences between successive elements).

Features 1, 10, 11 and 12 are time domain features while the rest are extracted in the frequency domain. MFCC is a widely used feature in many different areas of MIR, for our experiments we extracted the first 24 coefficients and their previously mentioned statistical properties. Features 2 ,5 through 8 provide various statistical measurements related to the frequency spectrum of a sound. Features 3 and 4 capture the human perception of loudness while auto correlation feature can be used to analyze reoccurring patterns (i.e beats, tempo) in a signal. We used two music datasets in our experiments. As our first dataset we chose the widely used genre benchmarking dataset GTZAN[13], it contains 10 genres with 100 clips each. In order to test the robustness of the proposed method on a larger dataset, and due to the lack sufficiently large datasets in the literature, we constructed our own dataset by extending the genre set of GTZAN to 15. The new dataset contains a carefully selected set of songs with no duplicates or ambiguous class labels. The genre labels are taken directly from official album/song information to ensure maximum accuracy of the labels. Details of both datasets are given in Table 2.

3.1   Experimental Setup

We used the WEKA[14] machine learning platform as the test bed for conducting our experiments. We chose the following 5 well known classifiers in our ensemble approach: K-Nearest Neighbour (IBK), Naive Bayes (NB), Decision Tree (J48), Logistic Regression (LR), and Sequential Minimal Optimization: Support Vector Machine (SMO).

As mentioned in section 2 we conducted our own experiments to choose the best binary multiclass decomposition scheme between OVA and OVO with the GTZAN dataset. Results are presented in table 3.

Table 3. Classifier performance between OVA and OVO

Classifier	One vs All	One vs One
IBK	54.70	**57.64**
NB	47.35	**56.74**
J48	48.82	**55.94**
SMO	50.52	**67.35**

From these results we can conclude that OVO scheme has a notable advantage over OVA. Therefore we primarily focus on OVO for implementing our proposed approach. There are many feature selection algorithms available in the literature which can be mainly classified into two groups: Filter methods and Wrapper methods [14]. Filter methods rely on general characteristics of the data; they attempt to measure the importance of each feature or feature subset using a score metric such as information gain, Chi-squared distributions or correlation coefficients, then choose ("Filter") the best set. In contrast, wrapper methods make use of a learning algorithm to search the feature space and evaluate the usefulness of features for the classification problem at hand, in other words the learning algorithm is "Wrapped" into the feature selection process. The later produce a more optimum set of features; however, they are much more computationally expensive and prone to over fitting. Sequential forward/backward selection and hill climbing methods are some of the search methods commonly used for searching the feature subsets in wrapper methods [14]. Since our primary aim is to investigate the effectiveness of class centric feature and classifier selection, we choose the more computationally efficient filter based approach. We used the following 4 widely used feature selection (FS) techniques [3]: Correlation based Feature Selection (CSF), Chi-square Feature Evaluation (CHI), Gain Ratio Feature Evaluation (GAIN), and Principal Component Analysis (PCA).

Table 4 list overall accuracies obtained for each classifier with and without feature selection on GTZAN dataset. This experiment was conducted to analyze the impact of different feature selection techniques on different classifiers, and choose the best technique for each classifier.

As seen in table 4, feature selection methods do have a significant impact on majority of the learning algorithms. These results also agree with the findings of [3]. One interesting observation is how PCA has negatively impacted the accuracy. Even though PCA can be used to reduce the dimensionality of data;

Table 4. Impact of feature selection on classifiers - HBA dataset

Classifier	No FS	CSF	CHI	GAIN	PCA
IBk	53.05	**59.60**	56.66	56.66	31.76
NB	58.94	**62.74**	57.64	57.64	54.31
J48	50.45	51.17	51.37	**51.37**	31.56
LR	65.00	**69.45**	68.63	68.12	67.56
DT	**41.05**	31.01	35.88	35.88	32.15
SMO	69.56	66.07	70.00	**70.00**	60.98

strictly speaking, PCA is not a feature selection technique, rather a feature extraction method that constructs new features through linear projection of high dimensional vector into a low dimensional vector while retaining characteristics of the dataset that contribute to most variation. Therefore even though PCA extracted features are optimal for reconstructing original data, they are not always better for classification [15]. We used these results as a guide for choosing the best feature selection technique for different classifiers in our ensemble approach.

3.2 Classifier Ensemble Approach

We adopt the classifier ensemble selection technique proposed in [[12]] under one-vs-one binary decomposition strategy to improve classification accuracy by narrowing down the selection criteria focusing on each class. Once the final classifier ensemble model has been constructed, classification for an unlabelled test instance is performed by choosing the class which generated majority votes among all $n(n-1)/2$ OVO problems. For each OVO problem, classifier ensembles are also evaluated based on majority voting. Results obtained for both datasets are listed in tables 5 and 6. The columns are labelled as follows:

- NoFS: Lists classification accuracies for each classifier on its own (without any feature selection).
- FS: Best accuracy obtained with feature selection for each classifier (not class centric).
- CCFS: Accuracies obtained for class centric feature selection through OVO decomposition
- CES: Accuracy of classifier ensemble selection (i.e conventional multiclass problem with classifier ensemble selection),
- CCFS+CES: Accuracy of class centric feature selection and classifier ensemble selection based on OVO decomposition (our proposed approach).

From these results we can observe that classification accuracies are improved across majority of the classifiers when class specific feature selection is performed, except for support vector machine. The implementation of SMO Support Vector Machine already performs OVO binary comparisons when handling multiclass problems. Therefore using a SMO classifier in an OVO setting has no extra advantage. Furthermore, a support vector machine is already a very efficient

Table 5. Comparison of classification accuracies between different classification techniques on the HBA Dataset

Classifier	No FS	FS	CCFS	CES	CCFS+CES
IBk	53.05	59.60	62.74		
NB	58.94	62.74	63.20		
J48	50.45	51.37	59.11	71.37	**77.88**
LR	65.00	69.45	71.30		
DT	41.05	35.88	38.80		
SMO	69.56	71.00	69.70	70.11	76.56

Table 6. Comparison of classification accuracies between different classification techniques on the GTZAN Dataset

Classifier	No FS	FS	CCFS	CES	CCFS+CES
IBk	54.70	64.41	66.70		
NB	57.05	60.88	61.80		
J48	49.41	49.70	57.11	65.60	**70.88**
LR	50.0	61.05	64.00		
DT	41.47	40.20	42.61		
SMO	66.17	72.05	69.70	70.11	71.0

and highly optimized multi class classifier; this is why adding SMO into the classifier ensemble didn't have a significant impact on the ensemble approach either.

Further analysis shows that classifier ensemble selection(CES) performs slightly higher (71.30% and 65.60%) than the best individual classifier: CF (69.45% and 64.41%) for HBA and GTZAN datasets respectively. Finally we can see a significant improvement when class specific ensemble classifier selection is performed. The improvement is about 6.5% for the HBA dataset and 5.28% for the GTZAN dataset. From these results we can conclude that the use of a class centric feature selection and classifier ensemble selection of otherwise weak classifiers can perform equally or better than a highly efficient classifier such as the support vector machine.

4   Conclusion

In this paper we proposed a class centric feature and classifier ensemble selection technique for music genre classification. We presented experiments to validate the selection the best multiclass decomposition technique and the best feature selection technique for music genre classification. Building upon these findings we tested our proposed method using two genre datasets and a set of weak classifiers and low level features. The promising results obtained through experiments validated our initial hypothesis that a class centric feature selection combined with a classifier ensemble selection can improve genre classification accuracy. In this work we tested our hypothesis using the most commonly used feature selection techniques (i.e: filter methods), aggregation of binary decomposition methods (i.e: majority voting) and classifier ensemble selection method (hill climbing forward selection by optimizing for accuracy). However better alternative techniques do exists and they may further improve performance, therefore need further investigation.

References

1. McKay, C., Fujinaga, I.: Musical Genre Classification: Is it worth pursuing and how can it be improved? In: Proc. ISMIR (2006)
2. Aucouturier, J., Pachet, F.: Improving timbre similarity: How high is the sky? Journal of Negative Results in Speech and Audio Sciences 1(1) (2004)
3. Doraisamy, S., Golzari, S., Norowi, N.M., Sulaiman, M.N., Udzir, N.I.: A Study on Feature Selection and Classification Techniques for Automatic Genre Classification of Traditional Malay Music. In: Proc. ISMIR (2008)
4. Yaslan, Y., Cataltepe, Z.: Audio Music Genre Classification Using Different Classifiers and Feature Selection Methods. In: Proc. ICPR (2006)
5. Soares, C., Williams, P., Gilbert, J.E., Dozier, G.V.: A Class-Specific Ensemble Feature Selection Approach for Classification Problems. In: Proc. ACMse (2010)
6. Zhang, T.: Semiautomatic approach for music classification. In: Proc. SPIE Conf. on Internet Multimedia Management Systems (2003)
7. Silla, C., Freitas, A.: Novel top-down approaches for hierarchical classification and their application to automatic music genre classification. In: Proc. IEEE SMC (2009)
8. Ariyaratne, H.B., Zhang, D.: A novel automatic hierarchical approach to music genre classification. In: Proc. ICME Workshop on Advances in Large Scale Multimedia Data Collection, Mining and Retrieval (2012)
9. Galar, M., Fernndez, A., Barrenechea, E., Bustince, H., Herrera, F.: An Overview of Ensemble Methods for Binary Classifiers in Multiclass Problems: Experimental Study on One-vs-One and One-vs-All Schemes. In: Pattern Recognition, vol. 44(8), pp. 1761–1776. Elsevier Science Inc. (2011)
10. Silla Jr., C.N., Koerich, A.L., Kaestner, C.A.A.: Feature Selection in Automatic Music Genre Classification. In: Proc. ISM (2008)
11. Rokach, L.: Taxonomy for Characterizing Ensemble Methods in classification Tasks: a review and annotated bibliography. In: Azen, S.P., Kontoghiorghes, E.J., Lee, L.C. (eds.) Computational Statistics & Data Analysis, vol. 53(12), pp. 4046–4072 (2009)
12. Caruana, R., Niculescu-Mizil, A., Crew, G., Ksikes, A.: Ensemble selection from libraries of models. In: Proc. ICML (2004)
13. Tzanetakis, G., Cook, P.: Musical Genre Classification of Audio Signals. IEEE Speech Audio Process 10(5), 293–302 (2002)
14. Witten, I., Frank, E., Hall, M.: Data Mining: Practical Machine Learning Tools and Techniques, 3rd edn. Morgan Kaufmann (2011)
15. Cevikalp, H.: Feature Extraction Techniques in High-dimensional Spaces: Linear and Nonlinear Approaches. Ph.D Thesis, Vanderbilt University (2005)

A Local Adaptation of the Histogram Radon Transform Descriptor: An Application to a Shoe Print Dataset

Makoto Hasegawa and Salvatore Tabbone

Université de Lorraine-LORIA, UMR 7503,
54506 Vandoeuvre-lès-Nancy, France
{makoto.hasegawa,tabbone}@loria.fr

Abstract. In this paper we propose a shape recognition approach applied to a dataset composed of 512 shoeprints where shapes are strongly occluded. We provide a local adaptation of the HRT (Histogram Radon Transform) descriptor. A shoeprint is decomposed into its connect components and describes locally by the local HRT. Then, following this description, we find the best local matching between the connected components and the similarity between two images is defined as mean of local similarity measures.

Keywords: Shape matching, local descriptor, histogram of Radon transform.

1 Introduction

In pattern recognition, image descriptions can be broadly categorized into statistical methods or structural methods. In the statistical method, many methods for a whole image have been proposed in the literature. The generic Fourier descriptor (GFD) proposed by Zhang and Lu [1] is a typical Fourier descriptor. The Fourier-Mellin transform (FMT) proposed Q. Chen et al. [2] is useful for many applications. Tabbone et al. propose methods called the histogram of Radon transform (HRT) [3] using the Radon transform. Recently, a local descriptor with vector of feature points in an image called SIFT is proposed by Lowe [4]. Structural methods[5] offer a good description thanks to the graph representation but they are not robust to noise.

In this paper, we consider a shoeprint dataset where shoeprints can be decomposed into connected components.The shoeprint can be decomposed into some connected components using the connectivity of 8 pixels around each pixel and each connected component is encoded using HRT for a local description. HRT has useful properties for shape rotation, shape scaling, and shape translation; then it is robust to geometric transformations of components. For a query shoe print image, we find the best local matching between the connected components of the query and the dataset and the similarity between two images is defined as mean of local similarity measures.

G.L. Gimel' farb et al. (Eds.): SSPR & SPR 2012, LNCS 7626, pp. 675–683, 2012.

We have carried out our experiments using a dataset composed of 512 shoeprints and we will show that our approach is competitive compared to well-known statistical descriptors (global HRT and FMT) and structural (graph edit distance) one.

2   HRT Descriptor

We recall the Radon transform definition in this section. Let a coordinate (x, y) in the two-dimensional $x-y$ plane described as $\mathbf{x}$, and an original image represented as $f(\mathbf{x})$. The Radon transform of $f(\mathbf{x})$ is defined as:

$$\mathcal{R}_f(\theta, \rho) = \int f(\mathbf{x})\delta(\mathbf{x} \cdot \xi - \rho)\mathrm{d}\mathbf{x}, \tag{1}$$

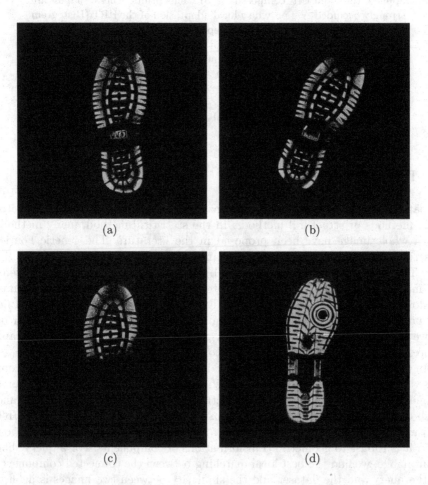

(a)   (b)

(c)   (d)

Fig. 1. (a) Full-print of "Shoe A"; (b) rotation; (c) occlusion. (d) Full-print of "Shoe B".

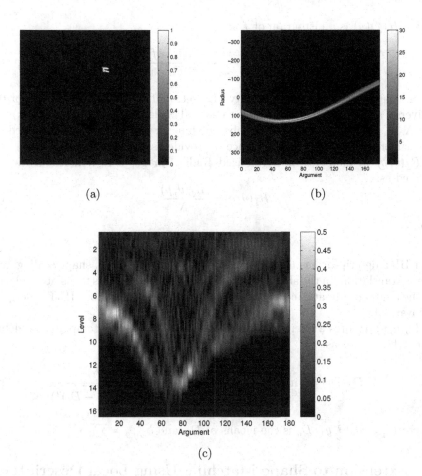

(a)   (b)

(c)

Fig. 2. (a) A connected component of "Shoe print A"; (b) and (c) respectively its Radon image and HRT descriptor

where $\xi = (\cos\theta, \sin\theta)$, and $\delta(\cdot)$ is a delta function. In other words, the Radon transform is the integral of $f(\mathbf{x})$ over lines

$$L_{\theta\rho} = \{\mathbf{x} \in R^2 | \mathbf{x} \cdot \xi = \rho\}, \tag{2}$$

where ρ is the distance between the origin and $L_{\theta\rho}$, the unit vector ξ and the angle θ describe the orientation of the line $L_{\theta\rho}$. The line integral is computed by a delta function $\delta(\cdot)$. An example of the Radon image using Fig. 2(a) is shown in Fig. 2(b); Fig. 2(a) is a connected component of "Shoe A" shown in Fig. 1(a).

The HRT descriptor is defined as a matrix of frequencies computed on the Radon image aggregated by the angle parameter of the Radon transform. The HRT descriptor $D(\theta, y)$ of $\dot{R}_f(\theta, \rho)$ for each orientation θ is:

$$D(\theta, y) = H(\dot{R}_f(\theta, \cdot))(y), \tag{3}$$

where $H(f)(y)$ is a histogram of f as:

$$H(f)(y) = \frac{\#\{x \in X | y = f(x)\}}{|X|}. \tag{4}$$

$\#$ in this equation is the cardinality of a set, and $|X|$ is the cardinality of the universal set. The universal set X is composed of elements in the case of $f(x) > 0$ as: $X = \{x | f(x) > 0\}$. The number of bins in the histogram is defined by experiments and we set it to 16 since it provide us the best performance.

$\dot{R}_f(\theta, \rho)$ in Eq.(3) is a normalized Radon image; here a normalization is defined as:

$$\dot{R}_f(\theta, \rho) = \frac{R_f(\theta, \rho)}{N}, \tag{5}$$

where

$$N = \max R_f(\theta, \rho). \tag{6}$$

Our HRT descriptor has useful properties for shape rotation, shape scaling, and shape translation. In the shape rotation, our HRT descriptor is translated relative to the angle coordinate; in the shape scaling and translation, our HRT descriptor is invariant.

$D_q(\theta, y)$ denotes a query descriptor, and $D_t(\theta, y)$ is for a template; we define a matching error as:

$$E(D_q, D_t) = \min_{\alpha \in [0, \pi)} \frac{\sum_{y, \theta}(D_q - \bar{D}_t)(D_q - \bar{D}_t)}{\sqrt{(\sum_{y, \theta}(D_q - \bar{D}_t)^2)(\sum_{y, \theta}(D_q - \bar{D}_t)^2)}}, \tag{7}$$

where $D_q = D_q(\theta, y)$, $\bar{D}_q$ is the means of D_q, and $\sum_{y, \theta} = \sum_y \sum_\theta$.

3   Extension to Shape Matching Using Local Descriptor

We decompose shapes on images into its connected components using the connectivity of 8 pixels around each pixel. Fig. 3 shows examples of the decomposition for shoeprint images. Each color of the connected component shows its component label.

Subsequently, each connected component is encoded using HRT, and we obtain two sets of descriptors of components, one is for a query, and the other is for a template.

We perform our local matching between the query descriptors and the template and find the closest one to each query component from the template components so that each best matching pair between the query and the template is defined. Matching error for each component pair is computed by Eq. (7); a mean value of the best matching pairs is the matching error between two images. Fig. 4 shows examples of our local shape matching between a query and a shoeprint image. We can remark that in some cases the matching between two connected components is not coherent and spatial constraints should be added to improve it.

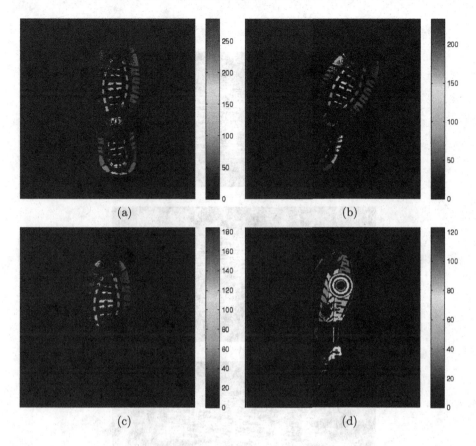

Fig. 3. Connected component decomposition

4   Experimental Results

The shoeprint dataset composed of 512 shoeprints; each shoeprint is a binary image with 512×512 pixels. Different example of shoes are shown in Fig. 5; each shoe has 8 images (4 images are full-prints; 4 images are strongly occluded: only toe, heel, light side, and left side). We will provide our dataset for a public common use.

4.1   Recognition Performance

Examples of matching score of our local descriptor HRT are shown in Table 1. We can see that shoes of the same class A are nearer than shoe B. The performance on the whole dataset is performed using the precision-recall curves defined as:

$$\text{Precision} = \frac{tp}{tp + fp}, \quad \text{Recall} = \frac{tp}{tp + fn}, \tag{8}$$

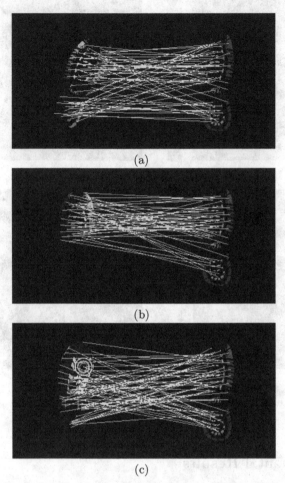

(a)

(b)

(c)

Fig. 4. Example of local shape matching using HRT local descriptors

Table 1. Matching errors using HRT local descriptor

		Template			
		Shoe A Fig.1(a)	Fig.1(b)	Fig.1(c)	Shoe B Fig.1(d)
Query	Shoe A Fig.1(a)	0.00	142.55	157.83	179.20
	Fig.1(b)	140.71	0.00	161.81	170.98
	Fig.1(c)	148.34	152.97	0.00	181.17
	Shoe B Fig.1(d)	182.62	179.15	196.05	0.00

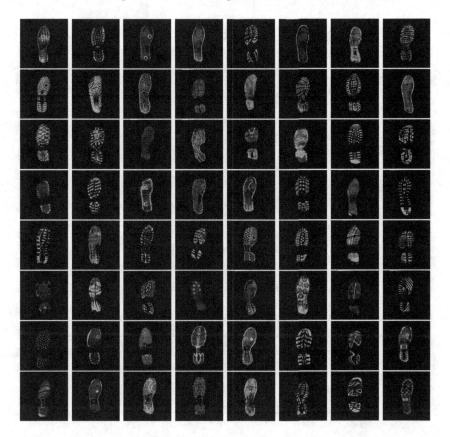

Fig. 5. Shoeprint dataset

where tp is the number of items correctly labeled as belonging to the positive class, fp is the number of items incorrectly labeled as belonging to the class, and fn is the number of items which are not labeled as belonging to the positive class but should have been.

We compare the precision-recall curves of our methods to the conventional HRT applied to a whole image, the Fourier-Mellin transform (FMT) and graph edit distance; the results are shown in Fig. 6. We can remark that global approaches are not suited for this kind of dataset and the performance provided by our local descriptor are globally good even if images are strongly occluded. For instance, as we can see in Table 1 the occluded shoeprint on Fig. 1(c) is nearer to Fig. 1(a) and 1(b) than Fig. 1(d).

4.2   Processing Time

Processing time for one matching using each method is shown in Table 2. Each method is performed by Intel Core i5-460 2.53 GHz CPU. HRT for local descriptor, Conventional HRT, and FMT are implemented using MATLAB and the

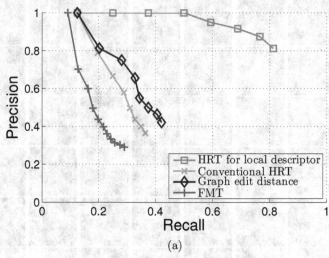

(a)

Fig. 6. Precision-recall curves of each method

Table 2. Processing time for one matching

Method	Processing time (Sec)
HRT for local descriptor	0.825
Conventional HRT	0.014
Graph edit distance	6.998
FMT	0.036

graph edit distance code provided in [5] is in C++. The processing time for our HRT local descriptor is high compared to the conventional HRT and FMT since we apply it locally.

5   Conclusion

We discuss a shape matching method using local descriptor. We apply HRT to connected components for local descriptions and find the best local matching between connect components. The obtained results are promising and could be improved by achieving a better decomposition of the image. Indeed, the performance of our local descriptor is strongly depend on the quality of the segmentation. In our case we use a simple decomposition based on the connected component. However, in some cases this decomposition is not appropriate. For instance, Fig. 7 show an original shape "Shoe B" and an connected component extracted on this image. This connected component has not been sufficiently decomposed, and in some cases a shoeprint is decomposed into one component only. This decomposition error leads to recognition errors and decreases the

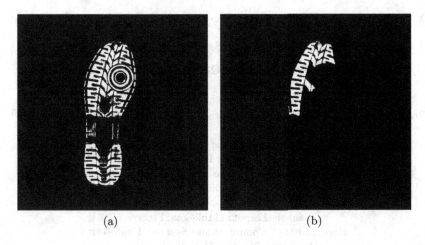

(a) (b)

Fig. 7. (a) Original shape "Shoe B"; (b) A large connected component

recognition performance. In this perspective, future investigations will be devoted to an appropriate shape decomposition method.

References

1. Zhang, D., Lu, G.: Shape-based image retrieval using generic Fourier descriptor. Sig. Proc.: Image Comm. 17(10), 825–848 (2002)
2. Sheng Chen, Q., Defrise, M., Deconinck, F.: Symmetric phase-only matched filtering of Fourier-Mellin transforms for image registration and recognition. IEEE Trans. PAMI 16(12), 1156–1168 (1994)
3. Tabbone, S., Ramos Terrades, O., Barrat, S.: Histogram of Radon transform. A useful descriptor for shape retrieval. In: International Conference on Pattern Recognition, Tampa, pp. 1–4 (2008)
4. Lowe, D.G.: Distinctive image features from scale-invariant keypoints. International Journal of Computer Vision 60(2), 91–110 (2004)
5. Riesen, K., Bunke, H.: Approximate graph edit distance computation by means of bipartite graph matching. Image Vision Comput. 27(7), 950–959 (2009)

A Multiple Classifier System for Classification of Breast Lesions Using Dynamic and Morphological Features in DCE-MRI

Roberta Fusco[1], Mario Sansone[1], Antonella Petrillo[2], and Carlo Sansone[3]

[1] Department of Biomedical, Electronic and Telecommunication Engineering
University Federico II of Naples, Italy
{roberta.fusco,mario.sansone}@unina.it
[2] Department of Diagnostic Imaging
National Cancer Institute of Naples 'Fondazione Pascale'
antonellapetrillo2@gmail.com
[3] Department of Computer and Systems Engineering
University Federico II of Naples, Italy
carlosan@unina.it

Abstract. In this paper we propose a Multiple Classifier System (MCS) for classifying breast lesions in Dynamic Contrast Enhanced-Magnetic Resonance Imaging (DCE-MRI). The proposed MCS combines the results of two classifiers trained with dynamic and morphological features respectively. Twenty-one malignant and seventeen benign breast lesions, histologically proven, were analyzed. Volumes of Interest (VOIs) have been automatically extracted via a segmentation procedure assessed in a previous study. The performance of the MCS have been compared with histological classification. Results indicated that with automatic segmented VOIs 90% of test-set lesions were correctly classified.

Keywords: breast DCE-MRI, multiple classification system, morphological and dynamic features.

1 Introduction

Breast cancer is the most common cancer among women in the Western world. To date it is the second leading cause of cancer death in women (after lung cancer) and is estimated to cause 15% of cancer deaths [2]. Therefore, screening for early diagnosis of breast cancer is of great interest.

The currently widespread screening method is RX mammography [1]. However, this method has some drawbacks: it uses ionizing radiation; it is not adequate for young women because of high density breasts; detection of breast lesions is difficult because of the lack of functional information, moreover no 3D information is available but the whole breast is projected on one or two planes. Although Magnetic Resonance Imaging (MRI) has some limitations such as long scanning time, cost, possible side effects of contrast media injection, the emerging methodology of Dynamic Contrast Enhanced-MRI (DCE-MRI) has

G.L. Gimel' farb et al. (Eds.): SSPR & SPR 2012, LNCS 7626, pp. 684–692, 2012.
© Springer-Verlag Berlin Heidelberg 2012

demonstrated a great potential in screening of high-risk women, in staging newly diagnosed breast cancer patients and in assessing therapy effects [2] thanks to its minimal invasiveness and to the possibility to visualize 3D high resolution dynamic (functional) information not available with conventional RX imaging.

Therefore MRI, and in particular DCE-MRI, is gaining popularity as an important complementary diagnostic tool for early detection of breast cancer [3].

In the analysis of breast lesions on MRI, radiologists agree that both *morphological* and *dynamic* features are important for distinguishing benign from malignant [6,5]. On the one hand, morphological features aim to quantify lesion margins characteristics, and are well assessed in the breast-MRI lexicon [6]: round shape and smooth margin for the benign lesions; more irregular shape for the malignant lesions. On the other hand, dynamic features, have shown a great potential in quantifying vascularity of tumors: malignant lesions usually show early enhancement with rapid wash out, whereas benign lesions typically show a slow increase followed by persistent enhancement [6].

Many recent works have attempted to take advantage of morphological features and dynamic information in a separate manner: dynamic information has been used for segmentation of volume of interests (VOIs) [13,12,11], while morphological features of the VOIs have been used for lesion classification [10,7,6]. For example, Nie et al. [4] demonstrated that quantitative analysis of morphology and texture features of breast lesions was feasible, and these features could be selected by artificial neural network to form a classifier for differential diagnosis. Agner et al. [9] showed that using a probabilistic boosting tree (PBT) classifier in conjunction with textural kinetic features good performances could be yielded but when the feature set included both textural kinetic and morphologic features the performances were lower. Tzacheva et al.[8] reported good performances using morphological features and MLP classifier on 14 breast lesions; Zheng et al. [10,18] has investigated the use of a feature set comprising dynamic, spatial, and morphological features with a linear classifier on 31 subjects.

To the best of our knowledge, a Multiple Classification System for classification of breast lesions using dynamic and morphological features in DCE-MRI has not been presented yet, although the idea of combining multiple classifiers is not new. For example, Keyvanfard et al. [21] proposed a multi classifier system composed of three classifiers that used dynamic features to classify breast lesion in DCE-MRI, but in their study morphological features were not used.

The aim of the present study is to propose a Multiple Classification System (MCS) for classification of breast lesions using both dynamic and morphological features in DCE-MRI.

The proposed MCS combines the results of two classifiers trained with dynamic and morphological features respectively. As classifiers, we used the best suited for the problem at hand, according to our previous studies [16].

Twenty-one malignant and seventeen benign breast lesions, histologically proven, were analyzed. Volumes of Interest (VOIs) have been both manually extracted by an expert radiologists and automatically extracted via a segmentation procedure assessed in a previous study. Both dynamic and morphological features were extracted. The performance of the MCS have been compared with histological classification.

The paper is organized as follows: in section 2.1 we describe the characteristics of the subjects enrolled in the study; in section 2.2 we give details on the data acquisition via DCE-MRI; in section 2.3 we show how the volumes of interest (VOIs) have been selected manually and automatically; in section 2.4 we illustrate the main characteristics of the features used; in section 2.5 we present the proposed multiple classifier system: in particular we summarise the characteristics of the classifiers used, together with the classifiers for dynamic features and morphological features respectively and the combination scheme. We subsequently report the performances of the system proposed in section 3 and finally we compare our results to the literature in section 4.

2   Materials and Methods

2.1   Patients Selection

38 women (average age 46 years, range 16-69 years) with benign or malignant lesions histopathologically proven were enrolled. 21 lesions were malignant and 17 were benign. The lesions were subdivided in two groups: training-test (12 benign and 16 malignant) and test-set (5 benign and 5 malignant).

2.2   Data Acquisition

The patients underwent imaging with a 1.5 T scanner (Magnetom Symphony, Siemens Medical System, Erlangen, Germany) equipped with breast coil. Turbo spin echo T2-weighted axial images (TR/TE: 4000/56 ms; flip angle: 180 degrees; field of view 340 x 340 mm x mm; matrix: 384 x 385; thickness: 2 mm; gap: 0; 56 slices spanning entire breast volume) and Turbo spin echo T1-weighted fut sat axial images (TR/TE: 564/12 ms; flip angle: 90 degrees; field of view 350 x 350 mm x mm; matrix: 512 x 256; thickness: 2 mm; gap: 0; 40 slices spanning entire breast volume) were acquired for morphological imaging.

DCE T1-weighted FLASH 3-D coronal images were acquired (TR/TE: 9.8 / 4.76 ms; flip angle: 25 degrees; field of view 330 x 247 mm x mm; matrix: 256 x 128; thickness: 2 mm; gap: 0; acquisition time: 56 s; 80 slices spanning entire breast volume). One series was acquired before and 9 series after intravenous injection of 2 ml/kg body weight of a positive paramagnetic contrast medium (Gd-DOTA, Dotarem, Guerbet, Roissy CdG Cedex, France). Automatic injection system was used (Spectris Solaris EP MR, MEDRAD, Inc.,Indianola, PA) and injection flow rate was set to 2 ml/s followed by a flush of 10 ml saline solution at the same rate.

2.3   VOI Segmentation

Manual. Manual ROI selection slice-by-slice was performed by an expert radiologist taking into account both morphological (Turbo spin echo T2- and T1 weighted images) and functional imaging (DCE T1 weighted images) based on the fat-suppressed image obtained subtracting the basal pre-contrast image from the 5-th post-contrast image. Per each patient all the slices including the lesion have been used. The segmentation was performed with OsiriX v.3.8.1 (fig. 1).

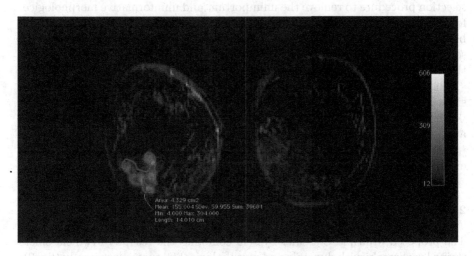

Fig. 1. Manual Segmentation

Automatic. In automatic selection the first step was the overall breast mask segmentation by means of Otsu thresholding of the parametric map obtained considering the pixel-by-pixel SOD followed by hole-filling and leakage removal by means of morphological operators as in a previous study [15]. Successively, **automatic** VOI segmentation has been obtained with pixel by pixel classification of dynamic features by using an MLP classifier.

The dynamic features used in the classification were sum of intensities difference (SOD), basal signal and relative enhancement slope calculated pixel by pixel on the breast mask: they were used as input of MLP classifier (learning rate = 0.3, momentum = 0.2, and a training time of 100 epochs). The classifier labeled each pixel as *suspicious* or *not suspicious*. The region of interest was obtained by the union of all suspicious pixels.

2.4   Morphological and Dynamic Features

Starting from our previous studies [16,17], we considered a feature set including 54 morphological features and 98 dynamic features, respectively.

The main categories of morphological features included areas, circularity, compactness, complexity, perimeter, radial length, smoothness, roughness, sphericity, eccentricity, volume, rectangularity, solidity, spiculation, convexity, curvature, edge [10,7,6]. For dynamic features the main categories included area, maximum intensity ratio, relative enhancement, relative enhancement slope, basal signal, perfusion index, sum of intensities difference (SOD), wash-in, wash-out, time to peak [13,12,11].

It is well known that training machine learning classifiers with large numbers of morphological features can lead to classifier overfitting, reduces the generalization capabilities of the classifiers and slows down the training process. Therefore, the number of morphological and dynamic features were reduced by a feature selection procedure to remove the unimportant and uninformative morphological features. To keep the loss of information to a minimum we tested Correlation-based Feature Selection (CFS) and Consistency feature Selection method with several search: the forward search,the backward search, the bidirectional search, the greedy search, feature ranking methods.

The dynamic features obtained by selection procedure were the sum of intensities difference (SOD), basal signal and relative enhancement slope [13,12,11].

The morphological features obtained by the selection procedure were instead area, eccentricity, compactness and perimeter.

The classifiers were trained with the morphological and dynamic feature extracted both by manual that automatic VOI segmentation.

2.5   VOI Classification

The proposed Multiple Classifier System combines the results of two classifiers trained separately with dynamic and morphological features respectively (fig. 2): in particular it was considered the weighted sum of probability of malignity and the probability of benignity of the two chosen classifiers as proposed in[20].

The choice of the classifier to be used was based on a previous study [16], where a Decision Trees (DT) and a Bayesian classifier gave us the best results when trained on morphological and dynamic features, respectively.

In order to combine the results of the two classifiers, each suspicious pixel within the VOI has been first classified as *benign* or *malignant* on the basis of the selected dynamic features. The whole VOI has been then classified as malignant if the number of malignant pixels n_m within the VOI was higher than that of benign pixel n_b within the same VOI.

The probabilities of malignant lesion (D_m) and benign lesion (D_b) were calculated as eq. 1:

$$D_m = \frac{n_m}{N}$$
$$D_b = \frac{n_b}{N} \tag{1}$$

where N is the total number of pixels in the lesion.

Morphological features were instead calculated for the whole VOI and were used to classify the lesion in malignant and benign. In this case the probability of malignity and benignity were M_m and M_b respectively.

Finally, the VOI was classified as malignant if $\alpha D_m + \beta M_m > \alpha D_b + \beta M_b$, where α and β were multiplicative coefficients ($\alpha + \beta = 1$) that must be suitably chosen in order to maximize the accuracy (fig. 2). In [20], a leave-one-out procedure is suggested.

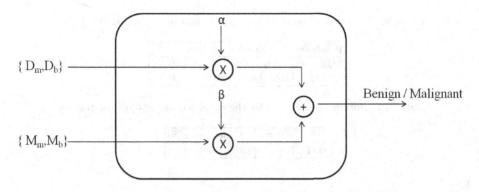

Fig. 2. Multiple Classifier System

3   Results

The results were reported for automatic VOI segmentation that have obtained the better findings. Table 1 shows the results obtained on the test set (10 patients) by the single classifiers that use dynamic and morphological features, respectively. Table 2 reports the performance obtained on the test set (10 patients) by our multiple classifiers system, using automatic VOI segmentation. The acronyms used in this table are the following: MC = Classifier trained with morphological feature; DC = Classifier trained with dynamic feature.

As far as the choice of optimal values for α and β is concerned we report in fig. 3 the percentage of correctly classified lesions vs. α. It is worth to notice that

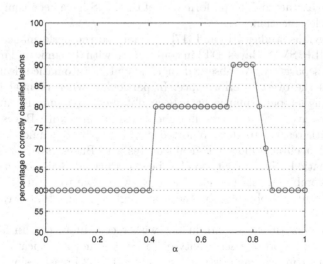

Fig. 3. Performances of the proposed MCS vs α

Table 1. Performance on the test set obtained by the single classifers

Classifer	Feature	accuracy[%]
Bayes	Dynamic	80.00
DT	Morphological	60.00

Table 2. Performance obtained by the proposed method on the test set

MC	DC	α	β	accuracy[%]
DT	BC	0.75	0.25	90.00

there is an interval ([0.7, 0.8]) of values in which high performance can be obtained. The best compromise among sensitivity, specificity and overall accuracy resulted in $\alpha = 0.75$.

4   Discussion and Conclusions

The aim of this study was to propose a Multiple Classifier System (MCS) for classifying breast lesions in Dynamic Contrast Enhanced-Magnetic Resonance Imaging (DCE-MRI).

The proposed MCS combines the results of two classifiers trained with dynamic and morphological features respectively.

Twenty-one malignant and seventeen benign breast lesions, histologically proven, were analyzed. Volumes of Interest (VOIs) have been both manually extracted by an expert radiologists and automatically extracted via a segmentation procedure assessed in a previous study [15]. Both dynamic and morphological features were extracted. The performance of the MCS have been compared with histological classification.

In our previous studies [16] and [17] we analysed the performance of several classifiers (MLP, SVM, Bayes, DT) in conjunction with dynamic and morphological features separately. We observed that manual or automatic selection of the VOI affected the overall performance. In particular, when manual ROIs have been used, higher performance have been obtained (dynamic features with a DT classifier gave 70% accuracy; morphological features with Bayes gave 90% accuracy) with respect to those reported in this paper (dynamic features with DT gave 60% accuracy; morphological features with Bayes gave 80% accuracy), where automatic ROIs segmentation has been performed. This notwithstanding, by suitably combining the two classifiers, we were able to obtain the same results (90% accuracy) obtained by the best single classifier that uses manually extracted VOIs.

The findings of this study are in line with recent literature. In fact, Wedegrtner et al [19] reported a sensitivity of 85% using morphological features and a receiver operating characteristic (ROC) curve for 62 breast lesions but without automatic classification; Tzacheva et al. [8] reported a sensitivity of 90%, a specificity of 91% and an accuracy of 91% using morphological features and

MLP classifier on 14 breast lesions. However, these result must be carefully considered because of the small number of patients and because they did not use an automatic segmentation step. Zheng et al [18] reported a sensitivity of 95% using a combination of temporal, spatial, and morphological attributes and a linear classifier for 31 subjects: even in this study the segmentation step was not completely automatic.

It should be noticed that although previous studies obtained in some cases higher accuracy in comparison to our findings, however, they did not employ a completely automatic multi-classifier system such as the system presented in the present study.

Moreover, further investigation is required for an optimal choice of α and β because as is clear from fig. 3 the specific value could affect the overall accuracy of the system.

In the future, our preliminary study will be extended on a larger number of patients, manual segmentation will be done by multiple readers and morphological, dynamic and texture [7] feature combination will be performed

References

1. Olsen, O., et al.: Screening for breast cancer with mammography. Cochrane Database of Systematic Reviews CD001877 (2001)
2. Lehman, C.D., et al.: MRI evaluation of the contralateral breast in women with recently diagnosed breast cancer. The New England Journal of Medicine 356, 1295–1303 (2007)
3. Schnall, M., et al.: Diagnostic architectural and dynamic features at breast MR imaging: Multicenter study. Radiology 238, 42–53 (2006)
4. Nie, K., et al.: Quantitative analysis of lesion morphology and texture features for diagnostic prediction in breast MRI. Acad Radiol. 15(12), 1513–1525 (2008)
5. Liney, G.P., et al.: Turnbull, Breast lesion analysis of shape technique: semiautomated vs. manual morphological de- scription. Journal of Magnetic Resonance Imaging 23, 493–498 (2006)
6. Ikeda, D.M., et al.: Development, standardization, and testing of a lexicon for reporting contrast-enhanced breast magnetic resonance imaging. Journal of Magnetic Resonance Imaging 13, 889–895 (2001)
7. McLaren, C.E., et al.: Prediction of malignant breast lesions from MRI features: a comparison of artificial neural network and logistic regression techniques. Acad. Radiol. 16(7), 842–851 (2009)
8. Tzacheva, A.A., et al.: Reast cancer detection in gadolinium-enhanced MR images by static region descriptors and neural networks. J. Magn. Reson. Imaging 17(3), 337–342 (2003)
9. Agner, S.C., et al.: Textural kinetics: a novel dynamic contrast-enhanced (DCE)-MRI feature for breast lesion classification. J. Digit. Imaging 24(3), 446–463 (2011)
10. Zheng, Y., et al.: Segmentation and classification of breast tumor using dynamic contrast-enhanced MR images. Med. Image Comput. Comput. Assist. Interv. 10(Pt 2), 393–401 (2007)
11. Degenhard, A., et al.: UK MRI Breast Screening Study. Comparison between radiological and artificial neural network diagnosis in clinical screening. Physiol. Meas. 23(4), 727–739 (2002)

12. Gibbs, P., et al.: Textural analysis of contrast-enhanced MR images of the breast. Magn. Reson. Med. 50(1), 92–98 (2003)
13. Schlossbauer, T., et al.: Classification of small contrast enhancing breast lesions in dynamic magnetic resonance imaging using a combination of morphological criteria and dynamic analysis based on unsupervised vector-quantization. Invest Radiol. 43(1), 56–64 (2008)
14. Juntu, J., et al.: Machine learning study of several classifiers trained with texture analysis features to differentiate benign from malignant soft-tissue tumors in T1-MRI images. J. Magn. Reson. Imaging 31(3), 680–689 (2010)
15. Fusco, R., et al.: Selection of Suspicious ROIs in Breast DCE-MRI. In: Maino, G., Foresti, G.L. (eds.) ICIAP 2011, Part I. LNCS, vol. 6978, pp. 48–57. Springer, Heidelberg (2011)
16. Fusco, R., et al.: Classification of breast lesions using dynamic and morphological features in DCE-MRI. In: Proceedings of Congresso Nazionale di Bioingegneria 2012, Rome, Italy, June 26-29 (2012)
17. Fusco, R., et al.: Segmentation and classification of breast lesions using dynamic features in Dynamic Contrast Enhanced-Magnetic Resonance Imaging. In: Proceedings of 25th IEEE International Symposium on Computer based Medical Systems (CBMS), Rome, Italy, June 20-22 (2012)
18. Zheng, Y., et al.: STEP: spatiotemporal enhancement pattern for MR-based breast tumor diagnosis. Med. Phys. 36(7), 3192–3204 (2009)
19. Wedeğartner, U., et al.: Differentiation between benign and malignant findings on MR-mammography: usefulness of morphological criteria. Eur. Radiol. 11(9), 1645–1650 (2001)
20. De Santo, M., et al.: Automatic Classification of Clustered Microcalcifications by a Multiple Expert System. Pattern Recognition 26, 1467–1477 (2003)
21. Keyvanfard, F., et al.: Specificity enhancement in classification of breast MRI lesion based on multi-classifier. Neural Computing and Applications (2012), doi:10.1007/s00521-012-0937-y

A Comparative Analysis
of Forgery Detection Algorithms

Davide Cozzolino, Giovanni Poggi, Carlo Sansone, and Luisa Verdoliva

Department of Electrical Engineering and Information Technologies
University Federico II of Naples
Via Claudio, 21, 80125 Naples, Italy
{davide.cozzolino,poggi,carlosan,verdoliv}@unina.it

Abstract. The aim of this work is to make an objective comparison between different forgery techniques and present a tool that helps taking a more reliable decision about the integrity of a given image or part of it. The considered techniques, all recently proposed in the scientific community, follow different and complementary approaches so as to guarantee robustness with respect to tampering of different types and characteristics. Experiments have been conducted on a large set of images using an automatic copy-paste tampering generator. Early results point out significant differences about competing techniques, depending also on complexity and side information.

Keywords: forgery detection, digital forensics, image tampering.

1 Introduction

With the ever increasing diffusion of simple and powerful software tools for digital source editing, image tampering is becoming more and more common, stimulating an intense quest for algorithms, to be used in the forensics field, which help deciding about the integrity of digital images. In particular, passive techniques draw the highest attention as they require no collaboration on the part of the user through some types of watermarks or signatures.

A large variety of approaches have been proposed in the literature, which take advantage of the inconsistencies (e.g., in the histogram or in the sampling grid) arising from the various tampering types. Following [1], they can be classified as pixel-based, format-based, camera-based, and phisics/geometric-based.

In extreme synthesis, pixel-based techniques analyze the correlation between pixels either directly in the spatial domain or in some transformed domain. Format-based methods, instead, exploit the usual adoption of some lossy compression scheme, like JPEG, which introduces some recognizable marks in the presence of manipulations. Camera-based techniques take advantage of features specific of any different camera models, and even of individual cameras, to use them as image signatures to exploit in forgery detection. Finally, phisics/geometric-based methods study higher-level inconsistencies between the imaged scene/object and the forgery source, such as illumination, object size, etc.

G.L. Gimel' farb et al. (Eds.): SSPR & SPR 2012, LNCS 7626, pp. 693–700, 2012.
© Springer-Verlag Berlin Heidelberg 2012

The many existing approaches, which comprise in turn a very large and ever growing number of individual detection techniques, testify both the interest towards this problem and its complexity. In fact, no ultimate solution exists to the image forgery detection problem. Each technique is based on some important hypotheses which limit its applicability, and therefore it is always possible to find cases where it fails. On the other hand, one should take for granted that a malicious tamperer, aware of the principles on which each technique works, will be able to trick it, given enough time and resources. Having a multiplicity of different tools is therefore essential to guarantee a high probability of detecting forgeries.

Given the great variety of approaches, comparing forgery detection algorithms turns out to be a very challenging task, although a much needed one. To increase confusion, many Authors, to support the claim of good performance of their proposed algorithm, provide results on a restricted set of images, without facing the problem of making an extensive comparison with other approaches in terms of key performance indices. So, the habit of proposing more and more new algorithms seems to prevail against the need of assessing the performance of the existing ones in an objective way.

In this work we perform an experimental performance comparison of a limited but significant set of forgery detection techniques. A similar work was presented in [2] where several techniques were considered, all belonging to the format-based category. Here, we consider four techniques chosen among the most popular and promising in the recent literature, belonging to different categories: pixel-based, format-based, and camera-based. As dataset, we used a set of images taken from a standard database [3] while forgeries have been obtained by using the copy-paste approach proposed in [4]. Tests have been carried out by varying the size and statistics of the tampered area, and results are reported in terms of the usual sensitivity and specificity parameters.

All algorithms are implemented in Matlab and integrated into an easy-to-use software tool, written in JAVA, easily extensible to include new techniques and functionalities. The tool allows one to process one or more images at once, select one or more detection algorithms, so as to allow comparison of results, and look for a forgery on the whole image or else only in a part of it.

In the following section we describe briefly the forgery detection algorithms under comparison, while in section three the experimental setting is presented and results are commented. Finally section 4 draws conclusions and outlines future work.

2   Forgery Techniques under Analysis

For this comparative study, in order to sample solutions as different as possible, we have chosen techniques that use various approaches: format based [5], pixel-based [6] and camera-based [7] [8]. In the following we will briefly describe the implemented algorithms.

Format-based methods take advantage of the specific format of images. Since most images are JPEG compressed, to detect a tampering it is possible to exploit the blocking effect introduced by JPEG, which gives rise to the so-called Block Artifact Grid (BAG).

In fact, manipulating images in this format causes an alteration of these artifacts, mainly in the case of *copy and paste* processing, since the BAG of the original image and that of the copied region very likely mismatch. In [5] a simple method is proposed to identify this type of forgery, named here Li-2009 after the name of the first Author and the year of publication, a convention followed for the other techniques as well. The basic idea is to extract the horizontal and vertical edges due to JPEG artifacts by means of a second order derivative followed by a thresholding operation in order to eliminate edges relative to signal discontinuities. A further enhancement is then realized to obtain the block artifact grid. If the image has been subject to a copy-paste processing a BAG mismatching can be detected when lines are present within a 8×8 block. The procedure delineated in [5] tries to determine this presence through summations along rows and columns both inside and at the boundaries of the block.

In [6] a pixel-based technique is proposed, Popescu-2005, relying on finding traces of resampling in the image. The idea is based on the observation that tampering may alter the underlying statistics. In fact, when an image is modified, operations like resizing, rotating and stretching must be typically performed, which require to resample the original image. This process introduces correlations that, once detected, can be considered as evidence of a digital tampering. The detection process is based on estimating, through the expectation/maximization algorithm, a set of periodic samples that are correlated to their neighbors.

A very powerful approach in detecting forgeries relies on artifacts introduced by the digital camera itself, and in particular the photo-response non uniformity (PRNU) which can be considered as a sort of intrinsic fingerprint of a specific digital camera. The PRNU arises from differences and imperfections in the silicon wafer used to manufacture the imaging sensor: these physical differences provide a unique sensor fingerprint which can be used for forgery detection. The Chen-2008 algorithm, proposed in [7], requires the preliminary estimation of the camera PRNU from a large number of images taken by the camera itself. Then, the PRNU of the image under investigation is estimated and compared with the reference. This step is quite challenging, since this fingerprint is much weaker than the image, therefore a denoising step is used, which removes much of the image content increasing the signal-to-noise ratio. In [9] we replaced the original denoising algorithm with state-of-the-art nonlocal filtering, obtaining a significant performance improvement. The PRNU comparison is carried out by sliding an analysis window of dimension 128×128 over the image: if the camera PRNU is present, the block (or more correctly its central pixel) is labeled as genuine, otherwise it is considered tampered. The test statistic used for detection is the normalized correlation value with a decision threshold selected so as to obtain the required false acceptance rate. A similar algorithm, Zhang-2010, has been recently proposed in [8]. It makes use of canonical correlation analysis (CCA) to measure the linear correlation between the two PRNU estimates. Only for heavily textured areas, identified in advance in the image, a Neyman-Pearson decision is used like in [7].

3   Experimental Results

We now study the performance of the selected forgery detection techniques through simulation experiments on a large number of tampered images.

Our test set is formed starting from 72 high-resolution photos of the Dresden Image Database [3]. Each image is then subject to copy-paste forgery by means of the Photoshop scripts of the University of Catania [4], which allows to superpose objects taken from a small library, scaled and translated at will, over the target image. Some photos have been used several times with different forgeries, reaching a total of 108 images, presenting one or more forgeries. Relatively large forgeries are considered, cover about 20% of the image area, because most techniques become too unreliable in the presence of small tampered regions. PRNU-based techniques, for example, discard altogether dubious objects smaller than 64×64 pixels, although a recent segmentation-based version [10] allows one to deal also with somewhat smaller objects. Fig.1 shows one of the original images together with its tampered version.

Fig. 1. An original test image (left) and its tampered version (right)

The selected techniques, as described in the literature, are not immediately comparable since they provide different types of results: some are pixel-based, some block-based, some require visual inspection. In order to provide homogeneous results, the image is divided in non-overlapping square blocks, of size 128×128 through 1024×1024 pixels, with decisions taken independently for each of them considered as a whole.

For each technique we compute on the entire database the quantities

TP (true positive): # forged blocks declared forged
FP (false positive): # genuine blocks declared forged
TN (true negative): # genuine blocks declared genuine
FN (false positive): # forged blocks declared genuine

Results are then given in terms of sensitivity, specificity and accuracy, computed as

$$\text{sensitivity} = \frac{TP}{TP + FN}$$

$$\text{specificity} = \frac{TN}{TN + FP}$$

$$\text{accuracy} = \frac{TN + TP}{TN + FP + TP + FN}$$

which measure, respectively, the ability to detect the presence of forgery, the ability to confirm the absence of forgery, and the overall classification accuracy, independent of the nature of the blocks.

Figures 2 through 5 report results for block-sizes 128×128, 256×256, 512×512, and 1024×1024.

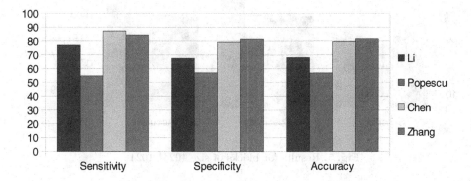

Fig. 2. Results for blocks of size 128×128

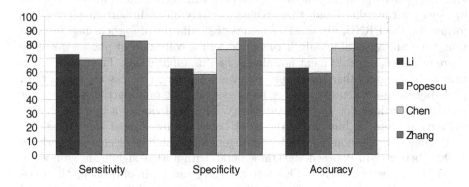

Fig. 3. Results for blocks of size 256×256

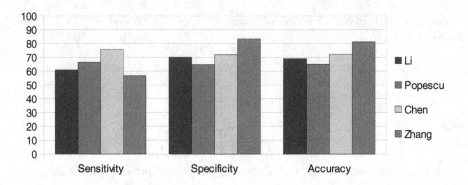

Fig. 4. Results for blocks of size 512×512

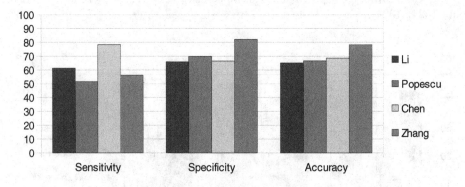

Fig. 5. Results for blocks of size 1024×1024

The first obvious consideration is that camera-based techniques, Chen-2008 and Zhang-2010, perform generally much better than the others, under all points of view. On the other hand, these techniques rely on an important piece of information, the PRNU, obtained by means of significant pre-processing. In addition, they are applicable only if the camera or a collection of photos taken by it are available, a requirement not always met in applications. The gap between camera-based and other techniques reduces when larger blocks are considered, probably for the increased number of blocks that are only partially forged, for which all decisions become quite arbitrary. This is obviously a limit of the experimental protocol, but also of the block-based nature of most algorithms.

Turning to the non-camera-based algorithms, Li-2009 seems to perform slightly better than Popescu-2005, in general, with a more significant gain when small blocks are considered. This latter quality can be important in the presence of small-size forgeries which are intrinsically more difficult to detect and therefore more challenging for a forgery detection algorithm.

As a visual example of results, Fig.6 shows the output provided for the image of Fig.1 by the Chen-2008 and Li-2009 methods, where only the 128×128 blocks considered tampered are shown. As expected, the camera-based Chen-2008 is clearly more accurate. In this case, however, the bad performance of Li-2009 is also a consequence of the low compression ratio of the original JPEG image, which therefore shows little or no trace of the block artifact grid needed to detect forgeries. This gives us the opportunity to underline once again that no single technique works in all cases and a judicious use of several complementary techniques is always recommended.

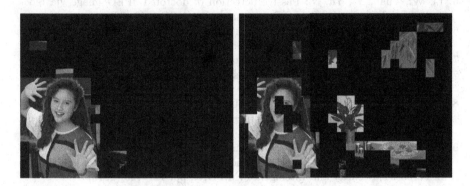

Fig. 6. Results of Chen-2008 (left) and Li-2009 algorithms for the image of Fig.1

4   Conclusions and Future Work

In this work we dealt with the challenging task of comparing several forgery detection techniques. Experimental results are quite reasonable, as performance seems to depend strongly on complexity and side information available. Obviously, we want is to extend this type of comparison including many more techniques, a work which is currently under way.

The major problems encountered, besides the implementation of the individual techniques, concern the interpretation of their results in terms of homogeneous and meaningful performance measures. Under this point of view, the block-based approach represents just a reasonable compromise, upon which we are already trying to improve.

Another important goal is to improve the overall detection reliability by implementing an information fusion level in which the output of many different detection algorithm is taken into account and properly combined. Some preliminary experiments have been conducted by using a simple weighted sum of the individual decisions [11]. Although only a very small set of detectors is available, results are already encouraging.

References

1. Farid, H.: Image forgery detection. IEEE Signal Processing Magazine 26, 16–25 (2009)

2. Battiato, S., Messina, G.: Digital Forgery Estimation into DCT Domain - A Critical Analysis. In: ACM Multimedia Workshop on Multimedia in Forensics, pp. 37–42 (2009)
3. Gloe, T., Böhme, R.: The Dresden Image Database for Benchmarking Digital Image Forensics Categories and Subject Descriptors. In: ACM Symposium on Applied Computing, pp. 1584–1590 (2010)
4. IPLab. Image Processing Laboratory - Forensic Database, http://iplab.dmi.unict.it/index.php?option=comdocman&Itemid=111
5. Li, W., Yuan, Y., Yu, N.: Passive detection of doctored JPEG image via block artifact grid extraction. Signal Processing 89(9), 1821–1829 (2009)
6. Popescu, A.C., Farid, H.: Exposing digital forgeries by detecting traces of resampling. IEEE Transactions on Signal Processing 53(2), 758–767 (2005)
7. Chen, M., Fridrich, J., Goljan, M., Lukas, J.: Determining Image Origin and Integrity Using Sensor Noise. IEEE Transactions on Information Forensics and Security 3, 74–90 (2008)
8. Zhang, C., Zhang, H.: Exposing Digital Image Forgeries by Using Canonical Correlation Analysis. In: International Conference on Pattern Recognition (ICPR), pp. 838–841 (2010)
9. Chierchia, G., Parrilli, S., Poggi, G., Sansone, C., Verdoliva, L.: On the influence of denoising in PRNU based forgery detection. In: 2nd ACM Workshop on Multimedia in Forensics, Security and Intelligence (MiFOR), pp. 117–122 (2010)
10. Chierchia, G., Parrilli, S., Poggi, G., Sansone, C., Verdoliva, L.: PRNU-based detection of small size image forgeries. In: 17th International Conference on Digital Signal Processing (DSP), pp. 1–6 (2011)
11. Kuncheva, L.I.: Combining Pattern Classifiers: Methods and Algorithms. Wiley Interscience (2004)

Low Training Strength High Capacity Classifiers for Accurate Ensembles Using Walsh Coefficients

Terry Windeatt and Cemre Zor

Univ Surrey, Guildford, Surrey, Gu2 7XH
t.windeatt'surrey.ac.uk

Abstract. If a binary decision is taken for each classifier in an ensemble, training patterns may be represented as binary vectors. For a two-class supervised learning problem this leads to a partially specified Boolean function that may be analysed in terms of spectral coefficients. In this paper it is shown that a vote which is weighted by the coefficients enables a fast ensemble classifier that achieves performance close to Bayes rate. Experimental evidence shows that effective classifier performance may be achieved with one epoch of training of an MLP using Levenberg-Marquardt with 64 hidden nodes.

Keywords: Ensembles, Multilayer Perceptrons, Boolean Function, Walsh Coefficients.

1   Introduction

For an ensemble of classifiers it is often useful to think of each base classifier as being controlled by two main parameters, the *capacity* and the *training strength* of the learning algorithm [1]. The term *capacity* refers to the flexibility of the classifier boundary. By *training strength* we mean the effort that is put into training the classifier. For an MLP, the capacity is the number of hidden nodes, and training strength is the number of epochs. In this paper we consider the trade-off between these two parameters, and what combination is suitable for a weighted majority vote.

The weighted vote is computed using Walsh coefficients. If each base classifier in an ensemble is given a binary decision, and if the problem is two-class, a Boolean mapping is defined. This mapping may be analysed using Walsh spectral coefficients. First order Walsh coefficients were shown to provide a measure of class separability for selecting optimal base classifiers in [2], in which it is also shown that this does not imply optimality of the ensemble. In contrast, in [3] it was shown that second order Walsh coefficients may be used to determine optimal ensemble performance. The motivation for using Walsh coefficients in ensemble design is fully explored in [4] and [2]. For further understanding of the meaning and applications of Walsh coefficients see [5] and [6].

To understand the computation of the weighted vote, the Tumer-Ghosh model [7] for ensemble classifiers will be described. This model defines Added Classification Error as the difference between classifier error and Bayes error, and provides a framework for understanding the reduction in error due to combining.

G.L. Gimel' farb et al. (Eds.): SSPR & SPR 2012, LNCS 7626, pp. 701–709, 2012.
© Springer-Verlag Berlin Heidelberg 2012

Section 2 explains the computation of the Walsh coefficients, and Section 3 discusses the relationship with the model of Added Classification Error. In Section 4, the weighted vote using Walsh coefficients is compared as the number of nodes and training epochs of MLP base classifiers are systematically varied.

2   Walsh Coefficients

Consider an ensemble framework, in which there are N parallel base classifiers, and X_m is the N-dimension vector representing the mth training pattern, formed from the decisions of the N classifiers. For a two-class supervised learning problem of μ training patterns, the target label given to each pattern X_m is denoted by $\Omega_m = \Phi(X_m)$ where $m = 1 \ldots \mu$, $\Omega_m \in \{1,-1\}$ and Φ is the unknown Boolean function that maps X_m to Ω_m. Thus the binary vector X_m represents the mth original training pattern

$$X_m = (X_{m1}, X_{m2}, \ldots, X_{mN})$$ (1)

where $X_{mi} \in \{1,-1\}$ is a vertex in the N-dimensional binary hypercube. The Walsh transform of Φ is derived from the mapping T_n and defined recursively as follows

$$T_n = \begin{bmatrix} T_{n-1} & T_{n-1} \\ T_{n-1} & -T_{n-1} \end{bmatrix} T_1 = \begin{bmatrix} 1 & 1 \\ 1 & -1 \end{bmatrix}$$ (2)

The first and second order spectral coefficients s_i and s_{ij} derived from (2) are defined in [5] as

$$s_i = \sum_{m=1}^{\mu} (X_{mi} \Omega_m)$$ (3)

$$s_{ij} = \sum_{m=1}^{\mu} (X_{mi} \oplus X_{mj}) \Omega_m$$ (4)

In (3) s_i represents the correlation between $\Phi(X_m)$ and X_{mi} and $s_{ij}(i, j = 1 \ldots N, i \neq j)$ in (4) represents correlation between $\Phi(X_m)$ and $X_{mi} \oplus X_{mj}$, where $\oplus$ is logic exclusive-OR.

Realistic learning problems are ill-posed [8], and therefore Φ may be partially specified, noisy and possibly contradictory. Relationships for computing spectral coefficients for partially specified Boolean functions, are proved in [9], for which the context is logic circuit design. The relevant ideas are presented here using different terminology, specifically minterms interpreted as patterns.

In [9], the concept of a standard trivial function Ψ is introduced. Each spectral coefficient gives a correlation value between the Boolean function Φ and Ψ. For first order coefficients, Ψ_i is the Boolean variable X_{mi} in (3) while for second order coefficients Ψ_{ij} is $X_{mi} \oplus X_{mj}$ in (4). Note in (4) $X_{mi} \oplus X_{mj} = 1$ implies pair of classifiers i and j disagree for pattern X_m and $X_{mi} \oplus X_{mj} = 0$ implies classifiers agree. For third order coefficients, Ψ_{ijk} is $X_{mi} \oplus X_{mj} \oplus X_{mk}$ and higher order follows, but in this paper we restrict ourselves to first and second order spectral coefficients.

The equations (3) and (4) require binary variables $\{1, -1\}$ but for computing coefficients it is notationally more convenient to use $\{0,1\}$. For $p, q \in \{0,1\}$ define n_{pq} to be the number of class p patterns of Boolean function Φ for which both Φ and Ψ have the logical value q. Then n_{11} is the number of class 1 patterns (true minterms in [9]) for which both Φ and Ψ that have the logical value 1. Similarly n_{00} is the number of class 0 patterns (false minterms in [9]) for which both Φ and Ψ have the logical value 0. Corresponding definitions follow for n_{01} and n_{10}. Now define d_1 and d_0 to be the number of unspecified patterns (don't care minterms) for which Ψ has the logical value 1 and 0 respectively. It is clear that the sum of all patterns of an N-dimensional Boolean function is given by

$$n_{11} + n_{00} + n_{01} + n_{10} + d_1 + d_0 = 2^N \qquad (5)$$

According to [9], all spectral coefficients s_l may be computed as

$$s_l = (n_{11} + n_{00}) - (n_{01} + n_{10}) \qquad (6)$$

where l may be i or ij. Substitution of (5) into (6) gives various equivalent formulae, but the advantage of (6) is that it is not necessary to include unspecified patterns d_1, d_0 explicitly in the computation.

3   Added Classification Error Model

Figure 1 shows the two class (ω_1, ω_0) model of Added Classification Error (E darkly shaded region) according to [7], which for simplicity is restricted to one dimension (x). The optimum (Bayes) boundary in Figure 1 is the loci of all points $\tilde{x} : P(\omega_1 \mid \tilde{x}) = P(\omega_0 \mid \tilde{x})$. The output of the classifier representing class ω_1 is given by $\hat{P}(\omega_1 \mid x) = P(\omega_1 \mid x) + \varepsilon_1(x)$ where $P, \hat{P}$ are the actual and estimated *a posteriori* probability distributions as shown in Figure 1, and $\varepsilon_1(x)$ is the difference between them. A similar equation is obtained for class ω_0 with $P(\omega_0 \mid x), \hat{P}(\omega_0 \mid x)$ and error $\varepsilon_0(x)$. In Figure 1 b is the amount that the *kth* classifier boundary (x_b) differs from the ideal Bayes boundary ($\tilde{x}$), and assuming that b is a Gaussian random variable with mean β and variance σ_b, in [7] it is shown that Added Classification Error for *kth* classifier is given by $E_k = \nabla P(\sigma_b^2 + \beta^2)$ where $\nabla P = 0.5(P'(\omega_1 \mid \tilde{x}) - P'(\omega_0 \mid \tilde{x}))p(\tilde{x})$ and P' indicates differentiation.

Figure 2 shows decision boundaries of *(i,j)th* classifiers for which it is assumed that the complexity is not sufficient to approximate the Bayes boundary, so that both classifiers under-fit. Note in Figure 2 that estimated probabilities $\hat{P}(\omega_0 \mid x)$ and $\hat{P}(\omega_1 \mid x)$ are omitted for clarity. Mutually exclusive areas under the probability distribution are labelled 1 – 8 in Figure 2, and denoting the number of patterns in area y by a_y, the contribution from classifiers i,j according to area is given in Table 1.

The model assumptions are discussed in [3], in which the expression for the difference in Added Classification Error of *ith* and *jth* classifiers $E_{ij} = E_i - E_j$ is derived

$$E_{ij} = E_i - E_j = 0.5(s_{ij} + \gamma) \tag{7}$$

where $\gamma = 1 - 2p_0$ and p_0 is the prior probability of class ω_0.

Averaging over all pairs of classifiers in (7) the mean difference in added error is given by

$$\Delta \bar{E} = \sum_{i,j,i \neq j} E_{ij} \tag{8}$$

Therefore from (7) and (8) we can approximate mean Added Error by subtracting γ and averaging over all pair-wise second order coefficients, call it S2M. In [3] it is shown that S2M is a good predictor of ensemble performance as number of epochs is increased. For the datasets in Section 4, optimal performance for majority vote occurs on average around 2-3 epochs.

The usual idea in weighted voting is to reward individual classifiers that perform accurately [10]. In this paper, a different approach is taken. For classifiers with lower training strength, it is expected that classifiers maybe unevenly spread around the optimal boundary. The idea is to give larger weight to pairs of classifiers with low

Added Error. The classifiers are chosen based on the product of first order coefficients as follows. The first order coefficients in (3) are decomposed into the contributions from the two classes $(n_{11} - n_{01})$ and $(n_{00} - n_{01})$ and the weight is proportional to their product. The weight of the *ith* classifier is given by

$$w_l = (n_{11} - n_{10})(n_{00} - n_{01}) \tag{9}$$

with negative weights in (9) set to zero. Considering Figure 1, classifiers close to the Bayes boundary will receive larger weight, but as they move further away, weight is decreased and becomes zero as n_{11} approaches n_{10} or as n_{00} approaches n_{01}. When classes are unbalanced, (9) tends to favour classifiers on either side of the Bayes boundary, in contrast to a weighting scheme based on training error. The weighting scheme using (9) is referred to as W1P in Section 4, and shown to reduce the mean Added Error given by (8).

4   Experimental Evidence

Natural two-class benchmark problems selected from [11] and [12] are shown in Table 2. The original features are normalised to mean 0 std 1, and for datasets with missing values the scheme suggested in [11] is used. Random perturbation of the MLP base classifiers is caused by different starting weights on each run. The number of hidden nodes and training epochs of homogenous (same number of nodes and epochs) MLP base classifiers are systematically varied over 1-5 epochs and 2-64 nodes. The experiments are performed with two hundred single hidden-layer MLP base classifiers, using the Levenberg-Marquardt training algorithm with default parameters. Combining uses majority (MAJ) or weighted vote. The random train/test split is 20/80 and experiments are repeated twenty times and averaged. Note that, for each dataset the class with most patterns is assigned ω_0 to give the same sign to γ in (7).

Bias/Variance will refer to 0/1 loss function using Breiman's decomposition [13], for which Bias plus Variance plus Bayes equals the base classifier error rate. Bias is intended to capture the systematic difference with Bayes, and requires Bayes probability. Patterns are divided into two sets, the Bias set containing patterns for which the Bayes classification disagrees with the ensemble classifier and the Unbias set containing the remainder. Bias is computed using the Bias Set and Variance is computed using the Unbias Set, but both Bias and Variance are defined as the difference between the probabilities that the Bayes and base classifier predict the correct class label. The Bayes estimation is performed for 90/10 split using original features, and a Support Vector Classifier (SVC) with polynomial kernel run 100 times. The polynomial degree and regularisation constant are varied, and lowest test error is given in Table 2.

Figure 3 gives mean results over seven datasets, which clearly indicates the overall trend. Figure 3 (a) (b) shows base and ensemble (MAJ) test error rates.

Figure 3 (c)-(f) shows difference between MAJ and various weighted vote schemes. Figure 3 (c) uses the first order Walsh coefficient (W1D) in (3), Figure 3 (d) is the proposed scheme (W1P) using (9), Figure 3 (e) uses the logarithmic weighting scheme used in Adaboost (ADA) [14]. Figure 3 (f) uses a trained linear perceptron (LIN) to learn the mapping. All the weighting schemes give a large improvement over MAJ at 1 epoch, the best being W1P, with a 13 percent improvement at 64 nodes. The best MAJ error occurs at 3-4 epochs, and here there is a small improvement W1P over MAJ of between 0.3 percent at 64 nodes and 1 percent at 4 nodes.

Fig. 4 shows various measures to help explain the results. Fig 4 (a) shows the mean second order coefficients (S2M), normalised by the total number of training patterns, and which is an estimate of the mean added error in (8). Figure 4 (b) is similar to (a) but shows coefficients weighted by (9) (for classifier i and j, weight is given by $(w_i + w_j)/2$). Figure 4 (c) – (f) show bias and variance for MAJ (Bias, Var) and W1P (BiasW, VarW). By comparing Figure 4 (a) and (b) the weighted coefficients (S2W) shows that weighted classifiers have reduced the Added Error. The Weighted bias (BiasW) in (d) is reduced in comparison with Bias in (c). For 64 nodes, the best weighted error rate is at 1 epoch, shown in (d), which is within 1 percent of Bayes rate. On the other hand, at 1 epoch Figure 4 (e) (f) show that weighted variance has increased, indicating that more diverse classifiers are weighted.

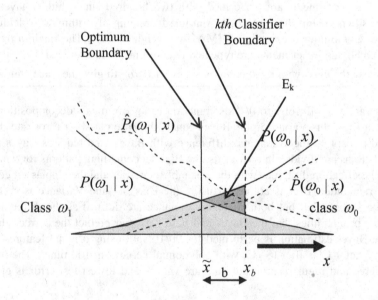

Fig. 1. Model of error region associated with *a posteriori* probabilities showing optimum (Bayes) boundary, kth classifier boundary with Added Classification Error (E_k)

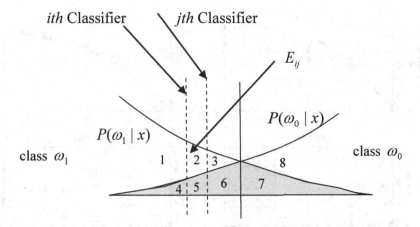

Fig. 2. Model showing pair of classifier boundaries and the difference in Added Classification Error between *ith* and *jth* classifiers E_{ij} (area 2)

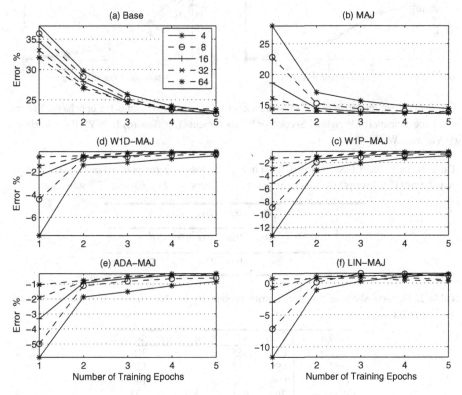

Fig. 3. Mean test errors over 2-class datasets for [4,8,16,32,64] nodes 1-5 epochs (a) Base Classifier (b) Majority Vote (c) –(f) Weighted votes with MAJ subtracted

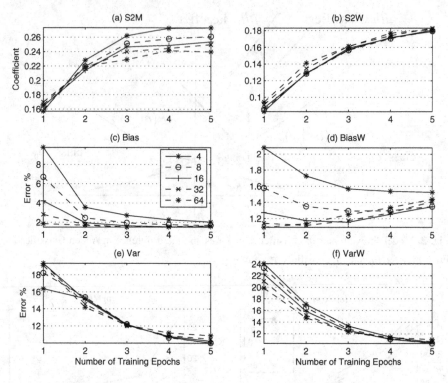

Fig. 4. (a) Mean measures over 2-class datasets for [4,8,16,32,64] nodes 1-5 epochs (a) Second order coefficients (b) Weighted Second order coefficients (c) Bias (d) Bias W1P (e) Variance (f) Variance W1P

Table 1. Areas under Distribution defined in Fig. **2**, showing corresponding number of class ω_1, ω_0 patterns (1st subscript) for which the pair of classifiers agree or disagree (2nd subscript)

	a_1	a_2	a_3	a_4	a_5	a_6	a_7	a_8
ω_1	n_{10}	n_{11}	n_{10}	n_{10}	n_{11}	n_{10}	n_{10}	
ω_0				n_{00}	n_{01}	n_{00}	n_{00}	n_{00}

Table 2. Datasets showing # patterns, prior probability ω_0, #continuous and discrete features and estimated Bayes error

DATASET	#pat	p_0	#con	#dis	%Bay
cancer	699	.655	0	9	3.1
card	690	.555	6	9	12.8
credita	690	.555	3	11	14.1
diabetes	768	.651	8	0	22.0
heart	920	.553	5	30	16.1
ion	351	.641	31	3	6.8
vote	435	.614	0	16	2.8

5   Conclusion

For two-class supervised learning problems, the spectral representation of the mapping between binary base classifier decisions and target class has been analysed with the help of the Tumer-Ghosh model of Added Classification Error. If the majority vote is weighted by the product of the class-dependent first-order coefficients, the ensemble has error rate that is close to optimal, even with fast inaccurate base classifiers.

References

[1] Smith, R.S., Windeatt, T.: A Bias-Variance Analysis of Bootstrapped Class-Separability Weighting for ECOC Ensembles. In: Proceedings of the 22nd International Conference on Pattern Recognition, Istanbul, Turkey (August 2010)

[2] Windeatt, T.: Vote Counting Measures for Ensemble Classifiers. Pattern Recognition 36(12), 2743–2756 (2003)

[3] Windeatt, T., Zor, C.: Minimising Added Classification Error using Walsh Coefficients. IEEE Trans. Neural Networks 22(8), 1334–1339 (2011)

[4] Windeatt, T.: Accuracy/ Diversity and ensemble classifier design. IEEE Trans. Neural Networks 17, 1194–1211 (2006)

[5] Hurst, L., Miller, D.M., Muzio, J.: Spectral Techniques in Digital Logic. Academic Press (1985)

[6] Beauchamp, K.G.: Walsh Functions and their Applications. Academic Press (1975)

[7] Tumer, K., Ghosh, J.: Error correlation and error reduction in ensemble classifiers. Connection Science 8(3), 385–404 (1996)

[8] Tikhonov, A.N., Arsenin, V.A.: Solutions of Ill-posed Problems. Winston & Sons, Washington (1977)

[9] Falkowski, B.J., Perkowski, M.A.: Effective Computer Methods for the Calculation of Rademacher-Walsh Spectrum for Completely and Incompletely Specified Boolean Functions. IEEE Trans. on Computer-Aided Design 11(10), 1207–1226 (1992)

[10] Kuncheva, L.I.: Combining Pattern Classifiers. Wiley (2004)

[11] Prechelt, L.: Proben1: A set of neural network Benchmark Problems and Benchmarking Rules. Tech Report 21/94, Univ. Karlsruhe, Germany (September 1994)

[12] Merz, C.J., Murphy, P.M.: UCI repository of ML databases,
http://www.ics.uci.edu/~mlearn/MLRepository.html

[13] Breiman, L.: Arcing Classifiers. The Annals of Statistics 26(3), 801–849 (1998)

[14] Freund, Y., Shapire, R.E.: A Decision-Theoretic Generalization of On-line Learning and its Application to Boosting. In: Vitányi, P.M.B. (ed.) EuroCOLT 1995. LNCS, vol. 904, pp. 23–37. Springer, Heidelberg (1995)

A Novel Shadow-Assistant Human Fall Detection Scheme Using a Cascade of SVM Classifiers*

Yie-Tarng Chen, You-Rong Lin, and Wen-Hsien Fang

Department of Electronic Engineering,
National Taiwan University of Science and Technology,
Taipei, Taiwan, R.O.C.
{ytchen,whf}@mail.ntust.edu.tw

Abstract. Visual recognition of human fall incidents in video clips has been an active research issue in recent years, However, most published methods cannot effectively differentiate between fall-down and fall-like incidents such as sitting and squatting. In this paper, we present a novel shadow-assistant method for detecting human fall. Normally, complex 3-D models are used to estimate the human height. However, to reduce the high computational cost, only the information of moving shadow is used for this context. Because the system is based on a combination of shadow-assistant height estimation, and a cascade of SVM classifiers, it can distinguish between fall-down and fall-like incidents with a high degree of accuracy from very short sequence of 1-10 frames. Our experimental results demonstrate that under bird's-eye view camera setting, the proposed system still can achieve 100% detect rate and a low false alarm rate, while the detection rate of other fall detection schemes have been dropped dramatically.

Keywords: fall detection, SVM.

1 Introduction

In recent years, visual recognition of human fall incidents in video clips has been an active research issue. In this paper, we consider the problem of using a mono un-calibrated camera to detect if senior citizens fall over, called fall-down incidents hereafter. Such incidents normally occur suddenly and take approximately 0.45 to 0.85 seconds. Both the posture and shape of the victim change rapidly, and he/she usually lies inactive on the floor. Hence, drastic changes in the posture, shape and height of the body are key features in human fall detection. However, modeling those features with low computational complexity is a not a trivial task, especially for accurate human height estimation.

A number of fall detection schemes have been proposed [4-5]. Simple features derived from shape analysis, such as the aspect ratio of the bounding box, the

---

* This work was supported by National Science Council of R.O.C. under contract NSC 100-222-E-011-134.

G.L. Gimel' farb et al. (Eds.): SSPR & SPR 2012, LNCS 7626, pp. 710–718, 2012.
© Springer-Verlag Berlin Heidelberg 2012

angle of the fall and a vertical projection histogram have been used for fall detection. Rougier [4] proposed a fall detection approach based on the Motion History Image (MHI) [3] and changes in body shape. Hidden Markov Models (HMMs) have also been utilized for fall detection. Hiseh [5] developed a triangulation-based skeleton extraction approach to analyze human movements; however, it is not designed specifically for detection fall-down incidents. No approach based on simple features can detect all kinds of human falls. Most video-based fall detection systems based on simple features suffer from high false alarm rates because they do not differentiate between fall-like and fall-down incidents. The high computational cost of human skeleton extraction discourages researchers from using it for real-time human fall detection. Hence, there is need for a reliable fall detection system, a combination of several approaches, which can increase the detection accuracy while still satisfying the real-time constraint.

1.1   A Motivation Example

Figure 1 illustrates the motivation for this paper, which attempted to differentiate the falling posture through the shadow information. We can observe a correlation between the height of standing, sitting down and falling postures and their relevant shadow areas. In particular, the shadow area approaches 0 for a falling posture. Hence, we attempt to investigate the possibility to utilize shadow information for human fall detection. However, shadow is not a stable image cue, especially, it is dependent on the capturing conditions. if a person stands just below a light source, where the projection angle of a light is vertical, the length of a person's shadow is still cannot be detected. Hence, there is need an intelligent combination of the shadow information with other approaches which can increase detection accuracy.

In this paper, we propose a real-time video-based human fall detection system which can support both bird-view and flat-view camera furnishing. The proposed system applies a novel shadow-assistant human height estimation scheme to differentiate between fall-down and fall-like incidents. Normally, complex 3-D models are required to estimate human height in bird-view camera finishing.

Fig. 1. The shadow (blue) and human foreground (red) for standing, sitting and falling

However, to reduce the high computational costs, the shadow-assistant approach instead, even though it provides less information than 3-D models, but satisfies the requirements of a real-time fall detection system. Furthermore, a fall into any angle can be detected. Specifically, the proposed approach can successfully detect fall toward camera with a low computational cost.

The contributions of this paper are two-fold:

1. We have address the use of shadow for visual recognition of human fall. Although shadows have been used to measure the height of a building on aerial images. To our knowledge, shadows are first introduced to human activity recognition in this paper.
2. We develop a fall detection system based on intelligent combination of shadow analysis,shape analysis and motion analysis, achieve 100% detect rate and a low false alarm rate from very short sequence of 1-10 frames while satisfying the real-time constraint.

The remainder of the paper is organized as follows. Section 2 presents foreground objects and moving shadows segmentation. In Section 3, we describe the novel shadow-assistant detection scheme. The results of experiments are detailed in Section 4 and Section 5 concludes the paper.

2   Foreground Objects and Moving Shadows Detection

Foreground object segmentation and moving shadows detection are important preprocessing steps in human fall detection schemes. We use the statistical background subtraction and shadow detection algorithm [1] developed by Horprasert et al.for this context. However, other foreground ground extraction and moving shadow detection algorithms still can be applied to the proposed fall detection scheme.

2.1   Background Modeling

In background modeling, we attempt to obtain a background model and its parameters by several selected images. Each pixel in the background model is assumed to be independent, and it can be represented as a tuple with four parameters The background is modeled statistically on a pixel by pixel basis. A pixel is modeled by a 4-tuple $< E_i, S_i, a_i, b_i >$. where S_i is a standard deviation of color value. It normalizes the pixel color in this work. It is given by

$$S_i = [\sigma_R(i), \sigma_G(i), \sigma_B(i)] \tag{1}$$

where $\sigma_R(i), \sigma_G(i), \sigma_B(i)$ are the standard deviations of the i-th pixel's red, green, blue values over N training frames. The expected color value of pixel i is given by

$$E_i = [\mu_R(i), \mu_G(i), \mu_B(i)] \tag{2}$$

where $\mu_R(i), \mu_G(i), \mu_B(i)$ denote arithmetic means of the i-th pixel's red, green, blue values over N training frames. Let $I_i = [I_R(i), I_G(i), I_B(i)]$ represent the pixel's RGB color value in current image. We want to measure the distortion of I_i from E_i by discomposing the distortion measurement into two components: brightness distortion α_i and and chromaticity distortion CD_i respectively, which are defined in Equations (3) and (4).

$$\alpha_i = \frac{\left(\frac{I_R(i)\mu_R(i)}{\sigma_R^2(i)} + \frac{I_G(i)\mu_G(i)}{\sigma_G^2(i)} + \frac{I_B(i)\mu_B(i)}{\sigma_B^2(i)} \right)}{\left(\left[\frac{\mu_R(i)}{\sigma_R(i)}\right]^2 + \left[\frac{\mu_G(i)}{\sigma_G(i)}\right]^2 + \left[\frac{\mu_B(i)}{\sigma_B(i)}\right]^2 \right)} \tag{3}$$

$$CD_i = \sqrt{\left(\frac{I_R(i) - \alpha_i\mu_R(i)}{\sigma_R(i)} \right)^2 \left(\frac{I_G(i) - \alpha_i\mu_G(i)}{\sigma_G(i)} \right)^2 \left(\frac{I_B(i) - \alpha_i\mu_B(i)}{\sigma_B(i)} \right)^2} \tag{4}$$

Furthermore, we consider the variation of the brightness and chromaticity distortions over space and time of the training background images. a_i represents the variation of the brightness distortion of i-th pixel, which is given by

$$a_i = \sqrt{\frac{\sum_{i=0}^{N} (\alpha_i - 1)^2}{N}} \tag{5}$$

b_i represents the variation of the chromaticity distortion of i-th pixel, which is given by

$$b_i = \sqrt{\frac{\sum_{i=0}^{N} (CD_i)^2}{N}} \tag{6}$$

2.2   Pixel Classification

Since the different pixels yield different distribution of α_i and CD_i. In order to use a single threshold for all of pixels, we re-scale α_i and CD_i as normalized brightness distortion $\hat{\alpha}_i$ and chromaticity distortion $\widehat{CD}_i$ respectively.

$$\hat{\alpha}_i = \frac{\alpha_i - 1}{a_i} \tag{7}$$

$$\widehat{CD}_i = \frac{CD_i}{a_i} \tag{8}$$

Each pixel $M(i)$ is classified into one of the four categories: B (Background), S (Shadow), H (Highlighted background), and F (Foreground object) by the following decision rule:

$$M(i) = \begin{cases} F: & \widehat{CD}_i > \tau_{CD} \ \text{ or } \ \hat{\alpha}_i < \tau_{alo}, \text{ else} \\ B: & \hat{\alpha}_i < \tau_{\alpha 1} \ \text{ and } \ \hat{\alpha}_i > \tau_{\alpha 2}, \text{ else} \\ S: & \hat{\alpha}_i < 0 \quad \text{ else} \\ H: & \text{otherwise} \end{cases} \tag{9}$$

where $\tau_{CD}, \tau_{alo}, \tau_{\alpha 1},$ and $\tau_{\alpha 2}$ are selected threshold values to determine the similarities of the chromaticity and brightness between the background image and the current observed image.

3   Human Fall Detection System

3.1   System Description

After Foreground object segmentation and object tracking, the system begins to perform fall detection. The proposed fall detection scheme as shown in Figure 2 consists of cascaded classifiers which integrates height estimation, posture analysis, shape analysis, and inactivity detection. We assume that a person is in an upright posture when first appearing in the video sequence and becomes immobile on the floor after a fall. All video sequences are captured by a stationary camera, which can be furnished in either a flat-view or a bird view.

Since Histograms of Oriented Gradients (HOG)[2] detectors have shown to give significantly high performance in upright human detection, first, an input image is sent to the HOG-based upright posture detector. Next, any non-upright human foreground is sent to a shadow-assistant falling posture detector as shown in Figure 3, which performs height estimation and shape analysis to detect a falling posture. Finally, we confirm a fall incident by monitoring the inactivity of the person by using a motion history image (MHI)[3].

For classification in the upright posture detection, the falling posture detector and inactivity detector, we use state-of-the-art machine learning techniques-support vector machines, which have been a popular approach for pattern classification and nonlinear regression because of its robustness even in the absence of a rich set of training examples. The virtual height, the HOG vector, and MHI vector are major features in the proposed fall detection scheme, which will be discussed in the following sections.

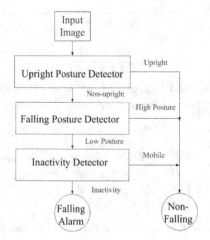

Fig. 2. The flowchart of the shadow-assistant fall detector

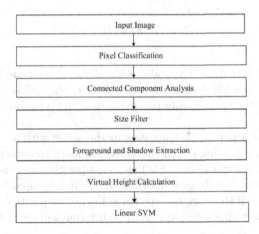

Fig. 3. The flowchart of the falling posture detection scheme

3.2 Upright Posture Detection

The Histogram of Oriented Gradient (HoG) human detector [2] is applied to detect the upright human posture. This detector combined locally normalized histogram of oriented Gradient (HoG) descriptor and a linear support vector machine classifier. The major idea behind the Histogram of Oriented Gradient descriptors is to describe local object appearance and shape by the distribution of intensity gradients or edge direction. The HOG upright posture detector consists of the following steps: gradient computation, orientation binning, block normalization and SVM classifiers. The input to the SVM upright pose classifier comes from a HOG vector in the foreground object. However, instead of using all pixels in an image to compute HOG, only pixels within a bounding box enclosed the foreground object. Similar to the SVM height classifier, The sign +1 of the SVM output was assigned to the upright posture and -1 to the non- upright posture. The details of the Histogram of Oriented Gradient (HoG) human detector can be found in [2].

3.3 Shadow-Assistant Falling Posture Detector

The virtual height, denoted as V_h, is defined as

$$V_h(B) = \begin{cases} \frac{HS_a}{WF_a} & \text{if } B \neq 0 \\ \frac{H}{W} & \text{if } B = 0 \end{cases} \tag{10}$$

where H and W are the height and the width of a bounding box respectively, S_a represents the areas of a shadow, and F_a represent area of a foreground object, which are given by:

$$S_a = \sum_{i=1}^{n} \delta(M(i) - S) \tag{11}$$

$$F_a = \sum_{i=1}^{n} \delta(M(i) - F) \tag{12}$$

If the shadow is occurring, both shadow analysis and shape analysis, such as the aspect ratio of the bounding box, are used to detect the falling posture, whereas, if the shadow is not available, only the shape analysis is considered to detect the falling posture.

We assume that a person first appears in the scene with an upright posture. We can use the first few images in the video sequence to detect the presence of the shadow. If the upright detector is flagged as positive and the shadow area is approximate to zero, the shadow information cannot be used for human falling detection. However, if the upright detector is flagged as positive, and the shadow area is larger than a threshold, the shadow information cannot be used to the human fall.

B is the index of the shadow-assistant function in the human fall detection system. If $B = 1$, the shadow-assistant function operates. On the other hand, if $B = 0$ the shadow information will not be considered in the proposed human fall detection scheme.

$$B = \begin{cases} 1 \text{ if } U_r(t) = 1 \text{ and } S(t) = 1 \\ 0 \text{ if } U_r(t) = 1 \text{ and } S(t) = 0 \end{cases} \tag{13}$$

where $U_r(t)$ is the upright detector for frame(t) and $U_r(t) = 1$ if the upright detector classifies frame(t) as an upright human posture and $U_r(t) = 0$ otherwise. $S(t)$ is the shadow index for frame(t) and $S(t) = 1$ if the area of the human shadow in frame(t) is larger than a predefined threshold and $S(t) = 0$ otherwise.

3.4 Inactivity Detector

Motion History Image (MHI) [3] is defined by a simple replacement and decay operators:

$$H_\tau(x, y, t) = \begin{cases} \tau & \text{if } D(x, y, t) = 1 \\ \max\left(H_\tau(x, y, t-1) - 1\right) & \text{otherwise} \end{cases}$$

where $D(x, y, t)$ is a binary image sequence indicating region of motion, generated by image differing.

Motion History Image (MHI) is a scalar-value image where more recently moving pixels are brighter. The Motion History Image can be used to represent how motion the image is moving.

We check the following condition to confirm the occurrence of a fall-down incident: the Motion History Image (MHI) of the human object is lower than the threshold T_p. This condition confirms that the person is inactive for a period of time after a fall.

4   Experiments and Results

4.1   Experimental Settings

In order to validate the efficacy of the shadow-assistant shape analysis for human fall detection, we perform experiments on five datasets: NTUST bird-view dataset, NTUST outdoor dataset, NTUST Lab dataset, ETHZ dataset, and Shoaib dataset. Since the page is limited, only the performance of NTUST bird-view dataset and Shoaib dataset are reported in this paper. We implemented the proposed shadow-based fall detection system on Intel's OpenCV library. The experiments were run on a PC with Windows XP, an Intel Pentium D 3.2GHz CPU and a 2 GB RAM. The performance metrics used in the fall detection experiments are the detection rate P_d and the false alarm rate P_n. The detection rate, i.e., the fraction of fall events detected correctly, is calculated as follows:

$$P_d = \frac{TP}{(TP + FN)} \qquad P_n = \frac{FP}{(TN + FP)}$$

where true positive (TP) and true negative (TN) are the counts of correct detection while false positive (FP) and false negative (FN) are the counts of incorrect prediction.

We compared the performance of the proposed fall detection system with that of existing approaches: approach 1: the proposed shadow assistant scheme, approach 2: combination of shape analysis and MHI [4], approach 3: Skeleton matching [5] and, approach 4: Shape analysis. shape analysis uses the aspect ratio of the bounding box as a feature and Support Vector Machine as a classifier. In skeleton matching, a skeletons is classified by near-neighbor scheme and the distance map is used as the distance function.

4.2   Experiments on NTUST Bird-View and Shoaib Data-Sets

NTUST bird-view dataset as shown in Figure 4 contains 6 activities performed by one actor in the NTUST design square, an outdoor environment and captured in a bird-view by a single stationary and un-calibrated camera furnished on 8 meters height. The dataset consists of 54 actions in 1800 frames, including 27 falling actions, 7 sitting-down on the ground, 5 sitting on a chair, 4 squat actions, 8 bend actions and 12 walking actions. Shoaib data-set contains one actor in a cluttered home environment.

The proposed shadow-assistant scheme outperforms other fall-detection approaches in both the NTUST bird-view dataset and Shoaib dataset as shown in Table 1 and Table 2 since the only light source in the NTUST bird-view dataset is the sunlight, a simple point light source, which can cast simple shadow contours. The detection rate of other human fall-down detection schemes drop significantly in bird-view dataset in compassion with they did in flat-view datasets since the actual height of a human posture in a bird-view image by cannot be estimated by conventional two-dimensional fall detection schemes. The average execution time of the shadow-assistant scheme takes 0.156 second per frame, which satisfies the real-time constraint.

Fig. 4. Parts of the test videos:(left three) walking, sitting-down, and falling down from brid-view dataset, and (right two) bending and falling down from Shoaib dataset

Table 1. Comparison of fall detection schemes for outdoor bird view data sets

approach	TP	FN	TN	FP	P_d	P_n
our approach	27	0	34	2	1.00	0.055
approach 2	20	7	25	11	0.74	0.31
approach 3	24	3	33	3	0.889	0.08
approach 4	25	2	31	5	0.926	0.138

Table 2. Comparison of fall detection schemes for Shoaib data sets

approach	TP	FN	TN	FP	P_d	P_n
our approach	10	0	24	4	1.00	0.166
approach 2	8	2	23	5	0.80	0.217
approach 3	4	6	26	2	0.40	0.077
approach 4	10	0	14	14	1.00	0.500

5   Conclusion

We have presented a novel shadow-assistant human fall detection system which can support different viewpoints. The robust human fall detection scheme relies on shadow and shape analysis to differentiate fall-down and fall-like incidents under different viewpoints. Our experiment results demonstrate that the proposed system can achieve a high detection rate and low false alarm rate while satisfying real-time constraints.

References

1. Horprasert, T., Harwood, D., Davis, L.: A Statistical Approach for Real-Time Robust Background Subtraction and Shadow Detection. In: IEEE International Conference on Computer Vision, ICCV 1999, Frame-Rate Workshop, pp. 1–19. IEEE Press, New York (1999)
2. Dalal, N., Triggs, B.: Histograms of Oriented Gradients for Human Detection. In: IEEE Computer Society Conference on Computer Vision and Pattern Recognition, CVPR 2005, pp. 886–893. IEEE Press, New York (2005)
3. Bobick, A., Davis, J.: The Recognition of Human Movement Using Temporal Template. IEEE Transactions on Pattern Analysis and Machine Intelligence 23, 257–267 (2001)
4. Rougier, C., Meunier, J., St-Arnaud, A., Rousseau, J.: Fall Detection from Human Shape and Motion History Using Video Surveillance. In: the 21st International Conference on Advanced Information Networking and Applications, Niagara Falls, Canada, pp. 875–880. IEEE Press, New York (2007)
5. Hsu, Y.T., Liao, H.Y., Chen, M.C.C., Hsieh, J.W.: Video-based Human Movement Analysis and Its Application to Surveillance Systems. IEEE Transactions on Multimedia 10, 372–392 (2008)

Analysis of Co-training Algorithm with Very Small Training Sets

Luca Didaci, Giorgio Fumera, and Fabio Roli

Department of Electrical and Electronic Engineering, University of Cagliari
Piazza d'Armi 09123, Cagliari, Italy
{luca.didaci,fumera,roli}@diee.unica.it
http://prag.diee.unica.it/

Abstract. Co-training is a well known semi-supervised learning algorithm, in which two classifiers are trained on two different views (feature sets): the initially small training set is iteratively updated with unlabelled samples classified with high confidence by one of the two classifiers. In this paper we address an issue that has been overlooked so far in the literature, namely, how co-training performance is affected by the size of the initial training set, as it decreases to the minimum value below which a given learning algorithm can not be applied anymore. In this paper we address this issue empirically, testing the algorithm on 24 real datasets artificially splitted in two views, using two different base classifiers. Our results show that a very small training set, even made up of one only labelled sample per class, does not adversely affect co-training performance.

Keywords: Semi-supervised learning, Co-training, Small sample size.

1 Introduction

Semi-supervised learning (SSL) methods are useful in many practical applications in which a small set of labelled samples L is available, but a large set of unlabelled samples U can be exploited to improve the performance of learning algorithms. Typical examples are text (e.g., Web page) classification, and biometric authentication. Co-training is a well known SSL algorithm originally proposed in [1], for binary classification problems in which two different views (feature spaces) X^1 and X^2 are available. Starting from a small training set $L = \{(x_i^1, x_i^2, y_i)\}_{i=1}^{n_L}$, where $x_i^1 \in X^1$, $x_i^2 \in X^2$ and $y \in \{-1, +1\}$, it consists of iteratively re-training a pair of classifiers $f^1 : X^1 \to Y$ and $f^2 : X^2 \to Y$, adding to L at each step the unlabelled samples from a given set $U = \{(x_i^1, x_i^2)\}_{i=1}^{n_U}$ that are classified with high confidence by one of the classifiers. Under the assumption of conditional independence between the views, given the class Y (i.e., $P(X^1, X^2|Y) = P(X^1|Y)P(X^2|Y)$), and of sufficiency of each view (i.e., the classes can be perfectly discriminated in each view, if there were a sufficient number of samples), it was shown that co-training allows both classifiers to attain the same performance as they were trained on a large set of labelled samples.

Several authors have theoretically or empirically investigated several aspects of co-training; for instance, how it works when the original assumptions do not hold (which

G.L. Gimel' farb et al. (Eds.): SSPR & SPR 2012, LNCS 7626, pp. 719–726, 2012.

can often happen in practice), or how it works under different and less restrictive assumptions. Balcan et al. [2] provide a PAC-style analysis, and showed that it is possible to relax the condition of independence between the views, if the classifiers are never "confident but wrong". Christoudias el al. [3] examined settings in which classifiers are not compatible due to noise in one view (i.e., even in a ideal situation with a very large training set, their decisions disagree). Didaci and Roli [4] investigated, form a Bayesian point of view, the consequence of the non-sufficiency of the views. Du et al. [5] investigated the possibility of predicting whether co-training will work or not for a given problem, and whether single-view problems can be artificially decomposed into two views to exploit co-training. Zhou et al. [6] considered the limit case when only one labelled sample (x, y) is available, and thus co-training can not be applied, and proposed a method for adding to (x, y) artificially labelled samples to enable the application of co-training.

In this work we address a different aspect of co-training behaviour that has been overlooked in the literature so far, namely, how it performs as the size of L decreases toward the minimum possible value, below which a given learning algorithm can not be applied anymore. As an example, to estimate the covariance matrix of a linear or quadratic Gaussian classifier, at least two samples are needed for each class. In this paper we address this issue empirically, using 24 single-view data sets from the UCI Repository [7], and two different base classifiers. Our goal is to provide a first answer to the questions of whether, and to what extent the performance of co-training is affected by the size of the initial training set L.

After a summary of the co-training algorithm and of previous works in Sect. 2, in Sect. 3 and 4 we present the results of our experiments. Conclusions are drawn in Sect. 5.

2 Background

2.1 The Co-training Algorithm

In this paper we consider the standard version of the co-training algorithm given in [1], which is reported as Algorithm 1. First, a subset U' of unlabelled samples is randomly selected from the available data set U. Then, the following steps are repeated for a predefined number of iterations. Two separate classifiers, f^1 and f^2, one for each view, are trained on the initial, small, training set L, and are then used to label the samples in U'. For each classifier, the p samples of class $+1$ and the n samples of class -1 that are labelled with the highest confidence among the ones of U' are added to L. Classifier confidence can be evaluated, for instance, as the estimated posterior probability, while p and n are chosen such that they are proportional to the (estimated) class priors. The selected $2p + 2n$ samples are then removed from U', and other $2p + 2n$ samples are randomly selected from U and added to U'. The reason of using a subset U' of the unlabelled samples U is that this forces f^1 and f^2 to select samples that are more representative of the underlying distribution, even if they may be not the ones labelled with highest confidence among the ones in U [1].

Previous works on co-training mainly considered Naive Bayes and decision trees as base classifiers. Nevertheless, to the purpose of this work, we point out that each base

---

Algorithm 1. Co-training algorithm

Input: L and U: sets of n_L labelled and n_U unlabelled samples, respectively, represented in two views X^1 and X^2; k: number of iterations; $n_{U'} < n_U$: number of samples to be drawn from n_U; n, p: number of pattern selected by each classifier at each step. n, p are proportional to priors.
Output: two trained classifiers $f^1 : X^1 \to Y$ and $f^2 : X^2 \to Y$.

    $U' \leftarrow$ a set of $n_{U'}$ samples randomly drawn from U
    for k iterations **do**
        Train a classifier f^1 on the X^1 view of L, and a classifier f^2 on the X^2 view of L
        for $i = 1, 2$ **do**
            Let f^i labels all samples in U'
            $U_i' \leftarrow$ the p samples labelled as $+1$ and the n ones labelled as -1 with higher confidence by f^i
            $L \leftarrow L \cup U_i', \;\; U' \leftarrow U' - U_i'$
        end for
        Randomly choose $2p + 2n$ samples from U, and move them to U'
    end for

---

classifier has an intrinsic limit to the minimum number of labelled samples of each class in the training set, that allows the corresponding learning algorithm to be applied. We denote these values as $|L^+|_{\min}$ and $|L^-|_{\min}$, respectively for $y = +1$ and $y = -1$. Usually $|L^+|_{\min} = |L^-|_{\min}$. In some cases $|L^+|_{\min} = |L^-|_{\min} = 1$ (e.g., a support vector machine), while in other cases both values can be greater than 1 (e.g., Gaussian classifiers).

2.2 Previous Works

Among previous works on co-training, we mention here the ones that are most related to this paper.

In [2] it was shown that the assumption of conditional independence given the class label can be relaxed, provided that the learning algorithm is never "confident but wrong", i.e., it never misclassifies samples with high confidence. This result could in principle be exploited to make co-training work even when the initial L is very small, as discussed in Sect. 4.

In [6], the limit case when only one labelled sample is available was considered, namely $|L| = 1$. This can happen in applications like content-based image retrieval, and online web-page recommendation. In this case the standard co-training algorithm can not be applied, since it requires a binary base classifier. The proposed solution is first to label and add to L some samples of U, using a different SSL method, such that both classes are represented, and then run co-training, starting from the updated L. The resulting performance of co-training was evaluated for some different sizes of L. However, sizes very close, or equal, to the minimum one required to run the considered base classifiers were not considered.

Finally, in [5] the possibility of predicting whether co-training will work for a given problem was investigated. To this aim, methods for estimating whether the original assumptions of [1] hold or not were devised, using the samples of L. The conclusion was that no reliable estimate can be obtained from a small L. The related problem of artificially decomposing single-view problems (with a small set of labelled samples) into two views, to exploit co-training, was addressed in the same work. No effective method was found to this aim. The reason is that this requires to find the "best" split of the original feature set, according to the co-training underlying assumptions. However, the validity of such assumptions can not be determined from a small L.

In our experiments we will exploit the methods of [5] to artificially split the considered data sets into two views, since for our purposes the best split can be estimated using the whole labelled data sets.

3   Experimental Setup

Co-training was implemented as in Algorithm 1, with $|U'| = 0.3|U|$, similarly to [1,5]. We chose the values of p and n, such that p/n is (approximately) equal to the estimated class priors. According to [2], to this end we chose the smallest possible p and n. We set the number k of co-training iterations in order to allow co-training to collect all samples in U. The exact value of k depends thus on the size of the data set, and on the values of n and p.

Two different base classifiers were used: Naive Bayes (NB), and K-nearest neighbors (K-NN), with $K = 1$. In the case of real-valued features, NB was implemented by subdividing their range into 10 bins of equal width. The experiments have been carried out on 24 single-view two-class data sets, previously used in [5]. They have been artificially subdivided into two views using the method proposed in [5], which aims at minimising the correlation between the corresponding feature subsets, given the class, and maximising the separability of classes in each view, to meet as much as possible the assumptions of [1].

Ten different runs of the experiments have been made. At each run, each data set was randomly subdivided into a labelled training set L, an unlabelled data set U, and a testing set. We considered different sizes of L, as explained below. The size of the testing set was 25% of the entire data set, whilst the remaining data was used as the set U.

The goal of our experiments was to analyse the behaviour of co-training, as the size of L decreases to $|L|_{\min} = |L^+|_{\min} + |L^-|_{\min}$. Note that, with the chosen base classifiers, $|L^+|_{\min} = |L^-|_{\min} = 1$. To this aim, we considered values of $|L|$ ranging from 2 (i.e., $|L^+| = |L^-| = 1$) to 50% of the entire data set. L was obtained by stratified sampling from the whole data set, i.e., $|L^+|$ and $|L^-|$ were chosen such that $|L^+|/|L^-|$ was (almost) equal to the original proportion between the two classes. When the size of L was reduced to the extent that the corresponding $|L^+|$ or $|L^-|$ (chosen as explained above) attained its lowest possible value (respectively, $|L^+|_{\min}$ and $|L^-|_{\min}$), then the most populated class was undersampled. Note that, in this case, L was not representative of the underlying class priors.

At each run, and for each given L, we run co-training and computed its testing set performance. We then checked whether co-training performance attained for $|L| = |L|_{\min}$ was better than the performance attained by the corresponding base classifier trained on the same L, without co-training. If not, we checked for which size of L (if any) co-training outperformed the base classifier trained on the same L, without co-training. We considered two performances significantly different, if the difference between their average values over the ten runs was higher than the sum of the corresponding standard deviations, divided by the square root of the number or runs.

4   Experimental Results

In Table 1 we report the characteristics of the data sets used in the experiments. Table 2 shows the comparison between co-training performance attained for $|L| = |L|_{\min}$, and the performance attained by the corresponding base classifier trained on the same L, without co-training, for both classifiers. The meaning of table entries is the following: 0: no statistically significant performance difference in both views; 1: co-training was outperformed by the base classifier in both views; 1*: co-training was outperformed by the base classifier in one view, no statistically significant performance difference in the

Table 1. Characteristics of the data sets used in the experiments. An asterisk after the data set name denotes that its classes have been merged into two artificial classes, as in [5]. The numbers between brackets in the "n. samples" column denote the number of samples per class. The column "views size" reports the number of features in each view, obtained with the method of [5].

ID	Dataset	n.features	n. samples	views size
1	Audiology*	55	200 [48, 152]	31/24
2	Automobile*	24	193 [130, 63]	22/2
3	Breast Cancer W.	8	699 [458, 241]	5/3
4	Winsconsin D.	30	569 [212, 357]	11/19
5	Winsconsin Progn. 1	33	194 [46, 148]	12/22
6	Winsconsin Progn. 2	32	198 [47, 151]	13/19
7	Contraceptive Method	9	1473 [629, 844]	2/7
8	Horse colic	5	368 [232, 136]	4/1
9	Credit Approval	15	653 [296, 357]	9/6
10	Dermatology*	33	366 [112,254]	17/16
11	Pima Indians Diabetes	8	768 [500, 268]	6/2
12	E.Coli*	7	336 [143/193]	3/4
13	Flags*	28	194 [134/60]	15/13
14	Heart (Cleveland)*	11	303 [164, 139]	5/6
15	Heart (LongBeach)*	4	200 [144, 56]	2/2
16	Heart-statlog	13	270 [150, 120]	4/9
17	Hepatitis_1	19	80 [13, 67]	12/7
18	Ionosphere	33	351 [225, 126]	6/27
19	Chess (King Rook vs King Pawn)	36	3196 [1669, 1527]	13/23
20	SolarFlare_2*	10	1389 [1321, 68]	8/2
21	Sonar Mines vs. Rocks	60	208 [97, 111]	34/26
22	Spambase	57	4601 [2788, 1813]	21/36
23	Splice2*	60	3186 [1532, 1654]	32/28
24	Tic-Tac-Toe	8	958 [626, 332]	3/2

Table 2. Comparison between the average co-training performance attained for $|L| = |L|_{\min}$, and the average performance attained by the corresponding base classifier trained on the same L, without co-training (see the text for the meaning of table entries)

ID	Dataset	Naive Bayes	K-NN
1	Audiology*	2	2*
2	Automobile*	2*	2*
3	Breast Cancer W.	2	2
4	Winsconsin D.	2	2
5	Winsconsin Progn. 1	2*	2
6	Winsconsin Progn. 2	2	2
7	Contraceptive Method	1*	1
8	Horse colic	0	1*
9	Credit Approval	2	0
10	Dermatology*	2	2
11	Pima Indians Diabetes	0	2*
12	E.Coli*	2	0
13	Flags*	2	2
14	Heart (Cleveland)*	2	0
15	Heart (LongBeach)*	2*	2
16	Heart-statlog	2	-
17	Hepatitis_1	2	2*
18	Ionosphere	2	0
19	Chess (King Rook vs King Pawn)	1	1*
20	SolarFlare_2*	2	2
21	Sonar Mines vs. Rocks	0	0
22	Spambase	2	0
23	Splice2*	0	1*
24	Tic-Tac-Toe	0	2

other view; 2: co-training outperformed the base classifier in both views; 2*: co-training outperformed the base classifier in one view, no statistically significant performance difference in the other view; -: co-training outperformed the base classifier in one view, and was outperformed by the base classifier in the other view.

When Naive Bayes was used, co-training outperformed in both views the base classifier trained only on L, in 14 data sets. It was instead outperformed in both views only once (Chess data set). The performance was similar in the remaining 5 data sets. When the K-NN classifier was used, results were similar: co-training was better than the base classifier, on both views, in 9 data sets; it was outperformed by the base classifier in one data set (Contraceptive Method); their performance was similar in the other 6 data sets.

We then evaluated co-training performance as above, for $|L| > |L|_{\min}$. The results (not reported here due to lack of space) showed that, when co-training outperformed in both views the base classifier for $|L| = |L|_{\min}$, the same happened for $|L| > |L|_{\min}$. A representative example of this behaviour is reported in Fig. 1 for the Hepatitis_1 data set. Note that co-training performance is almost constant for all the considered $|L|$ values.

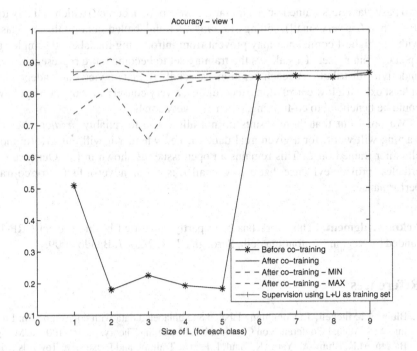

Fig. 1. Co-training performance on the Hepatitis_1 data set, as a function of $|L^+| = |L^-|$, using the NB classifier. "Before co-training": average performance of the base classifier trained on L. "After co-training": average co-training performance, starting from the same L. The minimum (MIN) and (MAX) co-training accuracy is also reported, over the ten runs. For reference, the average performance of the base classifier trained on $L + U$, using the true class labels of the samples in U, is also shown ("Supervision").

These results suggest that co-training performance seems not affected by the size of L, and that co-training can work (i.e., can outperform the base classifier trained only on L) also for very small $|L|$ values, including $|L| = |L|_{min} = 2$.

5   Conclusions

We addressed the issue of evaluating co-training performance as a function of the size of the labelled training set L, as it decreases to the minimum value below which the considered base classifier can not be applied anymore. Results attained on 24 real data sets, artificially splitted into two views, using two different base classifiers, showed that: (i) co-training performance seems not affected when L reduces to the smallest set of samples that allows the chosen learning algorithm to run; (ii) it can outperform the base classifier trained on L, for any size of L, and, in particular, also in the limit case $|L| = |L|_{min}$. In other words, co-training can work even with one sample per class. This behaviour could be explained by the results of [2], mentioned in Sect. 2.2. Even if the two base classifiers trained on the initial L have a poor accuracy (which is likely to happen, when L is very small), adding to L *only a few* unlabelled samples that are classified with the highest confidence may prevent from introducing mislabelled samples in the updated training set. This allows the training set to become more representative of the underlying distribution at each iteration, especially if the two views are independent, or at least exhibit a low correlation. Accordingly, a very conservative updating policy of L could be beneficial to co-training, when L is very small.

We point out that these results do not allow one to reliably *predict* whether co-training will work, for a given, real data set, i.e., whether it will outperform the base classifier trained on L. This remains an open issue, as shown in [5]. Our results nevertheless provide evidence that a very small L is not an adverse factor for co-training performance.

Acknowledgment. This work has been partly supported by the project CRP-18293 funded by Regione Autonoma della Sardegna, L.R. 7/2007, Bando 2009.

References

1. Blum, A., Mitchell, T.: Combining labeled and unlabeled data with co-training. In: Proceedings 11th Annual Conference on Computational Learning Theory, pp. 92–100. ACM (1998)
2. Balcan, M.F., Blum, A., Yang, K., Saul, L.K.: Co-Training and Expansion: Towards Bridging Theory and Practice. In: Weiss, Y., Bottou, L. (eds.) Advances in Neural Information Processing Systems 17, pp. 89–96. MIT Press (2005)
3. Christoudias, C.M., Urtasun, R., Kapoorz, A., Darrell, T.: Co-training with noisy perceptual observations. In: IEEE Conference on Computer Vision and Pattern Recognition, pp. 2844–2851 (2009)
4. Didaci, L., Roli, F.: A Bayesian Analysis of Co-Training Algorithm with Insufficient Views. In: Proc. 11th International Conference on Information Science, Signal Processing and their Applications, pp. 1141–1145. IEEE (2012)
5. Du, J., Ling, C.X., Zhou, Z.-H.: When Does Co-Training Work in Real Data? IEEE Transactions on Knowledge and Data Engineering 23(35), 788–799 (2011)
6. Zhou, Z.-H., Zhan, D.-C., Yang, Q.: Semi-Supervised Learning with Very Few Labeled Training Examples. In: Proc. AAAI, pp. 675–680 (2007)
7. Frank, A., Asuncion, A.: UCI Machine Learning Repository. University of California, School of Information and Computer Science, Irvine, CA (2010),
   http://archive.ics.uci.edu/ml

Classification of High-Dimension PDFs Using the Hungarian Algorithm

James S. Cope and Paolo Remagnino

Digital Imaging Research Centre, Kingston University, London, UK
{j.cope,p.remagnino}@kingston.ac.uk

Abstract. The Hungarian algorithm can be used to calculate the earth mover's distance, as a measure of the difference between two probability density functions, when the pdfs are described by sets of n points sampled from their distributions. However, information generated by the algorithm about precisely how the pdfs are different is not utilized. In this paper, a method is presented that incorporates this information into a 'bag-of-words' type method, in order to increase the robustness of a classification. This method is applied to an image classification problem, and is found to outperform several existing methods.

1   Introduction

For many machine learning problems, an object of data, e.g. an item to be classified, can be described as a single point within a feature space. Many different methods exist for classifying such objects, from simple methods, such as k-nearest-neighbour, to more sophisticated methods, such as support vector machines. However it is sometimes more appropriate to describe an object as a distribution within a feature space. A number of methods also exist for classifying data of this type. When described using histograms, the difference between two probability density functions (pdfs) can be calculated using bin-by-bin methods, such as the Jeffrey-divergence metric, however these methods encounter problems when the data has a high dimensionality, where a large number of bins makes the calculation expensive, whilst the sparse population of bins causes poor results. The earth mover's distance (EMD) [7] deals this by using signatures, and provides an accurate and intuitive measurement. These 'signatures' are weighted points within the feature space. This is akin to clustering data points drawn from the distribution, and weighting each cluster centroid by the number of points in the cluster. Another method is to use kernel density estimation to estimate a probability density function using points sampled from a distribution, and then to use this estimation to predict the probability of another sampling of points belonging to the same distribution. More recently, 'bag-of-words' methods have enjoyed increasing usage for this problem, particularly in the guise of 'bag-of-visual-words' [8] for image retrieval.

   In this paper we utilize information generated in the calculation of the earth mover's distance in order to allow for more robust classification of pdfs, combining this with the strengths of the 'bag-of-words' method.

G.L. Gimel' farb et al. (Eds.): SSPR & SPR 2012, LNCS 7626, pp. 727–733, 2012.

In section 2 we describe the EMD and 'bag-of-words' methods in more detail. In section 3 a new bag-of-words method is described. A comparison of this methods to other common methods, using empirical results, is given in section 4.

2 Background

2.1 The Hungarian Algorithm and the Earth Mover's Distance

The earth mover's distance (EMD) [7] is a measure of the difference between two pdfs. The analogy is that, to reform one mound of earth as another, the effort required would depend on the sum the of distances that each unit of dirt must be moved. Whilst bin-by-bin methods only consider the amount of 'earth' in each location, the EMD considers how far it must be moved. There are two forms of pdf descriptions that allow the EMD to be calculated, histogram binning, and the aforementioned signatures. Since the binning is analogous to using evenly spaced signatures, we need only consider the latter.

Whilst there may be many ways of reforming one pdf into another, the EMD is calculated as being the one that requires the minimum total movement. The standard way of determining this is to model it as the transportation problem. There are a number of methods for solving the transportation problem, but by reforming the data so that each signature has an equal weight, it becomes equivalent to the simpler assignment problem, which can be solved using the Hungarian algorithm [5]. Whilst the original Hungarian algorithm was $O(n^4)$, an $O(n^3)$ version has since been found by Edmond and Karp [4].

The EMD only uses the minimum cost calculated by the Hungarian algorithm, but in our usage here we will also record the corresponding mapping between signatures, as it provides not only a measurement of the difference between the pdfs, but also information about in what way they are different.

2.2 The Bag-of-Words Model

The 'bag-of-words' model was originally used for the retrieval of text documents [9]. The idea was to represent documents as the frequency of occurrence of different words, and to find similar documents by comparing these frequencies. In recent years this concept has been extended to allow for the classification of more general forms of data. Typically, a large number of points are sampled from the training distributions and then a clustering is performed on these. The cluster centroids are used as the 'codewords' in a 'dictionary' used to perform a quantization of the data, by assigning each data-point to its nearest 'codeword'. A set of points from a distribution can then be described as the frequency of occurrence of each 'codeword'. This concept has seen much use recently in the field of computer vision, for tasks such as image retrieval [1,8] and texture analysis [6,10].

3   Methodology

We define the problem as follows. An object, X, is described by a set of n data points, $X = \{\bar{x}_1, \bar{x}_2, ... \bar{x}_n\}$, sampled from a distribution. Each data point $\bar{x}$ is a feature vector, $\bar{x} = [x_1, x_2, ..., x_d]$, where d is the number of features. Given a number of different classes, where class i is described by another set of n data points, $C_i = \{\bar{y}_1, \bar{y}_2, ... \bar{y}_n\}$, drawn from all objects in the training set that belong to the class, we wish to determine the class to which object X most likely belongs. This is calculated using Bayes theorem.

The method involves first generating a set of codewords from the training set, suitable for representing the data. All points in the training and class objects are assigned to their nearest codeword. A mapping is calculated between the data points in each training object and its corresponding class object. For each pair of codewords and each class, the probability is calculated of a mapping having its training object point assigned to the first of these codewords and its class point assigned to the second. For classification, the same codeword assignments and mappings are performed, and the previously calculated probabilities are used to determine the class which the object belongs to.

3.1   Generating a Vocabulary

Within the literature there has been much discussion on the appropriate methods for generating, and ideal size of, the codeword dictionary. The simplest approach is choose points evenly distributed throughout the feature space. The main disadvantage of this is that large portions of the space may not be used, resulting in redundant codewords, whilst other, more useful areas may receive inadequate representation. Another simple method is to use randomly selected points from the training data as the codewords. This largely eradicates the above problems, although using the centroids from a clustering performed on the training data normally provides a better representation. Another approach is to perform a separate clustering for each class and combining the generated codewords. This ensures that each class has some appropriate codewords, but may result in very similar codewords in the combined dictionary. We found that a k-means clustering of the whole training set produces an appropriate dictionary for our method.

There is no consensus on the size of a dictionary, with suggestions varying greatly, but for this method we found that, with objects described using 1024 point, a dictionary of size 256 produced good results, with larger dictionaries providing little or no improvement. We call the ith codeword in the dictionary D_i.

3.2   Producing the Class Models

For each class, a class object is produced by randomly selecting n points from the class's example in the training set. For each training object, a mapping is found from its data points to those its class object using the Hungarian algorithm. This mapping pairs the points in one object to those in the other, such that the sum of the squared Euclidean distances between paired points is minimised. We define the point in the class object C_i to which point $\bar{x}$ is paired as $M(\bar{x}, C_i)$.

Each point in the training data is assigned to its nearest codeword. For each class i, for each pair of codewords, (D_a, D_b), the conditional probability is calculated of a point $\bar{x}$ in that class's training data being assigned to codeword D_a, given that the corresponding point in the class object has been assigned to D_b. This is calculated as follows:

$$P(\bar{x} \in D_a | M(\bar{x}, C_i) \in D_b) \tag{1}$$

$$= \frac{P(\bar{x} \in D_a, M(\bar{x}, C_i) \in D_b)}{P(M(\bar{x}, C_i) \in D_b)} \tag{2}$$

where

$$P(\bar{x} \in D_a, M(\bar{x}, C_i) \in D_b) = \sum_{\substack{T_{ij} \in D_a \\ M(T_{ij}, C_i) \in D_b}} \frac{1}{|T_i|} \tag{3}$$

$$P(M(\bar{x}, C_i) \in D_b) = \sum_{d=0}^{|D|} P(\bar{x} \in D_d, M(\bar{x}, C_i) \in D_b) \tag{4}$$

where T_{ij} is the jth point, $|T_i|$ is the total number of points in the training data for class i, $|D|$ is the number of codewords, and $\bar{x} \in D_a$ indicates that point $\bar{x}$ has been assigned to codeword D_a (likewise, $M(\bar{x}, C_i) \in D_b$ indicates that the point which $\bar{x}$ is paired with is assigned to codeword D_b).

Equation 3 calculates the probability of a point in D_a being mapped to a point in D_b as the fraction of training points for a class C_i for which this occurs. The probablity of a point, from any codeword, being mapped to one in D_b is then the sum of these for all codewords (equation 4).

3.3   Performing the Classification

To classify an object, we again assign all of the object's data points to determine to their nearest codewords. The object is mapped using the Hungarian algorithm to each of the class objects. We can then determine the class to which the object belongs using a Bayesian classifier.

$$c^* = \arg\max_i P(X|C_i)P(C_i) \tag{5}$$

$$P(C_i) = \frac{|T_i|}{\sum_j |T_j|} \tag{6}$$

$$P(X|C_i) = \prod_{\bar{x} \in X} P(\bar{x} \in D_a | M(\bar{x}, C_i) \in D_b) \tag{7}$$

4 Experiments

In this section the new algorithm is empirically evaluated by comparing it to a selection of other techniques. For these experiments we have 32 different classes, with 16 examples of each, performing a 16-fold cross validation. Each example's object has 1024 data-points. For the first method we use a dictionary of 64 codewords, and for the second method we use 16 clusters for each class.

To test the algorithm we apply it to a visual computing problem, the classification of plant species from images of their leaves. This is a problem which has received much interest recently [2]. For each leaf image in the database, we randomly select 1024 small windows. For each window we calculate 20 features based on the responses from different filters applied to all the pixels in the window. The set of features for each window becomes one of the objects data-points in a 20-dimension feature space.

4.1 Methods for Comparison

Three different methods are used for comparison:

1. Kernel Density Estimation - Kernel density estimation is used to predict the probability density function for each class. This estimate of the pdf is then used to calculate the likelihood of the object belonging to that class.

$$P(X|C_i) = \prod_{\bar{x} \in X} P(\bar{x}|C_i)$$

$$= \prod_{\bar{x} \in X} \sum_{\bar{y} \in C_i} \frac{\phi(||\bar{y} - \bar{x}||)}{|C_i|}$$

where $\phi(x)$ is a normal distribution function with mean, $\mu = 0$ and standard deviation, $\sigma = 0.1$. This kernel function was used as it appeared to give the best results for the dataset.

2. Earth Mover's Distance - For the we use the pure value calculated by the earth mover's distance instead of utilizing the mapping between objects. Each object is classified as belonging to the class whose object is closest to it according to the EMD metric.

3. Naive-Bayesian Bag-of-Words - For the bag of words method, we use the same codeword dictionary as for the new method, to allow fairer comparison. We use a Naive-Bayes classifier, as it is both one of the most common classifiers used for bag-of-words [3], and is similar to that used in the proposed method.

4.2 Results

Table 1 gives the results for the proposed method, using different numbers of data points, and different dictionary sizes. The overall results of the experiments are given in table 2.

Table 1. Results for the proposed method, varying object and dictionary size (in %)

$|D|$	$n = 256$	$n = 512$	$n = 1024$
16	67.97	73.05	75.39
32	75.39	80.66	81.64
64	84.77	85.35	88.09
128	86.13	90.04	90.06
256	90.02	91.02	92.97

Table 2. Overall results, using best parameter values for each method (in %)

Method	n		
	256	512	1024
Proposed Method	90.02	91.02	92.97
Kernel Density Estimation	69.73	73.83	77.73
Earth Mover's Distance	73.83	79.88	85.35
Bag-of-Words	77.15	79.30	80.27

As the results show, the new method both performed far better than the standard bag-of-words method. This is because when the difference between pdfs means that points are assigned to different codewords, the standard method considers only that these points are no longer assigned to the same codeword, whereas the new methods both consider where in the feature space those points may exist, given that particular class. The kernel density estimation and earth mover's distance methods both performed worse than the other methods. These methods both directly compare samplings from distributions, and so are susceptible to noise produced by the sampling. The bag-of-words methods eliminate much of this noise, by quantisation via assignment to codewords.

Given that the EMD must be calculated in performing the new method, it may be possible to improve the results by incorporating the EMD metric. In our experience, however, doing so produced no change in the results. As would be expected, increasing the number of points used to describe objects increases the quality of the classification, but the new method still performs better than the other methods when a smaller number of points are used, making it particularly suitable when larger samplings are not practicle.

5   Discussion

In this paper a new method for the classification of high-dimension probability density functions has been proposed. The method utilizes the Hungarian algorithm to calculate mapping between sets of points sampled from PDFs. This information is incorporated into a 'bag-of-words' type method by calculating the probabilities of a pair of corresponding data-points being assigned to particular pairs of 'codewords'. This allows for more robust classification than the traditional 'bag-of-words' method. For a visual object recognition problem the algorithm was found to perform significantly better than a number of existing techniques, achieving over 92% accuracy.

References

1. Chen, X., Hu, X., Shen, X.: Spatial Weighting for Bag-of-Visual-Words and Its Application in Content-Based Image Retrieval. In: Theeramunkong, T., Kijsirikul, B., Cercone, N., Ho, T.-B. (eds.) PAKDD 2009. LNCS, vol. 5476, pp. 867–874. Springer, Heidelberg (2009)
2. Cope, J.S., Corney, D.P.A., Clark, J.Y., Remagnino, P., Wilkin, P.: Plant species identification using digital morphometrics: A reviews. Expert Systems with Applications 39, 7562–7573 (2012)
3. Csurka, G., Dance, C.R., Fan, L., Willamowski, J., Bray, C.: Visual categorization with bags of keypoints. In: Workshop on Statistical Learning in Computer Vision, ECCV, pp. 1–22 (2004)
4. Edmonds, J., Karp, R.M.: Theoretical improvements in algorithmic efficiency for network flow problems. Journal of the ACM 19, 248–264 (1972)
5. Kuhn, H.W.: The Hungarian method for the assignment problem. Naval Research Logistics Quarterly 2, 83–97 (1955)
6. Leung, T., Malik, J.: Representing and recognising the visual appearance of materials using three-dimensional textons. International Journal Of Computer Vision 43, 7–27 (2001)
7. Rubner, Y., Guibas, L.J., Tomasi, C.: The earth mover's distance, multidimensional scaling, and color-based image retrieval. In: ARPA Image Understanding Workshop, pp. 661–668 (1997)
8. Sivic, J., Zisserman, A.: Google video: A text retrieval approach to object matching in videos. In: International Conference On Computer Vision, vol. 2, pp. 1470–1477 (2003)
9. Sparck-Jones, K., Needham, R.M.: Automatic term classifications and retrieval. Information Storage And Retrieval 4, 91–100 (1968)
10. Varma, M., Zisserman, A.: A statistical approach to material classification using image patch exemplars. IEEE Transactions on Pattern Analysis and Machine Intelligence 31, 2032–2047 (2009)

Face Recognition Using Multilinear Manifold Analysis of Local Descriptors

Xian-Hua Han and Yen-Wei Chen

Ritsumeikan University, 1-1-1, NojiHigashi, Kusatsu, Shiga, 525-8577, Japan

Abstract. In this paper, we propose to represent a face image as a local descriptor tensor and use a Multilinear Manifold Analysis (MMA) method for discriminant feature extraction, which is used for face recognition. The local descriptor tensor, which is a combination of the descriptor of local regions (K*K-pixel patch) in the image, can represent image more efficient than pixel-level intensity representation, and also than the popular Bag-Of-Feature (BOF) model, which approximately represents each local descriptor as a predefined visual word. Therefore it should be more effective in computational time than the BOF model. For extracting discriminant and compact features from the local descriptor tensor, we propose to use the proposed TMultilinear Manifold Analysis (MMA) algorithm, which has several benefits compared with conventional subspace learning methods such as PCA, ICA, LDA and so on: (1) a natural way of representing data without losing structure information, i.e., the information about the relative positions of pixels or regions; (2) a reduction in the small sample size problem which occurs in conventional supervised learning because the number of training samples is much less than the dimensionality of the feature space; (3) a neighborhood structure preserving in tensor feature space for face recognition and a good convergence property in training procedure. We validate our proposed algorithm on Benchmark database Yale and PIE, and experimental results show recognition rate with the proposed method can be greatly improved compared with conventional subspace analysis methods especially for small training sample number. ...

1 Introduction

Many face recognition techniques have been developed over the past few decades. One of the most successful and well-studied face recognition techniques is the appearance-based method [1,2]. When using appearance-based methods, an image of size $n_1 \times n_2$ pixels is usually represented by a vector in an $n_1 \times n_2$-dimensional space. In practice, however, these $n_1 \times n_2$-dimensional spaces are too large to allow robust and fast face recognition. Previous works have demonstrated that the face recognition performance can be improved significantly in lower dimensional linear subspaces [2-3]. Two of the most popular appearance-based face recognition methods include Eigenface [2] and Fisherface. Eigenface is based on Principal Component Analysis (PCA). PCA projects the face images along the directions of maximal variances. It also aims to preserve the Euclidean

G.L. Gimel' farb et al. (Eds.): SSPR & SPR 2012, LNCS 7626, pp. 734–742, 2012.

distances between face images. Fisherface is based on Linear Discriminant Analysis (LDA) [2]. Unlike PCA which is unsupervised, LDA is supervised. When the class information is available, LDA can be used to find a linear subspace which is optimal for discrimination. Recently there is considerable interest in geometrically motivated approaches to visual analysis. Therein, the most popular ones include Locality Preserving Projection (LPP) [3], Neighborhood Preserving Embedding (NPE) and so on, which can not only preserve the local structure between samples, and also obtain acceptable recognition rate for face recognition. In real application, all these subspace learning methods need to firstly reshape the 2D face image into 1D vector for analysis, which usually suffers "curse of dimension". Therefore, some researchers proposed to solve the "curse of dimension" problem with 2D subspace learning such as 2D-PCA, 2D-LDA [4] for analyzing directly on 2D image matrix, which was improved to be suitable to some extent. However, all of the conventional methods usually perform subspace analysis directly on the reshaped vector or matrix of pixel-level intensity, which would be unstable under illumination or pose variance.

In this paper, we propose to represent a face image as a local descriptor tensor, which is a combination of the descriptor of local regions (K*K-pixel patch) in the image, and more efficient than the popular Bag-Of-Feature (BOF) model [5] for local descriptor combination. In order to extract discriminant feature from the local regions, we explore an improved gradient (intensity-normalized gradient) of the face image, which is robust to illumination variance, and use histogram of orientation weighed with the improved gradient for local region representation. Furthermore, we propose to use a multilinear subspace learning algorithm for discriminant feature extraction from the local descriptor tensor of face images, which can preserve local sample structure in feature space. Compared with tensorfaces [6] method which also directly analyze multi-dimensional data, the proposed MMA uses supervised strategy, and thus can extract more discriminant features for distinguishing different objects (here facial images of different persons) and at the same time, can preserve samples' relationship of inner-person instead of only dimension reduction in tensorfaces. We validate our proposed algorithm on benchmark database Yale[2] and CMU PIE[7], and experimental results show recognition rate with our method can be greatly improved compared conventional subspace analysis methods especially for small training sample number.

The remaining parts of this paper are organized as follows. We introduce the local descriptor tensor for face images in section 2. Section 3 propose a Multilinear Manifold Analysis (MMA) for extracting discriminant feature for face representation. Finally, we report experiment setup and results in section 4, and give conclusion remarks in section 5.

2   Local Descriptor Tensor for Face Image Representation

In computer vision, local descriptors (i.e. features computed over limited spatial support) have proved well-adapted for matching and recognition tasks, as

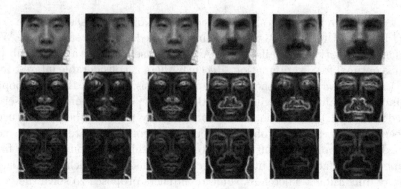

Fig. 1. Gradient image samples. Top row: Original face images; Middle row: the intensity-normalized gradient images; Bottom row: the conventional gradient images.

they are robust to partial visibility and clutter. The current popular one for local descriptor is SIFT feature, which is proposed by in [11] and is robust to small illumination variance. However with large illumination variance usually appeared in face recognition, it is still difficult to recognize correctly, and achieve acceptable recognition rate. Therefore, we proposed a histogram of orientation weighted with the improved gradient for local image representation. With the local descriptor, usually there are two types of algorithms for object recognition. One is to match the local point with SIFT feature in two images, and the other one is to use the popular Bag-Of-Feature model (BOF), which forms a frequency histogram of a predefined visual-words for all sampled region features [5]. For matching algorithm, it is usually not enough to recognize the unknown image even if there are several points well matched. How to combine more features is not unsolved still. The popular BOF model usually can achieve good recognition performance in most applications such as scene and object recognition. However, In BOF model, in order to achieve acceptable recognition rate it is necessary to sample a lot of points for extracting SIFT features (usually more than 1000 in an image), and compare the extracted local feature with the predefined visual-words (Usually more than 1000) to obtain the visual-word occurrence histogram. Therefore, BOF model need a lot of computing time to extract visual-words occurrence histogram. In addition, BOF model just approximately represent each local region feature as the predefined visual-words, and then, it maybe lose a lot of information and will be not efficient for image representation. Therefor, in this paper, we propose to represent a face image as a combined local descriptor tensor.

In our work, we combine two types of local features as a tensor for face image representation: SIFT feature and a intensity-Normalized Histogram of Orientation Gradient–NHOG.

(1) The SIFT descriptor computes a gradient orientation histogram within the support region. For each of 8 orientation planes, the gradient image is sampled over a 4 by 4 grid of locations, thus resulting in a 128-dimensional feature

vector for each region. A Gaussian window function is used to assign a weight to the magnitude of each sample point. This makes the descriptor less sensitive to small changes in the position of the support region and puts more emphasis on the gradients that are near the center of the region. To obtain robustness to illumination changes, the descriptors are made invariant to illumination transformations of the form $a\mathbf{I}(x) + b$ by scaling the norm of each descriptor to unity [8]. For representing the local region of a color image, we extract SIFT feature in each color component (R, G and B color components), and then can achieve a 128×3 2D tensor for each local region.

(2) In order to extract robust feature to illumination variance, we need to obtain the improved gradient. Given an image $\mathbf{I}$, we calculate the improved gradient (Intensity-normalized gradient) using the following Eq.:

$$\mathbf{I}_x(i,j) = \frac{\mathbf{I}(i+1,j) - \mathbf{I}(i-1,j)}{\mathbf{I}(i+1,j) + \mathbf{I}(i-1,j)}$$

$$\mathbf{I}_y(i,j) = \frac{\mathbf{I}(i,j+1) - \mathbf{I}(i,j-1)}{\mathbf{I}(i,j+1) + \mathbf{I}(i,j-1)} \tag{1}$$

$$\mathbf{I}_{xy}(i,j) = \sqrt{\mathbf{I}_x(i,j)^2 + \mathbf{I}_y(i,j)^2}$$

where $\mathbf{I}_x(i,j)$ and $\mathbf{I}_y(i,j)$ means the horizontal and vertical gradient in pixel position i,j, respectively, $\mathbf{I}_{xy}(i,j)$ means the global gradient in pixel position i,j. The idea of the normalized gradient is from χ^2 distance: a normalized Euclidean distance. For x-direction, the gradient is normalized by summation of the upper one and the bottom one pixel centered by the focused pixel; for y-direction, the gradient is normalized by that of the right and left one. With the intensity-normalized gradient, we can extract robust and invariant features to illumination changing in a local region of an image. Some examples with the intensity-normalized and conventional gradients are shown in Fig. 1. The local NHOG feature can be extracted as shown in Fig. 2. given a local region I^R in an face image, we firstly segment the region into 4 (2×2) patches,and in each patch, we extract a 20-bin histogram of orientation weighted by global gradient using the intensity-normalized gradients $\mathbf{I}_x^R$, $\mathbf{I}_y^R$ and $\mathbf{I}_{xy}^R$. Therefore, each region in a face image can be represent by 80-bin (20×4) histogram as shown in the left part of Fig. 2.

In order to efficiently represent a face image, we combine the extracted local SIFT or NHOG descriptors for face image representation. Firstly, we grid-segment an image, and can obtain $M2$ overlapping regions as shown in the right part of Fig. 2, and then in each region, we extract a L-dimension (128 for SIFT, 80 for NHOG) local feature (1D tensor). Furthermore we combine the $M2$ vectors (local descriptors) into a 2D tensor with of size $L \times M2$ in the space $R_{128or80} \bigotimes R_{M2}$ for representing a face image. The tensor NHPG feature extraction procedure of a face image is shown in Fig. 2.

3   Multilinear Manifold Analysis

In order to model N-D data without rasterization (2D is a special case), tensor representation is proposed and analyzed for feature extraction or modeling. In

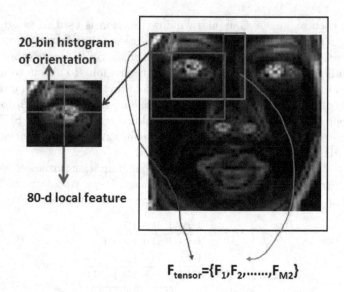

20-bin histogram of orientation

80-d local feature

$$F_{tensor}=\{F_1,F_2,\cdots\cdots,F_{M2}\}$$

Fig. 2. Extraction procedure of local descriptor tensor from a face image. The red rectangle in the right part of this figure is the first extracted region for calculating local descriptor (a 80-bin edge histogram); The green rectangle is the next extracted region after moving several pixels from the red one (predefined interval) along row, and continue this step until the end of row pixels. The purple rectangle is the first extracted region after moving several pixels for the red one along column, and then obtain next regions through moving pixel in row. The total number of extracted regions is $M2$.

this section, we propose a tensor supervised neighborhood embedding to not only extract discriminant feature but also preserve the local geometrical and topological properties in same category for recognition. The proposed approach decompose each model of tensor with objective function, which consider neighborhood relation and class label of training samples.

Suppose we have ND tensor objects $\mathcal{X}$ from C classes. The cth class has n^c tensor objects and the total number of tensor objects is n. Let $\mathcal{X}_{i_c} \in R^{N_1} \bigotimes R^{N_2} \bigotimes \cdots \bigotimes R^{N_L} (i_c = 1, 2, \cdots, n^c)$ be the ith object in the cth class. For a gray face image, we can directly represent it as pixel-level intensity tensor, where L is 2, N_1 is the row number, N_2 is the column number. We also can represent the face image as a feature-based tensor such as local descriptor feature tensor introduced in Sec. 2, where L is also 2, N_1 is the local feature dimension, N_2 is the sampled region number in an image. Then, we can build a nearest neighbor graph $\mathcal{G}$ to model the local geometrical structure and label information of $\mathcal{X}$. Let W be the weight matrix of $\mathcal{G}$. A possible definition of W is as follows:

$$W_{ij} = \begin{cases} exp^{-\frac{\mathcal{X}_i - \mathcal{X}_j}{t}} & \text{if sample } i \text{ and } j \text{ is in same class} \\ 0 & \text{otherwise} \end{cases} \quad (2)$$

Let $\mathbf{U}_d$ be the d-model transformation matrices(Dimension: $N_d \times D_d$). A reasonable transformation respecting the graph structure can be obtained by solving the following objective functions:

$$\min_{\mathbf{U}_1,\mathbf{U}_2,\cdots,\mathbf{U}_L} \sum_{ij} \|\mathcal{X}_{i\times1}\mathbf{U}_{1\times2}\mathbf{U}_2\cdots\times_L \mathbf{U}_L - \mathcal{X}_{j\times1}\mathbf{U}_{1\times2}\mathbf{U}_2\cdots\times_L \mathbf{U}_L\|W_{ij} \qquad (3)$$

The objective function incurs a heavy penalty if neighboring points $\mathcal{X}_i$ and $\mathcal{X}_j$ are mapped far apart. Therefore, minimizing it is an attempt to ensure that if $\mathcal{X}_i$ and $\mathcal{X}_j$ are "close", then $\mathcal{X}_{i\times1}\mathbf{U}_{1\times2}\mathbf{U}_2\cdots\times_L \mathbf{U}_L$ and $\mathcal{X}_{j\times1}\mathbf{U}_{1\times2}\mathbf{U}_2\cdots\times_L \mathbf{U}_L$ are "close" as well. Let $\mathcal{Y}_i = \mathcal{X}_{i\times1}\mathbf{U}_{1\times2}\mathbf{U}_2\cdots\times_L \mathbf{U}_L$ (Dimension: $D_1 \times D_2 \times \cdots \times D_L$) , and $(\mathbf{Y}_i)^d = (\mathbf{X}_{i\times1}\mathbf{U}_{1\times2}\mathbf{U}_2\cdots\times_L \mathbf{U}_L)^d$ (2D matrix, Dimension: $D_d \times (D_1 \times D_2 \times \cdots \times D_{d-1} \times D_{d+1} \times \cdots \times D_L)$) is the d-mode extension of tensor $\mathcal{Y}_i$. Let D be a diagonal matrix, $D_{ii} = \sum_j W_{ij}$. Since $\|\mathbf{A}\|^2 = tr(\mathbf{A}\mathbf{A}^T)$, we see that

$$\frac{1}{2}\sum_{ij} \|\mathcal{X}_{i\times1}\mathbf{U}_1\cdots\times_L \mathbf{U}_L - \mathcal{X}_{j\times1}\mathbf{U}_1\cdots\times_L \mathbf{U}_L\|W_{ij}$$

$$=\frac{1}{2}\sum_{ij} tr(((\mathbf{Y}_i)^d - (\mathbf{Y}_j)^d)((\mathbf{Y}_i)^d - (\mathbf{Y}_j)^d)^T)W_{ij}$$

$$=tr(\sum_i D_{ii}(\mathbf{Y}_i)^d((\mathbf{Y}_i)^d)^T - \sum_{ij} W_{ij}(\mathbf{Y}_i)^d((\mathbf{Y}_j)^d)^T)$$

$$=tr(\mathbf{U}_d^T(\sum_i D_{ii}((\mathcal{X}_{i\times1}\mathbf{U}_1\cdots\times_{d-1}\mathbf{U}_{d-1\times d+1}\mathbf{U}_{d+1}\cdots\times_L \mathbf{U}_L) \qquad (4)$$

$$(\mathcal{X}_{i\times1}\mathbf{U}_1\cdots\times_{d-1}\mathbf{U}_{d-1\times d+1}\mathbf{U}_{d+1}\cdots\times_L \mathbf{U}_L)^T$$

$$-\sum_{ij} W_{ij}((\mathcal{X}_{i\times1}\mathbf{U}_1\cdots\times_{d-1}\mathbf{U}_{d-1\times d+1}\mathbf{U}_{d+1}\cdots\times_L \mathbf{U}_L)$$

$$(\mathcal{X}_{j\times1}\mathbf{U}_1\cdots\times_{d-1}\mathbf{U}_{d-1\times d+1}\mathbf{U}_{d+1}\cdots\times_L \mathbf{U}_L)^T)\mathbf{U}_d)$$

$$=tr(\mathbf{U}_d^T(\mathbf{D}_d - \mathbf{S}_d)\mathbf{U}_d)$$

where $\mathbf{D}_d = \sum_i D_{ii}((\mathcal{X}_{i\times1}\mathbf{U}_1\cdots\times_{d-1}\mathbf{U}_{d-1\times d+1}\mathbf{U}_{d+1}\cdots\times_L \mathbf{U}_L) (\mathcal{X}_{i\times1}\mathbf{U}_1\cdots\times_{d-1}\mathbf{U}_{d-1\times d+1}\mathbf{U}_{d+1}\cdots\times_L \mathbf{U}_L)^T$ and $\mathbf{S}_d = \sum_{ij} W_{ij}((\mathcal{X}_{i\times1}\mathbf{U}_1\cdots\times_{d-1}\mathbf{U}_{d-1\times d+1}\mathbf{U}_{d+1}\cdots\times_L \mathbf{U}_L) (\mathcal{X}_{j\times1}\mathbf{U}_1\cdots\times_{d-1}\mathbf{U}_{d-1\times d+1}\mathbf{U}_{d+1}\cdots\times_L \mathbf{U}_L)^T$. Therefore the linear transformation $\mathbf{U}_d$ can be obtained by minimizing the objective function under constraint:

$$\mathbf{U}_d = \underset{\mathbf{U}_d^T\mathbf{D}_d\mathbf{U}_d=1}{\mathbf{argmin}} (\mathbf{U}_d^T(\mathbf{D}_d - \mathbf{S}_d)\mathbf{U}_d) \qquad (5)$$

In order to achieve the stable solution, we firstly regularize the symmetric matrix D as $D_{ii} = D_{ii} + \alpha$ (α is a small value). Finally, the minimization problem can be converted to solving a generalized eigenvalue problem as follows:

$$\mathbf{D}_d U_d = \lambda \mathbf{S}_d U_d \qquad (6)$$

After obtaining the MMA basis of each mode, we can project each tensor object into these MMA basis for each mode. For face recognition, the projection coefficients can represent the extracted feature vectors and can be used for classification using Euclidean distance or other similar measurement.

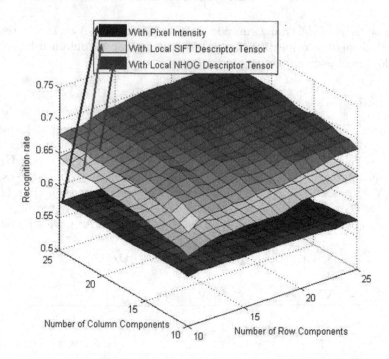

Fig. 3. Compared recognition rates using MMA for feature extraction with pixel-intensity tensor, local SIFT tensor and local NHOG tensor, respectively. X-axis denotes the number of retained row-mode components of the used tensor; Y-axis denotes the number of retained column-mode components of the used tensor.

4   Experimental Results

In this paper, we use the benchmark face dataset YALE, which includes 15 people and 11 facial images of each individual with different illuminations and expressions, and CMU PIE, which includes 68 people and about 170 facial images for each individual with 13 different poses, 43 different illumination conditions, and with 4 different expressions. For YALE dataset, we randomly select 2, 3, 4 and 5 facial images from each individual for training, and the remainders for test. We do 20 runs for different training number and average recognition rate. For comparison, we also do experiments using the proposed MMA analysis directly on the gray face image (pixel-level intensity, denoted MMA), and our proposed local descriptor tensor with SIFT descriptor (denoted MMA-SIFT) and intensity-Normalized Histogram of Orientation Gradient (denoted MMA-NHOG). Figure 3 gives the compared recognition rates after discriminant and compact feature extraction by the proposed MMA with the three types tensor (Pixel-intensity tensor, Local SIFT tensor and NHOG tensor), respectively. It is obvious from Fig. 3 that the proposed two local descriptor tensor representations for face image can achieve much higher recognition rates than those directly with pixel intensity tensor on any extracted feature number, and then the recognition

Table 1. Average recognition error rates (%) on YALE dataset with different training number

Method	2 Train	3 Train	4 Train	5 Train
PCA	56.5	51.1	57.8	45.6
LDA	54.3	35.5	27.3	22.5
Laplacianface	43.5	31.5	25.4	21.7
O-Laplacianface	44.3	29.9	22.7	17.9
TensorLPP	54.5	42.8	37	32.7
R-LDA	42.1	28.6	21.6	17.4
S-LDA	37.5	25.6	19.7	14.9
MMA	41.89	31.67	24.86	23.06
MMA-SIFT	35.22	26.33	22.19	20.83
MMA-NHOG	**29.74**	**22.87**	**18.52**	**17.44**

Table 2. Average recognition error rates (%) on YALE dataset with different training number

Method	PCA	LDA	LPP	MMA	MMA-NGOG
5 Train	75.33	42.8	38	37.66	**33.85**
10 Train	65.5	29.7	29.6	23.57	**22.06**

rates with the proposed NHOG feature, which is robust to large illumination variance, are better than those with SIFT feature, which just can deal with small illumination variation. In order to validate our proposed MMA algorithm with conventional subspace learning methods, we also give the compared results shown in Table 1 using MMA analysis with different tensors and the state-of-art subspace learning methods by He [3,9,10]. From Table 1, it is obvious that our proposed algorithm can obtain the best recognition performance especially using small training samples. For CMU PIE dataset, we randomly select 5 and 10 facial images from each individual for training, and the remainder for test. We also do 20 runs for achieving average recognition error rate. The compared recognition error rates between our proposed algorithms and the conventional subspace learning methods by He [3,9,10] are shown in Table 2.

5   Conclusions

In this paper, we proposed to represent a face image as a local descriptor tensor, which is a combination of the descriptor of local regions (K*K-pixel patch) in the image, and more efficient than the popular Bag-Of-Feature (BOF) model for local descriptor combination. Furthermore, we proposed to use Multilinear Manifold Analysis (MMA) for discriminant feature extraction from the local descriptor tensor of face images, which can preserve local sample structure in feature space. We validate our proposed algorithm on Benchmark database Yale and PIE, and experimental results show recognition rate with our method can be greatly improved compared conventional subspace analysis methods especially for small training sample numbers.

Acknowledgments. This work was supported in part by the Grant-in Aid for Scientific Research from the Japanese MEXT under the Grant No. 2430076, 24700179 and in part by the Research Matching Fund for Private Universities from MEXT (Ministry of Education, Culture, Sports, Science, and Technology).

References

1. Murase, H., Nayar, S.K.: Visual learning and recognition of 3-d objects from appearance. International Journal of Computer Vision 14(1), 5–24 (1995)
2. Turk, M., Pentland, A.: Eigenfaces for recognition. Journal of Cognitive Neuroscience 3(1), 71–86 (1991)
3. He, X., Yan, S., Hu, Y., Niyogi, P., Zhang, H.-J.: Face recognition using laplacianfaces. IEEE Transactions on Pattern Analysis and Machine Intelligence 27(3), 328–340 (2005)
4. Wang, X.-M., Huang, C., Fang, X.-Y., Liu, J.-G.: 2DPCA vs. 2DLDA: Face Recognition Using Two-Dimensional Method. In: International Conference on Artificial Intelligence and Computational Intelligence, vol. 2, pp. 357–360 (2009)
5. Csurka, G., Dance, C., Fan, L., Willamowski, J., Bray, C.: Visual categoraization with bags of keypoints. In: Proc. ECCV Workshop on Statistical Learning in Computer Vision, pp. 1–16.
6. Vasilescu, M.A.O., Terzopoulos, D.: Multilinear Analysis of Image Ensembles: TensorFaces. In: Heyden, A., Sparr, G., Nielsen, M., Johansen, P. (eds.) ECCV 2002, Part I. LNCS, vol. 2350, pp. 447–460. Springer, Heidelberg (2002)
7. Sim, T., Baker, S., Bsat, M.: The CMU Pose, Illumination, and Expression (PIE) Database of Human Faces. Robotics Institute, CMU-RI-TR-01-02, Pittsburgh, PA (2001)
8. Lowe, D.: Distinctive image features from scale-invariant keypoints. International Journal of Computer Vision 60(2), 91–110 (2004)
9. Cai, D., He, X., Hu, Y., Han, J., Huang, T.: Learning a Spatially Smooth Subspace for Face Recognition. In: CVPR 2007 (2007)
10. Cai, D., He, X., Han, J.: Spectral Regression for Efficient Regularized Subspace Learning. In: ICCV 2007 (2007)

A Genetic Inspired Optimization for ECOC

Miguel Ángel Bautista[1,2], Sergio Escalera[1,2], Xavier Baró[2,3], and Oriol Pujol[1,2]

[1] Dept. Matemàtica Aplicada i Anàlisi, Universitat de Barcelona, Gran Via 585, 08007 Barcelona, Spain
[2] Centre de Visió per Computador, Campus UAB, Edifici O, 08193 Bellaterra, Barcelona, Spain
[3] EIMS, Universitat Oberta de Catalunya, Rambla del Poblenou 156, 08018, Barcelona
{mbautista,opujol,sescalera}@ub.edu,
xbaro@uoc.edu

Abstract. In this work, we propose a novel Genetic Inspired Error Correcting Output Codes (ECOC) Optimization, which looks for an efficient problem-dependent encoding of the multi-class task with high generalization performance. This optimization procedure is based on novel ECOC-Compliant crossover, mutation, and extension operators, which guide the optimization process to promising regions of the search space. The results on several public datasets show significant performance improvements as compared to state-of-the-art ECOC strategies.

Keywords: Error-Correcting Output Codes, Genetic Optimization, Ensemble learning.

1 Introduction

A challenging task in Pattern Recognition is to develop efficient methodologies to process huge amount of data. Concretely, classification procedures present a lack of options when the number of categories is arbitrarily large. In this scope, the Error Correcting Output Codes (ECOC) framework has shown great performance results. At the ECOC *coding* step, a set of binary partitions of the original problem are encoded in a matrix of codewords (one code per class, univocally defined) which are learnt by binary classifiers. Then, at the ECOC *decoding* step a final decision is obtained by comparing the set of binary predictions with every class code, and choosing the class with the code at minimum 'distance'. Standard ECOC coding strategies need between N and $\binom{N}{2}$ classifiers to deal with a $N-$class problem (using the One vs. All and the One vs. One coding designs, respectively). This implies a scalability problem when dealing with a large number of classes. Recently, some works applied Genetic Algorithms (GA) to find a sub-optimal ECOC configuration. The underlying idea of GA is to reproduce the natural evolution by means of computer programs, using a chromosome based representation of the problems, and implementing from a functional point of view the processes involved in nature (crossover and mutation). Various works

G.L. Gimel' farb et al. (Eds.): SSPR & SPR 2012, LNCS 7626, pp. 743–751, 2012.

have treated the optimization of ECOC matrices with GA [2,6,5]. Nevertheless, they fail in taking into account the ECOC constraints, implying an unnecessary enlargement of the search space.

In this work, we propose a novel framework for treating the optimization of an ECOC matrix inspired on GA. In this framework the operators have been completely redefined in order to avoid non-valid individual generation, and thus, minimizing the search space in relation to previous works. In addition, the code length is reduced to be sub-linear in the number of categories, building both reduced and high-performance codes. This novel procedure is tested on several public datasets, obtaining significant performance improvements compared to state-of-the-art ECOC approaches.

The paper is organized as follows: Section 2 presents the novel genetic approach. Section 3 shows the experimental results and Section 4 concludes the paper .

2   ECOC-Compliant Genetic Algorithm

In this section we review the ECOC framework, its properties, and present the Genetic-ECOC.

2.1   ECOC Framework

The ECOC framework is composed of two different steps: *coding* and *decoding* [1]. At the coding step an ECOC coding matrix $M_{N \times n} \in \{-1, +1, 0\}$ is constructed, where N denotes the number of classes in the problem and n the number of bi-partitions defined to discriminate the N classes. In this matrix, the rows (also known as *codewords*) are univocally defined, since these are the identifiers of each category in the multi-class problem. On the other hand, the columns of M denote the set of bi-partitions, dichotomies, or meta-classes to be learnt by each base classifier h^j (also known as dichotomizer). Hence, classifier h^j is responsible for learning the bi-partition denoted on the j−th column of M [1]. From the learning point of view, the performance of the ECOC ensemble will increase as more bi-partitions are taken into account. However, by taking into account the problem idiosyncrasies the system is able to obtain great performance by using few bi-partitions.

At the decoding step a new sample s is classified according to the N possible categories. In order to perform the classification task, each dichotomizer predicts a binary value for s whether it belongs to one of the bi-partitions defined by the corresponding dichotomy. Once the set of predictions $x(s) \in \mathbb{R}^n$ is obtained, it is compared to the codewords of M using a distance metric δ, known as the decoding function.

---

[1] For notation purposes we will refer to the entry of M at the i-th row and the j-th column as $M_{i,j}$

2.2 ECOC Coding Matrix Properties

We define an ECOC coding matrix $M_{N \times n} \in \{-1, +1, 0\}$ to be constrained by,

$$\min(\delta_{AHD}(y^i, y^k)) \geq 1, \forall i, k : i \neq k, \ i, k \in [1, \ldots, N] \tag{1}$$

$$\min(\delta_{HD}(d^j, d^l)) \geq 1, \forall j, l : j \neq l, \ j, l \in [1, \ldots, n] \tag{2}$$

$$\min(\delta_{HD}(d^j, -d^l)) \geq 1, \forall j, l : j \neq l, \ j, l \in [1, \ldots, n] \tag{3}$$

where δ_{AHD} and δ_{HD} are the Attenuated Hamming Distance (AHD) and the Hamming Distance (HD) are defined as in [4].

2.3 Genetic Inspired ECOC Optimization

In this section we present the novel Genetic-ECOC.

Problem Encoding. In order to consider the ECOC properties and obtain smart heuristics to guide the optimization process, a novel representation of ECOC individuals is proposed. ECOC individuals are represented as structures $I = < M, C, H, P, E, \delta >$, where the fields are defined as follows,

- The **coding matrix**, $M_{N \times n} \in \{-1, +1, 0\}$ where $n \geq \lceil \log_2 N \rceil$. For the initial population we fix $n = \lceil \log_2 N \rceil$, where n can grow along generations.
- The **confusion matrix**, $C_{N \times N}$, over the validation subset. Let c^i and c^j be two classes of our problem, then the entry of C at the i-th row and the j-th column, defined as $C_{i,j}$, contains the number of examples of class c^i classified as examples of class c^j.
- The **set of dichotomizers** $H = < h^1, \ldots, h^n >$.
- The **performance of each dichotomizer**, $P \in \mathbb{R}^n$, $P = [p^1, \ldots, p^n]$. This vector contains the proportion of correctly classified examples over a validation subset for each dichotomizer in H.
- The **error rate**, E, over a validation subset. This scalar is the proportion miss-classified samples of the validation subset using the Loss-Weighted decoding [4]. Let the set of samples in the validation subset be $V = < (s_1, l(s_1)), \ldots, (s_v, l(s_v)) >$, then E is defined as,

$$E = \sum_{j=1}^{v} I(\Delta(M, x^{s_j}), l(s_j))/v, \tag{4}$$

$$\Delta(M, x) = \underset{i}{\arg\min} \ \delta(y_i, x), \ i \in \{1, \ldots, N\} \tag{5}$$

Fitness Function. The fitness function measures the environmental adaptation of each individual, and thus, is the one to be optimized. Individuals are evaluated according to the performance they obtain in the validation subset. Let E_{I_K} be the error rate of individual I_K and let n_{I_K} be the length of the coding matrix M of I_K, then, we define the fitness function as $F_f(I_k) = E_{I_k} + \lambda n_{I_k}$. [2]

---

[2] This expression (similar to the one showed by regularized classifiers), serves us to control the learning capacity of the ECOC matrix in order to not over-fit the training data.

ECOC Crossover and Mutation Operators In this section we introduce the novel ECOC crossover and mutation operators. These operators do not only take into account the restrictions of the ECOC framework (see Equations 1, 2, and 3) but also are carefully designed in order to avoid a premature convergence to local minima without generating non-valid individuals, and thus, converging to satisfying populations in fewer generations. In this sense, the crossover and mutation operators have two variants. The *Generic* one, which provides us with a tool to avoid premature convergence, and the *Specific* one, which guides the optimization to promising regions of the search space.

- **ECOC Crossover Algorithm**

Assume a N-class problem to be learnt and let I_F and I_M be two individuals encoded as shown in Section 2.3. Then, the crossover algorithm will generate a new individual I_S which coding matrix $M_{N\times n}^{I_S}$, $n = \min(n_{I_F}, n_{I_K})$ contains dichotomies of each parent. Therefore, the key aspect of this recombination is the selection of which dichotomies of each parent are suitable to be combined. We introduce a dichotomy selection algorithm that chooses those n dichotomies that hold the constraints shown in Equations 1, 2, and 3. The dichotomy selection algorithm generates a dichotomy selection order $\tau^I \in \mathbb{R}^n$ for each parent I. Moreover, the selection algorithm checks if the separation between codewords is congruent with the number of dichotomies left to be added. In this sense, the $(k - i)$-th extension dichotomy will be only added if it splits the existing codewords to define $|Y| = r \leq 2^{(k-i)}$ codes at $\delta_{AHD}(y^a, y^b) = 0\ \forall\ y^a, y^b \in Y :$ $a \neq b$, where k is the final length of the ECOC matrix. The Generic and Specific version of the ECOC crossover algorithm depend on how τ is defined. In the Generic version, τ is randomly generated, while in the Specific version τ is a classifier performance ranking.

In the crossover example shown in Figure 1 two individuals I_M and I_F are combined to produce a new offspring I_S. The crossover algorithm generates a dichotomy selection order τ for each parent. The first parent from which a dichotomy is taken is I_M, and d_3 is valid since $r \leq 2^{(3-1)} = 4$, and it only defines three codes without separation (y^1, y^2, and y^5). Once this step is performed, the parent is changed, and the following dichotomy will be extracted from I_F based on its selection order τ^{I_F}. In this case, d_4 is valid since $r \leq 2^{(3-2)} = 2$ and d_3 of I_M together with d_4 of I_F define only two equivalent codewords (y^1 and y^5). In the following iteration, the parent is changed again, and thus, I_M is used. Since $\delta_{AHD}(y^1, y^5) = 0$, d^1 can not be considered as an extension dichotomy, and therefore, the next dichotomy to use is d_2, which satisfies Equation 1 defining a valid ECOC coding matrix.

- **ECOC Mutation Algorithm**

Picture an individual I encoded as shown in Section 2.3 to be transformed by means of the mutation operator. This operator will select a set of positions $\mu =< M_{i,j}, \ldots, M_{k,l} >$, $i, k \in \{1, \ldots, N\}$, $j, l \in \{1, \ldots, n\}$ of M^I to be mutated. The value of these positions is changed constrained to values in the set $\{-1, +1, 0\}$.

```
    Data: I_F, I_M
    Result: I_S
 1  n := min(M^{I_F}, M^{I_M}) // Minimum code length among parents
 2  τ^{I_F} ∈ ℝ^n = selorder(I_F) // Dichotomy selection order of I_F
 3  τ^{I_M} ∈ ℝ^n = selorder(I_M);
 4  cp := I_F // Current parent to be used
 5  M^{I_S} := ∅ // Coding matrix of the offspring
 6  for i ∈ {1, ..., n} do
 7  │   for j ∈ {1, ..., n_{cp}} : τ_j^{cp} ≠ ∅ do
 8  │   │   f := 0 // Valid dichotomy search flag
 9  │   │   if calcRepetitions (M^{I_S}, d^{τ_j^{cp}}) ≤ 2^{(k-i)} then
10  │   │   │   d^i := d^{τ_j^{cp}} // Inheritance of dichotomies
11  │   │   │   h^i := h^{τ_j^{cp}} // Inheritance of dichotomizer
12  │   │   │   p^i := p^{τ_j^{cp}} // Inheritance of performance
13  │   │   │   τ_j^{cp} := ∅ // Avoid using a dichotomy twice
14  │   │   │   f := 1 // Valid dichotomy found
15  │   │   │   break;
16  │   │   end
17  │   end
18  │   if !f then
19  │   │   d^i := generateCol(M^{I_S}) // If non ECOC matrix can be built
20  │   │   h^i := ∅;
21  │   │   p^i := ∅;
22  │   end
23  │   if cp = I_F then
24  │   │   cp := I_M // Dichotomy inheritance parent switch
25  │   else
26  │   │   cp := I_F;
27  │   end
28  end
```

Algorithm 1. ECOC Crossover

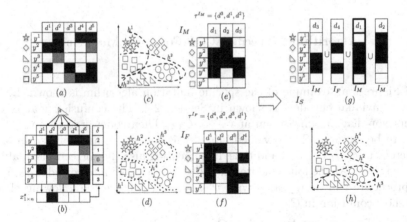

Fig. 1. (a). An example of an ECOC coding matrix. (b) Example of the decoding process. (c) Feature space and trained classifiers for parent I_M. (d) Feature representation and boundaries for parent I_F. (e) ECOC coding matrix of parent I_M. (f) Coding matrix of parent I_F. (g) ECOC coding matrix composition steps for the offspring I_S. (h) Feature space and inherited classifiers for I_S.

In the Generic version, the set of positions μ are those valued 0. Once μ is defined, the positions are randomly recoded to one of the three possible values in $\{-1, +1, 0\}$. In the Specific mutation algorithm, the set of positions μ is chosen taking into account the confusion matrix C. Once these classes are obtained, the algorithm will mutate the bits valued 0 of its codewords $\{y^i, y^j\}$ in order to increment the distance $\delta_{AHD}(y_i, y_j)$. The specific ECOC mutation algorithm is shown in Algorithm 2.

```
Data: I_T, mt_c
// Individual and mutation control value
Result: I_X
1  C_{N×N}^{I_T} // Confusion matrix of I_T
2  k := 0// Number of recoded bits of M^{I_T}
3  while k < mt_c do
4  │   (c^i, c^j) := argmax(C_{i,j} + C_{j,i}) ∀i,j : i ≠ j;
   │            i,j
5  │   for b ∈ {1, ..., n} do
6  │   │   if |y_b^i| + |y_b^j| ≤ 1  and  k < mt_c then
7  │   │   │   if y_b^i = 0 and y_b^j = 0 then
8  │   │   │   │   y_b^i := +1 // Invert both bits valued 0
9  │   │   │   │   y_b^j := -1;
10 │   │   │   else
11 │   │   │   │   if y_b^i = 0 then
12 │   │   │   │   │   y_b^i := -y_b^j // Invert bit valued 0
13 │   │   │   │   else
14 │   │   │   │   │   y_b^j := -y_b^i;
15 │   │   │   │   end
16 │   │   │   end
17 │   │   │   k := k + 1;
18 │   │   end
19 │   end
20 │   C_{i,j}^{I_T} := 0, C_{j,i}^{I_T} := 0;
21 end
```

Algorithm 2. Specific ECOC-Compliant Mutation

In Figure 2 an example of the specific mutation algorithm is shown. Let I_T be an individual encoded as shown in Section 2.3. The confusion matrix C_{I_T} has its non-diagonal maximum at $C_{4,3} + C_{3,4}$. Then codewords y^4 and y^3 are going to be mutated. The 0 valued bits of this codewords are changed in order to increment $\delta_{AHD}(y^4, y^3)$, and thus, incrementing also the correction capability between them. At the following iteration $C_{4,3}$ is not taken into consideration and the procedure will be repeated with y^5 and y^4 which are the following classes that show confusion in C.

Problem-Dependent Extension Operator. We propose an operator to extend ECOC designs based on the confusion matrix, focusing the extension of dichotomies on those categories which are difficult to be split. This methodology defines two types of extensions, the One vs. One extension (Generic extension) and the Sparse extension (Specific extension), which have the same probability of being executed along the optimization process. In the former, the ECOC coding

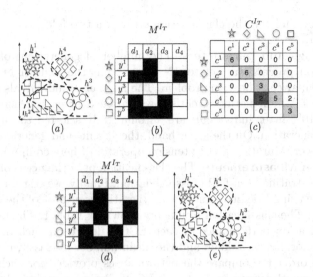

Fig. 2. Mutation example for a 5-class toy problem. (a) Feature space and trained dichotomizers for and individual I_T. (b) ECOC coding matrix of I_T. (c) Confusion matrix of I_T. (d) Mutated coding matrix. (e) Mutated feature space with trained dichotomizers.

matrix $M_{N \times n}$ will be extended with a dichotomy d^{n+1} which will be valued 0 except for those two positions d^i and d^j corresponding to the maximum confused classes $(c^i, c^j) = \underset{i,j}{\operatorname{argmax}}(C_{i,j} + C_{j,i})$, which will be inverse valued. The latter, follows the scheme in which two categories $\{c^i, c^j\}$ that maximize the confusion are discriminated.

3   Experimental Results

In order to present the results, we first discuss the data, methods, and evaluation measurements.

- **Data**: We consider five muti-class problems from the UCI Machine Learning Repository: Ecoli (8 classes), Vowel (11 classes), Yeast (10 classes), Shuttle (7 classes), and Glass (7 classes). In addition, we test our methodology over 4 challenging Computer Vision multi-class problems: 70 visual object categories from the MPEG dataset, 20 classes of the ARFace database, a real traffic sign categorization problem of 36 classes, and 7 handwritten music cleafs classes [2]. Computer Vision datasets are described using PCA keeping 99,9% of information.
- **Methods**: We compare the Genetic ECOC design with the One vs. All, One vs. One, Dense Random [1], Forest [8] and DECOC [7] designs. The ECOC base classifier is the libsvm implementation of a SVM with RBF kernel. The SVM ζ and γ parameters are tuned via Genetic Algorithms for all the

methods, minimizing the classification error of a two-fold evaluation over a training sub-set [2].

- **GA settings and parameters**: The number of generations of each GA optimization process was set to $3N$ where N is the number of classes of each particular classification problem. The number of individuals of the GA was set to $5N$. Furthermore, elitism was applied at each generation, and thus, the 10% fitter individuals are automatically selected to form part of the next generation. On the other hand, the specific and generic variants of the Crossover, Mutation and Extension operators where equiproportional.

- **Evaluation Measurements**: The classification performance is obtained by means of a stratified ten-fold cross-validation. Finally, we test for statistical significance using Friedman and Nemenyi statistics at 95% of the confidence interval [3]. The classification results are shown in Table 1. The table shows the classification performance of each ECOC design on each dataset, the average performance ranking, and the mean number of classifiers of the ensemble. In order to compare the performances provided for each strategy, Table 2 shows the mean rank of each ECOC design considering the 18 different experiments (9 dataset performances and 9 PC values).

We use the Nemenyi test to check if one of the techniques can be singled out. In our case with $k = 7$ ECOC approaches to compare and $N = 9 \cdot 2 = 18$ experiments, the critical value for a 90% of confidence is $CD = 1.415 \cdot \sqrt{\frac{56}{108}} = 1.0189$. Since none of the methods ranks intersect with the GA Inspired ECOC rank for $CD = 1.0189$, we can state that the proposed ECOC design significantly improves the rest of methods performances at 90% of confidence.

Table 1. Classification results and number of classifiers per coding design

Dataset	Compact ECOC		GA Ins. ECOC		D. Random ECOC	
	Perf.	Classif.	Perf.	Classif.	Perf.	Classif.
Ecoli	80.5±1.9	3	**81.4±1.3**	3.8	68.1±2.7	8
Vowel	48.6±3.5	3	54.4±4.3	3.2	42.8±1.1	7
Yeast	57.7±2.4	3	68.1±1.5	5.6	66.8±3.3	11
Shuttle	80.9±2.1	3	81.1±1.3	3.2	90.6±2.3	7
Glass	50.2±1.2	4	55.1±6.1	5	54.9±6.4	10
MPEG	90.8±4.1	6	95.3±3.2	6	83.3±1.0	36
ARFACE	61.5±3.2	5	86.3±1.2	6	73.0±1.3	20
TRAFFIC	81.2±1.2	3	96.3±2.4	4.2	82.3±1.1	7
CLEAFS	84.6±1.1	7	84.1±2.8	7	90.0±1.4	70
Mean Rank & #Class.	**5.6**	**4.2**	2.5	4.9	4.9	19.5

1vsAll		1vs1		DECOC		FECOC	
Perf.	Classif	Perf.	Classif.	Perf.	Classif.	Perf.	Classif.
75.5±1.8	8	79.2±1.8	28	69.4±1.3	7	75.2±3.5	21
53.8±6.2	7	**60.5±2.9**	15	55.1±2.5	6	43.9±2.1	15
80.7±2.2	11	78.9±1.2	28	66.7±1.3	10	68.1±1.3	30
90.6±1.1	7	86.3±1.1	21	77.1±1.4	6	80.3±1.5	18
47.1±1.3	10	52.4±2.8	45	55.8±2.2	9	**56.0±3.2**	27
91.8±2.6	36	90.6±2.1	630	86.2±4.2	35	**96.7±1.3**	105
84.0±3.3	20	**96.0±2.5**	190	82.7±2.1	19	81.6±0.4	57
80.8±1.2	7	84.2±2.8	21	96.9±2.4	6	**97.1±1.1**	18
87.8±2.4	70	**92.8±1.3**	2415	83.4±1.5	69	81.9±2.3	207
3.9	19.5	1.5	377	5.2	18.5	4.8	55.3

Table 2. Mean rank per coding design

Rank	Compact ECOC	GA ECOC	Dense ECOC
Perf. rank	5.6	2.5	4.9
Perf. per Class rank	1	2	5
Mean rank	3.3	**2.2**	4.9

Rank	1vsAll	1vs1	DECOC	FECOC
Perf. rank	3.9	**1.5**	5.2	4.8
Perf. per Class rank	4	7	3	6
Mean rank	3.9	4.2	4.1	5.4

4 Discussion and Conclusions

We presented the novel Genetic ECOC optimization procedure, which has been carefully defined in order to take into account the ECOC properties. New ECOC Crossover and Mutation operators have been defined to avoid non-valid coding matrix generation, reducing the search space and the number of individuals needed for convergence. Moreover, a new Extension ECOC operator has been proposed, which allows the ECOC design to take benefit from error correction in a problem dependent way. The methodology was tested on several public Machine Learning and Computer Vision datasets, obtaining significant performance improvements compared to state-of-the-art ECOC approaches using far less number of dichotomizers, which results in a much more efficient coding.

Acknowledgments. This work has been supported by projects TIN2009-14404-C01/C02 ,CONSOLIDER-INGENIO CSD 2007-00018, IMSERSO-Ministerio de Sanidad 2011 Ref. MEDIMINDER and RECERCAIXA 2011 Ref. REMEDI.

References

1. Allwein, E., Schapire, R., Singer, Y.: Reducing multiclass to binary: A unifying approach for margin classifiers. JMLR 1, 113–141 (2002)
2. Ángel Bautista, M., Escalera, S., Baró, X., Radeva, P., Vitriá, J., Pujol, O.: Minimal design of error-correcting output codes. PRL 33(6), 693–702 (2012)
3. Demsar, J.: Statistical comparisons of classifiers over multiple data sets. JMLR 7, 1–30 (2006)
4. Escalera, S., Pujol, O., Radeva, P.: On the decoding process in ternary error-correcting output codes. TPAMI 99(1) (2009)
5. Garcia-Pedrajas, N., Fyfe, C.: Evolving output codes for multiclass problems. TEC 12(1), 93–106 (2008)
6. Lorena, A.C., Carvalho, A.C.P.L.F.: Evolutionary design of multiclass support vector machines. JIFS 18, 445–454 (2007)
7. Pujol, O., Radeva, P., Vitrià, J.: Discriminant ECOC: A heuristic method for application dependent design of ecoc. TPAMI 28, 1001–1007 (2006)
8. Escalera, S., Pujol, O., Radeva, P.: Boosted landmarks and forest ECOC: Framework to detect and classify objects in clutter scenes. PRL 28(13), 1759–1768 (2007)

Author Index